Distance Formula

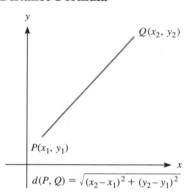

$$d(P, Q) = \sqrt{(x_2 - x_1)^2 + (y_2 - y_1)^2}$$

Slope of a Line

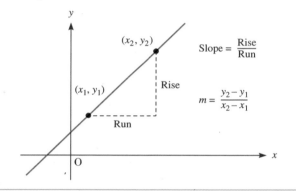

$$\text{Slope} = \frac{\text{Rise}}{\text{Run}}$$

$$m = \frac{y_2 - y_1}{x_2 - x_1}$$

Midpoint Formula

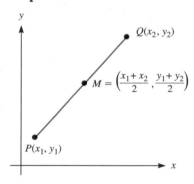

$$M = \left(\frac{x_1 + x_2}{2}, \frac{y_1 + y_2}{2}\right)$$

Vertical Line Test

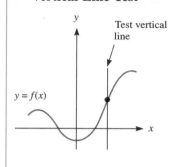
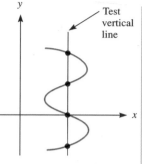

(a) Graph of a function   (b) Not the graph of a function

Vertical Translation

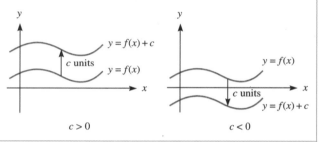

$c > 0$ $c < 0$

Horizontal Translation

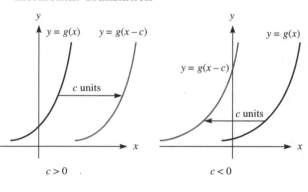

$c > 0$ $c < 0$

Algebra and Trigonometry and Their Applications

SECOND EDITION

Algebra and Trigonometry and Their Applications

SECOND EDITION

Larry Joel Goldstein

Wm. C. Brown Publishers

Dubuque, IA Bogota Boston Buenos Aires Caracas Chicago
Guilford, CT London Madrid Mexico City Sydney Toronto

Cover Image: A computer visualization of the flow of heat in the atmosphere.

Book Team

Developmental Editor   *Daryl Bruflodt*
Publishing Services Coordinator   *Julie Avery Kennedy*

 Wm. C. Brown Publishers

President and Chief Executive Officer   *Beverly Kolz*
Vice President, Publisher   *Earl McPeek*
Vice President, Director of Sales and Marketing   *Virginia S. Moffat*
Vice President, Director of Production   *Colleen A. Yonda*
National Sales Manager   *Douglas J. DiNardo*
Marketing Manager   *Julie Joyce Keck*
Advertising Manager   *Janelle Keeffer*
Production Editorial Manager   *Renée Menne*
Publishing Services Manager   *Karen J. Slaght*
Royalty/Permissions Manager   *Connie Allendorf*

A Times Mirror Company

Cover image courtesy of the National Center for Supercomputing Applications (NCSA)

Copyediting, design, and production by Publication Services, Inc.

Copyright © 1993 by Richard D. Irwin, Inc. and 1996 by Wm. C. Brown Communications, Inc. All rights reserved

Library of Congress Catalog Card Number: 95–76543

ISBN 0–697–26533–1   Student Edition
      0–697–26534–X   Instructor Edition

No part of this publication may be reproduced, stored in a retrieval system, or transmitted, in any form or by any means, electronic, mechanical, photocopying, recording, or otherwise, without the prior written permission of the publisher.

Printed in the United States of America by Times Mirror Higher Education Group, Inc., 2460 Kerper Boulevard, Dubuque, IA 52001

10 9 8 7 6 5 4 3 2 1

CONTENTS

PREFACE x
INDEX OF APPLICATIONS xxiv

CHAPTER 1 — REVIEW OF ELEMENTARY ALGEBRA

1.1 The Real Number System **2**
1.2 Exponents and Radicals **14**
1.3 Polynomials **29**
1.4 Rational Expressions **41**
1.5 Equations and Identities **53**
   Mathematics and the World around Us—Unbreakable Codes **62**
   Chapter Review **63**

CHAPTER 2 — GRAPHS AND FUNCTIONS

2.1 Two-Dimensional Coordinate Systems **69**
2.2 Graphs of Equations **75**
2.3 Straight Lines **86**
2.4 Inequalities **100**
2.5 The Concept of a Function **113**
2.6 Graphs of Functions **123**
2.7 Graphing Calculator Fundamentals **138**
2.8 Linear Functions **144**
   Chapter Review **148**
   Mathematics and the World around Us—Orbits of the Planets **149**

v

CHAPTER 3

APPLIED PROBLEMS, QUADRATIC EQUATIONS, AND INEQUALITIES

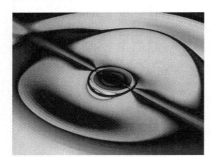

- 3.1 Solving Applied Problems **155**
- 3.2 Quadratic Equations **164**
- 3.3 Quadratic Functions and Optimization Problems **176**
- 3.4 Using Quadratic Equations to Solve Other Equations **186**
- 3.5 Equations and Inequalities Involving Absolute Value **195**
  Chapter Review **200**
  Mathematics and the World around Us—Fractals **201**

CHAPTER 4

POLYNOMIAL AND RATIONAL FUNCTIONS

- 4.1 Polynomial Functions of Degree Greater Than 2 **205**
- 4.2 Division of Polynomials **213**
- 4.3 Calculating Zeros of Polynomials **223**
- 4.4 Complex Numbers **231**
- 4.5 The Fundamental Theorem of Algebra and Descartes's Rule of Signs **239**
- 4.6 Rational Functions **245**
  Mathematics and the World around Us—Should Mathematical Results Be Secret? **258**
  Chapter Review **258**

CHAPTER 5

MORE ABOUT FUNCTIONS

- 5.1 Operations on Functions **266**
- 5.2 Inverse Functions **273**
- 5.3 Variation and Its Applications **284**
  Chapter Review **291**
  Mathematics and the World around Us—Predator-Prey Models **292**

Contents vii

CHAPTER 6

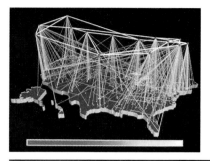

EXPONENTIAL AND LOGARITHMIC FUNCTIONS

6.1 Exponential Functions with Base a **296**
6.2 The Natural Exponential Function **304**
6.3 Logarithmic Functions **312**
6.4 Common Logarithms and Natural Logarithms **322**
6.5 Exponential and Logarithmic Equations **334**
    Mathematics and the World around Us—Population Models **338**
    Chapter Review **338**

CHAPTER 7

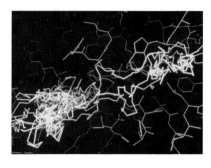

THE TRIGONOMETRIC FUNCTIONS

7.1 Angles and Radian Measure **344**
7.2 The Trigonometric Functions **354**
7.3 Trigonometric Functions of Acute Angles **363**
7.4 Applications of Trigonometry **371**
7.5 The Graphs of Sine and Cosine **378**
7.6 More about the Sine and Cosine Functions **388**
7.7 The Graphs of Tangent, Secant, Cotangent, and Cosecant **396**
7.8 Fundamental Identities for Trigonometric Functions **404**
    Mathematics and the World around Us—Can You Hear the Shape of a Drum? **410**
    Chapter Review **410**

CHAPTER 8

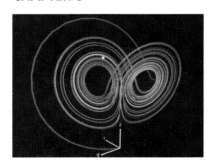

ANALYTIC TRIGONOMETRY

8.1 Trigonometric Identities **417**
8.2 The Addition and Subtraction Identities **424**
8.3 Multiple and Half-Angle Identities **437**
8.4 The Inverse Trigonometric Functions **450**
8.5 Solving Trigonometric Equations **456**
    Mathematics and the World around Us—Chaos Theory **461**
    Chapter Review **461**

Job satisfaction, 75
Labor negotiations, 157
Lawn service business, 533
Long distance savings, 103
Manufacturing, 122
Manufacturing cabinets, 565
Manufacturing calculators, 561
Manufacturing skis, 555
Manufacturing stereo receivers, 561
Manufacturing tires, 565
Manufacturing TV sets, 560
Market penetration, 75
Maximizing profit, 185, 213
Maximizing revenue, 2, 181, 185
Maximum affordable car loan, 649, 651
Merit salary plan, 644
Monthly payment on a mortgage, 648, 651
Motel management, 122
Multidivision corporations, 593
Multinational production, 593
Multiplier effect, 645
Mutual fund investment, 14, 16, 27
Occupancy rates of hotels, 153
Order fulfillment, 561

Parking fees, 138
Payout of an annuity, 650
Pension fund management, 163, 523, 642
Percent change in price, 13
Personnel management, 162
Pharmaceutical production, 558
Planning constructions, 561
Price markdown, 12, 157, 162
Price markup, 12
Production planning in a lumber mill, 561
Profit function, 40
Profit analysis, 112
Quantity discounts, 99
Rate of inflation, 16
Real estate sales, 217, 222, 636
Refinancing a mortgage, 651
Retail price determination, 13
Return on investment, 162
Revenue from sales, 100
Revenue function, 41
Revenue model, 171
Salary growth, 645
Salary of teachers, 73
Salary projection, 94

Sale of cold remedies, 211, 213
Sales commissions, 99, 152
Sales fluctuations, 394
Sales of a firm, 341
Sales of tennis rackets, 102
Sales projection, 56
Salvage value, 333
Satisfying final demand, 597
Saving for retirement, 651
Shipping automobiles, 560
Sports salaries, 644
Start-up business finance, 548
State income taxes, 135
Stock market investment, 528
Stock market trading, 112
Stockbroker fees, 102
Straight-line depreciation, 99, 158, 163
Tax audit, 664
Trade deficit, 175, 186
Transportation, 477
Unemployment model, 222
UPS shipping requirement, 34, 40
Utility pricing, 162
Yield curve, 327

CHEMISTRY

Acidity and alkalinity, 331, 332
Atomic numbers, 527
Boiling point of aluminum, 66
Carbon dating of an artifact, 69

Diameter of a DNA molecule, 24
Hydrogen ion concentration, 342
Manufacturing cosmetics, 159, 163
Mixing acid and water, 163

Mole of an element, 66
Radioactive decay of iodine 131, 300
Time of an explosion, 26

ENGINEERING

Accuracy of a speedometer, 290
Aerial navigation, 376, 508, 532
Air traffic control, 273
Air valve on a tire, 413
Architecture, 172
Binary representation of numbers, 319
Cable television network construction, 170
CD-ROMs, 327
Circuit board design, 75
Coefficient of linear expansion, 51
Communications engineering, 654, 656
Compression of data, 652
Computer networks, 295
Construction, 162
Construction of a music hall, 679
Construction of an office building, 227
Cost of a building, 230
Deflection of a beam, 212
Design of a cistern, 526
Design of packaging, 203
Design of a storage tank, 123

Design of a suspension bridge, 376
Design of a table leg, 377
Design of a window, 122, 185
Distance across a river, 372, 374
Earthquake magnitude, 331
Error-correcting codes, 603
Geometry of a suspension bridge, 679, 680, 720
Headlight of a car, 679
Height of a building, 373, 376
Hexadecimal numbers, 652
Highway safety, 99
Hyperbolic navigation, 696, 699
Intensity of an earthquake, 331
Intensity of sound, 331
Laying electric power cables, 123
Loudness of sound, 257, 293, 323, 330
Measurement error in manufacturing, 197
Military science, 375, 376, 477
Mining, 484
Nautical mile, 439

Navigation, 375, 377, 413, 434, 477, 483, 484, 512, 518
Noise in workplace, 330
Oil exploration, 636
Optimizing design of a box, 175
Optimizing design of a yard, 182
Optimizing production, 172
Pixels, 114
Power of a windmill, 290
Quality control in manufacturing, 619
Quality control in a lumber mill, 11, 13
Radioactive decay in nuclear power production, 303
Resistors in series, 51
Rotating spotlight, 404
Safe length of a beam, 290
Safe load, 294
Satellite power supply, 341
Speed determined from skid marks, 289
Stopping distance, 288
Supply of computer memory chips, 85

Support of a vertical pole, 370
Support of wood frame of a house, 377
Surface area of a storage tank, 120
Surveying, 377, 413, 414, 476, 477, 483
Surveying a building lot, 372, 374
Weight in an elevator, 556

ECOLOGY AND EARTH SCIENCE

Algae growth from pollution, 332
Height of clouds, 375
Industrial pollution, 323
Limited population growth, 312
Meteorology, 375
Oil spill, 272

Population models, 338
Population growth of Arizona, 342
Population growth of California, 332
Population growth of Hawaii, 333
Population growth of Kenya, 342
Population growth of Mexico City, 321

Population growth of Texas, 303
Population growth of world, 303
Population of Maryland, 341
Population of the United States, 332
Street noise, 342

GENERAL INTEREST

Age of Chinese artifacts, 332
Airplane travel, 548
Angle a ladder makes with the ground, 370
Archaeology, 326
Assignment of tasks, 654, 664
Baseball salaries, 175, 213
Bricklaying, 533
Chaos theory, 461
Coffee consumption, 26
Coin flipping, 662
College expenses, 16
College fund, 332, 645, 649
Computer art, 265
Counting license plates, 652
Design of a box, 230
Design of a flower box, 528
Design of a school test, 556
Distance and velocity, 4, 12
Diving, 375

Driving costs, 86
Driving safety, 210, 213
Gardening, 40
Geography, 483, 484
Grade point average, 42, 51
Home maintenance, 40
Home remodeling, 185
Hullian model of learning, 311
Inferring from data, 665
Interest on a certificate of deposit, 27
Installing a TV antenna, 376
Itemized deductions for health care, 99
Itemized deductions for taxes, 85, 98
Job planning, 192, 194
Marriage rate, 186
Maximizing volume of a box, 212
Movement of a phonograph record, 413
Navigation, 474

Nutrition, 556
Package design, 263
Personal computers for home use, 147
Pet rescue, 374
Population change, 645
Puzzle, 528
Ranching, 184
Rotation of a door, 353
Rotation of a Ferris wheel, 353
Savings account, 534
Seats in an auditorium, 634
Secrecy of mathematical results, 258
Spread of a rumor, 69, 312, 341
Tax cut, 13
Temperature conversion, 283, 527
Test score average, 113
Typing, 636
Unbreakable codes, 62

MEDICINE

Breast cancer, 619
Calcium injection, 20
Changes in blood pressure over time, 391, 394
Dosage of medication, 213, 230, 341
Ear injuries from loud noises, 331

New drug, 311
Pharmacology, 264
Physician visits, 213, 230
Radioactive decay in cancer treatment, 303
Spread of an epidemic, 308

Temperature during an illness, 200, 226, 230, 394, 403, 415
Treatments for a disease, 664
Urinary tract infections, 321
Viral infection, 615

PHYSICS

Acceleration due to gravity, 440
Atmospheric pressure, 332
Bouncing ball, 645
Boyle's law of gases, 294
Hearing the shape of a drum, 410
Centrifuge, 354

Charles's law of gas expansion, 99
Circular motion, 510, 518
Circular motion of a car tire, 413
Circular thermometer's movement, 353
Clock pendulum, 353
Convection in the atmosphere, 416

Damped motion of a spring, 403
Decay of a radioisotope, 321, 341
Decay of xenon-133, 321
Distance versus time, 95
Electrical current in a wire, 388, 396
Electrical attraction or repulsion, 291

Electrical resistance, 263, 286
Electron volt, 26
Escape velocity, 28
Falling objects, 28, 113
Force addition, 513
Forces on a structure, 509
Forces resolved into components, 508
Gas pressure, 289
Geometry of a reflecting telescope, 677
Illumination, 289
Inclined plane, 513, 518
Intensity of light, 333

Laser beam aimed at moon, 377
Measuring altitude, 175
Newton's law of gravitation, 24, 286, 289
Number of hours of daylight, 392
Oscillation of a weight on a spring, 200, 384, 388, 415
Parabolic mirror of a telescope, 679
Period of a pendulum, 289
Radiation energy, 289
Radioactive decay, 319
Reflection property of the parabola, 676
Rotation of a clock hand, 354

Rotation of wheel of exercise bike, 354
Speed of a tractor, 352
Speed of sound, 20, 28
Stretched spring, 285
Swimming pool leak, 66
Temperature conversion, 100, 152, 163
Temperature of a gas, 288
Universal gravitational constant, 26
Wavelength of a particle, 290
Wave motion, 396
Weight above the earth, 257

 SPORTS AND RECREATION

Area of a basketball floor, 556
Baseball, 28, 51, 66, 353
Bicycle race, 354
Boating, 533
Earned run average, 112, 257
Gambling, 26, 664
Genealogy, 645

Hockey, 556
Hockey team, 163
Photography, 374
Recreation, 41, 664
Roller coaster, 395
Skiing, 100

Skydiving, 307, 312
Speed of a sailboat, 2, 119
Swimming pool design, 533
Swing of a seesaw, 354
Theater tickets, 528, 534, 565
Travel, 528

CHAPTER 1

REVIEW OF ELEMENTARY ALGEBRA

Algebra is one of the most useful subjects you will ever study. It can be used to solve applied problems that arise in almost every field. If you are interested in business, architecture, medicine, law, biology, chemistry, agriculture, forestry, or journalism, you will find algebra a useful problem-solving tool.

Algebra has developed over many centuries, starting with the ancient Egyptians and Chinese, who developed mathematical tools to help in commerce, architecture, and agriculture more than four thousand years ago. Almost every civilization and culture on Earth has contributed to the development of algebra. The name **algebra** was derived from the title of a text written by the Arab mathematician al-Khwarizmi (780–850 A.D.).

Algebra is a prerequisite for learning most branches of modern mathematics, including calculus and statistics, fields many of you are destined to study.

Chapter Opening Image: *This abstract computer art, part of an animation by Brian D. Evans, represents music with changing colors.*

The branches of mathematics that rely on algebra are, in turn, the foundation for solving problems in almost all fields in which numerical problems arise.

To give you a preview of things to come later in the book, here are several applied problems that we will solve.

Problem 1 A model commonly used by yachtsmen states that the maximum speed of a sailboat (in knots) is 1.3 times the square root of the length of the boat at the water line. How long should a sailboat be if its maximum speed is to be 8 knots? (See Section 2.1 for the solution.)

Problem 2 A company manufactures upholstered chairs. If it manufactures x chairs, then the revenue $R(x)$ per chair is given by the function:
$$R(x) = 200 - 0.05x$$
How many chairs should the company manufacture to maximize total revenue? (See Section 3.5 for the solution.)

Problem 3 Suppose that a colony initially contains 5,000 bacteria and the number of bacteria doubles every 4 hours. How many bacteria are there after 10 hours? (See Section 5.1 for the solution.)

In this book we will develop the algebraic techniques to solve these and many other applied problems.

To start our study of algebra, this chapter will review some topics that you probably covered in earlier courses. Our discussion will provide a quick refresher.

1.1 THE REAL NUMBER SYSTEM

Arithmetic and geometric problems involve numbers like the following:
$$5, -18, 1.782, \frac{1}{3}, -\frac{7}{5}, \pi, \sqrt{2}, 97\%$$

These numbers are examples of **real numbers**. Such numbers are not new to you. You have been using them since your earliest days of grade school arithmetic. We will usually denote real numbers by letters, such as x, y, a, b, C, and S.

Several special types of real numbers should be mentioned here. The **counting numbers** are the numbers 1, 2, 3, ... These are the numbers we use to count people, cars, or bacteria. The **integers** are numbers such as:
$$\ldots, -3, -2, -1, 0, 1, 2, 3, \ldots$$

That is, the integers consist of the counting numbers, their negatives, and the number 0. A **rational number** is a real number that can be written as the quotient of two integers a/b with $b \neq 0$. Here are some examples of rational numbers:
$$\frac{5}{3}, -\frac{117}{89}, -\frac{45}{7}, \frac{5}{1} = 5$$

Every integer is a rational number and every rational number is a real number. The interrelationship between the integers, rational numbers, and real numbers is illustrated in Figure 1.

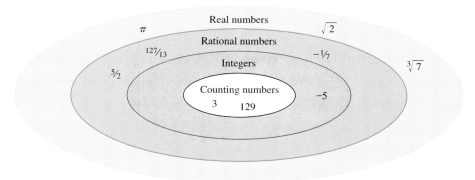

Figure 1

Arithmetic Operations on Real Numbers

There are four elementary arithmetic operations among real numbers: addition, subtraction, multiplication, and division. Throughout this book, we will be using the operations either on real numbers themselves or on variables that represent real numbers. Let's briefly review the most important properties of these operations.

The real number 0 is called the **additive identity** because it leaves a number unchanged under the operation of addition. That is,

$$a + 0 = 0 + a = a$$

for any real number a. The number 1 is called the **multiplicative identity** because it leaves a number unchanged under the operation of multiplication. That is,

$$1 \cdot a = a \cdot 1 = a$$

for every real number a.

The negative of a real number a is denoted $-a$ and is obtained by changing the sign of a. The number $-a$ is called the **additive inverse** of a because, when added to a, it gives the additive identity:

$$a + (-a) = 0$$

At left are some examples of additive inverses.

a	Additive Inverse of a
5	-5
3	-3
0	0
$-\frac{1}{2}$	$\frac{1}{2}$

A nonzero real number a has a **multiplicative inverse**, or **reciprocal**, denoted a^{-1}, with the property that the product of a and a^{-1} is the multiplicative identity. That is,

$$a \cdot a^{-1} = a^{-1} \cdot a = 1$$

Here are some examples of multiplicative inverses:

a	Multiplicative Inverse of a
2	$\frac{1}{2}$
-2	$\frac{1}{-2} = -\frac{1}{2}$
$\frac{1}{3}$	$\frac{1}{1/3} = 3$
1.2	$\frac{1}{1.2} = \frac{5}{6}$
0	does not have a multiplicative inverse

The next example provides some practice in translating phrases involving arithmetic operations into algebraic symbols.

EXAMPLE 1
Translating Words into Algebraic Language

Write the following phrases in algebraic language:
1. The sum of x and y.
2. Fifty percent of x.
3. One less than the product of x and z.
4. The additive inverse of a.
5. Five times the multiplicative inverse of b.

Solution

1. $x + y$
2. $0.5x$
3. $xz - 1$
4. $-a$
5. $5b^{-1}$

Throughout this book, we will use the arithmetic operations among real numbers to solve applied problems. The next example provides a taste of what is to come.

EXAMPLE 2
Distance = Rate × Time

A car is A miles west of Dayton. Going at a constant velocity and proceeding west, the car is B miles west of Dayton after t hours. Express the velocity of the car in terms of A, B, and t.

Solution

The velocity of the car is its rate of speed and we have the fundamental relationship:

$$\text{distance} = \text{rate} \times \text{time}$$

or

$$\text{rate} = \frac{\text{distance}}{\text{time}}$$

The car goes a total distance $B - A$ and it covers this distance in time t. So by the above formula, we have:

$$\text{velocity} = \frac{B - A}{t}$$

The following three rules of arithmetic are the basis for many of the simplifications that we can perform on algebraic expressions.

RULES OF ARITHMETIC

For any real numbers a, b, and c, the following relations are true:
1. Commutative laws
$$a + b = b + a$$
$$ab = ba$$

2. Associative laws
$$a + (b + c) = (a + b) + c$$
$$a(bc) = (ab)c$$

3. Distributive laws
$$a(b + c) = ab + ac$$
$$(a + b)c = ac + bc$$

EXAMPLE 3
Simplifying an Expression

Write the following real number without using parentheses:
$$3(x + y) + 4x$$

Solution

We apply the properties of real numbers, simplifying the expression using one law at a time.

$$\begin{aligned}
3(x + y) + 4x &= (3x + 3y) + 4x & \text{Distributive law} \\
&= 3x + 3y + 4x & \text{Associative law of addition} \\
&= 3x + 4x + 3y & \text{Commutative law of addition} \\
&= (3x + 4x) + 3y & \text{Associative law of addition} \\
&= 7x + 3y
\end{aligned}$$

Once you gain experience with algebraic simplification, you may do some (or all) of the intermediate steps in your head and perform the simplification in a single step.

Here are some useful properties of negatives that we will use throughout this book.

PROPERTIES OF NEGATIVES

$$-a = (-1)a$$
$$-(-a) = a$$
$$-(a + b) = -a - b$$
$$a - 0 = a$$
$$0 - a = -a$$
$$(-a)(-b) = ab$$
$$(-a)b = -ab = a(-b)$$

Subtraction of real numbers is defined in terms of addition and taking negatives: To subtract b from a, we add to a the negative of b. That is,
$$a - b = a + (-b)$$

6   CHAPTER 1   Review of Elementary Algebra

EXAMPLE 4
Simplifying an Expression Using Properties of Negatives

Write the following expressions without using parentheses.
1. $(-3x)(-4y)$
2. $x - (x - 2y)$
3. $3 - 5(x - 2)$

Solution

1. $(-3x)(-4y) = (3x)(4y) = 12xy$
2. $x - (x - 2y) = x + (-x - (-2y))$
$= x + (-x) + 2y$
$= 0 + 2y$
$= 2y$
3. $3 - 5(x - 2) = 3 - (5x + 5(-2))$
$= 3 - (5x + (-10))$
$= 3 - 5x - (-10)$
$= 3 - 5x + 10$
$= 13 - 5x$

The Order of Operations

In simplifying an arithmetic expression, you must perform arithmetic operations in the following order.

ORDER OF OPERATIONS

1. First, perform all multiplications and divisions, proceeding from left to right.
2. Next, perform all additions and subtractions, proceeding from left to right.
3. Simplify expressions in parentheses first, using rules 1 and 2. In the case of parentheses within parentheses, work from the innermost set outward.

EXAMPLE 5
Order of Operations

Calculate the following:
1. $5 \cdot 4 + 3 \cdot 7$
2. $5 \cdot 4 + 3 \cdot (7 - 2 \cdot 4)$

Solution

1. First, perform multiplications and divisions, proceeding from left to right:
$$20 + 21$$
Next, perform additions and subtractions from left to right:
$$41$$
2. First, evaluate the expression in parentheses. Use rules 1 and 2 first to do the multiplication, then to do the subtraction:
$$5 \cdot 4 + 3 \cdot (7 - 8) = 5 \cdot 4 + 3 \cdot (-1)$$

Next, perform the multiplications:
$$20 - 3$$
Finally, do the subtraction:
$$17$$

Using Calculators in Algebra

You may use a calculator to perform numerical calculations. In this book, we will give keystrokes to show how to perform various operations on a calculator. These keystrokes will apply to two general classes of calculators: scientific calculators with algebraic logic (almost all scientific calculators except those made by Hewlett-Packard) and graphing calculators (typified by the Texas Instruments TI-82).

Here are two important differences to watch out for between the classes of calculators:

1. A scientific calculator uses the $\boxed{=}$ key to display the answer to a calculation. On a graphing calculator, you would use the $\boxed{\text{ENTER}}$ key for the same purpose.[1] For instance, to add 2 and 3 on a scientific calculator, you would use the keystrokes:

   $\boxed{2}\boxed{+}\boxed{3}\boxed{=}$

   On a graphing calculator, the corresponding keystrokes would be

   $\boxed{2}\boxed{+}\boxed{3}\boxed{\text{ENTER}}$

   In each case, the calculator displays the answer 5.

2. To enter a negative number on a scientific calculator, you first enter the number and then press the $\boxed{+/-}$ key, which changes its sign. On a graphing calculator, you use the $\boxed{(-)}$ key to enter a minus sign. (Don't confuse this key with $\boxed{-}$, which indicates subtraction.) So, for example, to enter -5, the keystrokes for a scientific calculator are:

   For a graphing calculator, the corresponding keystrokes are:

EXAMPLE 6
Arithmetic Problems Using a Calculator

Give the keystrokes (for both a scientific and a graphing calculator) for calculating the following quantities:

1. $5 \cdot (3 + 7)$
2. $\dfrac{3}{4} + \dfrac{5}{8}$
3. $48\% - \dfrac{17}{9}$
4. $-8 + \dfrac{-1 + 2}{3 - 1}$

---

[1] Some calculators label this key $\boxed{\text{EXE}}$. Check which notation your graphing calculator uses. In what follows, we will assume that it is labeled $\boxed{\text{ENTER}}$. If your calculator has an $\boxed{\text{EXE}}$ key, you will need to translate the keystrokes given.

Solution

Calculation	Scientific Calculator	Graphing Calculator	Answer
$5 \cdot (3 + 7)$	5 × (3 + 7) =	5 × (3 + 7) ENTER	50
$\frac{3}{4} + \frac{5}{8}$	3 ÷ 4 + 5 ÷ 8 =	3 ÷ 4 + 5 ÷ 8 ENTER	1.375
$48\% - \frac{17}{9}$	.48 − 17 ÷ 9 =	.48 − 17 ÷ 9 ENTER	−1.4088888889
$-8 + \frac{-1+2}{3-1}$	In the fraction on the right, we want the calculator to perform the operations $-1 + 2$ and $3 - 1$ before performing the division. So we use parentheses around these calculations. 8 +/− + (1 +/− + 2) ÷ (3 − 1) =	In the fraction on the right, we want the calculator to perform the operations $-1 + 2$ and $3 - 1$ before performing the division. So we use parentheses around these calculations. (−) 8 + ((−) 1 + 2) ÷ (3 − 1) ENTER	−7.5

Sets

A **set** is a collection of objects. The collection of all real numbers is an example of a set. Sets are customarily denoted using capital letters, such as A, B, C, and so on. We will reserve the notation **R** to mean the set of real numbers.

An object in a set is called an **element** of the set and is said to belong to the set. Thus, $\frac{1}{3}$ and -3.75 are elements of the set **R**. If every element of the set A is also an element of the set B, then we say that A is a **subset** of B. For example, let A denote the set consisting of the numbers 0 and 1, and B the set consisting of all real numbers between 0 and 1 inclusive. Then A is a subset of B.

One way of describing a set is to list its elements within braces. For instance, the set A consisting of the integers 1, 2, 3, 4, and 5 can be written as:

$$A = \{1, 2, 3, 4, 5\}$$

Another method for describing sets is to give one or more conditions that specify the elements. This method is called **set-builder notation.** Using this notation, the set of nonzero real numbers x can be written as:

$$\{x \mid x \text{ real and } x \neq 0\} \quad \text{or} \quad \{x : x \text{ real and } x \neq 0\}$$

This notation is read: "The set of all x such that x is a real number not equal to 0."

The Number Line

We can represent the real numbers geometrically as points on a **number line**. To construct a number line, choose a point on the line to represent 0. This point is called the **origin**. Points to the right of 0 represent positive numbers and points to the left of 0 negative numbers. See Figure 2. Each point on the number line corresponds to one and only one real number. For example, Figure 3 shows a point labeled a that corresponds to the real number 3.3. A number line is also called a **one-dimensional coordinate system**.

Figure 2

Figure 3

Inequalities

The number line shows us how to compare real numbers with one another. If a and b are real numbers, we say that a is **greater than** b (and that b is **less than** a) provided that a lies to the right of b on the number line. (See Figure 4.) We indicate the relationships greater than and less than by using **inequality symbols:**

Inequality Symbol	Interpretation	Example
$a > b$	a is greater than b. That is, a lies to the right of b on the number line.	$3 > 1$
$a < b$	a is less than b. That is, a lies to the left of b on the number line.	$-3 < -1$
$a \geq b$	a is greater than or equal to b. That is, either a is greater than b or a is equal to b.	$5 \geq 5$, $\quad 0 \geq -2$
$a \leq b$	a is less than or equal to b. That is, either a is less than b or a is equal to b.	$-3 \leq -3$, $\quad -5 \leq 2$

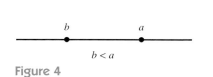

Figure 4

EXAMPLE 7 Expressing Algebraic Statements as Inequalities

Write the following statements as inequalities:

1. The profit of a company this year is at least equal to last year's profit.
2. The length of the new edition of a book is at most 20 percent longer than the previous edition.

Solution

1. Let P represent the profit this year and L the profit last year. The statement that P is at least equal to L can be written as the inequality:

$$P \geq L$$

2. Let x be the length of the previous edition and y the length of the new edition. Then 20 percent longer than the previous edition is equal to

$$x + 0.20x = 1.2x$$

The condition that the new edition is at most 20 percent longer than the previous edition may be written as the inequality:

$$y \leq 1.2x$$

Absolute Value

In many applications, it is necessary to consider the magnitude but not the sign of a real number. For dealing with such applications, it is convenient to define the notion of absolute value.

Definition Absolute Value Let a be a real number. The **absolute value** of a, denoted $|a|$, is equal to the distance from a to 0 on the number line.

The distance of a nonnegative number a from 0 is just a. However, if a is negative, then the distance of a to 0 is $-a$. That is, we have the following recipe for calculating the absolute value:

$$|a| = \begin{cases} a & \text{if } a \geq 0 \\ -a & \text{if } a < 0 \end{cases}$$

Because distances are always nonnegative, the absolute value of a number is nonnegative. To calculate the absolute value of a number, just disregard any negative sign. For example,

$$|5| = 5, \qquad |0| = 0, \qquad |-4| = 4$$

Absolute values occur in many applications, so it is important to be able to calculate with them. The next example gives some practice.

EXAMPLE 8
Calculating Absolute Values

Calculate the following:

1. $\left|-\dfrac{3}{8}\right|$
2. $|2|$
3. $\left|\dfrac{33}{12} - \pi\right|$

Solution

1. The number $-\dfrac{3}{8}$ is negative. To compute its absolute value, disregard the initial minus sign:

$$\left|-\dfrac{3}{8}\right| = \dfrac{3}{8}$$

2. Because 2 is positive, its absolute value is the number itself:

$$|2| = 2$$

3. We have

$$\dfrac{33}{12} - \pi = 2.75 - 3.14159\ldots = -0.39159\ldots$$

which is negative. To compute the absolute value, change the sign of the number:

$$\left|\dfrac{33}{12} - \pi\right| = -\left(\dfrac{33}{12} - \pi\right)$$
$$= \pi - \dfrac{33}{12}$$
$$= 0.39159\ldots$$

EXAMPLE 9
Accounting

A company is formulating a budget for the coming year. It orders each department to submit a budget that is within 3 percent of the budget for the preceding year and requests that there be no more than a $100,000 increase. Express in algebraic terms the conditions that the new budget must satisfy.

Solution

Let A represent the new budget and B the old budget for a particular department. The requirement that the new budget be within 3 percent of the old budget can be expressed by saying that the absolute value of the difference between the two budgets is at most $0.03B$; that is,

$$|A - B| \leq 0.03B$$

(The absolute value in the inequality measures the amount that the budget changes, either up or down.) The condition that the increase is no more than $100,000 can be expressed as:

$$A - B \leq 100{,}000$$

So the new budget must satisfy *both* of the inequalities:
$$|A - B| \le 0.03B, \qquad A - B \le 100{,}000$$

EXAMPLE 10
Manufacturing

A lumber mill cuts 8-foot studs to be used in house construction. The stud is rejected if its length differs from 8 feet by more than 0.25 inches. Express the condition for rejection of a stud in algebraic terms.

Solution

Let L be the actual length of a stud. The amount by which the stud differs from 8 feet (either more or less) equals $|L - 8|$. It is rejected if this quantity is more than 0.25. That is, the stud is rejected provided that:
$$|L - 8| > 0.25$$

The absolute value may be used to calculate the distance between points on the number line.

DISTANCE BETWEEN POINTS ON THE NUMBER LINE

Suppose that a and b are points on the number line. The distance $d(a, b)$ between them is equal to $|a - b|$.

EXAMPLE 11
Calculating Distances

Find the distance between the following pairs of points on the number line:
1. $a = 1, \quad b = 9$
2. $a = -1, \quad b = -5$

Solution

1.
$$d(a, b) = |a - b| = |1 - 9| = |-8| = 8$$
See Figure 5.

2.
$$d(a, b) = |a - b| = |-1 - (-5)| = |4| = 4$$
See Figure 6.

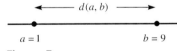
Figure 5

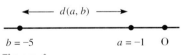
Figure 6

Exercises 1.1

Express each of the following in algebraic form.

1. The sum of twice x and y.
2. The difference of x minus y.
3. 30 percent of w.
4. 110 percent of z.
5. Five more than the product of x and z.
6. x divided by 4.
7. Twice the additive inverse of a.
8. Half the multiplicative inverse of b.
9. x is at most equal to y.
10. x is greater than one more than y.
11. x is less than 110 percent of y.
12. y is more than half of x.
13. The absolute value of x is at most 3.
14. The new budget x differs from the old one y by less than $50,000.

Write the following expressions without any parentheses and in as simple a form as possible.

15. $5x + 2x$
16. $1.1x + 3.2x$

17. $(a + b + c) + (2a + b)$
18. $(a - b) + (3a - 2b)$
19. $2(3x)$
20. $3(-5y)$
21. $(6x)(4y)$
22. $(-3)(-2x)$
23. $-(-5x)$
24. $-(-4 + 3x)$
25. $-(5 - 3y)$
26. $(-x)(2 + y)$
27. $7(3 - y)$
28. $x(3 + 2)$
29. $(2x + 3y) + (5x - 4y)$
30. $(9x - 12y) - (4x + y)$
31. $4 - (3 - x)$
32. $1 + (-2)(x - 1)$
33. $x - 5[3 - 2(x + 2)]$
34. $(x + 1)2 - (x - 1)4$
35. $4xy + 3(xy - 2yx)$
36. $x(y + 1) - y(1 - x)$

Fill in the blank with the correct symbol: $<$, $>$, or $=$.

37. $2.3 \_\_\_ 6.8$
38. $-6 \_\_\_ -12$
39. $0 \_\_\_ -1$
40. $56.7\% \_\_\_ 0.567$
41. $-9.0001 \_\_\_ -9$
42. $-3.216 \_\_\_ -3.220$
43. $\frac{3}{4} \_\_\_ \frac{5}{9}$
44. $(5 - 3) \_\_\_ (3 - 5)$
45. $(6.7 - 2.8) \_\_\_ (9.3 - 4.2)$
46. $\frac{45}{22} \_\_\_ 2.36$
47. $-1.732 \_\_\_ -\frac{4}{1.71}$
48. $-0 \_\_\_ 0$

Evaluate the following.

49. $\left|\frac{1}{4}\right|$
50. $|5 - 3|$
51. $|4 + 7|$
52. $|-5 + 0|$
53. $|-7 - 3|$
54. $|-8 + 2|$
55. $|3.28 - 3.28|$
56. $\left|-\frac{2}{3}\right|$
57. $|3.162|$
58. $|3.162 - \pi|$
59. Calculate the distance between -7 and 12 on the number line.
60. Calculate the distance between -3 and -11 on the number line.

Fill in the blank with the correct symbol: $<$, $>$, or $=$.

61. $|4| + |-6| \_\_\_ |4 + (-6)|$
62. $|7| - |9| \_\_\_ |7 - 9|$
63. $|-8| - |17| \_\_\_ |-8 - 17|$
64. $|-24| + |-18| \_\_\_ |-24 + (-18)|$

 Applications

65. **Distance-rate-time.** Suppose that a car goes a distance x miles at a velocity of 45 mph. Express the time taken in terms of x.
66. **Distance-rate-time.** Suppose that a cyclist goes a distance of x miles at 10 mph and y miles at 13 mph. Express the time taken in terms of x and y.
67. **Distance-rate-time.** Suppose that a car travels for h hours at a velocity of 60 mph. Express the distance traveled in terms of h.
68. **Distance-rate-time.** Suppose that a car travels for h hours at 50 mph, k hours at 10 mph, and 5 hours at 30 mph. Express the distance traveled in terms of h and k.
69. **Markup.** Suppose a retailer marks up prices of items by 35 percent over their wholesale price. If the wholesale price of an item is W, give an algebraic expression for the selling price.
70. **Markup.** A securities firm marks up the price of the bonds it buys by 5 percent. Suppose that it buys a bond for x dollars. At what price will it offer the bond for sale?
71. **Discount.** Suppose a retailer runs a sale in which each item is discounted 20 percent. If x is the price of an item before the sale, give an algebraic expression for the sale price.

72. **Discount.** A vacuum cleaner originally cost $75. During a sale, it is advertised at 30 percent off. What is the sale price?
73. **Change in price.** The **change in a price** equals the new price minus the old price. Suppose that the old price is $4 and the new price is x. Write an expression for the change in price.

74. **Percentage change in price.** The **percentage change in price** equals the quotient of the change in price divided by the old price. What is the percentage change in price of the product considered in the preceding problem?

75. **Budgeting.** A company is formulating a budget for the coming year. It orders each department to submit a budget that is within 2 percent of the budget for the preceding year and it requests that there be no more than a $50,000 increase. Express in algebraic terms the conditions that the new budget must satisfy.

76. **Taxes.** A politician promises that a tax cut will decrease the amount of taxes owed by each person by at least 20 percent. Express this promise in algebraic terms.

77. **Corporate downsizing.** A company orders each department to submit a budget that is at least 5 percent less than the budget of the preceding year, but with no more than a $50,000 decrease and no less than a $10,000 decrease. Express in algebraic terms the conditions that the new budget must satisfy.

78. **Price determination.** A retailer is holding a sale. She wishes to mark down each item by at least 10 percent, but with a price decrease of no more than $5. Express these conditions algebraically.

79. **Quality control.** A lumber mill cuts 10-foot studs to be used in office construction. A stud is rejected if its length differs from 10 feet by more than 0.15 inches. Express the condition for rejection of a stud in algebraic terms.

80. **Political polling.** A polling company estimates the number of voters likely to vote Democratic. It estimates that its poll has an error of no more than 5 percentage points. Express this condition in algebraic form.

81. **Depreciation.** In accounting, the cost of equipment is charged on a company's books in a number of annual payments. These charges are called **depreciation** and are meant to reflect how the equipment wears out or becomes obsolete. There are a number of methods used to calculate depreciation. The simplest method, called **straight-line depreciation**, charges equal amounts for each year of the equipment's lifetime. The total amount charged is the original cost of the equipment less the salvage value at the end of its lifetime. Suppose that a personal computer has a lifetime of 3 years, a purchase price of $1,200, and a salvage value of $300. What is the annual depreciation charge?

82. **Depreciation.** Suppose that an office copier has a lifetime of 7 years, a salvage value of S, and a cost of C. Give an algebraic expression for the annual depreciation charge.

Technology

Use a calculator to compute the value of the following expressions.

83. $5.77 - 3.88 + 4.19$
84. $100.01 + 358.77$
85. $78{,}391 \times 4{,}852$
86. $101{,}695 \div 5{,}901$
87. $58.99 \times 48 + 38.99 \times 1.01$
88. $119 \cdot (-38.79 + 3.05 \cdot (84.91 - 101))$
89. $\dfrac{4}{38} - 3 \cdot \dfrac{-12.9}{53.6}$
90. $5 - 3 \cdot (4 - 3 \cdot (3 - 1))$
91. $|5.37 \cdot 5.25 - 3.5 \cdot 26|$
92. $|13(2.11 - 4(3.61 - 4.88))|$
93. $x|x|$ where $x = -5.92$
94. $\dfrac{|x| + x}{x}$ for $x = -11.45$

Use your calculator to determine if the following inequalities are true.

95. $4.77(3.8 - 4.1) < 0.774$
96. $15.1 \geq 5.1 \cdot 3.04$
97. $1.1 + 2.1 + 3.1 < 4.1 + \dfrac{5.1}{0.85}$
98. $\left(4.88 - 3.1\left(1 - \dfrac{2}{0.88}\right)\right) < 1$

14   CHAPTER 1   Review of Elementary Algebra

99. The 12-month total return of science and technology mutual funds for the 12 months ending Oct. 31, 1993, was 33.15 percent. Suppose you invest $5,000 in a technology mutual fund. What would your shares be worth on Oct. 31, 1994, if the fund performance equaled the return of the preceding 12 months?

100. The 12-month total return of small company growth mutual funds for the 12 months ending Oct. 31, 1993, was 23.31 percent. Suppose you invest $15,000 in a technology mutual fund. What would your shares be worth on Oct. 31, 1994, if the fund performance equaled the return of the preceding 12 months?

In your own words

101. What is the difference between a real number and an integer?

102. Give three applied examples of quantities that can be represented by integers.

103. Give three applied examples of quantities that can be represented by real numbers that are not integers.

104. How many digits will your calculator carry out calculations to? In applications, will you meet any numbers that have more than this number of digits? Discuss the limitations this will impose on your calculations.

105. Describe how to calculate the absolute value of a number.

106. Give two examples of absolute values that arise in applied problems.

107. Give two applied examples of inequalities that you encountered in the last week.

108. Locate a number line in a magazine or newspaper. Describe how the number line is used.

1.2 EXPONENTS AND RADICALS

Many applications involve products in which a number is repeatedly multiplied by itself. Mathematicians have introduced a shorthand notation for such products. If n is a positive integer, then we write a^n to mean a product of n factors of a. That is, we have:

$$a^1 = a$$
$$a^2 = a \cdot a$$
$$a^3 = a \cdot a \cdot a$$
$$a^4 = a \cdot a \cdot a \cdot a$$

Here are some numerical examples of a^n for various values of a and n.

$$2^3 = 2 \cdot 2 \cdot 2 = 8$$
$$(-5)^3 = (-5) \cdot (-5) \cdot (-5) = -125$$
$$\left(\frac{1}{2}\right)^2 = \frac{1}{2} \cdot \frac{1}{2} = \frac{1}{4}$$

The expression a^n is read "a to the nth power" or simply "a to the nth." Computing a^n is called "raising a to the nth power." In an expression a^n, a is called the **base** and n is called the **exponent**.

Precedence of Powers

In performing calculations, standard algebra convention dictates that raising to a power has a higher precedence than addition, subtraction, multiplication, or division. That is, all raising to powers occurs before the other operations are performed. In the expression

$$\frac{5}{10} - 3 \cdot 2^3$$

we first perform all raising to powers, proceeding from left to right, to get the expression:

$$\frac{5}{10} - 3 \cdot 8$$

Next, we perform multiplications and divisions, proceeding from left to right, to obtain:

$$0.5 - 24$$

Finally, we perform additions and subtractions, proceeding from left to right, to obtain the result -23.5.

Exponents and Calculators

A scientific calculator uses the $\boxed{y^x}$[2] key to raise a number to a power. For example, to raise 6.13 to the power 4.1 on a scientific calculator, you would use the keystrokes:

$$\boxed{6.13}\,\boxed{y^x}\,\boxed{4.1}\,\boxed{=} \quad \text{Answer: } 1692.730991$$

On a graphing calculator, powers are indicated by the $\boxed{\wedge}$ key. (This notation for powers is common in computer usage.) The above calculation would be done using the keystrokes:

$$\boxed{6.13}\,\boxed{\wedge}\,\boxed{4.1}\,\boxed{\text{ENTER}} \quad \text{Answer: } 1692.730991$$

EXAMPLE 1
Calculating Powers Using a Calculator

Use a calculator to evaluate the following:

1. 1.55^3
2. $3 \cdot 4.1^5$
3. $500 \cdot \left(1 + \frac{0.06}{12}\right)^{5 \cdot 12}$
4. 5^{30}

Solution

Calculation	Scientific Calculator Keystrokes	Graphing Calculator Keystrokes	Answer
1.55^3	$\boxed{1.55}\,\boxed{y^x}\,\boxed{3}\,\boxed{=}$	$\boxed{1.55}\,\boxed{\wedge}\,\boxed{3}\,\boxed{\text{ENTER}}$	3.723875
$3 \cdot 4.1^5$	$\boxed{3}\,\boxed{\times}\,\boxed{4.1}\,\boxed{y^x}\,\boxed{5}\,\boxed{=}$	$\boxed{3}\,\boxed{\times}\,\boxed{4.1}\,\boxed{\wedge}\,\boxed{5}\,\boxed{\text{ENTER}}$	3475.68603
$500 \cdot \left(1 + \frac{0.06}{12}\right)^{5 \cdot 12}$	$\boxed{500}\,\boxed{\times}\,\boxed{(}\,\boxed{1}\,\boxed{+}\,\boxed{0.06}$ $\boxed{\div}\,\boxed{12}\,\boxed{)}\,\boxed{y^x}\,\boxed{(}\,\boxed{5}\,\boxed{\times}$ $\boxed{12}\,\boxed{)}\,\boxed{=}$ Note that it is necessary to enclose the product $5 \cdot 12$ within parentheses in order for the calculator to correctly evaluate the power.	$\boxed{500}\,\boxed{\times}\,\boxed{(}\,\boxed{1}\,\boxed{+}\,\boxed{0.06}$ $\boxed{\div}\,\boxed{12}\,\boxed{)}\,\boxed{\wedge}\,\boxed{(}\,\boxed{5}\,\boxed{\times}\,\boxed{12}$ $\boxed{)}\,\boxed{\text{ENTER}}$ Note that it is necessary to enclose the product $5 \cdot 12$ within parentheses in order for the calculator to correctly evaluate the power.	674.425076
5^{30}	$\boxed{5}\,\boxed{y^x}\,\boxed{30}\,\boxed{=}$ Note that, because the answer contains more than ten digits, the calculator automatically shifts to scientific notation[3] to display the answer.	$\boxed{5}\,\boxed{\wedge}\,\boxed{30}\,\boxed{\text{ENTER}}$ Note that, because the answer contains more than ten digits, the calculator automatically shifts to scientific notation to display the answer.	9.313225746E20

[2] Note that some calculators label this key $\boxed{x^y}$.
[3] See the end of this section for a discussion of scientific notation.

Compound Interest

One of the most important applications of exponents is to compound interest. At this point, let's derive the fundamental formula that underlies all such applications. Suppose that P dollars are deposited in a bank account paying interest at a rate r per year. (The value of r is usually expressed as a percentage.) At the end of a year, the interest earned is:

$$[\text{interest}] = [\text{rate}] \times [\text{amount}] = rP$$

So at the end of a year, the balance in the account is:

$$[\text{balance}] = [\text{old balance}] + [\text{interest}]$$
$$= P + rP$$
$$= P(1 + r)$$

That is, to get the new balance, multiply the old balance by $1 + r$. To get the balance after two years, multiply the balance after one year by $1 + r$ to obtain:

$$[\text{balance after two years}] = P(1 + r) \cdot (1 + r)$$
$$= P(1 + r)^2$$

In a similar fashion, we can obtain the following formula for the balance after n years:

$$[\text{balance after } n \text{ years}] = P(1 + r)^n$$

This formula assumes that the new balance is calculated once a year. In many applications, the new balance is calculated more than once a year, say semiannually (twice a year), quarterly (four times a year), or monthly (twelve times a year). If interest is compounded k times per year, the interest rate for each period (half-year, quarter, or month) is r/k. There are nk interest periods. Using the same reasoning used previously, we have the following formula:

COMPOUND INTEREST FORMULA

Suppose that interest is compounded k times per year and that the annual rate is r. Then the balance after n years is given by the formula:

$$[\text{balance after } n \text{ years}] = P\left(1 + \frac{r}{k}\right)^{nk}$$

EXAMPLE 2
Compound Interest

Suppose that a savings account earns interest at a 3 percent annual rate and that the interest is compounded monthly. The initial balance is $1,000. What is the balance in the account after 2 years?

Solution

Because the annual rate of interest is 3 percent, we have $r = 0.03$. Moreover, since the interest is compounded monthly, we have $k = 12$. The initial amount P is 1,000, and because we wish to find the balance after 2 years, we have $n = 2$. By the compound interest formula, the balance after 2 years is:

$$1{,}000\left(1 + \frac{0.03}{12}\right)^{2 \cdot 12} = 1{,}061.757044$$

(We used a calculator to do the arithmetic.) That is, after 2 years the balance is $1,061.76.

EXAMPLE 3
Rate of Inflation

If price inflation occurs at a constant rate r each year, then after n years an object with original price P will cost $P(1 + r)^n$. Suppose that inflation is 7 percent each year. A car currently costs $10,000. How many years before its price will increase by more than 50 percent?

Solution

By the given formula, with $r = 0.07$ and $P = 10,000$, the cost of the car after n years will be:

$$P(1 + r)^n = 10,000 \cdot (1.07)^n$$

Make a table listing the value of this expression for various values of n (see left margin). We see that the price will first increase by more than 50 percent (to more than $15,000) at the end of year 6.

n	$10,000 \cdot (1.07)^n$
1	$10,700.00
2	$11,449.00
3	$12,250.43
4	$13,107.96
5	$14,025.52
6	$15,007.30
7	$16,057.81

EXAMPLE 4
Expense of a College Education

The cost of a year at a certain college is currently $25,410. Assume that the cost increases 8 percent per year. What will be the cost of a four-year college education?

Solution

Use the inflation formula from the last example. Years are counted according to the number of years after the first year of inflation. That is, the cost of the second year involves one year of inflation, the cost of the third year involves two years of inflation, and so forth. Therefore, the inflation formula for n years corresponds to the cost of year $n + 1$, which is given by the formula:

$$[\text{cost of year } n + 1] = 25,410 \cdot (1 + 0.08)^n$$
$$= 25,410 \cdot (1.08)^n$$

Use this formula to calculate the cost of years 2, 3, and 4, and organize the data in the following table:

Year (n)	Formula for Cost	Cost
1		$25,410.00
2	$25,410 \cdot (1.08)^1$	$27,442.80
3	$25,410 \cdot (1.08)^2$	$29,638.22
4	$25,410 \cdot (1.08)^3$	$32,009.28
Total Cost		$114,500.30

Add the cost of each year to find the total cost for 4 years.

EXAMPLE 5
Mutual Fund Investment

In the year ending Oct. 31, 1993, natural resource mutual funds averaged a return of 15.42 percent. Suppose that you invest $5,000 in such a fund and suppose that its annual return over the next 5 years equaled the average return of 15.42 percent. What would your investment be worth at the end of 5 years?

Solution

The investment would return 15.42 percent compounded annually. By the compound interest formula, the value of the investment after 5 years is:

$$5,000 \cdot \left(1 + \frac{0.1542}{1}\right)^5 = 5,000 \cdot (1.1542)^5$$
$$= \$10,241.78$$

Square Roots and Other Radicals

As preparation for our discussion of rational exponents, let's review the basic facts about square roots and nth roots. We begin with the definition of square root.

Definition
Square Root

Suppose that a is a nonnegative real number. A **square root** of a is a real number b whose square is a. That is, b is a square root of a provided that $b^2 = a$.

For example, the number $a = 4$ has square roots $+2$ and -2, since $(+2)^2 = 4$ and $(-2)^2 = 4$. The number $a = 0$ has the square root 0, since $0^2 = 0$. On the other hand, the number -3 does not have a square root in the real number system, since the square of any real number b is nonnegative and hence cannot equal -3. For similar reasons, if a is any negative number, then a does not have a square root in the real number system.

Zero is the only real number that has only one square root, namely 0. However, each *positive* real number a has two square roots, one positive and the other negative. We define the symbol $\sqrt{a}$ to mean the positive square root. We call this the **principal square root** of a. We also set $\sqrt{0} = 0$. However, if a is negative, the symbol $\sqrt{a}$ is undefined in the real number system. Thus, for example, we have:

$$\sqrt{4} = 2, \qquad \sqrt{\frac{1}{16}} = \frac{1}{4}, \qquad \sqrt{0} = 0$$

$\sqrt{-1}$ is undefined in the real number system.

In the first three examples, the values of $\sqrt{a}$ are rational numbers. However, for many rational numbers a, the value of $\sqrt{a}$ is irrational. For example, it can be proved that $\sqrt{2}$ is irrational. The decimal expression of this real number is nonterminating and nonrepeating, and begins:

$$\sqrt{2} = 1.414213\ldots$$

Definition
nth Root

Let n be a positive integer and a be a real number. An **nth root** of a is a real number b whose nth power equals a. That is,

$$b^n = a$$

For example, a third root of 8 is 2, since $2^3 = 8$. Unlike square roots, a negative real number can have a *real* third root. The third root of -8 is -2, since $(-2)^3 = -8$.

Just as a negative number does not have a square root, a negative number does not have an nth root when n is even. Thus, for example, -2 does not have a fourth root or a sixth root. If a is positive and n is even, then a has two nth roots, which are negatives of each other. The positive value is denoted $\sqrt[n]{a}$ and is called the **principal nth root** of a. Thus, for example, the number 2 has two fourth roots, $\sqrt[4]{2}$ and $-\sqrt[4]{2}$. The numerical value of $\sqrt[4]{2}$ is approximately 1.189. The number 0 has just one nth root, namely 0.

In case n is odd, any real number a has a single nth root, which is denoted $\sqrt[n]{a}$.

We have:

$$\sqrt[3]{8} = 2, \qquad \sqrt[4]{\frac{1}{16}} = \frac{1}{2}, \qquad \sqrt[3]{-\frac{8}{27}} = -\frac{2}{3}$$

SECTION 1.2 Exponents and Radicals

If x is nonnegative, then $\sqrt{x^2}$ equals x. On the other hand, if x is negative, because $\sqrt{x^2}$ is the nonnegative number whose square is x^2, we have:

$$\sqrt{x^2} = -x, \quad x < 0$$

Putting the nonnegative and negative cases together, we have:

$$\sqrt{x^2} = \begin{cases} x & \text{if } x \text{ is nonnegative} \\ -x & \text{if } x \text{ is negative} \end{cases}$$

Recognize the right side of this equation as $|x|$. This gives the following important result:

$$\sqrt{x^2} = |x|$$

The symbols $\sqrt{a}$ and $\sqrt[n]{a}$ are called **radicals**. Manipulations involving radicals can usually be handled using the laws of exponents applied to rational exponents.

Rational Exponents Let a be a real number. To define a^r where r is a rational number, we consider the special case in which r is a number of the form $1/n$ for a positive integer n. In this case, we define $a^{1/n}$ by the formula:

$$a^{1/n} = \sqrt[n]{a}$$

assuming that the radical on the right side of the equation is defined. If the radical on the right side is not defined, then the power $a^{1/n}$ is undefined. Thus, for example,

$$2^{1/2} = \sqrt{2}$$
$$8^{1/3} = \sqrt[3]{8} = 2$$
$$(81)^{1/4} = \sqrt[4]{81} = 3$$

When r is a positive rational number

$$r = \frac{m}{n}$$

where m and n are positive integers and m/n is in lowest terms, we define a^r as:

$$a^r = a^{m/n} = \left(a^{1/n}\right)^m = \left(\sqrt[n]{a}\right)^m = \sqrt[n]{a^m}$$

This formula makes sense only if the radical is defined. Here are some examples of numbers raised to rational exponents:

$$4^{3/2} = \left(\sqrt{4}\right)^3 = 2^3 = 8$$
$$(-8)^{5/3} = \left(\sqrt[3]{-8}\right)^5 = (-2)^5 = -32$$
$$(-2)^{3/2} = \left(\sqrt{-2}\right)^3 = \text{undefined}$$

To calculate the value of a radical using a calculator, it is usually necessary to convert the radical to a rational exponent. For example, we can calculate $\sqrt[3]{4}$ as $4^\wedge(1/3)$. Because of the importance of square roots, however, most calculators have that particular radical available as a special key; in the case of square roots, it is not necessary to convert to a rational exponent.

EXAMPLE 6
The Speed of Sound

The speed of sound varies with the air temperature. At temperature T degrees Celsius, the speed of sound V is given in meters per second by the formula:

$$V = 331.5\left(\frac{T + 273}{273}\right)^{1/2} \text{ m/sec}$$

1. If the air temperature is 20°C, what is the speed of sound?
2. If the air temperature is 100°C, what is the speed of sound?
3. How does the speed of sound change as the temperature rises?

Solution

1. Replace T with 20 in the formula:

$$V = 331.5\left(\frac{20 + 273}{273}\right)^{1/2} \approx 343 \text{ m/sec}$$

2. Replace T with 100 in the formula:

$$V = 331.5\left(\frac{100 + 273}{273}\right)^{1/2} \approx 387.5 \text{ m/sec}$$

3. Compare the results of 1 and 2; it appears that as the temperature rises, the speed of sound increases. This is true in general (as we will see when we graph V versus T in the next chapter).

Negative and Zero Exponents

Let a be a nonzero number. We define the **zero power** of a by the formula:

$$a^0 = 1, \quad a \neq 0$$

Thus $3^0 = 1, (-7)^0 = 1, \left(\frac{2}{3}\right)^0 = 1$, and so forth.

Note, however, that the above definition excludes the case where a is equal to 0. That is, 0^0 is undefined.

Suppose that a is nonzero, r is a rational number for which a^r is defined, and $r > 0$. Then we define a^{-r} by the formula:

$$a^{-r} = \frac{1}{a^r}$$

Thus for example, we have:

$$2^{-3} = \frac{1}{2^3} = \frac{1}{8}$$

$$\left(\frac{1}{7}\right)^{-2} = \frac{1}{(1/7)^2} = \frac{1}{1/49} = 49$$

$$4^{-3/2} = \frac{1}{4^{3/2}} = \frac{1}{\left(\sqrt{4}\right)^3} = \frac{1}{2^3} = \frac{1}{8}$$

Rational exponents occur often in applications, as the following examples show.

EXAMPLE 7
Calcium Injection

In order to measure the rate at which calcium is absorbed by the body, a subject is injected with a radioactive isotope of calcium. At time t days after the injection, the amount of calcium remaining is given by an expression of the form

$$t^{-3/2} \quad (t > 0.5)$$

This expression assumes that the calcium is measured in appropriate units. How many units of calcium remain after 9 days?

Solution

The amount of calcium remaining after 9 days is obtained by evaluating the above formula for $t = 9$:

$$9^{-3/2} = \frac{1}{9^{3/2}}$$

$$= \frac{1}{\left(\sqrt{9}\right)^3}$$

$$= \frac{1}{3^3}$$

$$= \frac{1}{27}$$

That is, after 9 days there remains 1/27 unit of calcium.

The Laws of Exponents

In order to manipulate algebraic expressions involving exponents, it is critical that you understand the following rules.

LAWS OF EXPONENTS

Let a and b be real numbers and r and s rational numbers. Then the following laws of exponents hold, provided that all of the expressions appearing in a particular equation are defined.

1. $a^r \cdot a^s = a^{r+s}$
2. $(a^r)^s = a^{rs}$
3. $(ab)^r = a^r b^r$
4. $\dfrac{a^r}{a^s} = a^{r-s}$
5. $\left(\dfrac{a}{b}\right)^r = \dfrac{a^r}{b^r}$
6. $a^{-r} = \dfrac{1}{a^r}$
7. $\left(\dfrac{a}{b}\right)^{-r} = \left(\dfrac{b}{a}\right)^r$

We will assume that the above laws of exponents hold without providing any proofs.

EXAMPLE 8 Using Laws of Exponents

Use the laws of exponents to obtain equivalent expressions that involve no parentheses.

1. $(s^5)^2$
2. $(ab)^4$
3. $(x^2)^3 x^5$

Solution

1. By Law 2, we have:

$$(s^5)^2 = s^{5 \cdot 2} = s^{10}$$

2. By Law 3,
$$(ab)^4 = a^4 b^4$$

3. Combining Laws 2 and 1,
$$(x^2)^3 x^5 = x^{2 \cdot 3} x^5 = x^{6+5} = x^{11}$$

EXAMPLE 9
Eliminating Negative and Zero Exponents

Write the following expressions in a form that does not involve any negative or zero exponents.

1. $(x^3 y^{-2})^2$
2. $\left(\dfrac{x^{-5}}{x^{-3}}\right)^{-3}$

Solution

1. $(x^3 y^{-2})^2 = x^{3 \cdot 2} y^{(-2)2}$ *Laws 2 and 3*

$\qquad\qquad\quad= x^6 y^{-4}$

$\qquad\qquad\quad= x^6 \cdot \dfrac{1}{y^4}$ *Law 6*

$\qquad\qquad\quad= \dfrac{x^6}{y^4}$

2. $\left(\dfrac{x^{-5}}{x^{-3}}\right)^{-3} = \dfrac{x^{(-5) \cdot (-3)}}{x^{(-3) \cdot (-3)}}$ *Laws 2 and 5*

$\qquad\qquad\quad= \dfrac{x^{15}}{x^9}$

$\qquad\qquad\quad= x^{15-9}$ *Law 4*

$\qquad\qquad\quad= x^6$

EXAMPLE 10
Eliminating Negative Exponents and Radicals

Write the following expressions in a form that involves no negative exponents or radicals.

1. $(x^{1/2})^{-4}$
2. $\sqrt{\dfrac{x}{x^{-3/2}}}$

Solution

1. By Laws 2 and 6, we have:
$$(x^{1/2})^{-4} = x^{1/2 \cdot (-4)}$$
$$= x^{-2}$$
$$= \dfrac{1}{x^2}$$

2. By the definition of a rational exponent, we have:
$$\sqrt{\dfrac{x}{x^{-3/2}}} = \left(\dfrac{x}{x^{-3/2}}\right)^{1/2}$$
$$= \dfrac{x^{1/2}}{(x^{-3/2})^{1/2}}$$

$$= \frac{x^{1/2}}{x^{-3/4}}$$
$$= x^{(1/2)-(-3/4)}$$
$$= x^{5/4}$$

As a consequence of the laws of exponents, we deduce the following useful properties of radicals.

PROPERTIES OF RADICALS

1. $\sqrt[n]{a}\,\sqrt[n]{b} = \sqrt[n]{ab}$
2. $\sqrt[n]{\dfrac{a}{b}} = \dfrac{\sqrt[n]{a}}{\sqrt[n]{b}}, \quad b \neq 0$
3. $\sqrt[n]{a^n} = |a|, \quad n$ even
4. $\sqrt[n]{a^n} = a, \quad n$ odd

Note that these formulas hold only for values of a, b, and n for which all the radicals appearing are defined.

The next example illustrates the use of these formulas in manipulating radicals.

EXAMPLE 11
Simplifying Radical Expressions

Simplify the following expressions using the properties of radicals.

1. $\sqrt[3]{64y^3z^9}$
2. $\sqrt[3]{18x^4}\,\sqrt[3]{12x^2}$
3. $\dfrac{\sqrt[3]{54x^3}}{\sqrt[3]{16x^2}}$

Solution

1. Properties 1 and 4 of radicals give:
$$\sqrt[3]{64} \cdot \sqrt[3]{y^3} \cdot \sqrt[3]{z^9} = 4yz^3$$

2. Property 1 of radicals and Law 1 of exponents give:
$$\sqrt[3]{18x^4 \cdot 12x^2} = \sqrt[3]{3^2 \cdot 2 \cdot 3 \cdot 2^2 x^6}$$
$$= \sqrt[3]{2^3 3^3 x^6}$$
$$= 2 \cdot 3x^2$$
$$= 6x^2$$

3. Property 2 of radicals yields:
$$\sqrt[3]{\frac{54x^3}{16x^2}} = \sqrt[3]{\frac{27x}{8}} = \frac{3}{2}\sqrt[3]{x}$$

Scientific Notation

Powers of 10 are commonly used in scientific work, especially in describing numbers that are very large or very small. For example, the distance light travels in a year is approximately 10^{16} meters. The diameter of a DNA molecule is approximately 10^{-8} meters. To get a feel for these powers, review the following table containing the first few positive and negative powers of 10.

n	10^n	n	10^n
0	1	-1	0.1
1	10	-2	0.01
2	100	-3	0.001
3	1000	-4	0.0001
4	10000	-5	0.00001

Multiplication by 10^n is equivalent to shifting the decimal point n places. If n is positive, the shift is to the right, whereas if n is negative, the shift is to the left. Here are some examples:

$$15.354 \times 10^3 = 15{,}354 \text{ (shift right 3 places)}$$
$$15.354 \times 10^{-3} = 0.015354 \text{ (shift left 3 places)}$$

By shifting the decimal point an appropriate number of places, we can write any number as a power of 10 multiplied by a decimal, with exactly one nonzero digit to the left of the decimal point. For example,

$$15.354 = 1.5354 \times 10^1$$
$$0.0058723 = 5.8723 \times 10^{-3}$$
$$1{,}487 = 1.487 \times 10^3$$

This form of writing a decimal is called **scientific notation** and is commonly used in science. Its advantage is that very large and very small numbers can be expressed in a concise format. For instance, 1 followed by 100 zeros can be written in scientific notation as:

$$1.0 \times 10^{100}$$

A decimal point followed by 99 zeros and a 1 can be written in scientific notation as:

$$1.0 \times 10^{-100}$$

In scientific notation, the initial decimal is called the **mantissa** and the power of 10 is called the **exponent**. According to the laws of exponents, to multiply numbers written in scientific notation, we multiply mantissas and add the exponents. For example,

$$(2.0 \times 10^5) \times (3.5 \times 10^{-2}) = [(2.0) \times (3.5) \times (10^5 \times 10^{-2})] = 7.0 \times 10^3$$

Calculators allow you to enter numbers in scientific notation. Just enter the mantissa as usual, and then press the **Enter Exponent** key, often labeled $\boxed{\text{EE}}$, followed by the exponent. For example, to enter the number 5.78^{-5} on a graphing calculator, we would use the following keystrokes:

EXAMPLE 12
Gravitation

Newton's law of universal gravitation asserts that the force F of gravitational attraction between two bodies of masses m_1 and m_2 that are at a distance r from one another is given by the formula

$$F = G\frac{m_1 m_2}{r^2}$$

where G is the universal gravitational constant, which in the metric system is:
$$G = 6.670 \times 10^{-11}$$
In the metric system, F is measured in newtons, r in meters, and m_1 and m_2 in kilograms. The mass of the Earth is approximately 5.96×10^{24} kg. Determine the gravitational force exerted by the Earth on an asteroid of mass 10^{18} kg that lies at a distance of 10^{10} meters from the center of the Earth.

Solution

From the given formula,
$$F = G\frac{m_1 m_2}{r^2}$$
$$= (6.670 \times 10^{-11})\frac{(5.96 \times 10^{24}) \cdot (10^{18})}{(10^{10})^2}$$

Applying the laws of exponents to multiply the powers of 10, we have:
$$F = (6.670) \cdot (5.96) \cdot \frac{10^{-11+24+18}}{10^{20}}$$
$$= 39.75 \times \frac{10^{31}}{10^{20}}$$
$$= 39.75 \times 10^{31-20}$$
$$= 3.975 \times 10^{12} \text{ newtons}$$

In computer or calculator work, the multiplication sign and the 10 in scientific notation are replaced by the letter **E**, which stands for *exponent*. For instance, the number
$$1.275 \times 10^{-3}$$
is written for computer or calculator use as:
$$1.275\text{E} - 3$$

Exercises 1.2

Simplify the following expressions; write them in a form that does not involve any negative or zero exponents or any parentheses, and that contains as few multiplications as possible.

1. $a^{-5} \cdot a^2$
2. $t^{-7} \cdot t^7$
3. $q^3 \cdot q^{-6} \cdot q^7$
4. $m^{-3} \cdot m^{-5} \cdot m^{20}$
5. $(3x^5)(-4x^6)$
6. $(6y^{-4})(8y^{-10})$
7. $(4x^3 y^3)(9x^{-3} y^{-5})$
8. $(-12p^{-6}t^{11})(-5p^8 t^{-7})$
9. $(3ab^4)^2$
10. $(-5x^{-2} y^4)^{-3}$
11. $(2y)^4 (3x)^4$
12. $(2ab)^3 (3ab)^2$
13. $\dfrac{t^{-12}}{t^{-7}}$
14. $\dfrac{a^{10}}{a^8}$
15. $\dfrac{a^3 b^{-3}}{a^{-2} b}$
16. $\dfrac{10x^4 y^{-4}}{-2x^{-1} y^3}$
17. $(-3x^2 b^{-4})^{-2}$
18. $(-5p^{-3} q^4 r)^{-3}$
19. $\left(\dfrac{2a^3 b^{-2}}{3a^4 b^{-3}}\right)^3$
20. $\left(\dfrac{-4p^5 y^{-2}}{5p^{-1} q^6}\right)^{-4}$
21. $\dfrac{(4x^2 y^{-1} z^{-3})^{-2}}{(8xy^{-1} z^2)^{-1}}$
22. $\dfrac{(10^{-1} a^{-4} c^5)^{-2}}{(5^{-2} a^3 c^6)^{-3}}$
23. $(-2)^{-3} \cdot 5^2$
24. $-6^2 - 3^2$
25. $\dfrac{(-3)^2}{3^3}$
26. $(2^2 + 3^2)^2$
27. $(5^2 - 4^2)^2$
28. $(2^3 - 5^2)^2$
29. $\dfrac{(-2)^{-2} + (-2)^{-3}}{(-2)^5}$
30. $\dfrac{(-1)^5 + (-2)^{-3}}{(-2)^2}$

Simplify the following expressions so that the result contains no radicals or fractional exponents.

31. $\sqrt[3]{125}$

32. $\sqrt[5]{32}$
33. $8^{5/3}$
34. $81^{3/4}$

Write the following expressions in a form that involves no negative exponents or radicals. Assume that the values of all variables are positive.

35. $\sqrt[3]{y^2}$
36. $\sqrt[5]{t^3}$
37. $\sqrt[6]{y^{-17}}$
38. $\sqrt{(x^2-5)^{-1}}$
39. $(a^{4/5})^{-20}$
40. $(x^{-1/3})^{-12}$
41. $\sqrt{\dfrac{x}{x^{4/5}}}$
42. $\sqrt[3]{\dfrac{a}{a^{-2/3}}}$
43. $\sqrt[3]{a^{-18}}$
44. $\sqrt{p^4 q^8}$

Simplify the following expressions. Assume that the variables represent positive numbers.

45. $\sqrt{49x^2}$
46. $\sqrt{36t^2}$
47. $\sqrt{64(t+1)^2}$
48. $\sqrt{49x^6}$
49. $\sqrt{3x} \cdot \sqrt{6x}$
50. $\sqrt{8y} \cdot \sqrt{2y}$
51. $\sqrt{4x^2 y^4}$
52. $\sqrt[3]{64z^6}$
53. $\sqrt{98a^2 b^{-6}}$
54. $\sqrt[5]{-1 \cdot x^{10} y^{-15}}$
55. $\sqrt[3]{9x^2 y}\,\sqrt[3]{6xy^3}$
56. $\sqrt[4]{90ab^3}\,\sqrt[4]{16ab}$
57. $\dfrac{\sqrt[3]{54t}}{\sqrt[3]{2t}}$
58. $\dfrac{\sqrt{80a}}{\sqrt{5a}}$
59. $\dfrac{\sqrt{3(2x^2)^3}}{\sqrt{6x}}$
60. $\dfrac{\sqrt[3]{9xy}}{\sqrt[3]{3x^{-2}}}$
61. $\dfrac{\sqrt[3]{625 a^4 b^7}}{\sqrt[3]{5ab^2}}$
62. $\dfrac{\sqrt[3]{40x^5 y^2}}{\sqrt[3]{2xy}}$
63. $\sqrt{2(x+1)^2} \cdot \sqrt{4(x+1)}$
64. $\sqrt[3]{2(y-1)^2}\,\sqrt[3]{4(y-1)^6}$
65. $\sqrt[4]{\dfrac{243 a^6 b^{-13} c^{15}}{3 a^2 b^{-9} c^7}}$
66. $\sqrt[5]{\dfrac{160 x^9 y^{12}}{5xy^2}}$

Simplify the following expressions.

67. $\sqrt{\sqrt{\sqrt{\sqrt{65{,}536}}}}$
68. $\dfrac{(-1)^{n+3}}{(-1)^{n+1}}$
69. $\left(\sqrt[12]{\sqrt[6]{a^{80}}}\right)^9 \left(\sqrt[12]{\sqrt[6]{a^{80}}}\right)^9$
70. $\left[\dfrac{(5x^a y^b)^4}{(-5x^a y^b)^3}\right]^5$

Applications

Express the following in scientific notation.

71. **Coffee consumption.** 2,000,000,000 lb (the amount of coffee consumed annually in the United States)
72. **Light year.** 5,878,000,000,000 miles (the distance that light travels in one year. Also known as 1 light-year.)
73. **Universal gravitational constant.** 0.00000000006672 (in N-m^2/kg^2)
74. **Electron-volt.** 0.00000000000000000016 (in ft-lb)

Convert the following to decimal notation.

75. **Gambling.** $\$1.2 \times 10^{10}$ (the amount of money spent on lotteries in 1989)
76. **Time it takes a bullet cap to explode.** 10^{-6} sec
77. 5.78E$-$18
78. 7.02E16

Solve the following problems and give the answers in scientific notation.

79. **Astronomy.** It takes light 2,200,000 years to travel to Earth form the constellation Andromeda. How far is

it from Earth to Andromeda? (See Exercise 72 for the value of the speed of light.)

80. **Astronomy.** The sun is 93,000,000 miles from the earth. How long does it take for light to travel to Earth from the sun?

81. **Astronomy.** The moon travels about 1,500,078 miles in one orbit around the earth. About how far is the moon from the center of the earth? (Assume that the orbit of the moon is circular.)

82. **Imported oil.** The United States imported 8 million barrels of oil each day in 1992. How many barrels of oil did it import in the entire year?

83. **Mutual fund investment.** For the year ending Oct. 31, 1993, health and technology mutual funds returned an average of 3.68 percent a year. Suppose you invested $7,500 in such a fund that registered the same performance. What would be the value of your investment at the end of three years?

84. **Mutual fund investment.** On Nov. 11, 1993, the average return on the Dreyfus Short Term Income Fund for the preceding 30 days was 5.30 percent as an annual rate. Suppose that this return remains constant and that you invest $10,000. Assume that the interest is compounded monthly. What would be the value of your investment in 3 months?

85. **Compound interest.** Suppose that a savings account earns interest at a 4 percent annual rate and that the interest is compounded monthly. The initial balance is $5,000. What is the balance in the account after 2 years?

86. **Compound interest.** Suppose that a certificate of deposit earns interest at a 5 percent annual rate and that the interest is compounded monthly. The initial balance is $1,000. What is the balance in the account after 3 years?

87. **Inflation.** Suppose that inflation is 3 percent each year. A luxury car currently costs $40,000. How many years before its price will increase to more than $50,000?

88. **Inflation.** Suppose that inflation is 5 percent each year. A house currently costs $150,000. How many years before its price will increase to more than $225,000?

89. **Cost of college.** The cost of tuition at a certain college is currently $11,000 a year. Assume that the cost increases 8 percent each year. What will be the cost of a five-year stay at the college?

90. **Cost of college.** Suppose that the cost of tuition at a certain college is currently $10,000 each year. Assume the cost increases by 5 percent, 4 percent, and 3 percent in the second, third, and fourth years, respectively. What will be the cost of a four-year stay at the college?

91. **Effective rate of interest.** Suppose that an investment of initial amount P grows to value A after n years. The **effective rate of interest** r of the investment equals the rate of interest compounded annually that will produce the value of the investment in n years. The effective rate of interest can be calculated using the formula:
$$r = \left(\frac{A}{P}\right)^{1/n} - 1$$
Suppose that $2,000 grows to $3,500 after 10 years in a savings account, with interest compounded annually. What is the annual rate of interest?

92. **Effective rate of interest.** Refer to Exercise 91. A real estate investment of $2,000,000 grows to $3,000,000 in 2 years. What is the effective rate of interest on the investment?

93. **Effective rate of interest.** Refer to Exercise 91. A $100 investment in a promising biotechnology company is worth $50,000 after 10 years. What is the effective rate of interest?

94. **Effective rate of interest.** Refer to Exercise 91. Peter Minuit purchased Manhattan from Native Americans for $24 worth of jewelry in 1620. Suppose that the current value of all real estate in Manhattan is $5 trillion. What is the effective rate of interest on the investment? Who got the better side of the bargain?

95. **Calcium absorption.** Using the model of Example 7, determine the amount of calcium remaining after 16 days.

96. **Calcium absorption.** Using the model of Example 7, determine the amount of calcium remaining after 25 days.

97. **Baseball.** A baseball diamond is a square whose sides have length 90 ft. It is 60.5 ft from home plate to the pitcher's mound. (See figure.) About how far is it from the pitcher's mound to second base? (Use the Pythagorean equation for right triangles, $a^2 + b^2 = c^2$ or $c = \sqrt{a^2 + b^2}$, where c is the length of the hypotenuse and a and b are the lengths of the legs.)

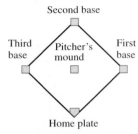

Exercise 97

98. **Geometry.** A circle with area 113.04 m² has a square inscribed inside it. (See figure.) Find the area of the square.

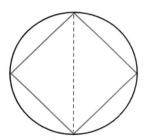

Exercise 98

The **escape velocity** V_0 of a projectile is the initial velocity needed in order for it to escape the gravitational pull of the planet. Escape velocity (in m/sec) is given by

$$V_0 = \sqrt{\frac{2GM}{R}}$$

where G is the universal gravitational constant 6.672×10^{-11} newtons · m²/kg², M is the mass of the planet in kilograms, and R is its radius in meters.

99. **Escape velocity.** The mass of the earth is 5.97×10^{24} kg and its radius is 6.37×10^6 m. What velocity is necessary for the rocket to escape the pull of the earth?

100. **Escape velocity.** The mass of Mars is 6.57×10^{23} kg and its radius is 3.45×10^6 m. What is the escape velocity for a rocket leaving Mars?

101. **Falling objects.** The time t, in seconds, it takes an object to fall a distance d, in feet, is given by $t = \frac{1}{4}\sqrt{d}$. Television station KTHI in Fargo, ND, has a transmitting antenna that is 2,063 ft high. A tool is dropped from the top by a repairperson. (See figure.) How long will it take to fall to the ground?

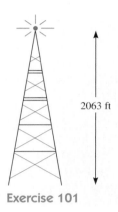

Exercise 101

102. **Heron's formula.** Heron's formula for the area A of a triangle (see figure) is given by

$$A = \sqrt{s(s-a)(s-b)(s-c)}$$

where $s = \frac{1}{2}(a + b + c)$, and a, b, and c are the lengths of the sides of the triangle. What is the area of a triangle with sides of lengths 5.5 ft, 7.2 ft, and 2.1 ft?

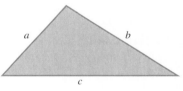

Exercise 102

103. **Speed of sound.** On a warm day at the beach the temperature is 25°C. What is the speed of sound on such a day? See Example 6.

104. **Speed of sound.** On a cold winter day the temperature is −3°C. What is the speed of sound on such a day? See Example 6.

Technology

Use your calculator to determine the values of the following expressions.

105. 15^5

106. 7^4

107. $5.11^{4.1}$

108. $119^{3.7}$

109. $12^{3/4}$

110. $81^{5/4}$

111. $\sqrt{5.897}$

112. $\sqrt{105,891}$

113. $\sqrt[3]{5.001}$

114. $\sqrt[3]{90.7}$

115. $41^{7/8}$

116. $105^{101/300}$

117. $(1.55\text{E}10)^{3/4}$

118. $(5\text{E}-12)^{5/3}$

119. $\sqrt{(8\sqrt{17-\pi})^2 - (-\sqrt{3}+\sqrt{\pi})^2}$

120. $\left[\dfrac{4\pi^2(1.069 \times 10^5 \times \frac{10,784}{2})}{(6.673 \times 10^{-8})(5.976 \times 10^{27})} \right]^{1/2}$

In your own words

121. Discuss the merits of scientific notation. When would you find scientific notation more convenient than standard decimal notation? When would you find it less convenient?

122. Consult your calculator manual to determine how to enter a number in scientific notation. Describe the procedure in your own words.

123. Consult your calculator manual to determine its upper and lower limits on a number in scientific notation. Test these limits by attempting to enter a number outside them. What happens?

124. Explain the concept of the square root of a number.

125. State the laws of exponents and give a concrete example of each law using rational exponents.

1.3 POLYNOMIALS

An **algebraic expression** is obtained by combining arithmetic operations and by forming radicals using real numbers and variables. Here are some examples of algebraic expressions involving the single variable x:

$$3x + 5, \qquad \frac{x^2 - 3}{\sqrt{x}}, \qquad \sqrt[3]{3x^2 - 3x + \frac{2}{\sqrt{x}}}$$

Here are some algebraic expressions involving several variables:

$$xy + y^2, \qquad -13xzy, \qquad \pi r^2 h$$

Polynomial Expressions in a Single Variable

A **monomial** in the variable x is an algebraic expression of the form

$$ax^n$$

where a is a constant and n is a nonnegative integer. The number a is called the **coefficient** of the monomial and n is called its **degree**. Some examples of monomials are:

$$5x^2, \qquad \frac{1}{3}x^9, \qquad 5x^0 \quad \text{or} \quad 5$$

To multiply two monomials in the same variable x, we apply the commutative and associative laws of multiplication along with the first law of exponents to obtain:

$$ax^n \cdot bx^m = abx^{m+n}$$

That is, to multiply monomials, multiply the coefficients and add the degrees. For example,

$$3x^2 \cdot 4x^5 = (3 \cdot 4)x^{2+5} = 12x^7$$

EXAMPLE 1
Multiplying Monomials

Calculate the product of the following monomials:

1. $(-5x^2) \cdot (10x^3)$
2. $\left(\dfrac{1}{2}x^{10}\right) \cdot (3x)$

Solution

1. $(-5x^2) \cdot (10x^3) = ([-5] \cdot 10) \cdot x^{2+3} = -50x^5$
2. $\left(\frac{1}{2}x^{10}\right) \cdot (3x) = \left(\frac{1}{2} \cdot 3\right) \cdot x^{10+1} = \frac{3}{2}x^{11}$

Definition
Polynomial Expression

A **polynomial expression** in x (**polynomial** for short) is a sum of monomials in x. That is, a polynomial in x is an expression of the form:

$$a_n x^n + a_{n-1} x^{n-1} + \cdots + a_1 x + a_0$$

where n is a nonnegative integer. The real numbers $a_n, a_{n-1}, \ldots, a_1, a_0$ are called the **coefficients** of the polynomial.

Here are some examples of polynomials in the variable x:

$$4x^3 - 2x^2 + x - 5, \qquad -5x + 3, \qquad x^2 + 1, \qquad -1$$

It is customary to omit any terms that have coefficient 0. For example, we write:

$$3x^2 - 1$$

rather than

$$3x^2 + 0x - 1$$

Note that all of the exponents appearing in a polynomial must be nonnegative integers. Thus, the following are *not* polynomials:

$$2x^{-2} + 1, \qquad 5x^{1/2} + x$$

The monomials that form a polynomial are called its **terms**. The nonzero term of highest degree is called the **leading term** of the polynomial, its coefficient is the **leading coefficient** of the polynomial, and its degree is the **degree of the polynomial**. The term corresponding to the zero power of the variable is called the **constant term**.

For example, the polynomial $-3x^2 + 2x + 1$ is of degree 2, has leading coefficient -3, has leading term $-3x^2$, and has constant term 1. The terminology is illustrated in Figure 1.

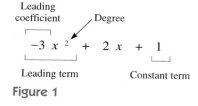

Figure 1

EXAMPLE 2
Identifying the Degree and Leading Term of a Polynomial

Determine the degrees, leading terms, and leading coefficients of the following polynomials:

1. $5x^3 - 3x + 1$
2. $x^{10} + 1$
3. $4 - 9x + 2x^2$

Solution

1. Degree = 3, leading term = $5x^3$, leading coefficient = 5.
2. Degree = 10, leading term = x^{10}, leading coefficient = 1.
3. Degree = 2, leading term = $2x^2$, leading coefficient = 2.

Two polynomials P and Q are said to be **equal** provided that the polynomials are of the same degree and that their corresponding coefficients are equal. The polynomial with all coefficients 0 is called the **zero polynomial** and is denoted 0. Mathematicians agree not to assign this polynomial a degree.

Special terms are used to describe polynomials of low degrees. A polynomial of degree 0 has the form a, where a is a nonzero constant and is called a **constant polynomial**. A polynomial of degree 1 has the form $ax + b$ where a and b are constants and $a \neq 0$. Such a polynomial is called **linear**. A polynomial of degree 2 has the form $ax^2 + bx + c$, where a, b, and c are constants. Such a polynomial is called **quadratic**. Polynomials of degrees three and four are called **cubic** and **quartic**, respectively.

Polynomials in Several Variables

The preceding discussion is concerned exclusively with monomials and polynomials in a single variable. Let us now consider monomials and polynomials in several variables. A monomial in several variables is a product of a constant and nonnegative integer powers of some variables. Here are examples of monomials in variables x, y, and z:

$$-3xyz, \qquad 14x^2y^3, \qquad 5xy^3z^9$$

The degree of a monomial in several variables is the sum of the exponents of the variables that appear in the monomial. For instance, the degree of xyz is $1 + 1 + 1 = 3$, and the degree of the monomial x^2yz^3 is $2 + 1 + 3 = 6$.

A polynomial in several variables is a sum of monomials in several variables. Here are some examples of polynomials in the variables x, y, and z:

$$-3xyz + 14x^2y^3 - 5xy^3z^9, \qquad x + y + z, \qquad xy + yz + xz^8$$

The degree of a polynomial in several variables is the highest degree of the monomials that appear. For instance, the degrees of the three polynomials above are 13, 1, and 9, respectively.

EXAMPLE 3 Degree of a Polynomial in Two Variables

What is the degree of the polynomial $3x^2y - 2xy^2 + x^3 - 4x^3y^2$?

Solution

The term with the highest degree is $-4x^3y^2$ with degree $3 + 2 = 5$. So the polynomial has degree 5.

Addition and Subtraction of Polynomials

Two terms of a polynomial are said to be **like**, or **similar**, if they have the same variables raised to the same powers. For example, the terms $-2xy^2$ and $4xy^2$ are similar, but the terms $4xy^2$ and $4x^2y$ are not similar.

Polynomials may be added and subtracted by combining similar terms.

EXAMPLE 4 Adding and Subtracting Polynomials

Suppose that $P = 2x^2 + 3x - 1$ and $Q = -3x^2 + 4x - 4$. Find $P + Q$ and $P - Q$.

Solution

$$P + Q = [2 + (-3)]x^2 + (3 + 4)x + [(-1) + (-4)]$$
$$= -x^2 + 7x - 5$$

$$P - Q = [2 - (-3)]x^2 + (3 - 4)x + [(-1) - (-4)]$$
$$= 5x^2 - x + 3$$

Multiplying Polynomials and Other Algebraic Expressions

To multiply two polynomials, we repeatedly apply the distributive and associative laws and the laws of exponents to multiply every term of the first polynomial by every term of the second polynomial. We then combine like terms to obtain the product. This procedure is illustrated in the following examples.

EXAMPLE 5
Multiplication of Polynomials

Calculate the following product:
$$3x^2(4x^3 - 3x^2 - 1)$$

Solution

By the distributive law, the desired product can be obtained by multiplying the monomial by each term of the polynomial and adding the resulting terms.
$$3x^2(4x^3 - 3x^2 - 1) = (3x^2 \cdot 4x^3) + [3x^2 \cdot (-3x^2)] + [3x^2 \cdot (-1)]$$
$$= 12x^5 - 9x^4 - 3x^2$$

EXAMPLE 6
More Multiplication of Polynomials

Determine the following product:
$$(x^2 - 1)(2x^2 - x - 3)$$

Solution

Apply the distributive law to the second factor, $2x^2 - x - 3$. Then, simplify the product.
$$(x^2 - 1)(2x^2 - x - 3) = x^2(2x^2 - x - 3) - 1(2x^2 - x - 3)$$
$$= 2x^4 - x^3 - 3x^2 - 2x^2 + x + 3$$
$$= 2x^4 - x^3 - 5x^2 + x + 3$$

Product Identities

Many products occur so often that it is useful to memorize them.

PRODUCT IDENTITIES

Square of a Binomial Sum: $(A + B)^2 = A^2 + 2AB + B^2$
Square of a Binomial Difference: $(A - B)^2 = A^2 - 2AB + B^2$
Product of a Binomial Sum and Difference:
$(A + B)(A - B) = (A - B)(A + B) = A^2 - B^2$
Product of Two Binomials: $(A + B)(C + D) = AC + AD + BC + BD$

The last formula is often referred to by the acronym **FOIL**, since the products on the right result from multiplying the First, Outside, Inside, and Last terms on the left.

In the above formulas, A and B may be replaced by any algebraic expression. For instance, suppose that we wish to calculate the product:
$$(3x^2 - 4y)(3x^2 + 4y)$$

We immediately recognize this as the product of a binomial sum and difference $(A - B)(A + B)$ with A replaced by $3x^2$ and B replaced by $4y$. So we apply the Product of a Binomial Sum and Difference formula. We obtain the result:
$$A^2 - B^2 = (3x^2)^2 - (4y)^2 = 9x^4 - 16y^2$$

We can also apply the various product identities with A and B replaced by expressions involving radicals, as the following example shows.

SECTION 1.3 Polynomials

EXAMPLE 7
Applying Product Identities

Determine the following products.

1. $\left(\sqrt{x} + \dfrac{1}{\sqrt{x}}\right)^2$
2. $\left(\sqrt{x-1} - 2\right)\left(\sqrt{x-1} + 2\right)$
3. $\left(2\sqrt{x} - 3x\right)\left(\dfrac{2}{x} + x^4\right)$

Solution

1. By the Square of a Binomial Sum formula, with A replaced by $\sqrt{x}$ and B replaced by $1/\sqrt{x}$, we have:

$$\left(\sqrt{x} + \dfrac{1}{\sqrt{x}}\right)^2 = \left(\sqrt{x}\right)^2 + 2 \cdot \left(\sqrt{x}\right) \cdot \left(\dfrac{1}{\sqrt{x}}\right) + \left(\dfrac{1}{\sqrt{x}}\right)^2$$

$$= x + \dfrac{2\sqrt{x}}{\sqrt{x}} + \dfrac{1}{x}$$

$$= x + 2 + \dfrac{1}{x}, \qquad x > 0$$

2. Apply the Product of a Binomial Sum and Difference formula, with A replaced by $\sqrt{x-1}$ and B replaced by 2. We then derive the product:

$$\left(\sqrt{x-1} - 2\right)\left(\sqrt{x-1} + 2\right) = \left(\sqrt{x-1}\right)^2 - 2^2$$

$$= (x - 1) - 4$$

$$= x - 5$$

3. Apply the Product of Two Binomials formula, with $A = 2\sqrt{x}$, $B = -3x$, $C = 2/x$, and $D = x^4$. In multiplying the first, outside, inside, and last terms, we use the laws of exponents:

$$\left(2\sqrt{x} - 3x\right)\left(\dfrac{2}{x} + x^4\right) = 4\dfrac{\sqrt{x}}{x} + 2x^4 \cdot \sqrt{x} - 6\dfrac{x}{x} - 3x^4 \cdot x$$

$$= \dfrac{4x^{1/2}}{x} + 2x^{1/2} \cdot x^4 - 6 - 3x^5$$

$$= \dfrac{4}{x^{1/2}} + 2x^{9/2} - 6 - 3x^5, \qquad x > 0$$

Applications of Polynomials

The following applied examples use the various algebraic operations among polynomials that we introduced above.

EXAMPLE 8
Compound Interest

Suppose that $50 is deposited in a savings account that pays compound interest at a rate r, compounded annually. Write the formula for the balance in the account after 2 years as a polynomial in r.

Solution

By the compound interest formula (page 16), the balance in the account after n years is:

$$50\left(1 + \dfrac{r}{1}\right)^{1 \cdot n} = 50(1 + r)^n$$

So the balance after 2 years is:
$$50(1 + r)^2$$
We can write $(1 + r)^2$ as a polynomial by multiplying $1 + r$ by itself. Using the square of a binomial formula, the result is:
$$(1 + r)^2 = 1 + r + r + r^2$$
$$= 1 + 2r + r^2$$
So the balance in the account after 2 years is given by the formula:
$$50(1 + r)^2 = 50(1 + 2r + r^2)$$
$$= 50 + 100r + 50r^2$$

EXAMPLE 9
UPS Shipping Requirement

United Parcel Service requires that the girth of a package (distance around a package horizontally) plus the height of the package must total no more than 130 inches. Suppose that a rectangular package has length equal to twice its width and height equal to one tenth the square of its width. Denote the width of the package by w.

1. Express the girth plus the height of the package as a polynomial in w.
2. Suppose that the width of the package is 10 inches. Is the package acceptable?
3. Suppose that the width of the package is 20 inches. Is the package acceptable?

Solution

1. The girth of the package is:
$$[\text{girth}] = 2 \times [\text{length}] + 2 \times [\text{width}]$$
$$= 2 \cdot (2w) + 2w$$
$$= 6w$$

$$[\text{girth}] + [\text{height}] = [\text{girth}] + \frac{1}{10}[\text{width}]^2$$
$$= \frac{1}{10}w^2 + 6w$$

2. If $w = 10$,
$$[\text{girth}] + [\text{height}] = \frac{1}{10}w^2 + 6w$$
$$= \frac{1}{10}(10)^2 + 6(10)$$
$$= 70$$

Because the number is less than 130, the package is acceptable.

3. If $w = 20$,
$$[\text{girth}] + [\text{height}] = \frac{1}{10}w^2 + 6w$$

$$= \frac{1}{10}(20)^2 + 6(20)$$
$$= 160$$

Because the number is more than 130, the package is not acceptable.

EXAMPLE 10
Determining a Revenue Function

A company produces sweatshirts with college insignias. If the price is x dollars for each sweatshirt, then the number actually sold (in thousands) is given by the expression:
$$N = 57 - x$$
How much revenue does the company get by selling sweatshirts at x dollars?

Solution

$$[\text{Revenue}] = [\text{number sold}] \times [\text{price}]$$
$$= Nx$$
$$= (57 - x)x$$
$$= 57x - x^2$$

Factoring

Factoring is the process of writing a polynomial as a product of polynomials. For example, the polynomial $x^2 - x$ can be written as the product $x(x-1)$. Factoring is an essential ingredient in the solution of many algebraic problems, such as finding the solutions to polynomial equations.

The simplest factors of a polynomial are monomial factors. In factoring a polynomial, we start by finding common monomial factors. This process, called removing the **greatest common factor**, is illustrated in the following example.

EXAMPLE 11
Finding Common Monomial Factors

Find common monomial factors and write each polynomial as a product of factors.
1. $3x^3 - 24x^2 - 18x$
2. $\frac{1}{2}x^2 - 4x$

Solution

1. Each term of the polynomial contains the factor $3x$, so we can write the polynomial in the form:
$$3x^3 - 24x^2 - 18x = 3x \cdot x^2 - 3x \cdot 8x - 3x \cdot 6$$
$$= 3x(x^2 - 8x - 6)$$

2. In factoring, it is always a good strategy to factor out fractional coefficients as part of a monomial. In this example, factoring out the fraction $\frac{1}{2}$ will result in all integer coefficients in the remaining factor. In addition, all terms have a common factor x. So we have:
$$\frac{1}{2}x^2 - 4x = \frac{1}{2}x \cdot x - \frac{1}{2}x \cdot 8$$
$$= \frac{1}{2}x(x - 8)$$

When factoring, always check your results by multiplication.

Here are some factorization formulas that can be used to solve some of the most commonly encountered factorization problems.

FACTORIZATION IDENTITIES

Square of a Binomial Sum: $A^2 + 2AB + B^2 = (A + B)^2$
Square of a Binomial Difference: $A^2 - 2AB + B^2 = (A - B)^2$
Difference of Squares: $A^2 - B^2 = (A + B)(A - B)$
Sum of Cubes: $A^3 + B^3 = (A + B)(A^2 - AB + B^2)$
Difference of Cubes: $A^3 - B^3 = (A - B)(A^2 + AB + B^2)$

The following example illustrates how these factorization identities can be used.

EXAMPLE 12
Applying Factorization Identities

Factor the following polynomials.
1. $4x^2 - 12x + 9$
2. $16x^6 - 25y^4$
3. $8x^4 - 8x$
4. $27x^6 + y^3$

Solution

1. The expression can be written:
$$4x^2 - 12x + 9$$
$$(2x)^2 - 2 \cdot 2x \cdot 3 + 3^2$$
$$A^2 - 2 \cdot A \cdot B + B^2$$
where
$$A = 2x \quad \text{and} \quad B = 3$$
So the Square of a Binomial Difference formula yields:
$$4x^2 - 12x + 9 = (2x - 3)^2$$

2. This expression is a difference of two perfect squares:
$$16x^6 - 25y^4$$
$$(4x^3)^2 - (5y^2)^2$$
Therefore, by the Difference of Squares formula, with A replaced by $4x^3$ and B replaced by $5y^2$,
$$16x^6 - 25y^4 = (4x^3 + 5y^2)(4x^3 - 5y^2)$$

3. Each term has a factor of $8x$. Before we attempt to apply any factorization formula, we factor out this common monomial factor:
$$8x^4 - 8x$$
$$8x(x^3 - 1)$$
We recognize the second factor as the difference of cubes. The Difference of Cubes formula yields the following factorization:
$$8x^4 - 8x = 8x(x - 1)(x^2 + x + 1)$$

4. By the Sum of Cubes formula, with $A = 3x^2$, and $B = y$,

$$\begin{aligned}27x^6 + y^3 &= (3x^2)^3 + y^3 \\ &= (3x^2 + y)\left((3x^2)^2 - 3x^2 y + y^2\right) \\ &= (3x^2 + y)(9x^4 - 3x^2 y + y^2)\end{aligned}$$

EXAMPLE 13
Factoring Polynomials

Factor the following polynomials into linear factors with integer coefficients, if possible.

1. $x^2 + 9x + 14$
2. $15x^2 + 2x - 1$
3. $x^2 + x + 1$

Solution

1. $x^2 + 9x + 14$. Suppose that we have a factorization of the form:

$$\begin{aligned}(ax + b)(cx + d) &= acx^2 + (ad + bc)x + bd \\ &= x^2 + 9x + 14\end{aligned}$$

Matching coefficients, we must have $ac = 1$, $ad + bc = 9$, and $bd = 14$. Use the first and last equations to narrow the choices. A quick check of the various possibilities shows that:

$$a = c = 1, \quad b = 7, \quad d = 2$$

So the factorization is:

$$(x + 7)(x + 2)$$

2. $15x^2 + 2x - 1$. We are looking for factors of the form:

$$\begin{aligned}(ax + b)(cx + d) &= acx^2 + (ad + bc)x + bd \\ &= 15x^2 + 2x - 1\end{aligned}$$

where a, b, c, d are integers. The product ac must equal 15 and the product bd must equal -1. This means that b and d must be 1 or -1, with opposite signs. Try a factorization of the form:

$$(\_ + 1)(\_ - 1)$$

The choices for the blanks are either $5x$ and $3x$ or x and $15x$. If we multiply out the corresponding products,

$$(3x + 1)(5x - 1) \quad \text{or} \quad (5x + 1)(3x - 1)$$

or

$$(x + 1)(15x - 1) \quad \text{or} \quad (15x + 1)(x - 1)$$

we see that the one that equals the desired product is:

$$(3x + 1)(5x - 1)$$

3. $x^2 + x + 1$. In trying out all possibilities for $a, b, c,$ and d in the Product of Two Binomials formula, we find no combination of linear factors that works. This quadratic polynomial cannot be factored into linear factors with integer coefficients.

Another technique of factoring is to group the terms so that you can apply the distributive law in reverse. The next example illustrates this technique.

EXAMPLE 14
Factoring by Grouping

Factor the following polynomials.
1. $6x^3 - 9x^2 + 4x - 6$
2. $t^3 + 2t^2 - t - 2$
3. $x^3 - xy^2 + 5x^2 - 5y^2$
4. $x^2 - 4xy + 4y^2 - 9$

Solution

1. Use parentheses to group the terms:
$$(6x^3 - 9x^2) + (4x - 6) = 3x^2(2x - 3) + 2(2x - 3)$$
$$= (3x^2 + 2)(2x - 3)$$

2. In this example, after factoring by grouping, one of the factors can be factored as a difference of squares:
$$t^3 + 2t^2 - t - 2 = (t^3 + 2t^2) + 1 \cdot (-t - 2)$$
$$= t^2(t + 2) - 1 \cdot (t + 2)$$
$$= (t^2 - 1)(t + 2)$$
$$= (t + 1)(t - 1)(t + 2)$$

3. In this example, we apply factoring by grouping to a polynomial in two variables:
$$(x^3 - xy^2) + (5x^2 - 5y^2) = x(x^2 - y^2) + 5(x^2 - y^2)$$
$$= (x + 5)(x^2 - y^2)$$
$$= (x + 5)(x + y)(x - y)$$

4. In this example, we recognize that the sum of the first three terms is a perfect square, so we group these terms together:
$$x^2 - 4xy + 4y^2 - 9 = (x^2 - 4xy + 4y^2) - 9$$
$$= (x - 2y)^2 - 9$$

Now apply the Difference of Binomial Squares formula to obtain the factorization:
$$(x - 2y - 3)(x - 2y + 3)$$

Exercises 1.3

Determine whether each of the following expressions is a polynomial. If not, explain why. If it is, state the leading coefficient and the degree of the polynomial.

1. $34x^5 - 12x^4 + x^2 - 3x + 4.5$
2. $-3x + \dfrac{5}{4}y^2 + 10$
3. $-7x^{-2} + 3x^{-1} + 5x^2$
4. $\dfrac{2}{3}x^3 + 5x^2 + x - 3 + x^{-1}$
5. $3.4y^3 - 5y^2 + x - 3 + x^{-1}$
6. $x^2 + z^3 + \dfrac{2}{z}$
7. $m^4 + \sqrt{m} - m^3 + 10$
8. $t^{1/2} + 4t^{2/3} - t^3 - 10t^4 + 4t^5$

Perform the indicated operations.

9. $(45x^3 - 23x^2 - 7x + 12) + (-28x^3 + 16x^2 + 10x - 13)$
10. $(-29x^3 - 10x^2 + 4x - 5) + (12x^3 + 18x^2 - 7x + 3)$
11. $(4x^2 - 2x^3y - 5x^2y^2 + 2xy^3) + (4x^3y - 6xy^3 + y^4)$

12. $(4a^2 - 3a + 9) - (a^2 - 3a + 9)$
13. $(3p^4 - 9p^3q^2 - 2q^4) - (-p^4 - 2p^2q^3 + q^4)$
14. $(6m^2n - 3mn^2 + 4n^3) + (2m^3 - 5m^2n - 8mn^2)$
15. $\left(4x\sqrt{y} - 3y\sqrt{x} + 2\sqrt{xy}\right) + \left(4\sqrt{xy} - 7x\sqrt{y} - 2y\sqrt{x}\right)$
16. $\left(\frac{2}{3}x^{-1} - \frac{1}{2}y\sqrt{x} - y\right) - \left(\frac{1}{2}x^{-1} - \frac{3}{4}y\sqrt{x} + 2y\right)$

Express the following products as polynomials.

17. $(4x + 3)(2x + 5)$ pg 32
18. $(a - 5)(9a - 2)$
19. $(9a + 2)(9a - 2)$
20. $(a - 2b)(a + 2b)$
21. $(a - 2)(a^2 + 2a + 4)$
22. $(m + n)(m^2 - mn + n^2)$
23. $(2a^2b^2 - 5ab)(3ab^2 + 2a^2b)$
24. $(2pq^2 + 10pq)(3p^2q - 5pq)$
25. $(5x - 3y)^2$
26. $(a + 7t)^2$
27. $(2ab + 3bc)^2$
28. $(a - b)(a + b)(a^2 + b^2)$
29. $(4x^2 - 7xy)^2$
30. $(3x + t)(3x - t)(9x^2 + t^2)$
31. $(a + b)^2(a - b)^2$
32. $(x + 2)^2(x - 2)^2$
33. $(a + b - y^3)(a + b + y^3)$
34. $(2x - y^2 + 5)(2x - y^2 - 5)$
35. $\left(\sqrt{2}xy - 3x\right)^2$
36. $\left(\sqrt{5}x + 2x^2\right)^2$
37. $\left(\sqrt{3}a + \sqrt{2}b\right)\left(\sqrt{3}a - \sqrt{2}b\right)$
38. $\left(\sqrt{7t} - \sqrt{5}\right)\left(\sqrt{7t} + \sqrt{5}\right)$
39. $\left(\frac{4}{5}x - \frac{3}{7}y\right)\left(\frac{4}{5}x + \frac{3}{7}y\right)$
40. $\left(\frac{4}{5}x + \frac{5}{6}y\right)^2$
41. $\left(\frac{2}{3}t^2 - \frac{3}{5}y^3\right)^2$
42. $(1.1x - 2y)^2$
43. $\left(\frac{1}{4}x^2 - y^2\right)\left(\frac{1}{2}x^2 - 4xy + \frac{1}{2}y^2\right)$
44. $(3x - 2y)^2(3x + 2y)$
45. Find a formula for $(A + B)^4$.
46. Find a formula for $(A - B)^4$.

Use the formulas found in the preceding exercises to do the following multiplications.

47. $(x + 8)^4$
48. $(x - 1)^4$
49. $(6x^2 + 1)^4$
50. $(3 - 4y^3)^4$

Factor the following expressions.

51. $-6x - 3x^2y + 9x$
52. $8a^2 - 2a + 12ab$
53. $x^2 - 4x - 21$
54. $x^2 - 11x + 30$
55. $(5x - 2)(x + 2) + (3x - 8)(5x - 2)$
56. $x\left(\sqrt{5} + \sqrt{7}\right) - y\left(\sqrt{5} + \sqrt{7}\right)$
57. $14a^2b - 12ab^2$
58. $21b^2 + 18xb$
59. $xy - zy - xw + zw$
60. $ab + ac - mb - mc$
61. $2x^3 + x^2 - 8x - 4$
62. $x^3 - 2x^4 + 8y^3 - 16xy^3$
63. $529 - 324x^2$
64. $49 - 9y^2$
65. $8ax - 12x^2 + 4a^2x - 20x$
66. $a^2 - ax - 6x^2$
67. $x^3 - 27$
68. $2z^3 - 16$
69. $6x^2 + 29x + 35$
70. $9a^2 + 3a - 42$
71. $0.25x^2 - 0.49y^2$
72. $\frac{1}{4}a^2b^2 - \frac{1}{9}c^2$
73. $-2t^2 + 11t - 12$
74. $-6m^2 - m + 1$
75. $27x^6 - 8y^3$
76. $2(x - 1)^2 + 3(x - 1) + 1$
77. $(a + 2)^3 + 8$
78. $2xy + 2y - x^2 - x$
79. $x^2 + 2xy + y^2 - 25$
80. $a^2 - b^2 + 6b - 9$

81. $3x^6 - 24y^6$

82. $x^6 - y^6$

83. $\frac{1}{6}x^2 - \frac{1}{72}x - \frac{1}{12}$

84. $0.06x^2 - 0.07x - 0.2$

Simplify the following expressions. Assume that the variables in the exponents represent rational numbers.

85. $(3x^a - 2y^c)^2$

86. $(x^a)^{a+b}(x^{-b})^{a-b}$

87. $(5t^n + 7)(4t^n - 5)$

88. $(x^n - y^n)(x^n + y^n)$

89. $(x^{m+n})^{m-n}(x^{m+n})^{m+n}$

90. $(x - 2)(x^2 + 2x + 4)(x + 2)(x^2 - 2x + 4)$

Factor the following expressions.

91. $a^{2n} - b^{2n}$

92. $10x^{2y} + x^y - 3$

93. $36t^{2n} - 60t^n + 25$

94. $a^{6n} - b^{3n}$

95. $1 + \frac{x^{12}}{1,000}$

96. $a^{2x} + 10a^x + 25 - 36y^{2x}$

Applications

97. **Compound interest.** Suppose that $100 is deposited in a savings account that pays compound interest at a rate r, compounded semiannually. Write the formula for the balance in the account after one year as a polynomial in r.

98. **Compound interest.** Suppose that $3,000 is deposited in a savings account that pays compound interest at a rate r, compounded annually. Write the formula for the balance in the account after three years as a polynomial in r.

99. **UPS shipping requirement.** Refer to Example 9. Suppose that a rectangular package has length equal to half its width and height equal to the square of its width. Denote the width of the package by w.
    a. Express the girth plus the height of the packages as a polynomial in w.
    b. Suppose that the width of the package is 10 inches. Is the package acceptable?
    c. Suppose that the width of the package is 20 inches. Is the package acceptable?

100. **UPS shipping requirement.** Refer to Example 9. Suppose that a rectangular package has length equal to twice its width and height equal to half the square of its width. Denote the width of the package by w.
    a. Express the girth plus the height of the package as a polynomial in w.
    b. Suppose that the width of the package is 10 inches. Is the package acceptable?
    c. Suppose that the width of the package is 20 inches. Is the package acceptable?

101. **Home maintenance.** Two rooms are being painted and carpeted. Room A measures 5 m by 4 m, and Room B measures 6 m by 4 m. Both rooms are 2.5 m high. The cost of paint is p dollars to cover a square meter, and the cost of carpet is c dollars per square meter. Find a polynomial expression for the cost of finishing (a) Room A, (b) Room B, and (c) both rooms.

102. **Gardening.** A gardener is preparing two areas for seeding. The first is a rectangle that measures 6 m by 7 m. The second is a circle that measures 8 m in diameter. It costs Z dollars per square meter for fertilizer and C dollars per meter for fencing. Find a polynomial expression for the cost of preparing (a) the first area, (b) the second area, and (c) both areas.

103. **Geometry.** One side of a rectangle is 3 m longer than the other side. Find a polynomial expression for the area of the rectangle in terms of the shorter side. Find an expression for the area in terms of the longer side.

104. **Geometry.** A square rug lies in the center of a rectangular room. There are two feet of uncovered floor on two sides of the rug and three feet of uncovered floor on the other two sides. Find a polynomial expression for the area of the room in terms of a side of the rug.

105. **Geometry.** A box is 10 ft longer than it is high and 3 ft wider than it is long. Find an expression for the volume of the box in terms of its height. Find an expression for the volume of the box in terms of its length.

In the following two exercises, an expression is given for the total revenue $R(x)$ from the sale of x units of a product and the total cost $C(x)$ of producing x units. Find an expression for the total profit $P(x)$, where
$$P(x) = R(x) - C(x)$$

106. **Profit function.** $R(x) = 9x - 2x^2$, $C(x) = x^3 - 3x^2 + 4x + 1$

107. **Profit function.** $R(x) = 100x - x^2$, $C(x) = \frac{1}{3}x^3 - 6x^2 + 89x + 100$

108. **Recreation.** To prepare a swimming pool for summer, the surface area must be painted and it must be filled with water. Two circular pools are being prepared: one is 4 ft deep and 10 ft in diameter, and the other is 7 ft deep and 20 ft in diameter. It costs P dollars to paint a square foot and W dollars to fill one cubic foot with water. Find a polynomial expression for the cost of preparing (a) the smaller pool, (b) the larger pool, and (c) both pools.

109. **Revenue function.** A company produces Washington Redskins sweatshirts. If the price is x dollars a sweatshirt, then the number actually sold (in thousands) is given by the expression

$$D = 78 - 1.5x$$

How much revenue does the company get by selling sweatshirts for x dollars?

110. **Revenue function.** A company sells CDs. If the price is x dollars a CD, then the number actually sold per week (in hundreds) is given by the expression

$$D = 6.3 - 0.25x$$

How much revenue does the company get by selling CDs for x dollars?

Technology

Use your calculator to determine the value of the following polynomials at the given value of the variable:

111. $5.1x^2 - 3x + 2$, $x = 1.98$

112. $-x^2 - 3.47x + 11.4$, $x = 5.9$

In your own words

113. Give three examples of polynomials.

114. Give three examples of algebraic expressions that are not polynomials.

115. Define the *degree* of a polynomial.

116. If you multiply a polynomial of degree 10 and a polynomial of degree 15, what is the degree of the resulting polynomial? Explain your answer.

117. Can you generalize the preceding exercise to determine the degree of the polynomial that results from multiplying a polynomial of degree m by a polynomial of degree n?

118. Explain the origin of the acronym FOIL for multiplication of two binomials.

119. Give a procedure for factoring a quadratic polynomial with integer coefficients.

1.4 RATIONAL EXPRESSIONS

A **rational expression** in the variable x is an algebraic expression of the form

$$\frac{A}{B}$$

where A and B are polynomials. We call A the **numerator** and B the **denominator**. For example, here are several rational expressions in the single variable x:

$$\frac{x}{x^2+1}, \quad \frac{5x^2+3x}{2x-1}, \quad \frac{\frac{1}{2}x^3}{(x-1)(x+1)}$$

A rational expression in several variables is a quotient of polynomials in those variables. Here are some examples of rational expressions in the variables x, y, and z:

$$\frac{x+y}{x}, \quad \frac{x^2 + y^2 + 3xyz}{2x - 3y}$$

We can evaluate a rational expression

$$R = \frac{A}{B}$$

for any value of the variables for which the denominator B is nonzero. (A zero denominator would lead to division by 0.) For example, consider the rational expression:

$$R = \frac{x+1}{(x-2)(x-1)}$$

For x equal to 0, the value of R is obtained by replacing x with 0 throughout the expression.

$$\frac{0+1}{(0-2)(0-1)} = \frac{1}{2}$$

However, the value of R at $x = 1$ is undefined, because the denominator has the value:

$$(1-2)(1-1) = 0$$

Similarly, the value of R at $x = 2$ is undefined.

EXAMPLE 1
Grade Point Averages

In many colleges and universities, a student earns 4 quality points for each credit hour of a grade of A, 3 quality points for each credit hour of a B, 2 quality points for a C, 1 quality point for a D, and 0 quality points for an F. The total number of quality points earned is given by the polynomial

$$4a + 3b + 2c + d$$

where the student has earned a credit hours of the grade A, b credit hours of B, c credit hours of C, and d credit hours of D. The student's grade point average, or GPA, is given by the rational expression

$$\text{GPA} = \frac{4a + 3b + 2c + d}{a + b + c + d + f}$$

where f is the number of credit hours earned of F. One semester, a student receives the following grades:

Course	Grade	Credit Hours
Chemistry	B	3
Trigonometry	A	4
Business	C	4
History	C	3
English	B	3

What is the student's GPA?

Solution

The number of credits of A is 4, the number of credits of B is $3 + 3 = 6$, and the number of credits of C is $4 + 3 = 7$. Substituting

$$a = 4, \quad b = 6, \quad c = 7, \quad d = 0, \quad f = 0$$

into the formula for GPA, we find that:

$$\text{GPA} = \frac{4a + 3b + 2c + d}{a + b + c + d + f}$$

$$= \frac{4 \cdot 4 + 3 \cdot 6 + 2 \cdot 7 + 0}{4 + 6 + 7 + 0 + 0}$$

$$= \frac{48}{17}$$

$$= 2.82$$

Reducing Rational Expressions

The procedure of reducing a rational expression to lowest terms is the technique of replacing a rational expression with an equivalent one, usually to obtain a simpler rational expression. We have the following general principle.

REDUCING RATIONAL EXPRESSIONS TO LOWEST TERMS

If A, B, and C are algebraic expressions, then we have the equivalence of algebraic expressions

$$\frac{AC}{BC} = \frac{A}{B}$$

where the equality is valid for all nonzero values of B and C. That is, we can cancel a common factor of the numerator and denominator. The process of removing the common factor C from the numerator and denominator is called **canceling**, or **reducing to lowest terms**.

EXAMPLE 2
Reducing a Rational Expression to Lowest Terms

Reduce the following rational expression to lowest terms:

$$\frac{x^5 - x^3}{x^2 + 3x + 2}$$

Solution

Factor the numerator and denominator:

$$\frac{x^3(x^2 - 1)}{(x + 1)(x + 2)} = \frac{x^3(x - 1)(x + 1)}{(x + 2)(x + 1)}$$

We see that the numerator and denominator have the common factor $x + 1$. Canceling this factor gives the equivalent expression:

$$\frac{x^3(x - 1)}{x + 2}, \qquad x \neq -1, -2$$

(The denominator of the original expression is 0 if x has either of the values -2 or -1, so to avoid division by 0, we must exclude these values.) After canceling the factor $x + 1$, the numerator and denominator have no factors in common, so the resulting rational expression is in lowest terms. By removing the common factor $x + 1$ from the numerator and denominator, the problem with division by 0 is eliminated for $x = -1$. However, this does not entitle you to say that the final expression equals the original expression when $x = -1$, since the original expression is undefined for this value.

EXAMPLE 3
Another Reduction

Reduce to lowest terms:
$$\frac{3x^2 + 3x}{x^6 - 1}, \quad x \neq 1, -1$$

Solution

We factor the numerator and denominator:
$$\frac{3x^2 + 3x}{x^6 - 1} = \frac{3x(x + 1)}{(x^3 - 1)(x^3 + 1)}$$
$$= \frac{3x(x + 1)}{(x - 1)(x^2 + x + 1)(x + 1)(x^2 - x + 1)}$$
$$= \frac{3x}{(x - 1)(x^2 + x + 1)(x^2 - x + 1)} \quad \text{Cancel } x + 1$$

The last expression is in lowest terms.

Addition and Subtraction of Rational Expressions

Let A/Q and B/Q be rational expressions with the same denominator. These expressions may be added and subtracted by adding numerators:
$$\frac{A}{Q} \pm \frac{B}{Q} = \frac{A \pm B}{Q}$$

For example,
$$\frac{3x - 1}{x^2 + 1} + \frac{2x + 5}{x^2 + 1} = \frac{(3x - 1) + (2x + 5)}{x^2 + 1}$$
$$= \frac{5x + 4}{x^2 + 1}$$

To add or subtract rational expressions that have *different* denominators, it is necessary to write them in equivalent forms that have the same denominators. To see how this may be done, consider the sum:
$$\frac{x}{x - 1} + \frac{x + 1}{x + 2}$$

Multiply the numerator and denominator of the first expression by $x + 2$ and the numerator and denominator of the second expression by $x - 1$. The sum is then equal to:
$$\frac{x}{(x - 1)} \cdot \frac{x + 2}{x + 2} + \frac{x + 1}{x + 2} \cdot \frac{x - 1}{x - 1} = \frac{x(x + 2)}{(x - 1)(x + 2)} + \frac{(x + 1)(x - 1)}{(x + 2)(x - 1)}$$

Note that the two expressions now have a common denominator, so they can be added by adding their numerators:
$$\frac{x(x + 2)}{(x - 1)(x + 2)} + \frac{(x + 1)(x - 1)}{(x - 1)(x + 2)} = \frac{x(x + 2) + (x + 1)(x - 1)}{(x - 1)(x + 2)}$$
$$= \frac{(x^2 + 2x) + (x^2 - 1)}{x^2 + x - 2}$$
$$= \frac{2x^2 + 2x - 1}{x^2 + x - 2}, \quad x \neq -2, 1$$

EXAMPLE 4
Adding Rational Expressions

Perform the indicated additions of rational expressions.

1. $\dfrac{1}{x} + \dfrac{1}{x-1}$
2. $\dfrac{2x}{x-1} + \dfrac{2}{x+1}$
3. $\dfrac{1}{x^2+x+1} + \dfrac{1}{x-1}$

Solution

1. Multiply numerator and denominator of the first expression by $x - 1$ and multiply the numerator and denominator of the second expression by x. This results in expressions with the common denominator $x(x - 1)$:

$$\dfrac{1}{x} + \dfrac{1}{x-1} = \dfrac{1}{x} \cdot \dfrac{x-1}{x-1} + \dfrac{1}{x-1} \cdot \dfrac{x}{x}$$

$$= \dfrac{(x-1) + x}{x(x-1)}$$

$$= \dfrac{2x-1}{x^2-x}, \qquad x \neq 0, 1$$

2. A common denominator is $(x - 1)(x + 1)$.

$$\dfrac{2x}{x-1} + \dfrac{2}{x+1} = \dfrac{2x}{(x-1)} \cdot \dfrac{x+1}{x+1} + \dfrac{2}{(x+1)} \cdot \dfrac{x-1}{x-1}$$

We can now add the rational expressions:

$$\dfrac{2x \cdot (x+1) + 2 \cdot (x-1)}{(x-1)(x+1)} = \dfrac{2x^2 + 2x + 2x - 2}{(x-1)(x+1)}$$

$$= \dfrac{2x^2 + 4x - 2}{x^2 - 1} \qquad (x \neq 1, -1)$$

3. A common denominator is $(x - 1)(x^2 + x + 1)$.

$$\dfrac{1}{x^2+x+1} + \dfrac{1}{x-1} = \dfrac{1}{x^2+x+1} \cdot \dfrac{x-1}{x-1} + \dfrac{1}{x-1} \cdot \dfrac{x^2+x+1}{x^2+x+1}$$

$$= \dfrac{(x-1) + (x^2+x+1)}{(x^2+x+1)(x-1)}$$

$$= \dfrac{x^2 + 2x}{x^3 - 1}, \qquad x \neq 1$$

EXAMPLE 5
Adding Three Rational Expressions

Write the following as a rational expression, reduced to lowest terms.

$$\dfrac{8y}{y^2-1} - \dfrac{2}{1-y} + \dfrac{4}{y+1}$$

Solution

The first denominator may be written in the form $(y + 1)(y - 1)$, and the second denominator in the form $-(y - 1)$. Therefore, the given rational expression can be written:

$$\dfrac{8y}{(y+1)(y-1)} - \dfrac{2}{-(y-1)} + \dfrac{4}{y+1} = \dfrac{8y}{(y+1)(y-1)} + \dfrac{2}{y-1} + \dfrac{4}{y+1}$$

A common denominator is $(y + 1)(y - 1)$.

$$\frac{8y}{(y - 1)(y + 1)} + \frac{2(y + 1)}{(y - 1)(y + 1)} + \frac{4(y - 1)}{(y + 1)(y - 1)}$$
$$= \frac{8y + 2y + 2 + 4y - 4}{(y - 1)(y + 1)}$$
$$= \frac{14y - 2}{(y - 1)(y + 1)}$$
$$= \frac{2(7y - 1)}{(y + 1)(y - 1)}, \qquad y \neq -1, 1$$

Here is another method for adding rational expressions. Write the rational expressions in the sum

$$\frac{A}{B} + \frac{C}{D}$$

over the common denominator BD:

$$\frac{A}{B} + \frac{C}{D} = \frac{AD}{BD} + \frac{BC}{BD} = \frac{AD + BC}{BD}$$

This formula can be used to easily compute sums of rational expressions. For example,

$$\frac{1}{x} + \frac{1}{x - 1} = \frac{1 \cdot (x - 1) + x \cdot 1}{x(x - 1)}$$
$$= \frac{x - 1 + x}{x^2 - x}$$
$$= \frac{2x - 1}{x^2 - x}, \qquad x \neq 0, 1$$

This agrees with the answer in part 1 of Example 4. Note that if we use the preceding formula for adding rational expressions, then it may be necessary to reduce the answer to lowest terms.

Multiplication and Division of Rational Expressions

Multiplication of rational expressions is carried out by multiplying numerators and multiplying denominators:

$$\frac{A}{B} \cdot \frac{C}{D} = \frac{AC}{BD}, \qquad B, D \neq 0$$

EXAMPLE 6
Multiplying Rational Expressions

Calculate the products.

1. $\dfrac{x}{x - 1} \cdot \dfrac{2}{x + 1}$

2. $\dfrac{x^2}{3x + 1} \cdot \dfrac{x}{5x + 1}$

Solution

1.
$$\frac{x}{x - 1} \cdot \frac{2}{x + 1} = \frac{2x}{(x + 1)(x - 1)}$$
$$= \frac{2x}{x^2 - 1}, \qquad x \neq 1, -1$$

2.
$$\frac{x^2}{3x+1} \cdot \frac{x}{5x+1} = \frac{x^2 \cdot x}{(3x+1)(5x+1)}$$
$$= \frac{x^3}{15x^2 + 8x + 1}, \qquad x \neq -\frac{1}{3}, -\frac{1}{5}$$

Division of rational expressions can be carried out by first forming the inverse of the denominator and then multiplying:

$$\frac{A/B}{C/D} = \frac{A}{B} \cdot \frac{D}{C} = \frac{AD}{BC}, \qquad B, C, D \neq 0$$

The next example shows how to apply this result.

EXAMPLE 7
Dividing Rational Expressions

Calculate the quotient:
$$\frac{x^2 - 1}{x} \div \frac{1}{x-1}$$

Solution

$$\frac{(x^2-1)/x}{1/(x-1)} = \frac{x^2-1}{x} \cdot \frac{x-1}{1}$$
$$= \frac{(x^2-1)(x-1)}{x}$$
$$= \frac{x^3 - x^2 - x + 1}{x}, \qquad x \neq 1, 0$$

The next example combines all of the operations on rational expressions that we have discussed to simplify a complex rational expression.

EXAMPLE 8
Reducing a Complex Rational Expression

Write the following as a rational expression in lowest terms:
$$\frac{a^{-1} + b^{-1}}{a^{-3} + b^{-3}}$$

Solution

The given expression can be written in the form:
$$\frac{\frac{1}{a} + \frac{1}{b}}{\frac{1}{a^3} + \frac{1}{b^3}}$$

Multiply both the numerator and denominator by $a^3 b^3$ to obtain the equivalent expression:

$$\frac{\left(\frac{1}{a} + \frac{1}{b}\right) \cdot (a^3 b^3)}{\left(\frac{1}{a^3} + \frac{1}{b^3}\right) \cdot (a^3 b^3)} = \frac{\frac{a^3 b^3}{a} + \frac{a^3 b^3}{b}}{\frac{a^3 b^3}{a^3} + \frac{a^3 b^3}{b^3}}$$
$$= \frac{a^2 b^3 + a^3 b^2}{b^3 + a^3}$$
$$= \frac{a^2 b^2 (b + a)}{b^3 + a^3}$$

And by the sum of cubes factorization formula, we can rewrite the expression in the form:

$$\frac{(a^2b^2)(a+b)}{(a+b)(a^2-ab+b^2)}$$

Simplifying this last expression to lowest terms gives us the result:

$$\frac{a^2b^2}{a^2-ab+b^2}, \quad a, b \neq 0, \quad a \neq -b$$

Rationalizing Denominators and Numerators

Expressions often involve radicals in the denominator. By multiplying numerator and denominator by a suitable common expression, you can often arrive at a form in which the radicals appear only in the numerator. For example, consider the expression:

$$\frac{1}{\sqrt{2}}$$

Here a radical appears in the denominator. To remove the radical from the denominator, multiply both numerator and denominator by $\sqrt{2}$ to obtain:

$$\frac{1}{\sqrt{2}} = \frac{1 \cdot (\sqrt{2})}{(\sqrt{2}) \cdot (\sqrt{2})} = \frac{\sqrt{2}}{2}$$

The process of removing a radical from a denominator is called **rationalizing the denominator.** This can be accomplished in most situations by multiplying both numerator and denominator by a suitable expression. Choosing the right expression to multiply by takes a bit of practice. As a guide, use the various product identities we have discussed in order to arrive at a product in the denominator that contains no radicals. For instance, suppose that the denominator is the expression:

$$1 + \sqrt{x}$$

The Product of a Sum and Difference formula suggests that this denominator may be rationalized by multiplying by the expression:

$$1 - \sqrt{x}$$

Indeed, the product of the two expressions is:

$$\left(1 + \sqrt{x}\right)\left(1 - \sqrt{x}\right) = 1^2 - \left(\sqrt{x}\right)^2 = 1 - x$$

As another example, suppose that the denominator of an expression is:

$$\sqrt[3]{x}$$

We can rationalize the denominator by multiplying both numerator and denominator by $x^{2/3}$. Indeed, by the laws of exponents, the new denominator is

$$\sqrt[3]{x} \cdot x^{2/3} = x^{1/3} x^{2/3} = x^{1/3+2/3} = x$$

The following example provides some practice in rationalizing denominators.

EXAMPLE 9
Rationalizing Denominators

Rationalize the denominator of the following expressions.

1. $\dfrac{1}{x + \sqrt{x}}$

2. $\dfrac{\sqrt[3]{x}}{\sqrt[3]{x} - 2}$

Solution

1. The Difference of Squares formula suggests that we multiply the numerator and denominator by the expression $x - \sqrt{x}$. Doing so gives us the following equivalent expression:

$$\frac{1}{x + \sqrt{x}} = \frac{1}{x + \sqrt{x}} \cdot \frac{x - \sqrt{x}}{x - \sqrt{x}}$$

$$= \frac{x - \sqrt{x}}{x^2 - \left(\sqrt{x}\right)^2}$$

$$= \frac{x - \sqrt{x}}{x^2 - x}$$

2. The Difference of Cubes formula, with $A = \sqrt[3]{x}$ and $B = 2$, suggests that we multiply the numerator and denominator by $A^2 + AB + B^2$. The denominator will then be:

$$A^3 - B^3 = \left(\sqrt[3]{x}\right)^3 - 2^3 = x - 8$$

Carrying out this calculation, we are able to rationalize the denominator:

$$\frac{\sqrt[3]{x}}{\sqrt[3]{x} - 2} = \frac{\sqrt[3]{x}}{\sqrt[3]{x} - 2} \cdot \frac{\sqrt[3]{x^2} + 2\sqrt[3]{x} + 2^2}{\sqrt[3]{x^2} + 2\sqrt[3]{x} + 2^2}$$

$$= \frac{x + 2\sqrt[3]{x^2} + 4\sqrt[3]{x}}{x - 8}$$

In calculus, it is often necessary to write expressions so that they have no radicals in the *numerator*. This process is called **rationalizing the numerator,** and is illustrated in the following example.

EXAMPLE 10
Rationalizing a Numerator

Write the expression

$$\frac{\sqrt{x + h} - \sqrt{x}}{h}$$

in a form that involves no radicals in the numerator.

Solution

The Difference of Squares formula suggests that we multiply both numerator and denominator by the expression:

$$\sqrt{x + h} + \sqrt{x}$$

This gives us the equivalent expression:

$$\frac{\left(\sqrt{x + h} - \sqrt{x}\right) \cdot \left(\sqrt{x + h} + \sqrt{x}\right)}{h \cdot \left(\sqrt{x + h} + \sqrt{x}\right)}$$

Applying the Product of a Sum and a Difference formula to the numerator, we have:

$$\frac{\left(\sqrt{x + h}\right)^2 - \left(\sqrt{x}\right)^2}{h\left(\sqrt{x + h} + \sqrt{x}\right)}$$

We can simplify the numerator to read:

$$\frac{(x+h) - x}{h\left(\sqrt{x+h} + \sqrt{x}\right)} = \frac{h}{h\left(\sqrt{x+h} + \sqrt{x}\right)}$$

$$= \frac{1}{\sqrt{x+h} + \sqrt{x}}, \quad h \neq 0$$

Thus,

$$\frac{\sqrt{x+h} - \sqrt{x}}{h} = \frac{1}{\sqrt{x+h} + \sqrt{x}}, \quad h \neq 0$$

EXAMPLE 11
Another Numerator to Rationalize

Rationalize the numerator of the expression:

$$\frac{\sqrt[3]{x+2} - \sqrt[3]{x}}{\sqrt[3]{x}}$$

Solution

We must come up with an expression whose product with the numerator has no radicals. When looking for such an expression, you should always keep in mind the various product or factoring formulas. In this case, the cube roots suggest the Difference of Cubes formula:

$$A^3 - B^3 = (A - B)(A^2 + AB + B^2)$$

If we set $A = \sqrt[3]{x+2}$ and $B = \sqrt[3]{x}$, then this formula reads:

$$\left(\sqrt[3]{x+2}\right)^3 - \left(\sqrt[3]{x}\right)^3 = \left(\sqrt[3]{x+2} - \sqrt[3]{x}\right)\left(\sqrt[3]{(x+2)^2} + \sqrt[3]{x(x+2)} + \sqrt[3]{x^2}\right)$$

$$2 = \left(\sqrt[3]{x+2} - \sqrt[3]{x}\right)\left(\sqrt[3]{(x+2)^2} + \sqrt[3]{x(x+2)} + \sqrt[3]{x^2}\right)$$

Using this formula, we can rationalize the desired numerator:

$$\frac{\sqrt[3]{x+2} - \sqrt[3]{x}}{\sqrt[3]{x}} = \frac{\sqrt[3]{x+2} - \sqrt[3]{x}}{\sqrt[3]{x}} \cdot \frac{\left(\sqrt[3]{(x+2)^2} + \sqrt[3]{x(x+2)} + \sqrt[3]{x^2}\right)}{\left(\sqrt[3]{(x+2)^2} + \sqrt[3]{x(x+2)} + \sqrt[3]{x^2}\right)}$$

$$= \frac{2}{\sqrt[3]{x}\left(\sqrt[3]{(x+2)^2} + \sqrt[3]{x(x+2)} + \sqrt[3]{x^2}\right)}$$

Exercises 1.4

Write the following rational expressions in lowest terms.

1. $\dfrac{4x^2 - 4}{8x^2 + 28x + 12}$

2. $\dfrac{x^2 - 9}{x^2 + 2x + 1}$

3. $\dfrac{a^4 + 2a^3 + a^2}{a^2 + 2a + 1}$

4. $\dfrac{t(t^2 + 7t + 12)}{t^2 + t - 12}$

5. $\dfrac{2r + 2s}{2(r^2 + s^2 + 2rs)}$

6. $\dfrac{(xy)^3 - 1}{xy - 1}$

7. $\dfrac{x^2 - 1}{x^3 - 1}$

8. $\dfrac{a^3 + b^3}{a^2 - b^2}$

Perform the indicated operations and write the results in lowest terms.

9. $\dfrac{3x - 5}{2x + 1} - \dfrac{5 - 3x}{2x + 1}$

10. $\dfrac{7a + 3}{2a + 7} - \dfrac{4a + 7}{2a + 7}$

SECTION 1.4 Rational Expressions

11. $\dfrac{3x-7}{x-3} + \dfrac{2x+3y}{3-x}$

12. $\dfrac{3a}{a-b} + \dfrac{2b+4}{b-a} + \dfrac{6a-b}{a-b}$

13. $\dfrac{2t^2-3t+5}{t^2+3t-5} - \dfrac{t^2-t+2}{5-3t-t^2}$

14. $\dfrac{x^2+6x+12}{x^2-3x-8} + \dfrac{2x^2-6x+4}{8+3x-x^2}$

15. $\dfrac{x-(3x+2)}{x^2-3x-8} + \dfrac{2x^2-6x+4}{8+3x-x^2}$

16. $\dfrac{a-4-(5-a)}{a^3+2a^2+a} - \dfrac{2a-4-(a-3)}{a^2-1}$

17. $\dfrac{t}{t^2+5t+6} - \dfrac{2}{t^2+3t+2}$

18. $\dfrac{t}{t^2+11t+30} - \dfrac{5}{t^2+9t+20}$

19. $\dfrac{6}{x+5} - \dfrac{2}{x-3} + \dfrac{4x-1}{x^2+2x-15}$

20. $\dfrac{4a}{a-b} + \dfrac{3b}{a+b} - \dfrac{2ab}{a^2-b^2}$

21. $\dfrac{x^2+5x+6}{x^2+8x+15} \cdot \dfrac{x^2+9x+20}{x^2+6x+8}$

22. $\dfrac{5x^2+30x+45}{6x^2-24} \cdot \dfrac{3x^2+12x+12}{10x^2-90}$

23. $\dfrac{t^3-1}{t^2-1} \div \dfrac{t^2+t+1}{t-1}$

24. $\dfrac{a^3+c^3}{4a-4c} \div \dfrac{a^3+a^3c^2}{2a^2-2c^2}$

Write the following expressions as rational expressions in the variables involved.

25. $\dfrac{ab^{-1}+ba^{-1}}{a^{-1}+b^{-1}}$

26. $\dfrac{2a^{-1}+2y^{-1}}{(x+y)y^{-1}}$

27. $\dfrac{\frac{1}{x}+\frac{1}{y}+\frac{1}{z}}{\frac{1}{xy}+\frac{1}{yz}}$

28. $\dfrac{\frac{x}{y}+\frac{y}{x}+2}{\frac{y+1}{x}+\frac{y+1}{y}}$

29. $\dfrac{\frac{1}{x+h}-\frac{1}{x}}{h}$

30. $\dfrac{\frac{1}{(x+h)^2}-\frac{1}{x^2}}{h}$

31. $\dfrac{\frac{1}{(x+h)^3}-\frac{1}{x^3}}{h}$

32. $\dfrac{(a+h)^3-a^3}{h}$

33. $\dfrac{\left[(x+x^{-1})^{-1}\right]^{-1}}{x+x^{-1}}$

34. $\dfrac{a+1-2a^{-1}}{a+4+4a^{-1}}$

35. $\dfrac{(a+h)^2-a^2}{h}$

36. $\dfrac{(x+h)^2-5(x+h)-(x^2-5x)}{h}$

37. $\dfrac{(x+h)^2+3(x+h)-(x^2+3x)}{h}$

38. $\dfrac{\frac{1}{2(x+h)+1}-\frac{1}{2x+1}}{h}$

Rationalize the denominators of the following expressions.

39. $\dfrac{1}{\sqrt{a}-\sqrt{b}}$

40. $\dfrac{3}{\sqrt{x}+\sqrt{h}}$

41. $\dfrac{\sqrt{a}+\sqrt{2}}{\sqrt{a}-\sqrt{b}}$

42. $\dfrac{2\sqrt{x}+3\sqrt{3}}{\sqrt{x}-\sqrt{3}}$

Rationalize the numerators of the following expressions.

43. $\dfrac{\sqrt{a+h}-\sqrt{a}}{h}$

44. $\dfrac{2\sqrt{x}+3\sqrt{y}}{x}$

45. $\dfrac{\sqrt{a}+\sqrt{2}}{\sqrt{a}-\sqrt{2}}$

46. $\dfrac{2\sqrt{x}+3\sqrt{3}}{\sqrt{x}-\sqrt{3}}$

47. $\dfrac{\sqrt{n+1}-\sqrt{n}}{\sqrt{n}}$

48. $\dfrac{\sqrt{n^2+1}-n}{n}$

49. $\dfrac{1-\sqrt{x}}{1-x}$

50. $\dfrac{\sqrt{4+h}-2}{h}$

Applications

51. **Baseball.** A baseball hitter's slugging percentage, P, is given by the rational expression

$$P = \dfrac{s+2d+3t+4h}{B}$$

where s = number of singles, d = number of doubles, t = number of triples, h = number of home runs, and B = number of at bats. In his career Hank Aaron had 2,294 singles, 624 doubles, 98 triples, and 755 home runs in 12,364 at bats. What was his slugging percentage?

52. **Electrical engineering.** The total resistance R of two resistors in parallel is given by the rational expression

$$R = \dfrac{1}{\dfrac{1}{R_1}+\dfrac{1}{R_2}}$$

where R_1 and R_2 are the resistances of the separate resistors. (See figure.)

a. Simplify this rational expression.

52   CHAPTER 1   Review of Elementary Algebra

b. What is the total resistance of a circuit with a 10-ohm and a 5-ohm resistor in parallel?

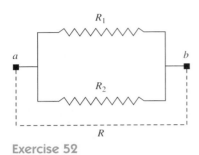

Exercise 52

53. **Physics of materials.** The **coefficient of linear expansion**, α, of a material describes how it expands as it is heated. (See figure.) The coefficient of linear expansion is given by the rational expression

$$\alpha = \frac{L_2 - L_1}{L_1(t_2 - t_1)}$$

where L_1 is the length of a material at the beginning temperature t_1, and L_2 is its length at temperature t_2.

a. Find the coefficient of linear expansion of aluminum if the length of a bar is 10 m at 10°C and 10.02 m at 100°C.

b. How long will the bar be at 250°C?

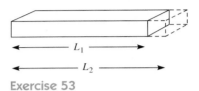

Exercise 53

54. **Physics of materials.** Refer to the preceding problem. Find the coefficient of linear expansion of steel if the length of a bar is 6 m at 20°C and 6.012 m at 200°C.

 Technology

Use a calculator to determine the values of the following rational expressions for the indicated values of the variable x.

55. $\dfrac{1}{x^3 + 10}$, $x = 5.7$

56. $\dfrac{x^2 + 1}{x^2 - 1}$, $x = -0.1$

57. $\dfrac{x^3 + 12x - 5}{3x^2 - x + 4}$, $x = 12.01$

58. $\dfrac{\frac{1}{x-1}}{x^{30} - 1}$, $x = \dfrac{2}{3}$

59. Use the formula for grade point average given in Example 1 of this section. Suppose a student gets the following grades for one term:

Course	Grade	Number of Credit Hours Earned in Course
Accounting	B	4
College Algebra	A	5
English	A	5
French	C	3
Phys. Ed.	F	1

Find the student's grade point average, or GPA.

60. Refer to the preceding exercise. Is it possible, by changing a single grade, to achieve a GPA above 3.7?

 In your own words

61. Describe in words the procedure for multiplying two polynomials.

62. In what ways are polynomial expressions and rational expressions similar? In what ways are they different?

63. Explain how a polynomial expression can be thought of as a special type of rational expression.

1.5 EQUATIONS AND IDENTITIES

An **equation** is a statement that two algebraic expressions are equal. Here are some examples of equations.

$$2 = 1 + 1$$
$$(x + 1)^2 = x^2 + 2x + 1$$
$$2x - 3 = 4$$
$$3x + 4y = 8$$

There are three types of equations: **identities, conditional equations,** and **inconsistent equations.** An **identity** is an equation that is true for all variable values. The second equation above is an example of an identity since it states an algebraic fact that remains true for all values of x.

A **conditional equation** is an equation that is true for only some variable values. In this chapter, most of our attention will be directed to conditional equations. The third equation above is a conditional equation: On the one hand, the equation is true when $\frac{7}{2}$ is substituted for x:

$$\begin{array}{c|c} 2x - 3 & 4 \\ 2\left(\frac{7}{2}\right) - 3 & \\ 7 - 3 & \\ 4 & \end{array}$$

On the other hand, suppose that $x = -1$ is substituted for x in the equation. The two sides are unequal, so the equation is not true if $x = -1$.

$$\begin{array}{c|c} 2x - 3 & 4 \\ 2(-1) - 3 & \\ -2 - 3 & \\ -5 & \end{array}$$

An **inconsistent equation** is an equation that states an impossibility, such as $x + 2 = x + 3$, which is not true for any value of x.

The process for finding the values of the variable(s) that make a conditional equation true is called **solving the equation.** The values of the variable(s) that make an equation true are called **solutions,** or **roots,** of the equation. The set of all of these values is called the **solution set** of the equation.

Methods for solving equations are one of the main topics of algebra. In this chapter, we are concerned with solving the simplest type of equation—linear. In later chapters, we learn to solve many other types of equations.

Equivalent Equations

To solve an equation, we can use any operation that changes an equation into an **equivalent equation,** one that has the same solutions as the original equation. For example, adding or subtracting the same number from both sides of an equation produces an equivalent equation, as does multiplication and division of both sides by the same nonzero number.

Consider the equation:

$$5x + 3 = 2x + 1$$

We can subtract $2x$ from both sides of the equation to obtain the equivalent equation:

$$(5x + 3) - 2x = (2x + 1) - 2x$$
$$3x + 3 = 1$$

Similarly, we can divide both sides of the equation

$$9x = 11$$

by 9 to obtain the equivalent equation:

$$\frac{9x}{9} = \frac{11}{9}$$
$$x = \frac{11}{9}$$

The following table lists some operations that lead to equivalent equations:

OPERATIONS THAT GENERATE EQUIVALENT EQUATIONS

1. Algebraic simplification:
   a. removing parentheses
   b. combining like terms
   c. performing indicated algebraic operations
   d. applying the laws of exponents
2. Interchange the two sides of the equation.
3. Add or subtract the same expression from both sides of the equation.
4. Multiply or divide both sides of the equation by the same nonzero number.

Linear Equations

The simplest equations in one variable are linear equations.

Definition
Linear Equation
A **linear equation** in one variable is an equation that is equivalent to an equation of the form

$$ax + b = 0$$

where a and b are constants and $a \neq 0$.

Here are some examples of linear equations:

$$-5x + 1 = 0$$
$$x + 3(2x - 4) = 1$$
$$5x + 2 = 3x - 1$$

By using one or more operations that replace a linear equation with an equivalent one, we may solve the equation. The procedure is illustrated in the following example.

EXAMPLE 1
Solving a Linear Equation

Solve the linear equation:

$$-2(3x + 7) + 4 = 3x - 2(x + 4)$$

SECTION 1.5 Equations and Identities

Solution

The first step in the solution of this equation is to simplify the expressions on both sides by removing the parentheses:

$$-6x - 14 + 4 = 3x - 2x - 8 \quad \text{Use distributive law}$$
$$-6x - 10 = x - 8 \quad \text{Combine like terms}$$

Next, we use various operations to arrive at an equivalent equation that has all terms involving x on one side of the equation and all terms not involving x on the other side of the equation.

$$-6x - 10 - x = x - 8 - x \quad \text{Subtract } x \text{ from both sides}$$
$$-7x - 10 = -8 \quad \text{Combine like terms}$$
$$-7x - 10 + 10 = -8 + 10 \quad \text{Add 10 to both sides}$$
$$-7x = 2$$
$$-\tfrac{1}{7} \cdot (-7x) = -\tfrac{1}{7} \cdot 2 \quad \text{Multiply both sides by } -\tfrac{1}{7}, \text{ using inverse}$$
$$\text{property and associative law}$$
$$x = -\tfrac{2}{7}$$

Thus, the solution is $-\tfrac{2}{7}$.

We now check that the solution works by substituting the value $-\tfrac{2}{7}$ for x in the equation:

$$-2(3x + 7) + 4 \quad \bigg| \quad 3x - 2(x + 4)$$
$$= -2\left[3\left(-\tfrac{2}{7}\right) + 7\right] + 4 \quad \bigg| \quad = 3\left(-\tfrac{2}{7}\right) - 2\left(-\tfrac{2}{7} + 4\right)$$
$$= -2\left[-\tfrac{6}{7} + 7\right] + 4 \quad \bigg| \quad = -\tfrac{6}{7} - 2\left[-\tfrac{2}{7} + \tfrac{28}{7}\right]$$
$$= -2\left[-\tfrac{6}{7} + \tfrac{49}{7}\right] + 4 \quad \bigg| \quad = -\tfrac{6}{7} - 2\left(\tfrac{26}{7}\right)$$
$$= -2\left(\tfrac{43}{7}\right) + \tfrac{28}{7} \quad \bigg| \quad = -\tfrac{58}{7}$$
$$= -\tfrac{58}{7}$$

EXAMPLE 2
Linear Equation with Fractional Coefficients

Solve the linear equation:

$$\tfrac{1}{3}x + \tfrac{3}{7} = \tfrac{1}{2}x + 4$$

Solution

It is usually helpful to eliminate fractions in an equation by first multiplying both sides by the least common denominator of all rational numbers that appear, in this case $3 \cdot 7 \cdot 2 = 42$. This gives:

$$42\left(\tfrac{1}{3}x + \tfrac{3}{7}\right) = 42\left(\tfrac{1}{2}x + 4\right) \quad \text{Multiply by least common denominator}$$
$$14x + 18 = 21x + 168$$

$$-150 = 7x \qquad \text{Move constant terms to one side and } x \text{ terms to the other}$$

$$-\frac{150}{7} = x \qquad \text{Multiply by } \tfrac{1}{7}$$

Check:

$$\frac{1}{3}x + \frac{3}{7} = \frac{1}{3} \cdot \left(-\frac{150}{7}\right) + \frac{3}{7} \qquad \frac{1}{2}x + 4 = \frac{1}{2} \cdot \left(-\frac{150}{7}\right) + 4$$

$$= -\frac{50}{7} + \frac{3}{7} \qquad\qquad\qquad = -\frac{75}{7} + \frac{28}{7}$$

$$= -\frac{47}{7} \qquad\qquad\qquad\qquad = -\frac{47}{7}$$

In Example 2, we eliminated fractions by multiplying both sides of the equation by the least common denominator. This process is called **clearing the denominators**. In an equation involving decimals, it is usually helpful to clear the decimals by multiplying both sides of the equation by an appropriate power of 10. For example, to clear the decimals from the equation

$$0.1x + 0.003 = 5.75$$

we would multiply both sides by 10^3 or 1,000, to obtain:

$$100x + 3 = 5,750$$

Equations Arising in Applications

In applied problems, we can often express real-world conditions as equations. By solving the equations, we can determine the value of the variable involved and thereby solve the applied problem.

EXAMPLE 3
Sales Projection

A manufacturer of medical electronics equipment estimates that the revenue R (in millions of dollars) from the sales of a new medical imaging device t years from the present is given by the equation

$$R = 230t + 50$$

How many years will it take for sales to pass the $1 billion level?

Solution

Since $1 billion equals 1,000 million dollars, sales of $1 billion corresponds to $R = 1,000$. So we must solve the equation:

$$1,000 = 230t + 50$$
$$1,000 - 50 = 230t$$
$$950 = 230t$$
$$t = \frac{950}{230} \approx 4.13$$

That is, in a little more than 4 years, sales will reach the $1 billion level.

EXAMPLE 4
Using a Calculator to Solve Linear Equations

Use a calculator to solve the equation:

$$5.88x - 3.19 = 4.31x + 11.95$$

Solution

Begin just as you would if you were solving the equation manually:

$$5.88x - 3.19 = 4.31x + 11.95$$
$$5.88x - 4.31x = 11.95 + 3.19$$
$$(5.88 - 4.31)x = 11.95 + 3.19 \quad \text{Distributive law}$$

Use a calculator to calculate the difference $5.88 - 4.31$ and the sum $11.95 + 3.19$ to obtain the equation:

$$1.57x = 15.14 \quad \text{Calculator}$$
$$x = \frac{15.14}{1.57}$$
$$\approx 9.643312102 \quad \text{Calculator}$$

Note that the first use of the calculator gave us exact results, whereas the second gave us only an approximation. Generally speaking, you should keep the full ten-digit calculator accuracy throughout a problem. If you require an answer to only two or three digits of accuracy, you should do your rounding only at the last step. This minimizes accumulated round-off errors throughout the problem.

EXAMPLE 5
Hourly Earnings

Based on data from the Department of Labor, hourly earnings E (in dollars per hour) in year x (with 1990 corresponding to 0) are given by the formula:

$$E = 11 + 0.27x$$

When are hourly earnings projected to be $13?

Solution

We must solve the equation:

$$E = 13$$
$$11 + 0.27x = 13$$
$$0.27x = 2$$
$$x = \frac{2}{0.27} = 7.4$$

That is, hourly earnings are projected to be $13 per hour sometime in 1997.

EXAMPLE 6
Income of Private Legal Offices

According to the Bureau of the Census, the income of private legal offices I (in thousands of dollars per year) in year x (with 1982 corresponding to 0) is given by the model:

$$I = 33.9 + 6.49x$$

In what year were the earnings $75,000?

Solution

We must solve the equation:

$$I = 75$$
$$33.9 + 6.49x = 75$$
$$6.49x = 41.1$$
$$x \approx 6.33$$

Because 1982 corresponds to $x = 0$, the \$75,000 income level was achieved in 1988.

Equations that Reduce to Linear Equations

Many equations can be simplified, resulting in equivalent equations. For example, we can multiply expressions, combine like terms, remove parentheses, and apply the laws of exponents. In many cases, such simplifications can result in a linear equation, as shown in the following example.

EXAMPLE 7
Reducing to a Linear Equation

Solve the following equation:
$$(3x + 1)(2x - 1) = (6x - 1)(x + 5)$$

Solution

The first step in the solution is to remove the parentheses by calculating the products on both sides of the equation.
$$6x^2 - x - 1 = 6x^2 + 29x - 5$$

At first glance, it looks as if this equation involves second powers of x and thus is not a linear equation. However, by subtracting $6x^2$ from both sides, we get a linear equation:
$$-x - 1 = 29x - 5$$
$$-30x = -4$$
$$x = \frac{2}{15}$$

The solution is $\frac{2}{15}$.

Check:

$$(3x + 1)(2x - 1) = \left(3\left(\frac{2}{15}\right) + 1\right)\left(2\left(\frac{2}{15}\right) - 1\right) \quad \Big| \quad (6x - 1)(x + 5) = \left(6\left(\frac{2}{15}\right) - 1\right)\left(\frac{2}{15} + 5\right)$$
$$= \left(\frac{21}{15}\right)\left(-\frac{11}{15}\right) \quad \Big| \quad = \left(-\frac{3}{15}\right)\left(\frac{77}{15}\right)$$
$$= -\frac{231}{225} \quad \Big| \quad = -\frac{231}{225}$$

Operations that Don't Lead to Equivalent Equations

As we showed in the preceding section, you may multiply both sides of an equation by a nonzero number and get an equivalent equation. However, note that *multiplying and dividing an equation by an algebraic expression does not necessarily lead to an equivalent equation*. The problem is that there may be values of the variable that make the numerator or denominator of the expression 0. If such values turn up as possible solutions of the equation, you must check them separately, since they may or may not actually be solutions. The method for handling such values is illustrated in the next three examples.

EXAMPLE 8
Dividing by an Expression that May Equal Zero

Determine all solutions to the following equation:
$$(x - 1)(x - 2) = (x - 1)(2x + 3)$$

Solution

We begin by dividing both sides by $x - 1$ and then performing allowable operations:

SECTION 1.5 Equations and Identities

$$\frac{(x-1)(x-2)}{x-1} = \frac{(x-1)(2x+3)}{x-1}$$

$$x - 2 = 2x + 3$$

$$x = -5$$

This value is a solution of the original equation, because:

$$[(-5) - 1][(-5) - 2] = [(-5) - 1][2(-5) + 3]$$

$$42 = 42$$

Dividing both sides by the factor $x - 1$ is allowed provided that $x - 1 \neq 0$, that is, provided that $x \neq 1$. The excluded value, namely $x = 1$, must be tested separately. Substituting $x = 1$ in the original equation yields:

$(x-1)(x-2)$	$(x-1)(2x+3)$
$(1-1)(1-2)$	$(1-1)(2(1)+3)$
0	0

We arrive at a true statement, so $x = 1$ is a solution of the original equation.

EXAMPLE 9
Extraneous Solutions

Solve the equation:

$$\frac{1}{(x-1)^2} = \frac{x}{(x-1)^2}$$

Solution

Multiply both sides of the equation by $(x - 1)^2$:

$$\frac{1}{(x-1)^2} \cdot (x-1)^2 = \frac{x}{(x-1)^2} \cdot (x-1)^2$$

$$1 = x$$

Thus, the only possible solution of the equation is $x = 1$. However, substitution of 1 for x in the equation leads to division by 0 on both sides of the equation, so $x = 1$ is not a solution. The given equation, therefore, has no solutions.

Note how multiplication by $x - 1$ in the last example introduced a false solution to the equation. Such a solution is called an **extraneous solution.**

Squaring both sides of an equation does not necessarily lead to an equivalent equation. For example, consider the equation:

$$x = 1$$

If we square both sides, we obtain the equation:

$$x^2 = 1$$

This equation has 1 and -1 as solutions, since $(-1)^2 = 1$. However, -1 is not a solution of the original equation! The operation of squaring both sides of the equation results in an equation that has all of the original solutions plus possibly some extraneous solutions.

Remember: Whenever you perform an operation that does not necessarily lead to an equivalent equation, the values you get may be extraneous. These values must be tested individually to determine whether or not they are solutions to the equation.

Taking square roots is another operation that requires care if you are to obtain an equivalent equation. Recall that the square root of x^2 equals $|x|$. If you forget to include the absolute value, then you will usually not get an equivalent equation and may lose solutions.

EXAMPLE 10
Taking Square Roots of Both Sides of an Equation

Determine all solutions to the following equation:
$$(5x + 1)^2 = x^2$$

Solution

Both sides of the equation are perfect squares, so it seems natural to take the square root of both sides. We can do so provided that we take the absolute values of both sides:
$$(5x + 1)^2 = x^2$$
$$\sqrt{(5x + 1)^2} = \sqrt{x^2}$$
$$|5x + 1| = |x|$$

Remove the absolute value signs by inserting $\pm$, the plus-or-minus sign:
$$5x + 1 = \pm x$$

So the solutions to the original equation are:
$$5x + 1 = x \quad \text{or} \quad 5x + 1 = -x$$

That is, x must have one of these values:
$$x = -\frac{1}{4}, \quad x = -\frac{1}{6}$$

EXAMPLE 11
An Equation with No Solutions

Determine the solutions of the equation:
$$5(x + 1)(x + 2) = 5x^2 + 15x - 12$$

Solution

Multiply out the left side of the equation:
$$5(x^2 + 3x + 2) = 5x^2 + 15x - 12$$
$$5x^2 + 15x + 10 = 5x^2 + 15x - 12 \quad \textit{Subtract } 5x^2 + 15x \textit{ from each side}$$
$$10 = -12$$

This last statement is a contradiction, so the given equation is equivalent to an equation that has no solutions. The given equation is inconsistent.

Exercises 1.5

Solve the following equations.

1. $-5x + 72 = 7$
2. $3t - 13 = -39$
3. $64 - 32t = 0$
4. $160 - 32t = 0$
5. $70 - 2x = 0$
6. $20 - 2x = 1$
7. $1 - 0.2y = 0.13$

8. $0.34t + 2 = 1.8$

9. $-\frac{2}{3}x - \frac{4}{5} = \frac{7}{12}$

10. $\frac{5}{6} - \frac{1}{3}t = \frac{1}{2}$

11. $-0.02x - 0.5 = 0$

12. $-1.5x + 150 = 0$

13. $97\% \cdot x = 4{,}850$

14. $5611.5 = 64.5\% y$

15. $y + 11\% y = 721.50$

16. $x + 8\% x = 216$

17. $5x - 17 - 2x = 6x - 1 - x$

18. $3x + 4 - x = 2(x - 2)$

19. $6 - \frac{3}{4}x = \frac{1}{2}x - 4$

20. $5x + \frac{3}{4} = \frac{1}{2}x - 3$

21. $5(2x - 3) = 12x - (x + 8)$

22. $2y - (3y + 2) = 5 - 4(3 - y)$

23. $(3x - 1) - (5x + 2) = 5 - [5x - (x + 2)]$

24. $6(x - 4) + 3x = 3 - [2x + 3(x - 5)]$

25. $\frac{1}{2}(5x + 3x - 2x) = \frac{3}{2}(x + 1)$

26. $\frac{1}{2}(6x + 2) = \frac{1}{3}(6 + 9x)$

27. $-3(2x-5)-[3+5(x-1)-3x] = 25-[5x+2-3(x+2)]$

28. $-4(3y + 1) - [4 - 2(2y - 5) - 3(y + 2)]$
$= 45 - [-2y + 4 - 4(y + 7)]$

29. $2.6t - 4.5t + 40 = -7.8t + 1.3t - 80$

30. $\frac{2}{3}y - \frac{1}{6}y - 2 = -\frac{3}{2}y + \frac{3}{4}y - 10$

31. $(x - 2)(x + 3) = (x + 5)(x - 7)$

32. $(y + 1)(y + 2) = (y - 2)(y + 6)$

33. $(2y + 1)(3y + 4) = (6y - 2)(y + 5)$

34. $(4t - 1)(2t - 5) = (t - 3)(8t + 3)$

Applications

35. **Hourly earnings.** Refer to Example 5. When will hourly earnings equal \$12.75?

36. **Hourly earnings.** Refer to Example 5. What were hourly earnings in 1991?

37. **Income of private law offices.** Refer to Example 6. When did the income of private law offices equal \$85,000 a year?

38. **Income of private law offices.** Refer to Example 6. What was the income of a private law office in 1993?

Technology

Use a calculator to solve the following linear equations.

39. $7.77x = 3.88x - 4.09$

40. $3105(1 - x) = 7301x$

41. $5E5x + 3.1E4 = 3.8594E5$

42. $4E10x + 1E10 = 3E10(2x - 1)$

43. $8.771 - 3.47(7.81 - 4.33x) = 11.98$

44. $8.983012x = 48.30927583x$

In your own words

45. Describe a procedure for solving a linear equation. Illustrate the procedure with an example.

46. Explain what an extraneous solution is. Give an example.

47. Give examples of three operations that lead to equivalent equations. Give an illustration of each.

48. Why does cubing both sides of an equation always lead to an equivalent equation but squaring does not?

49. In general, raising an equation to which powers always leads to equivalent equations? Raising to which powers does not always lead to equivalent equations?

Mathematics and the World around Us—Unbreakable Codes

When you read the word "codes," your first thoughts are probably of military intelligence operations, spies, and cloak-and-dagger intrigue. While these are all bound up with codes, a more current and personal connection is with the data that are stored in the millions of computers around the world. Such data include much sensitive information, including scientific secrets, business plans, financial records, medical files, and personnel records.

To make confidential data secure from electronic snooping, it is common to store the data in coded form. The most secure codes, only discovered in the early 1970s, are called **public key encryption systems;** they are as close to being unbreakable as any codes ever devised. What's more, they are based on a very simple mathematical principle: Factoring is much harder than multiplication.

To see this principle in action, use a stopwatch to calculate how long it takes you to multiply 94 times 87. Now try to factor the number 8,633. Unless you are a calculating prodigy, factoring this number will take you quite a while. (The answer is 89×97.) The point is that both 89 and 97 are primes and are big enough that if you use trial and error to factor the number, you won't readily discover them.

The difficulty we just demonstrated in factoring a four-digit number is magnified in attempting to factor larger numbers. For example, consider a number N that is the product of two primes pq that are each about 200 digits long. To factor such a number by trial and error would take even the fastest computer hundreds of billions of years! Yet multiplying p and q could be carried out by the same computer in a fraction of a second. For all practical purposes, it is impossible for someone to factor N using any known technique.

The idea of public key encryption is to publish a key that the public can use to send someone messages. This key is determined by the receiver, using his or her knowledge of p and q. She or he tells the world at large a recipe for using the key to encrypt a message. The recipe is set up so that decryption requires knowledge of p and q, so no one but the receiver can decrypt the message.

There are a number of recipes that can be used. They all involve a field of mathematics called **number theory**, but are simple enough for the novice to comprehend. If you are interested, see Martin Hellman, "Public Key Encryption Systems," *Scientific American,* 241, (2), (1979): pp. 146–157.

Public key encryption systems were recently in the headlines. In 1976, R. Rivest, L. Adelman, and A. Shamir, the inventors of public key encryption systems, created a key that consisted of 129 digits, obtained by multiplying two large primes. Then they encrypted a message using this key and posted a reward of $100 for anyone who could decrypt the message. In 1994 the reward was collected by A. Lenstra of Bellcore, the Bell Communications research division. It took 600 participants working for eight months on 1600 computers to crack this code. The novel method of collaboration involved exchange of data on Internet, the computer communications network.

Chapter Review

Important Concepts—Chapter 1

- algebra
- real numbers
- counting numbers
- rational numbers
- closure
- additive identity
- multiplicative identity
- additive inverse
- multiplicative inverse; reciprocal
- rules of arithmetic
- commutative laws
- associative law
- distributive law
- properties of negatives
- order of operations
- dividing by 0
- subtraction; difference
- quotient
- set
- R
- element
- subset
- set-builder notation
- number line
- origin
- one-dimensional coordinate system
- one-to-one correspondence
- greater than
- less than
- inequality symbols
- positive number
- nonnegative number
- absolute value
- distance between points on the number line
- base
- exponent
- compound interest formula
- square root
- principal square root
- nth root
- principal nth root
- radical
- rational exponent
- zero power
- negative exponent
- laws of exponents
- eliminating negative and zero exponents
- properties of radicals
- scientific notation
- mantissa
- algebraic expression
- monomial
- coefficient of a polynomial
- degree of a polynomial
- polynomial
- term of a polynomial
- leading term of a polynomial
- leading coefficient of a polynomial
- constant term of a polynomial
- equal polynomials
- zero polynomial
- constant polynomial
- linear polynomial
- quadratic polynomial
- cubic polynomial
- quartic polynomial
- polynomial in several variables
- addition and subtraction of polynomials
- similar polynomials
- multiplication of polynomials
- factoring polynomials
- product identities
- square of a binomial sum
- square of a binomial difference
- product of a binomial sum and difference
- product of two binomials
- factoring
- greatest common factor
- monomial factor
- factorization identities
- difference of squares
- sum of cubes
- difference of cubes
- factoring polynomials
- factoring by grouping
- rational expression
- numerator
- denominator
- reducing a rational expression to lowest terms
- canceling; reducing to lowest terms
- addition and subtraction of rational expressions
- multiplication and division of rational expressions
- rationalizing the denominator
- rationalizing the numerator
- equation
- identity
- conditional equation
- inconsistent equation
- solving an equation
- root; solution
- solution set
- equivalent equation
- operations that generate equivalent equations
- linear equation
- methods for solving a linear equation
- clearing the denominators
- equations that reduce to linear equations
- operations that don't lead to equivalent equations
- extraneous solution

Important Results and Techniques—Chapter 1

0 is the additive identity	$0 + a = a + 0 = a$	p. 3
1 is the multiplicative identity	$a \cdot 1 = 1 \cdot a = a$	p. 3
$-a$ is the additive inverse of a	$a + (-a) = 0$	p. 3
$\frac{1}{a}$ is the multiplicative inverse of a	$a \cdot \frac{1}{a} = 1$	p. 3
Commutative laws	$a + b = b + a$ $ab = ba$	p. 4
Associative laws	$a + (b + c) = (a + b) + c$ $a(bc) = (ab)c$	p. 5
Distributive laws	$a(b + c) = ab + ac$ $(a + b)c = ac + bc$	p. 5
Properties of negatives	$-a = (-1)a$ $-(-a) = a$ $-(a + b) = -a - b$ $a - 0 = a$ $0 - a = -a$ $(-a)(-b) = ab$ $(-a)b = -ab = a(-b)$	p. 5
Compound interest formula	Suppose that the interest on principal P is compounded k times per year and that the annual rate is r. Then the balance after n years is given by the formula: $\left[\begin{array}{c}\text{balance after} \\ n \text{ years}\end{array}\right] = P\left(1 + \frac{r}{k}\right)^{nk}$	p. 16
Laws of exponents	$a^r \cdot a^s = a^{r+s}, \quad \frac{a^r}{a^s} = a^{r-s}$ $(a^r)^s = a^{rs}$ $(ab)^r = a^r b^r$ $\left(\frac{a}{b}\right)^r = \frac{a^r}{b^r}$ $a^{-r} = \frac{1}{a^r}$ $\left(\frac{a}{b}\right)^{-r} = \left(\frac{b}{a}\right)^r$	p. 21
Properties of radicals	$\sqrt[n]{ab} = \sqrt[n]{a}\sqrt[n]{b}$ $\sqrt[n]{\frac{a}{b}} = \frac{\sqrt[n]{a}}{\sqrt[n]{b}}$ $\sqrt[n]{a^n} = \begin{cases} a & \text{if } n \text{ is odd} \\ |a| & \text{if } n \text{ is even} \end{cases}$	p. 23
Product and factorization identities	Product of a Binomial Sum and Difference: $(A + B)(A - B) = A^2 - B^2$	p. 32, 36
	Square of a Binomial Sum: $(A + B)^2 = A^2 + 2AB + B^2$	

Square of a Binomial Difference:
$$(A - B)^2 = A^2 - 2AB + B^2$$

Product of Two Binomials:
$$(A + B)(C + D) = AC + AD + BC + BD$$

Sum of Cubes:
$$A^3 + B^3 = (A + B)(A^2 - AB + B^2)$$

Difference of Cubes:
$$A^3 - B^3 = (A - B)(A^2 + AB + B^2)$$

Review Exercises—Chapter 1

Translate each of the following to algebraic expressions.

1. The square root of the sum of two numbers
2. The square root of the sum of the squares of two numbers
3. A number minus 37 percent of itself
4. The sum of two numbers divided by the difference of the same two numbers

Consider the following numbers:
-54.7, $\sqrt[3]{-25}$, $4^{5/2}$, 17, -4, $-\frac{1}{7}$, 6^0, -3^2, $9^{3/2}$, 67.2%, 0.892, 2.345, $\sqrt{3}$

5. List the numbers that are positive.
6. List the numbers that are real numbers.
7. List the numbers that are irrational.
8. List the numbers that are rational.

What law is exemplified by each of the following exercises?

9. $3(ab) = (3a)b$
10. $-9 + 0 = -9$
11. $x + (-x) = 0$
12. $a \cdot a^{-1} = 1$ for $a \neq 0$

Fill in the blank with the correct symbol: $<$, $>$, or $=$.

13. $0.019 \_\_ 0.02$
14. $78\% \_\_ 0.77$
15. $\sqrt{35} \_\_ 5.917$
16. $\sqrt[3]{16} \_\_ 2 \cdot \sqrt[3]{2}$

Perform the indicated operations and simplify.

17. $(2x - 3)(3x^2 + 4x + 1)$
18. $3(a^2 - 2)(a + 2) - (a^2 - 7a)$
19. $\dfrac{2}{t} - \dfrac{t}{3}$
20. $\dfrac{5}{y^2} + \dfrac{4}{y} - \dfrac{3}{y^3}$
21. $\dfrac{2a}{1 - 2a} - \dfrac{3}{4a^2 - 1} + \dfrac{3a}{2a + 1}$
22. $\left(\sqrt{7t} + \sqrt{5}\right)\left(\sqrt{7t} - \sqrt{5}\right)$
23. $\sqrt[3]{8x^2} + \sqrt[3]{x^5} + \sqrt{4x^3}$
24. $\dfrac{x + 4}{x^2 - x - 2} \cdot \dfrac{x - 2}{3x^2 - 4x + 1}$
25. $\dfrac{x}{x^2 + 17x + 72} - \dfrac{8}{x^2 + 15x + 56}$
26. $\dfrac{1}{(x + 1)^2} - \dfrac{1}{(x + 1)^3}$

Simplify the following expressions using the laws of exponents. Write the answers in a form that does not involve any negative or zero exponents.

27. $(9x^{-2})(-3x^{18})^{-4}$
28. $(8a^3b^5)(-4a^{-5}b^2)$
29. $\left(\dfrac{x^{-5}}{x^{-3}}\right)^{-2}$
30. $(10x^{-2}y^3)(25x^{-3}y^2)^{-1}$
31. $\left(\dfrac{27a^6b^{-5}c^2}{9a^{-4}b^3c^{-4}}\right)^{-3}$
32. $\dfrac{(2x^{-2}y^4z^{-5})^{-4}}{(3x^5y^4z)^{-2}}$

Write the following expressions in a form that involves no negative exponents or radicals.

33. $\sqrt{y^2}$
34. $\sqrt[3]{z^4}$
35. $\sqrt[4]{t^2}$
36. $\sqrt{(a^2 + b^2)^3}$
37. $\sqrt{\dfrac{x^{2/3}}{x^9}}$
38. $\sqrt[4]{a^{-8}b^{12}}$

Simplify the following expressions. Where possible, eliminate fractional or negative exponents, and use as few terms as possible.

39. $-8^{2/3}$
40. $16^{3/4}$
41. $25^{5/2}$
42. $27^{2/3} - \left(\dfrac{4}{9}\right)^{-3/2}$
43. $\dfrac{x^{-3} - y^{-3}}{x^{-2} - y^{-2}}$
44. $\dfrac{\frac{1}{a} + \frac{1}{b}}{a^2 + 2ab + b^2}$
45. $\dfrac{1}{x} + \dfrac{\frac{1}{x^3}}{1 - \frac{4}{x^4}}$
46. $\dfrac{ab - 1}{a^3b^3 - 1}$
47. $\sqrt{12x^2yz^3}$
48. $\sqrt[3]{\dfrac{27a^3}{64}}$

49. $\dfrac{\sqrt[3]{625a^4b^7}}{\sqrt[3]{5ab^2}}$

50. $\sqrt{(x+y)^5}\sqrt{(x+y)^8}$

51. $\sqrt[4]{(81a^4b^8)^3}$

52. $\sqrt[4]{\dfrac{160x^9y^{12}}{2\sqrt[5]{5xy^2}}}$

53. $\dfrac{\dfrac{2^{n+1}x^{n+1}}{(n+1)^5}}{\dfrac{2^n x^n}{n^5}}$

54. $\dfrac{\sqrt{(x-1)^{n/2}}(x-1)^{2n}}{(x-1)^{-n/4}}$

Factor if possible.

55. $8x^2 - 6 - 8x$
56. $49x^2 + 25y^2$
57. $1 + 125a^3$
58. $6x^{2a} - 5x^a - 1$
59. $x^{3a} + 27$
60. $7x^2y - 28y$
61. $4x^3 + 30x^2 + 14x$
62. $4a^6 + 108b^3$
63. $55 - 6x - x^2$
64. $a^3 - \dfrac{1}{8}$

65. $ax^2 + 2axy + ay^2 + bx^2 + 2bxy + by^2$
66. $x^2 - 10xy + 25y^2$
67. $36t^2 - 16m^2$
68. $a^{2x+4} - b^{6x+10}$
69. $\dfrac{1}{16}a^2 + \dfrac{1}{4}ab + \dfrac{1}{4}b^2 - \dfrac{1}{9}a^2b^2$
70. $x^2 - 7$ **Hint:** Let $7 = (\sqrt{7})^2$
71. $x^3 - 7$ **Hint:** Let $7 = (\sqrt[3]{7})^3$

Rationalize the denominators of the following expressions.

72. $\dfrac{5}{\sqrt{6}}$

73. $\dfrac{1 - \sqrt{a}}{1 + \sqrt{a}}$

74. $\dfrac{4}{\sqrt{3} - \sqrt{2}}$

75. $\dfrac{\sqrt[3]{18}}{\sqrt[3]{5}}$

Rationalize the numerators of the following expressions.

76. $\dfrac{\sqrt{2}}{2}$

77. $\dfrac{\sqrt[4]{t}}{5t}$

78. $\dfrac{\sqrt{3+h} - \sqrt{3}}{h}$

79. $\sqrt{n^2 - n} - n$

80. $\dfrac{\sqrt{x+3} - \sqrt{x-3}}{\sqrt{x+3} + \sqrt{x-3}}$

Simplify the following expressions.

81. $\sqrt{1 - a^2} + a^2(1 - a^2)^{1/2}$

82. $\left(x\sqrt{3} + x\sqrt{2}\right)^2$

83. $\dfrac{\sqrt{1+x}}{\sqrt{1-x}} - \dfrac{2(x-1)}{\sqrt{1-x^2}} - \dfrac{\sqrt{4-4x}}{\sqrt{1+x}}$

84. Simplify.

Solve the following equations.

85. $2(2x + 4 - 3x) = 8 - 2x$
86. $4(3x+5) - 25 - 6(4x-3) + 8x - 90 = 34 - 3x + 6 - 8(x + 4)$
87. $(2x-5)(3x+4) = (x+2)(6x-2)$
88. $\dfrac{x}{3} - \dfrac{3x-5}{2} = 8$

Applications

Convert to scientific notation.

89. **Chemistry.** 2467 (boiling point of aluminum in degrees Celsius)

90. **Baseball.** 0.00003 (the probability that Joe DiMaggio would get at least one hit in each of 56 consecutive games)

Convert to decimal notation.

91. **Unit conversion.** 6.48×10^{-2} (the number of grams in one grain)

92. **Astronomy.** 2.4×10^{13} (the distance of the star *Alpha Centauri* from Earth)

93. **Physics.** A tank contains water H meters deep. A hole is pierced in a wall of the tank h meters below the water surface. (See figure.) The stream of water coming from that hole will hit the floor x meters from the wall, where x is:

$$x = \sqrt{h(H - h)}$$

An above-ground swimming pool has water in it up to 3 m above the ground. A hole is pierced in the pool 2.5 meters above the ground. How far from the swimming pool will the water hit the ground?

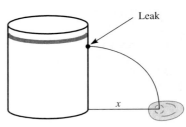

Exercise 93

94. **Chemistry.** One mole of any element contains 6.02×10^{23} atoms. Given the mass of one mole, as follows, find the mass of one atom of:

a. Hydrogen, 1.0079 g/mole

b. Aluminum, 26.98154 g/mole

c. Lead, 207.2 g/mole

95. **Geometry.** The hypotenuse of a triangle is twice as long as the shortest leg of the triangle. Find a formula for the length of the other leg.

96. **Geometry.** A circle is inscribed in a square. The area of the circle is 200 cm². (See figure.) Find the length of a side of the square.

97. **Geometry.** A 45°, 45°, 90° triangle has legs of length a. Find the length of the hypotenuse.

98. **Geometry.** A cylindrical can has a height twice its diameter and has a surface area of 300 square inches. Determine the dimensions of the can.

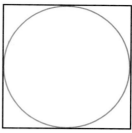

Exercise 96

Chapter Test

Solve the following problems.

1. Simplify the following expressions.
   a. $5(x - y) - 3(x + 2y)$
   b. $(-4)[-x + 2(x - 1)]$
   c. $\left(-\dfrac{1}{3}\right)^{-1}$

2. Simplify the following expressions.
   a. $(x^3)^7$
   b. $(x^4)^2 x^3 x^5$
   c. $\left(\dfrac{xy^2}{x^3 y}\right)^2$

3. Simplify the following expressions, eliminating all negative and fractional exponents.
   a. $\left(\sqrt{x}\right)^{12}$
   b. $\left(\dfrac{x^2}{y^{-1}}\right)^{-1}$
   c. $(x^{2/3})^4 x^{-1/3} \cdot \dfrac{1}{\sqrt[3]{x}}$

4. Determine the distance between the points 5 and -3 on the number line.

5. Replace the ___ with one of these symbols: $>$, $<$, or $=$
   a. -1 ___ -5
   b. $5(3 - 1)$ ___ 15
   c. $-(1 - 4)$ ___ 3

6. Simplify the following expressions.
   a. $(2x - 3y)^2$
   b. $(x - 7)(3x + 1)$
   c. $(x - 1)(x^2 + x + 1)$

7. Factor the following expressions.
   a. $x^2 + 6x + 9$
   b. $x^2 - 5x - 6$
   c. $x^3 y^3 - z^3$

8. Write the following rational expressions in lowest terms.
   a. $\dfrac{x^2 + 2x + 1}{x^2 - 1}$
   b. $\dfrac{3x^2 - 3x}{x^3 - 1}$

9. Simplify the following rational expressions and write in lowest terms.
   a. $\dfrac{x - 2}{x - 1} + \dfrac{3}{x + 1}$
   b. $\dfrac{x + 1}{x - 1} - 1$

10. Simplify the following rational expressions and write in lowest terms.
    a. $\dfrac{x^2}{x + 2} \cdot \dfrac{x^2 + 4x + 4}{x^5}$
    b. $\dfrac{x + 5}{x} \div x^3$

11. Suppose that $500 is deposited in a savings account paying interest at an annual rate r compounded semiannually. Express the amount in the account after one year as a polynomial in r.

12. Solve the following linear equation:
    $3(x - 1) = 2 - 4(-2x + 1)$

13. **Thought question.** Explain how scientific notation works. Give an example to show how scientific notation can result in simpler expressions for some numbers.

14. **Thought question.** Write a description of the method you would use to solve a linear equation. Illustrate your method with a concrete linear equation.

CHAPTER 2
GRAPHS AND FUNCTIONS

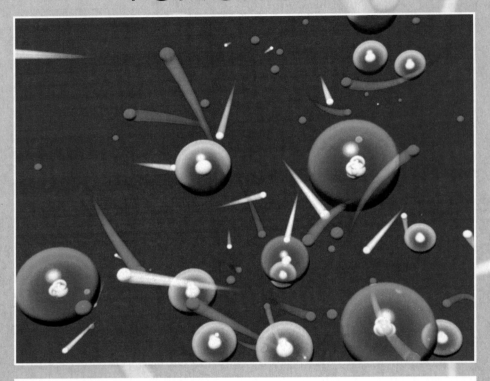

The concept of a function is a central idea in modern mathematics. It arose as a generalization of the concept of a formula, which expresses one variable in terms of another. Relationships among variables arise naturally in applications.

For example, the graph of Figure 1 expresses the relationship between elapsed time and the number of people who have heard a rumor. The graph shows how the rumor is slow to get started and then picks up steam as it spreads. As time passes, most of the population has heard the rumor, and the graph flattens out, indicating the slower spread of the rumor among the few remaining people. This graph is an example of a logistic curve, a curve that arises in many other mathematical models in such disparate fields as ecology and epidemiology.

Figure 2 shows the relationship between elapsed time and the amount remaining of a sample of radioactive carbon 14. The data contained in this graph

Chapter Opening Image: *Computer visualization of the expanding universe shortly after the big bang.*

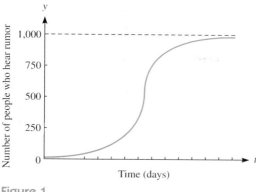

Figure 1

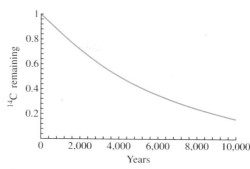

Figure 2

are used by archaeologists in applying the process of carbon dating to determine the ages of ancient artifacts such as mummies and papyrus scrolls.

In each of the two graphs above, there is a relationship between the variable time t and another variable, the number N of people having heard the rumor or the amount A of carbon 14 remaining. For each value of t, the relationships determine corresponding values of N or C. Such relationships among variables are examples of functions, the subject of this chapter.

2.1 TWO-DIMENSIONAL COORDINATE SYSTEMS

Coordinate Systems in a Plane

In Chapter 1, we showed how the points on the number line are represented by real numbers. Let's now consider a plane. How can we represent the points in a plane using numbers? It's really quite simple. Just use a pair of numbers (x, y) to represent a point. Here are the details for doing this.

Draw two perpendicular number lines in a plane, as shown in Figure 3. The perpendicular lines are called **axes**. Most often, the horizontal axis is labeled the **x-axis** and the vertical axis is labeled the **y-axis**. However, in applications, the axes may have various meanings. For example, suppose that we wish to explore the relationship between time t and the velocity v of a race car. In this case, we would label the horizontal axis t and the vertical axis v.

The point where the axes intersect is called the **origin**, and is customarily denoted by the letter O. (See Figure 4.)

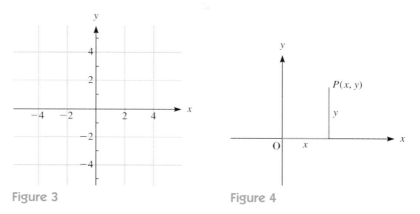

Figure 3   Figure 4

A pair of perpendicular number lines in the plane is called a **two-dimensional coordinate system** or a Cartesian coordinate system. Each point P in the plane may now be identified by a pair of real numbers (x, y): x is the directed horizontal distance from *P* to the origin, and *y* is the directed vertical distance from *P* to the origin. (See Figure 4.) **Directed distance** means that movements to the right correspond to positive values of *x*, whereas movements to the left correspond to negative values of *x*. Similarly, vertical movements upward correspond to positive values of *y* and vertical movements downward correspond to negative values of *y*.

For example, consider the point *P* in Figure 5. To get to this point from the origin, we must go 3 units to the right and 2 units upward. So the point corresponds to the pair (3, 2).

Similarly, consider the point *P* in Figure 6. To get to this point from the origin, we must go 2 units to the right and 4 units downward. Because downward motion is negative, the point corresponds to the pair (2, −4).

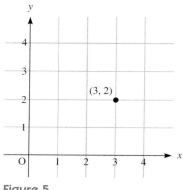

Figure 5

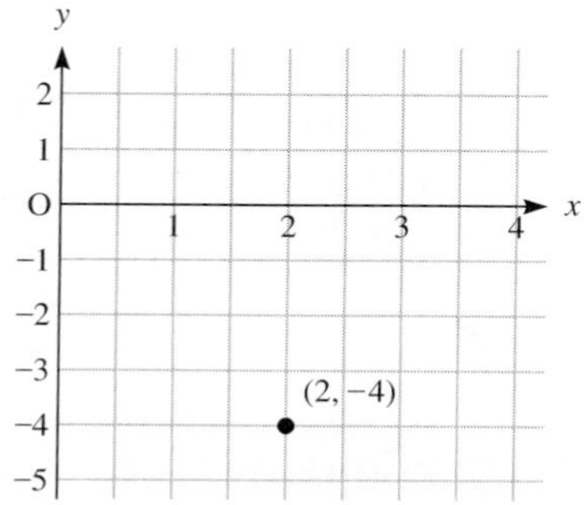

Figure 6

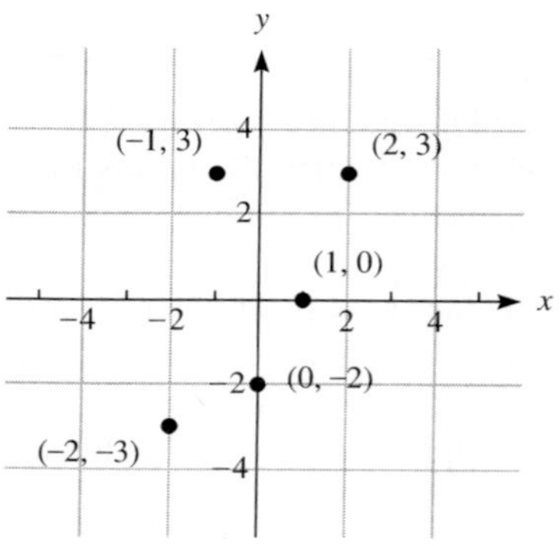

The number *x* is called the **x-coordinate**, or **abscissa**, of *P*. The number *y* is called the **y-coordinate**, or **ordinate**, of *P*.

Figure 7 shows the coordinate representations of various points in the plane. Here are the coordinates of some special points:

1. The origin has both coordinates 0: (0, 0).
2. A point on the *x*-axis has *y*-coordinate 0: (*x*, 0).
3. A point on the *y*-axis has *x*-coordinate 0: (0, *y*).

Figure 7

EXAMPLE 1
Plotting Survey Data

An airline does a survey of the travel habits of its customers and compares their annual income with the number of flights they take each year. Four of the survey participants give the following data:

Number of Flights per Year	Annual Income (Thousands)
10	38
12	42
7	28
15	45

Graph these data in a two-dimensional coordinate system in which the horizontal axis measures the number of flights per year and the vertical axis measures the annual income.

SECTION 2.1 Two-Dimensional Coordinate Systems

Solution

See Figure 8.

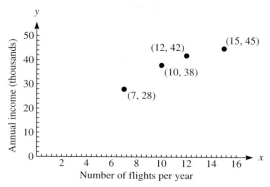

Figure 8

Warning: The coordinate representation (x, y) should not be confused with the open interval from x to y. Even though the same notation is used, it will be clear from context which is meant.

The coordinate axes divide the plane into four regions called **quadrants**. The quadrants are numbered counterclockwise starting from the positive x-axis, as shown in Figure 9. In a particular quadrant, the abscissa and ordinate have constant signs:

Quadrant	x-Coordinate	y-Coordinate
I	positive	positive
II	negative	positive
III	negative	negative
IV	positive	negative

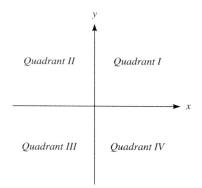

Figure 9

The Distance Formula

We can compute the distance between two points in terms of their coordinates using the following formula:

DISTANCE FORMULA

Let $P(x_1, y_1)$ and $Q(x_2, y_2)$ be two points. Then the distance between them is given by
$$d(P, Q) = \sqrt{(x_2 - x_1)^2 + (y_2 - y_1)^2}$$

Figure 10 illustrates the distance formula.

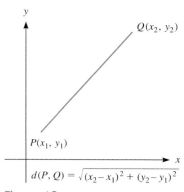

Figure 10

Verification Draw the line segment connecting P and Q and put it in a triangle with sides parallel to the horizontal and vertical axes, as shown in Figure 11. We see that the horizontal side has length $|x_2 - x_1|$ and that the vertical side has length $|y_2 - y_1|$. Because triangle PQR is a right triangle, we may apply the Pythagorean theorem to compute the length of the hypotenuse:

72 CHAPTER 2 Graphs and Functions

$$d(P, Q) = \sqrt{(RP)^2 + (RQ)^2}$$
$$= \sqrt{|x_2 - x_1|^2 + |y_2 - y_1|^2}$$
$$= \sqrt{(x_2 - x_1)^2 + (y_2 - y_1)^2}$$

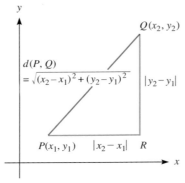

Figure 11

EXAMPLE 2
Calculating Distances

1. Determine the distance between the points $(-1, 3)$ and $(-2, -1)$.
2. Determine the closest distance between $(-1, 4)$ and the x-axis.

Solution

1. By the distance formula, the distance is equal to:

$$\sqrt{((-2) - (-1))^2 + ((-1) - 3)^2} = \sqrt{(-1)^2 + (-4)^2}$$
$$= \sqrt{1 + 16}$$
$$= \sqrt{17}$$

2. The point on the x-axis closest to $(-1, 4)$ is the point on the axis that has the same x-coordinate, namely $(-1, 0)$. The distance is shown in Figure 12. By the distance formula, the distance equals:

$$\sqrt{(-1 - (-1))^2 + (4 - 0)^2} = 4$$

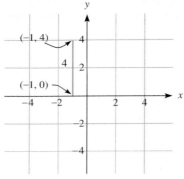

Figure 12

The Midpoint Formula

In many problems, especially those arising in geometry and physics, it is necessary to determine the coordinates of the midpoint of a line from the coordinates of the endpoints. The following formula asserts that the coordinates of the midpoint may be obtained by averaging the coordinates of both the endpoints. More precisely, we have the following result:

MIDPOINT FORMULA

The midpoint of the line segment with endpoints $P(x_1, y_1)$ and $Q(x_2, y_2)$ is the point

$$M\left(\frac{x_1 + x_2}{2}, \frac{y_1 + y_2}{2}\right)$$

SECTION 2.1 Two-Dimensional Coordinate Systems

Figure 13 illustrates the midpoint formula.

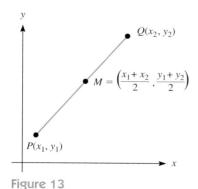

Figure 13

Verification Let's prove the result by establishing the equalities:

$$d(M, P) = \frac{1}{2}d(P, Q) \quad (*)$$

$$d(M, Q) = \frac{1}{2}d(P, Q) \quad (**)$$

These equalities state that M is equidistant from P and Q. Moreover, we can deduce from them that:

$$d(M, P) + d(M, Q) = \frac{1}{2}d(P, Q) + \frac{1}{2}d(P, Q)$$
$$= d(P, Q)$$

That is, the distance from P to M plus the distance from M to Q equals the distance from P to Q. From an elementary result in geometry, this implies that M lies on the line segment $\overline{PQ}$. To prove the equality (*), we apply the distance formula:

$$d(M, P) = \sqrt{\left(\frac{x_1 + x_2}{2} - x_1\right)^2 + \left(\frac{y_1 + y_2}{2} - y_1\right)^2}$$

$$= \sqrt{\left(\frac{x_2 - x_1}{2}\right)^2 + \left(\frac{y_2 - y_1}{2}\right)^2}$$

$$= \sqrt{\frac{(x_2 - x_1)^2 + (y_2 - y_1)^2}{4}}$$

$$= \frac{1}{2}\sqrt{(x_2 - x_1)^2 + (y_2 - y_1)^2}$$

$$= \frac{1}{2}d(Q, P) \quad \text{Distance formula}$$

$$= \frac{1}{2}d(P, Q) \quad \text{The distance from } P \text{ to } Q \text{ is equal to the distance from } Q \text{ to } P.$$

This proves the equality (*). The proof of the equality (**) is similar, and is omitted here. This completes the proof of the midpoint formula.

EXAMPLE 3
Calculating a Midpoint

Determine the midpoint of the line segment with endpoints $P(4, 1)$ and $Q(8, -5)$.

Solution

By the midpoint formula, the coordinates of the midpoint of the line segment are

$$\left(\frac{x_1 + x_2}{2}, \frac{y_1 + y_2}{2}\right) = \left(\frac{4 + 8}{2}, \frac{1 + (-5)}{2}\right) = (6, -2)$$

EXAMPLE 4
Interpolating Experimental Data

Suppose that the graph of salary versus years of teaching experience for a certain school district is a straight line. Further suppose that a teacher with 5 years of experience earns $31,000 a year, whereas a teacher with 9 years of experience earns $39,000 a year. How much does a teacher with 7 years of experience earn?

Solution

The graph of salary versus years of teaching experience is the straight line through the points (5, 31,000) and (9, 39,000). Because 7 is halfway between 5 and 9, the point with x-coordinate 7 is the midpoint of this line. By the midpoint formula, this point is:

$$\left(\frac{5+9}{2}, \frac{31{,}000 + 39{,}000}{2}\right) = (7, 35{,}000)$$

So a teacher with 7 years of experience earns $35,000 a year.

Exercises 2.1

Plot the following points on a Cartesian coordinate system.

1. a. $(3, 2)$
   b. $(3, -1)$
   c. $(0, 2)$
   d. $(-3, 0)$

2. a. $(0, -4)$
   b. $(-2, 3)$
   c. $(5, 0)$
   d. $(3, 4.5)$

3. For each point in Exercise 1, determine the quadrant in which it lies, or state the axis it lies on.

4. For each point in Exercise 2, determine the quadrant in which it lies, or state the axis it lies on.

Determine the distance between the points in Exercises 5–16.

5. $(2, 2)$ and $(-4, -3)$
6. $(6, 10)$ and $(-1, 5)$
7. $(0, -5)$ and $(2, -3)$
8. $(-3, -3)$ and $(3, 3)$
9. $(-2, 5)$ and $(4, 5)$
10. $(-3, 4)$ and $(-3, 8)$
11. $(2a, 3)$ and $(-a, 5)$
12. $(x, 0)$ and $(-2, -3)$
13. (a, b) and $(0, 0)$
14. $(\sqrt{3}, -2)$ and $(0, \sqrt{5})$
15. $(\sqrt{a}, \sqrt{b})$ and $(0, 0)$
16. $(a+b, a-b)$ and $(a-b, a-b)$
17. Determine the distance between $(-1, 4)$ and the y-axis.
18. Determine the distance between $(-2, 3)$ and the x-axis.
19. Determine the distance between the point (a, b) and the origin.
20. Determine the distance between the point (a, b) and the x-axis.

Determine the midpoints of the segments that have the following endpoints.

21. $(2, 2)$ and $(-4, -3)$
22. $(6, 10)$ and $(-1, 5)$
23. $(0, -5)$ and $(2, -3)$
24. $(-3, -3)$ and $(3, 3)$
25. $(-2, 5)$ and $(4, 5)$
26. $(-3, 4)$ and $(-3, 8)$
27. $(2a, 3)$ and $(-a, 5)$
28. $(x, 0)$ and $(-2, -3)$
29. (a, b) and $(0, 0)$
30. $(\sqrt{3}, -2)$ and $(0, \sqrt{5})$
31. $(\sqrt{a}, \sqrt{b})$ and $(0, 0)$
32. $(a+b, a-b)$ and $(a-b, a-b)$

33. Find the area of the triangle with vertices $A(2, 3)$, $B(-2, -3)$, and $C(5, -3)$.

34. Find the area of the parallelogram with vertices $P(-1, -7)$, $Q(-1, 2)$, $R(3, 5)$, and $S(3, -4)$.

35. Find the point on the y-axis equidistant from the points $(-2, 3)$ and $(-5, -7)$.

36. Find the point on the x-axis equidistant from the points $(-2, 3)$ and $(-5, -7)$.

Use the distance formula to determine whether each set of three points is on a straight line. (**Hint**: For three points to be on a straight line, the sum of the distances from the outer points to the middle one equals the distance between the outer points.)

37. $(-3, -4)$, $(2, 3)$, and $(4, 7)$
38. $(-3, -4)$, $(6, 1)$, and $(10, 6)$

In Exercises 39 and 40, use the distance formula to determine whether the three points are vertices of a right triangle. (**Hint**: A triangle is a right triangle provided that the Pythagorean theorem holds true.)

39. $(-4, 7)$, $(6, 3)$, and $(-8, -3)$
40. $(-3, 6)$, $(2, 4)$, and $(6, 14)$

41. Find an equation that must be satisfied by any point (x, y) that is equidistant from $(-2, 3)$ and $(5, -4)$.

42. Find an equation that must be satisfied by any point (x, y) that is always 5 units from the point (h, k).

Use the distance formula and/or the midpoint formula to prove each of the following statements. Locate the polygons on a coordinate system. Put one of the points at the origin to make the proof easier.

43. The midpoint of the hypotenuse of any right triangle is equidistant from each of the vertices of the triangle.

44. The diagonals of a rectangle have the same length.

Applications

45. Circuit board design. Assume that a two-dimensional coordinate system is imposed on a circuit board, with the origin in the lower-left corner of the board and the sides of the board parallel to the coordinate axes. Suppose that it is necessary to connect two transistors with coordinates (7.1 mm, 3.5 mm) and (4.2 mm, 1.9 mm). How long will a straight-line connection between the transistors be?

46. Job satisfaction. A business survey studies the relationship between job satisfaction and the number of years of education received. Job satisfaction is measured by scoring a questionnaire whose results range from 0 to 100. Here are several reports from the survey:

Job Satisfaction Score	Number of Years of Education
40	8
80	10
60	14
85	16

Graph these reports on a two-dimensional coordinate system.

47. Customer loyalty. A marketing survey measures the loyalty of customers against the number of years they have driven a particular brand of car. Here are several reports from the study:

Number of Years Customers Owned Current Car	Percentage of Customers Who Would Purchase Same Brand Again
1	60
2	52
3	44
4	65
5	69
6	71

Graph these reports on a two-dimensional coordinate system.

48. Market penetration. The market penetration of Chinese textile imports into the U.S. market was 10 percent in 1987 and 17.9 percent in 1993. Estimate the market penetration in 1990, assuming that market penetration grew according to a straight-line model.

Technology

Use your calculator to determine the following:

49. The distance between the points (5.81, 3.99) and (−1.001, 7.99).

50. The distance between the points $\left(\frac{1}{3}, \frac{2}{9}\right)$ and $\left(-\frac{3}{4}, \frac{5}{8}\right)$.

51. The midpoint of the line connecting (1,359, 1,781) and (1,597, 2,004).

52. The midpoint of the line connecting (0.00172, −0.0035) and (−0.0019483, 0).

In your own words

53. Think of an example of a real-life object that can be thought of as a two-dimensional coordinate system. How would you define distance in this system?

54. Explain the procedure for plotting points on a two-dimensional coordinate system.

55. State the distance formula. Use it to show how to calculate the distance between two particular points.

56. State the midpoint formula. Use it to show how to calculate the midpoint of a specific line from the coordinates of its endpoints.

2.2 GRAPHS OF EQUATIONS

In the preceding section, we learned to plot individual points in a Cartesian coordinate system, given their coordinates. In many applications, it is necessary to deal not with individual points, but with sets of points. This is the case, for example, when we study equations in two variables from a graphical viewpoint. In this section, we turn our attention to graphing sets of points and, in particular, to graphing equations.

76 CHAPTER 2 Graphs and Functions

Graphing Equations Suppose that we are given an equation in two variables x and y, such as:
$$3x + 5y = 7$$
or
$$3x^2 + 4y^3 = 1$$
or
$$y = 3x^4 + 1$$

A **solution** of such an equation is an ordered pair (c, d) for which the equation is satisfied when $x = c$ and $y = d$. For example, $(0, \frac{7}{5})$ is a solution of the first equation above; if $x = 0$ and $y = \frac{7}{5}$, the equation is satisfied:
$$3(0) + 5\left(\frac{7}{5}\right) = 7$$

Given an equation in x and y, we may plot all of its solutions as points in the plane. The set of points plotted will, generally speaking, form some sort of curve, which is called the **graph of the equation**. (See Figure 1.)

Graphing an equation is a way of representing the algebraic properties of the equation geometrically. In the remainder of this section, we discuss how to sketch the graphs of various equations and how to read information from graphs.

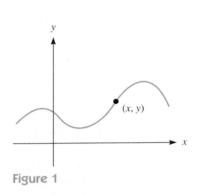
Figure 1

EXAMPLE 1
Points with Specific x- or y-Coordinates

Figure 2 shows the graph of an equation.
1. Determine all points on the graph for which $x = 1$.
2. Determine all points on the graph for which $y = 2$.

Solution
1. Draw a vertical line through the point $x = 1$ on the x-axis. (See Figure 3.) The points on this line all have x-coordinate 1. This line intersects the graph at the point $(1, 0)$, which is the only point on the graph for which $x = 1$.
2. Draw a horizontal line through the point $y = 2$. (See Figure 4.) The points on this line all have y-coordinate 2. This line intersects the graph at the points $(0, 2)$ and $(3, 2)$, which are the only points on the graph for which $y = 2$.

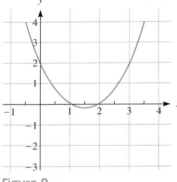
Figure 2

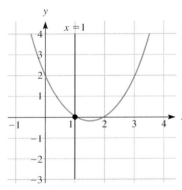
Figure 3

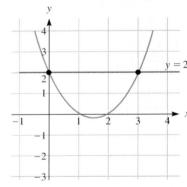
Figure 4

Graphing Linear Equations

The simplest equations in two variables have the form
$$cx + dy + e = 0$$
where c, d, and e are real numbers, with c and d not both zero. Such equations are called **linear equations in two variables**. The graphs of such equations may be described as follows:

GRAPH OF A LINEAR EQUATION IN TWO VARIABLES

1. The graph of a linear equation in two variables is a straight line.
2. Every straight line is the graph of some linear equation in two variables.

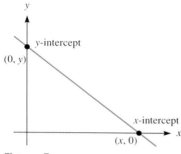
Figure 5

A point where the graph of an equation crosses the x-axis is called an ***x*-intercept**. A point where the graph of an equation crosses the y-axis is called a ***y*-intercept**.[1] (See Figure 5.) Note that the graph of an equation can, in general, have many x- and y-intercepts. (See Figure 6.)

A graph's intercepts often provide valuable geometric information that has meaning in the context of an application. For example, if we graph the price for a commodity (on the vertical axis) against its demand (on the horizontal axis), a horizontal intercept corresponds to the price at which 0 units of the commodity are sold.

Here's how to determine the intercepts of a graph.

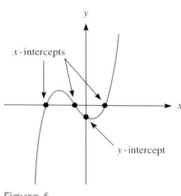
Figure 6

DETERMINING INTERCEPTS

1. x-intercept: Set $y = 0$ in the equation and solve for x. Each solution $x = a$ is an x-intercept.
2. y-intercept: Set $x = 0$ in the equation and solve for y. Each solution $y = b$ is a y-intercept of the graph.

The following example shows how to graph a linear equation by determining its intercepts.

EXAMPLE 2
Graphing a Line Using Intercepts

1. Determine the x- and y-intercepts of the graph of the linear equation:
$$2x + 3y = 1$$
2. Graph the equation.

Solution

1. To obtain the y-intercept, we substitute $x = 0$ in the given equation:
$$2(0) + 3y = 1$$
$$y = \frac{1}{3}$$

---

[1] If $(x, 0)$ is an x-intercept, we will also refer to the number x as an x-intercept. Similarly, the number y in $(0, y)$ is called a y-intercept.

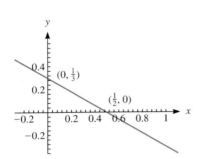

Figure 7

So the y-intercept is $\frac{1}{3}$. To obtain the x-intercept, we set y equal to 0 in the equation:

$$2x + 3(0) = 1$$
$$x = \frac{1}{2}$$

So the x-intercept is $\frac{1}{2}$.

2. To obtain the graph, use the intercepts to plot the points where the graph intersects the axes, and draw a line through the points. See Figure 7.

EXAMPLE 3
Vertical and Horizontal Lines

Graph the following linear equations.

1. $x = 1$ 2. $y = 3$

Solution

1. Because y does not appear in the equation, y may be any value, but the value of x is always 1. Therefore, the graph consists of all points with x-coordinate 1. The graph is a vertical line with x-intercept 1. See Figure 8.

2. Because x does not appear in the equation, x may be any value, but the value of y is always 3. Therefore, the graph consists of all points with y-coordinate 3. The graph is a horizontal line with y-intercept 3. See Figure 9.

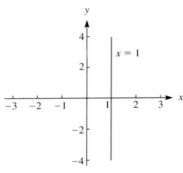

Figure 8

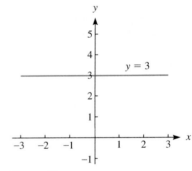

Figure 9

EXAMPLE 4

EPA Fines

The Environmental Protection Agency uses a formula to determine the level of fines levied for environmental pollution. Since 1984, fines levied by the agency have increased sharply. In year x (with 1984 corresponding to 0), the amount of fines levied y (in millions of dollars per year) is given by the model:

$$y = 4.64 + 8.83x$$

Graph this equation.

Solution

The y-intercept gives the point:

$$(0, 4.64 + 8.83 \cdot 0) = (0, 4.64)$$

The x-intercept is the point where:
$$0 = 4.64 + 8.83x$$
$$x = -\frac{4.64}{8.83} \approx -0.53$$

Plot these points and draw a graph through them. The graph is shown in Figure 10.

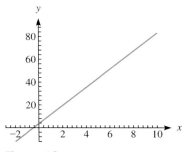

Figure 10

EXAMPLE 5
Demand for a Commodity

A supermarket observes the following relationship between the weekly sales of frozen orange juice q (measured in cans) and the price per can p (measured in dollars):
$$q = -3{,}000p + 12{,}000$$

1. Graph the linear equation representing the relationship.
2. Give an interpretation for the p-intercept of the graph.

Solution

1. In this example, the horizontal axis is labeled p and the vertical axis is labeled q. (See Figure 11.) The q-intercept is the value of q at which p equals 0. We obtain this value of q by setting p equal to 0 and solving the resulting equation:
$$q = (-3{,}000)0 + 12{,}000 = 12{,}000$$
So the q-intercept is 12,000. The p-intercept is the value of p for which the value of q is 0:
$$0 = -3{,}000p + 12{,}000$$
$$p = 4$$
So the p-intercept is 4. To graph the equation, we plot the two points at which the graph intersects the axes and draw a line through the plotted points. (See Figure 11.)

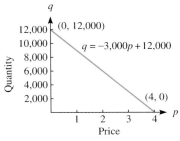

Figure 11

2. The p-intercept is the value of p at which the graph crosses the horizontal axis, that is, the point where the value of q is 0 and the value of p is 4. A value of zero for q means that no sales are being made. Because the value of p, price, is \$4 when q is zero, this graph can be interpreted to mean that the public will cease to buy orange juice when the price is \$4 per can.

EXAMPLE 6
Graphing an Absolute Value Equation

Determine the graph of the equation $y = |x + 1|$.

Solution

According to the definition of absolute value, we have

$$y = |x + 1| = \begin{cases} x + 1 & \text{if } x + 1 \geq 0 \\ -(x + 1) & \text{if } x + 1 < 0 \end{cases}$$

Therefore, the graph consists of two portions that correspond to values of x satisfying either $x \geq -1$ or $x < -1$, namely the graphs of the equations:

$$y = x + 1, \qquad x \geq -1$$

or

$$y = -x - 1, \qquad x < -1$$

Each of these equations is linear. However, the values of x are restricted. The graphs are shown in Figure 12.

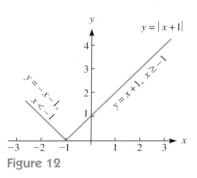

Figure 12

Graphing by Plotting Points

The simplest method for graphing an equation is to plot a number of points that lie on the graph and to connect the plotted points with a smooth curve. The next two examples illustrate this procedure.

EXAMPLE 7
Graphing a Quadratic Equation

Graph the equation:

$$y = x^2$$

Solution

We determine some points on the graph by using small values of x and determining the corresponding values of y, as shown in the following table. Because the equation expresses y in terms of x, we choose a representative set of values for x and calculate the corresponding values of y. In Figure 13, we plot the points and draw a smooth curve connecting the points. Note that as $|x|$ increases, so does y. Geometrically, this means that the graph rises as we proceed either right or left from the origin.

x	y	(x, y)
−3	9	(−3, 9)
−2	4	(−2, 4)
−1	1	(−1, 1)
0	0	(0, 0)
1	1	(1, 1)
2	4	(2, 4)
3	9	(3, 9)

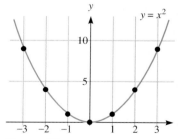

Figure 13

EXAMPLE 8
Graphing an Equation in Which x Is Expressed in Terms of y

Plot the graph of the following equation:

$$y^2 = x$$

Solution

In this example, because x is given in terms of y, it is simpler to choose representative values of y and compute the corresponding values of x. The results of the calculations are contained in the following table. These points are plotted and the resulting graph is sketched in Figure 14. Note that as $|y|$ increases, so does the value of x. Geometrically, this means that the graph heads to the right as we move either up or down from the origin.

y	x	(x, y)
−3	9	(9, −3)
−2	4	(4, −2)
−1	1	(1, −1)
0	0	(0, 0)
1	1	(1, 1)
2	4	(4, 2)
3	9	(9, 3)

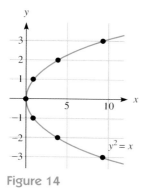

Figure 14

EXAMPLE 9
Graphing an Equation that Yields a Pair of Equations

Graph the equation:
$$x^2 + y^2 = 1$$

Solution

This equation does not allow immediate computation of either variable from the other. To obtain a set of points on the graph, it is easiest to first solve the equation for one variable in terms of the other. Solving for y in terms of x, we obtain:

$$y^2 = 1 - x^2$$
$$y = \pm\sqrt{1 - x^2}$$

Using this equation, it is a simple matter to compute a table of points on the graph. The table accompanying Figure 15 gives points corresponding to the following values of x:

$$x = 0, \quad \pm 0.2, \quad \pm 0.4, \quad \pm 0.6, \quad \pm 0.8, \quad \pm 1$$

The points are plotted in Figure 15. The graph of the equation with the $+$ sign is the upper semicircle and the graph of the equation with the $-$ sign is the lower semicircle.

x	$y = +\sqrt{1 - x^2}$	$y = -\sqrt{1 - x^2}$
−1	0	0
−0.8	0.6	−0.6
−0.6	0.8	−0.8
−0.4	0.91652	−0.91652
−0.2	0.97980	−0.97980
0	1	−1
0.2	0.97980	−0.97980
0.4	0.91652	−0.91652
0.6	0.8	−0.8
0.8	0.6	−0.6
1	0	0

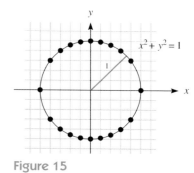

Figure 15

EXAMPLE 10
Graphing an Equation with an Undefined x-Value

Sketch the graph of the equation:
$$y = \frac{1}{x}, \quad x \neq 0$$

Solution

Whenever you sketch an equation by plotting points, it is necessary to plot a set of points that are representative of the various geometric features the graph possesses. Any kind of algebraic irregularity is likely to produce an interesting geometric feature that should be investigated in some detail. In this example, the expression $1/x$ is not defined for x equal to 0. This suggests that the graph of the equation has some unique behavior for points (x, y) when x is near 0. Therefore, in tabulating a set of points to plot, it is wise to use a number of points for which x is near 0. The table accompanying Figure 16 includes points for which x is equal to:

$$\pm 3, \quad \pm 2, \quad \pm 1, \quad \pm 0.5, \quad \pm 0.2$$

The points are plotted and the graph is sketched in Figure 16.

x	y	x	y
-3	$-\frac{1}{3}$	$\frac{1}{5}$	5
-2	$-\frac{1}{2}$	$\frac{1}{2}$	2
-1	-1	1	1
$-\frac{1}{2}$	-2	2	$\frac{1}{2}$
$-\frac{1}{5}$	-5	3	$\frac{1}{3}$

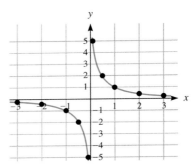

Figure 16

Based on the four examples just worked, we can state the following strategy for graphing an equation by plotting points:

GRAPHING AN EQUATION BY PLOTTING POINTS

1. Solve the equation for x in terms of y or for y in terms of x, whichever is easier.
2. If y is expressed in terms of x, choose a representative set of values for x; if x is expressed in terms of y, choose a representative set of values for y.
3. Calculate the other value corresponding to each of the representative values chosen.
4. Plot the points corresponding to the pairs (x, y) calculated.
5. Draw a smooth curve through the points, taking care, however, to make sure that the curve has breaks corresponding to undefined values of the variables.

To obtain an accurate graph, it is often necessary to plot a large number of points. The calculations involved in determining these points can be quite gruesome. Graphing calculators can alleviate some of the tedium by plotting the points for you. However, note that if you use a graphing calculator, it is almost always necessary to enter the equation solved for y in terms of x.

We should note, at this point, that plotting points will not always yield, unambiguously, the correct shape of a graph. Given a finite number of points, it is always possible to fit a number of different curves to the data. For instance, in Figure 17, we sketch several different curves that fit the same set of points. The only way to tell which of the graphs is correct is to further examine the algebraic properties of the equation that gave rise to the points. Let's now consider several pieces of information that allow us to interpret geometric properties of a graph and thereby supplement the graphing of points in determining the correct form of the graph.

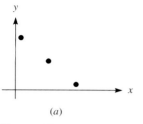 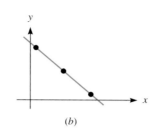 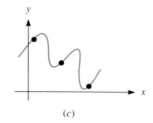 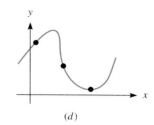

(a) (b) (c) (d)

Figure 17

Circles and Their Equations

Certain curves occur so often in applications that it is useful to know their equations in advance. Circles belong in this category.

A circle with center P and radius r is the set of all points whose distance from P equals r. (See Figure 18.) Let's determine the equation of such a circle. Let $Q(x, y)$ be a typical point on the circle and let $P(h, k)$ be the center. By the distance formula, the distance from P to Q equals

$$d(P, Q) = \sqrt{(x - h)^2 + (y - k)^2}$$

so that the condition

$$d(P, Q) = r$$

may be rewritten in the form:

$$\sqrt{(x - h)^2 + (y - k)^2} = r$$
$$(x - h)^2 + (y - k)^2 = r^2 \quad \text{Square both sides}$$

This proves the following result:

EQUATION OF A CIRCLE

Suppose that a circle has center (h, k) and radius r. Then the circle is the graph of the equation:

$$(x - h)^2 + (y - k)^2 = r^2$$

Figure 18

EXAMPLE 11 Graphing a Circle

Describe the graph of the equation:

$$(x - 1)^2 + (y + 5)^2 = 25$$

Then, draw the graph.

Solution

The equation may be written in the form:

$$(x - 1)^2 + (y - (-5))^2 = 5^2$$

This is the equation of a circle with center C at $(1, -5)$ and radius 5. We can draw the graph by first plotting the point $(1, -5)$ and then using a compass to draw the circle of the indicated radius with the plotted point as center (refer to Figure 19).

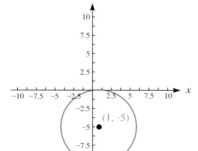

Figure 19

84 CHAPTER 2 Graphs and Functions

EXAMPLE 12
Completing the Square to Graph a Circle

Describe the graph of the equation:
$$4x^2 + 24x + 4y^2 - 8y = 41$$
Then, draw the graph.

Solution

The equation has terms x^2 and y^2 with the same coefficient. This suggests that the equation can be transformed into an equivalent equation that is the equation of a circle. We begin by dividing the equation by 4:
$$(x^2 + 6x) + (y^2 - 2y) = \frac{41}{4}$$

Now we make the expressions in parentheses into perfect squares by adding appropriate numbers to each expression in parentheses and to the other side of the equation. For the first, add $(\frac{6}{2})^2 = 9$ and for the second add $(-\frac{2}{2})^2 = 1$. (This procedure is called completing the square, and is discussed further when we cover quadratic equations in the next chapter.)
$$(x^2 + 6x + 9) + (y^2 - 2y + 1) = \frac{41}{4} + 9 + 1$$
$$(x + 3)^2 + (y - 1)^2 = \frac{81}{4}$$

To recognize this as the equation of a circle, we write it in the form
$$(x - (-3))^2 + (y - 1)^2 = \left(\frac{9}{2}\right)^2$$

This is the equation of a circle with center at $(-3, 1)$ and radius $\frac{9}{2}$. See Figure 20.

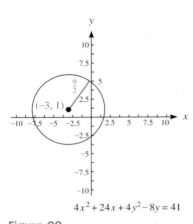

$4x^2 + 24x + 4y^2 - 8y = 41$

Figure 20

Exercises 2.2

1. Determine whether $(2, -1)$ is a solution of $-2x - 16y = 12$.
2. Determine whether $(5, -5)$ is a solution of $7x + 5y = 5$.
3. Determine whether $(1, -4)$ is a solution of $2x - 3y^2 = 46$.
4. Determine whether $(1, -1)$ is a solution of $2x^2 + y^2 = 3$.
5. Determine whether $(4, -2)$ is a solution of $-y = \sqrt{x}$.
6. Determine whether $(0, 2)$ is a solution of $y = \frac{1}{4}(x - 2)^2$.
7. Determine whether $(-2, 15)$ is a solution of $(x-3)^2 + 2(x-3) = y$.
8. Determine whether $(\frac{1}{8}, 16)$ is a solution of $x^{-2/3} = y^{1/2}$.

Sketch the graph of each equation by plotting a representative set of points. Use a calculator, where necessary, to calculate points on the graph.

9. $y = 2x + 3$
10. $y = 4 - x$
11. $y = -2$
12. $x = 4$
13. $2x - y = 4$
14. $x + y = 1$
15. $-4x - 3y = 12$
16. $2x + 5y = 10$
17. $y = x^2 + 1$
18. $y = 1 - x^2$
19. $x = y^2 + 3$
20. $x = -y^2$
21. $y = -\frac{1}{x}$
22. $y = \frac{2}{x}$
23. $y = |x|$
24. $y = |2x - 1|$
25. $y = x^3$
26. $x = y^3$
27. $x = |y|$
28. $x = |y + 3|$
29. $xy = 1$
30. $xy = -0.25$
31. $y = \frac{1}{x^2}$

32. $y = \dfrac{1}{|x|}$
33. $x^2 + y^2 = 4$
34. $x^2 + y^2 = 25$
35. $(x - 4)^2 + y^2 = 7$
36. $x^2 + (y + 1)^2 = 5$
37. $(x + 3)^2 + (y - 7)^2 = 13$
38. $(x - 6)^2 + (y + 2)^2 = 11$
39. $x^2 + y^2 + 12x + 19 = -1$
40. $x^2 + 6x + y^2 - 16y + 48 = 0$
41. $x^2 + x + y^2 - y - 5 = 0$
42. $x^2 + y^2 = 4y$

Find an equation in the standard form of a circle that satisfies the following conditions.

43. Center: $(-6, 1)$; radius: 2
44. Center: $(3, -4)$; radius: $\sqrt{3}$
45. Center at the origin; radius: $\sqrt{39}$
46. Center: $(0, a)$; radius: 0.1

Applications

47. **Demand for wheat.** The relationship between the demand for wheat y on the international market (in millions of tons per year) and the price of wheat x (in dollars per bushel) is estimated by a market analyst to be given by the model:

$$y = -3.11x + 8.9$$

   a. Sketch the graph of this equation.
   b. Interpret the x- and y-intercepts of the graph.

48. **Growth in sales.** The sales y of digital audio tape recorders (in millions of units) in year x after 1994 are estimated to be given by the model:

$$y = 2.1x + 0.15$$

   a. Sketch the graph of this equation.
   b. Interpret the y-intercept of the graph.
   c. Use the graph to estimate the number of digital audio tape recorders to be sold in 1998.

49. **Decrease in market share.** A company faced with foreign competition estimates that its market share y (measured as a percentage) in year x (measured with 1980 equal to 0) decreased according to the model:

$$y = -3x + 95$$

   a. Sketch the graph of this equation.
   b. Interpret the y-intercept of the graph.
   c. Use the graph to estimate the market share of the company in 1995.

50. **Supply of computer memory chips.** The relationship between the supply of SIMM computer memory chips y (in millions of units) and the price x (in dollars per megabyte) is given by the model:

$$y = 3.75x + 4.9$$

   a. Sketch the graph of this equation.
   b. Interpret the x- and y-intercepts of the graph.

51. **Income of private legal offices.** The income of private legal offices y (in thousands of dollars per year) in year x (with 1982 corresponding to 0) is given by the model:

$$y = 33.9 + 6.5x$$

   a. Sketch the graph of this equation.
   b. Interpret the y-intercept of the graph.

52. **Hourly earnings.** The hourly earnings y of U.S. workers in year x (with 1990 corresponding to 0) is given by the model:

$$y = 11 + 0.27x$$

   a. Sketch the graph of this equation.
   b. Interpret the y-intercept of the graph.

53. **Itemized deductions for state and local taxes.** Based on Internal Revenue Service data, the amount y of itemized deductions for taxes on a tax return that reports x dollars of income is given by the model:

$$y = 295.3 + 0.021689x$$

   a. Sketch the graph of this equation.
   b. Interpret the y-intercept of the graph.

86 CHAPTER 2 Graphs and Functions

54. Spread of an organism. A certain kind of dangerous organism is accidentally released over an area of 3 square miles. (See figure.) After time t (in years) it spreads over area A (in square miles) where A and t are related by the equation:

$$A = 1.8t + 3$$

a. Graph the linear equation representing the relationship between elapsed time and area.

b. Give an interpretation of the y-intercept.

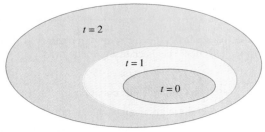

Exercise 55

55. Itemized deductions for health care. Based on Internal Revenue Service data, the amount y of itemized deductions for health care and health insurance on a tax return that reports x dollars of income is given by the model:

$$y = 49.86 + 0.077174x$$

a. Sketch the graph of this equation.

b. Interpret the y-intercept of the graph.

56. Driving costs. The cost C (in cents per mile) of driving a car is related to the number of years t (since 1980) by the equation

$$100C - 7t = 2,320$$

where $t = 0$ corresponds to 1980, $t = 1$ corresponds to 1981, and so on.

a. Graph the linear equation representing the relationship between cost and elapsed time. Assign time t to the x-axis and cost C to the y-axis.

b. Give an interpretation of the y-intercept.

Technology

Use your calculator to determine the point(s) on the graph of the indicated equation that corresponds to $x = 5$.

57. $y = 3.1x^2 - 4.75x + 11$

58. $y^2 = 3\sqrt{x}$

59. $y = \dfrac{1}{2x - 3}$

60. $2x^2 + 3y^2 = 100$

In your own words

61. Define the graph of an equation.

62. Can the graph of an equation have more than a single point with a particular x-coordinate? Explain, using examples.

63. Can the graph of an equation have more than a single point with a particular y-coordinate? Explain, using examples.

64. Explain how to find the x-intercepts of a graph.

65. Locate a graph from a magazine or newspaper and determine its x- and y-intercepts. Explain the applied significance of these points.

66. Locate a graph from a magazine or newspaper and determine three points on the graph. Explain the applied significance of the coordinates of these points.

67. How many x-intercepts can a straight line have? Explain, using examples.

68. How many y-intercepts can a straight line have? Explain, using examples.

2.3 STRAIGHT LINES

Straight lines are the simplest curves. In many applied situations, experiments or physical laws suggest that certain variables are related to one another by a linear equation. In such circumstances, the relationship between the variables can

Slope of a Straight Line

be depicted using a straight line graph. In this section, we take a closer look at straight lines and show how to find equations for them from various sets of data.

Let's consider a nonvertical straight line on which lie the distinct points (x_1, y_1) and (x_2, y_2). The difference $y_2 - y_1$ measures the vertical distance between the two points and is called the **rise**. The difference $x_2 - x_1$ measures the horizontal distance between the two points and is called the **run**. Geometric interpretations of the rise and run are shown in Figure 1. Because we assume that the two points are distinct and the line is nonvertical, the points must have different x-coordinates. That is, $x_2 - x_1 \neq 0$; the run is nonzero.

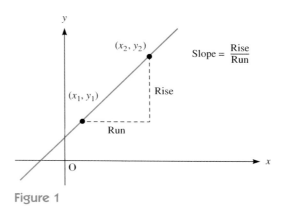

Figure 1

The ratio of the rise to the run measures the steepness of the line and is called the **slope**. That is, we have the following definition.

Definition
Slope of a Line

Suppose that L is a nonvertical line passing through the points (x_1, y_1) and (x_2, y_2). The **slope** of the line, denoted m, is defined as the ratio:

$$m = \frac{\text{rise}}{\text{run}} = \frac{y_2 - y_1}{x_2 - x_1}$$

For a vertical line, the slope is undefined.

EXAMPLE 1
Calculating Slope

For the straight lines in the following graphs, determine the slope.

1.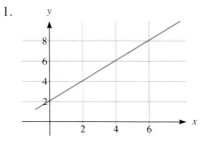

2.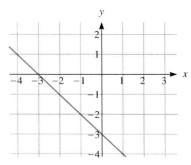

Solution

1. To calculate the slope, we need two points on the line. Examining the graph, we see that we can choose points $(2, 4)$ and $(4, 6)$. So the slope of the line is:

$$m = \frac{y_2 - y_1}{x_2 - x_1} = \frac{6 - 4}{4 - 2} = 1$$

2. Two points on the graph are $(0, -3)$ and $(-3, 0)$. Therefore, the slope is:

$$m = \frac{y_2 - y_1}{x_2 - x_1} = \frac{0 - (-3)}{-3 - 0} = -1$$

There's more to the definition of slope than meets the eye! As we will soon see, the concept of slope is one of the most important characteristics used in describing a straight line. However, before we describe any applications of slope, let's make several observations about the definition.

First, note that the value of the slope does not depend on which point is labeled with subscript 1 and which point is labeled with subscript 2.

$$\frac{y_2 - y_1}{x_2 - x_1} = \frac{-(y_1 - y_2)}{-(x_1 - x_2)} = \frac{y_1 - y_2}{x_1 - x_2}$$

Next note that the value of the slope does not depend on which points we choose as (x_1, y_1) and (x_2, y_2) provided, of course, that we choose points on the line. Suppose that the straight line has equation:

$$Ax + By + C = 0$$

In this equation, A and B can't both be 0. (Otherwise neither x nor y would be involved in the equation.) Because (x_1, y_1) and (x_2, y_2) are on the line, both points satisfy the equation of the line. That is,

$$Ax_1 + By_1 + C = 0$$
$$Ax_2 + By_2 + C = 0$$

Subtracting the first equation from the second, we have:

$$A(x_2 - x_1) + B(y_2 - y_1) = 0$$
$$A(x_2 - x_1) = -B(y_2 - y_1)$$
$$\frac{y_2 - y_1}{x_2 - x_1} = -\frac{A}{B}$$

The expression on the left side of the last equation is the slope. The expression on the right side depends only on the equation of the line and not on the particular points (x_1, y_1) and (x_2, y_2). That is, the value of the slope is independent of the points used to calculate it.

Next, let's emphasize that the slope is undefined for vertical lines. In the case of a vertical line, all points have the same x-coordinate. That is, x_1 equals x_2, so the formula used to define the slope would involve division by 0. Because that is an undefined operation, we leave as undefined the slope of a vertical line.

The Greek letter Δ (delta) is often used to denote the change in a variable. In computing the slope, the change in the y-variable from one point to the other is $y_2 - y_1$, or the rise. For this reason, the rise is sometimes denoted Δy (read "delta y"). Similarly, the run is sometimes denoted Δx (read "delta x"). In this notation, the slope is expressed as:

$$m = \frac{\Delta y}{\Delta x}$$

EXAMPLE 2
Calculating Slope from the Equation

Determine the slopes of the lines that have the following equations.
1. $y = 4x - 3$
2. $3x + 7y = -10$
3. $2y = 2x + 2y - 6$

Solution

1. To calculate the slope, we need two points on the line. To obtain such points, choose any two values for x, say $x = 0$ and $x = 1$, and determine the corresponding values for y:

$$y = 4 \cdot 0 - 3 = -3$$
$$(x_1, y_1) = (0, -3) \text{ on line}$$
$$y = 4 \cdot 1 - 3 = 1$$
$$(x_2, y_2) = (1, 1) \text{ on line}$$

We now calculate the slope:

$$m = \frac{y_2 - y_1}{x_2 - x_1} = \frac{1 - (-3)}{1 - 0} = 4$$

That is, the line has slope 4.

2. Again, we determine the slope by determining two points on the line that correspond to $x = 0$ and $x = 1$. The corresponding values of y are:

$$7y = -3x - 10$$
$$7y = -3 \cdot 0 - 10$$
$$y = -\frac{10}{7}$$
$$\left(0, -\frac{10}{7}\right) \text{ on line}$$
$$7y = -3 \cdot 1 - 10 = -13$$
$$y = -\frac{13}{7}$$
$$\left(1, -\frac{13}{7}\right) \text{ on line}$$

Therefore, the slope is:

$$m = \frac{y_2 - y_1}{x_2 - x_1} = \frac{\frac{-13}{7} - \frac{-10}{7}}{1 - 0} = -\frac{3}{7}$$

That is, the slope of the line is $-\frac{3}{7}$.

3. We simplify the equation by combining like terms:

$$2y = 2x + 2y - 6$$
$$0 = 2x - 6$$
$$x = 3$$

This is the equation of a vertical line. The slope of a vertical line is undefined.

CHAPTER 2 Graphs and Functions

EXAMPLE 3
Calculating Slope from Points

A line passes through the points $(-1, 5)$ and $(0, 4)$. Determine the slope of the line.

Solution
Use the definition of slope:
$$\text{Slope} = \frac{y_2 - y_1}{x_2 - x_1} = \frac{4 - 5}{0 - (-1)} = \frac{-1}{1} = -1$$

Next, let's develop an alternative method of computing the slope directly from the equation of a line. Suppose that the equation of a line is:
$$Ax + By + C = 0$$
where A and B are not both 0 (that is, the equation involves at least one of the x and y terms). If B is not 0, then we can solve the equation for y to obtain:
$$y = -\frac{A}{B}x + \left(-\frac{C}{B}\right)$$
As we showed earlier, the expression $-A/B$ equals the slope m of the line. Let's denote the expression $-C/B$ by b. Then the equation of the line may be written as:
$$y = mx + b$$
The number b has a geometric significance. Indeed, if we set x equal to 0, we see that $y = m(0) + b = b$. In other words, the point $(0, b)$ is on the line and b is its y-intercept. Since the equation $y = mx + b$ exhibits both the slope and the y-intercept of the line, it is called the **slope-intercept form of the equation of a line.**

The derivation of the slope-intercept form started from the assumption that $B \neq 0$. On the other hand, if $B = 0$, then the equation of the line reads:
$$Ax + C = 0, \qquad A \neq 0$$
$$x = -\frac{C}{A}$$
This is the equation of a vertical line.

We can summarize the above discussion in the following result:

EQUATIONS OF STRAIGHT LINES

Let $Ax + By + C = 0$ be the equation of a straight line. If $B \neq 0$, then solving the equation for y in terms of x gives an equivalent equation of the form $y = mx + b$, the slope-intercept form of the equation. Here m is the slope of the line and b is the y-intercept. If $B = 0$, then the line is vertical. In this case, solving the equation for x yields an equivalent equation of the form $x = a$.

EXAMPLE 4
Slope-Intercept Form

Determine the slope-intercept form, the slope, and the y-intercept of the line with equation $5x + 3y - 12 = 0$.

Solution
To determine the slope-intercept form, solve for y in terms of x.
$$5x + 3y - 12 = 0$$
$$3y = -5x + 12$$
$$y = -\frac{5}{3}x + 4$$

The last equation is in the slope-intercept form. The slope of the line equals the coefficient of x in the slope-intercept form. That is, the slope is $-\frac{5}{3}$. The y-intercept equals the constant term of the slope-intercept form, or 4.

Let's now give a geometric interpretation of the slope of a line. As we will see, this interpretation is useful in many applied problems.

GEOMETRIC INTERPRETATION OF SLOPE

Suppose that a line has slope m. Starting from any point on the line, if you move h units in the x-direction, then it is necessary to move mh units in the y-direction to return to the line. (See Figure 4.)

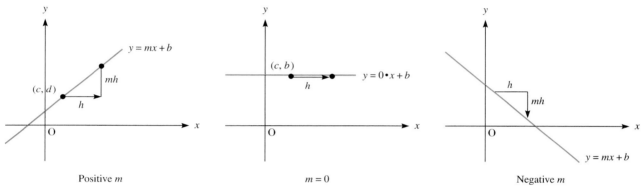

Figure 4

This fact is easy to prove from the definition of slope. However, we omit the proof here.

One consequence of the geometric definition of slope is the following:

RISING AND FALLING LINES AND SLOPE

1. A line with positive slope rises as you go from left to right.
2. A line with negative slope falls as you go from left to right.
3. A line with 0 slope remains level as you go from left to right.

EXAMPLE 5 Consider the line that has equation:
Application of Slope
$$y = -2x + 9$$

1. Suppose that x is increased by 5. What is the change in the value of y?
2. Suppose that x is decreased by 100. What is the change in the value of y?

Solution

1. The slope of the line is -2. By the geometric interpretation of slope, a change of h in x results in a change of mh in y. So an increase of 5 in x results in a change of
$$m \cdot 5 = -2 \cdot 5 = -10$$
in y. That is, y decreases by 10.

2. Again, as a consequence of the geometric interpretation of slope, a change of -100 in x results in a change of

$$mh = -2 \cdot (-100) = 200$$

in y. That is, y increases by 200.

EXAMPLE 6
Manufacturing

The cost C (in dollars) of manufacturing a quantity q television sets (in units) in a production run is given by:

$$C = 250q + 50{,}000$$

1. Give an interpretation for the slope of this linear equation.
2. Suppose that production is increased by 100 units. What will be the effect on production cost?

Solution

1. The slope of the line represented by the equation is 250. By the geometric interpretation of slope, if q is increased by 1, that is, if the company produces one more television set, then the value of C, the total cost of manufacturing, changes by:

$$250 \cdot 1 = 250$$

That is, the slope equals the cost of producing one additional television set. This additional cost is called the **marginal cost of production** and is used by economists in analyzing the microeconomics of production.

2. If production increases by $h = 100$ units, then the cost C changes by:

$$mh = 250 \cdot 100 = 25{,}000$$

That is, the cost increases by \$25,000.

EXAMPLE 7
Cost Equation

The developer of an office building determines that initial development costs are \$3,000,000 for land, permits, and site work. These costs don't vary with the size of the building, so they are called **fixed costs.** The construction of the building itself costs \$50 for each square foot of office space. This construction cost is called the **variable cost** of the building. Determine an equation that relates the total cost of the building to the number of square feet of office space it contains.

Solution

Let C denote the total cost of the building and S the number of square feet it contains. From the Geometric Interpretation of Slope, the slope of an equation expressing the relationship is 50 because an increase in S by 1 square foot causes C to increase by 50. The number 3,000,000 gives the constant term in the equation, because this number gives the cost when S is equal to 0. Therefore, using the slope-intercept form, we can write the equation:

$$C = 50S + 3{,}000{,}000$$

Finding Equations of Straight Lines

In solving problems, it is often necessary to determine equations of straight lines from given data. A number of different forms of equations for straight lines allow us to do just that. Let's now discuss these various equations. The first is the slope-intercept form, which we have been using in our discussion of slope.

The next formula allows us to determine the equation of a line given its slope and a point on the line.

POINT-SLOPE FORMULA

Suppose that a line has slope m and passes through the point (x_1, y_1). Then an equation for the line is

$$y - y_1 = m(x - x_1)$$

To verify the point-slope formula, first note that the point (x_1, y_1) is on the line determined by the equation. If we substitute x_1 for x and y_1 for y, we see that the equation is satisfied:

$$y_1 - y_1 = m(x_1 - x_1)$$
$$0 = 0$$

Writing the equation in the form

$$y - y_1 = mx - mx_1$$
$$y = mx + (y_1 - mx_1)$$
$$y = mx + b, \quad \text{where } b = y_1 - mx_1$$

This shows that the slope of the line is m.

EXAMPLE 8
Point–Slope Formula

Determine the equation of a line passing through the point $(3, -2)$ and having slope 4.

Solution

Using the point-slope formula, we have the equation:

$$y - (-2) = 4(x - 3)$$
$$y + 2 = 4x - 12$$
$$y = 4x - 14$$

EXAMPLE 9
Line Passing through Two Points

Determine the equation of a line passing through the points $(5, 3)$ and $(2, -2)$.

Solution

The slope of the line is:

$$m = \frac{3 - (-2)}{5 - 2} = \frac{5}{3}$$

Therefore, by the point-slope formula, the desired equation is:

$$y + 2 = \frac{5}{3}(x - 2)$$
$$y = \frac{5}{3}x - \frac{10}{3} - 2$$
$$y = \frac{5}{3}x - \frac{16}{3}$$

The following table summarizes the various equations of straight lines we have considered.

EQUATIONS OF LINES

1. General linear equations: $Ax + By + C = 0$, or $ax + by = c$.
2. Slope-intercept form: $y = mx + b$ (always a nonvertical line), where m is the slope and b is the y-intercept.
3. Point-slope formula: $y - y_1 = m(x - x_1)$ (always a nonvertical line), where m is the slope and (x_1, y_1) is any point on the line.
4. Vertical line: $x = a$, where a is the x-intercept.
5. Horizontal line: $y = b$, where b is the y-intercept.

EXAMPLE 10
A Linear Sales Model

A mathematical model of the effectiveness of advertising states that the graph of sales volume versus advertising expenditures is a straight line. Suppose that a company spends $3 million dollars on an advertising campaign and generates sales of $54 million. The next season, it spends $2 million and generates sales of $38 million.

1. Determine an equation that relates sales volume to advertising expenditures.
2. What should be the amount of advertising expenditures if the company wants to boost sales to $60 million?

Solution

1. We are given the two points (3, 54), (2, 38) on the straight line. So the slope of the line is:

$$m = \frac{54 - 38}{3 - 2} = 16$$

Therefore, the equation of the line is:

$$y - 38 = 16(x - 2)$$

where x denotes the advertising expenditure in millions and y the sales volume in millions. Solving for y in terms of x gives:

$$y = 16x + 6$$

2. If $y = 60$, then:

$$60 = 16x + 6$$
$$54 = 16x$$
$$x = 3.375$$

In other words, to generate sales of $60 million, advertising expenditures must be increased to $3.375 million.

EXAMPLE 11
Projecting Salaries

Starting salaries of programmers were $25,000 in 1985 and $35,000 in 1995.

1. Determine a mathematical model that describes the growth of starting salaries for programmers.
2. Project the starting salary of a programmer in 1998.

Solution

Use a coordinate system in which time corresponds to the x-axis and starting salaries correspond to the y-axis. Set up the coordinate system so that 1985 corresponds to 0 on the x-axis.

1. Use a mathematical model that assumes there is a straight-line relationship between starting salary and time. The following points are on the line:

$$(0, 25{,}000), \quad (10, 35{,}000)$$

The slope of the line is:

$$m = \frac{35{,}000 - 25{,}000}{10 - 0}$$
$$= 1{,}000$$

Therefore, by the point-slope formula, the equation of the line is:

$$y - 25{,}000 = 1{,}000(x - 0)$$
$$y = 1{,}000x + 25{,}000$$

The last equation is the desired mathematical model.

2. The year 1998 corresponds to $x = 13$. Substituting this value in the previous model, we have:

$$y = 1{,}000(13) + 25{,}000$$
$$= 38{,}000$$

EXAMPLE 12
Distance versus Time

After t hours of driving, a trucker's distance D, in miles, from his delivery point is given by the equation:

$$D = -50t + 500$$

Give a physical interpretation of the slope of the corresponding line.

Solution

The relationship connecting D and t is a linear equation expressed in slope-intercept form. The slope is -50 and the vertical-axis intercept (the D-intercept) is 500. According to the preceding discussion, for each increment of time Δt, the truck goes an additional distance ΔD. Moreover, the slope is the quotient:

$$\frac{\Delta D}{\Delta t} = \frac{\text{Change in distance}}{\text{Change in time}}$$

We know that distance = rate × time. Therefore, the ratio (Change in distance)/(Change in time) is equal to the **rate** at which the distance is changing per unit time. This rate is called the **velocity** of the truck and is expressed in units of miles per hour. The statement velocity = -50 means that the distance to the delivery point is decreasing at a rate of 50 miles per hour.

Parallel and Perpendicular Lines

The next two results describe the relationships between slopes of parallel and perpendicular lines.

SLOPES OF PARALLEL LINES

Two nonvertical lines are parallel if and only if they have the same slope and have different y-intercepts. See Figure 5.

SLOPES OF PERPENDICULAR LINES

Two lines with slopes m_1 and m_2 are perpendicular if and only if the product of the slopes is -1, that is, if and only if:

$$m_1 m_2 = -1, \quad \text{or} \quad m_1 = -\frac{1}{m_2}, \quad \text{where } m_2 \neq 0$$

(See Figure 6.)

The proof of this result is outlined in the exercises.

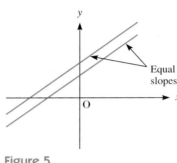

Figure 5   Figure 6

EXAMPLE 13
Perpendicular Lines

Determine whether the lines described by the following equations are perpendicular to one another.

$$-4x + 3y = 1, \qquad 3x + 4y = 17$$

Solution

Let's apply the preceding theorem. To do so, we must first determine the slopes of the lines. The slope-intercept forms of the equations are, respectively:

$$-4x + 3y = 1$$
$$3y = 4x + 1$$
$$y = \frac{4}{3}x + \frac{1}{3}$$

and

$$3x + 4y = 17$$
$$4y = -3x + 17$$
$$y = -\frac{3}{4}x + \frac{17}{4}$$

From these equations, we see that the slopes of the lines are $m_1 = \frac{4}{3}$ and $m_2 = -\frac{3}{4}$, respectively. We then have:

$$m_1 m_2 = \frac{4}{3} \cdot -\left(\frac{3}{4}\right) = -1$$

Therefore, by the Slopes of Perpendicular Lines theorem, the two lines are perpendicular.

EXAMPLE 14
Parallel Lines

Determine the equation of a line passing through the point $(-1, -2)$ and parallel to the line with equation:

$$y = 3x - 18$$

Solution

The line with the given equation has slope 3. By the Slopes of Parallel Lines theorem, a line parallel to that equation has the same slope, 3. Because the desired line must pass through the point $(-1, -2)$, its equation can be derived from the point-slope formula:

$$(y - (-2)) = 3(x - (-1))$$
$$y + 2 = 3x + 3$$
$$y = 3x + 1$$

EXAMPLE 15
Perpendicular Lines

Find the equation of a line passing through the point $(7, 4)$ and perpendicular to the line that has equation:

$$3x - 4y - 12 = 0$$

Solution

The given equation has slope-intercept form

$$y = \frac{3}{4}x - 3$$

so the slope is $\frac{3}{4}$. By the Slopes of Perpendicular Lines theorem, the slope m of a line perpendicular is equal to:

$$m = -\frac{1}{\frac{3}{4}} = -\frac{4}{3}$$

Because the perpendicular line must pass through $(7, 4)$, we can derive its equation using the point-slope formula:

$$y - 4 = -\frac{4}{3}(x - 7)$$
$$y = -\frac{4}{3}x + \frac{40}{3}$$

Exercises 2.3

Determine the slopes of the lines that have the following equations.

1. $y = 0.3x - 27$
2. $y = \frac{2}{3}x + \frac{4}{5}$
3. $-3y = 18x$
4. $x = 4y$
5. $x - 2y = 6$
6. $6x + 3y = 9$
7. $y + 6 = \frac{2}{3}(x - 3)$
8. $3(x - 3) - 4(y + 5) = 10(x + 7) - 5(y - 8)$
9. Consider the line with equation:
$$y = -12x + 34$$
 a. Suppose that x is increased by 3. What is the change in the value of y?
 b. Suppose that x is decreased by 50. What is the effect on y?

10. Consider the line with equation:
$$y = 3.4x - 15$$
 a. Suppose that x is increased by 200. What is the effect on y?
 b. Suppose that x is decreased by 39. What is the effect on y?

Determine the slope of the line passing through each pair of points.

11. $(0, 0)$ and $(3, 6)$
12. $(4, 0)$ and $(0, 2)$
13. $(1, -1)$ and $(3, -5)$
14. $(-2, -3)$ and $(-1, -6)$
15. $(2, 3)$ and $(-1, 3)$
16. $(-7, 3)$ and $(-7, 2)$
17. $(a, 2a+3)$ and $(a+h, 2(a+h)+3)$
18. $(k, -3k)$ and $(k, -3(k+h))$

Determine an equation of the line satisfying the given conditions.

19. Through point $(-1, 2)$ with slope -3.
20. Through point $(4, -2)$ with slope $-\frac{1}{2}$.
21. Through point $(2, -3)$ with slope undefined.

22. Through point $(2, -3)$ with slope 0.
23. With y-intercept $(0, -1)$ and slope $-\frac{1}{5}$.
24. With y-intercept $(0, 0)$ and slope -4.
25. Through points $(2, 3)$ and $(-1, 5)$.
26. Through points $(-2, -6)$ and $(4, -10)$.
27. Through points $(9, -3)$ and $(9, 7)$.
28. Through points $(4, -6)$ and $(3, -6)$.

Find the slope and y-intercept.

29. $y = -\frac{2}{3}x + 1$
30. $2y + 3x = 6$
31. $5x - 3y - 9 = 0$
32. $\frac{2}{3}x - \frac{1}{5}y - \frac{1}{10} = 0$
33. $2y + 3 = 6$
34. $-3x = 6$

Determine whether each pair of equations represents parallel lines, perpendicular lines, or neither.

35. $2x - 3y = 4$; $-2x + 3y = -8$
36. $y = \frac{4}{5}x - 15$; $4y - 5x + 5 = 0$
37. $2x - 3y = 56$; $x + 4y = 7$
38. $4x - 2y = 86$; $y = -3x + 7$
39. $x - 5y = 32$; $x + 11y = 4$
40. $x = -3$; $y = 0$
41. $x = -5y$; $x = 12$
42. $y - 4x = 34$; $y - 4x = -3$

Determine an equation of a line passing through the given point and parallel to the given line.

43. $(-3, -2)$; $2x + 2y = 10$
44. $(-3, 5)$; $9y = 3x + 1$
45. $(3, -1)$; $y = x$
46. $(5, -2)$; $x = 3$
47. $(0, 0)$; $4x - 6y = 2$
48. $(-3, 5)$; $9y = 18$

Determine an equation of a line passing through the given point and perpendicular to the given line.

49. $(0, -2)$; $y = \frac{2}{3}x + 1$
50. $(-1, 3)$; $2x - 4y = 6$
51. $(0, 0)$; $y = x$
52. $(4, 2)$; $-4x = -24$
53. $(-2, 5)$; $4y = 2$
54. $(-3, 5)$; $2x = 9y + 18$

Determine the equation of the line through P that is parallel to the line containing A and B. Determine also the equation of the line through P that is perpendicular to the line containing A and B.

55. $A(-2, 3)$, $B(4, 5)$; $P(4, -6)$
56. $A(-1, -8)$, $B(1, 10)$; $P(6, -2)$
57. Find an equation of the perpendicular bisector of the line segment with endpoints $(2, 3)$ and $(6, -7)$. (The perpendicular bisector is the line that passes through the midpoint of, and is perpendicular to, the line with the given end points.)
58. Find an equation of the perpendicular bisector of the line segment with endpoints $(0, 1)$ and $(-4, -5)$.
59. Find k so that the line containing $(-2, k)$ and $(3, 8)$ is parallel to the line containing $(5, 3)$ and $(1, -3)$.
60. Find k so that the line containing $(-2, k)$ and $(3, 8)$ is perpendicular to the line containing $(5, 3)$ and $(1, -3)$.
61. Prove that the graph of an equation in the form:
$$\frac{x}{a} + \frac{y}{b} = 1$$
has x-intercept a and y-intercept b.
62. **Slopes of Perpendicular Lines theorem. Outline of proof.** Consider the following figure showing two nonvertical lines L and M with equations $y = m_1 x + b_1$ and $y = m_2 x + b_2$. Lines L and M are perpendicular if and only if $\triangle PQR$ is a right triangle. Recall that $\triangle PQR$ is a right triangle if and only if $PR^2 + RQ^2 = PQ^2$; use this fact to complete the proof.

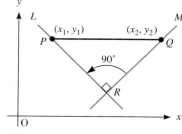

Exercise 62

Applications

63. **Hourly earnings.** The hourly earnings y of U.S. workers in year x (with 1990 corresponding to 0) is given by the model:
$$y = 11 + 0.27x$$
  a. Give an interpretation of the slope of the graph of this equation.
  b. By how much will income increase over five years?

64. **Itemized deductions for state and local taxes.** Based on Internal Revenue Service data, the amount y of itemized deductions for taxes on a tax return reporting x dollars of income is given by the model:
$$y = 295.3 + 0.021689x$$
  a. Give an interpretation of the slope of the graph of this equation.
  b. By how much will the deductions for state and local taxes decrease if total income decreases by $10,000?

65. **Itemized deductions for health care.** Based on Internal Revenue Service data, the amount y of itemized deductions for health care and health insurance on a tax return reporting x dollars of income is given by the model:
$$y = 49.86 + 0.077174x$$
   a. Give an interpretation of the slope of the graph of this equation.
   b. By how much will the deductions for health care increase if total income increases by $20,000?

66. **Income of private legal offices.** The income of private legal offices y (in thousands of dollars per year) in year x (with 1982 corresponding to 0) is given by the model:
$$y = 33.9 + 6.5x$$
   a. Give an interpretation of the slope of the graph of this equation.
   b. By how much will income increase over three years?

67. **Demand for orange juice.** A supermarket observes the following relationship between weekly sales of frozen orange juice q (measured in cans) and the price p (measured in dollars):
$$q = -3,000p + 12,000$$
   a. Give an interpretation of the slope of this line.
   b. What is the effect on sales of lowering the price $1?

68. **Charles's law.** In 1787, Jacques Charles, a French scientist, noticed that gases expand when heated and contract when cooled according to a linear equation, assuming the pressure remains constant.
   a. Suppose a particular gas has a volume $V = 500$ cc when $T = 27°C$ and a volume $V = 605$ cc when $T = 90°C$. Use the point-slope formula to find a linear equation relating V and T.
   b. At what temperature does $V = 0$ cc? This is an estimate of what is known as *absolute zero*, the coldest temperature possible.

69. **Highway safety.** Highway records show that about 34 percent of drivers aged 20 and about 22 percent of those aged 25 will be involved in at least one driving accident within a one-year period.
   a. Use the point-slope formula to find a linear equation that relates P to A, where P is the percentage of drivers who will have an accident at age A.
   b. Predict the percentage of those at age 30 who will have an accident.

70. **Quantity discounts.** A computer supply catalog recently listed boxes of ten $5\frac{1}{4}$-inch diskettes to be sold at $1.49 each if 10 diskettes are purchased. If 60 diskettes are purchased the price is reduced to $1.29 each.
   a. Use the point-slope formula to find a linear equation that relates P, the price per diskette, to N, the number purchased.
   b. Predict the price per diskette if 100 are purchased. The catalog lists the price at $1.19 each if 100 are purchased. How does this compare to your prediction?

71. **Straight-line depreciation.** A company buys an office machine for $9,700 on January 1 of a given year. The machine is expected to last for 6 years, at the end of which time its trade-in or salvage value will be $1,300. If the company figures the decline in value to be the same each year, then the book value or salvage value V after t years, where $0 \le t \le 6$, is given by:
$$V = \$9,700 - \$1,400t$$
Give an interpretation of the slope of this line.

72. **Cost of manufacturing.** A clothing manufacturer is planning to market a new type of sweatshirt to college students. The company estimates that fixed costs for setting up the production line are $43,000. Variable costs for producing each sweatshirt are $3.50. Determine an equation that relates total cost to the number of sweatshirts produced.

73. **Sales commissions.** A salesperson gets a salary of $24,000 a year plus a commission of 38 percent of sales. Determine an equation that relates salary to sales.

74. **Spread of an organism.** A certain kind of dangerous organism is accidentally released over an area of 3 square miles. After t years, it spreads over an area of A square miles, where A and t are related by the equation:
$$A = 1.8t + 3$$
   a. Give an interpretation of the slope of this linear equation.
   b. Suppose that time t is increased by 10 years. How large will the area A affected by the organism be?

75. **Driving costs.** The cost C, in cents per mile, of driving a car is related to the number of years t since 1980 by the equation:
$$100C - 7t = 2,320$$
where $t = 0$ corresponds to 1980, $t = 1$ corresponds to 1981, and so on.

a. Give an interpretation of the slope of this linear equation. Let time t be on the x-axis and cost C be on the y-axis.

b. Suppose that time is increased by 8 years. What will be the effect on the cost C?

76. **Revenue from sales.** The revenue R, in dollars, from the sale of q television sets from a production run is given by:
$$R = 409q$$

a. Give an interpretation of the slope of this linear equation.

b. Suppose that sales are increased by 2,000 units. What will the revenue be?

77. **Temperature conversion.** Fahrenheit temperatures F and Celsius temperatures C are related by the following linear equation:

$$F = \frac{9}{5}C + 32$$

a. Give an interpretation of the slope of this linear equation.

b. Suppose that a person's Celsius temperature rises 1°C during an illness. What is the change in the person's Fahrenheit temperature?

78. **Tail length of a snake.** A study has found that the total length L and the tail length t, both in millimeters, of females of the snake species *Lampropeltis polyzona* are related by the linear equation:
$$t = 0.143L - 1.18$$

a. Give an interpretation of the slope of this linear equation.

b. Suppose that one snake's tail is 80 mm longer than another's. What will be the effect on the total length of one snake over the other?

Technology

Use a calculator to approximate the slope and y-intercept of the following lines.

79. $y = -5.1x + 2.3$
80. $y = 1.17(x + 3.11) + 2x$
81. $5y + 3 = 4x - 7$
82. $5(y + 2.9) = 4(x - 1.1) + 31.7$

Use a calculator to determine the equation of the line passing through the following points.

83. $(5.8, 1.3)$ and $(-3.1, 8.9)$
84. $(-1, 0)$ and $(-3.01, 5.98)$
85. $(2, 11{,}000)$ and $(3, 40{,}500)$
86. $(0, 19{,}875)$ and $(500, 40{,}211)$

In your own words

87. Explain the geometric meaning of the slope of a line.

88. Explain the geometric meaning of the sign of the slope.

89. Why doesn't the formula that defines slope apply to a vertical line?

90. Suppose that you view a ski slope from the side, so that it looks like a straight line. What does the slope of that line tell you about the ski slope?

91. There is more than one way to define the distance function. For instance, consider the distance function $d(x, y) = |x_1 - x_2| + |y_1 - y_2|$.

a. How is this definition of distance different from the definition normally used?

b. Why do you think this definition is called the "Manhattan" distance function?

c. Describe another model that has another definition of distance.

2.4 INEQUALITIES

In Chapter 1, we introduced the inequality relations $<, >, \leq, \geq$ among real numbers. In many applications of algebra, inequalities are used extensively. In particular, we need inequalities and interval notation when we discuss domains of function in Section 2.5. In preparation, this section is devoted to inequalities that involve a single variable.

The fundamental rules for manipulating inequalities are summarized in the following result.

SECTION 2.4 Inequalities

PROPERTIES OF INEQUALITIES

Let a, b, and c be real numbers.
1. If $a < b$ and $b < c$, then $a < c$.
2. If $a < b$ and c is any real number, then $a + c < b + c$.
3. If $a < b$ and c is any real number, then $a - c < b - c$.
4. If $a < b$ and c is positive, then $ac < bc$.
5. If $a < b$ and c is negative, then $ac > bc$. That is, in multiplying an inequality by a negative number, it is necessary to reverse the inequality sign.

The above properties say that you can add or subtract the same quantity from both sides of an inequality, and you can multiply or divide both sides by a positive number. However, if you multiply or divide an inequality by a *negative* number, then you must reverse the direction of the inequality.

The above laws of inequalities are stated for the inequality symbol $<$. However, throughout their statement you may replace $<$ with any of the other inequality symbols $>$, $\leq$, $\geq$. (In Property 5, where two opposite symbols are used, you may replace the first symbol by any of the inequality symbols and the other by its opposite. That is, if the first symbol is replaced by $\leq$, then the second is replaced by $\geq$.)

Linear Inequalities

An inequality may involve one or more variables. For example, the inequality:

$$3x + 1 > -x - 4$$

involves the variable x. The inequality is true for some values of x. For instance, if we set $x = 4$, then we get the true inequality:

$$3(4) + 1 > -4 + 4$$
$$12 + 1 > 0$$
$$13 > 0 \quad \text{True}$$

On the other hand, the inequality is false for some values of x. For example, if we set $x = -5$, we obtain:

$$3(-5) + 1 > -(-5) + 4$$
$$-15 + 1 > 5 + 4$$
$$-14 > 9 \quad \text{False}$$

The process of determining the values of x that make the inequality true is called **solving the inequality**.

In many respects, solving an inequality is just like solving an equation. We use the laws of inequalities to obtain equivalent inequalities, that is, inequalities that have the same solutions. The next few examples provide some practice in solving simple inequalities that involve linear expressions.

> **EXAMPLE 1**
> Solving an Inequality
>
> Solve the inequality $3x + 1 > -x - 4$.
>
> **Solution**
>
> We perform allowable operations on the inequality as we would on an equation. First, add quantities to both sides to bring all terms involving x to one side of the inequality and all constant terms to the other side.

$$3x + x > -4 - 1$$
$$4x > -5$$

Next, multiply both sides of the inequality by $\frac{1}{4}$.

$$\frac{1}{4}(4x) > \frac{1}{4}(-5)$$
$$x > -\frac{5}{4}$$

At each step, the inequality is replaced by an equivalent one. So the solution of the original inequality is $x > -\frac{5}{4}$.

EXAMPLE 2 Solve the inequality $-2x + 3 > 5$.
Solving an Inequality

Solution

First, move $+3$ to the right side of the inequality by subtracting 3 from both sides:

$$-2x > 5 - 3$$
$$-2x > 2$$

Next, multiply both sides of the inequality by $-\frac{1}{2}$. Because $-\frac{1}{2}$ is negative, this operation requires that we reverse the direction of the inequality:

$$\left(-\frac{1}{2}\right) \cdot (-2x) < \left(-\frac{1}{2}\right) \cdot 2$$
$$x < -1$$

That is, the solution of the inequality is $x < -1$.

EXAMPLE 3 A local sporting goods store earns a profit of $30 per tennis racket. As part of
Merchandising a promotion, it spends $5,000 on advertising. How many tennis rackets must be sold if the profit is to be at least $10,000?

Solution

Let q denote the quantity of tennis rackets sold. Then the resulting profit is equal to $30q - 5{,}000$. The condition that the profit must be at least $10,000 can be expressed by the inequality:

$$\text{Profit} \geq 10{,}000$$
$$30q - 5{,}000 \geq 10{,}000$$

We can solve this last inequality for q:

$$30q \geq 15{,}000$$
$$q \geq 500$$

In other words, if the profit is to be at least $10,000, the quantity sold must be at least 500.

EXAMPLE 4 Stockbroker A charges $50 per trade plus $1.50 per share for shares trading at $50
Stockbroker Fees or more. For the same shares, Stockbroker B charges $20 per trade plus $2.10 per share. How many shares must you trade in order for the brokerage fees of Stockbroker B to be less than those of Stockbroker A?

SECTION 2.4 Inequalities

Solution

Suppose that x shares are traded. Let's first derive models for the cost of the trade with each of the stockbrokers:

$$\text{Stockbroker A: Cost} = \$50 \text{ per trade } + \$1.50 \text{ per share}$$
$$= 50 + 1.5x$$
$$\text{Stockbroker B: Cost} = \$20 \text{ per trade } + \$2.10 \text{ per share}$$
$$= 20 + 2.1x$$

We must determine the values of x for which:

$$\text{Cost with Stockbroker B} < \text{Cost with Stockbroker A}$$
$$20 + 2.1x < 50 + 1.5x$$

We now solve the inequality for x:

$$2.1x - 1.5x < 50 - 20$$
$$0.6x < 30$$
$$x < \frac{30}{0.6}$$
$$x < 50$$

Thus, if you trade fewer than 50 shares, Stockbroker B will be cheaper.

EXAMPLE 5
Long Distance Savings

Long distance discount Plan A costs $3 per month to enroll and $.18 per minute. Plan B has no enrollment fee but charges $.22 per minute. How many minutes of long distance per month is it necessary to use in order that Plan A be cheaper?

Solution

Suppose that you use x minutes of long distance per month. The cost models for each of the long distance plans are:

$$\text{Plan A: Cost} = \$3 \text{ per month } + \$.18 \text{ per minute}$$
$$= 3 + 0.18x$$
$$\text{Plan B: Cost} = \$0 \text{ per month } + \$.22 \text{ per minute}$$
$$= 0.22x$$

$$\text{Plan A} < \text{Plan B}$$
$$3 + 0.18x < 0.22x$$
$$3 < 0.22x - 0.18x$$
$$3 < 0.04x$$
$$\frac{3}{0.04} < x$$
$$75 < x$$

Plan A is cheaper if you use more than 75 minutes per month.

EXAMPLE 6
Break-Even Analysis

Suppose that it costs a toy company $5 to manufacture a doll. Fixed costs for the company (those costs that are independent of the number of dolls produced) are $20,000 per month. Each doll sells for $27. How many dolls must be manufactured in order for the company to make a profit?

Solution

In order for the company to make a profit, the following inequality must be satisfied:

$$\begin{bmatrix} \text{monthly} \\ \text{cost} \end{bmatrix} < \begin{bmatrix} \text{monthly} \\ \text{revenue} \end{bmatrix}$$

Let's derive mathematical models for both the cost and revenue. Suppose that x dolls are manufactured and sold. Then we have:

$$\begin{bmatrix} \text{monthly} \\ \text{cost} \end{bmatrix} = \begin{bmatrix} \text{fixed} \\ \text{cost} \end{bmatrix} + \begin{bmatrix} \text{manufacturing} \\ \text{cost} \end{bmatrix}$$

$$= 20{,}000 + \begin{bmatrix} \text{quantity} \\ \text{of dolls} \end{bmatrix} \times \begin{bmatrix} \text{cost to manufacture} \\ \text{one doll} \end{bmatrix}$$

$$= 20{,}000 + x \cdot 5$$

$$= 20{,}000 + 5x$$

$$\begin{bmatrix} \text{monthly} \\ \text{revenue} \end{bmatrix} = \begin{bmatrix} \text{quantity of} \\ \text{dolls sold} \end{bmatrix} \times \begin{bmatrix} \text{selling price of} \\ \text{one doll} \end{bmatrix}$$

$$= x \cdot 27$$

$$= 27x$$

$$\begin{bmatrix} \text{monthly} \\ \text{cost} \end{bmatrix} < \begin{bmatrix} \text{monthly} \\ \text{revenue} \end{bmatrix}$$

$$20{,}000 + 5x < 27x$$

$$20{,}000 < 27x - 5x$$

$$20{,}000 < 22x$$

$$x > \frac{20{,}000}{22}$$

$$x > 909.09$$

The company will make a profit provided that at least 910 dolls are sold each month.

EXAMPLE 7
Using a Calculator to Solve an Inequality

Use a calculator to solve the following inequality:

$$38.75x - 2.1 < -48.1(6.3x + 4.09)$$

Solution

Use your calculator to perform additions and multiplications just as if you were solving the inequality by hand computation:

$$38.75x - 2.1 < (-48.1) \cdot 6.3x + (-48.1) \cdot (4.09) \quad \textit{Distributive law}$$

$$38.75x - 2.1 < -303.03x - 196.729$$

$$38.75x + 303.03x < -196.729 + 2.1$$

$$341.78x < -194.629$$

$$x < \frac{-194.629}{341.78}$$

$$x < -0.56946$$

Note that all calculations are exact except for the final division, where we have approximated the result to five significant digits.

Many applications involve pairs of inequalities of the form
$$a < b \quad \text{and} \quad b < c$$
which can also be written as:
$$a < b < c$$
For instance, the interest rate I quoted on 30-year fixed-rate mortgages by different banks may be greater than or equal to 9.5 percent and less than 10.75 percent. This corresponds to the double inequality:
$$0.095 \leq I < 0.1075$$

The **solution set** of a double inequality consists of the numbers that satisfy both inequalities. That is, the solution set consists of the numbers in common to the two solution sets of the two inequalities.

For example, the solution set of the double inequality $-1 < x < 3$ consists of the numbers in common to the two solution sets: $\{x: x > -1\}$ and $\{x: x < 3\}$.

If A and B are sets, then the set that consists of the elements that A and B have in common is called the **intersection** of A and B, and is denoted by $A \cap B$. In terms of intersections, the solution set of the double inequality may be written as:
$$\{x: x > -1\} \cap \{x: x < 3\}$$

To solve a double inequality, we must solve two independent inequalities and form the intersection of the two solution sets. The following example illustrates the procedure.

EXAMPLE 8
A Three-Termed Inequality

Solve the inequality:
$$-5x < 3x + 1 < 2x + 4$$

Solution

We must determine the solution sets of the inequalities:
$$-5x < 3x + 1 \quad \text{and} \quad 3x + 1 < 2x + 4$$
The first inequality can be solved as follows:
$$-5x < 3x + 1$$
$$-5x - 1 < 3x$$
$$-1 < 3x + 5x$$
$$-1 < 8x$$
$$-\frac{1}{8} < x$$
The solution set of the first inequality is:
$$\left\{x: x > -\frac{1}{8}\right\}$$
The second inequality can be solved as follows:
$$3x + 1 < 2x + 4$$
$$3x - 2x + 1 < 4$$

106 CHAPTER 2 **Graphs and Functions**

$$x < 4 - 1$$
$$x < 3$$

The solution set of this inequality is:

$$\{x: x < 3\}$$

The solution set of the given double inequality consists of the set of numbers in common to the two solution sets just found, that is, the intersection:

$$\left\{x: x > -\frac{1}{8}\right\} \cap \{x: x < 3\}$$

This intersection consists of all numbers x that are greater than $-\frac{1}{8}$ and less than 3, so the solution set is:

$$\left\{x: -\frac{1}{8} < x < 3\right\}$$

In solving inequalities, it is often useful to break the solution into two or more cases. A solution must arise as a solution in one of the cases. If A and B are sets, then the set that consists of the elements in either A or B or both is called the **union** of A and B, and is denoted $A \cup B$. The next example illustrates an inequality whose solution can be expressed as a union.

EXAMPLE 9
Polynomial Inequality

Solve the inequality:

$$(x + 1)(x + 5) > 0$$

Solution

The inequality requires that the product of two factors be positive. Either both factors must be positive or both factors must be negative. It is simplest to determine this set using a picture such as Figure 1. We draw a number line for each of the factors and draw $+$ where the factor is positive and $-$ where the factor is negative. On a third coordinate axis, we indicate where both factors have the same sign. The solution set is:

$$\{x: x > -1\} \cup \{x: x < -5\}$$

This set consists of all x that are either greater than -1 or less than -5.

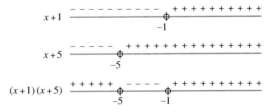

Figure 1

EXAMPLE 10
Rational Inequality

Solve the following inequality.

$$\frac{3x + 7}{4x - 5} > 0$$

Solution

In order for the inequality to hold, the numerator and denominator must have the same sign. Reason geometrically as in the previous example. (See Figure 2.) The solution is the set:

$$\left\{x: x > \frac{5}{4}\right\} \cup \left\{x: x < -\frac{7}{3}\right\}$$

Warning: You might be tempted to solve the inequality by multiplying both sides by $4x - 5$ to clear the denominator. However, this is not a useful way to proceed, because the inequality sign may face in either direction after multiplying, depending on the sign of $4x - 5$.

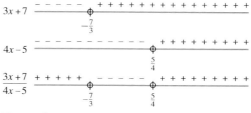

Figure 2

Intervals

The subset of the number line determined by a single or double inequality is called an **interval**. In the preceding examples, we determined solution sets of inequalities that turned out to be intervals, intersections of intervals, or unions of intervals. We classify intervals on the basis of whether their endpoints are included in the solution set, according to the following definition.

Definition
Intervals

Let a and b be real numbers. Then the **open interval** (a, b) is the set of real numbers:

$$(a, b) = \{x: a < x < b\}$$

The **closed interval** $[a, b]$ is the set of real numbers:

$$[a, b] = \{x: a \leq x \leq b\}$$

The **half-open intervals** $[a, b)$ and $(a, b]$ are defined, respectively, as the sets of real numbers:

$$[a, b) = \{x: a \leq x < b\}$$
$$(a, b] = \{x: a < x \leq b\}$$

An interval can be represented as a line segment on the number line. An open interval includes neither endpoint of the segment, a closed interval includes both endpoints, and a half-open interval includes one endpoint. The various types of intervals are illustrated in Figure 3. Note that the solid dot at the end of a line segment indicates that the endpoint is included. An open dot at the end of a line segment indicates that the endpoint is not included.

Intervals that extend indefinitely far to the right or left on the number line are called **infinite intervals.** The solution sets of the inequalities $x > a$ and $x < a$ are the open infinite intervals denoted (a, ∞) and $(-\infty, a)$, respectively. The solution sets of the inequalities $x \geq a$ and $x \leq a$ are $[a, \infty)$ and $(-\infty, a]$, respectively. The symbol ∞, or +∞, is read "infinity." The symbol −∞ is read "negative infinity." The infinity symbols indicate intervals extending infinitely far to the right and infinitely far to the left, respectively. The various types of infinite intervals are shown in Figure 4.

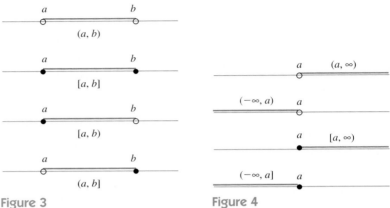

Figure 3

Figure 4

EXAMPLE 11
Sketching Intervals

Sketch the graphs of the following intervals.

1. $(-1, 3)$
2. $(-\infty, 5)$
3. $[0, 1]$
4. $[-2, \infty)$

Solution

1. This is an open interval, so neither endpoint is included. See Figure 5.
2. This is an open infinite interval. The right endpoint is not included and the interval extends infinitely far to the left. See Figure 6.
3. This is a closed interval. Both endpoints are included. See Figure 7.
4. This is a closed infinite interval. The left endpoint is included and the interval extends infinitely far to the right. See Figure 8.

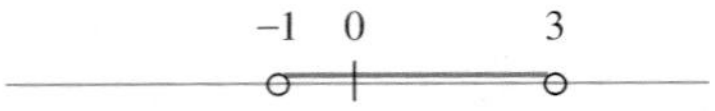

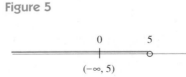

Figure 5

Figure 6

Figure 7

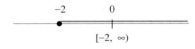

Figure 8

EXAMPLE 12
Graphing Intersections and Unions

Graph the following sets on the number line:

1. $(-1, 3) \cap (2, 4)$
2. $[0, 5] \cup [5, 7]$
3. $(0, 1) \cup (2, \infty)$

Solution

1. The graph consists of the points in common to the intervals $(-1, 3)$ and $(2, 4)$. In Figure 9, we have sketched both intervals. The points in common

are those that are shaded twice. These points compose the interval (2, 3). Note that the endpoints are not included, because 2 does not belong to the interval (2, 4) and 3 does not belong to the interval (−1, 3).

2. The graph consists of points that belong to either of the intervals [0, 5] or [5, 7]. In Figure 10, we have shaded each of these intervals. The graph of the union consists of the points that are shaded at least once. It includes the number 5, which is shaded twice. So the indicated union is the interval [0, 7].

3. The graph consists of points that belong to either of the intervals (0, 1) or (2, +∞). Figure 11 shows each of these intervals as shaded. The union consists of the points that are shaded at least once. In this case, the union cannot be represented as a single interval.

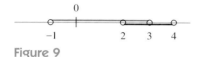
Figure 9

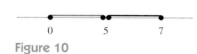
Figure 10

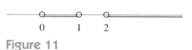
Figure 11

Solving Inequalities Using a Graphing Calculator

Many graphing calculators allow you to graph inequalities. Just enter the inequalities as if they were equations. For instance, suppose that you wish to graph the following inequality:

$$3x + 1 > 5x - 3$$

Call up the equation entry screen of your graphing calculator. (This is the screen that allows you to enter equations of the form Y=....) Enter the inequality just as if it were an equation. (You will need to locate the inequality symbol $>$ on your calculator. On the TI-82, you get it by pressing  and then selecting $>$ from the list of tests displayed.) Now press the key that shows the graph. The calculator will display a graph like the one shown in Figure 12. When the calculator graphs an inequality, it plots the point $(x, 1)$ if the inequality is true for x, and $(x, 0)$ if the inequality is false for x. In this example, the calculator shows the line:

$$y = 1 \quad (x < 2)$$

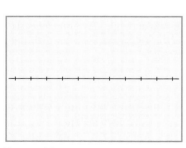
Figure 12

Be careful in interpreting calculator graphs. Depending on the x-range you select, you may see only part of the graph or nothing at all. For instance, if you selected the x-range to be from 3 to 10, then you will see the graph shown in Figure 13. Don't misinterpret this graph by saying that the inequality has no solutions!

You can determine the simultaneous solutions of several inequalities by writing the inequalities in sequence, each surrounded in parentheses. For example, to determine the graph of the inequality

$$1 < 2x + 1 < 3$$

Figure 13

we would write it as the simultaneous solutions of these two inequalities:

$$1 < 2x + 1 \quad \text{and} \quad 2x + 1 < 3$$

These would then be entered in the calculator as:

$$(1 < 2x + 1)(2x + 1 < 3)$$

The resulting graph is shown in Figure 14.

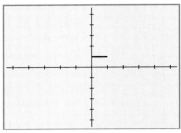

Figure 14

Polynomial Inequalities

An interesting application of intervals arises in solving polynomial inequalities, that is, inequalities of the form

$$p(x) \geq 0$$

or

$$p(x) \leq 0$$

where $p(x)$ is a polynomial in the variable x. Let's consider the particular case in which the polynomial may be factored into terms of the form

$$p(x) = A(x - a_1)(x - a_2) \cdots (x - a_n)$$

where the real numbers $a_1, a_2, \ldots, a_n$ are arranged in increasing size:

$$a_1 \leq a_2 \leq \cdots \leq a_n$$

Associated with such a polynomial, we have the following collection of intervals:

$$(-\infty, a_1), (a_1, a_2), \ldots, (a_{n-1}, a_n), (a_n, \infty)$$

These intervals are called **test intervals** for $p(x)$.

The factor $x - a_i$ is negative if x is less than a_i and positive if x is greater than a_i. Therefore, the product can change sign only at one of the numbers a_i and the product is of constant sign in each of the test intervals. We have the following result.

CONSTANT SIGN THEOREM

The polynomial $p(x) = A(x - a_1)(x - a_2) \cdots$ is of constant sign in each of the intervals $(-\infty, a_1), (a_1, a_2), \ldots$. That is, if c, d are two numbers in the same interval, then the numbers $p(c), p(d)$ are either both positive or both negative.

According to the theorem, if one number in a test interval is a solution to the inequality, then so is every other number in the test interval. Therefore, the solution set of a polynomial inequality consists of a union of test intervals. To determine which test intervals to include in the union, it suffices to test a single number from each interval. The procedure is illustrated by the following example.

EXAMPLE 13
Cubic Inequality

Solve the inequality $2x^3 + x^2 - x < 0$. Graph the solution.

Solution

We begin by factoring the polynomial:

SECTION 2.4 Inequalities

$$2x^3 + x^2 - x = x(2x^2 + x - 1) = x(2x - 1)(x + 1)$$
$$= 2(x - (-1))(x - 0)\left(x - \frac{1}{2}\right)$$

In the last factorization, the factors have been written so that the coefficient of x in each factor is 1 and the constants in the linear factors are in increasing order $(-1 < 0 < \frac{1}{2})$. The test intervals for this polynomial are:

$$(-\infty, -1), \quad (-1, 0), \quad \left(0, \frac{1}{2}\right), \quad \left(\frac{1}{2}, \infty\right)$$

We choose a number in each to act as a test case for that interval: $-2, -\frac{1}{2}, \frac{1}{4}, 1$. We calculate the value of the polynomial for each of these numbers and determine the sign of that value. If the value is less than zero, the corresponding interval is included in the solution set. The data are summarized in the following table and graphed in Figure 15.

Interval	Test Point x	Value of $2x^3 + x^2 - x$
$(-\infty, -1)$	-2	-10
$(-1, 0)$	$-\frac{1}{2}$	0.5
$\left(0, \frac{1}{2}\right)$	$\frac{1}{4}$	-0.15625
$\left(\frac{1}{2}, \infty\right)$	1	2

The first and the third intervals have test points that yield negative values for the polynomial. So the solution consists of the union of the first and third intervals:

$$(-\infty, -1) \cup \left(0, \frac{1}{2}\right)$$

The solution is graphed in Figure 16. You could also solve the inequality by using a graphing calculator, entering the inequality as:

$$2x\,\hat{}\,3 + x\,\hat{}\,2 - x < 0$$

The graph displayed would resemble the one shown in Figure 17.

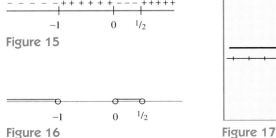

Figure 15

Figure 16

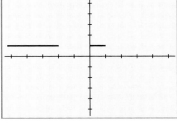

Figure 17

The method used in Example 13 may be used to solve any inequality of the form $p(x) > 0$, where $p(x)$ is a product of linear factors. Note that if we are given an inequality of the form $p(x) > c$, where c is nonzero, then, in order to apply the method, we must transform the inequality into one in which the right side is 0 and the left side is a product of linear factors.

Exercises 2.4

Translate each of the following statements into algebraic language.

1. c is positive.
2. t is negative.
3. a is nonnegative.
4. q is nonpositive.
5. -3 is less than or equal to m.
6. 9 is greater than b.
7. a is at most b.
8. p is at least q.

Write the solution set of the following inequalities in interval notation.

9. $-2 \leq x < 3$
10. $5 < t \leq 7$
11. $-4 \leq x \leq 5$
12. $-5 \leq t \leq -3$
13. $-2 \leq x < 3$
14. $5 \leq t \leq 17$
15. $x \geq -2$
16. $t < 3$

Write each of the following using set-builder notation.

17. $[-1, 3)$
18. $(-3, -1)$
19. $(-2, \infty)$
20. $(-\infty, 5]$
21. $(-\infty, 2]$
22. $[5, \infty)$

Solve the following inequalities and write the solution sets in interval notation.

23. $-2 \leq 5x + 3$
24. $3t + 2 < 17$
25. $-3 - 4x > 0$
26. $6 - 2t < 0$
27. $x + 3 \geq 2x - 1$
28. $5x - 8 < -2(x - 3)$
29. $3 - x \leq 4x + 7$
30. $10x - 3 \geq 13x - 8$
31. $3 - (4 - x) \leq 3x - (x + 5)$
32. $3 - x - x^2 \geq 2x - x(x + 1)$
33. $-0.3x > 0.2x + 1.3$
34. $-2.6x < 1.2 + 0.04x$
35. $-8 \leq 2x \leq 10$
36. $-9 < -3t \leq -6$
37. $-5 \leq 8 - 3x \leq 5$
38. $-7 \leq 2x - 3 \leq 7$
39. $-11 < 2x - 1 \leq -5$
40. $3 \leq 4y - 3 < 19$
41. $\frac{2}{5}x - 3 \leq \frac{4}{5}x - 13$
42. $\frac{3}{5}x \geq \frac{1}{3}x - \frac{1}{10}$
43. $-6 < \frac{2 - 3x}{2} \leq 13$
44. $1 \leq \frac{4 + 3t}{4} < 9$

45. $(3x - 2)^2 < 3(x - 2)(3x + 4)$
46. $4(x - 2)(x + 3) \geq (2x + 3)(2x - 3)$
47. $\frac{1}{2x + 3} < 0$
48. $\frac{-2}{4 - 5x} > 0$
49. $(x + 5)(x - 3) > 0$
50. $(x + 4)(x - 1) < 0$
51. $x^2 - 7x + 12 \geq 0$
52. $x^2 - x \leq 20$
53. $t^2 < 25$
54. $12x^2 > 4$
55. $36x^2 - 24x > 0$
56. $3x^2 + 6x + 3 > 0$
57. $(x + 3)(x - 2)(x + 1) > 0$
58. $(x - 4)(x - 5)(x + 2) \leq 0$
59. $x^3 + 3x^2 - 10x \leq 0$
60. $x^4 - 7x^3 + 12x^2 > 0$
61. $t^3 + 2t^2 - t - 2 < 0$
62. $2x^3 - x^2 - 8x + 4 \geq 0$
63. $\frac{x - 3}{x + 1} > 0$
64. $\frac{x + 5}{x - 2} < 0$
65. $\frac{2x - 3}{x - 4} \leq 0$
66. $\frac{3 - 2x}{3x + 5} \geq 0$

Applications

67. **Break-even analysis.** Suppose that it costs a clothing company $3 to manufacture a shirt. Fixed costs for the company (those costs that are independent of the number of shirts produced) are $75,000 per month. Each shirt sells for $18. How many shirts must be manufactured in order for the company to break even or make a profit?

68. **Stock trading.** Stockbroker A charges $20 per trade plus $1.00 per share. For the same shares, stockbroker B charges $10 per trade plus $1.55 per share. How many shares must you trade in order for the brokerage fees of stockbroker B to be less than those for stockbroker A?

69. **Sports.** A pitcher's earned run average, ERA, is given by $ERA = 9R/I$, where R is the number of earned runs given up while pitching I innings. A pitcher gives up 50 earned runs. What number of innings can be pitched so the pitcher's ERA is at most 3.00?

70. **Profitability analysis.** The following are expressions for total revenue and total cost from the sale of x units of

a product. $R(x) = 80x$ and $C(x) = 20x + 10{,}000$. Total profit $P(x)$ is given by $P(x) = R(x) - C(x)$.

a. Find the number of units x for which $P(x) > 0$.

b. Find the number of units x for which $P(x) \geq 0$.

71. **Education.** A student receives grades of 67, 83, and 78 on three math tests. What scores S can the student receive on the last test to obtain an average between 70 and 80?

72. **Falling objects.** The formula $h = -16t^2 + 64t + 960$ gives the height h of an object thrown upward from the roof of a 960 ft building at an initial velocity of 64 ft/sec.

a. For what times t will the height be greater than 992 ft?

b. For what times t will the height be less than 960 ft?

 Technology

Use a graphing calculator to solve the following inequalities.

73. $\dfrac{(x+1)(x-2)}{x+4} < 0$

74. $\dfrac{x+7}{(x-3)(x-2)} > 0$

75. $(x+3)(x-2)(x+1) > 0$

76. $(x-4)(x-5)(x+2) \leq 0$

77. $-11 < 2x - 1 \leq -5$

78. $3 \leq 4y - 3 < 19$

79. $-6 < \dfrac{2-3x}{2} \leq 13$

80. $1 \leq \dfrac{4+3t}{4} < 9$

 In your own words

81. Describe in words the procedure for solving a polynomial inequality by using test intervals.

82. Must a polynomial have opposite signs in consecutive test intervals? Explain your answer.

83. Describe in words the set of simultaneous solutions to three given inequalities.

2.5 THE CONCEPT OF A FUNCTION

Definition of a Function

The concept of a function was not formulated overnight. It took a number of reformulations and generalizations before the modern concept of a function arose. In order to understand functions, let's start with one of the older formulations. Suppose that a formula expresses the variable y in terms of the variable x. Then we say that y is a **function** of x. For example, consider the formula:

$$y = 3x^2$$

For each value of x, this formula determines a unique corresponding value of y. For instance, if $x = 1$, then the corresponding value of y is 3. We say that 3 is the **value** of the function at $x = 1$. It is customary to assign letters to denote functions. For instance, the above function can be denoted f and we can describe the function by the notation

$$y = f(x)$$

(read as "y equals f of x"). This notation indicates that the value of y is determined by the value of x that is used in the function f. We say that x is the **independent variable** and that y is the **dependent variable**.

Not every equation that relates x and y defines y as a function of x. For example, consider the equation $x = 3y^2$. Here, y is *not* a function of x, because for each positive value of x, there are two values of y that satisfy the equation. For instance, the value $x = 3$ corresponds to the two values $y = \pm 1$.

A function may have more than one independent variable. For example, the volume V of a right circular cylinder of radius r and height h is given by the following formula:
$$V = \pi r^2 h$$
This formula defines a function $V(r, h)$ of the two independent variables r and h. For positive r and h, the value of the function is obtained by the formula:
$$V(r, h) = \pi r^2 h, \qquad (r > 0, h > 0)$$
Note that the variables r and h are restricted to positive values because of the meaning of the variables as the radius and height of a cylinder, respectively.

The function f can be viewed as a correspondence that associates the real number x with the real number $3x^2$. In this view, the function is a device for associating elements of one set with elements of a second set. Namely, it associates with each real number x, the real number $3x^2$. The function V associates a real number with each point (r, h) in the first quadrant.

In many applications of mathematics, it is necessary to consider **correspondences** or **mappings** between sets that have nothing to do with numbers. For example, consider the set of all dots of light on a computer screen. Each dot of light is called a pixel. For each image portrayed on the screen, there is a correspondence that associates each pixel with its color. We can visualize the correspondence as shown in Figure 1. On the left is the set P of pixels on the screen and on the right is the set C of possible colors. The correspondence or mapping we have in mind associates each pixel p with a color c. We name the correspondence the letter S and we say that S associates c with p. We can write:
$$c = S(p)$$
Pictorially, we indicate that c is associated with p by drawing an arrow from p to $S(p)$, as in Figure 2. We can view the correspondence S as a general "formula" that tells you the color c for each pixel p. This "formula" is not an algebraic one such as $y = 3x^2$. Conceptually, however, associating a color with a pixel is no different from associating the value $3x^2$ with the value x.

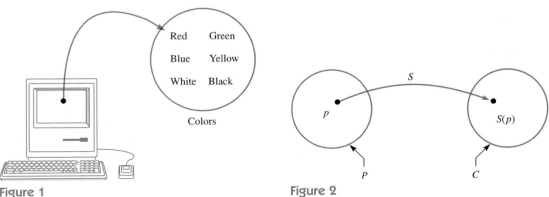

Figure 1

Figure 2

SECTION 2.5 The Concept of a Function

In a similar fashion, we can consider a correspondence F between the set of all people in your math class and the set of months of the year. The correspondence F associates with each person the month in which he or she was born. We might have, for example,

$$F(\text{Jane Smith}) = \text{December}$$

Again, the mapping F can be viewed as a rule that tells you the month of birth for each person in your math class.

These examples of correspondences lead us to the following definition.

Definition
Function as a Correspondence
Let A and B be sets. A **function**, or **mapping**, f from A to B is a correspondence that associates each element of A with one and only one element of B. If x is an element of A, then the element of B associated with x under the correspondence f is denoted $f(x)$ (read "f of x"), and is called the value of f at x.

For example, consider the function $f(x) = 3x^2$ from the set of real numbers A to the set of real numbers B. Here x can be any real number and the value of f, namely $3x^2$, is also a real number. The values of f at 0, 1, and -3 are:

$$f(0) = 3 \cdot 0^2 = 0$$
$$f(1) = 3 \cdot 1^2 = 3$$
$$f(-3) = 3 \cdot (-3)^2 = 27$$

You should view a function as a rule for associating each x in set A with one and only one value $f(x)$ in set B. You can also think of a function as a machine (say a computer) into which you feed a value x from A. The machine outputs a unique value $f(x)$ from B. See Figure 3.

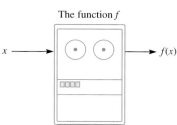

The function f

Figure 3

Definition
Domain of a Function
Suppose that f is a function from set A to set B. The set A is called the **domain** of f. That is, The **domain** of a function f consists of all x for which the function value $f(x)$ is defined.

If x is in the domain of f, then we say that f is **defined** at x. Similarly, if x is not in the domain of f, then we say that f is **undefined** at x.

The subset of B that consists of all function values $f(x)$ is called the **range** of f.

Examples of Functions, Domains, and Ranges

Let's now explore some of the simplest functions, namely those that have real values and are specified by formulas involving a single variable.

Suppose that a and b are real constants, with $a \neq 0$. We can define the function

$$f(x) = ax + b$$

where x is a real number. Such a function is called a **linear function**.

Similarly, a function of the form

$$f(x) = ax^2 + bx + c, \qquad a \neq 0$$

is called a **quadratic function**. A function of the form

$$f(x) = ax^3 + bx^2 + cx + d, \qquad a \neq 0$$

is called a **cubic function**. In general, a function in which $f(x)$ is specified as a polynomial expression in x is called a **polynomial function**. A function such as

$$f(x) = \frac{3x^2 + x - 1}{2x + 1}$$

which is given by a rational expression, is called a **rational function**.

In the following examples, we gain some experience in dealing with functions.

EXAMPLE 1
Calculating Domain and Range

Let f denote the linear function that associates each real number x with the real number $f(x) = 3x + 1$.

1. What is the domain of f?
2. Determine $f(2)$, $f(-1)$.
3. Determine $f(a)$, $f(a + 1)$, $f(2t - 1)$, where a and t are real numbers.
4. What is the range of f?

Solution

1. Any value of x is permissible in the formula $3x + 1$, so the domain consists of all real numbers, and can be written in interval notation as $(-\infty, \infty)$.
2. We have:
$$f(2) = 3(2) + 1 = 7$$
$$f(-1) = 3(-1) + 1 = -2$$
3. The function may be evaluated for any expression, as long as the value of the expression is in the domain. We assume a and t are real numbers. The expressions a, $a + 1$, and $2t - 1$ are real numbers and hence are in the domain. In each case, we substitute the given expression for x in the definition of the function. In the first case, we replace each occurrence of x by a:
$$f(a) = 3a + 1$$
In the second case, we replace each occurrence of x by $a + 1$:
$$f(a + 1) = 3(a + 1) + 1 = 3a + 4$$
In the third case, we replace each occurrence of x by $2t - 1$:
$$f(2t - 1) = 3(2t - 1) + 1 = 6t - 2$$
4. To determine the range, we must determine those real numbers c that are taken to be values of the function for some x, namely:
$$f(x) = c$$
This equation is equivalent to:
$$3x + 1 = c$$
$$x = \frac{1}{3}(c - 1)$$

That is, for this value of x, the function f equals the value c. Because c may be any real number, f takes all real numbers as values, and its range is all real numbers.

SECTION 2.5 The Concept of a Function

If we don't specify the domain of a function, we understand the domain to consist of all values of the variable for which the defining expression makes sense. We sometimes refer to this as the **natural domain** of x. For instance, because the expression $2x^3 - 1$ is defined for all real numbers x, the domain of the function

$$f(x) = 2x^3 - 1$$

consists of all real numbers x. Similarly, because the expression $1/x$ is defined for $x \neq 0$, the domain of the function

$$f(x) = \frac{1}{x}$$

consists of all nonzero real numbers x. This domain can be described in set-builder notation as $\{x: x \neq 0\}$, or in interval notation as:

$$(-\infty, 0) \cup (0, \infty)$$

The expression $\sqrt{x}$ is defined only for nonnegative real numbers. So the domain of the function $f(x) = \sqrt{x}$ consists of the nonnegative real numbers x. In set-builder notation, the domain can be described as $\{x: x \geq 0\}$. In interval notation, the domain can be written as:

$$[0, \infty)$$

The domain is often specified in parentheses after the expression defining the function. For example, the above three functions can be defined as follows:

$$f(x) = 2x^3 - 1 \qquad \text{(all real } x\text{)}$$

$$f(x) = \frac{1}{x} \qquad (x \neq 0)$$

$$f(x) = \sqrt{x} \qquad (x \geq 0)$$

In some cases, we wish to specify a domain that is smaller than the natural domain of the defining expression. This can be done by stating the conditions for specifying the domain in parentheses after the defining expression. For instance, the function

$$f(x) = 3x \qquad (3 < x \leq 5)$$

has as its domain the interval $(3, 5]$.

EXAMPLE 2
Evaluating a Function

Suppose that we are given the real-valued function:

$$f(x) = \sqrt{x - 1}$$

1. Determine the domain of f.
2. Determine the value of $f(10)$ and $f(-1)$.
3. Determine the value of $f(w^2)$.
4. Determine the range of f.

Solution

1. The expression $\sqrt{x - 1}$ produces a real number provided that:

$$x - 1 \geq 0, \qquad x \geq 1$$

So the domain is:

$$\{x: x \geq 1\}$$

or
$$[1, \infty)$$

2. For $f(10)$, we have:
$$f(10) = \sqrt{10 - 1} = \sqrt{9} = 3$$
For $f(-1)$, we have:
$$f(-1) = \sqrt{-1 - 1} = \sqrt{-2}$$
There is no real number whose square is -2 (because the square of any real number is nonnegative). So $\sqrt{-2}$ is undefined (at least as a real number). Thus, -1 is not in the domain and $f(-1)$ is undefined.

3. We substitute w^2 for x in the definition of the function:
$$f(w^2) = \sqrt{w^2 - 1}$$
The expression on the right gives the value of f at w^2.

4. As x takes values in the domain of f, the expression $x - 1$ assumes as values all nonnegative real numbers, so $\sqrt{x - 1}$ assumes the same set of values as the square root function, namely, all nonnegative real numbers. Hence the range of f is the set of all nonnegative real numbers.

EXAMPLE 3
Determining the Domain of a Quotient

Let h be the function defined by:
$$h(x) = \frac{\sqrt{x}}{(x - 1)(x - 2)}$$

1. What is the domain of h?
2. Determine $h(4)$.

Solution

1. The expression on the right requires a number of restrictions on x for it to be defined. First, the radical in the numerator is defined only for nonnegative numbers. And the expression is not defined at x for which the denominator is 0. This requires the exclusions $x \neq 1, 2$. Therefore,
$$\text{Domain of } h = \{x: x \geq 0 \text{ and } x \neq 1, 2\}$$
$$= [0, 1) \cup (1, 2) \cup (2, \infty)$$

2. We substitute the value 4 for x into the expression defining h:
$$h(4) = \frac{\sqrt{4}}{(4 - 1)(4 - 2)} = \frac{2}{3 \cdot 2} = \frac{1}{3}$$

Applications of Functions

As we have seen in Chapter 2, many applied problems give rise to equations that involve variables x and y. Many of these equations can be written in an equivalent form in which y is expressed as a function $f(x)$, say, $y = x^2 + 1$.

Any such equation can be regarded as defining a function that specifies the value of y in terms of the expression in x. In the examples that follow, we provide some practice in determining functions that arise in some typical applied situations.

EXAMPLE 4
Accrued Interest on a Municipal Bond

A bond that costs $1,000 is issued by a state bridge and turnpike authority, and it pays interest compounded semiannually at an interest rate r, to be determined. Rather than pay the interest each 6 months, the bridge and turnpike authority will accrue the interest. That is, the interest will be added to the amount owed, and at the next interest date, interest will be paid on the increased amount of debt. The entire amount owed will be paid at the end of 20 years. The law governing the issuance of bonds restricts the interest rate to be at most 10 percent.

1. Let $f(r)$ denote the amount owed on the bond after 20 years. Determine a formula for $f(r)$.
2. What is the meaning of the number $f(0.08)$?
3. What is the value of the bond after 20 years if the interest rate is 5 percent?
4. What is the domain of the function f?

Solution

1. The bond is like a savings account, with compound interest paid semiannually. By the compound interest formula, the value of an initial amount A after n interest years, with interest compounded k times per year, is:

$$A\left(1 + \frac{r}{k}\right)^{kn}$$

In this case, $A = 1{,}000$, $k = 2$, and $n = 20$. Therefore, the value $f(r)$ of the bond after 20 years is given by the formula:

$$f(r) = 1{,}000\left(1 + \frac{r}{2}\right)^{40}$$

2. $f(0.08)$ is the value of the function at $r = 0.08$. This is the value of the bond after 20 years if the annual interest rate is 8 percent.
3. If the interest rate is 5 percent, the value of the bond after 20 years is given by the function value $f(0.05)$. Numerically, this function value is:

$$f(0.05) = 1{,}000\left(1 + \frac{0.05}{2}\right)^{40}$$

We can evaluate this number using a calculator. Its value is 2,685.06. That is, if the annual interest rate is 5 percent, the bond will be worth $2,685.06 after 20 years.
4. The annual interest rate must be a positive number, that is, $r > 0$. Moreover, because by law the interest rate can be at most 10 percent, we must have $r \leq 0.1$. That is, r is subject to the restriction:

$$0 < r \leq 0.10$$

So the domain of the function f is the interval $(0, 0.1]$.

EXAMPLE 5
Speed of a Sailboat

A model commonly used by yachtsmen states that the maximum speed of a sailboat (in knots) is 1.3 times the square root of the length of the boat at the water line.

1. Express the maximum speed S of a sailboat as a function of the length l of the boat at the waterline (in feet).

2. What is the domain of the function S?
3. What is the maximum speed of a sailboat of length 20 feet?
4. How long would a sailboat be if its maximum speed is 8 knots?

Solution

1. From the verbal description of the model,
$$[\text{maximum speed}] = 1.3 \times [\text{square root of length at waterline}]$$
$$S = 1.3\sqrt{l}$$

2. Since the length of the boat must be positive and all lengths are theoretically possible, the domain of S is the interval $(0, \infty)$.

3. The maximum speed of a sailboat of length 20 feet is given by the function value $S(20)$.
$$S(20) = 1.3\sqrt{20} \approx 5.81 \text{ knots}$$

4. To achieve a maximum speed of 8 knots, we would need a boat of length l, where:
$$S(l) = 8$$
$$1.3\sqrt{l} = 8$$
$$\sqrt{l} = \frac{8}{1.3} \approx 6.1538$$
$$l \approx (6.1538)^2 \approx 37.870$$

That is, the length must be approximately 37.87 feet.

EXAMPLE 6
Storage Tank

Large tanks used to store industrial chemicals are built in a cylindrical shape 10 feet tall. The paint used for the tanks costs $50 per gallon, and one gallon is sufficient to cover 200 square feet of the tank's surface. Calculate the cost $C(r)$ of painting a tank as a function of its radius r. (Both the top and the sides of the tank must be painted.)

Solution

Let h denote the height of the tank (see Figure 4). From the formulas for the area of a circle and the surface area of a cylinder, the total area to be painted is:
$$\text{Area of top} + \text{Area of side} = \pi r^2 + 2\pi r h$$

Substituting $h = 10$, we have:
$$\text{Total area} = \pi r^2 + 2\pi r(10)$$
$$= \pi r^2 + 20\pi r$$

Because one gallon of paint covers 200 square feet, the number of gallons required to paint the entire cylinder is equal to the total area divided by 200:
$$\text{gallons} = \frac{\text{area}}{\text{gallons/area}}$$
$$= \frac{\pi r^2 + 20\pi r}{200}$$
$$= \frac{1}{200}\pi r^2 + \frac{1}{10}\pi r$$

We find the cost by multiplying this expression by 50 (the cost of one gallon) and arrive at an expression for the cost C in terms of the radius r:

$$\text{cost} = (\text{cost/gallon}) \cdot \text{gallons}$$

$$C(r) = 50 \cdot \left(\frac{1}{200}\pi r^2 + \frac{1}{10}\pi r\right)$$

$$= \frac{1}{4}\pi r^2 + 5\pi r$$

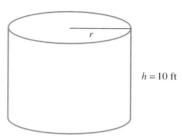

Figure 4

HISTORICAL NOTE

In the twentieth century, algebra has evolved in the direction of abstract algebra, in which mathematicians deduce properties of general algebraic structures that are defined by lists of axioms. This point of view has proven to be extremely fertile and has led to the application of algebra to many disciplines, including physics, chemistry, and engineering. One of the founders of the field of abstract algebra is the German mathematician Emmy Noether. Professor Noether's work was done in the early twentieth century at Gottingen University, the preeminent mathematics department in the world for 75 years. Noether was one of the first women mathematicians who actually lectured on her work. Notes of her lectures and those of her Gottingen colleague, Emil Artin, were developed into the famous, two-volume text *Modern Algebra*, by the Dutch mathematician van der Waerden. This text has been in use for almost 75 years, and has been the source from which many of the most prominent twentieth-century algebraists learned their abstract algebra.

Exercises 2.5

Let $f(x) = 4x - 5$. Find the following:

1. $f(0)$
2. $f(-2)$
3. $f(-2a)$
4. $f(a + h)$

Let $g(x) = 3x^2 + x - 1$. Find the following:

5. $g(10)$
6. $g(-1)$
7. $g(a - h)$
8. $g(-4t)$

For $f(x) = |5 - x^2|$, find each of the following:

9. $f(0)$
10. $f(-4)$
11. $f(1.2)$
12. $f(a + h)$

For $g(x) = \frac{1}{x^2} + \frac{1}{x} - 1$, find each of the following, if possible:

13. $g(0)$
14. $g(-2)$
15. $g\left(\frac{1}{3}\right)$
16. $g\left(\frac{1}{a}\right)$

Let $g(x) = 2x^3 - 3x^2$. Find each of the following:

17. $g(1)$
18. $g(-2)$
19. $g(2 - h)$
20. $g(a + h)$

Let $f(x) = (1 + x)^3$. Find each of the following:

21. $f(2)$
22. $f(-1)$
23. $f(a + h)$
24. $f\left(\dfrac{1}{t}\right)$

Let $g(x) = (2x + 1)^2$. Find each of the following:

25. $g(0)$
26. $g\left(-\dfrac{1}{2}\right)$
27. $g(a + h)$
28. $g(t + h)$

Let $f(x) = 12$. Find each of the following:

29. $f\left(-\dfrac{3}{2}\right)$
30. $f(0)$
31. $f(a + h)$
32. $f(t + h)$

Determine the domain and range of each function.

33. $f(x) = 3x + 7$
34. $g(x) = 5 - 4x$
35. $f(x) = 12$
36. $g(x) = -4$
37. $f(x) = \dfrac{1}{x - 3}$
38. $f(x) = -\dfrac{2}{3x + 7}$
39. $g(x) = \sqrt{3 - 2x}$
40. $g(x) = \sqrt{1 + 4x}$
41. $g(x) = x^3$
42. $g(x) = 4x^3 + 5$
43. $f(x) = |x|$
44. $g(x) = |5 - 3x|$
45. $g(x) = \sqrt{4 - x^2}$
46. $g(x) = \sqrt{3 - x^2}$
47. $g(x) = \sqrt{x - 1}$
48. $f(x) = \sqrt{x + 3}$

For each function, evaluate and simplify the expression:

$$\dfrac{f(a + h) - f(a)}{h}, \quad h \neq 0$$

49. $f(x) = 2x - 3$
50. $f(x) = 4 + 5x$
51. $f(x) = x^2 - 5x + 7$
52. $f(x) = x^2 + 3x$

Applications

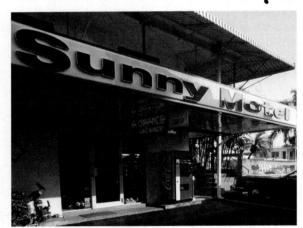

53. **Motel management.** The owner of a 30-unit motel, by checking records of occupancy, knows that when the room rate is $50 a day per unit, all units are occupied. For every increase of x dollars in the daily rate, there are x units vacant. Each unit occupied costs $10 per day to service and maintain. Let R denote the motel's total daily income. Determine a formula that expresses R as a function of x.

54. **Manufacturing.** A container company is constructing an open-top, rectangular, metal tank with a square base of length x that will have a volume of 300 cubic ft. Let S denote the surface area of the tank. Determine a formula that expresses S as a function of x.

55. **Manufacturing.** From a thin piece of cardboard that measures 20 in. by 20 in., square corners of length x are cut out so the sides can be folded up to make a box. Let V denote the volume of the box. Determine a formula that expresses V as a function of x.

56. **Architecture.** A Palladian window is a rectangle with a semicircle on top, as shown in the figure. Suppose the perimeter of a particular Palladian window is 28 ft. Let A denote the total area of the window. Determine a formula that expresses A as a function of x.

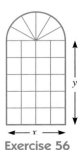

Exercise 56

57. **Storage tank.** A water storage tank is built in a cylindrical shape 30 feet tall. The paint used for the tanks costs $40 a gallon, and one gallon is sufficient to cover 200 square feet of the tank's surface. Determine a formula that expresses the cost C as a function of the radius r. (Both the top and the sides of the tank must be painted.)

58. **Electric power business.** A power line is to be constructed from a power station at point B to an island at point Q, which is directly 1 mile out in the water from a point A on the shore. See figure. Point A is 3 miles downshore from the power station at B. It costs $6,000 per mile to lay the line under the water and $4,000 per mile to lay the line underground. Suppose that x miles of power line are run diagonally from Q to the shoreline, with the remainder in a straight line to point B. Let C denote the total cost of the power line. Determine a formula that expresses C as a function of x.

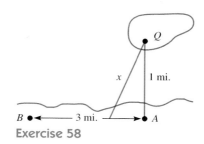

Exercise 58

 In your own words

59. Explain the concept of a function.
60. What is the domain of a function?
61. What is the range of a function?
62. Explain how to determine the value of the function $f(x) = 3x^2 - 2x$ for a particular value of x.
63. Give an example of a function defined by an equation.
64. Give an example of an equation that does not define a function.

2.6 GRAPHS OF FUNCTIONS

The graph of the equation $y = f(x)$ is called the **graph of the function f**. This graph is generally a curve that depicts the relationship between x and y. A point (x, y) is on the graph when x and y satisfy the equation—that is, if $y = f(x)$. So the graph consists of the points

$$(x, f(x))$$

where x is in the domain of f. See Figure 1.

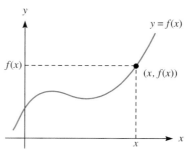

Figure 1

From the graph of a function, you can determine the function value $f(x)$ for any value of the independent variable x. Just start at x on the horizontal axis, and proceed vertically upward or downward until you get to the graph. The y-coordinate of the corresponding point is the value of $f(x)$. Once you are given the graph of a function, you know the function, that is, you can determine the function value (at least approximately) for any x. So a function and its graph are just two sides of the same coin.

The domain of a function can be visualized, in terms of its graph, as a subset of the x-axis. A point a is in the domain provided that there is a point on the graph directly above or below a. (See Figure 2.) Note that a_1 is in the domain but a_2 is not.

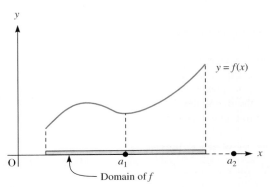

Figure 2

The range of a function can be visualized as a subset of the y-axis. A point b on the y-axis is in the range provided that there is a point on the graph at the same height as b. (See Figure 3.) Note that b_1 is in the range but b_2 is not.

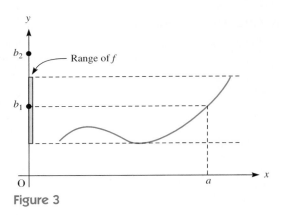

Figure 3

EXAMPLE 1
Determining Domain and Range from the Graph

Each of the following graphs is the graph of a function. From the graph, determine the domain and range of the corresponding function.

1.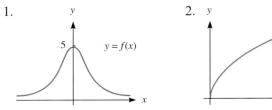

2.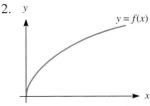

3.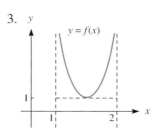

4.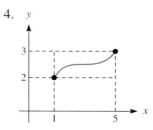

Solution

1. The domain consists of the values of x over which lie points on the graph. In this case, there is a point over every x value, so the domain is the infinite interval $(-\infty, \infty)$. The range consists of all values y on the y-axis for which there are points on the graph at the same height. In this case, the range is the interval $(0, 5]$.
2. Domain $= [0, \infty)$, range $= [0, \infty)$
3. Domain $= (1, 2)$, range $= [1, \infty)$
4. Domain $= [1, 5]$, range $= [2, 3]$

EXAMPLE 2
Graphing a Function

Graph the function:
$$f(x) = x^2 \quad (-2 \leq x \leq 2)$$
From the graph, determine the domain and range of f.

Solution

We graphed the equation
$$y = x^2$$
earlier in the chapter. Figure 4 shows this graph again. The graph of the function consists of the points with x-coordinates that satisfy the restriction $-2 \leq x \leq 2$. The domain and range are indicated on the x- and y-axes, respectively. The domain is the interval $[-2, 2]$ and the range is the interval $[0, 4]$.

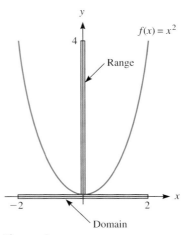

Figure 4

EXAMPLE 3 Graphing

Sketch the graph of the function:
$$f(x) = \sqrt{1 - x^2}$$

Solution

In order for $f(x)$ to be defined, the quantity within the radical must be nonnegative. That is,
$$1 - x^2 \geq 0$$
We can solve this polynomial inequality using the technique of Section 2.4. Write it in factored form:
$$(1 - x)(1 + x) \geq 0$$
Consulting Figure 5, we see that the solution of the inequality is:
$$-1 \leq x \leq 1$$

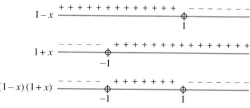

Figure 5

So the domain of the function is the closed interval $[-1, 1]$. To sketch the graph, we must graph the equation:
$$y = \sqrt{1 - x^2}$$
We could plot points, but there is a better method in this case. Note that the value of y is always ≥ 0. Square both sides of this equation to obtain:
$$y^2 = 1 - x^2, \qquad y \geq 0$$
$$x^2 + y^2 = 1, \qquad y \geq 0$$
The last equation has as its graph a circle with center $(0, 0)$ and radius 1. The inequality $y \geq 0$ restricts the graph to the upper semicircle. See Figure 6.

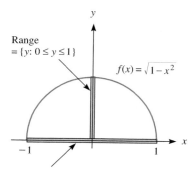

Range = $\{y: 0 \leq y \leq 1\}$

$f(x) = \sqrt{1-x^2}$

Domain of $f = \{x: -1 \leq x \leq 1\}$

Figure 6

With each x in the domain of f, the function associates a single real number $f(x)$. This means that, on a graph of $f(x)$, for a given value of x there is only one point (x, y). To put this another way, a vertical line passing through x on the horizontal axis intersects the graph in exactly one point, namely, the point $(x, f(x))$. See Figures 7(a) and 7(b). Not every vertical line must intersect the graph, of course—only vertical lines corresponding to x in the domain.

Vertical lines that pass through various points on the x-axis tell us much about a graph. If x is not in the domain of a function f, then there is no point of the form (x, y) on the graph, and thus a vertical line through x will not intersect the graph. See Figure 7(a). Furthermore, if a vertical line passing through x intersects the graph in more than one point, the graph does not have a single y associated with each x, and is not the graph of a function. See Figure 7(b).

SECTION 2.6 Graphs of Functions

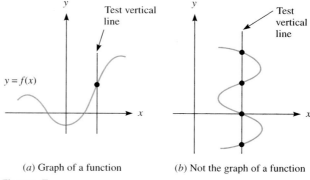

(a) Graph of a function   (b) Not the graph of a function

Figure 7

We can summarize these observations in the following test, which allows us to determine whether a graph in the xy-plane is the graph of a function.

VERTICAL LINE TEST

A set of points in the xy-plane is the graph of a function if and only if each vertical line intersects the graph at no more than one point.

EXAMPLE 4 Determine which of the graphs in Figure 8 are the graphs of functions.
Using the Vertical Line Test

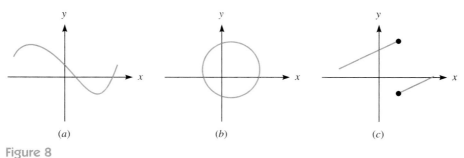

(a)   (b)   (c)

Figure 8

Solution

1. Any vertical line either does not intersect the graph or intersects it at a single point. See Figure 9(a). This graph is the graph of a function by the vertical line test.
2. In Figure 9(b), we have a vertical line that intersects the graph at more than one point. According to the vertical line test, the graph is not the graph of a function.
3. In Figure 9(c), we have a vertical line that intersects the graph more than once. According to the vertical line test, the graph is not the graph of a function.

128 CHAPTER 2 Graphs and Functions

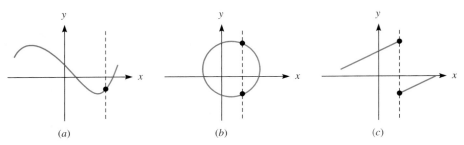

(a)　　　　　　　　(b)　　　　　　　　(c)

Figure 9

Horizontal and Vertical Translations, Scaling

Up to this point, we have described only one method for sketching the graphs of functions—by plotting points. This method is tedious and produces an accurate graph only if you use enough points and take care that these points represent all of the various geometric features of the graph. There are a number of additional methods for sketching graphs of functions. Some of these methods belong to the province of algebra, and others overlap into calculus. In the remainder of this section, we present several methods for sketching graphs using known graphs as a starting point.

The first method we present involves functions whose graphs are obtained by translating known graphs vertically or horizontally.

Suppose that f is a function and that c is a real number. The graph of f is the graph of the equation:

$$y = f(x)$$

Suppose that we translate this graph vertically c units (upward if $c > 0$ and downward if $c < 0$). Every y-value on the new graph is obtained by adding c to the corresponding value on the old graph. So if (x, y) is a point on the new graph, $(x, y - c)$ is a point on the old graph and (x, y) satisfies the equation:

$$y - c = f(x)$$

Or, solving for y, we have the equation

$$y = f(x) + c$$

This equation represents a function, which we will call g, defined by:

$$g(x) = f(x) + c$$

The function $g(x)$ can be evaluated for each x by adding c to the value of $f(x)$. As we have just seen, the graph of g can be obtained by vertically shifting the graph of f. See Figure 10. Rephrasing this geometric fact gives us the following result.

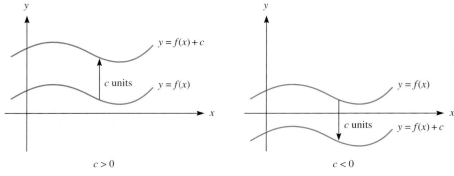

Figure 10

PRINCIPLE OF VERTICAL TRANSLATION

Let f be a function. Then the graph of the function g, defined by
$$g(x) = f(x) + c$$
is obtained by vertically translating the graph of f by c units. If $c > 0$, then the shift is upward. If $c < 0$, then the shift is downward.

EXAMPLE 5
Applying Translation

Sketch the graph of the function:
$$f(x) = \sqrt{1 - x^2} + 4$$

Solution

We have already seen (Example 3) that the graph of the function:
$$g(x) = \sqrt{1 - x^2}$$
consists of the upper half of the circle of radius 1 centered at the origin. According to the principle of vertical shifting, to get the graph of the function $f(x) + 4$, just shift the graph of $f(x)$ upward 4 units. See Figure 11.

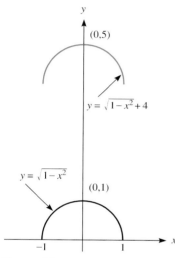

Figure 11

We can also sketch graphs by translating known graphs horizontally. Start with a function $f(x)$. Suppose that we shift the graph horizontally a directed distance of c units. Then if (x, y) is a point on the new graph, $(x - c, y)$ is a point on the old graph and satisfies the equation
$$y = f(x - c)$$
That is, we have the following result:

PRINCIPLE OF HORIZONTAL TRANSLATION

Let f be a function. Then the graph of the function defined by
$$g(x) = f(x - c)$$
is obtained by shifting the graph of f horizontally by c units. If $c > 0$, the shift is to the right. If $c < 0$, the shift is to the left. See Figure 12.

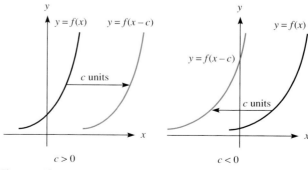

Figure 12

130   CHAPTER 2   Graphs and Functions

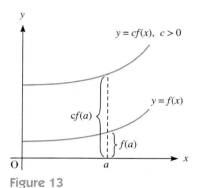

Figure 13

In addition to horizontal and vertical translation, we can also apply **scaling** to known graphs. Suppose that $f(x)$ is a given function and that c is a positive number. Consider the graph of the function $cf(x)$. A typical point on this graph is

$$(x, cf(x))$$

where x is in the domain of f. This point can be obtained by replacing the point $(x, f(x))$ with a point with the same x-coordinate but with the y-coordinate scaled to be c times as large. See Figure 13.

Scaling results from multiplying the y-values of a function by a positive number. If the positive number < 1, then each y-value decreases; in this case, the scaling is called **shrinking**. If the positive number > 1, then each y-value increases; in this case, the scaling is called **stretching**. We can summarize this discussion as follows:

PRINCIPLE OF SCALING

Let c be positive. To obtain the graph of the function $cf(x)$, scale the y-values of the graph of $f(x)$ by the factor c. If $0 < c < 1$, the y-coordinates are shrunk. If $c > 1$, the y-coordinates are stretched.

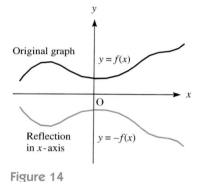

Figure 14

If we multiply a function f by -1, we obtain the function $-f$, whose graph is obtained by **reflecting** the graph of f in the x-axis. That is, the graph of $-f$ is the mirror image of the graph of f about the x-axis, as shown in Figure 14.

PRINCIPLE OF REFLECTION

The graph of $-f$ is obtained by reflecting the graph of f in the x-axis.

By combining scaling and reflection, we may obtain the graph of $cf(x)$ for any c, positive or negative: First scale by a factor of $|c|$, and then, if $c < 0$, reflect the resulting graph in the x-axis.

EXAMPLE 6
Applying Translation and Scaling

Sketch the graphs of the following functions.

1. $f(x) = (x - 2)^2$
2. $f(x) = (x + 3)^2$
3. $f(x) = 2x^2$
4. $f(x) = -2x^2$
5. $f(x) = (x - 1)^2 + 3$

Solution

1. By the principle of translation, the graph of $f(x)$ can be obtained by translating the graph of $g(x) = x^2$ two units to the right. See Figure 15(a).
2. By the principle of translation, the graph of $f(x)$ can be obtained by translating the graph of $g(x) = x^2$ three units to the left (because $c = -3$ is less than 0). See Figure 15(b).

3. The graph of $f(x) = 2x^2$ can be obtained by scaling the graph of $g(x) = x^2$ by 2. See Figure 15(c).
4. The graph of $f(x) = -2x^2$ can be obtained by first scaling the graph of $g(x) = x^2$ by $|-2| = 2$. Because the scaling factor -2 is less than 0, we must then reflect the graph in the x-axis, yielding the graph in Figure 15(d).
5. The graph of $f(x) = (x-1)^2 + 3$ can be obtained by translating the graph of $g(x) = x^2$ one unit to the right and three units upward. See Figure 15(e).

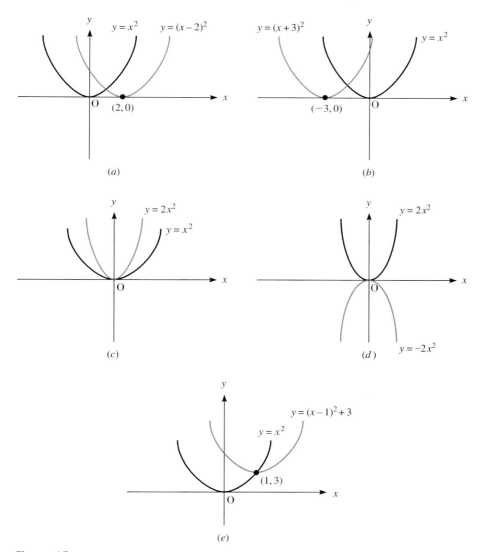

Figure 15

Increasing and Decreasing Functions

Up to this point in Section 2.6, we have concentrated on developing techniques for sketching graphs of functions. In many applied problems, however, we are given graphs of functions and must describe the features that they include and interpret these features in terms of the applications. One of the most important such features is whether a graph is increasing or decreasing for values of x in a particular interval. Specifically, we say that a function $f(x)$ is **increasing in the interval** $[a, b]$ provided that the values of $f(x)$ increase as x moves from left to right through the interval, that is, provided that $f(x_1) < f(x_2)$ whenever $a < x_1 < x_2 < b$. Figure 16 shows a graph that is increasing in an interval $[a, b]$. Similarly, we say that $f(x)$ is **decreasing in the interval** $[a, b]$ provided that the values of $f(x)$ are decreasing as x moves from left to right through the interval, that is, provided that $f(x_1) > f(x_2)$ whenever $a < x_1 < x_2 < b$. Figure 17 shows a graph that is decreasing on an interval $[a, b]$.

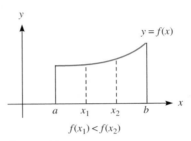

Figure 16 Figure 17

EXAMPLE 7
Income Function

Figure 18 shows the graph of the function that represents the income generated by a young biotechnology firm t years after its founding.

1. Determine the increase in revenue from year 1 to year 3.
2. In what intervals is the graph increasing and in what intervals is it decreasing?

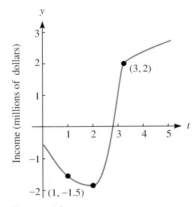

Figure 18

Solution

1. Reading the graph, we see that the ends of years 1 and 3 correspond to the points $(1, -1.5)$ and $(3, 2)$, respectively. The change from year 1 to year 3 is given by the difference in y-coordinates, namely $2 - (-1.5) = 3.5$. The positive sign of the difference indicates that revenue increased from year 1 to year 3 and that the amount of increase was $3.5 million.
2. Examining the graph, we see that it is decreasing in the interval $[0, 2]$ and increasing in the interval $[2, 5]$.

Symmetry of Graphs

We now introduce functions whose graphs exhibit various types of symmetry.

> **Definition**
> **Even and Odd Functions**
>
> Let f be a function. We say that f is an **even function** provided that
> $$f(-x) = f(x)$$
> for all x in the domain of f. We say that f is an **odd function** provided that
> $$f(-x) = -f(x)$$
> for all x in the domain of f.

EXAMPLE 8
Evenness and Oddness

Determine which of the following functions are even, which are odd, and which are neither.

1. $f(x) = x^3 - 2x$
2. $f(x) = \dfrac{1}{x^2 + 1}$
3. $f(x) = x^2 + 3$
4. $f(x) = (x + 1)^2$

Solution

1. We substitute $-x$ for x in the expression for f:
$$f(-x) = (-x)^3 - 2(-x)$$
$$= -x^3 + 2x$$
$$= -(x^3 - 2x)$$
We see that this last expression equals $-f(x)$. So f is an odd function.

2. Again we substitute $-x$ for x:
$$f(-x) = \frac{1}{(-x)^2 + 1} = \frac{1}{x^2 + 1}$$
Because the last expression is the same as the expression for $f(x)$, we see that $f(x)$ is an even function.

3. Again we substitute $-x$ for x in the expression for f:
$$f(-x) = (-x)^2 + 3 = x^2 + 3$$
But the last expression is the same as the expression for $f(x)$. In this case, $f(x)$ is an even function.

4. Proceeding as previously, we have:
$$f(-x) = (-x + 1)^2 = x^2 - 2x + 1$$
Because we have
$$f(x) = (x + 1)^2 = x^2 + 2x + 1$$
we see that $f(-x)$ is equal neither to $f(x)$ nor to $-f(x)$ for all x in the domain of f. Thus, f is neither even nor odd.

Suppose that $f(x)$ is an even function. If $(x, f(x))$ is a point on the graph of f, then so is $(-x, f(-x)) = (-x, f(x))$. These two points are located symmetrically with respect to the y-axis, as illustrated in Figure 19. In other words, an even function has a graph that is symmetric with respect to the y-axis.

134   CHAPTER 2   Graphs and Functions

As examples of graphs of even functions, consider the functions in parts 2 and 3 of Example 8. Their graphs are symmetric with respect to the y-axis. The graphs of these functions are shown in Figures 20 and 21.

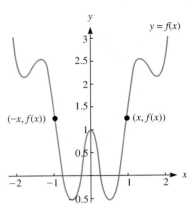

Figure 19

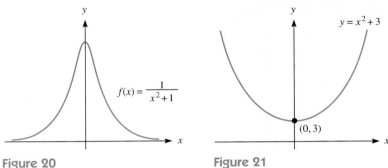

Figure 20   Figure 21

Suppose that f is an odd function. Then for each x in the domain of f, the points $(x, f(x))$ and $(-x, -f(x))$ are on the graph. These two points are located symmetrically with respect to the origin. That is, one can be obtained from the other by a 180° rotation about the origin. See Figure 22.

As an example of the graph of an odd function, consider the function f defined in part 1 of Example 8. Its graph is illustrated in Figure 23.

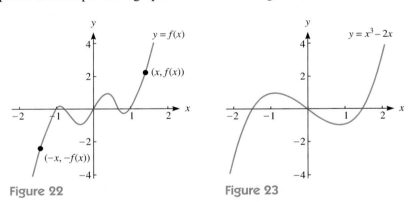

Figure 22   Figure 23

Piecewise-Defined Functions

All of the functions that we have considered so far have been defined using a single formula. However, it is possible to define a function using several formulas, one for each of several parts of the domain. For example, consider the function:

$$f(x) = \begin{cases} x & (x < 0) \\ -3x & (x \geq 0) \end{cases}$$

For $x < 0$, we use the first formula to determine $f(x)$; for $x \geq 0$, we use the second formula to determine $f(x)$. For instance, for $x = -3$ and $x = 2$, we have

$$f(-3) = -3, \qquad f(2) = -3(2) = -6$$

A function that is defined using several different formulas is said to be **piecewise-defined**. The graph of a piecewise-defined function is sketched by sketching separately the graph that corresponds to each formula. For example, the graph of the above function is shown in Figure 24.

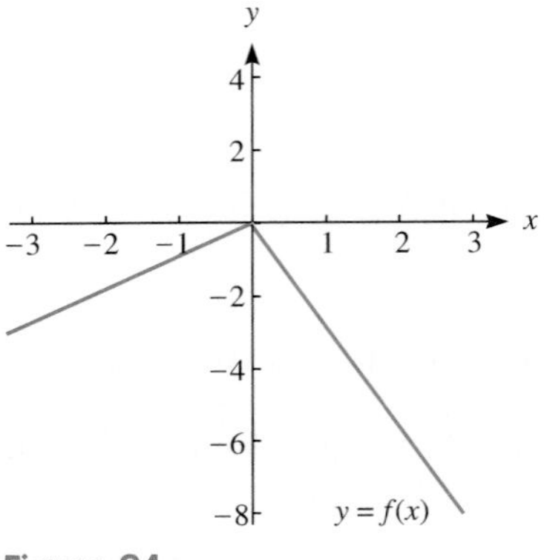

A piecewise-defined function that appears often in applied problems is the **greatest integer function**, denoted by $[x]$. This function is defined for all real numbers x and its value is given by:

Figure 24

$[x]$ = the greatest integer $\leq x$

Here are the values of the greatest integer functions for some typical values of x:

$$[5.1] = 5, \qquad [0.17] = 0, \qquad [111] = 111, \qquad [-1.1] = -2$$

Note that if $0 \leq x < 1$, the largest integer less than or equal to x is 0, so that:

$$[x] = 0 \qquad (0 \leq x < 1)$$

For $1 \leq x < 2$, the largest integer less than or equal to x is 1, so that:

$$[x] = 1 \qquad (1 \leq x < 2)$$

Similarly, for any nonnegative integer n, if $n \leq x < n+1$, the largest integer less than or equal to x is n, so that:

$$[x] = n \qquad (n \leq x < n+1)$$

The greatest integer function has a separate formula for each interval $n \leq x < n+1$. The graph of the function for x in this interval is a constant. The total graph resembles a series of steps, with a jump occurring at each integer. See Figure 25.

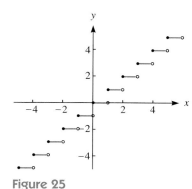

Figure 25

EXAMPLE 9
Cost Function

A cargo service charges by the weight of a package. It charges a flat fee of $2.00 plus $1.00 for each pound or fraction of a pound. Find a formula for the cost $C(x)$ of sending a package that weighs x pounds.

Solution

The function $C(x)$ is defined for $x > 0$. The $1 cost per pound can be expressed in terms of the greatest integer function as $[x]$ dollars. Indeed, for amounts less than 1 pound, $[x] = 0$; for weights at least 1 pound but less than 2 pounds, $[x] = 1$; and so forth. The function $[x] + 1$ counts 1 for each whole pound or fraction of a pound. Adding in the flat fee, the desired cost function is given by:

$$C(x) = 2 + ([x] + 1) = [x] + 3$$

EXAMPLE 10
State Income Taxes

Maryland imposes an income tax figured according to the following formula: 2 percent on the first $1,000 of income, 3 percent on the next $1,000, 4 percent on the next $1,000, 5 percent on the next $97,000, and 6 percent on any income above $100,000. Let $T(x)$ denote the amount of tax on an income of x dollars.

1. Determine the function $T(x)$.
2. What is the domain of T?
3. What is the tax on an income of $35,000?

Solution

1. We must figure the tax separately for each of the income intervals specified. For x between 0 and 1,000, the tax is 2 percent. That is,

$$T(x) = 0.02x, \qquad (0 \leq x \leq 1{,}000)$$

For income between $1,000 and $2,000, the tax equals the tax on the first $1,000 (or $20) plus 3 percent of the amount over $1,000. That is,

$$T(x) = 20 + 0.03(x - 1{,}000), \qquad (1{,}000 < x \leq 2{,}000)$$

For income between $2,000 and $3,000, the tax equals the tax on $2,000 (or $20 + 0.03(2{,}000 - 1{,}000) = 50$) plus 4 percent of the amount over $2,000. That is,

$$T(x) = 50 + 0.04(x - 2{,}000), \qquad (2{,}000 < x \leq 3{,}000)$$

For income between $3,000 and $100,000, the tax equals the tax on $3,000 (or $50 + 0.04(3,000 - 2,000) = 90$) plus 5 percent of the amount over $3,000. That is,

$$T(x) = 90 + 0.05(x - 3,000) \qquad (3,000 < x \le 100,000)$$

For income over $100,000, the tax equals the tax of $100,000 (or $4,940) plus 6 percent of the amount over $100,000. That is,

$$T(x) = 4,940 + 0.06(x - 100,000) \qquad (x > 100,000)$$

Combining the five different formulas for $T(x)$, we arrive at:

$$T(x) = \begin{cases} 0.02x & 0 \le x \le 1,000 \\ 20 + 0.03(x - 1,000) & 1,000 < x \le 2,000 \\ 50 + 0.04(x - 2,000) & 2,000 < x \le 3,000 \\ 90 + 0.05(x - 3,000) & 3,000 < x \le 100,000 \\ 4,940 + 0.06(x - 100,000) & x > 100,000 \end{cases}$$

2. The value of x can be any nonnegative number of dollars, so the domain of T consists of all nonnegative real numbers.
3. Since $35,000 lies between 3,000 and 100,000, we use the fourth formula for T:

$$T(35,000) = 90 + 0.05(35,000 - 3,000)$$
$$= 90 + 1,600$$
$$= \$1,690$$

Exercises 2.6

Determine whether each of the following is the graph of a function.

For each graph of a function f, find $f(-2)$, $f(0)$, $f(3)$, and $f(5)$.

1.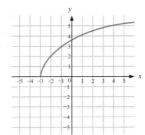

2.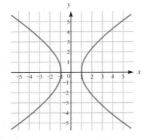

3.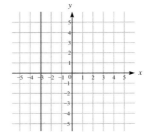

4.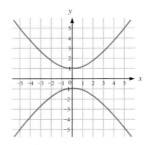

5.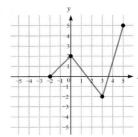

6.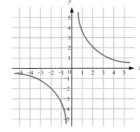

SECTION 2.6 Graphs of Functions

7.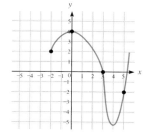

8.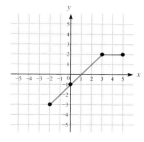

9–12. Determine the domain and range of the functions in Exercises 5–8.

Graph each function, and from the graph determine the domain and range.

13. $f(x) = 4 - 8x \quad (-1 \leq x \leq 1)$
14. $f(x) = 2x - 5 \quad (-3 \leq x \leq 3)$
15. $g(x) = x^2 \quad (-1 \leq x \leq 1)$
16. $g(x) = 3x^2 \quad (-1 \leq x \leq 1)$
17. $g(x) = x^2 - 3 \quad (-2 \leq x \leq 2)$
18. $f(x) = x^2 - 3$
19. $g(x) = x^3 - 1$
20. $g(x) = x^3 - 1 \quad (-2 \leq x \leq 2)$
21. $f(x) = \sqrt{4 - x^2} \quad (-2 \leq x \leq 2)$
22. $f(x) = \sqrt{-9 + x^2} \quad (|x| \geq 3)$
23. $f(x) = \dfrac{1}{x} \quad (x > 0)$
24. $f(x) = (x-1)^2 \quad (-1 \leq x \leq 3)$

Consider the following graph of a function $y = f(x)$ for Exercises 25–34. (See figure.) Graph each of the following:

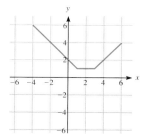

25. $y = f(x) + 1$
26. $y = f(x) - 1$
27. $y = f(x - 1)$
28. $y = f(x + 3)$
29. $y = f(x + 1) - 4$
30. $y = f(x - 2) + 3$
31. $y = -2f(x - 2) + 2$
32. $y = 2f(x) - 5$
33. $y = -\dfrac{1}{2}f(x) + 1$
34. $y = \dfrac{1}{2}f(x - 4)$

35. a. Graph $f(x) = \sqrt{x}$, and then use that graph to graph parts b–f.
    b. $g(x) = \sqrt{x - 3}$
    c. $g(x) = 2 + \sqrt{x + 1}$
    d. $g(x) = 4\sqrt{x}$
    e. $g(x) = -\sqrt{2x}$
    f. $g(x) = -1 - \dfrac{1}{2}\sqrt{x - 1}$

36. a. Graph $y = |x|$, and then use that graph to graph parts b–f.
    b. $y = |x + 3|$
    c. $y = |x - 1| - 2$
    d. $y = |-2x|$
    e. $y = -\dfrac{1}{2}|x| + 2$
    f. $2y - 6 = |x + 1|$

Determine whether each of the following functions is even, odd, or neither.

37. $f(x) = 3x - 2$
38. $f(x) = -5x$
39. $f(x) = 2x^3 - 2x^2 + x$
40. $f(x) = -2x^2$
41. $f(x) = \dfrac{3}{x^2}$
42. $f(x) = \dfrac{5}{x^3 - x}$
43. $f(x) = (x - 3)^2$
44. $f(x) = x^2 + 2x - 4$
45. $f(x) = \left(\dfrac{1}{x}\right)^{-3}$
46. $f(x) = \dfrac{x}{x + x^3}$
47. $f(x) = 4x^{2/3}$
48. $f(x) = (x^2 + 8)^{2/3}$

Graph the function.

49. $f(x) = \begin{cases} x & (x \leq 1) \\ -x & (x > 1) \end{cases}$

50. $f(x) = \begin{cases} -1 & (x < 0) \\ 0 & (x = 0) \\ 1 & (x > 0) \end{cases}$

51. $f(x) = \begin{cases} x - 1 & (x < -3) \\ x^2 & (-3 \leq x < 3) \\ 1 - x & (x \geq 3) \end{cases}$

52. $f(x) = \begin{cases} |x| & (x < 2) \\ -x^2 & (2 \leq x < 3) \\ 2x - 1 & (x \geq 3) \end{cases}$

53. $f(x) = \begin{cases} \dfrac{x^2 - 9}{x - 3} & (x \neq 3) \\ 2 & (x = 3) \end{cases}$

54. $f(x) = \begin{cases} \dfrac{x^2 - 9}{x + 3} & (x \neq -3) \\ 6 & (x = -3) \end{cases}$

55. $f(x) = \begin{cases} -x - 3 & (x \leq -2) \\ 3 - x^2 & (-2 < x < 2) \\ x - 3 & (x \geq 2) \end{cases}$

56. $f(x) = \begin{cases} -3 & (x \text{ not an integer}) \\ 3 & (x \text{ an integer}) \end{cases}$

The greatest integer function is given by $f(x) = [x]$, or, in computer language, $f(x) = \text{INT}(x)$. Use the graph of this function to construct the following graphs.

57. $g(x) = [x - 1]$
58. $g(x) = 2[x]$
59. $g(x) = \text{INT}(x)$
60. $f(x) = 3\text{INT}(x + 2) - 1$

61. We define a **decimal place function** as follows: Given a number x, we find its decimal notation. Then $f(x)$ is the digit in the fourth decimal place. For example,

$$f\left(\frac{2}{5}\right) = f(0.400000) = 0$$

$$f\left(\frac{18}{19}\right) = f(0.947368\ldots) = 3$$

$$f\left(-\frac{39}{14}\right) = f(-2.785714\ldots) = 7$$

and so on. Find $f(\frac{3}{4})$, $f(\frac{14}{23})$, $f(\frac{23}{13})$, and $f(\sqrt{2})$.

62. Refer to the preceding problem.
    a. Determine the domain and range of the function.
    b. Describe the graph of this function.

Applications

63. **Parking fees.** A parking garage charges $0.50 for the first hour or for part of the first hour, and $0.50 for each hour or partial hour thereafter, up to a maximum of $5.00 a day. Graph this function for 0 to 10 hours.

64. **Computer rental.** It costs $20 to connect to a mainframe computer and $20 per hour of CPU time. Graph this function for the use of 0 to 1 hour of CPU time.

In your own words

65. Explain how to sketch the graph of a function by plotting points. How should you go about selecting which points to plot?

66. Describe how to plot the graph of a linear equation using intercepts. Will this method work for all linear equations? Explain your answer.

2.7 GRAPHING CALCULATOR FUNDAMENTALS

Many hand-held calculators have the ability to graph equations. By using such calculators, you can apply technology to perform many of the graphical tasks of algebra with minimal effort. The following discussion is meant to apply to most graphing calculators. For your particular calculator, a particular keystroke described below may be a combination of keystrokes or a selection in a menu. Consult your calculator manual for details.

Fundamentals of Graphing Using a Calculator

Basic Calculator Techniques In order to graph a function using a graphing calculator, you must first enter the function. This is done by pressing the [Y=] key, which results in a display like the one shown in Figure 1. The screen shown allows you to enter up to four functions to graph simultaneously. (The exact number of allowable functions will depend on your calculator.) For instance, suppose that you wish to graph the following function:

$$y = x^2$$

You would move the cursor to the $Y_1 =$ line and type:

X^2

(To key this in, you will need to locate the key used to enter the variable X.)

You can edit the function or functions using the arrow keys to move the cursor, [Ins] to insert characters at the cursor, and [Del] to delete the character at the cursor.

Once the functions are entered to your satisfaction, you must set the x-range and y-range you wish to view. For the moment, let's use the default ranges of the calculator, which for most calculators are:

$$-10 \leq x \leq 10, \quad -10 \leq y \leq 10$$

```
:Y1 =
:Y2 =
:Y3 =
:Y4 =
```

Figure 1

SECTION 2.7 Graphing Calculator Fundamentals

Using the defaults means that we don't need to set the ranges and can proceed to view the graph.

To view the graph, press the [GRAPH] key. The calculator will display the graph of the specified equations in the specified ranges. You will see a graph like the one shown in Figure 2.

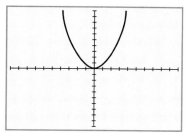

Figure 2

Setting Ranges To set ranges, press the [RANGE] key. You will see a display like that shown in Figure 3. Use the cursor-movement and editing keys to enter the desired minimum and maximum values of x (denoted Xmin and Xmax, respectively), and the desired minimum and maximum values of y (denoted Ymin and Ymax, respectively). For instance, to graph

$$y = x^2$$

in the range

$$-2 \leq x \leq 2, \qquad -4 \leq y \leq 4$$

we would set:

Xmin = -2

Xmax = 2

Ymin = -4

Ymax = 4

Pressing [GRAPH] again, we would see the graph shown in Figure 4.

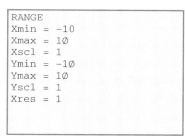

Figure 3

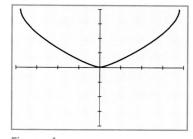

Figure 4

Tracing You can move a point along a displayed graph and have the calculator display the numerical coordinates of the point. This process is called **tracing** and is useful in calculating the coordinates of particular points that you can locate visually, such as the intersection point of two graphs.

To trace a graph, first display the graph and press the [TRACE] key. The calculator will display a point on the graph. On the bottom of the display will be the x and y-coordinates of the point. You can move the point along the curve using the left and right arrow keys. Each keystroke moves the point one dot to the left or right. Each time you move the point, the coordinates of the new point are displayed. (See Figure 5.)

If several curves are displayed, you can move the point from curve to curve using the up and down arrow keys.

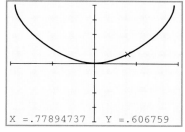

Figure 5

Zooming You can control the level of detail in a graph by **zooming**. Zooming out, you step away from the graph and thereby expand the range. Zooming in, you step toward the graph and thereby contract the range. A number of methods of zooming are listed in a menu that is displayed when you press the [ZOOM] key.

140 CHAPTER 2 Graphs and Functions

Perhaps the most useful method of zooming is by specifying a box on the graph. The portion of the graph in the box specifies the new range.

Here is how to zoom using a box: Press ZOOM. Select the option "Box" from the menu. The calculator will display the graph and a cursor. Using the arrow keys, move the cursor to one corner of the desired box and press the ENTER key. Then move the cursor to the opposite corner of the box. As you move the cursor, you will see the box expand or contract. (See Figure 6.) When the cursor indicates the corner you want, press the ENTER key. The calculator now displays the graph with the designated box as the range. (See Figure 7.)

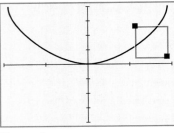

Figure 6   Figure 7

Graphing Calculator Explorations

The following examples illustrate some of the fine points of graphing functions on a calculator.

EXAMPLE 1
Graphing a Function

Use a graphing calculator to display the graph of:
$$y = x^3 - 3x - 2$$

Solution

Enter the function $Y_1 = X^3 - 3X - 2$ in the Y= menu. (See Figure 8.) Use the standard settings in the RANGE menu. Pressing GRAPH yields the graph of the function for $-10 \leq x \leq 10$. (See Figure 9.)

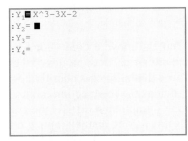 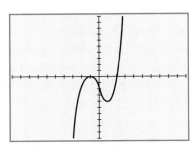

Figure 8   Figure 9

EXAMPLE 2
Graphing a Function with Limited Domain

Use a graphing calculator to display the graph of:
$$y = x^3 - 3x + 2 \qquad (-3 \leq x \leq 3)$$

Solution

We can express the limitation $-3 \leq x \leq 3$ in the range by placing the inequalities $(-3 \leq x)$ and $(x \leq 3)$ after the equation that defines the function. (See Figure 10.) Your calculator may have keys for the inequality symbols or it may allow you to enter them by choosing them from a menu of "tests." Consult your calculator

SECTION 2.7 Graphing Calculator Fundamentals

manual for details. Note that the minus sign in -3 is obtained by pressing the $\boxed{(-)}$ key, which is an initial minus sign. When we press $\boxed{\text{GRAPH}}$ we see the display in Figure 11. The domain of the displayed graph is $-3 \leq x \leq 3$, as required. Note that the calculator displays vertical lines that connect the points at the edges of the graph with the x-axis. These line segments should be ignored. They are not part of the function graph as defined mathematically.

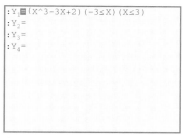
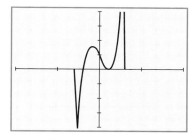

Figure 10   Figure 11

EXAMPLE 3
Graphing a Function That Involves a Rational Exponent

Use a graphing calculator to display the graph of:
$$y = x^{2/3}$$

Solution

Enter the function as:
$Y_1 = X\hat{\,}(2/3)$
Note that it is necessary to enclose the exponent in parentheses. The calculator exponentiation function $\hat{\,}$ is defined only for $X \geq 0$. Therefore, even though the function
$$y = x^{2/3} = \left(\sqrt[3]{x}\right)^2$$
is defined for all x, the graph you see will only include the portion for nonnegative values of x. See Figure 12. Note, however, that if you key in the function as $Y_1 = (X\hat{\,}2)(1/3)$, then the whole graph is shown. This is because $X\hat{\,}2$ is defined for all X and its value is nonnegative. So Y_1 is defined for all values of X.

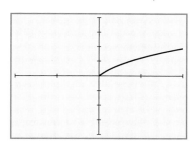

Figure 12

EXAMPLE 4
Graphing a Function with Limited Domain

Use a graphing calculator to display the graph of $y = \sqrt{x}$.

Solution

The function has a natural domain $\{x: x \geq 0\}$. Natural domains are built into the calculator and need not be entered separately. We enter the function as:
$Y_1 = \sqrt{X}$

The graph is shown in Figure 13.

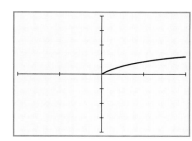

Figure 13

EXAMPLE 5
Graphing a Function

Use a graphing calculator to display the graph of:

$$y = \frac{x}{x-1} \quad (x \neq 1)$$

Solution

The exclusion of $x = 1$ is a natural range restriction, since the denominator is 0 for this value of x. This restriction is automatically handled by the calculator. It attempts to evaluate the function at $x = 1$ and finds that it cannot, so it omits this point in drawing the graph. The graph is shown in Figure 14. However, in drawing the graph, the calculator attempts to connect consecutive points. This leads to the vertical spike in the picture. Actually, this spike should not be included. The graph consists of two disconnected parts, one to the left of $x = 1$ and one to the right of $x = 1$. In studying graphs produced by a calculator, you will need to omit consideration of such spikes.

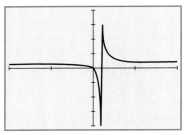
Figure 14

EXAMPLE 6
Determining When Two Functions Are Equal

Use a graphing calculator to determine the values of x for which $f(x) = g(x)$, where:

$$f(x) = x^2 + 2x - 4, \qquad g(x) = 2x - 1$$

Solution

Graph $f(x)$ as Y_1 and $g(x)$ as Y_2. With the standard range settings, the display resembles Figure 15. To determine the desired values of x, we must determine the x-coordinates of the points where the graphs intersect. We can do this using the trace feature. Trace along the graph of $f(x)$ until you get to an intersection, and note the x-coordinate shown at the bottom of the screen. From Figures 16 and 17, we see that:

$$x \approx -1.702128, \qquad 1.7021277$$

To obtain these values more accurately, we can use the box feature to enlarge the portion of the graph near an intersection, and use the trace to locate the intersection, as above. By enlarging the graph, each horizontal step in the trace is smaller, so the value of x can be determined with greater accuracy.

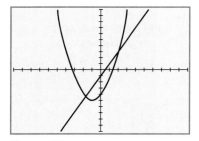
Figure 15

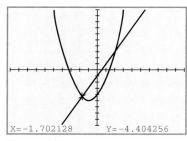
Figure 16

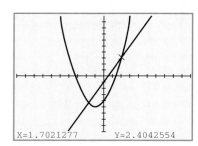
Figure 17

SECTION 2.7 Graphing Calculator Fundamentals

EXAMPLE 7
Determining the Domain and Range of a Function Using a Graphing Calculator

Use a graphing calculator to determine the domain and range of the function:

$$f(x) = 1 + \frac{x}{x^2 + 1} \qquad (x \geq 0)$$

Solution

Enter the function as shown in Figure 18. To make the features of the graph more visible, use the ranges $-10 \leq x \leq 30$, and $0 \leq y \leq 3$. The graph is shown in Figure 19. A point of the graph lies over each point on the x–axis. That is, the domain of the function consists of all nonnegative real x. (Note that although you allowed x to vary over the domain $-10 \leq x \leq 30$, the calculator correctly restricts this domain to $0 \leq x \leq 10$.) Examining the graph, we see that its points assume all y-values greater than or equal to 1 and less than or equal to 1.5. That is, the range is the interval $[1, 1.5]$.

Figure 18

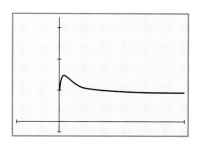

Figure 19

Exercises 2.7

In Exercises 1–8, use a graphing calculator to graph the following equations. Use the default values for the range of the viewing box.

1. $y = 3x - 5$
2. $y = -2x + 3.1$
3. $y = x^2 - x$
4. $y = 0.3x^2 + 0.151x$
5. $y = \dfrac{3}{x + 1}$
6. $y = \sqrt{3x - 1}$
7. $y = \dfrac{1}{x^2 + 1}$
8. $y = \dfrac{x - 1}{(x + 1)(x + 2)}$

9. Use the trace feature to determine the y-coordinate on the graph of $y = x^2 - 3x$ when x has the following values:
   a. $x = 1.2$
   b. $x = 2.1$
   c. $x = -1$
   d. $x = 0.001$

10. Use the trace feature to determine a point on the graph of $y = \dfrac{x}{x^2 + 1}$ at which $y = 0.1$.

11. Use the trace feature to determine a point on the graph in Exercise 7 for which $y = 0.3$.

12. Use the zoom feature to enlarge the graph in Exercise 8 in the range $-2 \leq x \leq 0, -10 \leq y \leq 10$.

13. By using the zoom and trace features, determine the points at which the graph of Exercise 7 crosses the horizontal axis.

14. By using the zoom and trace features, determine, to two significant digits, the points at which the graph of Exercise 8 crosses the horizontal axis.

15. Graph the equations $y = x^2$ and $y = -2x + 3$ on the same coordinate system. Choose the viewing box so that the two intersection points are visible.

144   CHAPTER 2   Graphs and Functions

16. Determine the coordinates of the intersection points of the graph in Exercise 15.

17. Graph the equations $y = 1/x$ and $y = 3x^3$ on the same coordinate system. Choose the viewing box so that the intersection points are visible.

18. Determine the coordinates of the intersection points of the graph in Exercise 17.

19. Graph the equations $y = x^2$, $y = 3x^2$, and $y = 0.6x^2$ on the same coordinate system. What do you observe about the points where they cross the horizontal axis? Can you make a conclusion about what happens to horizontal axis intersections when a graph is scaled?

20. On the same coordinate system, graph the equation $y = 5x + 3$, and the same graph translated 3 units horizontally and 2 units vertically.

21. On the same coordinate system, graph the equation $y = -x^2 + 1$, and the same graph translated -1 units horizontally and 3 units vertically.

22. On the same coordinate system, graph the equation $y = -x^2 + x - 1$, and the same graph translated -2 units horizontally and -1 units vertically.

Use a graphing calculator to determine the domains and ranges of the following functions:

23. $f(x) = 3x - 2$ $(0 \le x \le 2)$

24. $f(x) = 5 - \dfrac{3}{2}x$ $(x \ge 0)$

25. $f(x) = x^2 - 3x + 1$ $(1 \le x \le 3)$

26. $f(x) = x^3 - 2x$ $(0 \le x \le 3)$

27. $f(x) = \dfrac{x}{x^2 - 1}$

28. $f(x) = \dfrac{1}{x} + x$ $(0.5 \le x \le 2)$

29. $f(x) = \sqrt{1 - 4x^2}$

30. $f(x) = \dfrac{x}{x^2 + 1}$

31. $f(x) = |x(x + 1)|$

32. $f(x) = |x^2 - 2|$

Use a graphing calculator to determine the solutions of the following equations to one decimal place of accuracy.

33. $2x^2 = 5x - 14$

34. $18x^2 = 10x$

35. $13t^2 + 6t = 0$

36. $6x^2 + 5x = 21$

37. $10x^2 - 3x - 1 = 0$

38. $2 + \dfrac{3}{x^2} = \dfrac{1}{x}$

39. $3 - \dfrac{1}{x} + \dfrac{6}{x^2} = 0$

40. $\dfrac{2x - 3}{3x + 5} = \dfrac{x + 7}{3x - 4}$

41. $\dfrac{5 - 2x}{6x + 1} = \dfrac{3 + 2x}{5 - x}$

42. $x^2 + \sqrt{2}x - \sqrt{3} = 0$

43. $\sqrt{5}x^2 + 4x = \sqrt{2}$

44. $(y + 5)(y - 4) = (2y + 1)(3y - 4)$

45. $(2a - 1)(3a + 5) = (a - 3)(4a - 7)$

Suppose that $f(x) = 2 - x^2$. Graph the following along with f on the same coordinate system:

46. the graph of f translated to the right 4 units.

47. the graph of f translated to the left 6 units.

48. the graph of f translated up 1 unit.

49. the graph of f translated down 2 units.

50. the graph of f reflected in the x-axis.

51. the graph of f scaled by a factor of 3.

52. the graph of f scaled by a factor of 2 and then reflected in the x-axis.

53. the graph of f translated up 1 unit and right 2 units.

54. the graph of f translated down 2 units and reflected in the x-axis.

By graphing $f(x)$ and $f(-x)$ on the same coordinate system, determine which of the following functions f are even:

55. $f(x) = x^2 + \dfrac{1}{x^2}$

56. $f(x) = (x^3 + 1)^2 + (x^3 - 1)^2$

57. $f(x) = (2x + 1)^2$

58. $f(x) = (1 - x)^2$

In your own words

59. Give two advantages of using graphing calculators rather than manually plotting functions.

60. Describe three pitfalls to watch out for in graphing functions with a graphing calculator.

2.8 LINEAR FUNCTIONS

Among the simplest functions are the linear functions
$$f(x) = ax + b, \qquad a \ne 0$$
where a and b are real numbers. In this section, we explore these functions and present some models in which they are used.

SECTION 2.8 Linear Functions

The graph of the linear function above is the graph of the following linear equation:

$$y = ax + b$$

And as we saw in the preceding chapter, this graph is a nonvertical straight line with slope $m = a$. Moreover, because $a \neq 0$, the graph is not horizontal. That is, the graph of a linear function is a nonvertical, nonhorizontal straight line.

Because the graph of a linear function is nonhorizontal, all real values are assumed by the function. That is, the domain of a linear function consists of all real numbers.

In the preceding chapter, we explored straight lines and their equations. In the following examples, we apply what we developed there in order to solve various applications that involve linear functions.

EXAMPLE 1
Fitting a Linear Function to Data

Vehicles depreciate, or lose value, over time. Suppose that $D(x)$ denotes the depreciated value of a truck after x years, where the depreciated value is determined linearly, that is, the function $D(x)$ is a linear function. Suppose that the truck cost $50,000 when new, and its depreciated value is $10,000 after 5 years.

1. Determine the function $D(x)$.
2. What is the depreciated value of the truck after 3 years?

Solution

1. When $x = 0$, the value of D is 50,000 and when $x = 5$, the value of D is 10,000. That is, the graph is the line passing through the points (0, 50,000) and (5, 10,000). Let's apply the techniques of Section 2.3 to determine the equation of the line that passes through these points. The slope of the line is:

$$\frac{\text{difference in } y \text{ coordinates}}{\text{difference in } x \text{ coordinates}} = \frac{50,000 - 10,000}{0 - 5}$$

$$= -\frac{40,000}{5}$$

$$= -8,000$$

Because the slope is negative, the truck depreciates $8,000 each year. From the point-slope formula, the equation of the line is

$$y - 50,000 = -8,000(x - 0)$$

$$y - 50,000 = -8,000x$$

Solving this equation for y in terms of x gives us:

$$y = -8,000x + 50,000$$

The value of y is the value of the function D at x; that is,

$$D(x) = -8,000x + 50,000$$

2. The depreciated value of the truck after 3 years is equal to the function value $D(3)$.

$$D(3) = -8,000(3) + 50,000$$

$$= -24,000 + 50,000$$

$$= 26,000$$

After 3 years, the depreciated value of the truck is $26,000.

EXAMPLE 2
Graphing a Linear Function

Graph the linear function $g(x) = -3x + 15$.

Solution

The graph of the function is the graph of the linear equation:

$$y = -3x + 15$$

Let's use the intercept method for drawing the graph.

x-intercept: $y = 0$ y-intercept: $x = 0$

$-3x + 15 = 0$ $y = -3(0) + 15$

$3x = 15$ $= 15$

$x = 5$

x-intercept is 5 y-intercept is 15

Plot the points $(5, 0)$ and $(0, 15)$, and draw a line through them. This gives us the graph of the function. (See Figure 1.) Note that we include the entire line in the graph, not just the portion between the plotted points, because the domain of the function consists of all real numbers x.

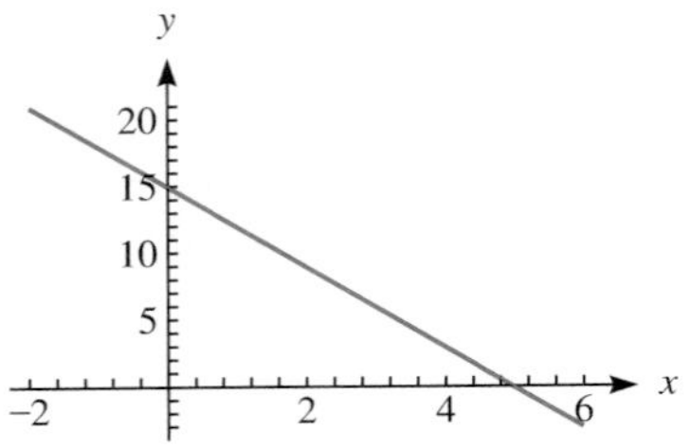

Figure 1

EXAMPLE 3
Application of Slope

The cost $C(x)$ of producing x cellular phones is given by the linear function $C(x) = 21x + 75{,}800$. What is the added cost of increasing production by 3,000 phones?

Solution

The graph of the function is the graph of the linear equation:

$$y = 21x + 75{,}800$$

This graph is a straight line. The equation is in slope-intercept form, so the slope is the coefficient of x, or 21. Increasing production by 3,000 phones increases y, the cost, by $21 \cdot 3{,}000$, or \$63,000.

EXAMPLE 4
A Piecewise Linear Function

Determine the piecewise linear function whose graph consists of the line segments from $(0, 5)$ to $(3, 2)$ and from $(3, 2)$ to $(5, 15)$. (See Figure 2.)

Solution

The desired function is defined by two linear functions; the first corresponds to the range $0 \leq x \leq 3$, and the second corresponds to the range $3 < x \leq 5$. To determine the linear functions, we use the given endpoints of the graph segment and the slope-intercept formula. The slope of the first segment is:

$$\frac{\text{difference in } y \text{ coordinates}}{\text{difference in } x \text{ coordinates}} = \frac{5 - 2}{0 - 3} = -1$$

So the line segment through the points $(0, 5)$ and $(3, 2)$ has equation:

$$y - 5 = (-1)(x - 0)$$

$$y = -x + 5$$

The slope of the second segment is

$$\frac{\text{difference in } y \text{ coordinates}}{\text{difference in } x \text{ coordinates}} = \frac{2 - 15}{3 - 5} = \frac{13}{2}$$

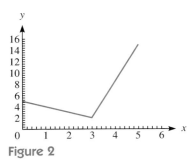

Figure 2

So the line segment through the points (3, 2) and (5, 15) has equation:
$$y - 2 = \frac{13}{2}(x - 3)$$
$$y - 2 = \frac{13}{2}x - \frac{39}{2}$$
$$y = \frac{13}{2}x - \frac{35}{2}$$

The desired function therefore is:
$$f(x) = \begin{cases} -x + 5 & (0 \le x \le 3) \\ \frac{13}{2}x - \frac{35}{2} & (3 < x \le 5) \end{cases}$$

Exercises 2.8

Determine the linear function $f(x)$ that satisfies the following conditions.

1. $f(0) = 0$, $f(2) = 4$
2. $f(0) = 0$, $f(3) = 15$
3. $f(0) = 3$, $f(-1) = 4$
4. $f(0) = -8$, $f\left(\frac{1}{2}\right) = 10$
5. $f(1) = 3$, $f(2) = 7$
6. $f(5) = 8$, $f(10) = 0$
7. $f(-3) = 4$, $f(-9) = 20$
8. $f\left(-\frac{1}{2}\right) = 8$, $f\left(-\frac{3}{2}\right) = 12$
9. x-intercept 4, y-intercept 10
10. x-intercept -1, y-intercept 1
11. x-intercept 12, $f(4) = 22$
12. y-intercept 10, $f(-5) = 9$

Graph the following linear functions.

13. $f(x) = 3x$
14. $f(x) = \frac{1}{2}x$
15. $f(x) = 2x - 1$
16. $f(x) = 4x + 3$
17. $f(x) = \frac{3}{2}x + \frac{7}{2}$
18. $f(x) = -x$
19. $f(x) = -2x$
20. $f(x) = -3x - 1$

Use the graph of the linear function $f(x) = x$ and the properties of translation to obtain the graphs of the following functions.

21. $f(x) = x - 1$
22. $f(x) = x + 1$
23. $f(x) = x - 3$
24. $f(x) = x - \frac{1}{2}$

Use the graph of the linear function $f(x) = -2x$ and the properties of translation and reflection to obtain the graphs of the following functions.

25. $f(x + 1)$
26. $f(x) + 1$
27. $f(x - 1)$
28. $f(x + 1) + 1$
29. $f(-x)$
30. $f(-x + 1)$

Applications

31. **Personal computers for home use.** The number of personal computers sold for home use (in millions) is denoted by y. The number sold in year x (with 1989 corresponding to 0) is given by the model:
$$y = \begin{cases} 4 & 0 \le x < 2 \\ 3 + \frac{x}{2} & 2 \le x \le 4 \end{cases}$$

a. Graph this function.

b. How many personal computers were sold for home use in 1990?

c. How many personal computers were sold for home use in 1992?

d. Interpret the graph in terms of the demand for home personal computers over the time period from 1989 through 1993.

148 CHAPTER 2 **Graphs and Functions**

32. **Health care employment.** The number y of workers in the health care field (in millions) in year x (with 1960 corresponding to 0) is given by the model:

$$y = 1.11 + 0.22x$$

 a. Graph this function.

 b. How many workers were in health care in 1965?

 c. How many workers were in health care in 1993?

 d. How many health care workers are added over each five-year period?

33. **Depreciation.** Suppose that $D(x)$ denotes the depreciated value of a building after x years, where the depreciated value is determined linearly. Assume that the building cost $3.7 million when new and its depreciated value after 30 years is $200,000.

 a. Determine the function $D(x)$.

 b. Graph the function $D(x)$.

 c. What is the significance of the y-intercept of the graph?

 d. What is the depreciated value of the building after 20 years?

 e. In how many years will the depreciated value of the building be $1 million?

34. **Cost function.** The cost $C(x)$ of producing x desks is given by the linear function $C(x) = 198x + 50,490$.

 a. What is the added cost of increasing production by 50 desks?

 b. How many more desks can be manufactured if an additional $105,000 can be allocated to production costs?

 In your own words

35. Is every straight line the graph of a linear function? Explain.

36. Explain the sort of modeling problem in which a linear function would be appropriate. Explain the kind of modeling problem in which a linear function would *not* be appropriate.

Chapter Review

Important Concepts—Chapter 2

- axes
- x-axis
- y-axis
- origin
- two-dimensional coordinate system
- directed distance
- coordinate representation of a point
- abscissa; x-coordinate
- ordinate; y-coordinate
- quadrant
- distance formula
- midpoint formula
- solution of an equation in two variables
- graph of an equation in two variables
- linear equation in two variables
- graph of a linear equation in two variables
- x-intercept
- y-intercept
- determining intercepts
- graphing an equation by plotting points
- equation of a circle
- completing the square
- rise
- run
- slope of a line
- slope-intercept form of the equation of a line
- equations of straight lines
- geometric interpretation of slope
- rising and falling lines and slope
- point-slope formula
- equations of lines
- rate

Chapter Review

Mathematics and the World around Us—Orbits of the Planets

Since humans evolved into thinking beings, they have observed the heavens with curiosity and emotion. In ancient times, people charted the motions of heavenly bodies and used the data to measure time and calibrate calendars. The ancient Greek philosophers proposed the theory that the earth was at the center of universe and that the sun and planets circled around the earth on spheres. This geocentric, or earth-centered, theory was based more on philosophy and ideology than on scientific observation. So important was the notion that the earth be at the center of the universe, that discrepancies between philosophy and observation were covered up with complex refinements in theory. This process led ultimately to Ptolemy's theory of planetary motion, which held that the sun and planets made "epicycles" on circular orbits around the earth. Even Ptolemy's theory did not lead to perfect predictions of events such as eclipses, however, and it could not account for the motion of comets and of the planet Mars.

In the late sixteenth century, the Polish astronomer Nicolaus Copernicus proposed a new theory in which the planets rotated in circular orbits about the sun. Copernicus was correct that the sun was at the center of things, but his circular orbit theory still made embarrassing errors in prediction.

Tycho Brahe and his protege, Johannes Kepler, observed the heavens with the naked eye and charted their observations for decades. It became clear that the orbits of the planets were not circles. However, the precise nature of the orbits was unclear until, in a flash of inspiration, Kepler decided to check whether the orbits might not be ellipses, rather than circles. And, indeed, they were. Based on the data he and Tycho Brahe had accumulated, Kepler formulated three laws of planetary motion, which are the basis of modern astronomy.

Kepler's three laws of planetary motion matched the observed data with uncanny accuracy. However, they had no theoretical underpinning. This was provided by Sir Isaac Newton in the early seventeenth century. Newton discovered the law of universal gravitation, which states that planets attract each other with a force that is directly proportional to the product of their masses and inversely proportional to the square of the distance between them. Newton used the "inverse square law" to mathematically deduce the equations of the planetary orbits. From the equations, Newton was able to deduce Kepler's three laws and prove that the orbits of the planets were elliptical.

Newton's mathematical techniques are among the most fundamental in the history of mathematics and science. They have been used for three centuries in exploring the physics of the world around us.

- velocity
- slopes of parallel lines
- slopes of perpendicular lines
- properties of inequalities
- solving an inequality
- solution set of an inequality
- intersection of sets
- three-termed inequality
- union of sets
- polynomial inequality
- rational inequality
- interval
- open interval
- closed interval
- half-open interval
- infinite interval
- test interval
- constant sign theorem
- function
- value of a function
- independent variable
- dependent variable
- correspondence
- mapping
- function as a correspondence
- domain of a function
- range of a function
- defined
- undefined
- linear function

- quadratic function
- cubic function
- polynomial function
- rational function
- natural domain
- graph of a function
- graphical interpretation of domain
- graphical interpretation of range
- vertical line test
- principle of vertical translation
- principle of horizontal translation
- principle of scaling
- shrinking a function
- stretching a function
- principle of reflection
- function increasing in an interval
- function decreasing in an interval
- even function
- odd function
- piecewise-defined function
- greatest integer function
- tracing
- zooming
- graphing a linear function
- piecewise linear function

Important Results and Techniques—Chapter 2

Distance formula	The distance between the points $P(x_1, y_1)$ and $Q(x_2, y_2)$: $$d(P, Q) = \sqrt{(x_1 - x_2)^2 + (y_1 - y_2)^2}$$	p. 71
Midpoint formula	The midpoint M of the line segment that connects the points (x_1, y_1) and (x_2, y_2): $$M = \left(\frac{x_1 + x_2}{2}, \frac{y_1 + y_2}{2}\right)$$	p. 72
Equation of a circle	Equation of a circle with center (h, k) and radius r: $$(x - h)^2 + (y - k)^2 = r^2$$	p. 83
Slope of a line	Slope m of a line that passes through the points (x_1, y_1) and (x_2, y_2): $$m = \frac{y_2 - y_1}{x_2 - x_1}, \quad x_2 - x_1 \neq 0$$	p. 87
Slope-intercept formula	Equation of a line with slope m and y-intercept b: $$y = mx + b$$	p. 90
Point-slope formula	Equation of a line that has slope m and passes through the point (x_1, y_1): $$y - y_1 = m(x - x_1)$$	p. 93
Slopes of parallel lines	If two lines are parallel, then they have the same slope: $$m_2 = m_1$$	p. 95
Slopes of perpendicular lines	If two lines are perpendicular, then their slopes satisfy the equation: $$m_1 m_2 = -1$$	p. 96

Review Exercises—Chapter 2

1. Determine the distance between the points $(-2, -7)$ and $(3, -4)$.
2. Determine the midpoint of the line segment that has the endpoints $(-2, -7)$ and $(3, -4)$.
3. Determine the slope of the line that passes through the points $(-2, -7)$ and $(3, -4)$.
4. Determine the equation of the line that passes through $(-2, -7)$ and $(3, -4)$.
5. Find the slope and y-intercept of $3x = 6y - 24$.
6. Determine the equation of the line that passes through $(5, -6)$ and has slope $-\frac{3}{2}$.

Determine whether each set of lines is parallel, perpendicular, or neither.

7. $3x + 6y = 3$, $\quad 6y = 10 - 4x$
8. $y = 0.32x + 16$, $1{,}000y = 3{,}125x + 7$
9. $12x - 45y = 17$, $77x - 76y = 87$
10. Find the point on the x-axis that is equidistant from $(7, 3)$ and $(1, -2)$.
11. Find the distance between $\left(2\sqrt{2}, \sqrt{5}\right)$ and $\left(3\sqrt{2}, -2\sqrt{5}\right)$.
12. Find the midpoint of the line segment that has endpoints $\left(2\sqrt{2}, \sqrt{5}\right)$ and $\left(3\sqrt{2}, -2\sqrt{5}\right)$.

Plot the graph of each equation.

13. $y = 4 - x^2$
14. $x = |y| - 3$
15. $xy = -2$
16. $y = -2x - 5$
17. $-3x + 6y = -12$
18. $x^2 + y^2 = 9$

Solve the following inequalities. Express the solutions in interval notation.

19. $3x - 4 < 5x + 6$
20. $5 - \dfrac{x - 3}{9} < \dfrac{2 + x}{6}$
21. $3 < 6 - \dfrac{1}{3}x \le 9$
22. $x^2 + 2x > 35$
23. $4x^2 - 5 < x$
24. $\dfrac{2x + 6}{3x - 1} > 1$

Graph each equation.

25. $x^2 + (y - 3)^2 = 9$
26. $(x + 2)^2 + (y + 3)^2 = 5$
27. $x^2 + y^2 - 4x + 2y - 7 = 0$
28. $x^2 + y^2 = 12y$
29. Determine the equation of a line that passes through $(-1, -5)$ and is perpendicular to the line $3x + 4y = 6$.
30. Determine the equation of a line that passes through $(-1, -5)$ and is parallel to the line $3x + 4y = 6$.
31. Determine the equation of a line that passes through $(-2, 3)$ and is perpendicular to the line that contains the points $(-3, -4)$ and $(-1, 2)$.
32. Determine the equation of a line that passes through $(-2, 3)$ and is parallel to the line that contains the points $(-3, -4)$ and $(-1, 2)$.

Determine whether each of the following is the graph of a function.

33.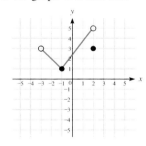

34.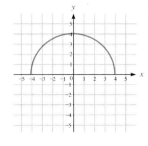

35.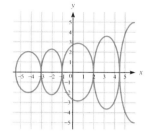

36.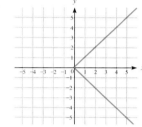

For $f(x) = x^2 + 3x - 4$, find each of the following:

37. $f(-1)$
38. $f(5)$
39. $f(0)$
40. $f(a + h)$
41. $f(a - 1)$
42. $f(0.75)$

For each function, evaluate and simplify the expression:

$$\dfrac{f(a + h) - f(a)}{h}, \quad h \ne 0$$

43. $f(x) = x^2 + 3x - 4$
44. $f(x) = 3\sqrt{x}$
45. $f(x) = x$
46. $f(x) = 1 - \dfrac{1}{x}$

Find the domain and range of each function.

47. $f(x) = -5x + 8$
48. $f(x) = x^3 - 1$
49. $f(x) = \sqrt{6 - x}$
50. $g(x) = \dfrac{x^2 - 1}{x + 1}$
51. $g(x) = 56.7$

152 CHAPTER 2 Graphs and Functions

52. $g(x) = -\dfrac{3}{x^2 + x - 30}$

53. $f(x) = \sqrt{64 - x^2}$

Find the domain of each function.

54. $f(x) = \sqrt{24 - 5x - x^2}$

55. $f(x) = 1 - \dfrac{1}{x}$

56. $g(x) = 3x^{45} - 2x^{34} + 6$

57. $f(x) = \sqrt{|x|(x+1)}$

58. $g(x) = \dfrac{3}{|2x - 6|}$

59. $f(x) = \sqrt{49x^2}$

60. $f(x) = \dfrac{1}{\sqrt{x - 1}}$

Consider the graph of $y = f(x)$ in the following figure. Graph each of the functions described in Exercises 61–64.

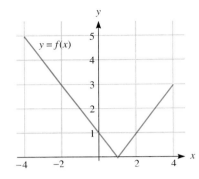

61. $y = f(x) - 3$

62. $y = f\left(\dfrac{1}{2}x\right) + 3$

63. $y = -f(x)$

64. $y = 2f(x - 1)$

Determine whether the graphs of the functions in Exercises 65–78 are odd, even, or neither.

65. $y = x^2 + 4$

66. $y = -x^{1/3}$

67. $y = \sqrt{-x^2 + 3}$

68. $y = -\dfrac{1}{2}x + 3$

69. $y = -x^2 + 4$

70. $x = y^{5/3}$

71. $f(x) = |4 - x|$

72. $f(x) = (x + 3)^2$

73. $f(x) = |x + 1| + |x - 1|$

74. $f(x) = -x^2 - 2$

75. $f(x) = (x^4 + 4)(x^3 - x)$

76. $f(x) = \dfrac{x}{x + 1}$

77. $f(x) = \sqrt{x^2 + 1}$

78. $f(x) = \sqrt[3]{x}$

Applications

79. **Temperature conversion.** Celsius temperatures C and Fahrenheit temperatures F are related by the linear equation:

$$C = \dfrac{5}{9}(F - 32)$$

a. Give an interpretation of the slope of this equation.

b. Suppose a person's Fahrenheit temperature dropped 1°F during an illness. What is the effect on the person's Celsius temperature?

80. **Sales commissions.** A salesperson gets a base salary of $25,000 plus a commission of 24 percent of sales that exceed $400,000.

a. Determine an equation that relates salary and sales.

b. Graph this equation.

Chapter Test

1. What is the distance between the two points $(4, -1)$ and $(-3, -2)$?

2. What is the midpoint of the line that connects the points $(3, 7)$ and $(4, 1)$?

3. Draw the graphs of the following equations:

   a. $2x + 4y = 8$

   b. $-3y = 12$

   c. $x + y = y - (y + 2)$

4. Find the slope and the y-intercept of the graph of the following equation:

$$3(y + 8) = -2(x - 1) + 3$$

5. Suppose that $y = -7x + 2$, and that the value of x is increased by 2. What is the change in the value of y?

6. Determine the equations of the following lines:

   a. passing through the points $(0, -1)$ and $(4, -2)$

   b. passing through $(6, -2)$ and having slope -3

   c. passing through $(5, 0)$ and perpendicular to the graph of $x + y = 1$

7. Solve the following inequalities:

   a. $3x + 1 \geq -x + 2$

   b. $-2 < 2x + 1 < 4$

   c. $x + 1 < -x + 3 < x + 5$

8. A market research company does research on the occupancy rates of hotels in resort areas. It determines that the occupancy rate is related to the tax rate on hotel rooms. When the tax rate is 9 percent, the occupancy rate is 65 percent; when the tax rate is 11 percent, the occupancy rate is 50 percent. Assume a linear model. Forecast the occupancy rate if the tax rate is 12 percent.

9. Suppose that $f(x) = 2x^2 - 3x + 1$. Determine the following:

   a. $f(-1)$

   b. $f(0)$

   c. The value(s) of x for which $f(x)$ equals 1.

10. What are the domains of the following functions?

    a. $f(x) = 5x^{10}$

    b. $f(x) = \dfrac{5}{x^{10}}$

    c. $f(x) = \sqrt{x+1}$

11. A carpet warehouse offers discounts to buyers in quantity. If x is the number of square yards of carpet, then $C(x)$ is the cost, where:

    $$C(x) = \begin{cases} 12x & (0 \leq x \leq 50) \\ 10x & (50 < x \leq 200) \\ 9x & (200 < x) \end{cases}$$

    Graph this function for $0 \leq x \leq 300$.

12. **Thought question.** State the definition of a function. Give an example of a function that is defined by one or more formulas.

13. **Thought question.** Suppose that we define $f(x)$ to be a real number whose square is x. Is this the definition of a function? Explain your answer.

CHAPTER 3
APPLIED PROBLEMS, QUADRATIC EQUATIONS, AND INEQUALITIES

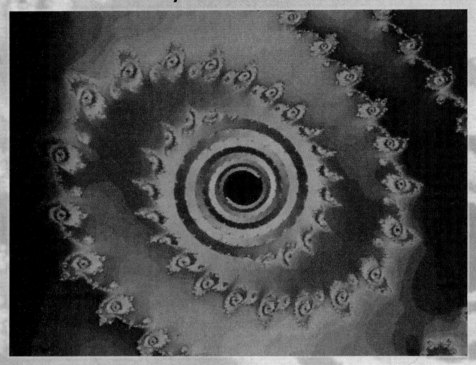

Problems in many disciplines, such as business, biology, chemistry, and sociology, require you to use variables to represent physical quantities: prices, blood pressures, temperatures, forces. Such problems often impose restrictions on the variables; some of these restrictions are specified as equations and others are specified as inequalities. Equations and inequalities, in addition to their importance in other disciplines, are the cornerstone on which algebra is built.

We begin this chapter by discussing techniques for attacking applied problems in general, using what we learned about linear equations in Section 1.5. Then we enlarge our technique by studying quadratic equations and quadratic functions. This is followed by a discussion of applied problems that these new techniques allow us to solve. Finally, we discuss equations and inequalities involving absolute values.

Chapter Opening Image: *Fractals, introduced in the early 1970s, consist of repeated patterns of increasingly smaller scale. They have been used to model many real-world phenomena.*

3.1 SOLVING APPLIED PROBLEMS

You can use equations to solve problems in many fields, including physics, chemistry, biology, engineering, business, psychology, political science, and computer science. Here is a concrete example of an applied problem and its solution. Let's work through it and then analyze the components of the solution to develop a general approach to problem solving.

EXAMPLE 1
Agricultural Management

A farm manager wishes to fence in a rectangular field twice as long as it is wide. Nine hundred feet of fencing will be used for the job. What are the dimensions of the field?

Solution

After reading the problem carefully, we analyze it to determine what information is given and what information is unknown. We are given the information that the field is twice as long as it is wide, and that the amount of fencing required to go around the perimeter is 900 feet. It is a good idea to draw a diagram that illustrates the given information. See Figure 1.

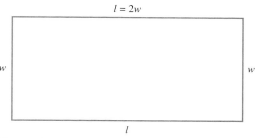

Figure 1

The next step in solving the problem is to translate the problem into algebraic language. First, we define variables to represent the unknown quantities. There are two such quantities, the length and the width, and we denote them with the variables l and w. From the given information, we know that the length is twice the width; we can express the length in terms of the width as follows:

$$l = 2w$$

From the given information, we also know the perimeter: $P = 900$.

From geometry, we have the following formula for the perimeter of a rectangle (see Figure 1):

$$P = 2l + 2w$$

The perimeter formula has three unknowns: P, l, and w. But we substitute $2w$ for l and 900 for P in the formula to obtain:

$$900 = 2(2w) + 2w$$
$$900 = 6w$$

This is a linear equation in one variable. We can solve it by dividing both sides of the equation by 6:

$$\frac{900}{6} = \frac{6w}{6}$$
$$w = 150$$

The solution of the problem does not end here. We have determined the value of the variable w, but we still have not solved the original problem—finding the dimensions of the field. We have determined only the width of the field. We can determine the length of the field by substituting the value for the width into the equation for length:

$$l = 2w = 2(150) = 300$$

The next step in solving a problem is to restate the answer in terms of the question asked, using appropriate units: The field is 300 feet long and 150 feet wide.

We should also check that the solution does indeed work by verifying the conditions set forth in the original problem. In this example, we check that the length is twice the width and that the perimeter of the field equals 900 feet:

$$l = 2w$$
$$300 = 2(150)$$
$$300 = 300 \quad ✓$$
$$900 = 2l + 2w$$
$$900 = 2(300) + 2(150)$$
$$900 = 900 \quad ✓$$

Strategy for Solving Applied Problems

Our solution to Example 1 involved a number of steps. By following similar sequences of steps, we can solve a wide variety of applied problems. Let's summarize these steps for future reference:

A STRATEGY FOR SOLVING APPLIED PROBLEMS

1. Read the problem carefully—several times if necessary.
   a. Determine what information is given.
   b. Determine what information is unknown, the information that you must find.
   c. Draw a sketch illustrating the given and unknown information.
2. Translate the problem into algebraic language.
   a. Represent the unknown quantities in the problem by variables. Choose one variable and express the other unknowns in terms of that variable.
   b. Express the given information in algebraic form. This may consist of given values or relationships between the variables in the form of equations or inequalities.
   c. State any relationships that are derived from mathematical formulas, such as geometric rules or physical laws.
3. Apply algebraic techniques to solve the equation.
4. Use the solution(s) of the equation and the relationships between the variables to determine the values of the other variables, if any.
5. Use the values of the variables to obtain a solution to the applied problem originally posed, and state the solution in terms of the problem. Be sure to use appropriate units.
6. Check the solution using the conditions of the original problem.

SECTION 3.1 Solving Applied Problems

Practice in Solving Applied Problems

The next six examples provide some practice in solving applied problems. Note that not all the problems make use of diagrams. To conserve space, some checks are omitted.

EXAMPLE 2
Price Markdown

A comparison shopper notes that the competition runs a sale in which a coat is marked down 20 percent to $72. What was the original price of the coat?

Solution

Let p denote the original price of the coat. From the data of the problem, we have:

$$[\text{original price}] - [20\% \text{ markdown}] = \$72$$

The amount of the 20 percent markdown equals $0.2p$. So the above word equation may be turned into the algebraic equation:

$$p - 0.2p = 72$$
$$0.8p = 72$$
$$p = \frac{72}{0.8} = 90$$

That is, the original price of the coat was $90.

EXAMPLE 3
Labor Negotiations

A union contract specifies that workers receive a 4 percent raise immediately and a 3 percent raise after a year. How much must a worker be currently earning each week if her weekly salary is to increase $40 over the life of the contract?

Solution

Let x denote the original salary. Then the salary at the end of the first year can be calculated using the model:

$$\text{salary after first raise} = \text{original salary} + 4\% \text{ of original salary} = x + 0.04x$$

The salary after the second raise can be calculated according to the model:

$$\text{salary after second raise} = \text{salary after first raise} + 3\% \text{ of salary after first raise}$$

Using the expression derived above for the salary after the first raise, we see that:

$$\text{salary after second raise} = [x + 0.04x] + 0.03[x + 0.04x]$$

We are given the information that the salary after the second raise is the original salary x plus $40. Thus, we have the equation:

$$x + 40 = [x + 0.04x] + 0.03[x + 0.04x]$$
$$x + 40 = x + 0.04x + 0.03x + 0.0012x$$
$$x + 40 = 1.0712x$$
$$0.0712x = 40$$
$$x = \frac{40}{0.0712} \approx 561.80$$

That is, the worker is currently earning $561.80 per week.

EXAMPLE 4
Inventory Management

A computer store stocks business computers and home computers. The business models generate a profit of 20 percent, whereas the home computers generate a profit of 35 percent. In a given month, the store sells $75,000 worth of computers. The profit is 28 percent. What is the dollar value of each type of computer sold?

Solution

The basic model for this situation is:

total profit = profit from business computers + profit from home computers

Let x denote the dollar amount of business computers sold. From the given data, we have:

$$\text{total profit} = 0.28 \cdot 75{,}000$$
$$\text{profit from business computers} = 0.20x$$

Moreover, because the total amount sold is $75,000, the total amount of home computers sold is $75{,}000 - x$, and we have:

$$\text{profit from home computers} = 0.35(75{,}000 - x)$$

Therefore, using the basic model, we have the equation:

total profit = profit from business computers
+ profit from home computers

$$0.28 \cdot 75{,}000 = 0.20x + 0.35(75{,}000 - x)$$
$$21{,}000 = 0.20x + 26{,}250 - 0.35x$$
$$-5{,}250 = -0.15x$$
$$x = \frac{5{,}250}{0.15} = 35{,}000$$

That is, $35,000 of business computers were sold. The amount of home computers sold was $75,000 − $35,000 = $40,000.

EXAMPLE 5
Straight-Line Depreciation

Depreciation is an accounting method for charging long-lasting assets (such as equipment and buildings) against the earnings of a company. The most common method for calculating depreciation is called **straight-line depreciation**, and works like this: Let L denote the useful life of the asset (in years), C the cost of the asset when new, and S the salvage value at the end of its useful life. Then the value V of the asset at the end of year N is given by the formula:

$$V = C - \frac{C - S}{L} N$$

Suppose that a truck has a useful life of 5 years and has a salvage value of $5,000. Its value at the end of the third year of service is $20,000. What was the cost of the truck when new?

Solution

Refer to the variables in the equation above. The following values are given:

$$L = 5 \text{ years}, \quad S = \$5{,}000, \quad N = 3, \quad V = \$20{,}000$$

and we wish to determine the value of C, the cost of the truck when new. Inserting the variable values into the above equation yields:

$$20{,}000 = C - \frac{C - 5{,}000}{5} \cdot 3$$

To solve this equation for C, let's first get rid of the fraction by multiplying both sides by 5:

SECTION 3.1 Solving Applied Problems

$$20{,}000 \cdot 5 = 5 \cdot C - (C - 5{,}000) \cdot 3$$
$$100{,}000 = 5C - (3C - 15{,}000)$$
$$100{,}000 = 5C - 3C + 15{,}000$$
$$100{,}000 - 15{,}000 = 5C - 3C$$
$$2C = 85{,}000$$
$$C = 42{,}500$$

That is, the cost of the truck when new was $42,500.

EXAMPLE 6
Mixture Problem

A cosmetic manufacturer requires 10,000 gallons of a mixture containing 30 percent isopropyl alcohol and 70 percent water. It has on hand two mixtures. Mixture A consists of 15 percent isopropyl alcohol and 85 percent water and mixture B consists of 40 percent isopropyl alcohol and 60 percent water. How many gallons of each should be mixed to obtain the desired mixture? (See Figure 2.)

Solution

Let x be the number of gallons of mixture A.

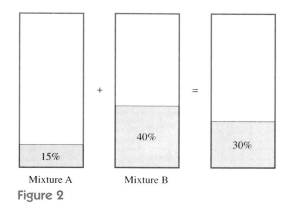
Figure 2

Then,

$$10{,}000 - x = \text{Number of gallons of mixture B}$$
$$\text{Alcohol in mixture A} = 0.15 \cdot x$$
$$\text{Alcohol in mixture B} = 0.4 \cdot (10{,}000 - x)$$

When the two mixtures are added together, they must contain 30 percent of 10,000, or 3,000, gallons of isopropyl alcohol.

Alcohol in mixture A + Alcohol in mixture B
= Alcohol in final mixture when A and B are combined

$$0.15x + 0.4(10{,}000 - x) = 3{,}000$$
$$0.15x + 4{,}000 - 0.4x = 3{,}000$$
$$-0.25x = -1{,}000$$
$$x = 4{,}000$$

The company should use 4,000 gallons of mixture A and $10{,}000 - 4{,}000 = 6{,}000$ gallons of mixture B.

160 CHAPTER 3 Applied Problems, Quadratic Equations, and Inequalities

Let's check this solution. The solution contains $4{,}000 + 6{,}000 = 10{,}000$ gallons. The amount of isopropyl alcohol in the mixture is

$$.15(4{,}000) + .4(6{,}000) = 600 + 2{,}400$$
$$= 3{,}000 \text{ gallons}$$

which is 30 percent of the mixture. The amount of water in the mixture is $10{,}000 - 3{,}000 = 7{,}000$ gallons, which is 70 percent of the mixture.

EXAMPLE 7
Financial Planning

The manager of the Amalgamated Industries Pension Fund wishes to make a one-year investment of $1,000,000 in a combination of stocks and bonds to yield interest income of $100,000 per year. The stocks yield 7 percent per year in interest and the bonds yield 12 percent per year. (Assume that interest is compounded annually.) How much should the manager invest in stocks and how much in bonds to achieve the desired total annual interest income of $100,000?

Solution

Let x denote the amount invested in stocks. Since the numbers given all have a large number of 0's, we may simplify writing by representing the number in the problem in $100,000 units. Since a total of $1,000,000 is to be invested, the amount to be invested in bonds is $10 - x$. From the statement of the problem, the total amount of investment income must be $100,000. This is a linear equation that we can easily solve:

$$\text{Interest from stocks} + \text{Interest from bonds} = 1$$
$$0.07x + 0.12(10 - x) = 1$$
$$0.07x + 1.2 - 0.12x = 1$$
$$-0.05x = -0.2$$
$$x = 4$$

That is, $400,000 should be invested in stocks. The amount to be invested in bonds is $1{,}000{,}000 - 400{,}000 = \$600{,}000$.

We must now check that these investments yield the desired total income:

$$\text{Interest from stocks} + \text{Interest from bonds} = 100{,}000$$
$$0.07(400{,}000) + 0.12(600{,}000) = 100{,}000$$
$$28{,}000 + 72{,}000 = 100{,}000$$

These investments in stocks and bonds do, indeed, produce the desired amount of income.

Solving for One Variable in Terms of Another

Often in an applied problem, we must solve an equation in two variables for one variable in terms of the other. We used this approach in Example 1. The next few examples give some practice in this important skill.

EXAMPLE 8
Solving for One Variable in Terms of Another

Solve the following equation for y in terms of x.

$$xy + y - 2x + 1 = 0$$

Solution

First, we use allowable operations to arrive at an equivalent equation with all terms involving y on one side of the equation and all terms that don't involve y on the other.

SECTION 3.1 Solving Applied Problems 161

$$xy + y = 2x - 1$$

We can factor y from both terms of the left side:

$$y(x + 1) = 2x - 1$$

Dividing both sides by $x + 1$ now yields:

$$y = \frac{2x - 1}{x + 1}, \qquad x \neq -1$$

Note that we must exclude the value $x = -1$, because this value would lead to a zero denominator.

EXAMPLE 9
Volume of a Cylinder

The volume V of a cylinder is given by the formula

$$V = \pi r^2 h$$

where r is the radius of the cylinder and h is its height. Express the radius in terms of the volume and the height. (See Figure 3.)

Solution

We divide both sides of the equation by πh to obtain:

$$r^2 = \frac{V}{\pi h}$$

We can solve for r by taking the square root of both sides. Remember that we must include the $\pm$ sign:

$$r = \pm \sqrt{\frac{V}{\pi h}}$$

Because r denotes the radius of a cylinder, its value must be positive, so we can ignore the negative value. The solution is:

$$r = \sqrt{\frac{V}{\pi h}}$$

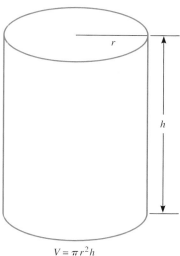

$V = \pi r^2 h$

Figure 3

EXAMPLE 10
Compound Interest

In Section 1.2, we introduced the following formula for compound interest: Suppose that interest on principal P is compounded k times per year and that the annual rate is r. Then the balance B after n years is given by the formula:

$$[\text{balance after } n \text{ years}] = P\left(1 + \frac{r}{k}\right)^{nk}$$

Express the interest rate r in terms of P, B, k, and n if the interest is compounded annually.

Solution

Because the interest is compounded annually, $k = 1$, and therefore:

$$B = P(1 + r)^n$$

We solve the equation by performing allowable operations on both sides:

$$(1 + r)^n = \frac{B}{P} \qquad \text{\textit{Divide both sides by } P}$$

$$[(1 + r)^n]^{1/n} = \left(\frac{B}{P}\right)^{1/n} \qquad \text{\textit{Raise to the power } } 1/n$$

The left side can be rewritten in the form:

$$(1 + r)^{n \cdot (1/n)} = (1 + r)^1 = 1 + r \qquad \text{\textit{Law of Exponents 2}}$$

CHAPTER 3 Applied Problems, Quadratic Equations, and Inequalities

Substituting $1 + r$ for the left side of the equation, we have:

$$1 + r = \left(\frac{B}{P}\right)^{1/n}$$

$$r = \left(\frac{B}{P}\right)^{1/n} - 1$$

Exercises 3.1

Solve the equation for the specified letter.

1. $F = pqV$, for V; for p
2. $A = \frac{1}{2}bh$, for h; for b
3. $2x + 2y + 5 = 0$, for y; for x
4. $\pi x + 2x + 2y = 24$, for y; for x
5. $k(36 - 2x - \pi x) + k\left(\frac{\pi}{4}\right)x = 0$, for x
6. $-6[2x - (100 + c)] = 0$, for x; for c
7. $A = \pi r^2$, for r
8. $P = ab^2$, for b
9. $P = ab^3$, for b
10. $V = \frac{4}{3}\pi r^3$, for r
11. $p + hdq + \frac{1}{2}dv^2 = k$, for d; for q
12. $A = P + Prt$, for P; for t
13. $S = \frac{a_1 - a_1 r^n}{1 - r}$, for a_1
14. $Q = \frac{k(t_2 - t_1)aT}{d}$, for t_2; for T
15. $m_1 u_1 + m_2 u_2 = m_1 v_1 + m_2 v_2$, for m_2; for v_1
16. $S = 2\pi rh + 2\pi r^2$, for h
17. $x^2 + y^2 = 4\lambda^2(x^2 + y^2)$, for λ
18. $\frac{2x}{16} = \frac{2(L - x)}{12\sqrt{3}}$, for x; for L
19. $k(4r^2 - 3x^2) = 0$, for x
20. $\frac{m - 3w^2}{4} = 0$, for w

Applications

21. **Price markdown.** A coat is marked down 25 percent to $69. What was the original price?

22. **Utility pricing.** A 7 percent raise in rates by the public utility company resulted in a $2.50 monthly increase. How much was the bill before the monthly increase? How much was the bill after the increase?

23. **Return on investment.** An investment grows to $1,635 after one year at an interest rate of 9 percent, compounded annually. What amount was originally invested?

24. **Return on investment.** An investment grows to $60,760 after one year at an interest rate of 8.5 percent, compounded annually. What amount was originally invested?

25. **Personnel management.** There are four employees in a small business—the president, the janitor, and two salespeople. The president earns ten times what the janitor earns. Each salesperson earns one and one-half times what the janitor earns. The total payroll is $161,000. What is the salary of each employee?

26. **Construction.** A rectangular lot is at the right-angle intersection of two roads. The lot has dimensions 200 ft by 80 ft. Both roads are to be widened by the same amount in such a way that the perimeter of the lot will be decreased to 512 ft. How much width was added to the roads? (See figure.)

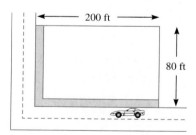

Exercise 26

27. **Compound interest.** An investment of $10,000 is made for 5 years at an interest rate that is compounded quarterly (4 times a year). It grows to $14,859.47. What was the interest rate?

28. **Compound interest.** $2,000 will grow to $3,338.68 in 5 years at an interest rate that is compounded daily. What is the interest rate?

29. **Manufacturing.** A cosmetic manufacturer requires 20,000 gallons of a mixture that is 35 percent isopropyl alcohol and 65 percent water. It has on hand two mixtures, mixture A, which is 15 percent isopropyl alcohol

and 85 percent water, and mixture B, which is 50 percent isopropyl alcohol and 50 percent water. How many gallons of each should be mixed to obtain the desired mixture?

30. **Chemistry.** A chemist has one solution of acid and water that is 65 percent acid and a second that is 20 percent. How many gallons of each should be mixed together to get 120 gallons of a solution that is 50 percent acid?

31. **Investing.** Two investments are made totaling $12,000. For a certain year these investments yield $885 in simple interest. Part of the $12,000 is invested at 6 percent and the rest is invested at 9 percent. How much is invested at each rate?

32. **Pension fund management.** The manager of a pension fund wants to invest $5,000,000 in a pension fund in a combination of stocks and bonds to yield an income of $330,000 per year. The stocks yield 6 percent per year and the bonds yield 8 percent per year. How much should the manager invest in stocks and how much in bonds?

33. **Sports.** In order to get a major league hockey team, a city's minor league team must average 6,000 per game in attendance. This year, the city's minor league team has played 9 of its 10 games and has averaged 5,800. How many people must attend the tenth game in order for the city to qualify for the major league team?

34. **Manufacturing.** A machine shop must average a weekly scrap rate of no more than 2 percent. Its scrap rates have been 1.5 percent, 2.4 percent, 2.1 percent, and 2.2 percent. If it produces the same number of parts each day, what must its scrap rate be on the fifth day to average 2 percent for the week?

35. **Straight-line depreciation.** A machine has a lifetime of 6 years, costs $6,000, and has a salvage value of $1,300.

   a. Find a formula for V, the depreciated value of the machine after n years. (The depreciated value equals the original cost minus the amount by which it has depreciated.)

   b. Find the depreciated value after 0 years, 1 year, 2 years, 3 years, 5 years, and 6 years.

36. **Biology.** It has been found in a study that the total length L and the tail length t of females of the snake species *Lampropeltis polyzona* are related by the linear equation

$$t = 0.143L - 1.18$$

   a. Such a snake is found with a total length of 455 mm. Use the equation to predict its tail length.

   b. Solve the equation for L. Determine whether the resulting equation is linear.

   c. Use the equation in part b to find the total length of a snake whose tail is 140 mm long.

37. **Temperature conversion.** The formulas for conversion between Celsius temperature C and Fahrenheit temperature F are given by the linear equations

$$F = \frac{9}{5}C + 32 \quad \text{and} \quad C = \frac{5}{9}(F - 32)$$

   a. Room temperature of 20°C corresponds to what Fahrenheit temperature?

   b. Water boils at 212°F. To what Celsius temperature does this correspond?

   c. At what temperature are the Fahrenheit and the Celsius temperatures the same?

164 CHAPTER 3 Applied Problems, Quadratic Equations, and Inequalities

In your own words

38. Describe an organized procedure for solving word problems. Illustrate each step with a problem of your choice.

39. What is the role of pictures or diagrams in the solution of word problems?

40. True or false: After you solve the equation that expresses a word problem, you have finished the solution. Explain your answer.

3.2 QUADRATIC EQUATIONS

In Sections 1.5 and 3.1, we studied linear equations, their solutions, and their applications. Let's now turn to the next important class of equations, polynomial equations of the second degree. In this section we learn three methods for solving such equations, and we discuss a number of their applications.

Definition
Quadratic Equation

A **quadratic equation** is an equation that is equivalent to an equation of the form

$$ax^2 + bx + c = 0$$

where a, b, and c are real numbers and $a \neq 0$.

Here are some examples of quadratic equations:
$$2x^2 + 3x + 1 = 0$$
$$x^2 - 3 = 0$$
$$x^2 = 9x + 10$$
$$x^2 - 5x = 0$$

The Method of Factoring

One method for solving quadratic equations involves collecting all terms on one side of the equation and factoring that side into a product of binomials. For instance, we can write the first equation in the form:

$$(2x + 1)(x + 1) = 0$$

If a product of real numbers is 0, then one of the factors must be 0. Therefore, one of the following equations must hold:

$$2x + 1 = 0 \quad \text{or} \quad x + 1 = 0$$

Solving each of these linear equations, we see that $x = -\frac{1}{2}$ or $x = -1$. Substituting these values of x into the original equation, we see that they are, indeed, solutions.

You can use the above method, called the **factoring method,** to solve any quadratic equation whose left side may be factored into a product of binomials.

 EXAMPLE 1
Factoring

Use the factoring method to solve the quadratic equation:
$$6x^2 + 13x - 5 = 0$$

Solution

We factor the left side:
$$(3x - 1)(2x + 5) = 0$$

SECTION 3.2 Quadratic Equations

To determine the solutions of the quadratic equation, we set each of the linear factors equal to 0.

$$3x - 1 = 0 \quad \text{or} \quad 2x + 5 = 0$$

We solve each equation separately to obtain the solutions:

$$x = \frac{1}{3} \quad \text{or} \quad x = -\frac{5}{2}$$

To check these solutions, we substitute their values into the given equation:

$$6 \cdot \left(\frac{1}{3}\right)^2 + 13 \cdot \left(\frac{1}{3}\right) - 5 \quad \Big| \quad 0$$
$$\frac{2}{3} + \frac{13}{3} - \frac{15}{3}$$
$$0$$

$$6 \cdot \left(-\frac{5}{2}\right)^2 + 13 \cdot \left(-\frac{5}{2}\right) - 5 \quad \Big| \quad 0$$
$$\frac{75}{2} - \frac{65}{2} - \frac{10}{2}$$
$$0$$

EXAMPLE 2
More Factoring

Solve the following quadratic equation using the factoring method:

$$2x^2 - 6x = x - 3$$

Solution

Collect all terms on the left side of the equation:

$$2x^2 - 7x + 3 = 0$$

We factor the quadratic polynomial on the left to obtain:

$$(2x - 1)(x - 3) = 0$$

and we set each linear factor equal to 0:

$$2x - 1 = 0 \qquad x - 3 = 0$$
$$x = \frac{1}{2} \qquad x = 3$$

The two solutions to the equation are $x = \frac{1}{2}$ and $x = 3$. We then check that the two solutions work:

$$2 \cdot \left(\frac{1}{2}\right)^2 - 6 \cdot \left(\frac{1}{2}\right) \quad \Big| \quad \frac{1}{2} - 3$$
$$\frac{1}{2} - \frac{6}{2} \qquad -\frac{5}{2}$$
$$-\frac{5}{2}$$

$$2 \cdot (3)^2 - 6 \cdot 3 \quad \Big| \quad 3 - 3$$
$$18 - 18 \qquad 0$$
$$0$$

Completing the Square If one side of a quadratic equation is a perfect square and the other is a constant, then the equation may be solved by taking the square roots of both sides. For instance, consider the equation:

$$x^2 = 4$$

Here, both x^2 and 4 are perfect squares, so we can solve the equation by taking square roots:

$$\sqrt{x^2} = \sqrt{4}$$
$$|x| = 2$$
$$x = \pm 2$$

Similarly, consider the equation

$$(x - 3)^2 = 10$$

Taking square roots of both sides gives:

$$|x - 3| = \sqrt{10}$$
$$x - 3 = \pm\sqrt{10}$$
$$x = 3 \pm \sqrt{10}$$

We can extend the above technique into a general method for solving quadratic equations: We transform one side of the equation into a perfect square and keep the other side constant. This method is called **completing the square.**

COMPLETING THE SQUARE

To turn the expression $x^2 + bx$ into a perfect square:

1. Divide the coefficient b of x by 2 to get $b/2$.
2. Square the result of part 1 to obtain $(b/2)^2$.
3. Add the result of part 2 to the original expression to obtain the perfect square:

$$x^2 + bx + \left(\frac{b}{2}\right)^2 = \left(x + \frac{b}{2}\right)^2$$

EXAMPLE 3
Completing the Square

Use the method of completing the square to solve the quadratic equation $x^2 + 6x + 5 = 0$.

Solution

First isolate all terms that involve x on the left side of the equation:

$$x^2 + 6x = -5$$

We want to turn the left side into a perfect square. This can be done by adding $\left(\frac{6}{2}\right)^2 = 9$ to both sides:

$$x^2 + 6x + 9 = -5 + 9$$
$$(x + 3)^2 = 4$$

SECTION 3.2 Quadratic Equations

$$x + 3 = \pm 2$$
$$x = -3 \pm 2 = -5, -1$$

So the given equation has the two solutions -5 and -1.

EXAMPLE 4
More Completing the Square

Use the method of completing the square to solve the following quadratic equation.

$$4x^2 + 12x + 5 = 0$$

Solution

First, we subtract the constant term from both sides of the equation:

$$4x^2 + 12x = -5$$

Next, we divide all terms by the coefficient of x^2:

$$x^2 + 3x = -\frac{5}{4}$$

Next, we complete the square by adding the square of half the coefficient of x, namely $(\frac{3}{2})^2 = \frac{9}{4}$, to each side:

$$x^2 + 3x + \frac{9}{4} = -\frac{5}{4} + \frac{9}{4}$$

$$x^2 + 3x + \frac{9}{4} = 1$$

$$\left(x + \frac{3}{2}\right)^2 = 1 \qquad \textit{Perfect square on the left}$$

Let's now take the square roots of both sides:

$$x + \frac{3}{2} = \pm 1$$

$$x = \pm 1 - \frac{3}{2}$$

$$x = 1 - \frac{3}{2} = -\frac{1}{2} \quad \text{or} \quad x = -1 - \frac{3}{2} = -\frac{5}{2}$$

The Quadratic Formula

A third method for solving quadratic equations is based on the **quadratic formula,** which allows us to calculate the solutions of any quadratic equation in terms of its coefficients.

QUADRATIC FORMULA

The solutions of the quadratic equation

$$ax^2 + bx + c = 0, \qquad a \neq 0$$

are given by the formula:

$$x = \frac{-b \pm \sqrt{b^2 - 4ac}}{2a}$$

The $\pm$ means that there are two solutions, one corresponding to the positive sign and one corresponding to the negative sign.

Proof Begin by writing the equation in a slightly altered form:
$$ax^2 + bx + c = 0$$
$$ax^2 + bx = -c$$
$$x^2 + \frac{b}{a}x = -\frac{c}{a}$$

Now complete the square on the left side by adding the square of half the coefficient of x, namely $(b/2a)^2$, to both sides of the equation:
$$x^2 + \frac{b}{a}x + \left(\frac{b}{2a}\right)^2 = -\frac{c}{a} + \left(\frac{b}{2a}\right)^2$$
$$\left(x + \frac{b}{2a}\right)^2 = -\frac{c}{a} + \frac{b^2}{4a^2}$$
$$= \frac{-4ac + b^2}{4a^2}$$

Taking the square roots of both sides, we have:
$$\left|x + \frac{b}{2a}\right| = \sqrt{\frac{b^2 - 4ac}{4a^2}}$$
$$x + \frac{b}{2a} = \pm\frac{\sqrt{b^2 - 4ac}}{2a}$$
$$x = -\frac{b}{2a} \pm \frac{\sqrt{b^2 - 4ac}}{2a}$$
$$= \frac{-b \pm \sqrt{b^2 - 4ac}}{2a}$$

The expression $b^2 - 4ac$, which occurs in the quadratic formula, is called the **discriminant.** The sign of the discriminant determines the nature of the solutions to the equation, as follows: If $b^2 - 4ac > 0$, then $\sqrt{b^2 - 4ac}$ is positive and the above formula gives two different real solutions for x. If $b^2 - 4ac = 0$, then $\sqrt{b^2 - 4ac} = 0$ and the above formula gives a single real solution for x. If $b^2 - 4ac < 0$, then $\sqrt{b^2 - 4ac}$ is not a real number (remember that a negative number does not have a real square root), and the above formula gives no real solutions for x. Summarizing the various cases, we have the following rules:

THE SOLUTIONS OF A QUADRATIC EQUATION

Consider the quadratic equation:
$$ax^2 + bx + c = 0, \quad a \neq 0$$
with discriminant $b^2 - 4ac$.

1. If $b^2 - 4ac > 0$, then the equation has the two different real solutions:
$$x = \frac{-b + \sqrt{b^2 - 4ac}}{2a}, \quad x = \frac{-b - \sqrt{b^2 - 4ac}}{2a}$$

2. If $b^2 - 4ac = 0$, the equation has the single real solution:
$$x = -\frac{b}{2a}$$

3. If $b^2 - 4ac < 0$, then the equation has no real solutions.

SECTION 3.2 Quadratic Equations

The three possibilities for the sign of the discriminant are illustrated in the following example.

EXAMPLE 5
Using the Quadratic Formula

Solve the following equations for real solutions using the quadratic formula.
1. $2x^2 - 3x - 4 = 0$
2. $9x^2 - 12x + 4 = 0$
3. $x^2 + x + 1 = 0$

Solution

1. From the equation $2x^2 - 3x - 4 = 0$,
$$a = 2, \quad b = -3, \quad c = -4$$
So the discriminant has the value:
$$b^2 - 4ac = (-3)^2 - 4(2)(-4) = 41$$
Because the discriminant is positive, the equation has two real solutions. They are given by the quadratic formula as:
$$x = \frac{-b \pm \sqrt{b^2 - 4ac}}{2a}$$
$$= \frac{-(-3) \pm \sqrt{41}}{2(2)}$$
$$= \frac{3 \pm \sqrt{41}}{4}$$
$$\approx 2.351, \quad -0.851$$

2. In the equation $9x^2 - 12x + 4 = 0$, we have $a = 9, b = -12, c = 4$, so the discriminant has the value:
$$b^2 - 4ac = (-12)^2 - 4(9)(4) = 0$$
So the equation has one real solution, that is, two identical real solutions, given by the quadratic formula as:
$$x = \frac{-b \pm \sqrt{b^2 - 4ac}}{2a}$$
$$= \frac{-(-12) \pm \sqrt{0}}{2(9)}$$
$$= \frac{12}{18}$$
$$= \frac{2}{3}$$

3. In the equation $x^2 + x + 1 = 0$, the discriminant has the value:
$$b^2 - 4ac = 1^2 - 4(1)(1) = -3$$
Because the discriminant is negative, the equation has no real solutions.

170   CHAPTER 3   Applied Problems, Quadratic Equations, and Inequalities

EXAMPLE 6
The Quadratic Formula Implemented on a Calculator

Use a calculator to solve the following quadratic equation:
$$2.1x^2 - 5.12x = 0.77x^2 + 4.11$$

Solution

First, write the equation in standard form by bringing all terms to one side of the equation:

$$2.1x^2 - 0.77x^2 - 5.12x - 4.11 = 0$$
$$(2.1 - 0.77)x^2 - 5.12x - 4.11 = 0 \quad \text{Distributive law}$$
$$1.33x^2 - 5.12x - 4.11 = 0 \quad \text{Using a calculator}$$

$$x = \frac{-(-5.12) \pm \sqrt{(-5.12)^2 - 4 \cdot 1.33 \cdot (-4.11)}}{2 \cdot 1.33} \quad \text{Quadratic formula}$$

We calculate the quantities on the right side of the last equation using a calculator, and we obtain two values of x, one corresponding to the $+$ in $\pm$ and one corresponding to the $-$. For example, to compute the value corresponding to the $-$ using a scientific calculator, we would use the following keystrokes:

Note that it is necessary to use parentheses around the quantity under the square root sign, the entire numerator, and the entire denominator in order to have the expression calculated correctly. As a result of the above keystrokes, the display reads:

$$-0.6819343932$$

The value corresponding to $+$ is:

$$4.531558453$$

Applications of Quadratic Equations

Quadratic equations occur in many applied problems. The next examples provide illustrations of how they are used.

EXAMPLE 7
Cable Television Network Construction

A local cable television company must run a cable to a house that is south and east of the nearest cable connection. The distance south is 50 yards. The direct route to the connection is 45 yards more than half the distance of the two-leg route. How far east is the cable connection?

Solution

Let x denote the distance east to the cable connection. See Figure 1. Then, we have:

$$\begin{array}{c}\text{distance of running cable}\\ \text{along two-leg route}\end{array} = \text{distance of east leg} + \text{distance of south leg}$$

$$= x + 50$$

The direct distance to the house is the hypotenuse of a right triangle with legs x and 50. Therefore, by the Pythagorean theorem, its length is:

$$\sqrt{(\text{east leg})^2 + (\text{south leg})^2} = \sqrt{x^2 + (50)^2}$$
$$= \sqrt{x^2 + 2{,}500}$$

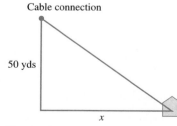

Figure 1

Because the direct route to the connection is 45 yards more than half the distance of the two-leg route, we have the equation:

$$\sqrt{x^2 + 2{,}500} = \frac{1}{2}(x + 50) + 45$$

$$2\sqrt{x^2 + 2{,}500} = x + 50 + 90 \quad \text{Multiply by 2}$$

$$2\sqrt{x^2 + 2{,}500} = x + 140$$

$$4(x^2 + 2{,}500) = (x + 140)^2 \quad \text{Square both sides}$$

$$4x^2 + 10{,}000 = x^2 + 280x + 19{,}600$$

$$3x^2 - 280x - 9{,}600 = 0$$

$$(3x + 80)(x - 120) = 0$$

$$x = -\frac{80}{3},\ 120$$

We can discard the negative solution, because x represents a distance and therefore is positive. We can check that the remaining solution is actually a solution by plugging it into the equation. (The calculations are omitted here.) Therefore, the cable connection is 120 yards east of the house.

EXAMPLE 8
Revenue Model

Suppose that if x cars are produced (in hundreds of thousands of cars), the selling price (in thousands of dollars) is $23 - 1.4x$. How many cars should be produced to generate revenue of $7.3014 billion?

Solution

The total revenue R generated by selling x cars is described by the model:

$$\text{Revenue} = \text{quantity} \times \text{price}$$

$$R = x(23 - 1.4x)$$

$$= 23x - 1.4x^2$$

We want to determine x so that the revenue is 7.3104 billion dollars. Because x is in units of hundred thousands and the price is in units of thousand dollars, their product is in units of hundred millions of dollars. In terms of these units, $7.3104 billion equals $73.104 hundred million. Therefore, we want to determine x so that it satisfies the equation:

$$23x - 1.4x^2 = 73.014$$

$$1.4x^2 - 23x + 73.104 = 0$$

$$x = \frac{-(-23) \pm \sqrt{(-23)^2 - 4(1.4)(73.014)}}{2(1.4)}$$

$$= \frac{23 \pm \sqrt{120.1216}}{2.8}$$

$$= \frac{23 \pm 10.96}{2.8}$$

$$= 4.3,\ 12.129$$

So there are two values of x that yield the desired revenue—either 430,000 cars, or 1,212,900 cars.

EXAMPLE 9
Cost Model

The cost C of a particular microprocessor t years after its introduction is projected using the model:
$$C = 900 - 120t + 0.1t^2 \quad (0 \le t \le 7)$$
How long will it take for the microprocessor to cost $100?

Solution

The value of C equals $100 when t satisfies the equation:
$$100 = 900 - 120t + 0.1t^2$$
$$0.1t^2 - 120t + 800 = 0$$
$$t = \frac{-(-120) \pm \sqrt{(-120)^2 - 4(0.1)(800)}}{2(0.1)}$$
$$= \frac{120 \pm \sqrt{14{,}080}}{0.2}$$
$$\approx 6.7041, \ 1{,}193.3$$

We can reject the second value of t as unrealistic because the model requires that $0 \le t \le 7$. So the microprocessor will cost $100 approximately 6.7 years after its introduction.

EXAMPLE 10
Architecture

An architect is designing a house that features Palladian windows, as shown in Figure 2. To be used for passive solar heating, each window, consisting of four rectangular panes and a semicircular pane, must have 2,500 square inches of glass. Determine the dimensions of each window.

Solution

The condition we seek to satisfy is:

Area of semicircular region + Area of rectangular region = 2,500

The semicircular region has diameter x and radius $x/2$. From the formula for the area of a circle, we have the following formula for a semicircle:
$$\frac{1}{2}\pi r^2 = \frac{1}{2}\pi \left(\frac{x}{2}\right)^2 = \frac{\pi}{8} x^2$$

The area of the entire rectangular region is:
$$2x \cdot x = 2x^2$$

Therefore, x must satisfy the equation:
$$\frac{\pi}{8} x^2 + 2x^2 = 2{,}500$$
$$\left(\frac{\pi}{8} + 2\right) x^2 = 2{,}500 \qquad \textit{Factoring out } x^2$$
$$x^2 = \frac{2{,}500}{(\pi/8) + 2}$$
$$x = \pm \sqrt{\frac{2{,}500}{(\pi/8) + 2}} \qquad \textit{Taking square roots}$$

Figure 2

SECTION 3.2 Quadratic Equations

Because x represents a physical length, we can ignore the negative value. So the width of the window is:

$$x = \sqrt{\frac{2{,}500}{(\pi/8) + 2}}$$

$$= \sqrt{\frac{20{,}000}{\pi + 16}} \text{ inches}$$

$$\approx 32.3241 \text{ inches}$$

The height of the rectangular portion of the window is twice this amount, or 64.6482 inches. The circumference of the semicircle is equal to:

$$\pi \cdot \text{radius} = \pi \frac{x}{2}$$

$$= \frac{\pi}{2}\sqrt{\frac{20{,}000}{\pi + 16}}$$

$$\approx 50.7746 \text{ inches}$$

EXAMPLE 11
Economics

The manufacturer of a very popular model of stereo compact disc system notes that the demand for the product fluctuates with the price. If the price is p dollars, then the market research department reports that monthly sales s can be predicted by the formula:

$$s = 5{,}000 - 10p$$

Suppose that the manufacturer wishes to achieve revenues of $600,000 in a given month. What should be the price of the compact disc players?

Solution

The revenues for the month are given by the product:

$$\text{Revenue} = \text{Sales} \cdot \text{Price}$$

Using the given expression for sales in terms of price, we have:

$$\text{Revenue} = (5{,}000 - 10p)p$$

To achieve a revenue of $600,000, the following condition must be fulfilled:

$$600{,}000 = (5{,}000 - 10p)p$$

$$-10p^2 + 5{,}000p - 600{,}000 = 0$$

$$p^2 - 500p + 60{,}000 = 0 \quad \text{Dividing by } -10$$

We could solve this equation by factoring, but it would take quite a while to work out. However, we can always solve a quadratic equation using the quadratic formula.

$$p = \frac{500 \pm \sqrt{(-500)^2 - 4(1)(60{,}000)}}{2(1)}$$

$$= \frac{500 \pm \sqrt{10{,}000}}{2}$$

$$= \frac{500 \pm 100}{2}$$

$$= 300, \ 200$$

174　CHAPTER 3　Applied Problems, Quadratic Equations, and Inequalities

In other words, the company can achieve the desired revenues by choosing one of the two prices, $300 or $200, for its compact disc player. Because both prices yield the desired revenue, the company can choose the price on the basis of some criterion other than revenue. The lower price should probably be chosen to generate consumer goodwill.

Exercises 3.2

Determine the real solutions of the following quadratic equations.

1. $x^2 - 3 = 0$
2. $5x^2 = 45$
3. $x^2 - 8 = 0$
4. $4x^2 - 3 = 0$
5. $3t^2 = -15$
6. $2a^2 = -11$
7. $(x - 2)^2 = 1$
8. $(x + 5)^2 = 20$
9. $(3x + 1)^2 = 9$
10. $(2x - 3)^2 = 5$

Solve the following equations, by completing the square.

11. $x^2 + 5x - 10 = 0$
12. $x^2 + 8x + 7 = 0$
13. $x^2 + 6 = 4x$
14. $x^2 = 3x - 5$
15. $3x^2 + 4x = 1$
16. $8x^2 = 12x + 8$
17. $7x + 10 = 3x^2$
18. $3y^2 - 10y - 1 = 0$

Use the quadratic formula to solve the following quadratic equations for real roots.

19. $3x^2 + 4x = 0$
20. $5x^2 = 8x$
21. $2x^2 - 4x + 5 = 0$
22. $7x^2 + 10x + 12 = 0$
23. $2x^2 + x = 4$
24. $2x^2 = 5x + 14$

Determine all solutions of the following equations in real numbers. Use any method you wish.

25. $18x^2 = 10x$
26. $13t^2 + 6t = 0$
27. $6x^2 + 5x = 21$
28. $10x^2 - 3x - 1 = 0$
29. $x^2 - 5x + 10 = 0$
30. $x^2 - 3x + 5 = 0$

Solve each formula for the indicated letter.

31. $G = \dfrac{n(n-1)}{2}$, for n
32. $D = \dfrac{n(n-3)}{2}$, for n
33. $H = -16t^2 + v_0 t + h$, for t
34. $s = 4.9t^2 + v_0 t$, for t
35. $S = 2\pi rh + 2\pi r^2$, for r
36. $S = 2\pi r^2 + \pi rs$, for r

Applications

37. **Compound interest.** What is the interest rate if interest is compounded annually and $1,200 grows to $1,323 in 2 years?

38. **Compound interest.** What is the interest rate if interest is compounded annually and $1,532 grows to $1,737.63 in 2 years?

39. **Compound interest.** What is the interest rate if interest is compounded semiannually and $2,500 grows to $2,924.65 in 2 years?

40. **Compound interest.** Assume $1,400 is invested at 10.5 percent for 2 years, and the interest is compounded quarterly. At what interest rate, compounded annually, would this same amount of money have to be invested in order for it to double?

41. **Geometry.** The length of each side of an equilateral triangle is a. (See figure.)

　a. Find the formula for the height h.

　b. Find a formula for the area A.

　c. Find a formula for the perimeter P.

　d. Assume the area of an equilateral triangle is $25\sqrt{3}$ cm². What is its perimeter?

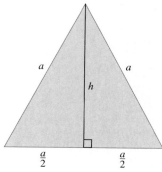

Exercise 41

42. **Manufacturing.** An open box is folded out of a 15-inch-by-25-inch piece of cardboard by cutting square corners in the box. (See figures.) The area at the bottom of the resulting box is 231 square inches. What is the length of each side of the square corners?

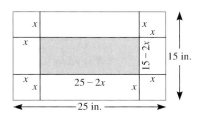

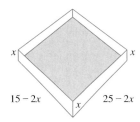

Exercise 42

43. **Demand for a product.** The manufacturer of a computer printer notices that demand for the product fluctuates with the price. If the price is p dollars, then the number N of printers sold per year is given by the formula:
$$N = 18{,}000 - 20p$$
Suppose the manufacturer wants to achieve revenues of $4,000,000 in a given year. What should the price of the printers be?

44. **Trade deficit.** The U.S. trade deficit D (in billions of dollars per year) at time x (in years since January 1992) is given by the model
$$D = 65 - 4.35x + 26x^2$$
(based on U.S. Department of Commerce data). Determine when the trade deficit was running at a rate of $75 billion per year.

45. **Baseball salaries.** The average salary S of a baseball player (in thousands of dollars per year) in year x (with 1982 corresponding to 0) is given by the model:
$$S = 246 + 64x - 8.9x^2 + 0.95x^3$$
   a. When was the average salary $246,000?
   b. What was the average salary in 1992?

46. **Measuring altitude.** One can use the boiling point T of water to estimate the elevation H at which the water is boiling. The quadratic equation that can be used is
$$H = 1{,}000(100 - T) + 580(100 - T)^2$$
where H is the elevation in meters, and T is the boiling point in degrees Celsius.
   a. The boiling point is 99.4°C. Find the elevation.
   b. What is the boiling point at sea level ($H = 0$)?
   c. Denver, Colorado, is 5,280 ft (1,609 m) above sea level. At what temperature does water boil?
   d. Mt. Elbert, outside Denver, is 14,431 ft (4,400 m) above sea level. A family drives up to the top of Mt. Elbert and boils some water for a tailgate party. How much lower is the boiling point than in Denver?

Technology

Solve the following equations using a calculator.

47. $5.81x^2 - 3.74x = 14.11$
48. $1\text{E}5x^2 - x - 1 = 0$
49. $x(x + 1) = 5{,}589$
50. $89{,}743x = 7{,}831x^2 + 983$

In your own words

51. Describe how to solve a quadratic equation using the method of factoring.
52. Describe how to solve a quadratic equation using the quadratic formula.
53. How do you decide which method to use in order to solve a quadratic equation?
54. Describe when extraneous solutions can arise.

CHAPTER 3 Applied Problems, Quadratic Equations, and Inequalities

3.3 QUADRATIC FUNCTIONS AND OPTIMIZATION PROBLEMS

A linear function

$$f(x) = ax + b, \quad a \neq 0$$

involves only the first power of x. Many mathematical models require higher powers of x. In this section, we introduce the polynomial functions, which are useful for such models.

**Definition
Polynomial Function** A **polynomial function** of the variable x and of degree n is a function of the form:

$$f(x) = a_n x^n + a_{n-1} x^{n-1} + \cdots + a_1 x + a_0, \quad a_n \neq 0.$$

The domain of a polynomial function consists of all real numbers. The range consists of a subset of the real numbers. (It may or may not consist of *all* real numbers.) We learn more about the range later in this section.

Here are some examples of polynomial functions:

$$f(x) = 5$$
$$f(x) = -0.1x + 0.001$$
$$f(x) = 3x^2 + 100x - 310$$
$$f(x) = -x^3 + x^2 - x + 1$$

As the first two examples illustrate, constant and linear functions are special cases of polynomial functions. Because we have already explored the properties of these functions and their graphs, we now turn to the next most complicated polynomial functions, namely those for which n equals 2.

**Definition
Quadratic Function** A **quadratic function** is a function defined by an expression of the form

$$f(x) = ax^2 + bx + c$$

where a, b, and c are real numbers and a is nonzero.

Here are some examples of quadratic functions:

$$f(x) = x^2$$
$$f(x) = \frac{1}{2}x^2 + 3x - 1$$
$$f(x) = 0.001x^2 + 5.07$$

Graphs of Quadratic Functions

Let's now explore the graphs of quadratic functions. We have already graphed the function $y = x^2$ and shown that its graph is bowl-shaped, opening upward, with the bottom of the bowl at the origin. (See Figure 1.)

The graphs of the quadratic functions

$$f(x) = ax^2, \quad a > 0$$

have the same general appearance as that of $f(x) = x^2$. The coefficient a controls the width of the opening of the graph. The smaller the value of a, the wider

SECTION 3.3 Quadratic Functions and Optimization Problems

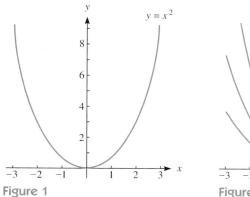

Figure 1

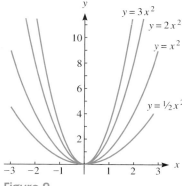

Figure 2

the opening. Figure 2 shows the graphs of such quadratic functions for various positive numbers a.

We obtain the graph of
$$f(x) = ax^2, \quad a < 0$$
by reflecting the graph of $f(x) = |a|x^2$ in the x-axis; it is a bowl-shaped curve that opens downward. (See Figure 3.)

We obtain the graph of
$$f(x) = a(x-h)^2 + k$$
by translating the graph of $f(x) = ax^2$; it is therefore a bowl-shaped graph with its bottom or top at (h, k). The graph opens upward if $a > 0$ and downward if $a < 0$. (See Figure 4.)

By completing the square, every quadratic function can be written in the form $f(x) = a(x-h)^2 + k$. This is called the **standard form** of the quadratic function. The bowl-shaped curve that arises as the graph of a quadratic function is called a **parabola**. The bottom or top of the bowl of a parabola is called its **vertex**. In the graph of this function, the vertex is at (h, k). The vertical line that passes through the vertex is called the **axis** of the parabola. (See Figure 5.)

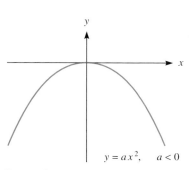

Figure 3

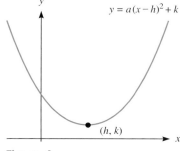

Figure 4

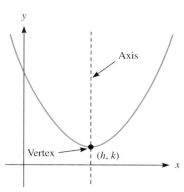

Figure 5

STANDARD FORM OF A QUADRATIC FUNCTION

The quadratic function $f(x) = ax^2 + bx + c$ can be written in the form:
$$f(x) = a(x - h)^2 + k$$
The graph of the function is a parabola with vertex (h, k). The parabola opens upward if $a > 0$ and downward if $a < 0$.

EXAMPLE 1
Standard Form of a Quadratic Function

Consider the quadratic function $f(x) = 2x^2 + 6x - 10$.
1. Write $f(x)$ in standard form.
2. Determine the vertex and the axis of the graph of $f(x)$.

Solution

1. Complete the square:
$$f(x) = 2x^2 + 6x - 10$$
$$= 2(x^2 + 3x) - 10$$
$$= 2\left(x^2 + 3x + \left(\frac{3}{2}\right)^2\right) - 10 - 2 \cdot \left(\frac{3}{2}\right)^2$$
$$= 2\left(x + \frac{3}{2}\right)^2 - \frac{29}{2}$$

This is of the standard form of a quadratic equation, with:
$$a = 2, \quad h = -\frac{3}{2}, \quad k = -\frac{29}{2}$$

2. The vertex is $(h, k) = \left(-\frac{3}{2}, -\frac{29}{2}\right)$. The axis is the vertical line that passes through the vertex and thus has equation $x = -\frac{3}{2}$.

EXAMPLE 2
Determining a Quadratic Function from Geometric Information

Determine the equation of a quadratic function whose graph passes through the point $(1, -2)$ and has vertex $(4, 5)$.

Solution

By the standard form of the quadratic function,
$$f(x) = a(x - 4)^2 + 5$$
for some value of a. Because the graph must pass through $(1, -2)$, we must have:
$$f(1) = -2$$
$$a(1 - 4)^2 + 5 = -2$$
$$9a + 5 = -2$$
$$a = -\frac{7}{9}$$
$$f(x) = -\frac{7}{9}(x - 4)^2 + 5$$

Because the function $y = ax^2$ is even, its graph is symmetric about the y-axis. Thus, the graph of $f(x) = a(x - b)^2 + k$, which is obtained by translating

SECTION 3.3 Quadratic Functions and Optimization Problems

the graph of $y = ax^2$, is symmetric about its axis. So to sketch the graph of a quadratic function, we can proceed as follows:

GRAPHING A QUADRATIC FUNCTION

1. Write the quadratic function in standard form: $f(x) = a(x - h)^2 + k$.
2. Plot the vertex (h, k).
3. Determine a point on the graph. The simplest such point to determine is usually the one that corresponds to $x = h + 1$.
4. Plot this point, and draw a smooth curve from the vertex through the point.
5. Complete the graph by drawing a symmetrical curve on the other side of the vertex.

EXAMPLE 3
Graphing Quadratic Functions

Graph the following quadratic functions.
1. $f(x) = 3x^2 - 12x - 10$
2. $f(x) = -2x^2 + 12x$

Solution

1. First, we write the function in standard form:
$$3x^2 - 12x - 10 = 3(x^2 - 4x) - 10$$
$$= 3(x^2 - 4x + 4) - 10 - 3 \cdot 4$$
$$= 3(x - 2)^2 - 22$$

The vertex is $(2, -22)$. When $x = 3$, the value of $f(x)$ is:
$$f(3) = 3 \cdot 1^2 - 22 = -19$$

So $(3, -19)$ is on the graph. Draw a smooth curve from $(2, -22)$ through $(3, -19)$. Now draw a symmetrical curve on the left side of the vertex. The completed graph is shown in Figure 6.

2. We put the function into standard form:
$$f(x) = -2x^2 + 12x$$
$$= -2(x^2 - 6x)$$
$$= -2(x^2 - 6x + 9) - (-2) \cdot 9$$
$$= -2(x - 3)^2 + 18$$

The vertex of the parabola is $(3, 18)$. Another point on the graph is $(4, 16)$. Draw a smooth curve from the vertex through $(4, 16)$ and complete the graph by drawing a symmetrical curve on the left side of the vertex. The completed graph is shown in Figure 7.

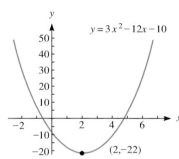
Figure 6

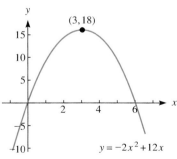
Figure 7

Applications of Quadratic Functions

The graph of a quadratic function is a parabola that opens either upward or downward. In the case of a graph that opens upward, the bottom point on the graph is called the **minimum point** of the function. Its y-value is the least value that the function assumes and is called the **minimum value** of the function. (See Figure 8.)

180 CHAPTER 3 Applied Problems, Quadratic Equations, and Inequalities

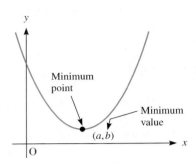

Figure 8

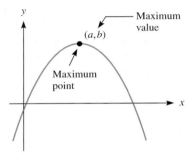

Figure 9

In the case of a graph that opens downward, the graph has a top point called the **maximum point** of the function. The y-value of the maximum point is the largest value that the function assumes and is called the **maximum value** of the function. (See Figure 9.)

The maximum or minimum point of the graph of a quadratic function is called the **vertex** or **extreme point** of the graph. The maximum or minimum value of the quadratic function is called the **extreme value**.

The extreme point of a quadratic function can easily be determined. Consider the quadratic function:

$$f(x) = ax^2 + bx + c$$

Assume that $a > 0$, so that the graph of f opens upward. The function then has a minimum point. Write the function in the form:

$$f(x) = a\left(x^2 + \frac{b}{a}x\right) + c$$

$$= a\left(x^2 + \frac{b}{a}x + \frac{b^2}{4a^2}\right) + \left(c - \frac{ab^2}{4a^2}\right)$$

$$= a\left(x + \frac{b}{2a}\right)^2 + \frac{4ac - b^2}{4a}$$

$$= a(x - h)^2 + k$$

where

$$(h, k) = \left(-\frac{b}{2a}, \frac{4ac - b^2}{4a}\right)$$

Recall that a perfect square is nonnegative. Because we are assuming that a is positive, the expression

$$a\left(x + \frac{b}{2a}\right)^2$$

is nonnegative. Therefore, for any value of x, we have:

$$f(x) \geq k$$

In other words, the value of $f(x)$ is always at least k. Moreover, for

$$x = -\frac{b}{2a}$$

the value of f is exactly k, because

$$f\left(-\frac{b}{2a}\right) = a\left(-\frac{b}{2a} + \frac{b}{2a}\right)^2 + k$$

$$= 0 + k$$

$$= k$$

We have proved that $f(x)$ is always at least k. So the minimum point of $f(x)$ is $(-b/2a, f(-b/2a))$.

If $a < 0$, the graph opens downward and f has a maximum point, the coordinates of which are given by the same formulas we just derived for the minimum point. The proof is similar to the one just given. We have, therefore, proved the following important result:

SECTION 3.3 Quadratic Functions and Optimization Problems

EXTREME POINT OF A QUADRATIC FUNCTION

Let $f(x) = ax^2 + bx + c$ be a quadratic function. The extreme point of its graph is:
$$(h, k) = \left(-\frac{b}{2a}, f\left(-\frac{b}{2a}\right)\right)$$
If $a > 0$, then the extreme point of $f(x)$ is a minimum point. If $a < 0$, then the extreme point of $f(x)$ is a maximum point.

We have shown that the extreme point of a parabola is at the vertex. Many applied problems can be reduced to determining the coordinates of the extreme point. Such problems are called **optimization problems** because, in the applied context, the maximum or minimum point represents some optimum condition, such as maximum profit, minimum cost, and so forth. In the next three examples, we illustrate a number of practical uses for quadratic functions.

EXAMPLE 4
Maximizing Manufacturing Revenue

A company manufactures upholstered chairs. If it manufactures x chairs, then the revenue $R(x)$ per chair is given by the function:
$$R(x) = 200 - 0.05x$$
How many chairs should the company manufacture to maximize total revenue?

Solution

We first find a function $T(x)$ that expresses the total revenue T received in terms of the number of chairs manufactured x. We have:
$$T(x) = \text{Number of chairs} \cdot \text{Revenue per chair}$$
Therefore,
$$T(x) = x \cdot R(x)$$
$$= x(200 - 0.05x)$$
$$= -0.05x^2 + 200x$$

Thus, we see that $T(x)$ is a quadratic function. Figure 10 shows the graph of this function. Note that the graph is a parabola opening downward. In economic terms, the graph says that, initially, as the number of chairs x increases, the total revenue $T(x)$ increases. At a certain level of production, the total revenue is maximized, and increasing the number of chairs further actually decreases total revenue. The maximum occurs for:
$$x = -\frac{b}{2a} = -\frac{200}{2(-0.05)} = 2{,}000$$
That is, the revenue is at a maximum when 2,000 chairs are manufactured.

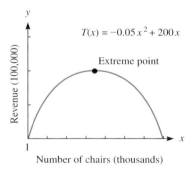

Figure 10

EXAMPLE 5
Maximizing Revenue When Faced with Variable Demand

Suppose that, on its Philadelphia-to-Chicago route, an airline is currently carrying 4,000 passengers each month and its fare is $300. The airline wishes to maximize its revenue by changing the fare, if warranted. Market research shows that for each $1 increase in the fare, the number of passengers per month will drop by 200. What is the fare that will maximize the airline's revenue?

Solution

Let x be the new fare. At this fare, the number of passengers is:

[number of passengers] = [original number] − 200 × $\begin{bmatrix} \text{number of dollars} \\ \text{fare increases} \end{bmatrix}$

$$= 4{,}000 - 200(x - 300)$$
$$= 4{,}000 - 200x + 60{,}000$$
$$= 64{,}000 - 200x$$

The new revenue R is given by the formula:

[Revenue] = [number of passengers] · [price per passenger]
$$= (64{,}000 - 200x)x$$
$$R = 64{,}000x - 200x^2$$

That is, R is a quadratic function of the fare x. Because the coefficient of x^2 is negative, the graph opens downward and the function has a maximum. This maximum occurs for $x = -b/2a = -64{,}000/[2(-200)] = 160$. That is, the maximum revenue occurs when the fare is set at $160. To get the maximum revenue, the airline should lower, not raise, its fares! The lower fares bring in more passengers, which increases revenue. This is amply illustrated by the periodic price wars that break out in the airline industry.

EXAMPLE 6
Architecture

Suppose that an architect wishes to design a house with a fenced backyard. To save fencing cost, he wishes to use one side of the house to border the yard and to use cyclone fence for the other three sides. The specifications call for using 100 ft of fence. What is the largest area that the yard can contain?

Solution

Let l denote the length of the yard and w denote its width. (See Figure 11.) Because the architect is required to use 100 ft of fence, the diagram in the figure shows that:

$$l + 2w = 100$$

The area of the yard, which is to be maximized, is given by the expression:

$$\text{Area} = lw$$

Figure 11

We can express the area as a function $A(w)$, as follows. Solve the first equation for l in terms of w:

$$l = 100 - 2w$$

Now substitute this expression for l into the expression for the area, to obtain:

$$\text{Area} = lw = (100 - 2w)w = -2w^2 + 100w$$

That is, the area function $A(w)$ is given by:

$$A(w) = -2w^2 + 100w$$

This is a quadratic function. The coefficient of the squared term, -2, is negative, so the extreme point is a maximum point. The value of w for which the maximum value of $A(w)$ occurs is then given by:

$$w = -\frac{b}{2a} = -\frac{100}{2(-2)} = 25 \text{ ft}$$

For this value of w, the corresponding value of l is

$$l = 100 - 2w = 100 - 2(25) = 50 \text{ ft}$$

SECTION 3.3 Quadratic Functions and Optimization Problems

In other words, if the area of the backyard is to be maximized, then the yard should be 25 feet wide and 50 feet long. The maximum area of the yard is
$$A(25) = -2(25)^2 + 100(25) = 1{,}250 \text{ ft}^2$$

EXAMPLE 7
Translating a Parabola Using a Calculator

Translate the graph of $f(x) = 2x^2 - 3x + 1$ two units to the left and 5 units down. Graph both the original and translated graphs on a single coordinate system.

Solution

The translated graph is the graph of:
$$f(x - (-2)) + (-5) = f(x + 2) - 5$$
$$= 2(x + 2)^2 - 3(x + 2) - 4$$

Using a graphing calculator, enter $f(x)$ for Y_1 and the above expression for Y_2. (See Figures 12 and 13.)

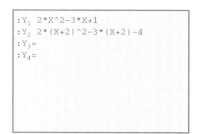

Figure 12

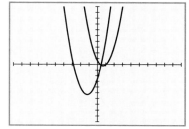

Figure 13

Exercises 3.3

Match each of the functions in Exercises 1–6 with its graph below.

1. $f(x) = 2x^2$
2. $f(x) = -x^2$
3. $f(x) = 2x^2 - x$
4. $f(x) = -(x + 1)^2 + 3$
5. $f(x) = (x - 3)^2 + 1$
6. $f(x) = x(x - 1)$

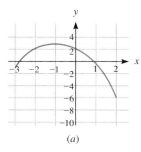

(a)

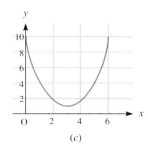

(c)

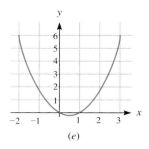

(e)

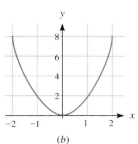

(b)

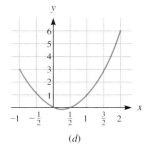

(d)

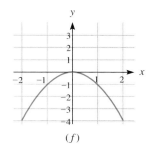

(f)

184   CHAPTER 3   Applied Problems, Quadratic Equations, and Inequalities

In Exercises 7–12, determine the equation of each parabola shown.

7.

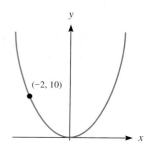

(−2, 10)

8.

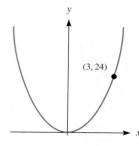

(3, 24)

9.

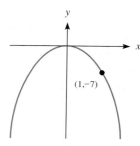

(1, −7)

10.

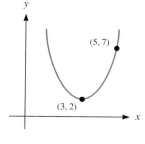

(5, 7)

(3, 2)

11.

(2, 4)

(−1, 1)

12.

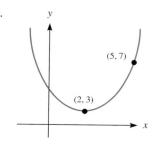

(5, 7)

(2, 3)

For each of the quadratic functions in Exercises 13–20:

a. Find the vertex.
b. Determine whether the graph opens upward or downward.
c. Graph the function.

13. $f(x) = -2x^2$

14. $f(x) = 2\frac{1}{4}x^2$
15. $f(x) = (x - 2)^2$
16. $f(x) = -2(x + 1)^2$
17. $f(x) = 2(x - 3)^2$
18. $f(x) = -1.5(x + 2)^2$
19. $f(x) = -(x + 3)^2 - 2$
20. $f(x) = \frac{1}{2}(x - 4)^2 - 3$

For each of the following quadratic functions:

a. Determine the standard quadratic form.
b. Find the vertex.
c. Determine whether the graph opens upward or downward.
d. Find the maximum or minimum value.
e. Graph the function.

21. $f(x) = x^2 + 6x + 5$
22. $f(x) = -13 - 8x - x^2$
23. $f(x) = -\frac{1}{4}x^2 - 3x - 1$
24. $f(x) = 2x^2 + x - 1$

Find the maximum or minimum value.

25. $f(x) = -3(x + 1)^2$
26. $f(x) = 4(x - 3)^2$
27. $f(x) = 5(x - 1)^2 + 3$
28. $f(x) = 12(x - 5)^2 + 2$
29. $f(x) = -2x^2 - x + 15$
30. $f(x) = x^2 - 6x + 9$
31. $f(x) = 3x^2 - x - 14$
32. $f(x) = -x^2 + 1$

Applications

33. **Ranching.** A rancher wants to build a rectangular fenced area next to a river, as shown in the figure, using 150 yd of fencing.

   a. Determine a formula in one variable for the area $A(x)$.
   b. At what value of x will the area be at a maximum?
   c. What is the largest area that can be enclosed?

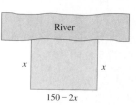

Exercise 33

SECTION 3.3 Quadratic Functions and Optimization Problems

34. Ranching. A rancher wants to enclose two rectangular areas next to a river, as shown in the figure, using 380 yd of fencing.

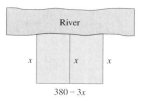

Exercise 34

a. Determine a formula in one variable for the area $A(x)$.
b. At what value of x will the area be at a maximum?
c. What is the largest area that can be enclosed?

35. Home remodeling. A carpenter is building a room with a fixed perimeter of 74 ft. What are the dimensions of the largest room she can build? What is the room's area?

36. Geometry. Of all the rectangles that have a perimeter of 98 ft, find the dimensions of the one with the largest area. What is its area?

37. Maximizing revenue. A clothing firm is coming out with a new line of suits. If it manufactures x suits, then the revenue $R(x)$ per suit is given by the function:
$$R(x) = 400 - 0.16x$$
How many suits should the company manufacture in order to maximize total revenue?

38. Maximizing profit. An electronics firm is marketing a new low-priced stereo. It determines that its total revenue from the sale of x stereos is given by the function:
$$R(x) = 280x - 0.4x^2$$

The firm also determines that its total cost of producing x stereos is given by:
$$C(x) = 5{,}000 + 0.6x^2$$
Total profit is given by:
$$P(x) = R(x) - C(x)$$

a. Find the total profit.
b. How many stereos must the company produce and sell in order to maximize profit?
c. What is the maximum profit?

39. Hotel management. By checking records of occupancy, the owner of a 30-unit motel knows that when the room rate is $50 a day per unit, all units are occupied. For every increase of x dollars in the daily rate, there are x units vacant. Each unit occupied costs $10 per day to service and maintain. Let R denote the total daily income.

a. Determine a formula in one variable for $R(x)$.
b. Find a value of x for which R is a maximum.
c. What is the maximum income?

40. Architecture. A Palladian window is a rectangle with a semicircle on top, as shown in the following figure. Suppose the perimeter of a particular Palladian window is to be 28 ft. Let A denote the total area of the window.

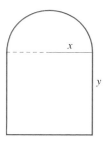

Exercise 40

a. Determine a formula in one variable for $A(x)$.
b. For what value of x will A be a maximum?
c. What is the maximum possible area?

41. Geometry. A 28-in. piece of string is to be cut into two pieces. (See figure.) One piece is to be used to form a circle and the other to form a square. Let A denote the total area.

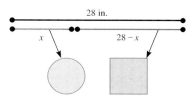

Exercise 41

a. Determine a formula for $A(x)$.

b. Find a value of x for which A is a maximum.

c. What is the maximum possible area?

42. **Farming.** A soybean farmer has 15 tons of beans on hand. If he sold them now, his profit would be $300 per ton. If he waits to sell, he can sell an additional 3 tons of beans each week, but he will lose $20 per ton for each week he delays to begin selling his beans.

    a. In how many weeks should he begin selling to maximize his profits?

    b. What would his maximum profit be?

43. **Trade deficit.** The annual trade deficit (in billions of dollars) for the preceding 12 months at time x (in years, with January 1992 corresponding to 0) is given by the model

$$y = 65 - 4.35x + 26.1x^2$$

a. What was the trade deficit for January 1992?

b. What was the trade deficit for January 1993?

c. In what month was the trade deficit the lowest?

d. What was the minimum trade deficit?

44. **Marriage rate.** The marriage rate (number of marriages per 1,000 people) in year x (with 1960 corresponding to 0) is given by the model:

$$y = 8.47 + 0.23x - 0.0065x^2$$

a. Graph this function.

b. What was the marriage rate in 1965?

c. What was the marriage rate in 1990?

d. In what year was the marriage rate the highest?

Technology

Use a graphing calculator to determine the maximum or minimum values of the following functions. Determine the value to two significant digits.

45. $y = 5.78x^2 + 3.11x$

46. $y = -12x^2 + 13x - 11$

47. $y = \dfrac{x^2}{21} + \dfrac{11x}{2} + \dfrac{5}{23}$

48. $y = 11(x - 3.4)(x + 7.9)$

49. $y = |x^2 - 3x + 4|$

50. $y = |1.1 - x^2|$

3.4 USING QUADRATIC EQUATIONS TO SOLVE OTHER EQUATIONS

In the preceding section, we learned to solve quadratic equations using any one of several methods. Moreover, we solved a number of applied problems that, when translated into algebraic language, resulted in quadratic equations. Many equations that are not quadratic can be reduced to quadratic equations and then solved. Some examples of such equations are:

$$x^4 - 7x^2 + 6 = 0$$

$$\left(\frac{x-3}{x+4}\right)^2 + 2\left(\frac{x-3}{x+4}\right) - 3 = 0$$

$$\sqrt{x-1} + \sqrt{x+2} = 5$$

This section is devoted to solving such equations.

SECTION 3.4 Using Quadratic Equations to Solve Other Equations

Equations That Reduce to Quadratic Equations

Many equations can be solved if we reduce them to quadratic equations through algebraic manipulation, such as clearing denominators of rational expressions or squaring to remove square roots. The next two examples illustrate such manipulation.

EXAMPLE 1
A Rational Equation

Determine all solutions of the following rational equation:
$$\frac{x}{x-1} = \frac{6x-5}{x+15}$$

Solution

When we are confronted with rational expressions in an equation, it is usually a good idea to rewrite the equation in order to clear the fractions. In this example, we can remove the rational expressions by multiplying both sides of the equation by $(x-1)(x+15)$:

$$\frac{x}{x-1}(x-1)(x+15) = \frac{6x-5}{x+15}(x-1)(x+15)$$
$$x(x+15) = (6x-5)(x-1), \qquad x \neq 1, -15$$
$$x^2 + 15x = 6x^2 - 11x + 5$$
$$5x^2 - 26x + 5 = 0$$

We now have a quadratic equation that we can solve using factoring:
$$(5x-1)(x-5) = 0$$
$$5x - 1 = 0 \quad \text{or} \quad x - 5 = 0$$
$$x = \frac{1}{5}, 5$$

The values of x for which the rational expressions in the equation involve division by 0 are 1 or -15. Neither of these values coincides with the values of x found from factoring. Therefore, the values $x = 5$ and $x = \frac{1}{5}$ are both solutions to our equation. Check the results by substituting these values for x in the original equation.

EXAMPLE 2
Another Rational Equation

Determine the solutions of the equation:
$$\frac{3}{x-5} + \frac{5}{x+5} = 4$$

Solution

The first step in the solution is to clear the fractions by multiplying both sides by the least common multiple of the denominators, namely $(x-5)(x+5)$. This gives the equation:

$$\frac{3}{x-5}(x+5)(x-5) + \frac{5}{x+5}(x+5)(x-5) = 4(x+5)(x-5)$$
$$3(x+5) + 5(x-5) = 4(x^2 - 25), \qquad x \neq 5, -5$$
$$8x - 10 = 4x^2 - 100$$
$$4x^2 - 8x - 90 = 0$$

We can divide both sides of this last equation by 2:
$$2x^2 - 4x - 45 = 0$$

This equation doesn't factor easily, so let's use the quadratic formula. The discriminant is:
$$b^2 - 4ac = (-4)^2 - 4(2)(-45) = 376$$
The quadratic formula yields:
$$x = \frac{-(-4) \pm \sqrt{376}}{2(2)}$$
$$= \frac{4 \pm 2\sqrt{94}}{4}$$
$$= \frac{2 \pm \sqrt{94}}{2}$$
$$\approx 5.848, -3.848$$

The values excluded when we multiplied by the denominators were $x = 5, -5$, which don't coincide with the solutions found by the quadratic formula. So our solutions are not extraneous and are solutions to the original equation.

EXAMPLE 3
A Radical Equation

Determine all solutions of the following radical equation:
$$3x + 2\sqrt{x} = 1$$

Solution

When confronted with an equation containing a radical, we should attempt to transform the equation so that the radical disappears. In this case, it is simplest to isolate the radical on one side of the equation and to square both sides.

$$2\sqrt{x} = 1 - 3x$$
$$(2\sqrt{x})^2 = (1 - 3x)^2 \quad \text{Squaring both sides}$$
$$4x = 1 - 6x + 9x^2$$
$$9x^2 - 10x + 1 = 0$$

We now have a quadratic equation that we can solve by factoring:
$$(9x - 1)(x - 1) = 0$$
$$9x - 1 = 0 \quad \text{or} \quad x - 1 = 0$$
$$x = \frac{1}{9}, 1$$

Note that we squared both sides of the equation. It is necessary to check whether this operation introduced any extraneous solutions. Substituting $x = \frac{1}{9}$ into the given equation yields:

$$3x + 2\sqrt{x} \quad \Big| \quad 1$$
$$= 3 \cdot \frac{1}{9} + 2\sqrt{\frac{1}{9}}$$
$$= \frac{1}{3} + 2 \cdot \frac{1}{3}$$
$$= 1$$

SECTION 3.4 Using Quadratic Equations to Solve Other Equations

So $\frac{1}{9}$ is a solution. Next, we substitute $x = 1$ into the given equation:

$$\begin{array}{c|c} 3x + 2\sqrt{x} & 1 \\ = 3 \cdot 1 + 2\sqrt{1} & \\ = 3 + 2 & \\ = 5 & \end{array}$$

The two sides are unequal, so $x = 1$ is an extraneous solution that arose from squaring the original equation, and hence must be discarded.

The Method of Substitution

We can often rewrite an equation in a solvable form by changing the variable. For example, the equation

$$x^4 - 4x^2 + 4 = 0$$

is a fourth degree equation. However, if we introduce the new variable

$$y = x^2$$

then we can rewrite the equation in terms of the new variable as:

$$y^2 - 4y + 4 = 0$$

We can solve this quadratic equation using the methods demonstrated in the preceding section. For instance, the equation can be factored into

$$(y - 2)^2 = 0$$

which gives the solution $y = 2$. To solve the original equation, we must relate the solution in y back to the original variable x. To do this, substitute $y = 2$ into the equation that gives the relationship between the variables:

$$y = x^2$$
$$2 = x^2$$
$$x = \pm\sqrt{2}$$

These are the possible solutions of the given equation. By substituting these values into the original equation, we check that both values are solutions. The method used to solve the equation is called the **method of substitution**.

We can summarize this method as follows:

METHOD OF SUBSTITUTION FOR SOLVING AN EQUATION

1. Make an appropriate substitution $y =$ [expression in x]
2. Solve the resulting equation for y.
3. For each solution $y = a$ in step 2, solve the equation:
   $$a = [\text{expression in } x]$$
4. Check all values of x to determine which are solutions of the original equation.

The following two examples provide further practice in applying the method of substitution.

CHAPTER 3 Applied Problems, Quadratic Equations, and Inequalities

EXAMPLE 4
Method of Substitution

Use the method of substitution to solve the equation
$$(x^2 + 3x)^2 - 7(x^2 + 3x) + 6 = 0$$

Solution

In the equation $(x^2 + 3x)^2 - 7(x^2 + 3x) + 6 = 0$, we make the substitution:
$$y = x^2 + 3x$$

We arrive at the equation
$$y^2 - 7y + 6 = 0$$

and factor it to obtain solutions:
$$(y - 6)(y - 1) = 0$$
$$y = 6, 1$$

That is,
$$x^2 + 3x = 6 \quad \text{or} \quad x^2 + 3x = 1$$

Using the quadratic formula, we have the solutions:
$$x = \frac{-3 \pm \sqrt{33}}{2} \quad \text{and} \quad \frac{-3 \pm \sqrt{13}}{2}$$

Let's check these values in the original equation. In the case of the values $x = (-3 \pm \sqrt{33})/2$, we find that:
$$x^2 + 3x = 6$$
$$(x^2 + 3x)^2 - 7(x^2 + 3x) + 6 = 0$$
$$(6)^2 - 7(6) + 6 = 0$$

In a similar fashion, for the values $x = (-3 \pm \sqrt{13})/2$, we have
$$x^2 + 3x = 1$$
$$(x^2 + 3x)^2 - 7(x^2 + 3x) + 6 = 0$$
$$1^2 - 7(1) + 6 = 0$$

Thus, all four values of x are solutions of the given equation.

EXAMPLE 5
Substitution That Leads to Extraneous Solutions

Solve the following equation:
$$\left(\frac{x-3}{x+4}\right)^2 + 2\left(\frac{x-3}{x+4}\right) - 3 = 0$$

Solution

In this example, we make the substitution:
$$y = \frac{x-3}{x+4}$$

In terms of y, the equation becomes:
$$y^2 + 2y - 3 = 0$$

This equation can be factored:
$$(y + 3)(y - 1) = 0$$
$$y = -3, 1$$

SECTION 3.4 Using Quadratic Equations to Solve Other Equations

The first value, $y = -3$, gives:

$$y = \frac{x-3}{x+4} = -3$$

The second value, $y = 1$, gives:

$$y = \frac{x-3}{x+4} = 1$$

Provided that $x \neq -4$, the first equation is equivalent to:

$$x - 3 = -3(x+4)$$
$$4x = -9$$
$$x = -\frac{9}{4}$$

Provided that $x \neq -4$, the second equation is equivalent to:

$$x - 3 = x + 4$$

which leads to the contradiction:

$$0 = 7$$

This last equation is inconsistent. So the only solution of the given equation is $x = -\frac{9}{4}$. Here is a check that this value is a solution:

$$\left(\frac{x-3}{x+4}\right)^2 + 2\left(\frac{x-3}{x+4}\right) - 3 = \left(\frac{-\frac{9}{4}-3}{-\frac{9}{4}+4}\right)^2 + 2\left(\frac{-\frac{9}{4}-3}{-\frac{9}{4}+4}\right) - 3$$

$$= \frac{\left(-\frac{21}{4}\right)^2}{\left(\frac{7}{4}\right)^2} + 2\left(\frac{-\frac{21}{4}}{\frac{7}{4}}\right) - 3$$

$$= 9 + 2(-3) - 3$$

$$= 0 \qquad\qquad 0$$

So $-\frac{9}{4}$ is a solution of the equation.

EXAMPLE 6

Raising an Equation to a Power

Solve the equation:

$$\sqrt[3]{7x^2 - 15} = -2$$

Solution

When we encounter a radical equation, we should attempt to remove the radical by raising both sides of the equation to an appropriate power. In this case, we raise both sides to the third power:

$$\left[\sqrt[3]{7x^2 - 15}\right]^3 = (-2)^3$$
$$7x^2 - 15 = -8$$
$$7x^2 = 7$$
$$x^2 = 1$$
$$x = \pm 1$$

Let's check these solutions:

192 CHAPTER 3 Applied Problems, Quadratic Equations, and Inequalities

$$\sqrt[3]{7(\pm 1)^2 - 15} = -2$$
$$\sqrt[3]{-8} = -2$$
$$-2 = -2$$

So the solutions are $x = \pm 1$.

EXAMPLE 7
Several Radicals in One Equation

Solve the following equation:
$$\sqrt{x - 1} + \sqrt{x + 2} = 5$$

Solution

As in the preceding example, we must remove the radicals. To do so, we move one radical to the right and square both sides:

$$\sqrt{x - 1} = 5 - \sqrt{x + 2}$$
$$(\sqrt{x - 1})^2 = (5 - \sqrt{x + 2})^2$$
$$x - 1 = 5^2 - 2(5\sqrt{x + 2}) + (\sqrt{x + 2})^2$$
$$x - 1 = 25 - 10\sqrt{x + 2} + (x + 2)$$
$$\sqrt{x + 2} = \frac{28}{10}$$
$$x + 2 = \left(\frac{28}{10}\right)^2 = \frac{784}{100}$$
$$x = \frac{584}{100} = \frac{146}{25}$$

To check this solution, we substitute it into the given equation:

$$\sqrt{x - 1} + \sqrt{x + 2} = \sqrt{\frac{146}{25} - 1} + \sqrt{\frac{146}{25} + 2}$$
$$= \sqrt{\frac{121}{25}} + \sqrt{\frac{196}{25}}$$
$$= \frac{11}{5} + \frac{14}{5}$$
$$= 5 \qquad\qquad 5$$

More Applied Problems

In the previous section, we learned to solve equations that could be reduced to quadratic equations. Here we illustrate how such equations arise in compound interest and in job planning.

EXAMPLE 8
Compound Interest

A savings account contains $700. At what rate of interest, compounded annually, will the account contain $1,000 after two years?

Solution

According to the compound interest formula, the balance A after two years is given by the formula

$$A = P\left(1 + \frac{r}{k}\right)^{kn}$$

where P is the initial principal, r is the annual interest rate, n is the number of years the principal is invested, and k is the number of times per year interest is compounded. In this case,

$$P = 700, \quad A = 1{,}000, \quad k = 1, \quad n = 2$$

so we have the equation:

$$1{,}000 = 700(1 + r)^2$$

$$(1 + r)^2 = \frac{10}{7}$$

Taking square roots of both sides gives us:

$$1 + r = \pm\sqrt{\frac{10}{7}}$$

$$1 + r \approx \pm 1.195$$

$$r \approx -1 + 1.195, \, -1 - 1.195$$

$$r \approx 0.195, \, -2.195$$

Because r is the rate of interest of a savings account, we can ignore the negative value. So r is approximately 19.523 percent.

EXAMPLE 9
Job Planning

Maria and Jose have a painting business. Maria can paint a particular room in one hour less than Jose. Together they can paint the room in 5 hours. How long would it take Jose to paint the room alone?

Solution

For each of the painters, we have the relationship:

$$\text{Rate of working} \times \text{Time worked} = \text{Fractional part of job done}$$

Let t denote the total amount of time (in hours) that Jose requires to paint the room by himself. Then his rate of working is $1/t$ rooms per hour. By the above equation, in five hours the part of the job he accomplishes equals $5/t$. Maria paints the room in one hour less than Jose, or $t - 1$ hours. So Maria's rate of working is $1/(t - 1)$ rooms per hour. Therefore, in 5 hours, the part of the job she accomplishes equals

$$5 \cdot \frac{1}{t - 1} = \frac{5}{t - 1}$$

Together, Maria and Jose can paint the entire room in 5 hours, which we can express as:

$$\text{Jose's fractional part} + \text{Maria's fractional part} = \text{Whole job}$$

$$\frac{5}{t} + \frac{5}{t - 1} = 1$$

We can clear the fraction in this equation by multiplying by the least common denominator, $t(t - 1)$. This gives us:

$$5(t - 1) + 5t = t(t - 1)$$

$$10t - 5 = t^2 - t$$

$$t^2 - 11t + 5 = 0$$

By the quadratic formula, we have

$$t = \frac{11 \pm \sqrt{(-11)^2 - 4(1)(5)}}{2}$$

$$= \frac{11 \pm \sqrt{101}}{2}$$

$$= 10.5249, \ 0.4751$$

We reject the solution $(11 - \sqrt{101})/2$ because it leads to a negative proportion, $1/(t - 1)$, for Maria's work. So t has the value $(11 + \sqrt{101})/2$. That is, it takes Jose about 10.5 hours to paint the room by himself.

Exercises 3.4

Solve the following equations.

1. $2 + \dfrac{3}{x^2} = \dfrac{1}{x}$
2. $3 - \dfrac{1}{x} + \dfrac{6}{x^2} = 0$
3. $\dfrac{2x - 3}{3x + 5} = \dfrac{x + 7}{3x - 4}$
4. $\dfrac{5 - 2x}{6x + 1} = \dfrac{3 + 2x}{5 - x}$
5. $x - \dfrac{1}{x} = 17$
6. $m + \dfrac{2}{m} = 23$
7. $x = 7\sqrt{x} + 30$
8. $x - 13\sqrt{x} + 12 = 0$
9. $\sqrt{x + 5} = 3$
10. $\sqrt{x + 10} = 4$
11. $\sqrt{x + 7} + 5 = x$
12. $\sqrt{y + 5} = y - 1$
13. $\sqrt{x + 3} = x$
14. $\sqrt{x^2 + 5} = 2x^2$
15. $x^3 + 3x^2 - x - 3 = 0$
16. $3y^3 - 4y^2 - 12y + 16 = 0$
17. $12x^3 - 12x^2 = 0$
18. $4x^3 - 4 = 0$
19. $5x^4 - 5 = 0$
20. $2x - \dfrac{432}{x^2} = 0$
21. $\dfrac{1}{5} + \dfrac{1}{7} = \dfrac{2}{t}$
22. $\dfrac{2}{x} - \dfrac{3}{x} = 1$
23. $\dfrac{x - 1}{3} + \dfrac{2x + 1}{4} = \dfrac{1}{2}$
24. $\dfrac{3x + 2}{3} - \dfrac{x + 5}{2} = \dfrac{x + 1}{3}$
25. $\dfrac{10}{x} - \dfrac{1}{x} = 3 - \dfrac{7}{x}$
26. $\dfrac{16}{t} - \dfrac{16}{t + 5} = \dfrac{8}{t}$
27. $\dfrac{2}{a} + \dfrac{1}{a + 2} = \dfrac{4}{a^2 + 2a}$
28. $\dfrac{2}{y - 2} + \dfrac{2y}{y^2 - 4} = \dfrac{3}{y + 2}$
29. $\sqrt{x^2 + 3x - 2} = x + 3$
30. $\sqrt{2x^2 + 7x + 10} = x + 4$
31. $t^{-1/4} = 2$
32. $s^{-1/3} = -\dfrac{1}{3}$
33. $\sqrt[3]{3x + 7} = 4$
34. $\sqrt[3]{x - 3} = -2$
35. $\sqrt{t - 1} + \sqrt{t + 4} = 5$
36. $\sqrt{m + 10} = 10 + \sqrt{m - 10}$
37. $x^4 + 5x^2 - 36 = 0$
38. $3x^4 + 2 = 7x^2$
39. $(x^2 - 3x)^2 - 2(x^2 - 3x) - 8 = 0$
40. $(t^2 - 7t)^2 = 2(t^2 - 7t) + 48$
41. $(\sqrt{y} - 10)^2 + 30 = 11(\sqrt{y} - 10)$
42. $(3 - \sqrt{t})^2 - 7(3 - \sqrt{t}) + 18 = 0$
43. $\left(\dfrac{x}{x - 1}\right)^2 + \left(\dfrac{x}{x - 1}\right) - 20 = 0$
44. $\left(\dfrac{2x + 3}{x + 1}\right) + 12 = \left(\dfrac{2x + 3}{x + 1}\right)^2$
45. $7z^{1/2} - 3 = 2z$
46. $3x^{2/3} = 20 - 7x^{1/3}$
47. $\sqrt[3]{(x^2 + 2x)^2} - \sqrt[3]{x^2 + 2x} - 2 = 0$
48. $(\sqrt{2x + 1} - 2)^2 - 4(\sqrt{2x + 1} - 2) = 5$
49. $\sqrt{3 + 2x} - (3 - 2x)^{1/2} = \sqrt{2x}$
50. $\sqrt[6]{2x + 2} = \sqrt[3]{3x - 1}$

Applications

51. **Job planning.** A painter can paint a room in 3 hours. It takes a trainee 5 hours to paint the same room. How long would it take if they worked together?

52. **Job planning.** A can mow the lawn in $\frac{1}{3}$ hr, B can do it in $\frac{1}{4}$ hr, and C can mow it in $\frac{2}{5}$ hr. How long would it take them if they all worked together?

SECTION 3.5 Equations and Inequalities Involving Absolute Value

53. **Job planning.** Marcia can deliver the papers on her route in 2 hours. Louis can deliver the same route in 3 hours. How long would it take them to deliver the papers if they worked together?

54. **Job planning.** A can do a job in 0.4 hr. When A works with B, the job takes 0.25 hr. After C is hired, all three can do the job in 0.18 hr. How long would it take B, working alone, to do the job?

55. **Job planning.** It takes Madonna 9 hours longer to build a wall than Sean. If they work together, they can build the wall in 20 hours. How long would it take Sean to build the wall by himself?

56. **Geometry.** A right triangle contains as many square centimeters in area as it has centimeters in perimeter. Its longer leg is twice as long as its shorter leg. What is the length of the shorter leg?

57. **Investment analysis.** At the beginning of the year, $2,000 is deposited in a bank. At the beginning of the next year, $1,200 is deposited in another bank at the same interest rate. At the beginning of the third year, there is a total of $3,573.80 in both accounts. If interest is compounded annually, what is the interest rate?

 Technology

Solve the following equations using a calculator. (A certain amount of algebraic manipulation by hand may be necessary.)

58. $5.75\sqrt{x + 3.9} = 3 + 2.1\sqrt{x}$

59. $9.1\sqrt{x} - 3.4 = 4.89(x - 2.1)$

60. $\dfrac{3.11}{x - 4.8} = \dfrac{2.1x + 4.69}{x + 2.75}$

61. $\dfrac{5}{(x - 1)(x - 5)} = 17 + \dfrac{1}{x}$

 In your own words

62. Give an example of an equation that can be solved by using the method of substitution. Use this equation to describe the steps of the method.

63. The method of substitution for reducing a polynomial equation to a quadratic equation does not work for a polynomial that has an x^3 term. Why?

3.5 EQUATIONS AND INEQUALITIES INVOLVING ABSOLUTE VALUE

As we saw in Section 1.1, the absolute value can be used to calculate the distance between points on the number line. In many problems, it is necessary to consider points for which distances from other points are restricted by inequalities, for example, the points that have a distance from 5 that is less than 2. The algebraic statement of such problems gives rise to inequalities that involve absolute values. Such inequalities also arise when we deal with numerical approximations, say, when we approximate the area of a field within a measurement error of at most 1 square foot. In this section, we discuss techniques for solving equations and inequalities that involve absolute values.

Equations Involving Absolute Values

The main point to remember when we deal with absolute values is that they have a two-part definition. Most absolute value problems must be broken into two cases: the case that the expression within the absolute value bars is positive or zero, and the case that it is negative. The next two examples demonstrate how this is typically handled.

EXAMPLE 1
Absolute Value Equation

Solve the equation:
$$|2x - 1| = 5$$

Solution

Because $|2x - 1|$ equals either $2x - 1$ or $-(2x - 1)$, the equation is equivalent to:
$$2x - 1 = 5 \quad \text{or} \quad -(2x - 1) = 5$$

The first equation has the solution 3 and the second has the solution -2. So the given equation has two solutions, 3 and -2.

EXAMPLE 2
Quadratic Absolute Value Equation

Determine all solutions to the equation:
$$x|x| = 3x - 1$$

Solution

We break the solution into two cases.

Case 1: $x \geq 0$

In this case, we have $|x| = x$ and the equation reads:
$$x \cdot x = 3x - 1$$
$$x^2 - 3x + 1 = 0$$

This quadratic equation can be solved using the quadratic formula:
$$x = \frac{3 \pm \sqrt{5}}{2}$$

Both solutions are positive and thus satisfy the condition in Case 1. So the solutions in Case 1 are:
$$x = \frac{3 + \sqrt{5}}{2} \quad \text{or} \quad x = \frac{3 - \sqrt{5}}{2}$$

Case 2: $x < 0$

In this case, we have $|x| = -x$, so that the equation reads:
$$x(-x) = 3x - 1$$
$$-x^2 - 3x + 1 = 0$$
$$x^2 + 3x - 1 = 0$$

Solve this equation by using the quadratic formula to obtain the solutions:
$$x = \frac{-3 + \sqrt{13}}{2} \quad \text{or} \quad x = \frac{-3 - \sqrt{13}}{2}$$

Because $x = \dfrac{-3 + \sqrt{13}}{2}$ is positive, this solution violates our assumption that x is negative. The only solution in case 2 is
$$x = \frac{-3 - \sqrt{13}}{2}$$

Taking into account both cases, we see that the given equation has the three solutions:
$$x = \frac{3 + \sqrt{5}}{2}, \frac{3 - \sqrt{5}}{2}, \frac{-3 - \sqrt{13}}{2}$$

SECTION 3.5 Equations and Inequalities Involving Absolute Value

Inequalities Involving Absolute Values

Let's turn to inequalities that involve absolute values. Here are some examples of such inequalities:

$$|2x + 5| < 4$$
$$|x| > 6$$

The key to solving inequalities involving absolute values is the following fundamental result:

FUNDAMENTAL ABSOLUTE VALUE INEQUALITY

Let a be a positive real number. Then x satisfies the inequality $|x| < a$ if and only if $-a < x < a$.

Verification Recall that $|x|$ is the distance of x from the origin. The inequality $|x| < a$ says that the distance from x to the origin is less than a. However, the numbers that satisfy this property are exactly those that satisfy the inequality:

$$-a < x < a$$

The next several examples illustrate how the Fundamental Absolute Value Inequality can be used to solve inequalities involving absolute values.

EXAMPLE 3 Absolute Value Inequality

Solve the inequality $3|2 - 5x| + 1 < 10$.

Solution

We can simplify the inequality by subtracting 1 from each side and then dividing each side by 3. This gives the inequality:

$$|2 - 5x| < 3$$

Use the Fundamental Absolute Value Inequality to convert this last inequality to a double inequality:

$$-3 < 2 - 5x < 3$$
$$-5 < -5x < 1 \quad \text{Subtract 2 from both sides}$$
$$1 > x > -\frac{1}{5} \quad \text{Multiply by } -\frac{1}{5}$$
$$-\frac{1}{5} < x < 1 \quad \text{Rewrite the inequality}$$

The solution set of the given inequality is the interval $(-\frac{1}{5}, 1)$.

EXAMPLE 4 Measurement Error

Suppose that a paper cutter is cutting 36-inch square sheets of Christmas wrapping paper. Suppose that the cutter can make an error of 0.25 inch in cutting either the length or the width.

1. By how much can the area of the paper be in error?
2. How many sheets of paper can the cutter waste in a run of 1,000 sheets?

Solution

1. Let L be the actual length of one side of the paper. Then, from the given information, we see that:

$$|L - 36| \leq 0.25$$

We solve this inequality using the Fundamental Absolute Value Inequality:
$$-0.25 \leq L - 36 \leq 0.25$$
$$36 - 0.25 \leq L \leq 36 + 0.25$$
$$35.75 \leq L \leq 36.25$$

Because the area of a square sheet of paper is L^2, we must find an inequality for L^2. To do so, square each member of the last inequality to obtain:
$$35.75^2 \leq L^2 \leq 36.25^2$$
$$1{,}278.0625 \leq L^2 \leq 1{,}314.0625$$

The ideal area of the sheet is $36^2 = 1{,}296$. The error in the area can be as much as the larger of these two values:
$$1{,}296 - 1{,}278.0625 = 17.9375$$
$$1{,}314.0625 - 1{,}296 = 18.0625$$

That is, the error in the area can be as much as 18.0625 square inches.

2. The cutter can waste as much as 18.0625 in.² per sheet. In a run of 1,000 sheets, this can amount to as much as:
$$\frac{18.0625 \times 1{,}000}{1{,}296} = 13.94 \text{ sheets}$$

That is, the cutter can waste almost 14 sheets per run of 1,000 sheets.

In the preceding examples, we considered inequalities in which an absolute value was less than a constant. In some applications, it is necessary to consider inequalities of the form
$$|x| > a$$
where a is a positive number. This inequality states that the distance from x to the origin is greater than a. The values of x for which this holds fall into two categories: Either $x > a$ or $x < -a$. That is, the solution set of the inequality is:
$$\{x : x < -a\} \cup \{x : x > a\}$$
The graph of this union is shown in Figure 1.

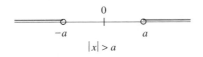

Figure 1

EXAMPLE 5
Solution Set Is a Union

Solve the inequality:
$$|4x - 1| > 1$$
Graph the solution set.

Solution

According to the preceding discussion, this inequality can be replaced by the two inequalities:
$$4x - 1 > 1 \quad \text{or} \quad -(4x - 1) < 1$$
Solving these inequalities, we have:

SECTION 3.5 Equations and Inequalities Involving Absolute Value

Figure 2

$$4x - 1 > 1 \qquad\qquad 4x - 1 < -1$$
$$4x > 1 + 1 \qquad\qquad 4x < -1 + 1$$
$$4x > 2 \qquad\qquad 4x < 0$$
$$x > \frac{1}{4} \cdot 2 \qquad\qquad x < 0$$
$$x > \frac{1}{2}$$

The solution set consists of those x that satisfy either of the above inequalities. That is, the solution set is the following:

$$\left\{x : x > \frac{1}{2}\right\} \cup \{x : x < 0\} = \left(\frac{1}{2}, \infty\right) \cup (-\infty, 0)$$

The graph of this union is sketched in Figure 2.

We can summarize the two types of absolute value inequalities as follows:

INEQUALITIES INVOLVING ABSOLUTE VALUE

Inequality	Solution	Distance Interpretation	Solution in Interval Notation
$\|x\| < a$	$-a < x < a$	The set of x for which the distance from x to the origin is less than a	$(-a, a)$
$\|x\| > a$	$x > a$ or $x < -a$	The set of x for which the distance from x to the origin is greater than a	$(-\infty, -a) \cup (a, \infty)$

Exercises 3.5

Solve each equation or inequality. Where appropriate, describe your answers using interval notation.

1. $|x| = 3$
2. $|x| = 6$
3. $|x + 3| = 2$
4. $|5 - x| = 7$
5. $|3x| = 12$
6. $4|x - 1| = 16$
7. $|2x - 3| = 15$
8. $|5 + 6x| = 41$
9. $|x| \leq 23$
10. $|x| > 23$
11. $|x| > 14$
12. $|x| < 14$
13. $|x - 2| > 4$
14. $|x + 9| \geq 7$
15. $|6 - 3x| < 9$
16. $\left|1 - \frac{2}{3}x\right| \leq 7$
17. $\left|\frac{3x - 2}{4}\right| \leq 5$
18. $\left|\frac{6 + 7x}{4}\right| < 26$
19. $4|5 + 6x| - 7 < 19$
20. $6|2 - 5x| + 12 > 36$
21. $|2x + 1| \leq x - 4$
22. $|3 - 2x| \geq 1 + 2x$
23. $|2x - 1| < -2$
24. $|3x + 1| > 0$

Each of the following is a solution set of an absolute value inequality. Find the inequality.

25. $[5 - r, 5 + r]$
26. $\left(-2, \frac{2}{3}\right)$
27. $(-\infty, -4] \cup [8, +\infty)$
28. $\left(-\infty, -\frac{3}{4}\right) \cup \left(\frac{13}{4}, +\infty\right)$

Applications

29. **Estimation of profit.** A company's profits P for a certain year are estimated to satisfy the inequality:
$$|P - \$3{,}000{,}000| \le \$225{,}000$$
Determine the interval over which the company's profits will vary.

30. **Temperature during an illness.** During a strange illness, a patient's temperature T varied according to the inequality:
$$|T - 98.6°| \le 2$$
Determine the interval over which the patient's temperature varied.

31. **Oscillation of a weight on a spring.** A weight is bobbing on a spring in such a way that its distance d, in inches, from the top satisfies the inequality:
$$|d - 10| < 4$$
Determine the interval over which the distance varies.

Technology

Solve the following inequalities using a calculator.

32. $15.1\,|x - 2.5| < 4.98$
33. $3.75\,|2x + 1| < 5.2$
34. $|4.1x - 12.8| > 1$
35. $|-100x + 260| > 300$

In your own words

36. Give a geometric interpretation of the solutions of the inequality $|x| < a$.
37. Give a geometric interpretation of the solutions of the inequality $|x| > a$.

Chapter Review

Important Concepts—Chapter 3

- applied problem
- strategy for solving an applied problem
- straight-line depreciation
- quadratic equation
- factoring method for solving a quadratic equation
- completing the square
- quadratic formula
- discriminant of a quadratic equation
- solutions of a quadratic equation
- polynomial function
- quadratic function
- standard form of a quadratic function
- parabola
- vertex of a parabola
- axis of a parabola
- graphing a quadratic function
- minimum point of a graph of a function
- minimum value of a function
- maximum point of a graph of a function
- maximum value of a function
- extreme point of a quadratic function
- extreme value of a quadratic function
- optimization problem
- method of substitution for solving an equation
- fundamental absolute value inequality

Mathematics and the World around Us—Fractals

A **fractal** is a geometric shape that is built out of repeated copies of a basic shape that is moved around and scaled. For example, consider the shape in the figure at left. It consists of repetitions of a triangle, and is obtained by starting with a single triangle and adding scaled copies of it, then scaled copies of those, and so forth, eventually including an infinite number of triangles, each successively smaller.

Fractals were first introduced by the IBM mathematician Benoit Mandelbrot. They have been used to model the shapes of many irregular patterns in nature, including clouds, mountains, landscapes, and even smoke patterns.

The idea of fractals is that, by specifying a simple pattern and rule for creating repetitions, you can create a geometric pattern with infinite detail. Because the instructions for creating the repetitions are specified by a mathematical rule, the patterns created have geometric properties that can be analyzed and recreated by a computer. However, since a fractal can have any level of complexity, it can be used to describe incredibly complicated patterns.

In the early 1990s fractals were first used for approximating pictures stored on a computer. By storing a basic shape and a rule for repetition, a complex picture can be stored in an extremely compact form. This is called *fractal image compression*.

Fractals are more than just an interesting piece of mathematical trivia; in the hands of scientists in many fields, fractals appear to be a new and highly useful way of describing patterns that appear in a large number of natural phenomena.

For more information about fractals, see *Chaos—Making a New Science* by James Gleick, Penguin Books, 1987, pp. 83–118, and *Fractals Everywhere* by Michael Barnsley, Academic Press, 1988.

Important Results and Techniques—Chapter 3

| **Quadratic formula** | Consider the quadratic equation $$ax^2 + bx + c = 0, \quad a \neq 0$$ with discriminant $b^2 - 4ac$.

1. If $b^2 - 4ac > 0$, then the equation has the two different, real solutions: $$x = \frac{-b + \sqrt{b^2 - 4ac}}{2a}$$ and $$x = \frac{-b - \sqrt{b^2 - 4ac}}{2a}$$
2 If $b^2 - 4ac = 0$, the equation has the single real solution: $$x = -\frac{b}{2a}$$
3. If $b^2 - 4ac < 0$, then the equation has no real solutions. | p. 167 |

CHAPTER 3 Applied Problems, Quadratic Equations, and Inequalities

Quadratic function	$f(x) = ax^2 + bx + c, \quad a \neq 0$	p. 176
Optimizing a quadratic function	The maximum or minimum point of the quadratic function $f(x) = ax^2 + bx + c, \quad a \neq 0$ has x-coordinate $-\dfrac{b}{2a}$ and y-coordinate $f\left(-\dfrac{b}{2a}\right).$	p. 181
Absolute value inequalities	1. The inequality $\|x\| < a$ has solution $-a < x < a$. 2. The inequality $\|x\| > a$ has the solution $x < -a$ or $x > a$	p. 199

Review Exercises—Chapter 3

Solve these equations and inequalities. Where appropriate, describe your answers using interval notation.

1. $4x^2 + 5x = 1$
2. $4x^2 = 3 + x$
3. $10t^2 - t = 100$
4. $1 - \dfrac{1}{6x} + \dfrac{2}{3x^2} = 0$
5. $\dfrac{5}{2x+6} = \dfrac{1-2x}{4x} + 2$
6. $2x^2 + 10x = 14$
7. $(2a - 3)^2 = (a + 1)^2$
8. $y^2 + \sqrt{3}y - 2 = 0$
9. $\sqrt{x^2 + 7} = x - 1$
10. $15 - 3\sqrt[3]{2x + 1} = 0$
11. $x^2 - 9x^{3/2} + 20x = 0$
12. $\sqrt{\dfrac{x+1}{4x-1}} = \dfrac{1}{2}$
13. $\dfrac{2}{x-2} + \dfrac{2x}{x^2-4} = \dfrac{3}{x+2}$
14. $\dfrac{5}{y^2+5y} - \dfrac{1}{y^2-5y} = \dfrac{1}{y^2-25}$
15. $\dfrac{4x-20}{x-9} - \dfrac{16}{x} = \dfrac{144}{x^2-9x}$

16. $\dfrac{6t}{t-5} - \dfrac{300}{t^2+5t+25} = \dfrac{2{,}250}{t^3-125}$
17. $m^4 - 7m^2 + 10 = 0$
18. $6t^{-2} = 6t^{-1} + 36$
19. $-3\left(\dfrac{x+2}{x-3}\right)^2 - 4\left(\dfrac{x+2}{x-3}\right) + 4 = 0$
20. $t^{2/3} = 16$
21. $x^3 + 4x^2 - 9x - 36 = 0$
22. $x - 10\sqrt{x} + 16 = 0$
23. $|2x - 5| = 7$
24. $|2x - 5| < 7$
25. $|3x + 12| \geq 9$
26. $\left|1 - \dfrac{2}{x}\right| < 3$

Solve for the indicated letter.

27. $Q = mn^2$, for n
28. $Q = mn^3$, for n
29. $v = \dfrac{1}{2}\sqrt{1 + \dfrac{T}{L}}$, for L
30. $r = \sqrt[3]{\dfrac{3w}{4\pi d}}$, for w
31. $S = \dfrac{n(n+1)}{2}$, for n
32. $S = -16t^2 + v_0 t$, for t

For each of the following quadratic functions:

a. Determine the standard form.
b. Find the vertex.
c. Determine whether the graph opens upward or downward.
d. Find the maximum or minimum value.
e. Graph the function.

33. $f(x) = x^2 - 3x$
34. $f(x) = 2(x - 1)^2 + 4$
35. $f(x) = 2x^2 + 12x + 2$
36. $f(x) = \frac{1}{2}x^2 + 2x + 5$

Find the maximum or minimum value.

37. $f(x) = 2x^2 + 17x - 35$
38. $f(x) = \frac{1}{4}x^2 - 20x + \frac{2}{3}$

Applications

39. **Design of packaging.** A rectangular box with a volume of 560 cubic feet is to be constructed so that it has a square base and top, each with sides of length x. The cost per square foot for the bottom is $0.35, for the top is $0.20, and for the sides is $0.11. Let C denote the total cost of the box. Determine C as a function of x.

40. **Investment analysis.** Suppose $50,000 is invested at 8.4 percent. After 5 years, how much is in the account if interest is compounded (a) annually, (b) semiannually, (c) monthly, (d) daily? Assume there are 360 days in a year.

41. **Investment analysis.** A principal of $50,000 is invested for 2 years at an interest rate that is compounded annually. It grows to $59,405. What is the interest rate?

42. **Bank promotion.** To promote business, bank A offers to give a grandfather clock worth $595 to anyone who deposits over $5,000 at 6 percent for 10 years, where interest is compounded annually. Bank B offers no free gift, but compounds interest quarterly, also at 6 percent. Which bank has the better deal for a $5,000 deposit? Which has the better deal for a $10,000 deposit?

Chapter Test

1. Solve the following quadratic equation by factoring: $2x^2 + 5x - 3 = 0$

2. Use completing the square to solve the following quadratic equation: $x^2 - 14x = 7$

3. Use the quadratic formula to solve the following quadratic equations:
   a. $3x^2 - x - 2 = 0$
   b. $9x^2 + 6x + 1 = 0$

4. Solve the equation: $\dfrac{3x-1}{x+1} + 2 = x$

5. Solve the equation: $\sqrt{x+2} = x$

6. Solve the following equation for r in terms of a: $a = 4\pi r^2 + 3$

7. Solve the following inequalities:
   a. $|x - 2| < 3$
   b. $2|x + 5| > 4$

8. Determine the minimum value of the function $f(x) = (x+3)(x-1)$.

9. Determine the intercepts of the function $f(x) = 3x^2 + 5x - 2$

10. A small business is marketing a new tape recorder. It determines that its total revenue from the sale of x tape recorders (in hundreds) is given by the function
    $$R(x) = 44x - 2x^2$$
    (in hundreds of dollars). The firm also determines that its total cost of producing x tape recorders (in hundreds) is given by
    $$C(x) = 200 - 36x$$
    (in hundreds of dollars). Total profit is given by:
    $$P(x) = R(x) - C(x)$$
    a. Find the total profit.
    b. How many tape recorders must the company produce and sell to maximize profit?
    c. What is the maximum profit?

11. **Thought question.** Explain what is meant by the term *maximum point* of a graph.

12. **Thought question.** Does every graph have a maximum point? Explain your answer.

13. **Thought question.** Give an illustration of how the absolute value function is used in an applied setting.

CHAPTER 4
POLYNOMIAL AND RATIONAL FUNCTIONS

As we have seen in the preceding chapter, linear and quadratic polynomials can be used to model a variety of interesting situations and, in some cases, lead to optimization problems. In this chapter, we learn more about the properties of polynomials of higher degree. Knowing the zeros of a polynomial helps us to sketch its graph. Moreover, there is a fundamental connection between zeros of a polynomial and long division. So our study of polynomials concentrates on their zeros and leads us beyond the real number system, into the complex numbers.

Chapter Opening Image: *A black hole is a collapsed star that has a gravitational field so strong that it traps all light the star emits. This image is a computer simulation of a black hole being distorted by a gravitational field in its vicinity.*

4.1 POLYNOMIAL FUNCTIONS OF DEGREE GREATER THAN 2

As we have already seen, the graph of a linear function is a straight line and the graph of a quadratic function is a parabola. Many applied problems, especially those in science and engineering, require us to deal with polynomial functions of degree 3 or higher. In this section, we develop some elementary methods for sketching the graphs of such functions.

Graphs of the Functions $y = ax^n$

Among the most elementary polynomial functions are those of the form

$$y = ax^n$$

where n is a positive integer and a is a real constant. We begin our study of the graphs of polynomial functions by examining the graphs of these functions.

Consider the function

$$f(x) = x^3$$

To get a feel for the behavior of the function, consider the following table of values:

x	f(x)
0	0
0.2	0.008
0.5	0.125
1	1
2	8
3	27

We see that as x increases, so does $f(x)$. In fact, as x increases without bound, the value of $f(x)$ does the same. So we draw the graph heading steadily upward without bound as x increases.

Next, we note that the function is odd:

$$f(-x) = (-x)^3 = -x^3 = -f(x)$$

Therefore, its graph is symmetric with respect to the origin, so the portion of the graph for $x < 0$ is obtained by reflecting the portion for $x > 0$ in the origin. Using a graphing calculator, we can display the graph, which is shown in Figure 1.

Starting from the graph in Figure 1, we can sketch the graphs of many other functions using the geometric transformations of scaling, reflection, horizontal translation, and vertical translation. The next example illustrates some of the possibilities.

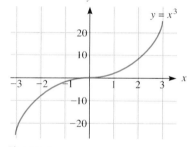

Figure 1

EXAMPLE 1 Graphing Cubics

Sketch the graphs of the following functions.

1. $f(x) = 3x^3$
2. $f(x) = -x^3$
3. $f(x) = (x - 1)^3$
4. $f(x) = (x + 2)^3 + 3$
5. $f(x) = -2(x - 1)^3 - 4$

Solution

1. To sketch the graph of this function, we scale the graph of Figure 1 using a factor of 3. See Figure 2.
2. To graph this function, we reflect the graph of Figure 1 in the x-axis because the scaling factor is negative. See Figure 3.

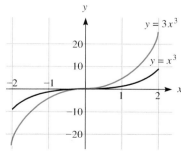

Figure 2

206 CHAPTER 4 Polynomial and Rational Functions

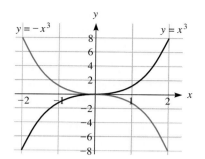
Figure 3

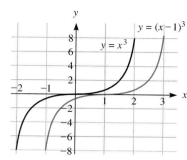
Figure 4

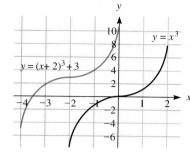
Figure 5

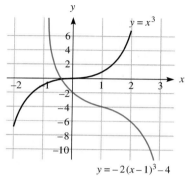
Figure 6

3. We translate the graph of Figure 1 by 1 unit to the right. See Figure 4.
4. We translate the graph of Figure 1 by 2 units to the left and 3 units upward. See Figure 5.
5. We perform the following geometric transformations to Figure 1: First, we scale the graph using a factor of 2. Next, we reflect the graph in the x-axis. Finally, we translate the graph 1 unit to the right and 4 units downward. See Figure 6.

Let's consider the graphs of the functions:

$$y = x^n, \quad n \geq 2$$

When n is odd, the function is odd, just as we observed in Example 1, in which $n = 3$. For n odd, the graph has the same general shape as the graph $y = x^3$. In Figure 7, we show the graphs corresponding to several small, odd values of n.

If n is even, the function

$$y = x^n$$

is even, like the function $y = x^2$. In this case, the graph is symmetric with respect to the y-axis and has the general shape of a parabola. In Figure 8, we have sketched the graphs that correspond to several small, even values of n. Note that as n increases, the graph tends to be flatter for x between -1 and 1 and to rise more steeply for $x < -1$ and $x > 1$.

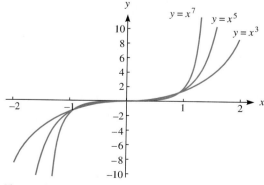
Figure 7

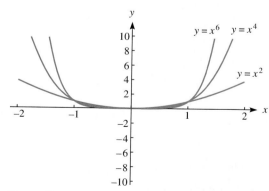
Figure 8

SECTION 4.1 Polynomial Functions of Degree Greater than 2

By using scaling, reflection, and translation, and by referring to the basic graphs in Figures 7 and 8, we can sketch the graphs of all functions of the form:

$$y = ax^n, \quad n \geq 2$$

Graphs of Other Polynomial Functions

Let's now turn our attention to graphing more general polynomial functions. A full discussion of this subject belongs in calculus, where the tools necessary to describe the various specific geometric features of these graphs are developed. However, using a few fundamental ideas, we can make some inroads into the problem of sketching graphs of polynomial equations, and we can come up with rough sketches of such graphs.

Definition
Zero of a Polynomial

Let $f(x)$ be a polynomial function. A real number α_0 for which

$$f(\alpha_0) = 0$$

is called a **zero** of f.

Suppose that x_0 is a real zero of f. Then the point $(x_0, 0)$ is an x-intercept of the graph of f. (See Figure 9.) In general, a polynomial function $f(x)$ has a number of zeros, which can be found as solutions to the polynomial equation $f(x) = 0$. Later in the chapter, we discuss these zeros more fully and develop techniques for determining them.

Figure 10 shows the graph of a polynomial function of degree greater than 2. Notice that the graph consists of a number of peaks and valleys. A point on the graph that is at either the top of a peak or the bottom of a valley is called a **maximum point** or a **minimum point**. Each of these is called a **turning point** or an **extreme point**. Using calculus, it is possible to prove that a polynomial of degree n has at most $n - 1$ extreme points.

Note that in the interval between two consecutive extreme points, the graph is either increasing or decreasing. If the graph changed from increasing to decreasing or vice-versa, then there must be an extreme point within the interval. (The proof of this fact is part of calculus.)

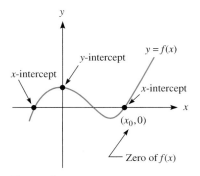

Figure 9

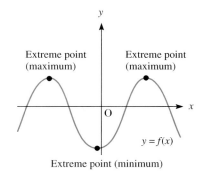

Figure 10

Here is another result that is useful in sketching graphs of polynomials (again, the proof is derived from calculus):

INTERMEDIATE VALUE THEOREM

Suppose that $f(x)$ is a polynomial with real coefficients and that a and b are real numbers with $a < b$. Suppose that one of $f(a)$ and $f(b)$ is positive and the other is negative. Then $f(x)$ has a real zero in the interval (a, b).

As a consequence of the Intermediate Value theorem, we conclude that if a and b are two consecutive real zeros of f, then for x between a and b, the values of f are all of one sign, either all positive or all negative. (If not, then by the Intermediate Value theorem, there would be a zero between the supposedly consecutive zeros.) We can determine which sign prevails by evaluating f for a **test value** lying between a and b. Using the signs of f in each of the intervals determined by consecutive real zeros allows us to produce a rough sketch of the graph, as shown in the following two examples.

EXAMPLE 2
Graphing a Polynomial Function

Sketch the graph of the function:
$$f(x) = 3(x + 2)(x - 1)(x - 4)$$

Solution

The zeros of f are determined as the solutions of the equation
$$3(x + 2)(x - 1)(x - 4) = 0$$

A product can be zero only if one of the factors is 0. In this case, that means that either $x + 2 = 0$, $x - 1 = 0$, or $x - 4 = 0$. That is, the zeros of x are -2, 1, and 4. These three real zeros divide the x-axis into four intervals, namely:

$$(-\infty, -2), \quad (-2, 1), \quad (1, 4), \quad (4, +\infty)$$

In each of these intervals, we choose a test value and determine the sign of f at the test value. The results are summarized in the following table:

Interval	Test Value x	f(x)	Sign of f
$(-\infty, -2)$	-3	$3(-1)(-4)(-7)$	$-$
$(-2, 1)$	0	$3(2)(-1)(-4)$	$+$
$(1, 4)$	2	$3(4)(1)(-2)$	$-$
$(4, +\infty)$	5	$3(7)(4)(1)$	$+$

Now we can sketch the graph. We begin by plotting the points corresponding to the x-intercepts, namely the points $(-2, 0)$, $(1, 0)$, $(4, 0)$. Next, we consult the table of test intervals. The sign of f in a test interval determines whether the graph is above or below the x-axis in that interval. For the first interval listed, the value of f is negative. So the graph lies below the x-axis throughout the interval. As x becomes large, either positively or negatively, the values of f become arbitrarily large in absolute value. (This is a general property of polynomial functions. We omit the proof here.) Because f is negative in the first interval, the values must get arbitrarily larger in the negative direction as x becomes larger in the negative direction. This allows us to sketch the graph of f for x in the first interval.

In the second interval, x is positive. At the left and right endpoints of the interval, the value of $f(x)$ is 0. This means that the graph must have a turning

SECTION 4.1 Polynomial Functions of Degree Greater than 2

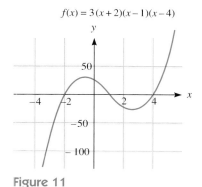

$f(x) = 3(x+2)(x-1)(x-4)$

Figure 11

point somewhere in the interval. Knowing this allows us to make a sketch of the graph for x in the second interval. (We have drawn the graph with one turning point in this interval. It might have more than one, but without using calculus, we have no way of knowing.)

In the third interval, f is negative. Because again both endpoints give the value 0 for $f(x)$, we can use the same reasoning as above to sketch the graph corresponding to x in this interval.

Finally, the last interval includes values of x that are arbitrarily large. For these values, $f(x)$ grows arbitrarily large. And because the sign of $f(x)$ is positive, the growth is in the positive direction. The final sketch of the graph is shown in Figure 11. You can check the above reasoning by using a graphing calculator to display the graph.

EXAMPLE 3
Graphing a Fourth-Degree Polynomial

Sketch the graph of the function:
$$f(x) = x^4 - 4x^2$$

Solution

Factor the expression on the right to obtain
$$f(x) = x^2(x^2 - 4) = x^2(x+2)(x-2)$$
Examining the factors on the right, we see that the zeros of f are -2, 0, and 2. These divide the x-axis into four intervals, as shown in the following table. As in the preceding example, we choose a test value in each interval and determine the sign of f at the test value. The results are summarized in the table:

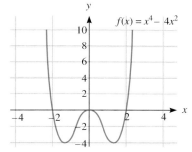

$f(x) = x^4 - 4x^2$

Figure 12

Interval	Test Value x	$f(x)$	Sign of f
$(-\infty, -2)$	-3	$(-3)^2(-1)(-5)$	$+$
$(-2, 0)$	-1	$(-1)^2(1)(-3)$	$-$
$(0, 2)$	1	$(1)^2(3)(-1)$	$-$
$(2, +\infty)$	3	$(3)^2(1)(5)$	$+$

In Figure 12, we sketch the graph of f by first plotting the zeros and then drawing a section of the graph that corresponds to each of the intervals in the table of test intervals. The accuracy of the graph is enhanced by noting that the function $f(x)$ is even, so that the graph is symmetric with respect to the y-axis.

Let's now illustrate how graphing calculators can be applied to solve problems involving polynomials.

EXAMPLE 4

Using a Graphing Calculator to Graph a Fourth-Degree Polynomial

Use a graphing calculator to graph the following fourth-degree polynomial function and to determine the number of maximum and minimum points.
$$f(x) = x^4 + x^3 - 2x^2 + x - 1$$

Solution

The graph is shown in Figure 13. There is a single minimum point and no maximum point. Note that, by using the graphing calculator to produce the graph, there is no need to factor the polynomial, as we did in the preceding example. Factoring can often be a difficult or impossible task.

In the preceding chapter, we learned to solve optimization problems, which involve determining the maximum or minimum point of a quadratic function. Many applied problems require us to determine maximum or minimum points of other functions. (Calculus teaches techniques for solving such problems.) Using a

Figure 13

EXAMPLE 5
Solving an Optimization Problem by Using a Graphing Calculator

Determine the maximum and minimum points of the function $f(x) = x^3 - 3x + 5$.

Solution

Graph the function to obtain the display in Figure 14. The graph has one maximum point and one minimum point. To determine the coordinates of these points, we use the trace function. To obtain greater accuracy, we use the [Box] command to enlarge the areas of the graph around the maximum and minimum points. In Figure 15, we trace the curve near the minimum point, the coordinates of which are approximately (1.0094183, 3.0002669). In Figure 16, we trace the curve near the maximum point, the coordinates of which are approximately (−0.9894737, 6.9996688).

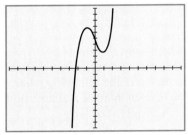

Figure 14

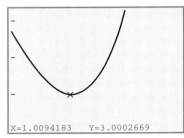

Figure 15

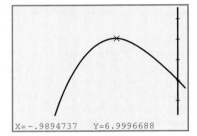

Figure 16

Applications   Let's now examine some polynomials that arise from real-world data.

EXAMPLE 6
Driving Safety

The Wall Street Journal studied the Department of Transportation's fatal accident reports and mileage statistics for 1990. The newspaper reported that fatalities in motor vehicle accidents at first decrease as driver age increases, and then increase as drivers reach old age. If x denotes the age of the driver and y denotes the number of fatalities caused by drivers of age x per hundred million miles driven, then x and y are related by the model:

$$y = A(x - 51.5)^4 + 3.5, \qquad A = 1.3101 \times 10^{-5}$$

Graph this model for $18 \leq x \leq 85$.

Solution

Key the model into your calculator and graph the function. (Because the model is complex and the microprocessor in your calculator is not all that powerful, graphing will take some time. Be patient.) The graph is shown in Figure 17.

Figure 17

EXAMPLE 7
Sale of Cold Remedies

Based on data reported in the *Wall Street Journal*, the sales of cold and cough remedies throughout a flu season can be described by a polynomial model. Let x denote the number of weeks after September 27 and let y denote the total sales of cold and cough remedies, in millions of dollars. Then x and y are related by the model:

$$y = 11.25 + 0.9597x + 0.5039x^2 - 0.04133x^3 + 0.0007916x^4$$

Graph this equation for $0 \leq x \leq 28$. (The flu season lasts 28 weeks.)

Solution

Key the model into your calculator. The graph is shown in Figure 18.

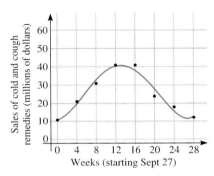

Figure 18

EXAMPLE 8
Coffee Consumption

Daily coffee consumption in the United States has varied considerably over the years. A model describing the number of cups per day consumed by the average adult in year x (with 1955 corresponding to $x = 0$) can be described by the model:

$$y = 2.76775 + 0.0847943x - 0.00832058x^2 + 0.000144017x^3$$

1. Graph the equation to show daily coffee consumption from 1955 through 1992.
2. In what year was daily coffee consumption least?
3. In what year was coffee consumption greatest?

Solution

1. Using a graphing calculator, we arrive at the graph in Figure 19.
2. By magnifying the graph, we find that the minimum point occurs approximately at $x = 32.5$. So daily coffee consumption was lowest in 1987.

   By magnifying the graph, we find that the maximum point occurs approximately at $x = 6$. So daily coffee consumption was highest in 1961.

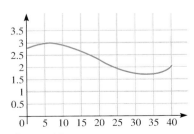

Figure 19

Exercises 4.1

Match the following functions with the corresponding graph.

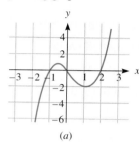
(a)

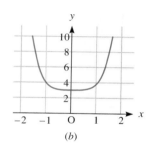
(b)

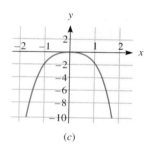
(c)

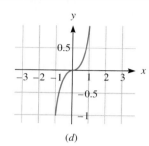
(d)

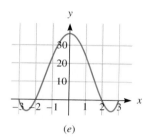
(e)

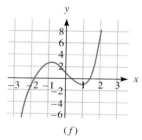
(f)

1. $f(x) = x^3$
2. $f(x) = x^4 + 3$
3. $f(x) = x(x+1)(x-2)$
4. $f(x) = x^3 - 3x + 1$
5. $f(x) = -x^4$
6. $f(x) = (x^2 - 4)(x^2 - 9)$

Sketch graphs of each of the following functions. Do not use a graphing calculator.

7. $f(x) = \frac{1}{2}x^3$
8. $f(x) = 4x^3$
9. $f(x) = (x-2)^3 + 1$
10. $f(x) = -(x+1)^3 - 4$
11. $f(x) = x(x-1)(x+1)$
12. $f(x) = (x+3)(x-1)(x+2)$
13. $f(x) = 0.25x^4$
14. $f(x) = -x^4$
15. $f(x) = -x^5$
16. $f(x) = -x^6$
17. $f(x) = x^3 - 3x + 22$
18. $f(x) = x^3 - x^2 - 2x$
19. $f(x) = x^4 - 2x^2$
20. $f(x) = x^4 - 6x^2$
21. $f(x) = (x-2)^2(x+3)$
22. $f(x) = x^2(x+2)(x^2+1)$

Applications

You may use a graphing calculator to solve the following problems.

23. **Maximizing volume.** From a thin piece of 8-inch by 8-inch cardboard, square corners are cut out so that the sides can be folded up to make a box (see figure). Let V denote the volume of the box.

Exercise 23

a. Determine a polynomial function in one variable for $V(x)$.
b. Graph the function on the interval $[0, 4]$.
c. Use the graph to estimate the maximum value of the function and the value of x at which it occurs.

24. **Deflection of a beam.** A beam rests at two points, and a concentrated load is applied to its center. Let y denote the deflection of the beam at a distance of x units measured from the beam to the left of the load. The deflection depends on the elasticity of the board, the load, and other physical characteristics. Suppose that, under certain conditions, y is given by:

$$y = \frac{1}{12}x^3 - \frac{1}{16}x$$

a. Graph the function over the interval [0, 10].

b. Use the graph to estimate the minimum value of the function and the value of x at which it occurs.

25. **Maximizing profit.** The following are total revenue and total cost functions:

$$R(x) = 100x - x^2; \quad C(x) = \frac{1}{3}x^3 - 6x^2 + 89x + 100$$

a. Total profit is given by $P(x) = R(x) - C(x)$. Find $P(x)$.

b. Graph $R(x)$, $C(x)$, and $P(x)$ using the same set of axes.

c. Estimate the value or values of x at which $P(x)$ is at a maximum. Estimate the maximum value(s) of $P(x)$.

26. **Medical dosage.** The function

$$N(t) = -0.045t^3 + 2.063t + 2$$

gives the body's concentration, in parts per million, of a certain dosage of medication after time t.

a. Graph the function.

b. Estimate the maximum value of the function and the time t at which it occurs.

c. The *minimum effective dosage* of the medication occurs for values of t for which $N(t) \geq 2$. Use the graph to estimate the times at which the medication has its minimum effective dosage.

27. **Driving fatalities.** Refer to Example 6. At what age is a driver least likely to cause a fatality?

28. **Driving fatalities.** Refer to Example 6. At what age after 70 is a driver as likely to cause a fatality as an 18-year-old?

29. **Sale of cold and cough remedies.** Refer to Example 7. At what point in the flu season do the sales of cold and cough remedies equal $15 million a week?

30. **Sale of cold and cough remedies.** Refer to Example 8. What is the maximum amount of cold and cough remedies sold?

Technology

Graph each of the following functions using a graphing calculator. Use the resulting graph to estimate the x-intercepts.

31. $f(x) = x^3 - 3x^2 - 144x - 140$
32. $f(x) = x^4 - 2x^3$
33. $f(x) = 6x^5 - 24x^3$
34. $f(x) = x^2(x + 5)(x^2 + 3x + 5)^2$

35. **Doctor's visits as a function of age.** The number y of doctor's visits per year by a patient of age x is given by the model:

$$y = 6.95 - 0.3x + 0.0083x^2 - 0.00002x^3$$

a. Graph this function.

b. At what age are doctor's visits minimized?

36. **Baseball salaries.** The average salary S of a baseball player (in thousands of dollars per year) in year x (with 1982 corresponding to 0) is given by the model:

$$S = 246 + 64x - 8.9x^2 + 0.95x^3$$

a. Graph this function.

b. By how much did baseball salaries increase from 1990 to 1993?

In your own words

37. Explain what is meant by a zero of a function. Use a specific function to illustrate your explanation.

38. Describe the graph of the function $f(x) = x^n$ for even n.

39. Describe the graph of the function $f(x) = x^n$ for odd n.

40. What is a maximum point of a function?

41. What is a minimum point of a function?

4.2 DIVISION OF POLYNOMIALS

Long Division of Polynomials

We can divide two polynomials to produce a quotient and a remainder. As an illustration, consider the quotient:

$$\frac{x^2 + 3x + 1}{x}$$

In order to divide a polynomial by a monomial, we divide each term of the polynomial by the monomial:

$$\frac{x^2+3x+1}{x} = \frac{x^2}{x} + \frac{3x}{x} + \frac{1}{x} = x+3+\frac{1}{x}$$

We can express this in terms of polynomials by multiplying each term by the denominator x to obtain

$$x^2 + 3x + 1 = x(x+3) + 1$$

The polynomial $x+3$ is the quotient and 1 is the remainder.

We can calculate the quotient and remainder using the process of **long division of polynomials**, which is included in introductory algebra courses. The next example recalls this procedure.

EXAMPLE 1
Long Division

Calculate the quotient and remainder when $f(x) = 2x^3 + 5x^2 - 10x - 7$ is divided by $g(x) = 2x - 1$.

Solution

Because we seek the quotient and remainder for the division $f(x)/g(x)$, we organize the calculation as a long division problem of $f(x)$ divided by $g(x)$.

$$
\begin{array}{r}
x^2 + 3x - \frac{7}{2} \\
2x-1 \,\overline{\smash{)}\, 2x^3 + 5x^2 - 10x - 7} \\
\underline{2x^3 - x^2} \\
6x^2 - 10x \\
\underline{6x^2 - 3x } \\
-7x - 7 \\
\underline{-7x + \frac{7}{2}} \\
-\frac{21}{2}
\end{array}
$$

(Quotient above; Remainder $-\frac{21}{2}$)

Therefore, in this example, the quotient $q(x)$ and the remainder $r(x)$ are:

$$q(x) = x^2 + 3x - \frac{7}{2}, \qquad r(x) = -\frac{21}{2}$$

Note that the degree of the denominator $g(x)$ is 1 and the degree of the remainder is 0, which is less than the degree of $g(x)$.

The division algorithm states that a computation like the one just carried out can always be carried out for the quotient $f(x)/g(x)$.

> **DIVISION ALGORITHM FOR POLYNOMIALS**
>
> Suppose that $f(x)$ and $g(x)$ are polynomials, and that $g(x)$ is not the zero polynomial. Then there are polynomials $q(x)$ and $r(x)$ such that:
>
> 1. $f(x) = g(x)q(x) + r(x)$
> 2. Either $r(x)$ is the zero polynomial or the degree of $r(x)$ is less than the degree of $g(x)$.
>
> The polynomial $q(x)$ is the quotient and the polynomial $r(x)$ is the remainder when $f(x)$ is divided by $g(x)$.

SECTION 4.2 Division of Polynomials

The quotient and remainder in the division algorithm can be determined using long division, as in Example 1.

Factoring and Zeros Let's now consider the special case of a polynomial $f(x)$ that is divided by a linear polynomial $x - a$, in which the leading coefficient is 1. By the division algorithm, we can express the division in the form

$$f(x) = (x - a)q(x) + r(x)$$

where $q(x)$ is the quotient and $r(x)$ is the remainder. From the statement of the division algorithm, either $r(x)$ is the zero polynomial or the degree of $r(x)$ is less than the degree of $x - a$, which is 1. In either case, $r(x)$ is a constant polynomial.

We can determine the value of this constant by replacing x in the above equation by a. We then obtain:

$$f(a) = (a - a)q(a) + r(a)$$
$$= 0 + r(a)$$
$$= r(a)$$

Thus, the value $r(a)$ is equal to $f(a)$. But because $r(x)$ is a constant polynomial, this means that $r(x)$ is equal to $f(a)$. In other words, we have the following theorem:

REMAINDER THEOREM

Let $f(x)$ be a polynomial. Then the remainder after dividing $f(x)$ by $x - a$ is the constant polynomial $f(a)$. That is,

$$f(x) = (x - a)q(x) + f(a)$$

where $q(x)$ is the quotient.

As the following example shows, using the Remainder theorem to determine the remainder is usually more efficient than using long division.

EXAMPLE 2
Calculating the Remainder

Suppose that:
$$f(x) = -x^4 + 5x^2 - 2$$

What is the remainder when $f(x)$ is divided by $x - 3$?

Solution

By the Remainder theorem, the remainder is the constant polynomial $f(3)$, which is:

$$-(3)^4 + 5(3)^2 - 2 = -81 + 45 - 2 = -38$$

That is, the remainder is the constant polynomial $r(x) = -38$.

Recall that a zero of f is a number a for which $f(a) = 0$. Let's use the Remainder theorem to obtain some information about the zeros of f. Because the remainder is 0, we have:

$$f(x) = (x - a)q(x)$$

That is, f has $x - a$ as a factor. The same argument works in reverse: If f has $x - a$ as a factor, then
$$f(x) = (x - a)q(x)$$
so that the remainder of f on dividing by $x - a$ is 0. Thus, we have the following fundamental connection between factoring and zeros:

FACTOR THEOREM

Let $f(x)$ be a polynomial. Then $x - a$ is a factor of $f(x)$ if and only if a is a zero of $f(x)$.

We used a special case of this theorem when we studied the method of factoring for solving quadratic equations. In this method, we factored the quadratic polynomial into linear factors and set each factor equal to 0. Each linear factor of the quadratic polynomial corresponds to one solution of the quadratic equation.

EXAMPLE 3
Applying the Factor Theorem

Suppose that $f(x)$ is a cubic polynomial with zeros $\frac{1}{2}$, 1, and 3. Further, suppose that $f(0)$ is equal to 4. Determine $f(x)$.

Solution

Because $\frac{1}{2}$, 1, and 3 are all zeros of $f(x)$, the Factor theorem asserts that $f(x)$ has as factors each of the polynomials $x - \frac{1}{2}$, $x - 1$, and $x - 3$. Therefore, $f(x)$ has as a factor the polynomial:
$$\left(x - \frac{1}{2}\right)(x - 1)(x - 3)$$
And because $f(x)$ is a cubic polynomial,
$$f(x) = c\left(x - \frac{1}{2}\right)(x - 1)(x - 3)$$
for some constant c. We are also given the fact that $f(0) = 4$. To use this fact, we substitute 0 for x into the above equation for $f(x)$ to obtain:
$$4 = f(0) = c\left(0 - \frac{1}{2}\right)(0 - 1)(0 - 3)$$
$$4 = -\frac{3}{2}c$$
Solving this equation for c gives us:
$$c = -\frac{8}{3}$$
Finally, substituting this back into the formula for $f(x)$ gives us:
$$f(x) = -\frac{8}{3}\left(x - \frac{1}{2}\right)(x - 1)(x - 3)$$
This answer is acceptable as it stands. Or, if you wish, you can multiply out the right-hand side to obtain the cubic polynomial expression for $f(x)$. However, this expression is preferable since it is clear just by looking at it that it satisfies the conditions specified in the problem.

If a polynomial f has different zeros $a_1, a_2, \ldots, a_k$, then, by the Factor theorem, f is divisible by:

$$(x - a_1)(x - a_2) \cdots (x - a_k)$$

In particular, because this last product has degree k, the degree of f must be at least equal to k, the number of different zeros. That is, we have the following:

NUMBER OF ZEROS THEOREM

A polynomial of degree n has at most n different zeros.

EXAMPLE 4
Using the Factor Theorem in Modeling

During a recession, housing sales typically fall and then increase during the following recovery. Suppose that the sales reported by a local board of realtors x months after the beginning of a recession can be approximated by a quadratic function $f(x)$. Suppose that 230 sales are recorded in month 3 and in month 9, and that the number of sales at the beginning of the recession is 400 houses per month. Determine the function $f(x)$.

Solution

Because $f(x)$ has the value 230 for $x = 3$ and for $x = 9$, we can write

$$f(x) = 230 + g(x)$$

where $g(x)$ is a quadratic polynomial that is 0 for $x = 3$ and $x = 9$. By the Factor theorem, we can write

$$g(x) = A(x - 3)(x - 9)$$

for some real number A. That is, we have:

$$f(x) = 230 + A(x - 3)(x - 9)$$

Furthermore, we know that $f(0) = 400$, so we can solve for A and substitute that value back into the equation:

$$400 = f(0) = 230 + A(0 - 3)(0 - 9)$$
$$400 = 230 + 27A$$
$$A = \frac{170}{27}$$

$$f(x) = 230 + \frac{170}{27}(x - 3)(x - 9)$$
$$= 400 - \frac{680}{9}x + \frac{170}{27}x^2$$

Synthetic Division

In the preceding discussion, we showed the significance of dividing a polynomial $f(x)$ by polynomials of the form $x - a$. The calculational procedure we have shown for performing this division is often tedious and time-consuming. There is a simple algorithm, called **synthetic division**, for carrying out such calculations in a much simpler and faster way.

Let's begin by working out an example and showing how the efficient organization of long division leads naturally to synthetic division. Consider the problem of dividing

$$3x^5 - 2x^4 - 5x^3 + x^2 - x + 4$$

by $x - 2$. Here is the traditional method of performing the division.

$$
\begin{array}{r}
3x^4 +4x^3 +3x^2 +7x +13 \\
x-2 \overline{\smash{\big)}\, 3x^5 -2x^4 -5x^3 +x^2 -x +4} \\
\underline{3x^5 -6x^4} \\
4x^4 -5x^3 \\
\underline{4x^4 -8x^3} \\
3x^3 +x^2 \\
\underline{3x^3 -6x^2} \\
7x^2 -x \\
\underline{7x^2 -14x} \\
13x +4 \\
\underline{13x -26} \\
30
\end{array}
$$

To simplify this computation, the first thing to note is that the calculations are completely contained in the various coefficients. It is not really necessary to include the variables at all. Moreover, because we will always be dividing by a polynomial of the form $x - a$, we can omit the coefficient 1, which corresponds to x. This leaves a computation of the following form:

$$
\begin{array}{r}
3 4 3 7 13 \\
-2 \overline{\smash{\big)}\, 3 -2 -5 1 -1 4} \\
\underline{3 -6} \\
4 -5 \\
\underline{4 -8} \\
3 1 \\
\underline{3 -6} \\
7 -1 \\
\underline{7 -14} \\
13 4 \\
\underline{13 -26} \\
30
\end{array}
$$

Note that this array of numbers still contains a lot of duplicate information. Consider the first step of the division, in which we obtain the first 3 in the quotient (top row). This is just a duplicate of the first 3 in the dividend. (The dividend is in the second row, which is the first row inside the division sign.) Moreover, the third row duplicates the 3 again. Let's simplify the computation by writing the 3 only twice, in the quotient and in the dividend. The third row then consists of the single number -6, which is computed as 3 (from the quotient) multiplied by the -2 in the divisor (which is left of the division sign). The next entry in the quotient, 4, is calculated by subtracting -6 from -2. But we only write 4 once, in the quotient. The fourth row is determined by bringing down the -5 from the dividend, and the fifth row is the product of 4 (from the quotient) and -2 (from the dividend). And so forth. This abbreviated computation is shown below:

$$
\begin{array}{r|rrrrrr}
 & 3 & 4 & 3 & 7 & 13 & \\
\hline
-2 & 3 & -2 & -5 & 1 & -1 & 4 \\
 & -6 & & & & & \\
\hline
 & & -5 & & & & \\
 & & -8 & & & & \\
\hline
 & & & 1 & & & \\
 & & & -6 & & & \\
\hline
 & & & & -1 & & \\
 & & & & -14 & & \\
\hline
 & & & & & 4 & \\
 & & & & & -26 & \\
\hline
 & & & & & & 30 \\
\end{array}
$$

Rather than subtracting numbers at each stage, let's perform addition. This can be done by changing the sign of the -2 in the divisor. The multiplications will then result in the negatives of the previous results, so we can add the entries at each step. The computation now looks like this:

$$
\begin{array}{r|rrrrrr}
 & 3 & 4 & 3 & 7 & 13 & \\
\hline
2 & 3 & -2 & -5 & 1 & -1 & 4 \\
 & & 6 & & & & \\
\hline
 & & -5 & & & & \\
 & & & 8 & & & \\
\hline
 & & & 1 & & & \\
 & & & & 6 & & \\
\hline
 & & & & -1 & & \\
 & & & & & 14 & \\
\hline
 & & & & & 4 & \\
 & & & & & & 26 \\
\hline
 & & & & & & 30 \\
\end{array}
$$

Let's now compress the entire display into three lines, as follows:

$$
\begin{array}{r|rrrrrr}
 & 3 & 4 & 3 & 7 & 13 & \\
\hline
2 & 3 & -2 & -5 & 1 & -1 & 4 \\
 & & 6 & 8 & 6 & 14 & 26 & 30 \\
\end{array}
$$

Finally, let's move the quotient from the top row to the bottom row, and combine it with the remainder, 30, in a single row. Moreover, tradition dictates that we turn the division sign inside out. (Of course, this doesn't affect the calculation.)

$$
\begin{array}{r|rrrrrr}
2 & 3 & -2 & -5 & 1 & -1 & 4 \\
 & & 6 & 8 & 6 & 14 & 26 \\
\hline
 & 3 & 4 & 3 & 7 & 13 & 30 \\
\end{array}
$$

Note how simply the bottom row can be calculated: The initial 3 in the quotient is a copy of the 3 in the dividend in the first row. Then compute 6 as 2 (divisor) multiplied by 3 (quotient). Compute 4 by adding -2 and 6. Now compute 8 as 2 (divisor) multiplied by 4 (quotient). And so forth. The quotient corresponds to the terms of the last row, except for the last, which is the remainder.

This procedure can be used to calculate the quotient of any polynomial $f(x)$ divided by $x - a$.

SYNTHETIC DIVISION

To divide $f(x)$ by $x - a$ using synthetic division:

1. Write the coefficients of f across the top row. Make sure to include a coefficient for each power of x. For powers that don't explicitly appear, write 0.
2. Start with the leftmost column. Bring down the leading coefficient into the third row.
3. Multiply the new entry in the third row by a.
4. Move one column to the right and put the result of step 3 in the second row.
5. Add rows 1 and 2 of the current column and put the result in the third row.
6. Repeat steps 3–5 for each of the columns in turn.
7. The numbers in the third row are the coefficients of the quotient, except for the rightmost number, which is the remainder.

The following examples illustrate the mechanics of synthetic division and show how it can be applied to get information about a polynomial $f(x)$.

EXAMPLE 5
Synthetic Division

Use synthetic division to determine the quotient and remainder when $f(x) = 4x^5 - 2x^4 + x^3 - 7x^2 + 3$ is divided by $x - 2$.

Solution

We use the algorithm for synthetic division. Note that we have included a zero to represent the zero coefficient of the x term, because there must be one column corresponding to each power of x in the dividend. If a power of x is missing, we must use 0 as a placeholder in the corresponding column. Be careful—leaving out a 0 placeholder is an easy mistake to make. Because we are dividing by $x - 2$, we change the sign of the -2 and put 2 on the left side of the top row.

$$
\begin{array}{r|rrrrrr}
2 & 4 & -2 & 1 & -7 & 0 & 3 \\
  &   & 8 & 12 & 26 & 38 & 76 \\
\hline
  & 4 & 6 & 13 & 19 & 38 & 79
\end{array}
$$

The coefficients of the quotient are given by the first five entries in the third row of the table, making the quotient

$$q(x) = 4x^4 + 6x^3 + 13x^2 + 19x + 38$$

The last entry in the table, 79, gives the remainder. That is, we have:

$$4x^5 - 2x^4 + x^3 - 7x^2 + 3 = (4x^4 + 6x^3 + 13x^2 + 19x + 38)(x - 2) + 79$$

EXAMPLE 6
Using Synthetic Division to Evaluate a Polynomial

Use synthetic division to determine $f(-3)$, where
$$f(x) = -x^4 - 5x^3 + 4x^2 - 9x + 10$$

Solution

By the Remainder theorem, the value of $f(-3)$ is the remainder when $f(x)$ is divided by $x - (-3)$. We use synthetic division to determine the value of this remainder.

$$\begin{array}{r|rrrrr}
-3 & -1 & -5 & 4 & -9 & 10 \\
 & & 3 & 6 & -30 & 117 \\
\hline
 & -1 & -2 & 10 & -39 & 127
\end{array}$$

The last entry in the table, the remainder, provides the value of $f(-3)$. That is,
$$f(-3) = 127$$

EXAMPLE 7
Manufacturing Cellular Phones

The cost of producing x cellular phones is given by the function:
$$C(x) = 0.000001x^3 + 0.01x^2 + 23x + 575{,}000$$

Use synthetic division to determine the cost of manufacturing 10,000 cellular phones.

Solution

The cost of manufacturing 10,000 cellular phones equals $C(10{,}000)$. By the Remainder theorem, this number is the remainder in dividing $C(x)$ by $x - 10{,}000$. And this remainder can be obtained by synthetic division (a calculator was used to determine the entries in the synthetic division table):

$$\begin{array}{r|rrrr}
10{,}000 & 0.000001 & 0.01 & 23 & 575{,}000 \\
 & & 0.01 & 200 & 2{,}230{,}000 \\
\hline
 & 0.000001 & 0.02 & 223 & 2{,}805{,}000
\end{array}$$

That is, the total cost of producing 10,000 cellular phones is $2,805,000.

Exercises 4.2

For Exercises 1–6, a polynomial $f(x)$ and a divisor $g(x)$ are given. Calculate the quotient $q(z)$ and the remainder $r(x)$ by long division.

1. $f(x) = x^5 - 2x^4 + x^3 - 5$,
   $g(x) = x - 2$

2. $f(x) = 2x^5 - 3x^3 + 2x^2 - x + 3$,
   $g(x) = x + 1$

3. $f(x) = 3x^4 - x^3 + 2x - 6$,
   $g(x) = 3x - 2$

4. $f(x) = 2x^5 - 3x^3 + 2x^2 - x + 3$,
   $g(x) = 4x + 5$

5. $f(x) = x^5 - 2x^4 + x^3 - 5$,
   $g(x) = x^2 - 2$

6. $f(x) = 2x^5 - 3x^3 + 2x^2 - x + 3$,
   $g(x) = x^2 + 1$

7. Suppose that $f(x)$ is a quadratic polynomial with zeros -1 and 2, and suppose that $f(3) = -10$. Determine f.

8. Suppose that $f(x)$ is a quadratic polynomial with zeros 5 and -2, and suppose that $f(1) = 7$. Determine f.

9. Suppose that $f(x)$ is a cubic polynomial with zeros -1, 0, and 1, and suppose that $f(2) = 24$. Determine f.

10. Suppose that $f(x)$ is a cubic polynomial with zeros -1, 0, and 2, and suppose that $f(1) = -24$. Determine f.

Use synthetic division to find the quotient and remainder when $f(x)$ is divided by $g(x)$.

11. $f(x) = 4x^5 + 2x^3 - x + 5$,
    $g(x) = x + 2$

12. $f(x) = 2x^5 - 3x^3 + 2x^2 - x + 3$,
    $g(x) = x + 1$

13. $f(x) = 2x^3 + x^2 - 13x + 6$,
    $g(x) = x - 2$

14. $f(x) = 6x^4 - 5x^3 + 9x^2 + 2x - 10$,
    $g(x) = x - 1$
15. $f(x) = x^3 + 125$,
    $g(x) = x + 5$
16. $f(x) = x^3 - 1$, $\quad g(x) = x - 1$
17. $f(x) = 3x^4 - 4x^3 - 19x^2 + 8x + 12$,
    $g(x) = x - \frac{2}{3}$
18. $f(x) = 4x^5 + 2x^3 - x + 5$,
    $g(x) = x + \frac{3}{4}$
19. $f(x) = x^6 - y^6$,
    $g(x) = x - y$
20. $f(a) = a^5 + b^5$,
    $g(a) = a + b$

Use synthetic division to determine the given function values.

21. $f(x) = 2x^4 - x^3 - 3x^2 + 2$;
    find $f(1), f(-2), f(4)$
22. $f(x) = x^4 - 3x^3 + 5x - 2$;
    find $f(-2), f(0), f(2)$
23. $f(x) = x^3 - 4x^2 + 9$;
    find $f(-1), f(3), f(5)$
24. $f(x) = 2x^3 - 5x^2 + x - 3$;
    find $f(-2), f(1), f(4)$
25. $f(x) = 2x^3 - 3x^2 + 5x - 1$;
    find $f(3), f(-5), f(11), f\left(\frac{2}{3}\right)$
26. $f(x) = x^5 - 3x^4 + 2x^2 - x - 5$;
    find $f(-3), f(20), f\left(-\frac{1}{2}\right)$

Use synthetic division to determine whether each number is a zero of the given polynomial.

27. $-1, 1$;
    $f(x) = x^4 + 2x^3 + 3x^2 - 2$
28. $-1, 2$;
    $f(x) = x^4 - 2x^3 + x^2 - x - 2$
29. $\frac{1}{3}, -5$;
    $f(x) = 6x^3 + 31x^2 + 4x - 5$
30. $3, -\frac{1}{2}$;
    $f(x) = x^3 + \frac{17}{2}x^2 + 19x + \frac{15}{2}$

In the following exercises, a polynomial $f(x)$ and a divisor $g(x)$ are given. Calculate the quotient $q(x)$ and the remainder $r(x)$.

31. $f(x) = (x + 1)^2 - 2(x + 1) + 2$,
    $g(x) = x + 1$
32. $f(x) = (x - 2)^3 + 7(x - 2) + 4$,
    $g(x) = x - 2$
33. $f(x) = (x + 1)^5 4(x - 5)^3 +$
    $(x - 5)^4 3(x + 1)^2$,
    $g(x) = x + 1$
34. $f(x) = (2x^2 - 5x - 3)^3$,
    $g(x) = 2x + 1$

Applications

35. **Realty sales model.** During a 12-month period, the monthly dollar volume of homes sold is given by a quadratic function $f(x)$, where x is the number of months since the beginning of the period. Determine $f(x)$ if the sales are $5.5 million at the beginning of the period, and $8 million in the fourth and eighth months.

36. **Unemployment model.** A model for the number of unemployed in an urban area is given by a cubic function. Suppose that the number equals 370,000 in months 3, 5, and 10 of a survey year and is 450,000 at the end of the year. Determine the cubic polynomial that describes the model.

37. **Evaluating a cost function.** The cost of producing x T-shirts is given by the function $C(x) = 0.00002x^3 - 0.001x^2 + 2.5x + 5,000$. Use synthetic division to determine the cost of producing 5,000 T-shirts.

38. **Evaluating a profit function.** The profit from the sale of x boxes of chocolate is given by the function:
    $$P(x) = -0.0000001x^3 + 0.0001x^2 + 0.25x - 30,000$$
    Use synthetic division to determine the profit from the sale of 15,000 boxes.

Technology

39. Let $f(x) = x^3 - 5x + 1$, $\quad g(x) = x^2 - 2x$.
    a. Calculate $q(x), r(x)$ using the division algorithm.
    b. Use a graphing calculator to graph $f(x)/g(x)$ and $q(x)$ on the same coordinate system.

c. Zoom out the graph in step b several times. Describe what you see.

40. Repeat the preceding exercise with $f(x) = x^4 - 2x^3 + 10$, $g(x) = 3x^2 - x$.

41. Can you generalize the graphical phenomenon exhibited in the preceding two exercises? Test your generalization using four more sets of polynomials.

In your own words

42. State the Factor theorem. Illustrate it with a specific polynomial.

43. State the Remainder theorem. Illustrate it with a specific polynomial.

44. Suppose that the polynomials f and g have rational coefficients. Must all the coefficients of the quotient and remainder of f divided by g be rational? Explain your answer.

45. Suppose that the polynomials f and g have integer coefficients. Must all the coefficients of the quotient and remainder of f divided by g be integers? Explain your answer.

46. Suppose someone tells you that a certain polynomial has 300 zeros. What can you say about its degree? Explain your answer.

4.3 CALCULATING ZEROS OF POLYNOMIALS

In the preceding section, we discussed the connection between the zeros of a polynomial function and its factorization. Let's now take up the problem of approximating the zeros of a polynomial and of obtaining information about the zeros without doing any calculation at all.

Approximating Real Zeros

For many applications, it is sufficient to determine a real zero with an accuracy of a certain number of decimal places. Here we present a method for determining such approximations based on the Intermediate Value theorem, stated in Section 4.1. This theorem states that if there are two points at which the graph of a polynomial lies on opposite sides of the x-axis, then the graph crosses the x-axis at some point in between those two points. See Figure 1.

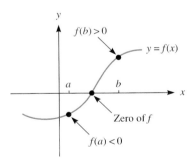

Figure 1

EXAMPLE 1
Isolating a Zero

Show that the polynomial
$$f(x) = 3x^3 - x + 1$$
has a zero in the interval $(-1, 0)$.

Solution
We note that:
$$f(-1) = 3(-1)^3 - (-1) + 1 = -1 < 0$$
$$f(0) = 3(0)^3 - 0 + 1 = 1 > 0$$
Therefore, $f(-1)$ and $f(0)$ have opposite signs. The Intermediate Value theorem states that the given polynomial has a zero in the interval $(-1, 0)$.

We can place a real zero in an interval as small as we wish by applying the Intermediate Value theorem to successively smaller intervals. Suppose that we determine from using the Intermediate Value theorem that $f(x)$ has a zero within a certain interval. We can divide this interval into two subintervals of equal length. For one of these subintervals, $[a, b]$ the values of $f(a)$ and $f(b)$ are of opposite sign. So there is a zero x in the interval, $a < x < b$. We can break this subinterval into two equal parts and apply the same procedure again. In this way, we can approximate the zero to any desired degree of accuracy. This method of approximating real zeros is called the **method of bisection**. It can be carried out with manual calculation, but it is most easily implemented using a scientific calculator.

The next example illustrates the same general method, except that each interval is subdivided into ten equal subintervals, rather than two. Using ten subintervals allows determination of one additional decimal place of accuracy with each repetition.

EXAMPLE 2
Approximating a Zero to Specified Accuracy

Determine the real zero of $f(x) = x^4 - 2x^2 + 3x - 1$ that lies between 0 and 1. Calculate the zero to within 0.001 accuracy.

Solution

We begin by tabulating the values of $f(x)$ for x between 0 and 1, at intervals of 0.1. We have used the computer program Mathematica® to prepare the following table. You can verify the entries using your calculator. According to the shaded rows, the zero lies between 0.4 and 0.5, because the value of $f(x)$ changes sign in this interval.

x	f(x)
0	−1
0.1	−0.7199
0.2	−0.4784
0.3	−0.2719
0.4	−0.0944
0.5	0.0625
0.6	0.2096
0.7	0.3601
0.8	0.5296
0.9	0.7361
1.0	1.0

We now tabulate the values of the function for x between 0.4 and 0.5 at intervals of 0.01. We arrive at the following table. The shaded rows show that the zero lies between 0.45 and 0.46.

x	f(x)
0.4	−0.0944
0.41	−0.0779424
0.42	−0.061683
0.43	−0.045612
0.44	−0.029719
0.45	−0.0139938
0.46	0.00157456
0.47	0.0169968
0.48	0.0322842
0.49	0.047448
0.5	0.0625

Finally, we tabulate the function values for x between 0.45 and 0.46 at intervals of 0.001. The following table shows that the zero lies between 0.458 and 0.459. To three significant digits, the zero is therefore 0.459.

SECTION 4.3 Calculating Zeros of Polynomials

x	f(x)
0.45	−0.0139938
0.451	−0.01243
0.452	−0.0108679
0.453	−0.00930727
0.454	−0.00774819
0.455	−0.00619065
0.456	−0.00463462
0.457	−0.0030801
0.458	−0.00152706
0.459	0.0000244838
0.46	0.00157456

Approximating Real Zeros Using a Graphing Calculator

The above method for approximating the zeros of a polynomial requires a fair amount of calculation—we must evaluate the polynomial ten times to get an extra digit of accuracy.

The next example illustrates another method for approximating the real zeros of a polynomial (or of any function, for that matter) using a graphing calculator.

EXAMPLE 3
Approximating Real Zeros Graphically

Approximate the zeros of the polynomial function:
$$f(x) = x^4 - x^3 - x^2 + 2x - 2$$

Solution
Begin by graphing the function using the standard range settings $-10 \leq x \leq 10$. (See Figure 2.) It appears that the only possible zeros lie in the interval $-2 \leq x \leq 2$. So change the range to this interval and regraph. (See Figure 3.) We now see that there is a zero in the interval $-2 \leq x \leq -1$ and another in $1 \leq x \leq 2$. Use a box to zoom in on these ranges for each of these intervals and use the trace; we find that these zeros are approximately -1.418504 and 1.418974. See Figures 4 and 5.

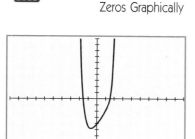

Figure 2

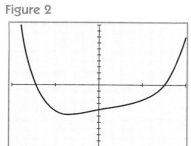

Figure 3

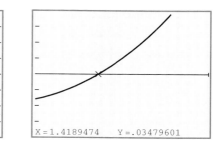

Figure 4: X = −1.418504 Y = .06600443

Figure 5: X = 1.4189474 Y = .03479601

Applications

Here are some applications requiring estimation of the zeros of polynomials.

EXAMPLE 4
Break-Even Analysis

If $R(x)$ denotes the revenue from selling x units of a commodity and $C(x)$ denotes the cost of producing the units, then a value of x is called a **break-even point** if $R(x) = C(x)$, that is, if revenue equals cost. Suppose that the revenue and cost functions for selling large-screen television sets are:

$$R(x) = 2,000x$$

and

$$C(x) = -0.0001x^3 + 0.005x^2 + 700x + 75,000$$

Determine the smallest break-even point to two significant digits.

Solution

The condition $R(x) = C(x)$ gives the equation:

$$2,000x = -0.0001x^3 + 0.005x^2 + 700x + 75,000$$
$$0 = 0.0001x^3 - 0.005x^2 + 1,300x - 75,000$$

We can solve this equation using either bisection or a graphing calculator. To two significant digits, the only real solution of the equation is $x = 58$. That is, the break-even point is reached by selling 58 televisions.

EXAMPLE 5 Temperature Variation during an Illness

The temperature T, t days after someone comes down with the flu, is given by the function:

$$T(t) = 98.6 + 2t^2(3 - t) \qquad (0 \leq t \leq 3)$$

Determine the times when the temperature reaches 101°F.

Solution

The times we want to find satisfy the equation:

$$98.6 + 2t^2(3 - t) = 101$$
$$2t^2(3 - t) - 2.4 = 0$$
$$-2t^3 + 6t^2 - 2.4 = 0$$

Using the method of bisection, we see that there are two real zeros with approximate values 0.73 and 2.85. So the temperature hits 101°F twice, once on its way up and once on its way down. Note that at 3 days, the temperature is 98.6°F, or normal body temperature.

Rational Zeros of Polynomials with Integer Coefficients

The methods presented thus far determine the approximate values of the zeros of a polynomial function. Let's now consider some methods that can be used to determine the exact values of the zeros (at least in some instances).

Consider a polynomial with integer coefficients, such as:

$$f(x) = 3x^2 + 5x + 4$$

The simplest zeros to look for are the rational ones, that is, the zeros of the form c/d, where c and d are integers and d is nonzero. We can determine all such zeros by examining the first and last coefficients, in this case 3 and 4. Namely, c must be a factor of 3 and d must be a factor of 4. This is the content of the following useful result:

RATIONAL ZERO THEOREM

Let

$$f(x) = a_n x^n + a_{n-1} x^{n-1} + \cdots + a_1 x + a_0, \qquad a_0 \neq 0, a_n \neq 0$$

be a polynomial with integer coefficients. Suppose that c/d is a rational zero of $f(x)$, where c and d are integers, d is nonzero, and the fraction c/d is in lowest terms. Then c is a factor of the constant coefficient a_0 and d is a factor of the leading coefficient a_n.

The proof of this theorem belongs in either a course in abstract algebra or a course in the theory of numbers, and so it is omitted here.

When we use the Rational Zero theorem, the problem of determining the rational zeros of $f(x)$ is reduced to testing a finite number of possibilities. We factor the leading and constant coefficients of $f(x)$ and form all possible fractions c/d, and then we test all c/d by substituting them for x in the function. The next two examples illustrate how to organize the calculations.

EXAMPLE 6
Rational Zeros

Find all rational zeros of the polynomial:
$$x^3 - 2x^2 - 2x - 3$$

Solution

In this case, the leading and constant coefficients are 1 and -3. By the Rational Zero theorem, if c/d is a rational zero in lowest terms, then d is a factor of 1 and c is a factor of -3. So d is either 1 or -1 and c is either 1, -1, 3, or -3. This gives us four choices for the zero: 1, -1, 3, or -3. We can test each of these possibilities by substituting them for x in the polynomial:
$$1^3 - 2(1)^2 - 2(1) - 3 = -6$$
$$(-1)^3 - 2(-1)^2 - 2(-1) - 3 = -4$$
$$(-3)^3 - 2(-3)^2 - 2(-3) - 3 = -42$$
$$(3)^3 - 2(3^2) - 2(3) - 3 = 0$$

We see that 3 is the only rational zero of the polynomial.

EXAMPLE 7
Determine Rational Zeros

Determine all rational zeros of the polynomial:
$$6x^3 - 13x^2 + 9x - 2$$

Solution

In this case, the leading coefficient is 6 and the constant coefficient is -2. By the Rational Zero theorem, if c/d is a rational zero in lowest terms, then d is a factor of 6 and c is a factor of -2. That is, d is one of the numbers $\pm 1, \pm 2, \pm 3, \pm 6$ and c one of the numbers $\pm 1, \pm 2$. Therefore, c/d is one of the numbers $\pm 1, \pm \frac{1}{2}, \pm \frac{1}{3}, \pm \frac{1}{6}, \pm 2, \pm \frac{2}{3}$. There are twelve possibilities to check. We can proceed as in the previous example, by evaluating the polynomial at each number and determining the numbers for which the polynomial equals 0. Another possibility is to use synthetic division to determine for which values of c and d the polynomial is divisible by $x - c/d$. We omit the arithmetic here. The results show that the zeros of the polynomial are $\frac{1}{2}, \frac{2}{3}$, and 1.

EXAMPLE 8
Architecture

The cost of building an office building of n floors is:
$$100n^3 + 3{,}000n^2 + 50{,}000$$

Suppose that the total cost of construction is $450,000. How many stories is the building?

Solution

We must solve the following equation for n:
$$100n^3 + 3{,}000n^2 + 50{,}000 = 450{,}000$$
$$100n^3 + 3{,}000n^2 - 400{,}000 = 0$$
$$n^3 + 30n^2 - 4{,}000 = 0$$

Let's apply the Rational Zero theorem to the polynomial:
$$f(x) = x^3 + 30x^2 - 4{,}000$$

We are looking for a positive zero. Because the leading coefficient is 1, the Rational Zero theorem asserts that the zero must be a factor of 4,000. If we factor 4,000, we see that
$$4{,}000 = 2^5 5^3$$

So the factors of 4,000 are obtained by multiplying a certain number of factors of 2 by a certain number of factors of 5. The first few of these are 1, 2, 4, 5, 8, 10, 16, 20, 32, ... Using a calculator, we can test these factors one by one to determine which is a zero of $f(x)$. After a bit of calculation, we see that the only one that works is 10. So 10 is a solution of the equation and the building has 10 stories.

Bounds on the Zeros of Real Polynomials

As we have just seen, the rational zeros of a polynomial with integer coefficients can be determined by examining a finite (and possibly long) list of possibilities. The number of possibilities can be narrowed by using the following result, which we cite without proof:

UPPER AND LOWER BOUNDS FOR ZEROS OF A POLYNOMIAL

Let $f(x)$ be a polynomial with real coefficients and a positive leading coefficient.

1. Let $a > 0$ be chosen so that the third row in synthetic division of $f(x)$ by $x - a$ has all positive or zero entries. Then all real zeros of $f(x)$ are less than or equal to a. The number a is called an **upper bound** for the zeros of $f(x)$.
2. Let $b < 0$ be chosen so that the third row in the synthetic division of $f(x)$ by $x - b$ has alternating signs. (A zero coefficient is always assumed to cause an alternation in sign.) Then all real zeros of $f(x)$ are greater than or equal to b. The number b is called a **lower bound** for the zeros of $f(x)$.

The next example shows you how this last result can be applied to obtain information about the zeros of a real polynomial.

EXAMPLE 9
Calculating Upper and Lower Bounds for Zeros

Let:
$$f(x) = x^5 - x^3 + 2x^2 - 2x + 3$$
Determine upper and lower bounds for the real zeros of $f(x)$.

Solution

In a search for an upper bound for the zeros, we can try the various positive integers in turn and perform synthetic division. For the integer $a = 1$, we have:

$$\begin{array}{r|rrrrrr} 1 & 1 & 0 & -1 & 2 & -2 & 3 \\ & & 1 & 1 & 0 & 2 & 0 \\ \hline & 1 & 1 & 0 & 2 & 0 & 3 \end{array}$$

Because all entries in the last row are nonnegative, we see that all zeros of f are ≤ 1. To obtain a lower bound, we test negative integers and look for third rows with alternating signs (possibly with 0's in between). For $a = -1$, we have:

$$\begin{array}{r|rrrrrr} -1 & 1 & 0 & -1 & 2 & -2 & 3 \\ & & -1 & 1 & 0 & -2 & 4 \\ \hline & 1 & -1 & 0 & 2 & -4 & 7 \end{array}$$

The signs don't alternate. Next, we try $a = -2$:

$$\begin{array}{r|rrrrrr} -2 & 1 & 0 & -1 & 2 & -2 & 3 \\ & & -2 & 4 & -6 & 8 & -12 \\ \hline & 1 & -2 & 3 & -4 & 6 & -9 \end{array}$$

The signs alternate, so all zeros of f are ≥ -2.

Exercises 4.3

Show that each of the following polynomials has a zero in the given interval.

1. $f(x) = x^3 + x + 1$; $(-1, 0)$
2. $f(x) = x^3 - x + 1$; $(-2, -1)$
3. $f(x) = x^3 - 3x^2 + 4x - 5$; $(2, 3)$
4. $f(x) = x^3 + x - 3$; $(1, 2)$
5. $f(x) = x^3 + 2x^2 + 8x - 2$; $(0, 1)$
6. $f(x) = x^3 - 3x^2 - 5$; $(3, 4)$
7. $f(x) = x^4 - x^2 - 3$; $(-1, -2)$ and $(1, 2)$
8. $f(x) = x^4 - x^3 - 1$; $(-1, 0)$ and $(1, 2)$

Approximate the real zeros of each of the following functions to an accuracy of 0.1 using the method of bisection.

9. $f(x) = x^3 + x + 1$
10. $f(x) = x^3 - x + 1$
11. $f(x) = x^3 - 3x^2 + 4x - 5$
12. $f(x) = x^3 + x - 3$
13. $f(x) = x^3 + 2x^2 + 8x - 2$
14. $f(x) = x^4 - x^2 - 3$
15. $f(x) = x^4 - x^3 - 1$
16. $f(x) = x^4 - 3x^3 - 2x - 3$
17. $f(x) = x^5 - 3x^4 + 5x^3 - 7x^2 + 6$
18. $f(x) = -x^5 + 4x^4 - 2x^3 + x^2 - 4x + 1$

Determine all rational zeros of the following polynomials.

19. $9x^3 - 18x^2 + 11x - 2$
20. $4x^3 - x^2 - 100x + 25$
21. $3x^3 - 2x^2 - 3x + 2$
22. $x^3 - 3x^2 - 3x - 4$
23. $4x^4 - 21x^2 - 25$
24. $81x^4 - 16$

Use synthetic division to determine upper and lower bounds for the real zeros of $f(x)$. Use this information to display the graph on a graphing calculator, showing all real zeros.

25. $f(x) = 2x^4 - x^3 + 3x^2 - 5$
26. $f(x) = 3x^3 + 4x^2 + 12$
27. $f(x) = x^5 + 3x^4 - x^2 + 14$
28. $f(x) = 2x^4 - 3x^2 - 9x + 1$
29. $f(x) = 16x^{12} - 11x^{10} + 2x^9 - 3x^8 + 4x - 6$
30. $f(x) = 4x^4 + x^3 - 6x^2 + 10$
31. $f(x) = x^4 - 25$
32. $f(z) = x^6 - 1$

Find the exact values of all the zeros.

33. $x^3 - 10x + 3$
34. $x^3 + 2x^2 + 9$
35. $x^3 + 6x^2 + 7x - 2$
36. $x^3 + 4x^2 + 7x + 6$
37. $x^4 + 3x^3 - 2x^2 - 2x + 1$
38. $6x^4 + x^3 - 8x^2 - x + 2$
39. $6x^4 + x^3 + 4x^2 + x - 2$
40. $4x^4 - 4x^3 - 5x^2 + x + 1$
41. $4x^5 - 24x^4 + 25x^3 + 39x^2 - 38x - 24$
42. $x^5 + x^4 - 9x^3 - 5x^2 + 16x + 12$
43. $3x^4 - 39x^2 + 18$
44. $x^4 - 10x^2 + 23$

Applications

You can use a scientific or graphing calculator to solve the following applied problems.

45. **Cost of a building.** The cost of building an office building of n floors is:
$$100n^3 + 3{,}000n^2 + 50{,}000$$
   a. Suppose that the total cost of construction is $3,487,500. How many stories is the building?
   b. Suppose that the total cost of construction is about $776,000. Use the method of bisection to estimate how many stories are in the building. Give your answer correct to the nearest 1.

46. **Volume of a box.** An open box of volume 108 in.^{3} can be made from a 10-in.-by-15-in. piece of cardboard by cutting a square from each corner and folding up the sides (see figure). What is the length of a side of the squares? If appropriate, use the method of bisection to find an approximate answer.

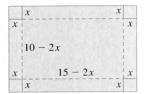

Exercise 46

47. **Break-even points.** We are given the following total revenue and total cost functions:
$$R(x) = 100x - x^2 \qquad C(x) = \tfrac{1}{3}x^3 - 6x^2 + 89x + 100$$
Find the break-even point(s).

48. **Medical dosage.** The function
$$N(t) = 0.045t^3 + 2.063t + 2$$
gives the bodily concentration in parts per million of a certain dosage of medication after time t, in hours. Find all times in the interval [0, 10] for which the concentration is 4 parts per million.

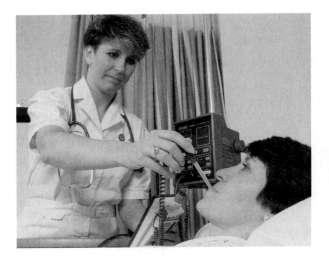

49. **Temperature during an illness.** A patient's temperature T during an illness is given by:
$$T(t) = 98.6 - t^2(t - 4)$$
Find the times t at which the patient's temperature was 100°F.

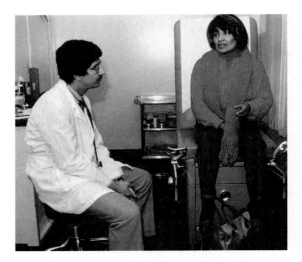

50. **Physician visits.** The number of physician visits y per year by a person of age x is given by the model:
$$y = 6.95 - 0.3x + 0.0083x^2 - 0.000052x^3$$

a. Graph this equation.

b. Determine x so that the number of visits to a doctor per year is 5.

c. Beyond age 20, how old must a person be before the number of visits to a physician per year equals that required by a 10-year-old?

In your own words

51. Describe the method of bisection. Illustrate it with a particular polynomial.

52. Describe how to determine the zeros of a function using a graphing calculator.

53. Compare and contrast the method of bisection and the graphing calculator method for finding zeros of a function.

54. Why might you want to find bounds on the zeros of a polynomial?

55. Suppose you wish to approximate the zero of a polynomial using a graphing calculator. To what extent does the accuracy of the procedure depend on the fineness of the dot image, that is, the resolution of the display?

4.4 COMPLEX NUMBERS

As comprehensive as the set of real numbers appears to be, certain elementary arithmetic operations are not possible using real numbers. For example, there is no real number whose square is -1. To put this another way, it is impossible to compute $\sqrt{-1}$ within the real numbers because the square of any real number is nonnegative, and so cannot be -1. In fact, the real numbers don't have a square root of *any* negative number. This serious shortcoming of the real numbers has motivated mathematicians to create a larger number system, called the **complex numbers**, in which taking square roots is always possible.

Definition of Complex Numbers

We introduce the number i, which is a square root of -1. A typical complex number is of the form $a + bi$, where a and b are real numbers. Let's record this definition for future reference.

Definition
Complex Numbers

The **imaginary unit,** denoted i, is a number whose square is -1. That is,
$$i^2 = -1, \quad i = \sqrt{-1}$$
A **complex number** is a number of the form
$$a + bi$$
where a and b are real numbers and i is the imaginary unit.

Here are some examples of complex numbers:
$$2 + 3i, \quad \frac{1}{2} - 4i, \quad 5 + 0i, \quad \sqrt{2}i$$

If $a + bi$ is a complex number, then a is called its **real part** and b is called its **imaginary part.**

Definition
Equality of Complex Numbers

Two complex numbers $a + bi$ and $c + di$ are equal if and only if $a = c$ and $b = d$.

The real number a is the same as the complex number $a + 0i$, so every real number is also a complex number. Figure 1 shows the relationships between the set of complex numbers and the various number systems we introduced earlier.

CHAPTER 4 Polynomial and Rational Functions

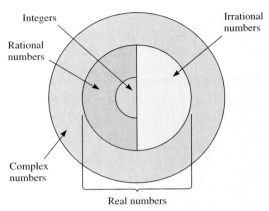

Figure 1

Arithmetic Involving Complex Numbers

Addition and subtraction of complex numbers is defined by adding real and imaginary parts.

For example, we have:

$$(5 - 3i) + (4 + 2i) = (5 + 4) + (-3 + 2)i = 9 - i$$

To multiply complex numbers, first use FOIL, and then simplify, using the fact that $i^2 = -1$. For example,

$$(2 + 4i)(3 - 7i) = 2 \cdot 3 + 2 \cdot (-7i) + (4i)(3) + (4i)(-7i)$$
$$= 6 - 14i + 12i - 28i^2$$
$$= 6 - 2i - 28(-1)$$
$$= 6 - 2i + 28$$
$$= 34 - 2i$$

EXAMPLE 1
Arithmetic of Complex Numbers

Determine the following complex numbers:

1. $2(3 + i) - i(2 - 4i)$
2. $(1 + i)^2$

Solution

1. $2(3 + i) - i(2 - 4i) = (6 + 2i) + (-2i + 4i^2)$
   $= (6 + 2i) + (-2i - 4)$ *Because $i^2 = -1$*
   $= 2 + 0i$

2. $(1 + i)^2 = (1 + i)(1 + i)$
   $= 1 + i + i + i^2$
   $= 1 + 2i - 1$
   $= 2i$

EXAMPLE 2
Complex Products

Compute the product:

$$(x - i)(x + i)$$

Solution

Apply the distributive law:

$$(x - i)(x + i) = x^2 + ix - ix - i^2$$
$$= x^2 - (-1) \quad \text{Using } i^2 = -1$$
$$= x^2 + 1$$

The preceding example can also be read in reverse, not as a multiplication example, but as a factorization. Starting with our solution in that example, a sum of squares,

$$x^2 + 1$$

we can arrive at the factorization:

$$(x - i)(x + i)$$

In a similar fashion, we can factor the sum of squares

$$x^2 + y^2$$

into the linear factors:

$$(x - yi)(x + yi)$$

To verify this fact, we just multiply the linear factors as in Example 2. Let's record this fact for future reference:

FACTORING A SUM OF SQUARES

$$x^2 + y^2 = (x + yi)(x - yi)$$

Note that the sum of two squares is not factorable using the real numbers, yet we can factor it using complex numbers.

Algebraic Properties of Complex Numbers

Addition and multiplication of complex numbers have the same fundamental properties as addition and multiplication of real numbers. Namely, they obey the commutative, associative, and distributive laws: If α, β, and γ are complex numbers, then:

$$\alpha + \beta = \beta + \alpha \quad \text{Commutative law of addition}$$
$$(\alpha + \beta) + \gamma = \alpha + (\beta + \gamma) \quad \text{Associative law of addition}$$
$$\alpha\beta = \beta\alpha \quad \text{Commutative law of multiplication}$$
$$\alpha(\beta\gamma) = (\alpha\beta)\gamma \quad \text{Associative law of multiplication}$$
$$\alpha(\beta + \gamma) = (\alpha \cdot \beta) + (\alpha \cdot \gamma) \quad \text{Distributive law}$$

Complex numbers have an **additive identity,** namely 0 (or $0 + 0i$), with the property:

$$\alpha + 0 = \alpha$$
$$a + bi + 0 = a + bi$$

A complex number $\alpha = a + bi$ has an **additive inverse,** namely $-\alpha = -a - bi$, with the property:

$$\alpha + (-\alpha) = 0$$
$$(a + bi) + (-a - bi) = 0$$

Addition and subtraction of complex numbers is consistent with the corresponding operations for real numbers. Indeed, if we consider two real numbers a and c, then:

$$(a + 0i) + (c + 0i) = (a + c) + 0i + 0i = a + c$$
$$(a + 0i) - (c + 0i) = (a + 0i) + (-c - 0i) = (a - c) + 0i - 0i = a - c$$

The number 1 (or $1 + 0i$) is the **multiplicative identity:**

$$\alpha \cdot 1 = \alpha$$
$$(a + bi) \cdot 1 = a + bi$$

Just as with the real number system, the only way for a product of complex numbers to be zero is for at least one of the factors to be zero.

$$\alpha\beta = 0 \quad \text{implies that} \quad \alpha = 0 + 0i \quad \text{or} \quad \beta = 0 + 0i$$

This fact is of great use in determining solutions of polynomial equations in the complex number system.

Powers of i The complex number i has the property that its square is -1. Using this fact and the laws of exponents, we can easily calculate higher **powers of i.** Here are the first few:

$$i^3 = i^2 \cdot i = (-1) \cdot i = -i$$
$$i^4 = i^2 \cdot i^2 = (-1)(-1) = 1$$

Using the values of the first four powers of i, we can now compute higher powers, as follows:

$$i^5 = i^4 \cdot i = 1 \cdot i = i$$
$$i^6 = i^4 \cdot i^2 = 1 \cdot (-1) = -1$$
$$i^7 = i^4 \cdot i^3 = 1 \cdot (-i) = -i$$
$$i^8 = i^4 \cdot i^4 = 1 \cdot 1 = 1$$

Note that the powers of i repeat each time the exponent is increased by 4. Moreover, if the exponent is divisible by 4, then the power of i is 1. We can use these facts to rapidly calculate any power of i.

EXAMPLE 3 Calculate:
Calculating a Power of i

$$i^{57}$$

Solution

First, divide the exponent by 4 and write it in the form:

$$57 = 14 \cdot 4 + 1$$

Substitute this expression for the exponent of i:

$$i^{57} = i^{14 \cdot 4 + 1}$$
$$= i^{14 \cdot 4} \cdot i^1$$
$$= (i^4)^{14} \cdot i$$
$$= (1)^{14} i$$
$$= i$$

Square Roots of Negative Real Numbers

Mathematicians introduced the complex numbers to create a number system that contains square roots of negative numbers. Suppose that p is a positive real number. Then, as we have observed, there is no real number whose square is $-p$. However, there is a complex number with this property, namely:

$$\sqrt{p}\, i$$

For we have

$$(\sqrt{p}\, i)^2 = (\sqrt{p})^2 i^2$$
$$= p(-1)$$
$$= -p$$

The quantity $\sqrt{p}\, i$ is called the **principal square root** of $-p$ and is written $\sqrt{-p}$.

EXAMPLE 4   Square Roots of a Negative Number

Determine two square roots of -4.

Solution

One square root of -4 is given by the principal square root:

$$\sqrt{-4} = \sqrt{4}\, i = 2i$$

A second square root is given by the negative of this quantity, $-2i$, because

$$(-2i)^2 = (-2)^2 i^2 = 4 \cdot (-1) = -4$$

So two square roots of -4 are $\pm 2i$.

HISTORICAL NOTE

Square roots of negative numbers were used for centuries in calculating solutions to equations of degrees 2, 3, and 4. However, mathematicians regarded them with great suspicion. They realized, of course, that they were not real numbers but were not quite sure of the legitimacy of working with them. Accordingly, square roots of negative numbers were called **imaginary numbers** or just **imaginaries**. The German mathematician Karl Friedrich Gauss, in 1801, gave a formal definition of the complex numbers that provided a logical foundation for imaginary numbers. He established the basis for legitimate proofs of theorems that involve imaginary numbers.

Quadratic Equations That Have Complex Solutions

The quadratic equation,

$$ax^2 + bx + c = 0$$

has solutions given by the quadratic formula:

$$x = \frac{-b \pm \sqrt{b^2 - 4ac}}{2a}$$

As we have seen, when the discriminant $b^2 - 4ac$ is negative, the equation has no real solutions. However, in this case, the formula gives two distinct **complex solutions,** that is, the quadratic equation has **complex roots.**

EXAMPLE 5
Quadratic Equation with Complex Solutions

Determine all solutions of the equation $x^2 + x + 1 = 0$.

Solution

By the quadratic formula, the solutions are given by:

$$x = \frac{-b \pm \sqrt{b^2 - 4ac}}{2a}$$

$$= \frac{-1 \pm \sqrt{1^2 - 4 \cdot 1 \cdot 1}}{2 \cdot 1}$$

$$= \frac{-1 \pm \sqrt{-3}}{2}$$

$$= \frac{-1 \pm \sqrt{3}i}{2}$$

Conjugates

Each complex number has a conjugate, defined as follows:

Definition
Conjugate of a Complex Number

Let $a + bi$ be a complex number. Its conjugate is the number $a - bi$ and is denoted $\overline{a + bi}$.

Here are some examples of conjugates:

$$\overline{2 + 3i} = 2 - 3i, \qquad \overline{5} = 5, \qquad \overline{\frac{1}{2}i} = -\frac{1}{2}i, \qquad \overline{-8 - 7i} = -8 + 7i$$

Suppose that α is a nonzero complex number. Then α has a **multiplicative inverse** α^{-1}, a complex number that satisfies the equation:

$$\alpha \alpha^{-1} = 1$$

This is exactly analogous to the definition of the multiplicative inverse of a real number.

We can calculate the multiplicative inverse using conjugates. For example, suppose we wish to compute:

$$(2 + i)^{-1}$$

Just as in working with negative exponents of a real number, we write this number as:

$$\frac{1}{2 + i}$$

Now multiply the numerator and denominator by the conjugate of the denominator:

$$\frac{1}{2+i} = \frac{1}{2+i} \cdot \frac{2-i}{2-i}$$
$$= \frac{2-i}{2^2+1^2}$$
$$= \frac{2-i}{5}$$
$$= \frac{2}{5} - \frac{1}{5}i$$

The multiplicative inverse of $2+i$ is $\frac{2}{5} - \frac{1}{5}i$. We can check the solution using the definition of multiplicative inverse.

$$(2+i)\left(\frac{2}{5} - \frac{1}{5}i\right) = \frac{4}{5} - \frac{2}{5}i + \frac{2}{5}i - \frac{1}{5}i^2$$
$$= \frac{4}{5} + \frac{1}{5}$$
$$= 1$$

We can perform **division of complex numbers** in terms of multiplicative inverses, using the equation:

$$\frac{\alpha}{\beta} = \alpha \cdot \beta^{-1}, \qquad \beta \neq 0$$

EXAMPLE 6
Inversion and Division of Complex Numbers

Express the following complex numbers in the form $a + bi$.
1. $(5 - 4i)^{-1}$
2. $\dfrac{3+i}{1+2i}$

Solution

1. We can write the complex number as $\dfrac{1}{5-4i}$. Multiply the numerator and denominator by the conjugate $5+4i$:

$$\frac{1}{5-4i} = \frac{1}{5-4i} \cdot \frac{5+4i}{5+4i} = \frac{5+4i}{5^2+4^2} = \frac{5}{41} + \frac{4}{41}i$$

2. We can calculate the quotient by multiplying numerator and denominator by the conjugate of the denominator:

$$\frac{3+i}{1+2i} = \frac{3+i}{1+2i} \cdot \frac{1-2i}{1-2i}$$
$$= \frac{3 \cdot 1 + (-2)i^2 + (1 - 2 \cdot 3)i}{1^2 + 2^2}$$
$$= \frac{5 - 5i}{5}$$
$$= 1 - i$$

Exercises 4.4

Express the following in terms of i.
1. $\sqrt{-3}$
2. $\sqrt{-4}$
3. $\sqrt{-81}$
4. $\sqrt{-27}$
5. $\sqrt{-98}$

6. $-\sqrt{-18}$
7. $-\sqrt{-49}$
8. $-\sqrt{-125}$
9. $4 - \sqrt{-60}$
10. $6 - \sqrt{-84}$
11. $\sqrt{-4} + \sqrt{-12}$
12. $-\sqrt{-76} + \sqrt{-125}$

Calculate the following powers of i.

13. i^7
14. i^{11}
15. i^{24}
16. i^{35}
17. i^{42}
18. i^{64}
19. i^9
20. $(-i)^{71}$

Write the following complex numbers in the form $a + bi$:

21. $7 + i^4$
22. $-18 + i^3$
23. $i^4 - 26i$
24. $i^5 + 37i$
25. $i^2 + i^4$
26. $5i^5 + 4i^3$
27. $i^5 + i^7$
28. $i^{84} - i^{100}$
29. $1 + i + i^2 + i^3 + i^4$
30. $i - i^2 + i^3 - i^4 + i^5$
31. $5 - \sqrt{-64}$
32. $\sqrt{-12} + 36i$

33. $\dfrac{8 - \sqrt{-24}}{4}$
34. $\dfrac{9 + \sqrt{-9}}{3}$
35. $\dfrac{\sqrt{-16}}{\sqrt{-25}}$
36. $\dfrac{\sqrt{-9}}{\sqrt{-36}}$
37. $(3 + 2i) + (2 + 4i)$
38. $(2 - 5i) + (3 + 6i)$
39. $(3 + 4i) + (3 - 4i)$
40. $(2 + 5i) + (-2 - 5i)$
41. $(9 + 12i) - (7 + 8i)$
42. $(10 - 4i) - (6 + 2i)$
43. $6i - (7 - 4i)$
44. $45 - (23 + 5i)$
45. $(1 - 3i)(2 + 4i)$
46. $(-2 + 3i)(6 - 7i)$
47. $(2 - 5i)(2 + 5i)$
48. $(-5 + 7i)(-5 - 7i)$
49. $2i(4 - 3i)$
50. $5i(-8 + 6i)$
51. $(5 - 2i)^2$
52. $(3 + i)^2$
53. $\dfrac{1 + 2i}{3 - i}$
54. $\dfrac{2 + i}{2 - i}$
55. $\dfrac{\sqrt{3} - i}{\sqrt{3} + i}$
56. $\dfrac{\sqrt{2} + i}{\sqrt{2} - i}$
57. $\dfrac{5 - 3i}{i}$

58. $\dfrac{\sqrt{2} - i}{i}$
59. $\dfrac{i}{1 - i}$
60. $\dfrac{4}{3 + 10i}$
61. $\dfrac{2 + i}{(2 - i)^2}$
62. $\dfrac{5 - i}{(5 + i)^2}$
63. $(1 - 2i)^{-1}$
64. $(1 + i)^{-1}$
65. $\dfrac{(1 + i)(2 - i)}{(4 - 2i)(5 - 3i)}$
66. $\dfrac{(2 - 3i)(5 - 6i)}{(9 + 2i)(4 + 3i)}$
67. $\dfrac{1 + i}{1 - i} + \dfrac{2 - i}{2 + i}$
68. $\dfrac{5 - 2i}{3 + 2i} - \dfrac{7 - i}{4 + i}$
69. $(1 + i)^{-2}$
70. $(\sqrt{3} - 2i)^{-2}$

Determine two square roots of each of the following.

71. -5
72. -9
73. -64
74. -17

Factor the following polynomials, using complex numbers where necessary.

75. $x^2 + 4$
76. $x^2 + 25$
77. $x^2 + 3$
78. $x^2 + 5$
79. $a^2 + b^2$
80. $x^2 + 16y^2$

In your own words

81. Why should we use complex numbers?
82. Explain how some polynomials cannot be factored into factors that have real coefficients, but can be factored into factors that have complex coefficients. Illustrate with a specific polynomial.
83. Describe in words a procedure for determining whether a quadratic equation has nonreal solutions.
84. What is the conjugate of the conjugate of a complex number?

4.5 THE FUNDAMENTAL THEOREM OF ALGEBRA AND DESCARTES'S RULE OF SIGNS

As we have seen in the preceding section, quadratic polynomials can have complex numbers as zeros. The same is true for polynomials of higher degrees. In this section, we show how to get information about zeros of these polynomials.

One of the most basic results of algebra states that a nonconstant polynomial can be factored into linear factors with complex coefficients. More specifically, this result is stated as follows:

FUNDAMENTAL THEOREM OF ALGEBRA

Let $f(x)$ be a polynomial with leading coefficient a and positive degree n. Then $f(x)$ can be written in the form

$$f(x) = a(x - \alpha_1)(x - \alpha_2) \cdots (x - \alpha_n)$$

where

$$\alpha_1, \alpha_2, \ldots, \alpha_n$$

are complex numbers and are zeros of $f(x)$.

Any proof of the Fundamental Theorem of Algebra involves mathematical ideas that are significantly beyond the scope of this book. Although we won't prove this result here, let's illustrate it using several examples.

EXAMPLE 1
Factoring a Polynomial into Linear Factors

Factor the following polynomial into linear factors with complex coefficients:

$$f(x) = 6x^4 + 6x^3 + 18x^2$$

Solution

We begin by noting that we can factor out $6x^2$ from each term to obtain:

$$6x^2(x^2 + x + 3)$$

To factor the second quadratic polynomial, we use the quadratic formula to determine its zeros:

$$\frac{-1 \pm \sqrt{1^2 - 4(1)(3)}}{2} = \frac{-1 \pm \sqrt{-11}}{2} = \frac{-1 \pm \sqrt{11}i}{2}$$

Therefore, the factorization of the given polynomial is:

$$6x^4 + 6x^3 + 18x^2 = 6x^2\left(x - \frac{-1 + \sqrt{11}i}{2}\right)\left(x - \frac{-1 - \sqrt{11}i}{2}\right)$$

Zeros and Their Multiplicities

In the factorization of a polynomial $f(x)$ given by the Fundamental Theorem of Algebra, the zeros

$$\alpha_1, \alpha_2, \ldots, \alpha_n$$

may not all be different. The number of times a particular zero appears among the factors is called the **multiplicity** of the zero. For many purposes, it is convenient to count a zero of multiplicity m as m zeros. In this case, we say that zeros are counted according to their multiplicities. For example, the polynomial

$$f(x) = (x - 1)^3$$

has the single zero 1 of multiplicity 3, which is counted as three zeros.

Suppose that the polynomial $f(x)$ has degree n. When zeros are counted according to their multiplicities, the factors listed above consist of n zeros. That is, we have the following result.

NUMBER OF ZEROS OF A POLYNOMIAL

Suppose that $f(x)$ is a polynomial of positive degree n. If zeros are counted according to their multiplicities, then $f(x)$ has exactly n zeros.

EXAMPLE 2
Determining Multiplicities of Zeros

Determine the multiplicities of the zeros of the polynomial:
$$f(x) = x^5(x-1)^6(x+7)^2(x+i)^2(x-i)^2$$

Solution

The zeros can be read off from the distinct factors appearing in the factorization of $f(x)$. The multiplicities can be read off from the exponents, which indicate the number of times the particular factor is present. Here are the results:

$\alpha_1 = 0$ *Multiplicity = 5*
$\alpha_2 = 1$ *Multiplicity = 6*
$\alpha_3 = -7$ *Multiplicity = 2*
$\alpha_4 = -i$ *Multiplicity = 2*
$\alpha_5 = i$ *Multiplicity = 2*

Polynomials with Real Coefficients

You may have already observed that if a quadratic polynomial (with real coefficients) has nonreal roots, then the roots that are complex numbers are conjugates of one another. For example, consider the polynomial
$$f(x) = x^2 + x + 1$$
By the quadratic formula, the zeros are:
$$\frac{1}{2} + \frac{\sqrt{3}}{2}i, \quad \frac{1}{2} - \frac{\sqrt{3}}{2}i$$
These two complex numbers are conjugates of one another. This is a general fact:

CONJUGATE ROOT THEOREM

The nonreal zeros of a polynomial with real coefficients occur in complex-conjugate pairs.

This result can be used to determine information about the zeros of a polynomial without determining their precise values, as the following example illustrates.

EXAMPLE 3
Zeros of a Fifth-Degree Polynomial

How many real zeros can a fifth-degree polynomial with real coefficients have?

Solution

By the Conjugate Root theorem, the nonreal zeros occur in pairs. The number of pairs can be 0, 1, or 2. (Any more pairs would result in more than five zeros.) These three possibilities correspond to the following numbers of real zeros:

SECTION 4.5 The Fundamental Theorem of Algebra and Descartes's Rule of Signs

zero pairs: $5 - 0 = 5$ real zeros
1 pair: $5 - 2 = 3$ real zeros
2 pairs: $5 - 4 = 1$ real zero

EXAMPLE 4
Determining Complex Zeros

Determine the zeros of the polynomial:

$$x^6 - 1$$

Solution

We proceed by using the sum and difference of cubes factorization formulas from Chapter 1:

$$\begin{aligned} x^6 - 1 &= (x^3)^2 - 1 \\ &= (x^3 - 1)(x^3 + 1) \\ &= (x - 1)(x^2 + x + 1)(x + 1)(x^2 - x + 1) \end{aligned}$$

From this factorization, we see that the zeros are solutions of the equations:

$$x - 1 = 0, \qquad x + 1 = 0$$
$$x^2 + x + 1 = 0, \qquad x^2 - x + 1 = 0$$

The first two equations give the zeros $1, -1$. To solve the last two equations, we use the quadratic formula:

$$x^2 + x + 1 = 0$$

$$x = \frac{-1 \pm \sqrt{(1)^2 - 4(1)(1)}}{2(1)}$$

$$= \frac{-1 \pm \sqrt{3}i}{2}$$

In a similar fashion, the last equation yields the zeros

$$x = \frac{1 \pm \sqrt{3}i}{2}$$

So the six zeros of the given polynomial are:

$$\pm 1, \quad \frac{1 \pm \sqrt{3}i}{2}, \quad \frac{-1 \pm \sqrt{3}i}{2}$$

The Fundamental Theorem of Algebra says that a polynomial with real coefficients can be factored in the form

$$f(x) = a(x - \alpha_1)(x - \alpha_2) \cdots (x - \alpha_n)$$

where the complex numbers

$$\alpha_1, \alpha_2, \ldots, \alpha_n$$

are the zeros of $f(x)$. According to the Conjugate Root theorem, the nonreal zeros occur in complex-conjugate pairs. If such a pair of zeros is

$$\beta, \overline{\beta}$$

then the product

$$(x - \beta)(x - \overline{\beta})$$

is contained in the factorization of $f(x)$ and is equal to:
$$x^2 - (\beta + \overline{\beta})x + \beta\overline{\beta}$$
The coefficients of this polynomial are real because, if we let $\beta = a + bi$, then:
$$\beta + \overline{\beta} = (a + bi) + (a - bi) = 2a$$
and
$$\beta\overline{\beta} = (a + bi)(a - bi) = a^2 + b^2$$
And both of the expressions on the right of these two equations are real numbers. This shows that the factor corresponding to a complex-conjugate pair of zeros, namely
$$(x - \beta)(x - \overline{\beta})$$
is a real polynomial. On the other hand, if a zero α is real, the factor $(x - \alpha)$ has real coefficients. This proves the following important result about the factorization of polynomials with real coefficients:

POLYNOMIAL FACTORIZATION OVER THE REAL NUMBERS

Let $f(x)$ be a polynomial of positive degree with real coefficients. Then $f(x)$ can be factored into a product of linear and quadratic polynomials that have real coefficients, where the quadratic factors have nonreal zeros.

EXAMPLE 5 Factor the following polynomial into polynomials that have real coefficients:
Factoring a Polynomial
into Real Factors
$$x^4 + x^3 + 5x^2 + 4x + 4$$

Solution

Note that the coefficient 4 appears twice. This suggests that we group terms together so that we can factor out a 4. In addition to grouping the last two terms together, we can rewrite the middle term as $x^2 + 4x^2$. So we can factor two groups of terms:
$$(x^4 + x^3 + x^2) + (4x^2 + 4x + 4) = x^2(x^2 + x + 1) + 4(x^2 + x + 1)$$
$$= (x^2 + 4)(x^2 + x + 1)$$
Notice that the zeros of each of the two factors are nonreal complex numbers, because their respective discriminants are negative (-16 for the first and -3 for the second). Therefore, the polynomials cannot be factored further into real polynomials. Note, however, that by using the quadratic formula to determine the zeros of each of the factors, we see that the given polynomial can be factored into complex polynomials:
$$(x + 2i)(x - 2i)\left(x - \frac{-1 + \sqrt{3}i}{2}\right)\left(x - \frac{-1 - \sqrt{3}i}{2}\right)$$

Descartes's Rule of Signs It is possible to derive information about the zeros of a polynomial $f(x)$ that has real coefficients by examining the number of changes of sign in the sequence of coefficients. Consider the polynomial:
$$5x^6 - 3x^4 + x^3 + 0.5x^2 - 10x + 4$$

The sequence of signs in this polynomial is

$$+ \quad - \quad + \quad + \quad - \quad +$$

The number of sign changes in this sequence is four: the change from the first sign to the second sign, from the second sign to the third, from the fourth to the fifth, and from the fifth to the sixth. Note that no change in signs is recorded for terms that have zero coefficients. The number of changes in the sequence of signs is related to the number of real zeros of the polynomial by the following famous result of Descartes.

DESCARTES'S RULE OF SIGNS

Let $f(x)$ be a polynomial with real coefficients and with a nonzero constant term.

1. The number of positive real zeros is at most equal to the number of changes M in the sequence of signs of the coefficients of $f(x)$ and differs from M by an even integer.
2. The number of negative real zeros is at most equal to the number of changes N in the sequence of signs of the coefficients for $f(-x)$ and differs from N by an even integer.

Note that in counting the changes in sign in the sequence of coefficients, zero coefficients can be skipped.

EXAMPLE 6
Number of Real Zeros

Apply Descartes's Rule of Signs to determine the possible number of positive and negative real zeros of the polynomial

$$5x^6 - 3x^4 + x^3 + 0.5x^2 - 10x + 4$$

Solution

The polynomial has a nonzero constant term, so we can apply Descartes's Rule of Signs. As we have already seen, the sequence of signs for this polynomial is $+ \ - \ + \ + \ - \ +$, and there are four changes in this sequence, so $M = 4$. By Descartes's Rule of Signs, the polynomial has at most four positive real zeros, and the actual number differs from 4 by an even integer. Thus, the number of positive zeros is either 4, 2, or 0. To analyze the possibilities for negative zeros, we form $f(-x)$:

$$5(-x)^6 - 3(-x)^4 + (-x)^3 + 0.5(-x)^2 - 10(-x) + 4$$
$$= 5x^6 - 3x^4 - x^3 + 0.5x^2 + 10x + 4$$

The sequence of signs for $f(-x)$ is $+ \ - \ - \ + \ + \ +$. There are two changes in sign, so $N = 2$. Descartes's Rule of Signs then states that the polynomial has at most two negative real zeros, and the number of negative real zeros is either 2 or 0.

Thus, Descartes's Rule of Signs shows that there are six possibilities for the number of positive and negative real zeros of the polynomial, as the following table shows:

Positive Real Zeros	Negative Real Zeros	Nonreal Zeros	Total Zeros
4	2	0	6
4	0	2	6
2	2	2	6
0	0	6	6
0	2	4	6
2	0	4	6

EXAMPLE 7
Study of Real Zeros

Use Descartes's Rule of Signs to analyze the zeros of the equation:
$$f(x) = x^3 + 2x^2 - x + 1$$

Solution

There are two changes in the sequence of signs of $f(x)$. By Descartes's Rule of Signs, this means that there are at most two positive zeros. Moreover, the number of positive zeros differs from 2 by an even integer. This means that the number of positive zeros is either 2 or 0.

To analyze the negative zeros, we form $f(-x)$:
$$f(-x) = (-x)^3 + 2(-x)^2 - (-x) + 1 = -x^3 + 2x^2 + x + 1$$

There is a single change in sign. Descartes's Rule of Signs tells us that there is at most one negative zero. Because the actual number of negative zeros differs from 1 by an even integer, there is only one possibility: a single negative zero. Combining the various possibilities, we see that there are either 2 positive zeros and 1 negative zero, or no positive zeros, 1 negative zero, and 2 nonreal zeros.

Descartes's Rule of Signs assumes that the constant term of the polynomial is nonzero. However, this is not much of a restriction when it comes to analyzing zeros, because a polynomial with a zero constant term has a power of x as a factor. And by factoring out this power of x, we can arrive at a polynomial with a nonzero constant term to which Descartes's rule can be applied. For example, consider the polynomial:
$$f(x) = x^5 - 3x^3 + 12x^2$$
It has a zero constant term but can be written in the form:
$$f(x) = x^2(x^3 - 3x + 12)$$
The first factor on the right, x^2, corresponds to a zero of 0 with a multiplicity of 2. We can investigate the remaining zeros by applying Descartes's Rule of Signs to the polynomial factor, $x^3 - 3x + 12$.

Exercises 4.5

Factor each of the following polynomials into linear factors with complex coefficients.

1. $f(x) = 6x^4 + 5x^3 + 5x^2$
2. $f(x) = 3x^6 + 6x^5 - 3x^4$
3. $f(x) = (x^2 + 1)(x^2 + 2x + 3)$
4. $f(x) = (x^2 + 4)(x^2 - 4x - 5)$
5. $f(x) = x^3 - 1$
6. $f(x) = x^3 + 8$
7. $f(x) = x^3 + 3x^2 + 5x + 15$ (**Hint:** Consider factoring by grouping.)
8. $f(x) = x^3 - 5x^2 + 7x - 35$ (**Hint:** Consider factoring by grouping.)

9. $f(x) = (x^2 + x + 3)^2$
10. $f(x) = (x^2 - x - 5)^2$

Determine the zeros and their multiplicities of the following polynomials.

11. $f(x) = 3x(x - 2)^2(2x - 5)^3$
12. $f(x) = (2x - 3)(x - 4)^2(x + 1)^2$
13. $f(x) = (x^2 - 4)^2(x + 2)^3$
14. $f(x) = (2x^2 - 5x - 3)^3$
15. $f(x) = (x - 2)^3(x + 1)$
16. $f(x) = (x^2 - x - 6)^2(x^2 + 2x - 3)^2$
17. $f(x) = x(x - 3)^2(x^2 + 1)$
18. $f(x) = x^2(x + 5)(x^2 + 3x + 5)^2$
19. $f(x) = x^5(x^2 + x - 1)^3$
20. $f(x) = x^4(x^2 + 5)^3(x^2 - x)$

Factor each polynomial into polynomials that have real coefficients and then into polynomials that have complex coefficients. Then, find the zeros.

21. $x^6 - 64$
22. $t^6 - 1$
23. $x^3 - 1$
24. $x^3 + 1$
25. $m^3 + 8$
26. $x^3 - 8$
27. $y^4 - 4$
28. $y^4 - 9$
29. $r^7 + 2r^5 + r^3$
30. $p^6 + 2p^4 + p^2$
31. $x^4 - x^3 + 3x^2 - 4x - 4$
32. $x^4 - 2x^3 + 7x^2 - 18x - 18$
33. $x^4 - 2x^2 - 3$
34. $x^4 - 2x^2 - 35$
35. $r^3 - r^2 - 8r + 12$
36. $t^3 - 4t^2 - 12t + 48$

Use Descartes's Rule of Signs to analyze the number of positive and negative zeros of each polynomial.

37. $5x^4 + x^3 - 2x^2 - 3x + 5$
38. $8x^6 - 3x^5 - 2x^3 + x^2 - 3x + 8$
39. $11x^{10} + 8x^3 + x^2 - x + 2$
40. $x^8 - x^7 + 3x^6 + 2x^2 - 1$
41. $5x^4 + x^3 - 2x^2 - 3x + 5$
42. $8x^6 - 3x^5 - 2x^3 + x^2 - 3x + 8$
43. $x^5 - x^4 - 3x^3 + 2x^2 - x - 1$
44. $10x^4 - 3x^3 + 2x^2 + x + 4$
45. $11x^{10} + 8x^3 + x^2 - x + 2$
46. $x^8 - x^7 + 3x^6 + 2x^2 - 1$
47. $10x^6 - 3x^5 + 2x^4 + x^3 + 4x^2$
48. $x^8 - x^7 - 3x^6 + 2x^5 - x^4 - x^3$

For each polynomial, certain zeros are given. Find the remaining zeros.

49. $x^3 + 5x^2 + 9x + 5$;
$-2 - i$
50. $3x^3 + 2x^2 + 3x + 2$;
i
51. $x^4 - 3x^3 - 2x^2 + 2x$;
$2 \pm \sqrt{2}$
52. $x^4 - 4x^3 + 8x^2 - 8x - 5$;
$1 + 2i$
53. $x^4 - 6x^2 - 8x - 3$;
-1 is a zero of multiplicity 3
54. $x^4 - 5x^3 + 6x^2 + 4x - 8$;
2 is a root of multiplicity 3
55. $x^5 - 12x^3 + 46x^2 - 85x + 50$;
$-5, 1 + 2i$
56. $12x^5 + 4x^4 + 7x^3 + 14x^2 - 34x + 12$;
$-\sqrt{2}i, -\frac{3}{2}$

4.6 RATIONAL FUNCTIONS

Recall that a rational function is a function given by an expression of the form

$$f(x) = \frac{p(x)}{q(x)}$$

where $p(x)$ and $q(x)$ are polynomials with real coefficients. The domain of such a function consists of all real numbers x for which $q(x)$ is nonzero. In this section, we learn to sketch the graphs of rational functions. The significant new feature is the behavior of the functions for x near a zero of the denominator, $q(x)$.

Vertical Asymptotes

We begin with a graph of a typical rational function.

EXAMPLE 1
Graph of a Rational Function

Sketch the graph of the rational function

$$f(x) = \frac{1}{x^2}$$

Solution

First note that $f(x)$ is not defined for $x = 0$, the value at which the denominator is 0. Values of $f(x)$ for x near 0 therefore display special characteristics. Note that

$$f(-x) = \frac{1}{(-x)^2} = \frac{1}{x^2}$$

so $f(x)$ is even and its graph is symmetric with respect to the y-axis. Because the graph is symmetric with respect to an axis, we can restrict ourselves at first to the positive values of x. Here is a table of the values of $f(x)$ at some representative values that decrease through positive numbers and get increasingly close to 0. We say that these values of x **approach zero from the right.**

x	$f(x)$
1	1
0.1	100
0.01	10,000
0.001	1,000,000

The data in the table suggest that as x approaches zero from the right, the values of $f(x)$ increase without bound. Geometrically, this means that, as x approaches 0 from the right, the graph of $f(x)$ rises without bound and approaches arbitrarily close to the y-axis without ever touching it. We say that the y-axis is a **vertical asymptote** of the graph. Using the data in the table, we can make a sketch of the graph for $x > 0$. By symmetry, the portion of the graph corresponding to $x < 0$ can be obtained by reflection in the y-axis. (See Figure 1.)

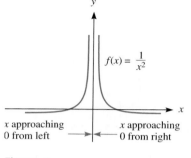

Figure 1

Suppose that the rational function $f(x)$ is undefined for $x = a$ (that is, the value a would make the denominator of the function equal to zero). As we did in Example 1 for $a = 0$, we can examine the behavior of $f(x)$ as x approaches any value a from the right. If the values of $f(x)$ increase without bound as x approaches a from the right, we say that $f(x)$ **approaches positive infinity from the right,** and we write $f(x) \to +\infty$ as $x \to a^+$. (See Figure 2(a).)[1] If, as x approaches a from the right, the values of $f(x)$ decrease without bound (say, $-1, -100, -10,000$, and so on), then we say that $f(x)$ **approaches negative infinity from the right,** and we write $f(x) \to -\infty$ as $x \to a^+$. (See Figure 2(b).)[2]

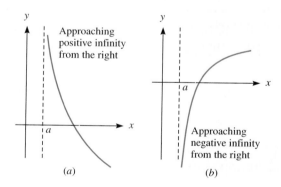

Figure 2

In a similar fashion, we can consider x approaching a from the left, that is, through values smaller than a. If the values of $f(x)$ increase without bound when x approaches a from the left, we say that $f(x)$ **approaches positive infinity from**

---

[1] In calculus, we describe this situation by saying that the limit of f as x approaches a from the right equals $+\infty$. There we use the limit notation $\lim\limits_{x \to a^+} f(x) = +\infty$.

[2] In calculus, this situation is described using the limit notation $\lim\limits_{x \to a^+} f(x) = -\infty$.

the left, and we write $f(x) \to +\infty$ as $x \to a^-$. (See Figure 3(a).)[3] If the values of $f(x)$ decrease without bound, when x approaches a from the left (see Figure 3(b)), then we say that $f(x)$ **approaches negative infinity from the left,** and we write $f(x) \to -\infty$ as $x \to a^-$. (See Figure 3(b).)[4]

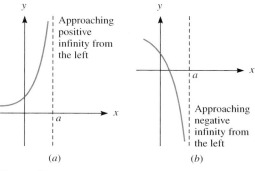

Figure 3

Definition **Vertical Asymptote**	If, as x approaches a, the graph of a rational function $f(x)$ approaches positive infinity or negative infinity from either the right or the left, then we say that the line $x = a$ is a vertical asymptote of the graph. An asymptote that is not one of the axes is indicated by a dashed line on the graph.

Graphs of rational functions (assumed to be written in lowest terms) have a vertical asymptote corresponding to each zero of the denominator. It is necessary to allow x to approach such values from both the right and the left to determine the behavior of the graph in the vicinity of the asymptote. The next two examples illustrate how to do this.

EXAMPLE 2
Graph with Vertical Asymptote

Sketch the graph of the rational function

$$y = \frac{1}{x + 1}$$

Solution

The function is undefined for $x = -1$, the value at which the denominator equals 0. This means that the line $x = -1$ is a vertical asymptote of the graph. To determine the behavior of the graph in the vicinity of this asymptote, we allow x to approach -1 from both the right and the left. As x approaches -1 from the right, $x > -1$, so the value of $f(x)$ is positive. Moreover, as x approaches -1 from the right, the values of $f(x)$ increase without bound. That is, they approach positive infinity. This information allows us to sketch the graph for $x > -1$. We first sketch the asymptote $x = -1$. Then we draw in the portion of the graph to the right of the asymptote. See Figure 4. Next, we consider values of x when x approaches -1 from the left. These values for x are less than -1, so $x + 1$ is negative, as are the values of $f(x)$. Moreover, as x approaches -1 from the left, the values of $f(x)$ decrease without bound. That is, $f(x)$ approaches negative infinity. See Figure 4.

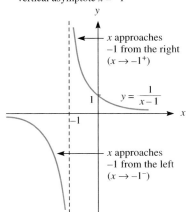

Figure 4

[3] In calculus, this situation is described using the limit notation $\lim_{x \to a^-} f(x) = +\infty$.

[4] In calculus, this situation is described using the limit notation $\lim_{x \to a^-} f(x) = -\infty$.

Horizontal Asymptotes

In drawing an accurate sketch of a graph, it is helpful to indicate how the function behaves for values of x that are very large in either the positive or the negative direction. These values lie far to the right or left on the number line.

If x increases without bound through positive numbers, then we say that x **approaches positive infinity.** We can visualize this concept by imagining the value of x as represented by a point on the number line moving indefinitely far to the right. If x decreases without bound through negative numbers, then we say that x **approaches negative infinity.** We can visualize this concept by imagining the value of x as represented by a point on the number line moving indefinitely far to the left.

As x approaches either positive or negative infinity, the values of a rational function $f(x)$ may exhibit several different behaviors. On the one hand, the value of $f(x)$ may approach positive infinity or negative infinity. These possibilities are shown in Figures 5(a) and (b), respectively.

On the other hand, the values of $f(x)$ may approach a real number c. Two ways in which this can occur are illustrated in Figure 5(c).

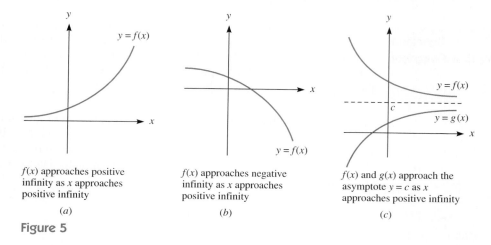

$f(x)$ approaches positive infinity as x approaches positive infinity
(a)

$f(x)$ approaches negative infinity as x approaches positive infinity
(b)

$f(x)$ and $g(x)$ approach the asymptote $y = c$ as x approaches positive infinity
(c)

Figure 5

Definition
Horizontal Asymptote

Suppose that the values of the rational function $f(x)$ approach a real number c as x approaches either positive or negative infinity. We then say that the line $y = c$ is a **horizontal asymptote** of the graph of $f(x)$.

As x approaches positive infinity, so do all positive powers of x. However, negative powers of x approach 0 because the reciprocal of a large number is small.

As x approaches negative infinity, negative powers of x still approach 0. However, a positive, even power of x approaches positive infinity and a positive, odd power of x approaches negative infinity. A positive, even power of a large negative number is a large positive number; a positive, odd power of a large negative number is a large negative number.

Using these facts, we can determine the behavior of a rational function as x approaches positive infinity or negative infinity. The next example illustrates how this can be done.

EXAMPLE 3
Graph with Horizontal Asymptotes

Determine the horizontal asymptotes, if any, of the rational function:

$$f(x) = \frac{x^2}{3x^2 - 4x - 1}$$

Solution

Divide the numerator and denominator by the largest power of x in the denominator, namely x^2, to obtain the following expression for $f(x)$:

$$f(x) = \frac{\frac{x^2}{x^2}}{\frac{3x^2 - 4x - 1}{x^2}} = \frac{1}{3 - \frac{4}{x} - \frac{1}{x^2}}$$

As x approaches either positive infinity or negative infinity, the terms

$$\frac{4}{x}, \quad \frac{1}{x^2}$$

approach 0 and the value of $f(x)$ approaches:

$$\frac{1}{3 + 0 + 0} = \frac{1}{3}$$

That is, the graph has as a horizontal asymptote at the line $y = \frac{1}{3}$. This asymptote is indicated by a dashed line, along with a sketch of the graph, in Figure 6.

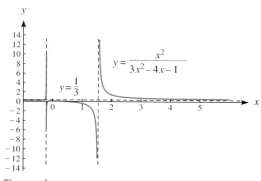

Figure 6

Note that whereas a graph can never cross a vertical asymptote, it is perfectly possible for a graph to cross a horizontal asymptote. In fact, some graphs oscillate an infinite number of times around a horizontal asymptote, approaching closer with each oscillation.

Oblique Asymptotes

A line is said to be oblique if it is nonhorizontal and nonvertical. In our discussion above, we introduced both horizontal and vertical asymptotes. We also have the concept of an **oblique asymptote**, which is a line that a graph approaches as x approaches either positive infinity or negative infinity. (See Figure 7.)

Here is an example to illustrate how oblique asymptotes arise. Consider the rational function $f(x) = (x^2 + 3x + 1)/x$. By performing the indicated division, we arrive at the following alternate expression for $f(x)$.

$$f(x) = x + 3 + \frac{1}{x}$$

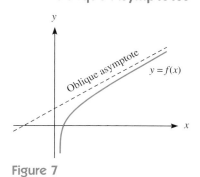

Figure 7

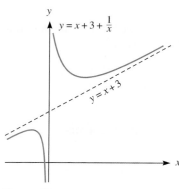

Figure 8

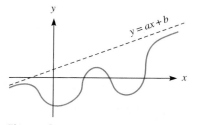

$y = ax + b + \dfrac{r(x)}{q(x)}$
degree of $r(x)$ < degree of $q(x)$

Figure 9

It is clear that as x approaches either positive infinity or negative infinity the term $1/x$ approaches 0, and so the function value approaches the value of the linear function $x + 3$. Geometrically, this means that the graph of $f(x)$ approaches the graph of $x + 3$ as x approaches either positive or negative infinity. That is, the line $y = x + 3$ is an oblique asymptote of the graph. (See Figure 8.)

In a similar fashion, suppose that we are given a rational function $f(x) = p(x)/q(x)$, where the degree of $p(x)$ exceeds the degree of $q(x)$ by 1. Then we can perform division to write the rational function in the form

$$f(x) = ax + b + \dfrac{r(x)}{q(x)}$$

where the degree of $r(x)$ is less than the degree of $q(x)$. This assures us that $r(x)/q(x)$ approaches 0 as x approaches $+\infty$ or $-\infty$. In this case, the line $y = ax + b$ is an oblique asymptote of the graph of $f(x)$. (See Figure 9.)

We can summarize our discussion of horizontal and oblique asymptotes as follows:

ASYMPTOTES OF RATIONAL FUNCTIONS

Let $f(x) = p(x)/q(x)$ be a rational function, where $p(x)$ is a polynomial of degree m and with leading coefficient c, and where $q(x)$ is a polynomial of degree n and with leading coefficient d.

1. If $m < n$, then the x-axis is a horizontal asymptote of the graph of $f(x)$; that is, $f(x)$ approaches 0 as x approaches either positive infinity or negative infinity.
2. If $m = n$, then the line $y = c/d$ is a horizontal asymptote of the graph of $f(x)$; that is, $f(x)$ approaches c/d as x approaches either positive infinity or negative infinity.
3. If $m = n + 1$, then we can write $f(x)$ in the form
$$f(x) = ax + b + \dfrac{r(x)}{q(x)}$$
where $r(x)$ has degree less than n. In this case, $y = ax + b$ is an oblique asymptote of the graph of $f(x)$. That is, $f(x)$ approaches either positive or negative infinity along the path of a line, the oblique asymptote.
4. If $m > n$, then the graph of $f(x)$ has no horizontal asymptotes. Moreover, as x approaches positive infinity or negative infinity, $f(x)$ approaches either positive infinity or negative infinity.

Examples of Asymptotes

Let's now work out some examples that use everything we have learned about sketching graphs of rational functions.

EXAMPLE 4
Asymptotes of Rational Functions

Describe the behaviors of the following rational functions as x approaches positive infinity and negative infinity.

1. $f(x) = \dfrac{1}{2x - 1}$
2. $f(x) = \dfrac{2x^3 - 1}{(x - 1)(x + 1)}$
3. $f(x) = \dfrac{(x - 1)(3x - 5)}{(2x + 7)(x + 1)}$

Solution

1. In this example, the degree of the numerator is less than the degree of the denominator. By the first rule for the asymptotes of rational functions, the graph has the x-axis as a horizontal asymptote. That is, as x approaches positive or negative infinity, the values of $f(x)$ approach 0.
2. In this case, the degree of the numerator exceeds the degree of the denominator by one, so the graph has no horizontal asymptotes but has an oblique asymptote. According to the third rule, the oblique asymptote is determined by dividing the numerator by the denominator and writing the result in the form $ax + b + r(x)/q(x)$:

$$\frac{2x^3 - 1}{x^2 - 1} = 2x + \frac{2x - 1}{x^2 - 1}$$

   This equation shows that the oblique asymptote is $y = 2x$.
3. In this case, the degree of the numerator equals the degree of the denominator. So by the second rule, the graph has a horizontal asymptote. The leading coefficient of the numerator is 3 and the leading coefficient of the denominator is 2, so the horizontal asymptote is the line $y = \frac{3}{2}$. As x approaches either positive infinity or negative infinity, the value $f(x)$ approaches $\frac{3}{2}$.

Sketching Graphs of Rational Functions

Here is a general, organized approach to graphing rational functions.

GRAPHING RATIONAL FUNCTIONS

Suppose that $f(x) = p(x)/q(x)$ is a rational function. Follow these steps to sketch the graph of $f(x)$:

1. Write the given expression of the rational function in lowest terms. Restrict the domain to exclude the zeros of any canceled factor.
2. Determine the vertical asymptotes by calculating the real zeros of the denominator. For each zero a, the line $x = a$ is a vertical asymptote.
3. Determine the horizontal and oblique asymptotes. If there are none, determine the behavior of $f(x)$ as x approaches both positive infinity and negative infinity.
4. Determine the zeros of $f(x)$ by determining the zeros of the numerator. These are the x-intercepts of the function.
5. Divide the x-axis into intervals determined by the real zeros of $f(x)$ and by the vertical asymptotes. For each interval, choose a test value and determine the sign of $f(x)$ in that interval. Then use the signs to determine whether the graph is above or below the x-axis in each interval.
6. Draw the horizontal and vertical asymptotes and plot the zeros.
7. Draw the portion of the graph that corresponds to each interval using the information developed in steps 1–6.

The next example illustrates how to apply this general approach to sketching a graph with vertical asymptotes.

EXAMPLE 5
A Graph with Horizontal and Vertical Asymptotes

1. Determine all asymptotes of the rational function:
$$f(x) = \frac{(x-2)(x-4)}{(x-1)(x-3)}$$
2. Sketch the graph of the function.

Solution

1. There is a vertical asymptote corresponding to each zero of the denominator. There are two such zeros, namely 1 and 3, so the vertical asymptotes are the lines $x = 1$ and $x = 3$. Because the numerator and denominator have the same degree, there is a horizontal asymptote. In finding the horizontal asymptote, use the expanded form of $f(x)$:
$$f(x) = \frac{x^2 - 6x + 8}{x^2 - 4x + 3}$$
The leading coefficients of the numerator and denominator are each 1, so the horizontal asymptote is the line $y = \frac{1}{1} = 1$.
2. To sketch the graph, we must first determine the zeros of $f(x)$. These are the zeros of the numerator. In this case, the numerator is factored for us, so we can read off the zeros directly, namely 2 and 4. The zeros and the vertical asymptotes determine a division of the x-axis into intervals:
$$(-\infty, 1), (1, 2), (2, 3), (3, 4), (4, +\infty)$$
In each of these intervals, the value of $f(x)$ has a constant sign. Let's choose a test point in each interval and determine the corresponding sign. The results can be summarized in the following table:

Interval	Test Point	Sign of $f(x)$
$(-\infty, 1)$	0	+
$(1, 2)$	$\frac{3}{2}$	−
$(2, 3)$	$\frac{5}{2}$	+
$(3, 4)$	$\frac{7}{2}$	−
$(4, +\infty)$	5	+

Finally, we turn to the graph itself. First, we draw asymptotes and plot the zeros. We now plot the section of the graph lying in each interval. For the leftmost interval, $(-\infty, 1)$, the graph approaches the horizontal asymptote $y = 1$ from above, as x approaches negative infinity. On the right side of the interval, the graph must approach the vertical asymptote $x = 1$. The graph must head upward as it does so, because otherwise it would cross the x-axis within the interval, and there are no zeros of $f(x)$ within the interval. In the second interval, $(1, 2)$, the function value is negative. This means that the graph must head downward toward the asymptote on the left. Moreover, the graph crosses the x-axis at $x = 2$. In the third interval, $(2, 3)$, the graph is positive. So the graph must increase as it approaches the asymptote $x = 3$ on the right. In the fourth interval, $(3, 4)$, the function value is negative. This means that the graph must head downward toward the asymptote on the left. Moreover, the graph crosses the x-axis at $x = 4$. In the fifth interval, $(4, +\infty)$, the graph rises from the x-axis on the left and must approach the horizontal asymptote $y = 1$ from below, as x approaches positive infinity. The completed graph is sketched in Figure 10.

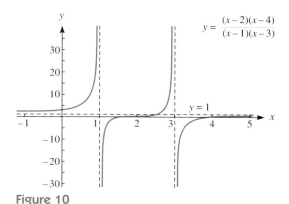

Figure 10

EXAMPLE 6
Inventory Management

The cost $C(x)$ of managing an inventory of x cases of a certain brand of canned beans is determined by a team of financial analysts to be given by the function:

$$C(x) = \frac{10,000}{x} + 3x$$

Sketch the graph of $C(x)$.

Solution

Because x represents a number of cases of canned goods, x must be positive. So the domain of the function $C(x)$ is the set of all positive integers. If we write the expression for $C(x)$ in the form

$$\frac{3x^2 + 10,000}{x}$$

we see that $C(x)$ is a rational function. The only zero of the denominator is 0, so there is a single vertical asymptote, namely $x = 0$. Moreover, because the degree of the numerator is one more than the degree of the denominator, there are one oblique asymptote and no horizontal asymptotes. If we write

$$C(x) = 3x + \frac{10,000}{x}$$

then we see that the line $y = 3x$ is an oblique asymptote. We now sketch the graph, beginning with the single vertical asymptote, $x = 0$. Because $C(x)$ is positive throughout the domain, the graph must rise as it approaches the asymptote from the right. We draw the oblique asymptote and show the graph approaching this asymptote as x approaches positive infinity. The completed sketch is shown in Figure 11.

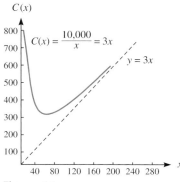

Figure 11

In case the numerator and denominator of a rational function have a common factor, the rational function is not defined at the zeros of the common factor, and the graph has "holes" in it corresponding to the values of x at which the function is undefined. The next example illustrates this phenomenon.

EXAMPLE 7
Rational Function with Cancellation

Sketch the graph of the function:

$$f(x) = \frac{x^2 - 25}{x - 5}$$

Solution

Because

$$\frac{x^2 - 25}{x - 5} = \frac{(x + 5)(x - 5)}{x - 5}$$

we see that $f(x)$ has the equivalent expression $f(x) = x + 5$ as long as x is not equal to a zero of the common factor $x - 5$. That is, we have

$$f(x) = x + 5, \quad x \neq 5$$

The graph of this function is shown in Figure 12. The empty circle indicates an interruption in the graph at the point where $f(x)$ is undefined.

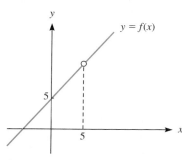

Figure 12

Graphing Rational Functions with a Graphing Calculator

A graphing calculator makes graphing even very complicated rational functions easy. You may then wonder why it is important to be able to sketch these graphs by hand, or to understand the algebra, if the technology will do all of the graphing for you. The answer to this is that sometimes a graph on a calculator does not look anything like the graph of the function.

For instance, look at Figure 13. This is the graph of

$$f(x) = \frac{(x - 2)(x - 4)}{(x - 1)(x - 3)}$$

displayed on a graphing calculator, using the x-range $-1 \leq x \leq 1$. The graph of this function was also sketched in Figure 10, in Example 5. The two graphs do not look similar at all. The reason is that the x range $-1 \leq x \leq 1$ does not show all of the vertical asymptotes. Without a knowledge of asymptotes and rational functions, you might be fooled into thinking Figure 13 is what the graph looks like.

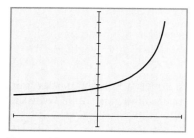

Figure 13

Here is another example of how you could get a misleading graph from a graphing utility. Figure 14 was produced on a graphing calculator. It is the graph of:

$$f(x) = \frac{x^2 - 25}{x - 5}$$

Again, notice that this is not the same graph that was shown in Figure 12, in Example 7. The reason is that the graphing calculator can only evaluate the function at a certain number of points, and it uses these points to plot the graph. It happened that $x = 5$ was not one of the points evaluated, so the graph does not show the hole that we know is there. Even using a graphing utility's zoom feature does not guarantee that it will show these holes. Figure 15 shows the same function graph in the range $4.95 \leq x \leq 5.05$. Even this close-up view does not show the hole. Without some understanding of the concepts involved, the technology could mislead us.

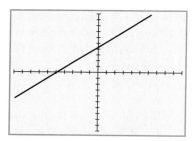

Figure 14

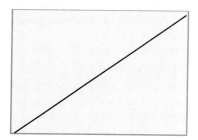

Figure 15

SECTION 4.6 Rational Functions

EXAMPLE 8
Graphing a Rational Function Using a Calculator

Determine the asymptotes of the function $f(x) = (x + 1)/(x^2 - x - 4)$.

Solution

Graph the function for $-3 \leq x \leq 3$. See Figure 16. We see that there is a single horizontal asymptote, namely $y = 0$, and two vertical asymptotes. Using the trace and observing the value of the x-coordinate, we determine that the vertical asymptotes are $x = a$ and $x = b$, where $a \approx -1.55$, and $b \approx 2.68$.

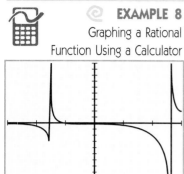

Figure 16

EXAMPLE 9
Graphing Calculator Solution to an Optimization Problem

In Example 6, we graphed the function

$$C(x) = \frac{10{,}000}{x} + 3x$$

which gives the cost of managing an inventory of x cases of canned beans. We saw that this function had a single minimum point. Approximate the minimum value of this function.

Solution

First use a graphing calculator to graph the function within the range $0 \leq x \leq 200$, $0 \leq y \leq 3{,}000$. See Figure 17. Use the trace to locate this point. Move from left to right and locate the point at which the y-coordinate stops decreasing. This point is approximately $(56.842105, 346.45224)$. Therefore, the minimum value of the function is approximately 346.45.

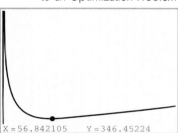

Figure 17

Exercises 4.6

Match the functions in Exercises 1–6 with their graphs.

1. $f(x) = \dfrac{x}{(x-1)(x+2)}$

2. $f(x) = \dfrac{1}{x+3}$

3. $f(x) = \dfrac{x^2 + 1}{x}$

4. $f(x) = \dfrac{1}{x(x+1)(x+2)}$

5. $f(x) = \dfrac{x^2}{x^2 + 1}$

6. $f(x) = \dfrac{1}{x^2 + 1}$

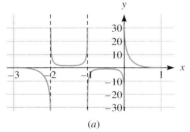

(a)

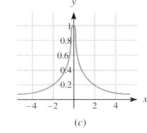

(c)

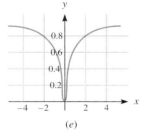

(e)

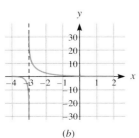

(b)

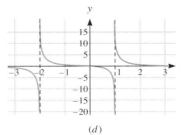

(d)

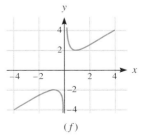

(f)

The graphs of rational functions $f(x)$ and $g(x)$ are shown in figures (*a*) and (*b*). Sketch the graphs of the functions in Exercises 7–12 without using a graphing calculator.

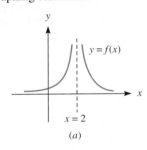

(a)

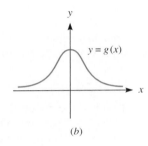

(b)

7. $f(x + 1)$
8. $f(x) + 3$
9. $f(x - 2) + 3$
10. $g(x + 1) + 1$
11. $g(x - 4)$
12. $g(x) - 2$

For each function, do the following:

a. Determine all the asymptotes.
b. Sketch the graph of the function.

You may use a graphing calculator.

13. $f(x) = \dfrac{2}{x^2}$
14. $f(x) = \dfrac{1}{x^3}$
15. $f(x) = -\dfrac{1}{x^3}$
16. $f(x) = -\dfrac{3}{x^2}$
17. $f(x) = \dfrac{1}{x + 1}$
18. $f(x) = \dfrac{1}{x - 1}$
19. $f(x) = -\dfrac{2}{x + 2}$
20. $f(x) = -\dfrac{2}{(x - 3)^2}$
21. $f(x) = \dfrac{1}{(x + 3)^2}$
22. $f(x) = \dfrac{x + 1}{2 - x}$
23. $f(x) = \dfrac{3 - x}{x + 2}$
24. $f(x) = \dfrac{2x + 3}{16 - 5x}$
25. $f(x) = \dfrac{8 - 2x}{10 + 3x}$
26. $f(x) = \dfrac{1}{x^2 + 1}$
27. $f(x) = -\dfrac{1}{x^2 + 3}$
28. $f(x) = \dfrac{1}{x^2 - 4}$
29. $f(x) = -\dfrac{1}{x^2 - 9}$
30. $f(x) = \dfrac{2x}{x^2 - x - 6}$
31. $f(x) = \dfrac{x - 3}{2x^2 + 5x - 3}$
32. $f(x) = \dfrac{x^2 - 9}{x + 2}$
33. $f(x) = \dfrac{x^2 - 4}{x - 3}$
34. $f(x) = \dfrac{x^2 - 1}{x^2 + x - 6}$
35. $f(x) = \dfrac{x^2 - 9}{x^2 - x - 2}$
36. $f(x) = \dfrac{x^2 - x - 2}{x + 2}$
37. $f(x) = \dfrac{x^2 - 2x - 8}{x + 4}$
38. $f(x) = x + \dfrac{4}{x}$
39. $f(x) = x + \dfrac{1}{x}$
40. $f(x) = \dfrac{2x^2 - x - 3}{3x^2 - 2x - 8}$
41. $f(x) = \dfrac{x^2 + 2x - 3}{3x^2 - 1}$
42. $f(x) = \dfrac{x^3 + 1}{x}$
43. $f(x) = \dfrac{x^3 - 2}{4x}$
44. $f(x) = \dfrac{x^2 - 1}{x - 1}$
45. $f(x) = \dfrac{x^2 - 9}{x + 3}$
46. $f(x) = \dfrac{(x + 2)(x^2 - 9)}{x + 2}$
47. $f(x) = \dfrac{(x - 1)(3 - x)^2}{x - 1}$
48. $f(x) = \dfrac{x^2 + 2x - 15}{x + 5}$
49. $f(x) = \dfrac{x^2 + x - 2}{x - 1}$
50. $f(x) = \dfrac{x^2 - 4x + 3}{x^2}$
51. $f(x) = \dfrac{4x^2}{x^2 - 1}$
52. $f(x) = \dfrac{x^3 - 2x^2 - 8x}{x^2 - 9}$
53. $f(x) = \dfrac{x^3 + 2x^2 - 3x}{x^2 - 25}$
54. $f(x) = \dfrac{x^3 - 4x^2 + x + 6}{x^2 + x - 2}$

Applications

You may use a graphing calculator to solve the following problems.

55. **Earned-run average.** The earned-run-average, or ERA, of a pitcher is given by $ERA = 9R/I$, where $R =$ the number of earned runs allowed and $I =$ the number of innings pitched.

    a. Suppose $R = 1$. Graph the function over the interval $(0, 20]$.
    b. Suppose $R = 2$. Graph the function over the interval $(0, 20]$.
    c. Suppose $R = 5$. Graph the function over the interval $(0, 20]$.

56. **Distance as a function of time.** The distance from Dallas, Texas, to El Paso, Texas, is 617 miles. A car makes the trip in t hours, at various speeds r, in miles per hour.
   a. Find a rational function for t in terms of r.
   b. Graph the function.

57. **Loudness of Sound.** A person is sitting at a distance d from a stereo speaker (see figure). The loudness L of the sound varies inversely as the square of the distance d according to the equation

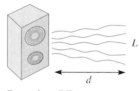

Exercise 57

$$L = \frac{k}{d^2}$$

where k depends on the original sound intensity. Graph this function assuming $k = 0.1$.

58. **Inventory management.** The cost $C(x)$ of managing an inventory of x television sets is determined by a team of financial analysts to be given by the function:

$$C(x) = 5x + \frac{50,000}{x} + 22,500$$

Sketch the graph of this function over the interval $(0, +\infty)$.

59. **Weight above the earth.** A person's weight W_h at a height h above sea level satisfies the equation

$$W_h = \left(\frac{r}{r+h}\right)^2 W_0$$

where W_0 is the person's weight at sea level, h is the height, in miles, of the person above sea level, and r is the radius of the earth, in miles. Assume that $r \approx 4{,}000$ miles.
   a. Suppose a person weighs 200 lb at sea level. Find a rational function for the person's weight h miles above the earth.
   b. Graph this function on the interval [0, 16,000].
   c. At what height above the earth's surface would the 200 lb person weigh 100 lb?

In your own words

60. Write a detailed procedure for determining the vertical asymptotes of a rational function.

61. Write a detailed procedure for determining the horizontal asymptotes of a rational function.

62. Given an applied example of a rational function with a horizontal asymptote. Explain the applied significance of the horizontal asymptote.

63. Give an applied example of a rational function with a vertical asymptote. Explain the applied significance of the vertical asymptote.

Mathematics and the World around Us—Should Mathematical Results Be Secret?

Mathematicians usually work in the privacy of their offices, developing new mathematical ideas and applying existing results to a host of real-world problems. Usually, their work does not have the immediacy and urgency of work in other fields, such as medicine or economics. An exception is the field of cryptography, where the work of mathematicians gives rise to debates on ethics, morality, and public policy.

Cryptography has its origins in coded messages of war. And some of the greatest intellectual triumphs of the field have occurred during wartime. During World War II, for example, England assembled a group of Cambridge professors at a small town called Bletchley. Their purpose was to decode ciphers used by the Nazi regime. (One member of this group, Alan Turing, later became famous when he founded the modern field of computer science.)

The most sophisticated code then in use was named ENIGMA and was used for communications between members of the German general staff. The Bletchley group succeeded in decoding ENIGMA (with the help of a stolen enciphering machine from a Czech defector). Throughout the war, English Prime Minister Winston Churchill and his staff received messages from the German high command almost as quickly as German commanders in the field. The fact that ENIGMA was cracked was one of the most closely guarded secrets of the war and was not revealed until several decades after peace was declared. Of such perceived importance was the secret, that Churchill allowed the city of Coventry, England, to be bombed, even though he had enough advance knowledge of the raid to evacuate the city in time.

Should results in cryptography be kept secret? Most mathematicians are used to working quite openly, and they freely lecture all over the world and circulate preprints and reprints of their papers to any interested colleagues.

Alan Turing

Chapter Review

Important Concepts—Chapter 4

- zero of a polynomial
- Intermediate Value theorem
- extreme point; turning point
- maximum point
- minimum point
- long division of polynomials
- division algorithm for polynomials
- Remainder theorem
- Factor theorem
- Number of Zeros theorem
- synthetic division
- bisection method for approximating real zeros of a polynomial
- Rational Zero theorem
- upper and lower bounds for zeros of a polynomial
- complex number
- imaginary unit
- real part
- imaginary part
- equality of complex numbers

Winston Churchill

But should this apply to results in cryptography? A policy debate on this question has been raging throughout the West for the last several decades.

In the 1970s, the National Security Agency, the U.S. government agency charged with creating and breaking ciphers, attempted to start a program of self-censorship by mathematics journals. They asked that all papers in cryptography be submitted to them for clearance so that no national security interests would be violated by publishing them. After a spirited debate within the mathematical and scientific community, this program was abandoned.

With the advent of unbreakable codes (see Chapter 1, Mathematics and the World around Us), codes of incredible complexity could be easily implemented on personal computers. The U.S. government declared it illegal to export cryptography technologies. For the most part, they were able to bar export of computer chips that implemented cryptographic algorithms. However, the algorithms themselves are easily sent by electronic mail. Throughout the 1980s, they spread like wildfire all over the world, in spite of U.S. restrictions.

The impact of the cryptography programs has been mixed. On the one hand, during the 1993 battle between Boris Yeltsin and the Russian parliament, dissidents all over Russia used a program imported from the United States to encrypt exchanges. Many of the dissidents credit their success to the encrypted messages. However, the same program was used to encrypt the computer diary of a convicted California child pornographer and the police were unable to decrypt it to obtain information about a child pornography ring.

Is it illegal to send such encryption programs over computer networks and thereby aid in their export? As of this writing, a federal grand jury in California is examining this question. (See *Wall Street Journal*, April 28, 1994, p. 1.) Even if it is decided that laws were violated, should such laws be on the books?

- addition and multiplication of complex numbers
- additive identity for complex numbers
- additive inverse of a complex number
- multiplicative identity for complex numbers
- powers of i
- principal square root of a negative real number
- factoring a sum of squares
- complex roots of a quadratic equation
- imaginary number; imaginary
- complex solutions
- conjugate of a complex number
- multiplicative inverse of a complex number
- division of complex numbers
- Fundamental Theorem of Algebra
- multiplicity of a zero
- number of zeros of a polynomial
- Conjugate Root theorem
- polynomial factorization over the real numbers
- Descartes's Rule of Signs
- values of a function approach zero from the right (or left)
- vertical asymptote
- values of a function approach positive infinity from the right (or left)
- values of a function approach negative infinity from the right (or left)
- horizontal asymptote
- oblique asymptote
- asymptotes of rational functions
- graphing rational functions

Important Results and Techniques—Chapter 4

Synthetic division	To divide $f(x)$ by $x - a$ using synthetic division: 1. Write the coefficients of f across the top row. Be sure to include a coefficient for each power of x. For powers that don't explicitly appear, write 0. 2. Start with the leftmost column. Bring down the leading coefficient into the third row. 3. Multiply the new entry in the third row by a. 4. Move one column to the right, and put the result of step 3 in the second row. 5. Add rows 1 and 2 of the current column and put the result in the third row. 6. Repeat steps 3–5 for each of the columns in turn. 7. The numbers in the third row are the coefficients of the quotient, except for the rightmost number, which is the remainder.	p. 220
Upper and lower bounds for zeros of a polynomial	Let $f(x)$ be a polynomial with real coefficients and with a positive leading coefficient. 1. Let $a > 0$ be chosen so that the third row in synthetic division of $f(x)$ by $x - a$ has all positive or zero entries. Then all real zeros of $f(x)$ are less than or equal to a. The number a is called an **upper bound** for the zeros of $f(x)$. 2. Let $b < 0$ be chosen so that the third row in the synthetic division of $f(x)$ by $x - b$ has alternating signs. (We assume a zero coefficient always causes an alternation of signs.) Then all real zeros of $f(x)$ are greater than or equal to b. The number b is called a **lower bound** for the zeros of $f(x)$.	p. 228
Complex numbers	Let $\alpha = a + bi$ and $\beta = c + di$ be complex numbers. $$i^2 = -1$$ $$\alpha + \beta = (a + c) + (b + d)i$$ $$\alpha \cdot \beta = (ac - bd) + (ad + bc)i$$	p. 231–232

Conjugate of a complex number	$\overline{a+bi} = a-bi$	p. 236
Descartes's Rule of Signs	The number of positive real zeros is at most equal to the number of sign changes M of $f(x)$, and differs from M by a multiple of 2. The number of negative real zeros is at most equal to the number of sign changes N of $f(-x)$, and differs from N by a multiple of 2.	p. 243
Asymptotes of rational function	If $f(x) = p(x)/q(x)$, where $p(x)$ is a polynomial of degree m and has leading coefficient c, and $q(x)$ is a polynomial of degree n and has leading coefficient d, then the following rules apply: 1. If $m < n$, then the x-axis is a horizontal asymptote of the graph of $f(x)$, and $f(x)$ approaches 0 as x approaches either positive infinity or negative infinity. 2. If $m = n$, then the line $y = c/d$ is a horizontal asymptote of the graph of $f(x)$, and $f(x)$ approaches c/d as x approaches either positive infinity or negative infinity. 3. If $m = n + 1$, then we can write $f(x)$ in the form $f(x) = ax + b + r(x)/q(x)$, where $r(x)$ has degree less than n. In this case, $y = ax + b$ is an oblique asymptote of the graph of $f(x)$. 4. If $m > n$, then the graph of $f(x)$ has no horizontal asymptotes. Moreover, as x approaches positive infinity or negative infinity, $f(x)$ approaches either positive infinity or negative infinity.	p. 250
Graphing rational functions	1. Determine any common factor of the numerator and denominator. The zeros of this common factor are points at which the rational function is undefined. Write the given expression of the rational function in lowest terms. Restrict the domain to exclude the zeros of any canceled factor. 2. Determine the vertical asymptotes by calculating the real zeros of the denominator. For each zero a, the line $x = a$ is a vertical asymptote. 3. Determine the horizontal and oblique asymptotes. If there are none, determine the behavior of $f(x)$ as x approaches both positive infinity and negative infinity.	p. 251

4. Determine the zeros of $f(x)$ by determining the zeros of the numerator. These are the x-intercepts of the function.
5. Divide the x-axis into intervals determined by the real zeros of $f(x)$ and by the vertical asymptotes. For each interval, choose a test value and determine the sign of $f(x)$ in that interval. Then use the signs to determine whether the graph is above or below the x-axis in each interval.
6. Draw the horizontal and vertical asymptotes and plot the zeros.
7. Draw the portion of the graph that corresponds to each interval using the information developed in steps 1–6.

Review Exercises—Chapter 4

In the following exercises, a polynomial $f(x)$ and a divisor $g(x)$ are given. Calculate the quotient $q(x)$ and the remainder $r(x)$.

1. $f(x) = x^5 + 3x^4 - 2x^2 + 6x - 3$
   $g(x) = x - 1$
2. $f(x) = x^5 + 3x^4 - 2x^2 + 6x - 3$
   $g(x) = x^3 - x^2 + 4$
3. Suppose that $f(x)$ is a quadratic polynomial with zeros $-2 + \sqrt{3}$ and $-2 - \sqrt{3}$ and assume that $f(-1) = 5$. Determine $f(x)$.
4. Suppose that $f(x)$ is a quadratic polynomial with zeros $\dfrac{1+3i}{2}$ and $\dfrac{1-3i}{2}$ and suppose that $f(-1) = 5$. Determine $f(x)$.

Use synthetic division to find the quotient and the remainder when $f(x)$ is divided by $g(x)$.

5. $f(x) = 4x^5 + 6x + 1$
   $g(x) = x + 2$
6. $f(x) = x^4 - 2x^3 - 5x^2 - x + 1$
   $g(x) = x + 1$
7. $f(x) = \dfrac{1}{3}x^2 + \dfrac{5}{6}x - 3$
   $g(x) = 2x + 9$
8. $f(x) = x^4 - 2x^3 - 2x - 1$
   $g(x) = x + i$

Use synthetic division to determine the given function values.

9. $f(x) = x^4 - 2x^3 - 2x - 1$; find $f(0)$, $f(-1)$, and $f(i)$.
10. $f(x) = x^5 + x^4 + 2x^3 - 2x^3 - 2x^2 + x - 1$; find $f(-1)$, $f(2)$, and $f(-i)$.

Use synthetic division to determine whether each number is a zero of the given polynomial.

11. $2, 3, -2$; $f(x) = x^3 - 9x^2 + 23x - 15$
12. $-1, i, 2$; $f(x) = x^4 - x^3 - 7x^2 - 7x - 2$

Factor each polynomial $f(x)$. Then use the result of the factoring to solve the equation $f(x) = 0$.

13. $f(x) = x^3 - 125$
14. $f(x) = x^3 - 6x^2 + 3x + 10$

For each polynomial, determine the zeros and their multiplicities.

15. $f(x) = 5x^3(x - 4)^2(2x + 3)$
16. $f(x) = (3x^3 - 6x^2 + 3x)^2$

Use Descartes's Rule of Signs to analyze the zeros of each polynomial.

17. $6x^4 - 30x^3 - 32x^2 - 23 + 2$
18. $3x^6 - 5x^4 + 2x^3 - x - 4$

For each polynomial, certain zeros are given. Find the remaining zeros.

19. $f(x) = 2x^4 + 4x^3 + 3x^2 - 6x - 9$; $-1 - i\sqrt{2}$
20. $f(x) = x^3 - x^2 - 7x + 3$;   3
21. $f(x) = x^4 + 2x^3 + x^2 + 60x + 144$;
    -3 is a root of multiplicity 2
22. $f(x) = x^4 - 4x^3 + 18x^2 + 8x - 40$; $2 + 4i$

Find the rational solutions of the following equations.

23. $6x^4 - 30x^3 - 32x^2 - 23x + 2 = 0$

24. $3x^6 - 5x^4 + 2x^3 - x - 4 = 0$
25. $8x^5 + 3x^4 - x^3 + 5x^2 - 3 = 0$
26. $x^5 - 2x^4 + x^3 - 1 = 0$

Solve the following equations.

27. $(x - 1)(x^2 + x + 1) = 0$
28. $x^3 + 31x = 10x^2 + 30$
29. $5x^2 = x^3 + 5x + 3$
30. $12x^6 - 52x^5 + 91x^4 - 82x^3 + 40x^2 - 10x + 1 = 0$
31. $x^3 + 2 = 2x^2 + 7x$
32. $x^4 + x^2 + 1 = 0$

Show that each of the following polynomials has a zero in the given interval.

33. $f(x) = x^3 - 2x - 5$; $(2, 3)$
34. $f(x) = x^4 - x^3 - 2x^2 - 6x - 4$; $(-1, 0)$

Use the method of bisection to approximate the real zeros of each of the following polynomial functions in the indicated interval with an accuracy of 0.0001.

35. $f(x) = x^3 - 2x - 5$; $(2, 3)$
36. $f(x) = x^4 - x^3 - 2x^2 - 6x - 4$; $(-1, 0)$

Sketch graphs of each of the following functions.

37. $f(x) = (x^2 - 1)(x^2 - 4)$
38. $f(x) = 4x^2 - x^3$
39. $f(x) = (x + 4)(x - 1)(x - 3)$
40. $f(x) = -(x - 2)^3 - 3$
41. $f(x) = (x + 3)^2(x - 2)$
42. $f(x) = 2x^4 + 8$
43. $f(x) = -\dfrac{2}{x^2}$
44. $f(x) = \dfrac{1}{x + 1}$
45. $f(x) = \dfrac{8}{x^2 - 4}$
46. $f(x) = \dfrac{x^2 + 2x - 3}{x^2 - x - 2}$
47. $f(x) = 2x + \dfrac{2}{x}$
48. $f(x) = \dfrac{1}{x^2 + 2}$
49. $f(x) = \dfrac{x^3 + 12x^2 + 4x + 48}{x^3 + 2x^2 - 3x}$
50. $f(x) = \dfrac{x^2 - 5}{x - 4}$
51. $f(x) = \dfrac{(x - 2)(x^2 - 1)}{x - 2}$
52. $f(x) = \dfrac{x^2 - 4}{x - 2}$
53. $f(x) = \dfrac{x^3 + 2}{x}$
54. $f(x) = \dfrac{x^3 + x^2 - 6x}{x^2 + x - 2}$

Applications

55. **Package design.** An open box of volume 132 cubic inches can be made from a piece of cardboard that is 10 in. by 15 in. by cutting a square from each corner and folding up the sides. What is the length of a side of the squares? Use the method of bisection to find an approximate answer, if appropriate.

56. **Electrical resistance.** Suppose three resistors are connected in parallel, as shown in the following figure. The result of the three resistances is an equivalent resistance R, and is given by:
$$\dfrac{1}{R} = \dfrac{1}{R_1} + \dfrac{1}{R_2} + \dfrac{1}{R_3}$$
Suppose $R = 1$, the second resistance is two more than the first, and the third resistance is three more than the first. (See figure.)

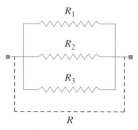

Exercise 56

a. Find a polynomial equation in R_1 that can be used to find the resistances.

b. Find each of the three resistances. **Remember:** Resistance is always nonnegative. Use a calculator to approximate the values.

57. Inventory management. The cost $C(x)$ of managing an inventory of x pool tables is determined by a team of financial analysts to be given by the function:

$$C(x) = 10x + \frac{4,000}{x} + 1,600$$

Sketch the graph of this function over the interval $(0, +\infty)$.

58. Pharmacology. There are many rules for determining the dosage of medication for a child based on the corresponding adult dosage. The following are two such rules:

$$\text{Young's Rule: } C = \frac{A}{A + 12}D$$

$$\text{Cowling's Rule: } C = \frac{A + 1}{24}D$$

where A = the age of the child, in years, D = the adult dosage, and C = child's dosage. (**Warning!** Do not apply these formulas without consulting a physician. These are estimates that may not apply to a particular medication.) An adult dosage of a liquid medication is 5 cc (cubic centimeters).

a. Sketch a graph of Young's Rule for values of A in the interval [0, 18].

b. Sketch a graph of Cowling's Rule for values of A in the interval [0, 18].

c. For what ages A are the child's dosages given by the two rules the same?

Chapter Test

1. Find the quotient and the remainder when $f(x) = 10x^3 + 3x - 1$ is divided by $2x - 1$.

2. Use synthetic division to determine the quotient and the remainder when $f(x) = x^5 - 5x^4 + 3x^3 - x + 17$ is divided by $x - 1$.

3. Suppose that $f(x) = -3x^5 + 2x^4 + x^3 - 2x^2 - 4x + 5$. Use synthetic division to calculate the value of $f(2)$.

4. Suppose that $f(x)$ is a quadratic polynomial with zeros $\frac{1}{2}$ and $-\frac{2}{3}$. Furthermore, assume that $f(-1) = 5$. Determine $f(x)$.

5. Determine all zeros of the following polynomial and state the multiplicity of each zero: $f(x) = x^4(x - 2)^3(x^2 + 1)^2$.

6. Use the method of bisection to determine the zero of $f(x) = x^5 - x + 1$ in the interval $(-2, 0)$ to within 0.001 accuracy.

7. Use Descartes's Rule of Signs to analyze the zeros of the polynomial $8x^5 + 3x^4 - x^3 + 5x^2 - 3$.

8. Calculate the following:

   a. $(5 - 2i) + (3 + 4i)$

   b. i^{15}

   c. $(10 - 3i)(5 - 8i)$

   d. $\dfrac{3 - i}{1 + i}$

9. Determine the zeros of the function $f(x) = 2x^4 - x^2 - 10$.

10. Determine the zeros of the function $f(x) = (x^3 - 1)^2(x^2 + 1)$.

11. Sketch the graph of the function $f(x) = x + 3/x$. Determine all asymptotes.

12. **Thought question.** Explain how you would determine the asymptotes of the graph of a rational function.

13. **Thought question.** Explain how you would estimate the zeros of a polynomial. (You may use any of the methods described in this chapter.)

CHAPTER 5
MORE ABOUT FUNCTIONS

In the preceding chapters, we explored the properties of many of the functions one encounters in mathematical modeling, including linear, quadratic, polynomial, and rational functions. As preparation for our study of other important functions, in this chapter we develop various facts about functions in general, namely the various algebraic operations on functions, inverse functions, and the applications of functions to variation.

Chapter Opening Image: *"In my work, I am trying to bridge the gap between the seemingly disparate worlds of sonic and visual art—trying to create a coherent, integrated, 'visual music.' To this end, I explore the world of mathematics for raw material. Like a sculptor searching for good marble, I search for mathematical formulae that offer aesthetic possibilities. With materials chosen, I then chisel and constrain the numeric marble, mapping the results into colors and musical pitches."—Brian D. Evans, creator of the above image.*

5.1 OPERATIONS ON FUNCTIONS

It is possible to perform algebraic operations on functions similar to the operations we discussed for algebraic expressions, namely addition, subtraction, multiplication, and division. We can also perform a unique operation called composition of functions. In this section, we introduce these operations and demonstrate some of their applications.

Sums and Differences of Functions

Let f and g be functions with domains A and B, respectively. We define the **sum** of f and g, denoted $f + g$, to be the function such that:

$$(f + g)(x) = f(x) + g(x)$$

That is, the value of $f + g$ at x is obtained by adding the value of f at x to the value of g at x. The domain of $f + g$ consists of those x that belong to both A and B.

EXAMPLE 1
The Sum of Two Functions

Suppose that:

$$f(x) = \frac{x}{x-1}, \qquad g(x) = \frac{2}{x-2}$$

Determine the sum $f + g$.

Solution

To compute the sum function, we just add the rational expressions for f and g:

$$(f + g)(x) = \frac{x}{x-1} + \frac{2}{x-2}$$

$$= \frac{x(x-2) + 2(x-1)}{(x-1)(x-2)}$$

$$= \frac{x^2 - 2}{(x-1)(x-2)}$$

Because f is defined for $x \neq 1$ and g is defined for $x \neq 2$, the sum is defined for $x \neq 1$ or 2.

The graph of $f + g$ can be described simply in terms of the separate graphs of f and g. For each x in the domain of $f + g$, the distance of the point

$$\bigl(x, (f + g)(x)\bigr)$$

from the x-axis is obtained by adding the corresponding y-coordinates of the graphs of f and g, namely $f(x)$ and $g(x)$. (See Figure 1.)

The definition of the **difference** of f and g is parallel to the definition of the sum of f and g:

$$(f - g)(x) = f(x) - g(x)$$

That is, the value of the difference $f - g$ at x equals the value of f at x minus the value of g at x.

The domain of $f - g$ consists of those x that belong to the domains of both f and g.

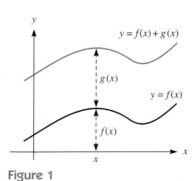

Figure 1

Sums and differences of functions often appear in applications. For example, suppose that $R(x)$ and $C(x)$ denote the revenue and cost, respectively, of producing x bushels of corn for a certain agricultural corporation. Then the difference

$$R(x) - C(x)$$

represents the profit from producing x bushels of corn.

Products and Quotients of Functions

The **product** of f and g, denoted fg, is the function defined by the formula:
$$(fg)(x) = f(x)g(x)$$
Like the sum and the difference, the product is defined for values of x that are in the domains of both f and g.

EXAMPLE 2 Product of Two Functions

Suppose that:
$$f(x) = \sqrt{x}, \qquad g(x) = \frac{x-1}{\sqrt{x}}$$
Determine the product function fg.

Solution
$$(fg)(x) = (\sqrt{x})\left(\frac{x-1}{\sqrt{x}}\right) = x - 1$$

Because f is defined for nonnegative x and g is defined for positive x, the product is defined where both f and g are defined, namely for positive x. Note that, although the expression simplified for the product is defined for all real numbers x, the product fg is defined only for positive values of x.

A very common mistake is to compute a product, simplify it, and then substitute a value for x that is not in the domain of the product. This sort of error can lead to mathematical nonsense, statements such as $0 = 1$, for example. To avoid such an error, be careful to note the domains of the functions you compute.

The **quotient** of f divided by g, denoted f/g, is the function defined by the formula:
$$\left(\frac{f}{g}\right)(x) = \frac{f(x)}{g(x)}$$
For the expression on the right to be defined, x must be in both the domain of f and the domain of g, and $g(x)$ must be nonzero. (Division by 0 is undefined.)

EXAMPLE 3 Determining a Quotient Function

Suppose that:
$$f(x) = \frac{x}{x^2 + 1}, \qquad g(x) = (x-1)(x-2)$$
Determine an expression for the quotient f/g.

Solution

The quotient is defined by the formula:
$$\left(\frac{f}{g}\right)(x) = \frac{x}{(x^2 + 1)(x - 1)(x - 2)}, \qquad x \neq 1, 2$$
Note that f and g are both defined for all x. However, because the denominator $g(x)$ is equal to 0 for $x = 1, 2$, these values must be excluded from the domain of the quotient.

In working with quotients, you must exercise the same care as in working with products with regard to the domain. Although you may be able to simplify the quotient to an expression that has a larger domain, the quotient is defined only for the domain that consists of values that are common to the domains of both the numerator and denominator, and for which the denominator is nonzero.

Operations on Functions Using a Graphing Calculator

If you are using a graphing calculator or graphing software, you will find that most models are able to graph sums, differences, products, and quotients of functions.

For example, suppose you are using the TI-82 and want to graph the functions $f(x) = 2x$ and $g(x) = x^2$ and their sum $f(x) + g(x)$. Begin by setting the domain and range for the graph, $-2 \leq x \leq 2$, and $-1 \leq y \leq 4$. Then, using the [Y=] key, enter $2x$ for Y_1 and x^2 for Y_2. The separate graph of each function is shown in Figure 2.

Next, let's graph the sum. Return to the [Y=] menu. Enter $Y_1 + Y_2$ for Y_3. To enter Y_1 within an equation, press [2nd] [Vars] [ENTER]. Then select Y_1 from the menu displayed and press [ENTER]. In the same way, enter Y_2. Once $Y_1 + Y_2$ is entered as Y_3, deselect the graphs of Y_1 and Y_2 (that is, make these graphs invisible). To do this, position the cursor on the $=$ in the equation definition and press [ENTER]. Finally, press [GRAPH]. The graph should look like Figure 3.

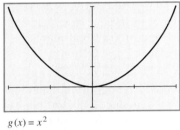

$g(x) = x^2$

Figure 2

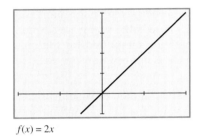

$f(x) = 2x$

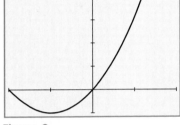

Figure 3

EXAMPLE 4
Graphing a Product of Functions

Let $f(x) = x^3$ and $g(x) = 1/x$ $(x \neq 0)$. Use a graphing calculator to display the graph of $f(x)g(x)$.

Solution

Set Y_1 equal to $f(x)$, Y_2 equal to $g(x)$, and Y_3 equal to $Y_1 * Y_2$. (See Figure 4.) Now deselect the graphs of Y_1 and Y_2 using the procedure described above. Follow a similar procedure to turn off the display for Y_2. The final display consists only of the graph of Y_3, as shown in Figure 5. This is the graph of $f(x)g(x)$. Be careful—this graph has a hole at $x = 0$, because the product is not defined there. This hole is very hard to see, but you should note its existence from the definition of the functions.

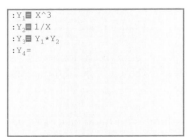

Figure 4

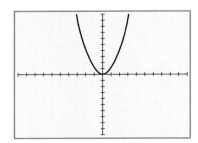

Figure 5

SECTION 5.1 Operations on Functions

EXAMPLE 5
Graphing a Quotient of Functions

Let $f(x) = \sqrt{x}$ and $g(x) = x^2 + 1$. Display the graph of $f(x)/g(x)$.

Solution

The domain of $f(x)$ is $x \geq 0$ and the domain of $g(x)$ is all real x. Moreover, because $g(x) \neq 0$ for any real x, the denominator is never 0 and the domain of the quotient is $x \geq 0$. Set Y_1 equal to $f(x)$, set Y_2 equal to $g(x)$, and set Y_3 equal to Y_1/Y_2. (See Figure 6.) To make the features of the graph more evident, choose the range $0 \leq y \leq 2$. The graph of $f(x)/g(x)$ is the graph of Y_3, and is shown in Figure 7.

Figure 6

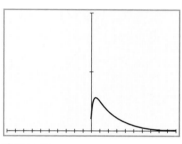

Figure 7

Composition of Functions

In many applications, it is necessary to substitute an expression for the variable of a function $g(x)$. Such substitutions are conveniently described as follows:

Definition
Composition of Functions

Suppose that f and g are functions. The composite

$$g \circ f$$

is the function defined by the formula:

$$(g \circ f)(x) = g(f(x))$$

That is, to compute the value of the composite at x, we first evaluate f at x and then use that value to evaluate g at $f(x)$. (See Figure 8.) We refer to $g \circ f$ as the **composite** of g by f.

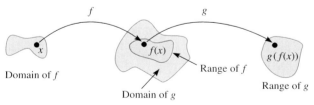

Figure 8

The machine concept of functions gives us a convenient way of viewing a composite. Suppose we view f as one machine and g as a second. Then the machine that corresponds to the composite is a serial combination of the first machine followed by the second.

Suppose that x is a value in the domain of f. For the value $g(f(x))$ to make sense, the value $f(x)$ must be contained in the domain of g. This is necessary if we are to be able to evaluate $g(z)$ at the value $z = f(x)$.

CHAPTER 5 More about Functions

In forming composites, it helps to think in terms of an "inner function" and an "outer function." In forming the expression

$$g(f(x))$$

$f(x)$ is the inner function and $g(x)$ is the outer function.

EXAMPLE 6
Calculating a Composite

Suppose that:

$$f(x) = x + 1, \qquad g(x) = x^2$$

Determine the composite $g \circ f$.

Solution

To form the composite of g by f, we substitute the expression for f wherever x appears in the expression for g. We have $g(x) = x^2$. Therefore, to obtain an expression for the composite, we replace x with $x + 1$ in the expression for g. We have:

$$\begin{aligned}(g \circ f)(x) &= g(f(x)) \\ &= g(x + 1) \\ &= (x + 1)^2\end{aligned}$$

EXAMPLE 7
Calculating a Composite

Suppose that:

$$f(x) = \frac{1}{\sqrt{x}}, \qquad g(x) = x^2 + 1$$

Determine the composite of g by f.

Solution

The expression for the composite of g by f is obtained by substituting $f(x)$ for x in the expression for g:

$$\begin{aligned}(g \circ f)(x) = g(f(x)) &= g\left(\frac{1}{\sqrt{x}}\right) \\ &= \left(\frac{1}{\sqrt{x}}\right)^2 + 1 \\ &= \frac{1}{x} + 1\end{aligned}$$

Note that the order of forming composites is important. For instance, if we let f and g be as in the last example, we can form the composite of f by g as follows:

$$(f \circ g)(x) = f(g(x)) = f(x^2 + 1) = \frac{1}{\sqrt{x^2 + 1}}$$

Thus, in this case, $f \circ g$ is not equal to $g \circ f$.

EXAMPLE 8
Writing a Function as a Composite

Let h be the function defined by

$$h(x) = \sqrt[3]{x^2 + 5}$$

Express h as the composite of two functions.

Solution

Suppose that we define:

$$g(x) = \sqrt[3]{x}, \qquad f(x) = x^2 + 5$$

SECTION 5.1 Operations on Functions

Then, we have:
$$g(f(x)) = h(x)$$
In other words, h is the composite of g by f. Note that there is no unique way to express a function as a composite. For example, we can use other functions for f and g in Example 8, such as $g(x) = \sqrt[3]{(x+5)}$ and $f(x) = x^2$. However, the functions we chose were the most obvious ones.

Composition of Functions on a Graphing Calculator

Now let's graph a composite of functions on a graphing calculator. Consider the functions:
$$f(x) = \frac{1}{\sqrt{x}} \quad \text{and} \quad g(x) = x^2 + 1$$

The composite function we are looking for is $g(f(x))$. To graph this, first enter $f(x)$ in Y_1 in the [Y=] menu. Now move to Y_2 and enter $Y_1\hat{\ }2 + 1$. Then graph Y_2. The results should resemble Figure 9.

The power of a graphing utility is the ease with which we can graph many functions. For instance, by entering $g(x)$ in Y_1 and $1/\sqrt{Y_1}$ in Y_2 and graphing Y_2, we have the graph of $f(g(x))$ as shown in Figure 10. By looking at the two graphs, we clearly see that $f(g(x)) \neq g(f(x))$, as we observed above.

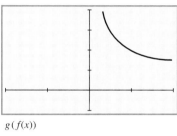

$g(f(x))$
Figure 9

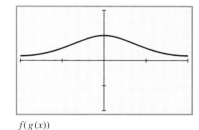

$f(g(x))$
Figure 10

EXAMPLE 9
Graphing a Composite of Functions

Let $f(x) = x - 1/x$ and $g(x) = 1/x$. Display the graph of $g(f(x))$.

Solution
$$g(f(x)) = 1/f(x)$$
So set Y_1 equal to $f(x)$ and Y_2 equal to $1/Y_1$. (See Figure 11.) The graph of Y_2 is the graph of the composite $g(f(x))$. (See Figure 12.) To make the features of the graph more visible, we have chosen the range $-3 \le x \le 3$. Moreover, we have turned off the graph of Y_1.

Figure 11

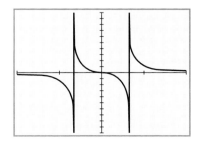

Figure 12

Exercises 5.1

For Exercises 1–4, find the domain of each of the following:

f, g, $f + g$, $f - g$, fg,

ff, f/g, g/f, $f \circ g$, $g \circ f$

Then, find:

$(f + g)(x)$, $(f - g)(x)$, $(fg)(x)$,

$\left(\dfrac{f}{g}\right)(x)$, $(f \circ g)(x)$

1. $f(x) = x - 4$, $g(x) = x + 5$
2. $f(x) = x^2 - 4$, $g(x) = 2x + 3$
3. $f(x) = 2x^2 + x - 3$, $g(x) = x^3$
4. $f(x) = \sqrt{x}$, $g(x) = x^2$

Let $f(x) = x^2 - 1$ and $g(x) = 2x + 3$. Find each of the following.

5. $(f - g)(3)$
6. $(f + g)(-1)$
7. $(f + g)(x)$
8. $(f - g)(x)$
9. $(fg)(3)$
10. $\left(\dfrac{f}{g}\right)(-1)$
11. $\left(\dfrac{g}{f}\right)(-1)$
12. $(fg)(x)$
13. $\left(\dfrac{f}{g}\right)(-1.5)$
14. $\left(\dfrac{f}{g}\right)(x)$
15. $\left(\dfrac{g}{f}\right)(x)$
16. $(f \circ f)(x)$
17. $(f \circ g)(x)$
18. $[f \cdot (f \circ f)](x)$
19. $(g \circ g)(x)$
20. $(g \circ f)(x)$

In each of the following, find $f(x)$ and $g(x)$ such that $h(x) = (f \circ g)(x)$. Answers may vary, but try to choose the most obvious answer.

21. $h(x) = (4x^3 - 1)^5$
22. $h(x) = \sqrt[3]{x^2 + 1}$
23. $h(x) = \dfrac{1}{(x + 5)^4}$
24. $h(x) = \dfrac{1}{\sqrt{7x + 2}}$
25. $h(x) = \dfrac{x^3 + 1}{x^3 - 1}$
26. $h(x) = |4x^2 - 3|$
27. $h(x) = \left(\dfrac{1 + x^3}{1 - x^3}\right)^4$
28. $h(x) = \left(\sqrt{x} + 3\right)^4$
29. $h(x) = \sqrt{\dfrac{x + 2}{x - 2}}$
30. $h(x) = \sqrt{2 + \sqrt{2 + x}}$
31. $h(x) = (x^5 + x^4 + x^3 - x^2 + x - 2)^{63}$
32. $h(x) = (x - 7)^{2/3}$

For Exercises 33–44, refer to the graph of the function $f(x)$ shown below. In each exercise, draw the graph of the indicated function.

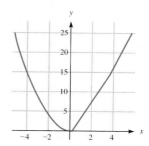

33. $f(x) + 1$
34. $f(x) + 3$
35. $f(x + 2)$
36. $f(x + 1)$
37. $f(x - 1)$
38. $f(x - 5)$
39. $-f(x)$
40. $-f(x - 1)$
41. $f(-x)$
42. $-f(-x + 1)$
43. $f(-x + 2)$
44. $1 - f(x)$

Applications

45. **Ecology.** A tanker has an oil spill. Oil is moving away from the tanker in a circular pattern such that the radius of the spill is increasing at the rate of 10 ft/hr. (See figure.)

    a. Find a function $r(t)$ for the radius in terms of the time t.

    b. Find a function $A(r)$ for the area of the oil spill in terms of the radius r.

    c. Find $(A \circ r)(t)$. Explain the meaning of this function.

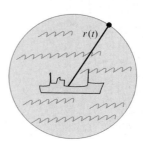

Exercise 45

SECTION 5.2 Inverse Functions 273

46. **Air traffic control.** An airplane at the end of the runway is 200 ft from the control tower. It takes off at a speed of 180 ft/sec. (See figure.)

a. Find a formula for a, the distance the plane has traveled down the runway, in terms of the time t the plane travels. That is, find an expression for $a(t)$.

b. Let d be the distance of the plane from the control tower. Find a formula for d in terms of the distance a. That is, find an expression for $d(a)$.

c. Find $(d \circ a)(t)$. Explain the meaning of this function.

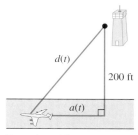

Exercise 46

 Technology

Use a graphing calculator to graph the following on a single coordinate system: $f(x)$, $g(x)$, $f(x)/g(x)$. What do you observe about the graph of the quotient at zeros of the denominator?

47. $f(x) = x + 1$, $g(x) = x$
48. $f(x) = x^2 - x$, $g(x) = (x + 1)^2$
49. $f(x) = x^2 - 4x + 4$, $g(x) = x - 2$

50. $f(x) = \sqrt{x + 1}$, $g(x) = x^2$

Graph $f(g(x))$, where f and g are given by the following:

51. $f(x) = x - 4$, $g(x) = x + 5$
52. $f(x) = x^2 - 4$, $g(x) = 2x + 3$
53. $f(x) = 2x^2 + x - 3$, $g(x) = x^3$
54. $f(x) = \sqrt{x}$, $g(x) = x^2$

In your own words

55. Explain why the composition of two even functions is even.
56. Explain why the sum of two even functions is even.
57. Explain why the sum of two odd functions is odd.
58. Explain why the composition of two odd functions is odd.
59. Explain why the product of an even function and an odd function is odd.

5.2 INVERSE FUNCTIONS

One important task of algebra is defining functions that are suitable for solving a wide variety of real-world problems. Throughout this book, we introduce new functions required to describe various events and processes. One general method for constructing new functions from known ones is by using inverse functions. In this section, we define inverse functions and develop an initial facility in dealing with them. Many of the functions we encounter later in the book, such as the logarithmic functions, are constructed as inverses of known functions. Related to inverse functions is the idea of one-to-one functions.

One-to-One Functions

Suppose that f is a function with domain A and range B. Then, for every element c in B, there is some x in the domain A such that $f(x) = c$. However, there may be more than one such x for which f has the value c. For instance, consider the function $f(x) = x^2$. In this case, there are two values of x for which $f(x) = 4$, namely $x = 2$ and $x = -2$. On the other hand, for some functions, each value c in the range corresponds to exactly one value of x for which $f(x) = c$. Let's assign such functions a name:

Definition
One-to-One Function
We say that f is a **one-to-one function** provided that for each c in the range, there is exactly one x in the domain such that $f(x) = c$.

When the domain and range are both subsets of the set of real numbers, we can use a simple geometric test to determine whether a function is one-to-one. On the graph of $f(x)$, the values of x for which $f(x) = c$ are the y-coordinates of the points at which the graph intersects the horizontal line $y = c$. Generally, there may be several such values of x, that is, many such points of intersection. (See Figure 1(a).) However, if $f(x)$ is a one-to-one function, then *every* line $y = c$ intersects the graph of $f(x)$ in at most one point. (See Figure 1(b).)

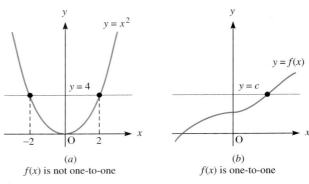

(a) $f(x)$ is not one-to-one

(b) $f(x)$ is one-to-one

Figure 1

Thus, we have the following geometric criterion for one-to-one functions:

HORIZONTAL LINE TEST FOR ONE-TO-ONE FUNCTIONS

A function f is one-to-one if and only if each horizontal line intersects the graph of f in at most one point.

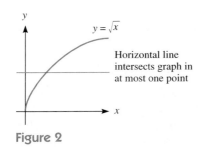

Horizontal line intersects graph in at most one point

Figure 2

As an example of a one-to-one function, consider the function $f(x) = \sqrt{x}$ ($x \geq 0$). Its graph is shown in Figure 2. Note that if a horizontal line intersects the graph, then it intersects it in precisely one point. (Of course, not every horizontal line intersects the graph.) So the function f is one-to-one.

EXAMPLE 1
Determining if a Function Is One-to-One

Determine which of the following functions are one-to-one.

1. $f(x) = -2x + 5$
2. $f(x) = \sqrt{1 - x^2}$ $(0 \leq x \leq 1)$
3. $f(x) = x^2$

SECTION 5.2 Inverse Functions

Solution

1. Suppose that we are given a value y in the range of f (which is the set of all real numbers). A particular value of x satisfies $f(x) = y$ exactly if:
$$-2x + 5 = y$$
Solving this equation for x in terms of y, we obtain:
$$x = -\frac{1}{2}(y - 5)$$
This equation shows that for a particular value of y, there is exactly one value of x such that $f(x) = y$. Thus, f is one-to-one. The graph of f is shown in Figure 3. From the horizontal line test, we obtain a second proof that the function is one-to-one.

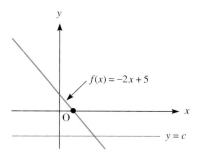

Figure 3

2. Again, suppose that y is in the range of f. The condition $f(x) = y$ is equivalent to:
$$\sqrt{1 - x^2} = y$$
$$1 - x^2 = y^2 \quad \text{and } x \text{ is in the interval } [0, 1]$$
$$x^2 = 1 - y^2$$
$$x = \pm\sqrt{1 - y^2}$$
At first glance, it appears that for a given value of y there are two possible corresponding values of x, one for each choice of sign. However, because x is in the interval $[0, 1]$, the negative sign cannot hold, and there is a single choice for x, namely:
$$x = \sqrt{1 - y^2}$$
Because a single value of x corresponds to a given value of y, the function f is one-to-one. In Figure 4, we sketch the graph of f. It is easy to verify that the horizontal line test holds in this case.

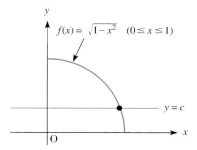

Figure 4

3. Figure 5 shows the graph of f. It is clear that the horizontal line test fails; the figure shows a horizontal line that intersects the graph in two places. This is reflected in the algebra as follows: If we solve the equation $f(x) = y$ for x in terms of y, we have:
$$y = x^2 \qquad x = \pm\sqrt{y}$$
For all $y \neq 0$, the numbers $\sqrt{y}$ and $-\sqrt{y}$ are different (one is positive and the other is negative). Therefore, for $y \neq 0$, there are two distinct values of y corresponding to each x. This is confirmed by examining the graph.

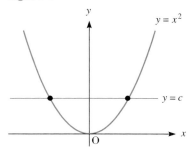

Figure 5

Inverse of a Function

Let $y = f(x)$ be a one-to-one function; the function assumes each value of y exactly once. So we can create a new function that associates to each value of y the corresponding value of x. For example, let's consider the function
$$y = f(x) = -2x - 5$$
which is defined for any real value of x. We have shown that this function is one-to-one and assumes as values all real y. Suppose we are given a value of y. What value of x corresponds to y? To find out, we solve the equation for x in terms of y:
$$-2x = y + 5$$
$$x = -\frac{1}{2}(y + 5)$$

That is, if y is any real number, then the given function f has value y at the x-value $-\frac{1}{2}(y+5)$. The correspondence that associates y with its x-value $-\frac{1}{2}(y+5)$ is called the **inverse function** of f, and is denoted f^{-1} (read "f inverse"). That is, we have:

$$f^{-1}(y) = -\frac{1}{2}(y+5), \qquad \text{where } y \text{ is any real number}$$

More generally, suppose that f is a one-to-one function with domain A and range B. For each y in B, there is exactly one x in A for which $f(x) = y$. The correspondence that associates to y its corresponding x value is a function with domain B and range A. This function is called the inverse function of f and is denoted f^{-1}. (Note that -1 here is not an exponent, but a symbol that represents the inverse function.)

We can explain the inverse function graphically, as follows. Recall that we can represent a function from A to B as an arrow from an element x of A to the element $y = f(x)$ in B. The inverse function is represented by the arrow in the opposite direction, from y to x. (See Figure 6.)

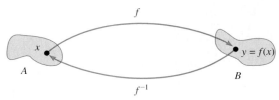

Figure 6

Examples of Inverse Functions

We can condense the preceding discussion into the following method for determining the inverse of a one-to-one function.

DETERMINING THE INVERSE OF A FUNCTION

1. Determine that $f(x)$ has an inverse by confirming that it is a one-to-one function.
2. Write the equation $f(x) = y$.
3. Solve the equation for x in terms of y:

$$x = \text{expression in } y$$

4. Write:

$$f^{-1}(y) = \text{expression in } y$$

5. Replace y throughout with x to obtain the inverse function:

$$f^{-1}(x) = \text{expression in } x$$

EXAMPLE 2
Calculating Inverses

Determine the inverse functions of the following functions. Determine the domain and range of each inverse function.

1. $f(x) = 3x + 5$
2. $f(x) = \sqrt{x}, \qquad x \geq 0$

Solution

1. The graph of f is a nonvertical, straight line. By the horizontal line test, f is one-to-one and so has an inverse function. The domain and range of f both consist of all real numbers. Therefore, the domain and range of f^{-1} are also the set of all real numbers. To obtain a formula for f^{-1}, we write:

$$y = 3x + 5$$

We now solve this equation for x in terms of y:

$$y - 5 = 3x$$

$$x = \frac{1}{3}(y - 5)$$

Finally, we interchange x and y to obtain:

$$y = \frac{1}{3}(x - 5)$$

$$f^{-1}(x) = \frac{1}{3}(x - 5)$$

2. Figure 7 shows the graph of $y = \sqrt{x}$. By inspecting the graph, we see that it passes the horizontal line test and so is one-to-one. Therefore, f has an inverse function. Both the domain and range of f are the set of all nonnegative real numbers. Therefore, both the domain and range of the inverse function are also the set of all nonnegative real numbers. We now obtain a formula for the inverse function by first solving the equation for x in terms of y:

$$y = \sqrt{x}$$

$$x = y^2, \qquad y \geq 0$$

Interchanging x and y, we have:

$$y = x^2, \qquad x \geq 0$$

$$f^{-1}(x) = x^2$$

That is, the inverse function is $f^{-1}(x) = x^2$, where $x \geq 0$.

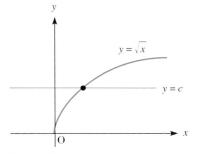

Figure 7

The following facts are derived directly from the definition of inverse functions.

INVERSE FUNCTIONS

1. The inverse function f^{-1} is defined if and only if the function f is one-to-one.
2. The domain of the inverse function is the range of f.
3. The range of the inverse function is the domain of f.
4. The inverse function satisfies the property:

$$f^{-1}(y) = x \quad \text{if and only if} \quad f(x) = y$$

Suppose that we start with a number x in the domain of the function f. The function value $f(x)$ lies in the range of f (the domain of the inverse function). The definition of the inverse function states that:

$$f^{-1}(f(x)) = x$$

Similarly, suppose that y is a number in the domain of f^{-1} (in other words, in the range of f). Then the definition of the inverse function states that:

$$f(f^{-1}(x)) = x$$

The last two formulas can be phrased in terms of composition of functions:

$$(f^{-1} \circ f)(x) = x, \quad \text{where } x \text{ is in the domain of } f$$
$$(f \circ f^{-1})(x) = x, \quad \text{where } x \text{ is in the domain of } f^{-1}$$

These formulas state important relationships between a function and its inverse. We use these formulas in many special cases to perform algebraic manipulations in expressions that involve both a function and its inverse.

The Graph of the Inverse Function

From the definition of the inverse function, we can see that the point (a, b) is on the graph of f if and only if (b, a) is on the graph of the inverse of f. Indeed, we have:

$$b = f(a) \quad \text{if and only if} \quad a = f^{-1}(b)$$

To better understand the relationship of the points (a, b) and (b, a), think of the line $y = x$ as a mirror that reflects images from one side of it to the other side, preserving all distances. Then the point (a, b) is the reflection of the point (b, a), and vice versa. (See Figure 8.) This fact provides us with a simple method for sketching the graph of the inverse function.

GRAPH OF AN INVERSE FUNCTION

The graph of the inverse function $f^{-1}(x)$ is the reflection of the graph of f in the line $y = x$. (See Figure 9.)

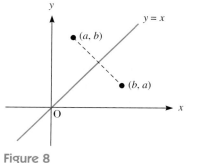

Figure 8

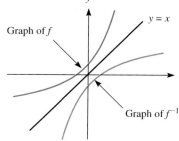

Figure 9

EXAMPLE 3
Graphing Inverse Functions

Sketch the graph of the inverse function of $f(x)$.

1. $f(x) = -2x + 5$
2. $f(x) = \sqrt{1 - x^2}$ $(0 \leq x \leq 1)$

Solution

1. We first sketch the graph of $f(x)$, as in Figure 10. To obtain the graph of the inverse, we reflect this graph in the line $y = x$.
2. We proceed as above. First we graph f, and then we reflect that graph in the line $y = x$. Note that, in this case, the reflected graph is the same as the original. This graph geometrically confirms a fact we discovered by computation in Example 1 of this section—the inverse of the function

$$f(x) = \sqrt{1 - x^2} \quad (0 \leq x \leq 1)$$

is the function $f(x)$ itself. (See Figure 11.)

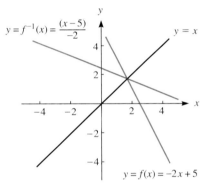

Figure 10

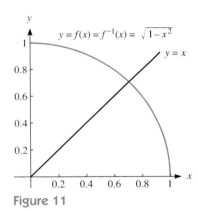

Figure 11

EXAMPLE 4
Using a Graphing Calculator to Check that a Function Is One-to-One

Determine if the following functions are one-to-one using a graphing calculator.

1. $f(x) = x^2 + 2x$ $(-1 \leq x \leq 1)$
2. $f(x) = x + \dfrac{1}{x}$ $(x \neq 0)$

Solution

1. The graph of $f(x)$ is shown in Figure 12. From the horizontal line test, we see that the function is not one-to-one.
2. The graph of $f(x)$ is shown in Figure 13. From the horizontal line test, we see that the function is not one-to-one.

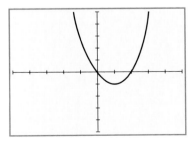

Figure 12

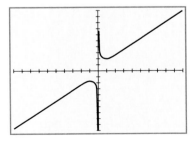

Figure 13

EXAMPLE 5
Graphing an Inverse Function with a Graphing Calculator

Graph the function $f(x) = 3x + 1$ and its inverse function $g(x) = (x - 1)/3$ on the same coordinate system, along with the graph of $y = x$.

Solution

Graph $f(x)$ as Y_1 and $g(x)$ as Y_2, and then graph $Y_3 = X$. The three graphs are shown in Figure 14. Note that the graphs of Y_1 and Y_2 are reflections of each other in the line Y_3. This demonstrates graphically that the functions are inverses of one another.

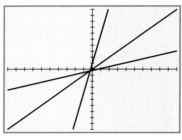

Figure 14

EXAMPLE 6
Using a Graphing Calculator to Prove that Functions Are Inverses of One Another

The inverse of
$$f(x) = x^2 - 4 \qquad (x \geq 0)$$
is:
$$g(x) = \sqrt{x + 4} \qquad (x \geq -4)$$

Use a graphing calculator to prove that:
$$g(f(x)) = x \qquad (x \geq 0)$$

Solution

Set Y_1 equal to $f(x)$ and Y_2 equal to $\sqrt{Y_1 + 4} = g(f(x))$. Finally, set Y_3 equal to the function X restricted to the domain $X \geq 0$. We key in:

$Y_3 = X \qquad (X \geq 0)$

(See Figure 15.) Graphing Y_2 and Y_3, we see that the two graphs are the same. (See Figure 16.) This provides a graphical demonstration that:
$$g(f(x)) = x \qquad (x \geq 0)$$

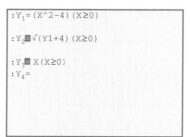

Figure 15

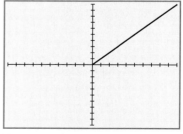

Figure 16

SECTION 5.2 Inverse Functions

EXAMPLE 7
Finding Domain and Range of an Inverse Function with a Graphing Calculator

Use a graphing calculator to determine the domain and range of the inverse function of $f(x)$:

$$f(x) = \frac{x-1}{x+1} \quad (x > -1)$$

Solution

The graph of f is shown in Figure 17, using the domain $-1 \leq x \leq 5$. From the graph, we see that f is one-to-one and therefore has an inverse function. The range of the inverse is the domain of f, that is, $\{y: y > -1\}$. The domain of the inverse is the range of f, which, as the figure shows, is the set of all real numbers.

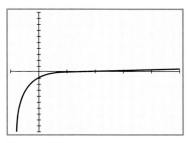

Figure 17

EXAMPLE 8
Evaluating an Inverse Function with a Graphing Calculator

Suppose that:

$$f(x) = \frac{1}{x^2 + 1} \quad (x \geq 0)$$

Use a graphing calculator to determine $f^{-1}(0.25)$.

Solution

The graph of f is shown in Figure 18. If $y = f(x)$, then $x = f^{-1}(y)$. To determine $f^{-1}(0.25)$, we need to determine the value of x for which $f(x) = 0.25$. To do this, we use the trace to move along the curve until we arrive at a point for which the y-coordinate, that is, the value of f, is equal to 0.25. The corresponding x-value is approximately 1.73.

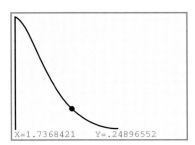

Figure 18

Exercises 5.2

The following are graphs of functions. Determine which functions have inverse functions. For those that do, draw the graph of the inverse function.

1.
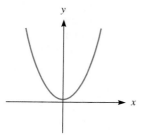

2.
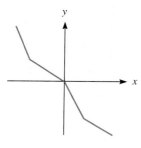

3.
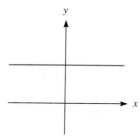

4.
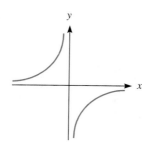

5.
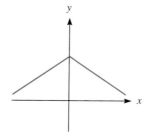

6.
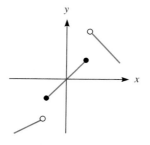

Graph each of the following functions. Determine which functions have an inverse. For those that do, find a formula for $f^{-1}(x)$ and draw its graph.

7. $f(x) = 5 - 8x$
8. $f(x) = 4$
9. $f(x) = -8$
10. $f(x) = \frac{x-2}{4}$
11. $f(x) = x^2 - 3$
12. $f(x) = 3x^2$
13. $f(x) = \sqrt{4 - x^2}$ $(0 \le x \le 2)$
14. $f(x) = \sqrt{1 - x^2}$ $(-1 \le x \le 1)$
15. $f(x) = \sqrt{4 - x^2}$ $(-2 \le x \le 2)$
16. $f(x) = \sqrt{4 - x^2}$ $(-2 \le x \le 0)$

Determine whether these functions are inverses of each other by calculating $(f \circ g)(x)$ and $(g \circ f)(x)$.

17. $f(x) = \frac{2x}{3}$, $g(x) = \frac{3x}{2}$
18. $f(x) = x + 5$, $g(x) = x - 5$
19. $f(x) = 2x + 1$, $g(x) = \frac{x-1}{2}$
20. $f(x) = \frac{4x}{3} + \frac{16}{3}$, $g(x) = \frac{3x}{4} - 4$
21. $f(x) = x^3 - 5$, $g(x) = \sqrt[3]{x + 5}$
22. $f(x) = \sqrt[7]{x + 1}$, $g(x) = x^7 - 1$
23. $f(x) = \sqrt{x - 2}$, $g(x) = x^2 + 2$
24. $f(x) = \sqrt[4]{x}$ $(x \ge 0)$, $g(x) = x^4$
25. $f(x) = \frac{1}{x + 1}$, $g(x) = \frac{1-x}{x}$
26. $f(x) = \frac{x^2 + 1}{x^2 - 1}$, $g(x) = \frac{2x - 5}{3x + 4}$
27. $f(x) = 3$, $g(x) = \frac{1}{3}$
28. $f(x) = -4$, $g(x) = 3x - 1$

For each of the following functions, find a formula for $f^{-1}(x)$.

29. $f(x) = 2x - 3$
30. $f(x) = 7x + 2$
31. $f(x) = \frac{2x}{3} + \frac{3}{5}$
32. $f(x) = 1.6x - 6.4$
33. $f(x) = x^2$ $(x \ge 0)$
34. $f(x) = 4x^2$ $(x \le 0)$
35. $f(x) = \sqrt{x + 2}$ $(x \ge -2)$
36. $f(x) = \sqrt{7x - 2}$ $\left(x \ge \frac{2}{7}\right)$
37. $f(x) = x^5$
38. $f(x) = x^3 - 2$

39. $f(x) = \dfrac{2}{x}$

40. $f(x) = -\dfrac{4}{x}$

41. $f(x) = \dfrac{2x-3}{3x+1}$

42. $f(x) = \dfrac{7x+2}{8x-3}$

43. $f(x) = \sqrt{25-x^2}$ $(-5 \le x \le 0)$

44. $f(x) = \sqrt{49-x^2}$ $(0 \le x \le 7)$

45. $f(x) = \sqrt[3]{x+8}$

46. $f(x) = \sqrt[5]{6-3x}$

47. For $f(x) = 23{,}457x - 3{,}456$, find $f \circ f^{-1}(739)$ and $(f^{-1} \circ f)(5.00023)$.

48. For $f(x) = 677x^3$, find $(f \circ f^{-1})(958.34)$ and $(f^{-1} \circ f)\left(\dfrac{765}{577}\right)$.

49. Let $f(x) = mx + b$, $m \ne 0$. Find a formula for $f^{-1}(x)$.

50. For what integers n does $f(x) = x^n$ have an inverse?

Applications

51. **Temperature conversion.** We have seen the following formulas for conversion between Fahrenheit and Celsius temperatures several times before in the text:
$$F = \dfrac{9}{5}C + 32, \qquad C = \dfrac{5}{9}(F-32)$$
Show that these functions are inverses of one another.

52. **Area of a circle.** Find a formula for the area A of a circle in terms of its radius r. Then find a formula for its radius r in terms of the area A. Show that the functions defined by these formulas are inverses of one another.

Technology

For each of the following functions $f(x)$, use a graphing calculator to determine whether or not it has an inverse function. If the inverse exists, calculate $f^{-1}(2)$.

53. $f(x) = x^2 - 3.5x + 1$

54. $f(x) = -3.9x^3 + 9.6$

55. $f(x) = \dfrac{x}{4x+3}$ $(x > 0)$

56. $f(x) = \dfrac{1}{x^2 - 17}$

57. $f(x) = |3x - 11|$

58. $f(x) = \sqrt{x^2 - x}$ $(x \ge 1)$

Use a graphing calculator to determine whether or not the following functions are one-to-one.

59. $f(x) = x^2 + 2x$ $(0 \le x \le 2)$

60. $f(x) = 1 - x^2$ $(0 \le x \le 5)$

61. $f(x) = \dfrac{2}{x}$ $(x > 0)$

62. $f(x) = \dfrac{x}{1-x}$

In your own words

63. Write a paragraph in which you define the concept of a one-to-one function. Provide examples of functions that are one-to-one and functions that are not one-to-one.

64. Explain the procedure for determining the inverse of a one-to-one function.

65. What happens when you attempt to apply the procedure you described in Exercise 64 to a function that is not one-to-one?

66. Determine whether or not an even function has an inverse. Explain.

67. Find examples of odd functions that are inverses of one another. Explain why your examples fulfill the stated conditions.

68. Suppose that f and g both have inverses. Show that $(f \circ g)^{-1}(x) = (g^{-1} \circ f^{-1})(x)$.

69. The function f is said to be increasing over an interval provided that, if $a < b$ with a and b in the interval, $f(a) < f(b)$. Prove that if f is increasing over an interval, then it has an inverse.

70. Does an inverse function always exist for a decreasing function? Explain your reasoning.

5.3 VARIATION AND ITS APPLICATIONS

In many applications, an equation expresses one variable in terms of other variables. Such an equation expresses a **functional relationship** between the variables. In this section, we explore some of the most common types of functional relationships.

Direct Variation

Let's begin with the simplest type of functional relationship between two variables:

> **Definition**
> **Direct Variation**
>
> We say that **y varies directly as x** or that **y is proportional to x**, provided that
> $$y = kx$$
> for some constant k. The constant k is called the **proportionality factor**.

When y varies directly as x, any increase in x results in a proportional increase in y, and any decrease in x results in a proportional decrease in y. Moreover, in this case, y is a linear function of x with constant term zero.

EXAMPLE 1
Direct Variation

Suppose that y varies directly as x. Suppose also that the value of y is 100 when $x = 4$.

1. Determine the functional relationship between y and x.
2. What is the value of y when $x = 5$?

Solution

1. Because y varies directly as x, we have
$$y = kx$$
for some positive constant k. We can determine the constant k by using the given data. Substitute the values $y = 100$ and $x = 4$ into this equation to obtain:
$$100 = k(4)$$
$$k = 25$$
So the functional relationship between x and y is given by:
$$y = 25x$$

2. Using the functional relationship just derived, we see that, when $x = 5$,
$$y = 25x = 25(5) = 125$$

EXAMPLE 2
Hooke's Law

Hooke's law from physics states that the distance a spring stretches from its resting position is directly proportional to the amount of force applied. Suppose that a spring is stretched 8 in. by a force of 50 lb. How far will the spring be stretched when a force of 60 lb is applied? (See Figure 1.)

Solution

Denote the force acting on the spring by F and the distance the spring stretches by x. Then Hooke's law states that F varies directly as x. Therefore, we have

$$F = kx$$

for some constant k. Inserting the given data into this equation, we find that:

$$F = kx$$
$$50 = k(8)$$
$$k = \frac{25}{4}$$
$$F = \frac{25}{4}x$$

Therefore, when F is equal to 60 pounds, we have:

$$60 = \frac{25}{4}x$$
$$\frac{60 \cdot 4}{25} = x$$
$$x = \frac{48}{5} \text{ inches}$$

When a force of 60 pounds is applied, the spring is stretched $\frac{48}{5}$ or 9.6 inches.

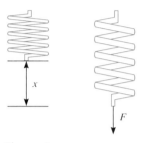

Figure 1

In many applications, functional relationships of the form

$$y = kx^m$$

are required, where m and k are constants. This is a generalization of the formula for direct variation. In the case of a functional relationship of this form, we say that **y varies directly as x^m**, or that **y is proportional to x^m**.

Inverse Variation

In other applications, as one variable increases, a related variable decreases. A relationship of this type can be described as follows:

Definition
Inverse Variation

We say that **y varies inversely as x^m** varies or that **y is inversely proportional to x^m**, provided that

$$y = \frac{k}{x^m}$$

for some constant k.

Here are some examples of functional relationships in which y varies inversely with some power of x:

$$y = \frac{5}{x^2}, \qquad y = \frac{0.3}{x}, \qquad y = -\frac{5}{x^4}$$

EXAMPLE 3
Inverse Variation

Suppose that y varies inversely with x. Furthermore, suppose that when $x = 0.1$, the value of y is 20. Determine y as a function of x.

Solution

Because y varies inversely as x, we have

$$y = \frac{k}{x}$$

for some constant k. We can determine the value of k from the given data by substituting $x = 0.1$ and $y = 20$ into the equation to obtain:

$$20 = \frac{k}{0.1}$$

$$k = 2$$

Replacing k with 2, we have

$$y = \frac{2}{x}$$

EXAMPLE 4
Electrical Resistance

The electrical resistance of a wire of a certain length varies inversely with the square of the radius of the wire. Suppose that a certain wire has a radius of 0.05 in. and that the resistance is 1,000 ohms. What will the resistance be if the radius of the wire is 0.1 in.? See Figure 2.

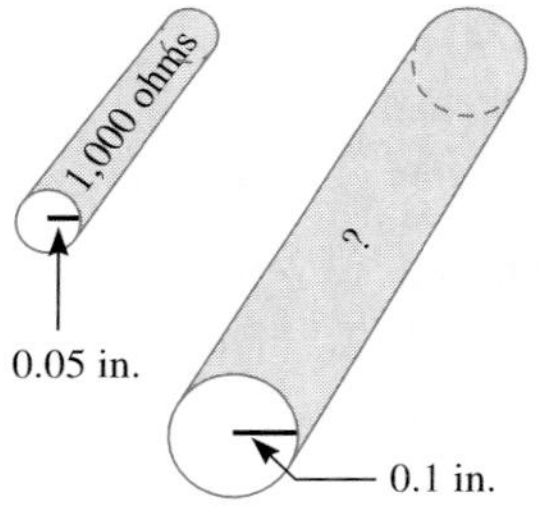

Solution

Let R denote the resistance and r the radius of the wire. Then we have

$$R = \frac{k}{r^2}$$

for some constant k. To determine k, we substitute $R = 1{,}000$ and $r = 0.05$ into the equation:

$$1{,}000 = \frac{k}{0.05^2}$$

$$k = 2.5$$

Therefore, we have the functional relationship:

$$R = \frac{2.5}{r^2}$$

In particular, if $r = 0.1$ in., then:

$$R = \frac{2.5}{0.1^2} = 250 \text{ ohms}$$

Figure 2

EXAMPLE 5
Newton's Law of Gravitation

The force of gravitational attraction between two bodies varies inversely with the square of the distance between them. Suppose that when two space rocks are 100 miles apart, the force of gravitational attraction between them is 50 lb. What is the gravitational force between them when the rocks are 30 miles apart? See Figure 3.

Solution

Let F denote the force of gravitational attraction and d the distance between the rocks. Newton's law of gravitation states that

$$F = \frac{k}{d^2}$$

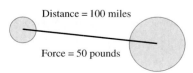

Figure 3

for some constant k. Substituting $F = 50$ and $d = 100$ into this equation, we have:

$$50 = \frac{k}{100^2}$$

$$k = 500{,}000$$

This means that the gravitational force between the rocks is given by the functional relationship:

$$F = \frac{500{,}000}{d^2}$$

To determine the amount of the force when the rocks are 30 miles apart, we substitute $d = 30$ into this equation to obtain:

$$F = \frac{500{,}000}{30^2} = 555.56 \text{ lb}$$

Other Forms of Variation

Other types of variation correspond to relationships that occur in applications. A commonly encountered form of variation is joint variation, which is defined as follows:

Definition **Joint Variation**
We say that y **varies jointly as x and t** or that **y is proportional to x and t**, provided that

$$y = ktx$$

for some constant k.

EXAMPLE 6
Joint Variation

Suppose that y varies jointly as x and t. What is the effect if both x and t are increased by 50 percent?

Solution

From the information given, we can write:

$$y = kxt$$

If x and t are each increased by 50 percent, then x is replaced by $1.5x$ and t is replaced by $1.5t$. This means that y is replaced by:

$$k(1.5x)(1.5t) = 2.25(kxt)$$

But kxt is the original value of y. So the original value of y is multiplied by 2.25.

Other types of variation are obtained by combining the types of variation already introduced. For example, to say that y varies jointly as a and b and inversely as the square root of c means that y is defined by a relationship of the form

$$y = k\frac{ab}{\sqrt{c}}$$

for some constant k. Similarly, to say that y varies jointly with the square of a and the fourth power of b means that y satisfies a relationship of the form

$$y = ka^2b^4$$

for some constant k. Other types of variation are defined similarly.

Exercises 5.3

1. y varies directly as x, and the value of y is 0.04 when $x = 2.8$.
   a. Determine y as a function of x.
   b. What is the value of y when $x = 350$?

2. y varies directly as x, and the value of y is 5.6 when $x = 3.2$.
   a. Determine y as a function of x.
   b. What is the value of y when $x = 0.64$?

3. y varies directly as the square of x, and $y = 0.8$ when $x = 0.2$.
   a. Determine y as a function of x.
   b. What is the value of y when $x = 3.7$?

4. y varies inversely as x, and $y = 2.6$ when $x = 1.2$.
   a. Determine y as a function of x.
   b. What is the value of y when $x = 0.64$?

5. y varies inversely as the square of x, and $y = 2$ when $x = 0.7$.
   a. Determine the relationship between y and x, also known as an equation of variation.
   b. What is the value of y when $x = 0.1$?

6. y varies jointly as x and z, and $y = 3.4$ when $x = 4$ and $z = 5$.
   a. Determine the equation of variation between y, x, and z.
   b. What is the value of y when $x = 8.4$ and $z = 505$?

7. y varies jointly as x and z and inversely as w. It is known that $y = 52$ when $x = 2$, $z = 4$, and $w = \frac{1}{2}$.
   a. Determine the relationship between y, x, z, and w.
   b. What is the value of y when $x = 12$, $z = 24$, and $w = 15$?

8. y varies jointly as x and the square of z, and $y = 152.1$ when $x = 1.8$ and $z = 6.5$.
   a. Determine the relationship between y, x, and z.
   b. What is the value of y when $x = 0.0034$ and $z = 23.4$?

9. y varies jointly as x and z and inversely as the square of w. It is known that $y = 22$ when $x = 14$, $z = 10$, and $w = 12$.
   a. Determine the relationship between y, x, z, and w.
   b. What is the value of y when $x = 12.4$, $z = 1,400$, and $w = 5$?

10. y varies jointly as x and z and inversely as the product of w and p. It is known that $y = 16$ when $x = 23$, $z = 31$, $w = 42$, and $p = 19$.
    a. Determine the relationship between y, x, z, w, and p.
    b. What is the value of y when $x = 12.4$, $z = 1,400$, $w = 35.8$, and $p = 200$?

11. y varies inversely as the square of x and directly as the cube of z. It is known that $y = 3$ when $x = 4$ and $z = 6$.
    a. Determine the relationship between y, x, and z.
    b. What is the value of y when $x = 6$ and $z = 3$?

12. y varies directly as the square root of x and inversely as the product of the cube of z and the square of w. It is known that $y = 25$ when $x = 16$, $z = 2$, and $w = 40$.
    a. Determine the functional relationship between y, x, z, and w.
    b. What is the value of y when $x = 25$, $z = 4$, and $w = 20$?

Exercises 13–16 each present a situation involving variation. Determine a functional relationship, list the variation constant, and describe the variation.

13. The area A of a circle and its radius r.

14. The area A of a triangle and its base b and height h.

15. The distance d a car travels at a constant speed of 65 mph in t hours.

16. The simple interest I on a principal of P dollars at interest rate r for t years.

Applications

17. **Temperature of a gas.** The temperature of a gas varies jointly as its pressure P and volume V. A tank contains 100 liters of oxygen under 15 atmospheres (atm) of pressure at 20°C. If the volume of the gas is decreased to 75 liters and the pressure is increased to 22 atm, what is the temperature of the gas?

18. **Stopping distance.** The stopping distance of a car on a certain surface varies directly as the square of the velocity before the brakes are applied. A car traveling 55 ft/sec needs 79 ft to stop. What is the stopping distance for a car on the same road surface at a velocity of 95 ft/sec?

SECTION 5.3 Variation and Its Applications

19. **Water in a carrot.** The amount of water A in a raw carrot varies directly as the carrot's weight w. Suppose that a 25-gram carrot contains 22 grams of water. How much water does a 32-gram carrot contain?

20. **Radiation energy.** The amount of energy E emitted by radiation varies directly as the wavelength L of the radiation. Suppose that one type of x-ray has a wavelength of 10^{-6} cm and emits 2×10^{-10} erg. How much energy is emitted from infrared light, which has a wavelength of 7×10^{-4} erg?

21. **Illumination.** The illumination I from a light source varies inversely as the square of the distance d from the source. A flashlight is shining at a painting on a wall 8 ft away. At what distance should the flashlight be placed so the amount of light is doubled? (See figure.)

Exercise 21

22. **Period of a pendulum.** The period of a pendulum varies directly as the square root of the length of the pendulum. A 3.15-m pendulum has a period of 3.56 sec. What is the period of a 10-m pendulum?

23. **Speed from skid marks.** Police have used the formula $s = \sqrt{30fd}$ to estimate the speed s (in mph) that a car was traveling if it skidded d feet. The variable f is the coefficient of friction determined by the kind of road (concrete, asphalt, gravel, tar) and whether the road was wet or dry. The following are some values of f:

	Concrete	Tar
Wet	0.4	0.5
Dry	0.8	1.0

   a. Determine a functional relationship between s and d on a dry tar road. At 40 mph, about how many feet will a car skid? A car leaves a skid mark about 141 ft long. How fast was it going?

   b. Determine a functional relationship between s and d on a wet concrete road. At 55 mph, about how many feet will a car skid? A car leaves a skid mark about 208 ft long. How fast was it going?

24. **Force of attraction.** In Newton's law of gravitation, the force F with which two masses attract each other varies directly as the product of the masses M and m, and varies inversely as the square of the distance d between the masses. What happens to the force of attraction when the distance between the two masses is increased from 3 ft to 12 ft? (See figure.)

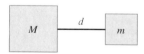

Exercise 24

25. **Gas pressure.** The pressure P of a given quantity of gas varies directly as the absolute temperature T and inversely as the volume V. At what temperature T, with $V = 200$ in.^{3}, will the pressure be three times that which exists when $T = 300°$ and $V = 500$ in.^{3}?

26. **Kepler's law** states that the time *t* for a planet to make an orbit around the sun is directly proportional to the three-halves power of the planet's mean distance *d* from the sun. Mean distances from the sun are 93 million miles for Earth and 141 million miles for Mars. In days, how long will it take Mars to make one orbit around the sun? (Assume it takes the earth 365 days to orbit the sun.)

27. **Accuracy of a speedometer.** If a car's speedometer is not accurate, its accuracy varies directly with the speed of the car. A speedometer reads 38 mph when it is actually going 42 mph. How fast is the car going when the speedometer reads 60 mph?

28. **Safe length of a beam.** The safe length *S* of a wooden beam varies jointly as the width *w* and as the square of the height *h*, and varies inversely as the length *L*. An old house has beams that are 3 in. by 10 in. by 18 ft long. A remodeler wants to replace these with beams that are 2 in. wide and that have the same length. What new height would the beams have to be to have twice the safe length as the old beams? (See figure.)

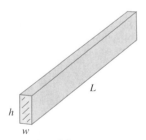

Exercise 28

29. **Wavelength of an electronic particle.** The wavelength *w* of a particle is inversely proportional to its momentum *M*, which is its mass *m* times its velocity *v*. The mass of an electron is 9.19×10^{-28} g. Traveling at one-tenth the speed of light (3×10^3 m/sec), its wavelength is 2.5×10^{-11} m. What is the wavelength of a 150-g baseball traveling at 40 m/sec? Give your answer using scientific notation.

30. **Power of a windmill.** Within certain limitations, the power *P*, in watts, generated by a windmill is proportional to the cube of the velocity *V*, in mph, of the wind, as given by the equation:

$$P = 0.015V^3$$

(The constant 0.015 is a typical value that does not apply in all situations.) (See figure.)

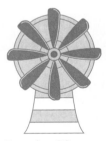

Exercise 30

a. How much power would be generated by a continuous 6 mph wind?

b. How much power would be generated by a continuous 3 mph wind?

c. By what fraction is the power changed if the wind speed is cut in half?

d. How fast would the wind speed need to be to produce 120 watts of power?

(Note: The fact that, by this law, half the wind speed generates only $\frac{1}{8}$ the power is one of the frustrations of recent attempts to use windmills as an alternative energy source. Also, the functional relationship does not apply for higher and higher wind speeds. That is, the design of a windmill may not triple the power; consider the effect of a tornado, for example.)

31. **Electrical attraction or repulsion.** The force E of electrical attraction or repulsion between two charged particles varies jointly as the magnitudes C_1 and C_2 of the two charges, and varies inversely as the square of the distance d between them. Suppose the force of repulsion between two electrons 1×10^{-8} cm apart, each of charge -1.6×10^{-19} coulomb, is 2.3×10^{-13} dyne. What is the force of attraction between an electron and a proton if the proton's charge is 1.6×10^{-19} coulomb and the two particles are 8×10^{-9} cm apart? Give your answer using scientific notation.

In your own words

32. Explain direct variation of one variable with respect to another. Use a specific example in your explanation.

33. Explain inverse variation of one variable with respect to another. Use a specific example in your explanation.

34. Explain joint variation. Use a specific example in your explanation.

Chapter Review

Important Concepts—Chapter 5

- sum of two functions
- difference of two functions
- product of two functions
- quotient of two functions
- composition of functions
- one-to-one function
- horizontal line test for one-to-one functions
- inverse function
- determining the inverse of a function
- graph of an inverse function
- direct variation
- y varies directly as x; y is proportional to x
- proportionality factor
- y varies directly as x^m; y is proportional to x^m
- inverse variation
- y varies inversely as x^m; y is inversely proportional to x^m
- joint variation
- y varies jointly as x and t; y is proportional to x and t

Important Results and Techniques—Chapter 5

Operations on functions	Let f and g be functions.	p. 266–269
	The sum of f and g: $(f + g)(x) = f(x) + g(x)$	
	The difference of f and g: $(f - g)(x) = f(x) - g(x)$	
	The product of f and g: $(fg)(x) = f(x)g(x)$	
	The quotient of f and g: $\left(\dfrac{f}{g}\right)(x) = \dfrac{f(x)}{g(x)}, \quad g(x) \neq 0$	
	The composition of f and g: $(f \circ g)(x) = f(g(x))$	

Mathematics in the World around Us—Predator-Prey Models

Mathematical models are used to study the interaction between predators and prey. When there are many predators, the prey population decreases. However, if there are too few prey, the predators have insufficient food and some of them starve to death. But as the number of predators decreases, the prey population has a chance to recover.

Under what conditions will the predators avoid killing off the prey population and thereby doom themselves to extinction? Or, to put it differently, under what conditions can the predators and prey coexist so that their populations oscillate, but neither group ever is extinguished? Such questions belong to the field of mathematical biology. Founded in the 1920s by the mathematicians Lotka and Volterra, mathematical biology has become an important field of research.

The original model of predator-prey interaction is governed by a pair of differential equations, called the Lotka-Volterra equations, that are based on simple proportions: At any given moment, let the number of prey be $N(t)$ and the number of predators be $P(t)$. Then the rate of change in the number of prey at time t is affected by two forces. The prey are reproducing at a rate proportional to the number of prey, and they are being eaten at a rate proportional to the number of contacts between predator and prey, which is proportional to the product $N(t)P(t)$. So the rate at which the number of prey is changing at any given time t is

$$aN(t) - bN(t)P(t)$$

for constants a and b. Similarly, at any given time t, the predators are dying off in proportion to their numbers, and they are increasing at a rate proportional to the number of contacts between predator and prey. So the rate at which the number of predators is changing is

$$-cP(t) + dN(t)P(t)$$

where c and d are constants. In a course in differential equations, you can learn how to solve the Lotka-Volterra equations.

Since Lotka and Volterra's original work, many more sophisticated models have been developed that take into account multiple predators, interchangeable prey, and other environmental factors.

Inverse of a function	An inverse exists provided that f is one-to-one. An inverse exists if the graph passes the horizontal line test.	p. 274, 278
	If f^{-1} is the inverse function, then $f(f^{-1}(y)) = y$, $\quad f^{-1}(f(x)) = x$ for all x in the domain and for all y in the range of f.	

Variation	y varies directly with x: $y = kx$	p. 284
	y varies inversely with x: $y = \dfrac{k}{x}$	p. 285
	y varies jointly with x and z: $y = kxz$	p. 287

Review Exercises—Chapter 5

For $f(x) = x^2 - 1$ and $g(x) = 2x + 3$, find each of the following.

1. The domain of f.
2. The domain of g.
3. a. The domain of $\dfrac{f}{g}$.
   b. $\left(\dfrac{f}{g}\right)(x)$
4. a. The domain of $\dfrac{g}{f}$.
   b. $\left(\dfrac{g}{f}\right)(x)$
5. a. The domain of $f + g$.
   b. $(f + g)(x)$
6. a. The domain of $f \circ g$.
   b. $(f \circ g)(x)$
7. a. The domain of $g \circ f$.
   b. $(g \circ f)(x)$

Find $(f \circ g)(x)$ and $(g \circ f)(x)$.

8. $f(x) = x^2 - 2x + 1$, $g(x) = \dfrac{3}{x^2}$
9. $f(x) = x^3 + 2$, $g(x) = \sqrt[3]{x - 2}$
10. $f(x) = x^3$, $g(x) = x^{11}$
11. $f(x) = \dfrac{1}{2 - x}$, $g(x) = \dfrac{2x - 1}{x}$
12. $f(x) = \dfrac{2x + 7}{3x - 4}$, $g(x) = \dfrac{5}{x^2}$

In each of the following, find $f(x)$ and $g(x)$ such that $h(x) = (f \circ g)(x)$. Answers may vary, but try to choose the most obvious answer.

13. $h(x) = (3x^2 - 5x + 1)^{11}$
14. $h(x) = \sqrt[5]{\dfrac{x^3 + 1}{x^3 - 1}}$
15. $h(x) = (x^3 + 2)^5 + (x^3 + 2)^3 - 2(x^3 + 2)^2 - 7$
16. $h(x) = \left|\dfrac{x^2 - 5}{x^2 + 5}\right|$
17. Find the domain of $f \circ g$ and of $g \circ f$ if:
    $$f(x) = \dfrac{1}{x - 4} \text{ and } g(x) = 4$$
18. For the following functions, find $(f \circ g)(x)$ and $(g \circ f)(x)$.
    $$f(x) = \begin{cases} x^2 & (-3 \leq x \leq 0) \\ 1 - x & (0 < x \leq 2) \\ \dfrac{1}{x - 2} & (2 < x) \end{cases}$$
    $$g(x) = \begin{cases} 2x - 7 & (-3 \leq x \leq 0) \\ x^3 & (0 < x \leq 2) \\ \dfrac{x + 3}{x} & (2 < x) \end{cases}$$

Determine whether these functions are inverses of each other by calculating $(f \circ g)(x)$ and $(g \circ f)(x)$.

19. $f(x) = -\dfrac{4}{5}x$, $g(x) = -\dfrac{5}{4}x$
20. $f(x) = 2x - 5$, $g(x) = \dfrac{5}{2} + \dfrac{1}{2}x$
21. $f(x) = x^2$, $g(x) = \sqrt{x}$
22. $f(x) = x^5$, $g(x) = \sqrt[5]{x}$

For each of the following, find a formula for $f^{-1}(x)$.

23. $f(x) = 9x - 14$
24. $f(x) = \dfrac{3x - 4}{4 - 2x}$
25. $f(x) = \sqrt{16 - x^2}$ $(0 \leq x \leq 4)$
26. $f(x) = x^3 - 1$
27. For $f(x) = 45{,}677x^5 + 0.0457$, find:
    $$(f \circ f^{-1})(78.8999)$$
28. Refer to the preceding problem. Determine:
    $$(f^{-1} \circ f)(-2{,}344{,}789)$$

Applications

29. **Loudness of sound.** Suppose you are sitting at a distance d from a stereo speaker. You and the speaker are outside. The loudness L of the sound is inversely proportional to the square of d. What happens to the sound if you move three times the distance d from the speaker?

30. **Safe load.** The safe load S for a rectangular beam of fixed length varies jointly as the width w and the square of the height h.

   a. Determine a relationship between S, w, and h.

   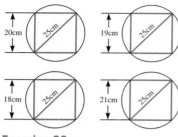

   Exercise 30

   b. Which will give the strongest single beam that can be cut from a cylindrical log of diameter 25 cm? (See figure.)

31. **Boyle's law.** Boyle's law asserts that the volume V of a gas at a constant temperature is inversely proportional to its pressure P.

   a. Determine a functional relationship between V and P.

   b. A tank contains 16 cubic feet of oxygen at a pressure of 50 pounds per square inch. What volume of oxygen will be occupied if the pressure is changed to 15 pounds per square inch?

Chapter Test

1. Suppose that $f(x) = \sqrt{2x-1}$ and $g(x) = x^2 - 3$. Determine the following functions and state the domain of each:

   a. $f + g$
   b. fg
   c. $\dfrac{f}{g}$
   d. $f \circ g$

2. Suppose that $f(x) = x^2 - 1$, $x \geq 0$. Determine $f^{-1}(x)$.

3. Does the function $f(x) = x^2$ have an inverse? If it does, find it; if it doesn't, explain why.

4. Are the following functions one-to-one? For those that are, find the inverse function. For those that are not, explain why.

   a. $f(x) = 5x - 1$
   b. $f(x) = x^4 + x^2$
   c. $f(x) = \dfrac{x}{x+1}$, $x \neq -1$
   d. $f(x) = \dfrac{1}{x^2}$

5. Suppose that a spring obeys Hooke's law. It takes 5 kg to stretch it 10 cm. Suppose that a force of 35 kg is applied. By how much does the spring stretch?

6. Suppose that the variable y varies inversely with x. When y is 4, the value of x is 13. Express y as a function of x.

7. Suppose that f and g are inverses of one another. What is the value of $g(f(4)) - f(g(4))^2$?

8. **Thought question.** Define the notion of a one-to-one function. Give examples of a function that is one-to-one and a function that is not one-to-one.

9. **Thought question.** Describe a procedure for determining the inverse of a one-to-one function. Will this procedure work for *all* one-to-one functions? Explain your answer.

CHAPTER 6
EXPONENTIAL AND LOGARITHMIC FUNCTIONS

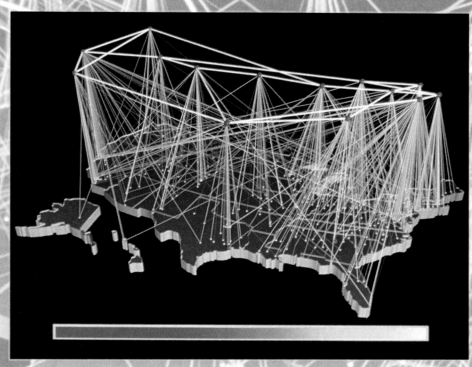

Applied mathematics uses various functions that describe common situations. In Chapter 2, we introduced linear functions that could be either increasing or decreasing. However, in many mathematical models, the simple increase or decrease portrayed by linear functions doesn't accurately reflect what is taking place. To portray more complex models, we need other kinds of functions, the first of which are introduced in this chapter. Here, we add to our repertoire the exponential and logarithmic functions, which are used for describing real-world situations such as compound interest, radioactive decay, bacterial growth, earthquakes, hearing ability, and the transmission of light to ocean depths.

Chapter Opening Image: *This data visualization illustrates the rapid growth of computer networking from 1990 to 1992; during this two-year period, traffic growth exceeded tens of billions of bytes per day.*

6.1 EXPONENTIAL FUNCTIONS WITH BASE a

Many models make use of functions in which the variable is in the exponent, as in the function
$$f(x) = a^x$$
where a is a fixed real number.

Calculating a^x with a Calculator

To calculate a^x, it is most convenient to use a scientific calculator. Such calculators have a key, usually labeled [x^y] or [^], for calculating powers. To calculate a^x, first enter the value of a. Then, press the [x^y] or [^] key. Next, enter the value of x. Finally, press the [ENTER] or [=] key. For example, to calculate $2.57^{3.4875}$, first key in the number 2.57. Then, press the [x^y] or [^] key. Next, key in the exponent 3.4875. Finally, press the [ENTER] or [=] key. The approximate value is 26.893.

Laws of Exponents

When we introduced the laws of exponents in Section 1.2, the exponents were rational numbers. The laws of exponents continue to hold true for all real exponents. Let's record them here for reference:

LAWS OF REAL EXPONENTS

Let a and b be positive real numbers and let x and y be any real numbers. Then the following laws of exponents hold:

1. $a^x a^y = a^{x+y}$
2. $(a^x)^y = a^{xy}$
3. $(ab)^x = a^x b^x$
4. $\dfrac{a^x}{a^y} = a^{x-y}$

Definition Exponential Function with Base a

Let a be a positive real number other than 1. The **exponential function with base** a is the function
$$f(x) = a^x, \qquad a \neq 1 \text{ and } a > 0$$

Here are some examples of exponential functions with various bases:
$$f(x) = 2^x$$
$$g(x) = 3^x$$
$$h(x) = \sqrt{5}^x$$
$$k(x) = \left(\frac{1}{2}\right)^x$$

Note that we excluded the possibility $a = 1$; the constant function $f(x) = 1^x = 1$ is not considered an exponential function.

EXAMPLE 1
Calculating Values of Exponential Functions

Let $f(x) = 2^x$ be the exponential function with base 2. Calculate the following:

1. $f(1)$
2. $f(2)$
3. $f(3)$
4. $f(-1)$
5. $f(-2)$
6. $f(0)$
7. $f\left(\dfrac{1}{2}\right)$

SECTION 6.1 Exponential Functions with Base a

Solution

1. $f(1) = 2^1 = 2$
2. $f(2) = 2^2 = 4$
3. $f(3) = 2^3 = 8$
4. $f(-1) = 2^{-1} = \dfrac{1}{2^1} = \dfrac{1}{2}$
5. $f(-2) = 2^{-2} = \dfrac{1}{2^2} = \dfrac{1}{4}$
6. $f(0) = 2^0 = 1$
7. $f\left(\dfrac{1}{2}\right) = 2^{1/2} = \sqrt{2}$

Graphs of Exponential Functions

As we have seen with polynomial and rational functions, the graph of a function provides valuable information about the function, such as where it increases or decreases, what its zeros are, what its asymptotes are, and so on. With this in mind, let's now learn to sketch the graphs of exponential functions. The following example discusses the graph of a particular exponential function.

EXAMPLE 2
Graphing an Exponential Function

Sketch the graph of the function:
$$f(x) = 2^x$$

Solution

We can record the results of Example 1 in the table at left. In Figure 1, we plot the points listed in the table and connect them with a smooth curve. Note that as x increases, the graph rises very sharply. As x approaches negative infinity, the graph has the negative x-axis as an asymptote.

For $a > 1$, the graph of $f(x) = a^x$ has the same general shape as the graph in Figure 1. Namely, as x increases, so does the value of $f(x)$, and as x approaches negative infinity, the graph approaches the negative x-axis as an asymptote. Furthermore, the larger the value of a, the steeper the rate at which the graph rises. In Figure 2, we show the graphs of several exponential functions that correspond to assorted values of a, all larger than 1.

In the next example, we consider a particular exponential function for which $0 < a < 1$.

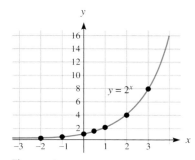

x	$f(x)$
-2	$\dfrac{1}{4}$
-1	$\dfrac{1}{2}$
0	1
$\dfrac{1}{2}$	$\sqrt{2}$
1	2
2	4
3	8

Figure 1

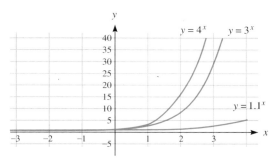

Figure 2

EXAMPLE 3
Graphing a Decreasing Exponential

Sketch the graph of the exponential function

$$f(x) = \left(\frac{1}{2}\right)^x$$

Solution

To get a sense of the shape of the graph, we plot points that correspond to some representative values of x. The calculations require the use of the laws of real exponents. For instance,

$$f(-1) = \left(\frac{1}{2}\right)^{-1} = \frac{1}{\frac{1}{2}} = 2$$

The accompanying table summarizes various values of the function. You should verify these values as an exercise.

x	$f(x)$
-3	8
-2	4
-1	2
0	1
1	$\frac{1}{2}$
2	$\frac{1}{4}$
3	$\frac{1}{8}$

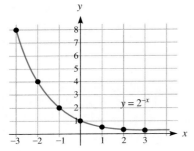

Figure 3

In Figure 3, we plot the points that correspond to the table entries and draw a smooth curve through the points. Note that as x increases, the graph of $f(x)$ decreases. Moreover, as x approaches positive infinity, the graph approaches the positive x-axis as an asymptote. As x decreases and approaches negative infinity, function values increase without bound and approach infinity.

For $0 < a < 1$, the graph of $f(x) = a^x$ has the same general shape as the graph shown in Figure 3. In Figure 4, we sketch the graphs of several exponential functions with base values a that are less than 1 and greater than 0. Note that the smaller the value of a, the steeper the graph.

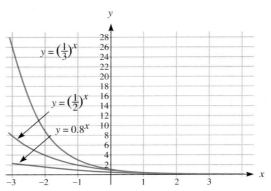

Figure 4

Applications of Exponential Functions

Exponential functions abound in applications, such as radioactive decay, bacteria growth, and the spread of epidemics. Such functions typically arise multiplied by a constant, in the form

$$f(x) = A \cdot b^{kx}$$

where A is the constant, b is a positive real number, and k is a real number that may be positive or negative.

The following examples illustrate some of the instances in which such functions arise.

EXAMPLE 4 Compound Interest

Suppose that P dollars are deposited in a bank account that pays annual interest at a rate r, and the interest is compounded n times per year. After a length of time t, in years, the amount $A(t)$ in the account is given by the formula:

$$A(t) = P\left(1 + \frac{r}{n}\right)^{nt}$$

Suppose that the interest rate is 10 percent, the amount is compounded annually, and the initial deposit is $10,000.

1. Write the function $A(t)$ in terms of an exponential function.
2. Determine the amount in the account after 6 months, 1 year, 2 years, and 10 years.
3. Graph the function $A(t)$.

Solution

1. In this example,

$$P = 10{,}000, \qquad r = 0.1, \qquad n = 1$$

So in this case, the compound interest formula is

$$A(t) = 10{,}000\left(1 + \frac{0.1}{1}\right)^{1 \cdot t}$$

$$A(t) = 10{,}000 \cdot (1.1)^t$$

2. The desired amounts are given, respectively, by the function values $A(0.5)$, $A(1)$, $A(2)$, $A(10)$. Computing these values, we have:

$$A(0.5) = 10{,}000 \cdot (1.1)^{0.5}$$
$$= 10{,}000 \cdot \sqrt{1.1}$$
$$= \$10{,}488.09$$
$$A(1) = 10{,}000 \cdot (1.1)^1 = \$11{,}000.00$$
$$A(2) = 10{,}000 \cdot (1.1)^2 = \$12{,}100.00$$
$$A(10) = 10{,}000 \cdot (1.1)^{10} = \$25{,}937.43$$

To calculate these values, we use a calculator.

3. We know the general shape of the graph of the exponential function $f(t) = 1.1^t$ from Figure 2. The graph of $A(t)$ is obtained by scaling the graph of $f(t)$ by a factor of 10,000. From part 2, we have several points we can plot. The graph must pass through these points and increase rapidly as the value of t increases. From the context of the application, only nonnegative values of t are relevant, so we limit the domain of the graph to $t > 0$. A sketch of the graph is given in Figure 5.

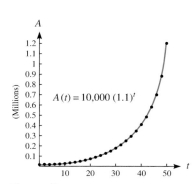

Figure 5

EXAMPLE 5
Radioactive Decay

Radioactive decay is described by an exponential function. Suppose that A_0 denotes the amount of a radioactive substance present at time 0, and suppose that $Q(t)$ represents the quantity present at time t. As time passes, the amount of radioactive material decays, with the amount approaching 0 as time approaches infinity. Suppose H denotes the time it takes for a quantity of the radioactive material to decay to half its original amount. The time H depends on the particular radioactive material and is called the **half-life** of the material. The function $Q(t)$ is then given by the expression

$$Q(t) = A_0 2^{-t/H}$$

Radioactive iodine (^{131}I) is used in various medical diagnostic tests. Its half-life is 8 days.

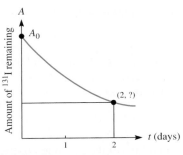

Figure 6

1. Write the formula for $Q(t)$ for radioactive iodine in terms of an exponential function.
2. What proportion of the original radioactive iodine is present 48 hours after it is originally swallowed? (See Figure 6.)

Solution

1. We are given the information that $H = 8$ days. Because we are not given the original amount of radioactive iodine, we leave it in the formula as A_0. We then have:

$$Q(t) = A_0 2^{-t/8}$$
$$= A_0 2^{-1/8 \cdot t}$$
$$= A_0 \left(2^{-1/8}\right)^t$$
$$= A_0 (0.917)^t$$

The function $Q(t)$ is a constant, A_0, multiplied by an exponential function, a^t, where a is equal to 0.917.

2. When t equals 48 hours, the amount of radioactive iodine remaining is $Q(2)$, which is equal to:

$$Q(2) = A_0 2^{(-1/8)2} = A_0 2^{-1/4} = 0.840896 A_0$$

That is, after 48 hours, about 84 percent of the radioactive iodine remains.

EXAMPLE 6
Bacterial Growth

Colonies of certain bacteria exhibit exponential growth. At time 0 hours, suppose that there are N_0 bacteria. Then $N(t)$, the number of bacteria at time t, in hours, is given by

$$N(t) = N_0 2^{t/D}$$

where D denotes the time, in hours, that it takes the colony to double in size.

1. Write $N(t)$ in terms of an exponential function.
2. Suppose that the colony initially contains 5,000 bacteria and the doubling time is 4 hours. How many bacteria are there after 10 hours?
3. Consider again the bacteria colony in part 2. How long will it be before the colony contains 160,000 bacteria?

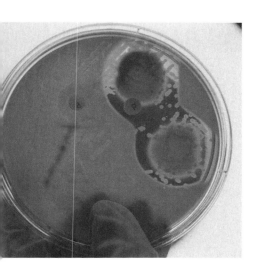

Solution

1. Use the laws of exponents to write:

$$N(t) = N_0 2^{t/D}$$
$$= N_0 2^{(1/D)t}$$
$$= N_0 \left(2^{1/D}\right)^t$$

Let $a = 2^{1/D}$. Then the formula takes the form:

$$N(t) = N_0 a^t$$

2. We are given the information that the doubling time, D, is 4 hours, and $N_0 = 5{,}000$. So the formula for $N(t)$ is:

$$N(t) = 5{,}000 \cdot 2^{t/4}$$

The number of bacteria after 10 hours is given by the value $N(10)$:

$$N(10) = 5{,}000 \cdot 2^{10/4} = 5{,}000 \cdot 2^{5/2} = 5{,}000 \cdot \sqrt{2^5} = 28{,}284$$

Note that the last number was rounded to the nearest integer. The formula for the number of bacteria, like most mathematical descriptions of physical phenomena, is only approximate; answers must be interpreted to conform to physical reality, in this case, the fact that there are no fractional bacteria.

3. We must determine the value of t for which:

$$N(t) = 160{,}000$$

Use the equation for $N(t)$ from part 2. We have:

$$5{,}000 \cdot 2^{t/4} = 160{,}000$$
$$2^{t/4} = 32$$
$$2^{t/4} = 2^5$$

For this equation to hold, the exponents must be equal. (Look at the graph of 2^x in Figure 1; it is constantly increasing, so that no two values of x have the same value for y.) Thus, we have the equation:

$$\frac{t}{4} = 5$$
$$t = 20$$

After 20 hours, the bacteria colony has 160,000 bacteria.

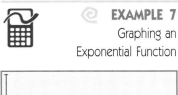

EXAMPLE 7
Graphing an Exponential Function

Consider the function:

$$f(x) = 3 \cdot 1.1^{-3x} + 1$$

1. Use a graphing calculator to display the graph of $f(x)$.
2. Determine the asymptotes of the graph.

Solution

1. Because the exponential 1.1^{-3x} decreases so quickly as x increases, we can use the range $0 \leq x \leq 10$ (rather than $-10 \leq x \leq 10$) to get a reasonable idea of the shape of the graph. See Figure 7.
2. From the graph, we see that the line $y = 1$ is a horizontal asymptote and that there are no vertical asymptotes.

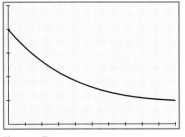

Figure 7

Exercises 6.1

1. Let $f(x) = 3^x$
   a. Find $f(1)$, $f(2)$, $f(3)$, $f(0)$, $f(-1)$, $f(-2)$, $f(-3)$, and $f\left(\dfrac{1}{2}\right)$.
   b. Sketch the graph of $f(x) = (3)^x$.

2. Let $f(x) = \left(\dfrac{1}{3}\right)^x$
   a. Find $f(1)$, $f(2)$, $f(3)$, $f(0)$, $f(-1)$, $f(-2)$, $f(-3)$, and $f\left(\dfrac{1}{2}\right)$.
   b. Sketch the graph of $f(x) = \left(\dfrac{1}{3}\right)^x$.

Sketch the graphs of the following functions.

3. $f(x) = 4^x$
4. $f(x) = 8^x$
5. $f(x) = \left(\dfrac{1}{4}\right)^x$
6. $f(x) = \left(\dfrac{1}{5}\right)^x$
7. $f(x) = \left(\dfrac{3}{4}\right)^x$
8. $f(x) = \left(\dfrac{2}{3}\right)^x$
9. $f(x) = (2.4)^x$
10. $f(x) = (3.6)^x$
11. $f(x) = (0.38)^x$
12. $f(x) = (0.76)^x$
13. $y = 5^x$
14. $y = 3^x$
15. $y = 10^x$
16. $y = 2.7^x$
17. $f(x) = 2^{-x}$
18. $f(x) = 2^{x-1}$
19. $f(x) = 2^{-(x-3)}$
20. $f(x) = 2^x - 4$
21. $f(x) = 2^{|x|}$
22. $f(x) = 2^x + 2^{-x}$
23. $f(x) = 2^{-x^2}$
24. $f(x) = |2^x - 4|$
25. $f(x) = 2^x - 2^{-x}$
26. $f(x) = 2^{-(x+3)^2}$
27. $f(x) = 1 - 2^{-x^2}$
28. $f(x) = x + 2^{-x^2}$

Match the functions in Exercises 29–38 with their corresponding graphs in Figures (a)–(j).

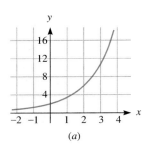
(a)

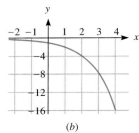
(b)

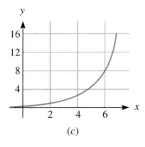
(c)

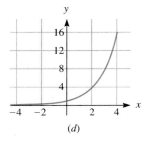
(d)

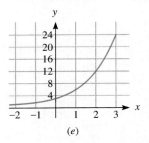
(e)

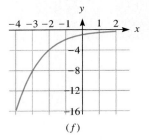
(f)

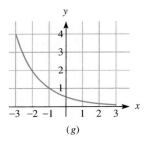
(g)

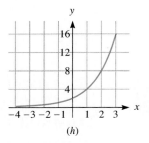
(h)

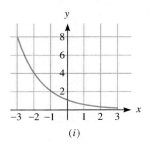
(i)

SECTION 6.1 Exponential Functions with Base a

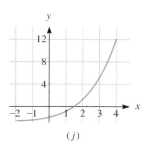

(j)

29. $f(x) = 2^x$
30. $f(x) = 2^{-x}$
31. $f(x) = 2^{x+1}$
32. $f(x) = 2^x + 1$
33. $f(x) = 2^{-(x+1)}$
34. $f(x) = 2^{x-3}$
35. $f(x) = 2^x - 3$
36. $f(x) = 3 \cdot 2^x$
37. $f(x) = -2^x$
38. $f(x) = -2^{-x}$

Applications

39. **Compound interest.** Suppose that a bank account pays 8 percent interest, the amount is compounded annually, and the initial deposit is $10,000. Let $A(t)$ denote the amount in the account after t years.

    a. Write the function $A(t)$ in terms of an exponential function.

    b. Determine the amount in the account after 1, 2, 3, and 10 years.

    c. Graph the function $A(t)$.

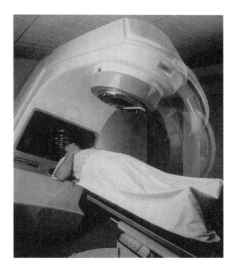

40. **Radioactive decay in cancer treatment.** A radioactive decay function is given by
$$Q(t) = A_0 2^{-t/H}$$
where A_0 denotes the amount of a radioactive substance present at time $t = 0$, and H is the half-life. Radioactive cobalt 60 (Co^{60}) has a half-life of 5.3 years. It is used for the chemotherapy treatment of cancer tumors.

    a. Write the formula for $Q(t)$ for radioactive cobalt in terms of an exponential function.

    b. Suppose a chemotherapy device contains 4 grams of Co^{60}. That is, A_0 is 4 grams. How much will be present after 1 day ($\frac{1}{365}$ yr)? after 1 year? 2 years? 10.6 years? 20 years?

41. **Radioactive decay in nuclear power production.** Plutonium, a common byproduct of nuclear reactors and ingredient of nuclear bombs, is a great concern to those who oppose the building of such reactors. The half-life of plutonium is 23,105 years.

    a. Write the formula for the quantity of plutonium, $Q(t)$, in terms of an exponential function.

    b. A nuclear reactor contains 400 kilograms of plutonium. What proportion of this original amount of plutonium is present after 1 year? 2 years? 200 years?

42. **Population growth** The population of the world was about 5 billion on January 1, 1987, and the population doubles every 43 years. Suppose we let $t = 0$ correspond to 1987, and suppose we consider the world population N_0 to be 5 billion at $t = 0$. The population after t years is $N(t)$, where
$$N(t) = N_0 2^{t/D}$$
and where D is the doubling time.

    a. Write $N(t)$ in terms of an exponential function.

    b. What will the world population be in 43 years? in 86 years? in 1997? in 2000?

    c. In how many years will the population of the world be 6 billion?

43. **Population growth.** The population of Texas was 16,370,000 on January 1, 1985, and it doubles every 23 years. Suppose we let $t = 0$ correspond to 1985, and suppose we consider the Texas population N_0 to be 16,370,000 at $t = 0$. The population after t years is $N(t)$, where
$$N(t) = N_0 2^{t/D}$$
and D is the doubling time.

    a. Write $N(t)$ in terms of an exponential function.

    b. What will the population of Texas be in 23 years? in 46 years? in 1995? in 2000?

44. **Investment performance.** An investment grows exponentially in such a way that it doubles every 10 years. An initial amount of N_0 dollars at time $t = 0$ will grow to an amount $N(t)$ in t years, where

$$N(t) = N_0 2^{t/D}$$

and D is the doubling time.

a. Write $N(t)$ in terms of an exponential function, assuming the initial amount invested is $100,000.
b. What will the amount of the investment be after 5 years? 10 years? 20 years? 32 years?
c. How many years will it take for the amount of the investment to be $1 million?

Technology

Use a graphing calculator to graph the following functions for $0 \leq x \leq 5$. Experiment with the y-range to include points that correspond to all values of x.

45. $f(x) = 3^x$
46. $f(x) = 2.5 \cdot 2^x$
47. $f(x) = 8.71 \cdot 2^{5x}$
48. $f(x) = -3.1 \cdot 2^{-1.1x}$
49. $f(x) = 2^x - 2^{-x}$
50. $f(x) = 2^{2^x}$
51. Determine all points at which the graphs of $f(x) = 2^x$ and of $f(x) = 3^x - 2.5^x$ intersect. Approximate coordinates to two significant digits.
52. Graph the function $f(x) = 1{,}000(1 - 2^{-x})$. By zooming out, determine any asymptotes of the graph.

In your own words

53. Describe the sort of physical situation that would best be modeled by an exponential function of the form $f(x) = a^x$, $a > 1$.
54. Describe the sort of physical situation that would best be modeled by an exponential function of the form $f(x) = a^x$, $0 < a < 1$.
55. In the laws of real exponents, explain why the base must be a positive number. Why is this restriction not necessary for integer exponents?

6.2 THE NATURAL EXPONENTIAL FUNCTION

The Number e

Any positive real number $a \neq 1$ can serve as the base for an exponential function. The most important value for a is a particular real number designated e. The number e is an irrational number, the value of which is approximately 2.718281. Although this number may seem to come out of nowhere, it arises quite naturally in many applications of mathematics in the biological and physical sciences as well as in business. To introduce the number e, let's consider a problem in compound interest.

Suppose that P dollars are deposited for t years in a bank account that yields interest at an annual rate r, and suppose the interest is compounded n times per year. As we have already seen, the value $A(t)$ of the account at time t is given by the compound interest formula:

$$A(t) = P\left(1 + \frac{r}{n}\right)^{nt}$$

Suppose that the initial deposit is $1, the length of time is one year, and the rate of interest is 100 percent. In this case, the compound interest formula reads:

$$A(1) = 1 \cdot \left(1 + \frac{1}{n}\right)^{n \cdot 1} = \left(1 + \frac{1}{n}\right)^n$$

SECTION 6.2 The Natural Exponential Function

Let's consider the effect of different periods of compounding on the amount after 1 year. This means we compute the value of $A(1)$ for various values of n. We compute assorted values on a calculator and list them in the following table.

n	Amount from Compounding the Interest n Times a Year
1	2
2	2.25
3	2.370370
4	2.441406
10	2.5937425
50	2.691588
100	2.7048138
1,000	2.7169239
10,000	2.7181459
100,000	2.718268
1,000,000	2.7182805
10,000,000	2.7182817

Note that as the frequency of compounding increases without bound, the amount in the account at the end of the year approaches an amount approximately equal to 2.718. This is rather surprising. Your intuition might suggest that as the money is compounded more and more frequently (every day, every hour, every second), the amount in the account would grow without limit. However, this is just not the case. As the frequency of compounding increases without bound, the amount in the account approaches a specific real number, denoted e, whose decimal expansion is given by 2.718281.... Let's record these results for future reference.

THE NUMBER e

The quantity $\left(1 + \dfrac{1}{n}\right)^n$ approaches e as n becomes arbitrarily large.

$$e \approx 2.718281$$

Calculations Involving e   Algebraic expressions involving e^x can be manipulated using the fact that e^x is just an ordinary power to which the laws of exponents apply. The next example provides some practice in carrying out such algebraic manipulations.

EXAMPLE 1   Simplify the following expressions that involve e.

Calculating with e
1. $e^3 e^{-1} e^{-5}$
2. $(e + 1)^2$
3. $(e^x)^5$
4. $(e^x + e^{-x})^2$
5. $(e^x + 1)(3e^x - 4)$
6. $\sqrt{e^x}$

Solution

1. By the laws of exponents, we have:
$$e^3 e^{-1} e^{-5} = e^{3+(-1)-5} = e^{-3}$$

2. Use the formula for the square of a binomial to obtain:
$$(e+1)^2 = e^2 + 2 \cdot e \cdot 1 + 1^2$$
$$= e^2 + 2e + 1$$

3. By the laws of exponents,
$$(e^x)^5 = e^{x \cdot 5} = e^{5x}$$

4. Multiply the binomials and apply the laws of exponents to obtain:
$$(e^x + e^{-x})^2 = (e^x)^2 + 2e^x e^{-x} + (e^{-x})^2$$
$$= e^{2x} + 2e^0 + e^{-2x}$$
$$= e^{2x} + e^{-2x} + 2$$

5. Multiply, and apply the laws of exponents:
$$(e^x + 1)(3e^x - 4) = e^x \cdot 3e^x - 4e^x + 3e^x - 4$$
$$= 3e^{2x} - e^x - 4$$

6. Recall that taking the square root is equivalent to raising to the power $\frac{1}{2}$:
$$\sqrt{e^x} = (e^x)^{1/2} = e^{x \cdot 1/2} = e^{x/2}$$

You can do numerical calculations with e just as you would with any other decimal number. Most scientific calculators have an $\boxed{e^x}$ key for calculating powers of e. For example, to calculate e^2 on a graphing calculator, you would use the keystrokes: $\boxed{e^x}\,\boxed{2}\,\boxed{\text{ENTER}}$. On most scientific calculators, you need to use the reverse order: $\boxed{2}\,\boxed{e^x}$.

EXAMPLE 2
Calculators and e

Calculate $2e^{-4.78x}$ for $x = 3.5$.

Solution

Here are the required keystrokes for a graphing calculator. Note that we enclose the entire exponent within parentheses.

$\boxed{2}\,\boxed{\times}\,\boxed{e^x}\,\boxed{(}\,\boxed{(-)}\,\boxed{4.78}\,\boxed{\times}\,\boxed{3.5}\,\boxed{)}\,\boxed{\text{ENTER}}$

The answer displayed is $1.0846342480\text{E} - 7$.

The Natural Exponential Function

The exponential function with base e is the most important exponential function, because it plays a role in both the sciences and calculus. To indicate its importance, let's assign it a name as follows:

Definition
Natural Exponential Function

The **natural exponential function** is the function $f(x)$, defined as:
$$f(x) = e^x$$

In the preceding section, we considered applications that involved exponential functions of the form
$$f(x) = a^x$$

10. $(2e^{-x})^4$
11. $(e^x - e^{-x})^2$
12. $(1 + 2e^x)^2$
13. $(e^x - 1)^3$
14. $(e^x + e^{-x})^3$
15. $(e^{x-2})(4e^x + 5)$
16. $(e^x + e^{-x})(e^x - e^{-x})$
17. $(e^x - 3)(e^x + 3)$
18. $(e^x - 1)(e^{2x} + e^x + 1)$
19. $\sqrt{e^{2x}}$
20. $\sqrt[3]{e^{6x}}$
21. $\dfrac{(1 + e^x)(1 + e^{-x}) - (1 + e^x)(1 - e^{-x})}{(1 + e^x)^2}$

22. $\dfrac{(e^x + e^{-x})(e^x + e^{-x}) - (e^x - e^{-x})(e^x - e^{-x})}{(e^x + e^{-x})^2}$

Determine

$$\dfrac{f(x + h) - f(x)}{h}, \quad h \neq 0$$

and simplify the resulting expression, where:

23. $f(x) = e^x$
24. $f(x) = \dfrac{e^x - e^{-x}}{2}$
25. $f(x) = \dfrac{e^x + e^{-x}}{2}$
26. $f(x) = e^{-3x}$

Applications

27. **Psychology.** The Hullian Model of Learning asserts that the probability $P(t)$ of mastering a certain concept after t learning trials is given by

$$P(t) = 1 - e^{-kt}$$

where k is a constant, the value of which depends on the difficulty of the learning. An educational psychologist analyzes the amount of time it takes to learn a concept in mathematics, and finds that $k = 0.18$.

a. What is the probability of learning the concept after 1 trial? 2 trials? 3 trials? 5 trials? 13 trials?

b. Sketch a graph of the function $P(t)$.

28. **Advertising.** A company introduces a new product with a trial run in a city. It advertises the product on TV and gathers data concerning the percentage $P(t)$ of the people in the city who buy the product after it is advertised a certain number of times t. It determines that the **limiting effect** of the advertising is 75 percent.

This means that no matter how many times the company advertises its product, the percentage of people who buy the product never reaches or exceeds 75 percent, although the percentage gets closer and closer to 75 percent. The company determines that $P(t)$ can be described in terms of exponential functions

$$P(t) = 0.75(1 - e^{-kt})$$

where $k = 0.05$.

a. What percentage of the population buys the product after 1 advertisement? After 2 ads? 10 ads? 20 ads? 50 ads?

b. Sketch a graph of $P(t)$.

29. **Advertising.** Repeat Exercise 28, assuming that the limiting effect is 50 percent and that $k = 0.08$.

30. **Advertising.** Repeat Exercise 28, assuming that the limiting effect is 100 percent and $k = 0.06$.

31. **Use of a new drug.** Even after a new medication has been extensively tested and approved by the FDA, it takes time for doctors to accept and use it. First, they have to become aware of it, and then they must try it on their patients. The usage approaches a limiting value of 100 percent, or 1, as a function of t, in months. The percentage $P(t)$ of doctors who use the medication can be described in terms of exponential functions, as follows,

$$P(t) = 100(1 - e^{-kt})$$

where $k = 0.3$.

a. What percentage of the doctors have accepted the medication after 1 month? 2 months? 5 months? 10 months? 1 year?

b. Sketch a graph of $P(t)$ for $t \geq 0$.

312 CHAPTER 6 Exponential and Logarithmic Functions

32. **Skydiving.** The velocity of a particular skydiver $v(t)$ can be described in terms of the exponential functions
$$v(t) = v_T(1 - e^{-kt})$$
where k is a positive constant. Suppose that, for a certain skydiver, the terminal velocity v_T is 160 miles per hour and k is equal to 0.2.

a. What is the velocity of the skydiver 1 second after jumpoff? 2 sec? 5 sec? 10 sec? 20 sec?

b. Sketch a graph of the function for $t \geq 0$.

Technology

Use a calculator to determine the values of the following expressions.

33. **Limited population growth.** A herd of caribou on a Canadian island has its growth limited by the environment and by predators. The limiting population is 500. Suppose that the number of caribou at time t years after they were first studied is:
$$N(t) = \frac{500}{1 + 400e^{-t}}$$

a. What is the population after 10 years? 20 years? 30 years? 100 years?

b. Sketch the graph of $N(t)$, where $0 \leq t \leq 30$.

34. **Spread of a rumor.** In a college with a population of 800, a group of students spread the rumor that if they study algebra from a certain book, they will graduate with honors. The number of people $N(t)$ who have heard the rumor after t minutes is given by
$$N(t) = \frac{P}{1 + Ae^{-kt}}$$
where A and k are constants. In this situation, $P = 800$, $A = 132$, and $k = 0.4$.

a. How many people have heard the rumor after 1 minute? 2 minutes? 6 minutes? 12 minutes? 30 minutes? 1 hour?

b. Sketch the graph of $N(t)$ for $0 \leq t \leq 180$.

Calculate the numerical values of the following.

35. $e^{0.1654} + e^{-2.34}$

36. $2e^{-4.6} - e^{5.71}$

Use a graphing calculator to graph the following functions in the domain $-4 \leq x \leq 4$. Experiment with the y-range so that a point is included for each given x-value.

37. $f(x) = e^x$

38. $f(x) = 5e^{3x}$

39. $f(x) = e^{-0.1x}$

40. $f(x) = \frac{e^x - e^{-x}}{e^x + e^{-x}}$

41. $f(x) = \frac{1}{e^x + 1}$

42. $f(x) = 5 - 5e^{-x}$

In your own words

43. What is the difference between an exponential function and a natural exponential function?

44. What sort of natural exponential functions can be used to model increasing quantities?

45. What sort of natural exponential functions can be used to model decreasing functions?

46. What happens to the values of e^{-x} as x becomes a very large positive number?

6.3 LOGARITHMIC FUNCTIONS

Inverse functions provide us with a useful way of creating new functions from given ones. In this section, we introduce the logarithmic functions by taking the inverse of the exponential function.

Definition of Logarithmic Functions

By examining the graph, we see that the exponential function
$$f(x) = a^x, \quad a \neq 1 \quad \text{and} \quad a > 0$$
is one-to-one. Therefore, it has an inverse function that depends on the value of a. The inverse function is denoted $\log_a x$, and is called the **logarithm with base a**. We restate the definition of an inverse function for this particular instance:

SECTION 6.3 Logarithmic Functions

Definition
Logarithmic Function with Base a

Let x be a positive real number. Then:
$$\log_a x = y \quad \text{if and only if} \quad a^y = x$$
Or, in words, $\log_a x$ is the exponent to which a must be raised to obtain x.

For example, because $3^2 = 9$, we have:
$$\log_3 9 = 2$$
That is, 2 (or $\log_3 9$) is the exponent to which 3 must be raised to obtain 9. Similarly, because
$$2^{10} = 1{,}024$$
we have:
$$\log_2 1{,}024 = 10$$
That is, 10 (or $\log_2 1{,}024$) is the exponent to which 2 must be raised to obtain 1,024.

A Word about Notation: In writing logarithmic functions, it is common to omit the parentheses if no confusion results. For example, it is common to write $\log_a x$ instead of $\log_a(x)$. However, in cases where we take the logarithm of a complicated expression, it is best to use the parentheses notation.

The domain of $\log_a x$ is the range of a^x, that is, all positive real numbers. The range of $\log_a x$ is the domain of a^x, that is, all real numbers.

Properties of Logarithms

For each fact about raising a number to a power, there is a corresponding fact about logarithms. For example, the property $a^1 = a$ gives us:
$$\log_a a = 1$$

Similarly, the property $a^0 = 1$ yields:
$$\log_a 1 = 0$$

The fact that $\log_a x$ and a^x are inverse functions of one another yields the following formulas:

FUNDAMENTAL PROPERTIES OF LOGARITHMS

1. $a^{\log_a x} = x, \quad x > 0$
2. $\log_a a^x = x$

Note that these two properties of logarithms are special cases of the formulas of Section 5.2 for the composite of a function and its inverse function. Indeed, if we set $f(x) = a^x$ and $g(x) = \log_a x$, then, because f and g are inverses of one another, we have:
$$f(g(x)) = f(\log_a x) = a^{\log_a x} = x$$
$$g(f(x)) = g(a^x) = \log_a(a^x) = x$$

The two Fundamental Properties of Logarithms are very useful in simplifying expressions that involve logarithmic functions, as the following example illustrates.

314 CHAPTER 6 Exponential and Logarithmic Functions

EXAMPLE 1
Calculations that Involve Logarithms

Calculate the values of the following expressions:

1. $\log_{10} 10^{5.781}$
2. $\log_{10} 1{,}000$
3. $\log_2 \frac{1}{8}$
4. $4^{\log_4 11.1}$
5. $\log_a a^{3x}$
6. $a^{\log_a x^3}$

Solution

1. Property 2 states that $\log_a a^x = x$, so the value of the expression is $x = 5.781$.
2. Because $1{,}000 = 10^3$, we see that $\log_{10} 1{,}000 = \log_{10} 10^3 = 3$, by Property 2.
3. Using Property 2, we have:
$$\log_2 \frac{1}{8} = \log_2 2^{-3}$$
$$= -3$$
4. Property 1 states that $a^{\log_a x} = x$, so the value of the expression is $x = 11.1$.
5. By property 2, the value of the expression is $3x$.
6. By property 1, the value of the expression is x^3.

The Graph of $\log_a x$

Because $\log_a x$ is the inverse of the function a^x, the graph of $\log_a x$ can be obtained by reflecting the graph of a^x in the line $y = x$ (See Section 5.2). In Figure 1, we sketch the graphs of the two functions for $a > 1$, and the graph of the line $y = x$.

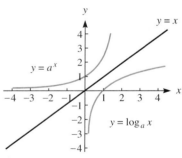

Figure 1

By using translation and scaling, we can sketch the graphs of many functions related to logarithmic functions, as the next example shows.

EXAMPLE 2
Graphing Logarithmic Functions

Sketch the graphs of the following functions.

1. $y = \log_2 x$
2. $y = \log_2(x - 1)$
3. $y = \log_2 x + 3$

SECTION 6.3 Logarithmic Functions

Solution

1. The function $y = \log_2 x$ is the inverse of the function $y = 2^x$. Figure 2 shows the graph of $y = 2^x$ in black. To obtain the graph of the inverse function, we reflect this graph in the line $y = x$. The graph of the inverse is shown in color. The y-axis is a vertical asymptote of the colored graph, and the graph increases without bound as x approaches positive infinity.

2. According to our discussion about translating graphs of functions, the graph of $y = \log_2(x - 1)$ can be obtained by translating the colored graph in Figure 2 one unit to the right. The translated graph is sketched in Figure 3. Note that the vertical asymptote is now the line $x = 1$.

3. According to our discussion on translating graphs of functions, the graph of $y = \log_2 x + 3$ can be obtained by translating the graph $\log_2 x$ three units upward. The translated graph is shown in Figure 4.

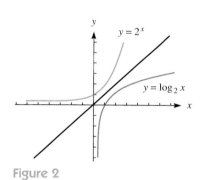

Figure 2

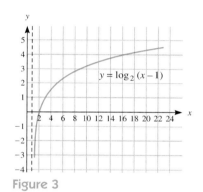

Figure 3

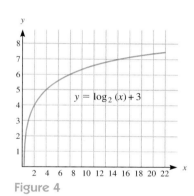

Figure 4

Properties of $\log_a x$

The following result is helpful in simplifying expressions that involve logarithmic functions:

LAWS OF LOGARITHMS

The function $\log_a x$ has the following properties where $x, y > 0$ and where a is a positive number other than 1.

1. The logarithm of a product is the sum of the logarithms:
$$\log_a(xy) = \log_a x + \log_a y$$

2. To compute the logarithm of x raised to a power, multiply the logarithm of x by the exponent:
$$\log_a x^y = y \log_a x$$

3. The logarithm of a quotient is the difference of the logarithms:
$$\log_a\left(\frac{x}{y}\right) = \log_a x - \log_a y$$

Verification of the Laws of Logarithms

1. By Fundamental Property 1, we have:
$$x = a^{\log_a(x)}$$
$$y = a^{\log_a(y)}$$

Multiplying these expressions gives:
$$xy = a^{\log_a(x)} \cdot a^{\log_a(y)}$$
$$= a^{\log_a x + \log_a y}$$

The last equation states that $\log_a x + \log_a y$ is the exponent to which a must be raised to obtain the value xy. This means that:
$$\log_a x + \log_a y = \log_a xy$$

2. Again, we use Fundamental Property 1 to obtain $x = a^{\log_a x}$, and we raise both sides of the equation to the power y:
$$x^y = (a^{\log_a(x)})^y = a^{y \log_a(x)}$$

This equation states that $y \log_a x$ is the exponent to which a must be raised to obtain x^y. This means that:
$$y \log_a x = \log_a x^y$$

3. Write the expression $\dfrac{x}{y}$ in the form
$$xy^{-1}$$
and apply what we just proved in parts 1 and 2:

$$\log_a\left(\frac{x}{y}\right) = \log_a(xy^{-1}) \quad \quad \textit{Laws of exponents}$$
$$= \log_a(x) + \log_a(y^{-1}) \quad \textit{Part 1}$$
$$= \log_a(x) + (-1)\log_a(y) \quad \textit{Part 2}$$
$$= \log_a(x) - \log_a(y)$$

The laws of logarithms can be used to simplify expressions that involve logarithmic functions, as the following two examples illustrate.

EXAMPLE 3
Simplifying Logarithmic Expressions

Write the following expressions using a single logarithm.
1. $\log_5[(x-1)(x-2)] - \log_5(x-2)$
2. $\log_{10} \sqrt{x} + \log_{10} x^{3/2}$

Solution

1. By Law 1 of logarithms, we have:
$$\log_5[(x-1)(x-2)] - \log_5(x-2)$$
$$= [\log_5(x-1) + \log_5(x-2)] - \log_5(x-2)$$
$$= \log_5(x-1)$$

SECTION 6.3 Logarithmic Functions

2. By Law 2 of logarithms, we have:
$$\log_{10} \sqrt{x} + \log_{10} x^{3/2} = \log_{10} x^{1/2} + \log_{10} x^{3/2}$$
$$= \frac{1}{2} \log_{10} x + \frac{3}{2} \log_{10} x$$
$$= 2 \log_{10} x$$

EXAMPLE 4
Further Practice with the Laws of Logarithms

Write the following expressions using a single logarithm.
1. $3 \log_a x + 4 \log_a y + \frac{1}{2} \log_a z$
2. $\log_2 \frac{x}{y} - \log_2 \frac{y^2}{z}$

Solution

1. By applying Law 2 of logarithms, we can write the given expression in the form:
$$\log_a x^3 + \log_a y^4 + \log_a z^{1/2}$$
We can now apply Law 1 of logarithms, to combine the three terms:
$$\log_a (x^3 y^4 z^{1/2})$$

2. By Law 3 of logarithms, we have:
$$\log_2 \frac{x}{y} - \log_2 \frac{y^2}{z} = \log_2 \left[\frac{x}{y} \div \frac{y^2}{z} \right]$$
$$= \log_2 \left[\frac{x}{y} \cdot \frac{z}{y^2} \right]$$
$$= \log_2 \left[\frac{xz}{y^3} \right]$$

Equations that Involve Logarithms

We can solve equations that involve logarithms and exponentials of a variable by using the following approach:

SOLVING EXPONENTIAL AND LOGARITHMIC EQUATIONS

1. Solving an equation that involves a variable exponent:
   a. Transform the equation so that it has a single exponential on one side of the equation and a number on the other side.
   b. Take logarithms of both sides of the equation.
   c. Apply the laws of logarithms to solve for the variable.
2. Solving an equation that involves a logarithm of a variable:
   a. Transform the equation so that it has a single logarithm on one side of the equation and a number on the other side.
   b. Apply the exponential function to both sides of the equation.
   c. Apply the laws of logarithms to solve for the variable.

The next example illustrates how to apply this approach.

EXAMPLE 5
Solving Logarithmic Equations

Solve the following equations for x.

1. $\log_5(3x - 1) = 2$
2. $7^{2x} = 4$

Solution

1. When faced with an equation involving logarithms, the best strategy is to use the properties of logarithms to eliminate them from the equation. In this case, we can raise 5 to the quantity on both sides of the equation:

$$5^{\log_5(3x-1)} = 5^2$$

By Fundamental Property 1, the left side of the equation equals $3x - 1$ and the equation reads:

$$3x - 1 = 5^2$$
$$3x - 1 = 25$$
$$x = \frac{26}{3}$$

2. Apply the function $\log_7$ to both sides of the equation:

$$\log_7 7^{2x} = \log_7 4$$
$$2x \log_7 7 = \log_7 4$$
$$2x = \log_7 4$$
$$x = \frac{\log_7 4}{2}$$

EXAMPLE 6
Solving More Logarithmic Equations

Solve the following equation:

$$\log_2(x + 1) + \log_2(x - 1) = 3$$

Solution

First, we can combine the two terms on the left using the first fundamental property of logarithms, and then eliminate the logarithm:

$$\log_2[(x + 1)(x - 1)] = 3$$
$$\log_2(x^2 - 1) = 3$$
$$2^{\log_2(x^2-1)} = 2^3$$
$$x^2 - 1 = 8$$
$$x^2 = 9$$
$$x = \pm 3$$

We now test these potential solutions to determine whether they satisfy the equation. Substituting $x = 3$ into the equation gives:

$$\log_2(x + 1) + \log_2(x - 1) \quad 3$$
$$\log_2(3 + 1) + \log_2(3 - 1)$$
$$\log_2 4 + \log_2 2$$
$$2 + 1$$
$$3$$

So $x = 3$ is a solution to the equation. Note, however, that for $x = -3$, the expression $\log_2(x + 1)$ is not even defined; therefore, $x = -3$ is not a solution.

EXAMPLE 7
Binary Representation of Numbers

A **binary number** is a number composed of a sequence of the digits 0 and 1, such as 1001 or 111101000111. It is possible to represent any decimal integer as a binary number. For example, 189 can be written as the binary number 10111101. (At the moment, we won't worry about how to determine this binary representation.) The electronic circuits of a computer work with binary numbers, so the binary representation of integers has great significance. Suppose that x is a positive integer. It can be shown that the number of digits $B(x)$ in the binary representation of x is given by the formula:

$$B(x) = \text{INT}(\log_2 x) + 1$$

where INT denotes the greatest integer function, that is, INT(x) is equal to the greatest integer that is less than or equal to x. How many binary digits are required to represent the integer 32,767? How many are needed to represent the integer 32,768?

Solution

To solve this problem, we must first obtain some feeling for the values assumed by the function $\log_2 x$. We begin by tabulating the various values of the powers of 2 and the corresponding values of $\log_2$, as in the following table. According to the table, the value of $\log_2 32,767$ is an integer at least 14 and less than 15, and therefore equals 14. The value of $B(32,767)$ is equal to $14 + 1 = 15$. That is, it takes 15 binary digits to represent the decimal number 32,767. Also according to the table, the value of $\log_2 32,768$ is exactly 15, so the value of INT($\log_2 32,768$) is 15 and the value of $B(32,768)$ is $15 + 1 = 16$. That is, it takes 16 binary digits to represent the decimal number 32,768.

x	$\log_2 x$	x	$\log_2 x$
1	0	1,024	10
2	1	2,048	11
4	2	4,096	12
8	3	8,192	13
16	4	16,384	14
32	5	32,768	15
64	6	65,536	16
128	7	131,072	17
256	8	262,144	18
512	9	524,288	19

The next example shows how the properties of logarithms can be applied in dealing with the exponential functions that arise in describing radioactive decay.

EXAMPLE 8
Radioactive Decay

A physicist measures the decay of a radioactive isotope. She finds that after 20 days, only 40 percent of the original amount of the isotope remains. Determine the half-life of the isotope.

Solution

Let A_0 denote the original quantity of the radioactive isotope, let $Q(t)$ denote the quantity left after time t, and let H denote the half-life of the isotope. In Example

5 of Section 6.1, we stated the result from physics that:
$$Q(t) = A_0 2^{-t/H}$$
In this example, we are given the information that
$$Q(20) = 0.4A_0$$
Insert the value 20 for t in the first formula, and insert the value $0.4A_0$ for $Q(20)$:
$$0.4A_0 = A_0 2^{-20/H}$$
$$0.4 = 2^{-20/H}$$
To solve this equation, take $\log_2$ of both sides to obtain:
$$\log_2(0.4) = \log_2(2^{-20/H})$$
$$\log_2(0.4) = -\frac{20}{H}$$
$$(\log_2(0.4))H = -20$$
$$H = -\frac{20}{\log_2(0.4)}$$

This gives an expression from which a numerical value for the half-life H can be computed. We show how to compute the numerical value for H in the next section.

Exercises 6.3

Determine the equivalent exponential equation.

1. $\log_2 8 = 3$
2. $\log_5 3{,}125 = 5$
3. $\log_3 243 = 5$
4. $\log_9 729 = 3$
5. $\log_2 \frac{1}{8} = -3$
6. $\log_5 \frac{1}{25} = -2$
7. $\log_x 3 = -12$
8. $\log_4 Q = t$

Determine the equivalent logarithmic equation.

9. $4^2 = x$
10. $20^0 = 1$
11. $10^{-3} = 0.001$
12. $27^{2/3} = 9$
13. $\sqrt{4} = 2$
14. $\sqrt[3]{27} = 3$
15. $x^{-3} = 0.01$
16. $a^b = c$

Simplify the following expressions.

17. $\log_x x$
18. $\log_e e$
19. $\log_{10} 10$
20. $\log_{23} 23$
21. $\log_m 1$
22. $\log_{10} 1$
23. $\log_e 1$
24. $\log_k 1$
25. $25^{\log_{25} 4}$
26. $9^{\log_9 3x}$
27. $\log_{10} 10^{-3}$
28. $\log_8 8^k$
29. $e^{\log_e 0.8241}$
30. $10^{\log_{10} 25}$
31. $\log_2 8$
32. $\log_3 9^2$
33. $7^{\log_7(2x-1)}$
34. $\log_5 5^{(3x+5)}$
35. $\log_m m^{x^4}$
36. $t^{\log_t(x^2+3)}$

Sketch graphs of the following functions.

37. $f(x) = \log_3(x)$
38. $f(x) = \log_3(x+1)$
39. $f(x) = \log_4(x)$
40. $f(x) = \log_4(x-1)$
41. $f(x) = \log_3(x) - 2$
42. $f(x) = \log_4(x) + 3$
43. $f(x) = \log_5(x)$
44. $f(x) = \log_{10}(x)$
45. $f(x) = \log_2(x-3)$
46. $f(x) = \log_2(x+3)$
47. $f(x) = \log_2 |x|$
48. $f(x) = |\log_2(x)|$
49. $f(x) = \log_2(x^2)$
50. $f(x) = \log_2(\sqrt{x})$
51. $f(x) = \log_2\left(\frac{1}{x}\right)$

52. $f(x) = \dfrac{1}{\log_2 x}$

Write the following expressions using a single logarithm.

53. $\dfrac{1}{3}[2\log_{10} 8 - 6\log_{10} 3] - 2\log_{10} 2 + \log_{10} 3$

54. $\dfrac{1}{2}[\log_{10} 6 + \log_{10} 4 - \log_{10} 12 + \log_{10} 2] + 2\log_{10} 5$

55. $\dfrac{1}{2}\log_a 36 - \dfrac{1}{3}\log_a 8$

56. $2\log_a 25x - \log_a 125x$

57. $2\log_b \sqrt{bx} - \log_b x$

58. $\dfrac{1}{2}\log_a 8x^2 - \dfrac{1}{6}\log_a 8x^3$

59. $\log_a(x^2 - 9) - \log_a(x - 3)$

60. $\log_b(x^2 - 1) - [\log_b(x + 1) + \log_b(x - 1)]$

61. $\log_3(2x - 3) - \log_3(2x^2 - x - 3) + \log_3(3x + 3)$

62. $\log_a(3x^2 - 5x - 2) - \log_a(x^2 - 4) - \log_a(3x + 1)$

63. $\log_a(x^3 - y^3) - \log_a(x^2 + xy + y^2)$

64. $\log_a(m + n) + \log_a(m^2 - mn + n^2)$

In Exercises 65–74, express the given logarithm in terms of $\log x$, $\log y$, and $\log z$.

65. $\log_b\left(\dfrac{x\sqrt{y}}{z^2}\right)$

66. $\log_c\left(x^3 \sqrt[4]{\dfrac{y}{z}}\right)$

67. $\log_2(x^2 y^2 z^4)$

68. $\log_5 x^3(yz)^4$

69. $\log_3 \dfrac{x^4 y}{z^2}$

70. $\log_2 \dfrac{1}{x^2 y z^3}$

71. $\log_{10} \dfrac{x^2 y^{2/3}}{z^4}$

72. $\log_a \sqrt{\dfrac{x}{y^2 z^3}}$

73. $\log_{10} \sqrt{x^3 \sqrt{yz^5}}$

74. $\log_6 \dfrac{x^3 y^2 z^{-1}}{xy^4 z^{-2}}$

Solve each of the following equations.

75. $\log_x 125 = 3$

76. $\log_x 10 = 2$

77. $3.4^x = 1$

78. $56^x = 0$

79. $\log_3 81 = x$

80. $\log_7 x = -2$

81. $\log_{13} x = 3$

82. $\log_x 256 = 4$

83. $\log_x \dfrac{1}{36} = -2$

84. $\log_5 \dfrac{1}{25} = x$

85. $\log_x 3 = 2$

86. $\log_5 1 = x$

87. $5^{x^2 - 5} = 14$

88. $2^{5x+7} = 16$

89. $5^{3x-1} = 125^{2x}$

90. $4^{x^2} = 8^{2/3} \cdot 2^{-3x}$

91. $b^x = 5$

92. $\dfrac{1}{8} = 4^{x-3}$

93. $5^x = 7$

94. $4^{3x+5} = 13$

95. $16^{-x} = 1{,}024$

96. $8^{2x-5} = 2^{x+3}$

97. $4^{2x-1} + 8 = 40$

98. $2^{2-3x} + 2 = 18$

99. $5^{2x}(25^{x^2}) = 125$

100. $\left(\dfrac{1}{25}\right)^{x+3} = 0.2^{x+9}$

101. $24^{\log_{24}(x^2 - 8)} = 1$

102. $5^{-2\log_5 x} = 4$

103. Suppose that $f(x) = 3^x$. Determine $f^{-1}(x)$.

104. Suppose that $f(x) = \log_2 x$. Determine $f^{-1}(x)$.

105. Suppose that $f(x) = 2^x + 2^{-x}$, $x \geq 0$. Prove that $f(x)$ is one-to-one. Determine $f^{-1}(x)$.

106. Suppose that $f(x) = 2^{x^2}$, $x \geq 0$. Prove that $f(x)$ is one-to-one. Determine $f^{-1}(x)$.

107. Prove that $\log_a \dfrac{x + \sqrt{x^2 - 1}}{x - \sqrt{x^2 - 1}} = 2\log_a\left(x + \sqrt{x^2 - 1}\right)$.

Applications

108. **Population growth of Mexico City.** The population of Mexico City was 14.5 million in 1980 and 21.5 million in 1989. Find the doubling time of the population. Assume the population grows exponentially.

109. **Decay of xenon-133.** Xenon-133 is a radioactive chemical. It is known that 1,000 grams of this chemical will decay to 246.6 grams in 10 days. What is its half-life?

110. **Decay of a radioisotope.** A physicist measures the decay of a radioisotope. He finds that, after 33 days, only 40 percent of the original amount remains. Determine the half-life of the isotope.

111. **Urinary tract infections.** The bacteria *Escherichia coli* is a common cause of urinary tract infections. A population of 10,000,000 bacteria will grow to 10,352,650 in 1 minute. Find the doubling time of the population. Assume the growth is exponential.

Technology

Graph the following functions with the aid of a graphing calculator.

112. $f(x) = 5 \log_{10} x$
113. $f(x) = \log_{10}(x + 1)$
114. $f(x) = x \log_{10} x$
115. $f(x) = x^2 \log_{10} x$
116. $f(x) = (\log_{10} x)^2$

In your own words

117. Give the definition of the logarithm to the base a.
118. Describe how the function $f(x) = \log_{10} x$ increases as x increases. Give some examples of how $f(x)$ compares with x for several large values of x.
119. Explain why $\log_a x$ is not defined for negative values of x.

6.4 COMMON LOGARITHMS AND NATURAL LOGARITHMS

Rather than a single function, $\log_a x$ is really an infinite family of functions, one for each value of a. It is very inconvenient to have calculations stated in terms of many different logarithmic functions. For instance, calculations that involve the strength of earthquakes and intensity of light are most conveniently phrased in terms of the function $\log_{10} x$, whereas calculations that involve information theory are most conveniently phrased in terms of the function $\log_2 x$. In this section, we examine more closely the two logarithmic functions that are most commonly used, namely $\log_{10} x$ and $\log_e x$.

Common Logarithms

The decimal number system is based on the number 10: place values are determined by powers of 10. Therefore, it should come as no surprise that logarithms to base 10 play a special role in applications.

Definition
Common Logarithms
Logarithms to base 10 are called **common** (or **Briggsian**) **logarithms**, and the function $\log_{10} x$ (also written $\log x$) is called the **common logarithmic function.**

Here are some special values of $\log_{10} x$:

$$\log_{10} 10 = 1$$
$$\log_{10} 100 = \log_{10} 10^2 = 2$$
$$\log_{10} 1{,}000 = \log_{10} 10^3 = 3$$
$$\log_{10} 0.01 = \log_{10} 10^{-2} = -2$$

Before the common availability of scientific calculators, values of common logarithms were determined from tables. However, there is rarely a need for such tables today.

Scientific calculators have a key for calculating values of the common logarithmic function. This key is usually labeled [log] or [log₁₀]. To compute the value of

SECTION 6.4 Common Logarithms and Natural Logarithms

$\log_{10} x$ on a calculator that uses algebraic logic, press the [log] key, and then enter the value of x, then press [=] or [ENTER].

EXAMPLE 1
Calculators and Logarithms

Calculate the following using a graphing calculator.
1. $\log_{10} 50.39$
2. $\log_{10} \dfrac{5{,}000}{4.1}$

Solution

1.  [log] [50.39] [ENTER] The answer is 1.7023.
2. [log] [(] [5,000] [÷] [4.1] [)] [ENTER] The answer is 3.08619.

HISTORICAL NOTE

Before the invention of computers, common logarithms were used to perform arithmetic calculations. By the laws of logarithms, the logarithm of a product is the sum of the logarithms. So to multiply two numbers, one could first determine their logarithms, add them, and then determine the number for which the logarithm was the sum. Values of common logarithms were determined from tables. It was commonplace for most textbooks in the sciences to include a table of common logarithms to aid in doing the computations in the text. For more exacting work, book-length tables of logarithms, carried out to many decimal places, were considered a standard tool for anyone who needed to perform lengthy scientific calculations, especially those involving powers.

Common logarithms were invented by the English mathematician Henry Briggs in the seventeenth century. In some old books, common logarithms are called Briggsian logarithms. Although common logarithms are an anachronism for computational purposes today, it is impossible to overemphasize the great advance in calculation they afforded to scientists of the seventeenth through nineteenth centuries. Important calculations in astronomy, physics, and chemistry became possible only after tables of logarithms became available.

So important were tables of logarithms to calculation, that when the U.S. government's Works Projects Administration was looking for jobs for unemployed scientists and mathematicians during the Great Depression, it commissioned a new set of logarithm tables, carried out to 14 decimal places.

EXAMPLE 2
Loudness of Sound

The loudness D of a sound, as measured in decibels, can be calculated in terms of the sound's intensity I using the formula

$$D(I) = 10 \log\left(\dfrac{I}{I_0}\right)$$

where I is measured in watts per square centimeter and I_0 is the minimum perceptible intensity, set by international standard at 10^{-16} watts per square centimeter.

1. Determine the number of decibels in a human conversation if the intensity of the sound of a human voice is 10^{-10} watts per square centimeter.
2. An antinoise ordinance bans noise of more than 80 decibels after midnight. What is the intensity of an 80-decibel sound?

324   CHAPTER 6   Exponential and Logarithmic Functions

Solution

1. The number of decibels in a human conversation is equal to:
$$D(10^{-10})$$
And by the given formula, this quantity is equal to:
$$10 \log\left(\frac{10^{-10}}{10^{-16}}\right) = 10 \log(10^{-10-(-16)})$$
$$= 10 \log(10^6)$$
$$= 10 \cdot 6$$
$$= 60 \text{ decibels}$$

2. The intensity I of an 80-decibel sound satisfies the equation:
$$80 = 10 \log\left(\frac{I}{I_0}\right)$$
$$8 = \log\left(\frac{I}{I_0}\right)$$
$$10^8 = 10^{\log(I/I_0)}$$
$$10^8 = \frac{I}{I_0}$$
$$I = 10^8 I_0$$
$$I = 10^8 \, 10^{-16}$$
$$= 10^{-8} \text{ watts per square centimeter}$$

Natural Logarithms

The number e is important because mathematical formulas (particularly in calculus) that involve exponential and logarithmic functions assume an especially simple form when expressed in terms of base e. We have already explored such exponential functions. Let's now discuss the corresponding logarithmic functions.

Definition
Natural Logarithmic Function

Logarithms to base e are called **natural logarithms**. The function
$$f(x) = \log_e(x), \qquad x > 0$$
is called the **natural logarithmic function,** and is denoted $\ln x$.

The notation ln is shorthand for the Latin term for "natural logarithm." This notation has been in use for several hundred years and has become traditional. As with the common logarithmic function, it has become traditional to omit the parentheses, where possible, in writing the natural logarithmic function. That is, we write $\ln x$ rather than $\ln(x)$, and we write $\ln e^x$ rather than $\ln(e^x)$. We use this notation so long as it does not result in confusion. However, in taking the natural logarithm of complicated expressions, we always use the parentheses for clarity.

To calculate values of the natural logarithmic function, we can use a scientific calculator, which has a key labeled "ln" for this purpose. For calculators such as the TI-82 that use algebraic notation, calculate $\ln x$ by pressing the [ln] key, keying in the value of x, and then pressing the [ENTER] key.

We can apply to the natural logarithmic function the properties we previously established for logarithmic functions in general:

SECTION 6.4 Common Logarithms and Natural Logarithms

PROPERTIES OF NATURAL LOGARITHMS

1. $\ln 1 = 0$
2. $\ln e = 1$
3. $e^{\ln x} = x$
4. $\ln e^x = x$

The general laws of logarithms proved in Section 6.3 are also valid in the special case of the natural logarithmic function.

LAWS OF NATURAL LOGARITHMS

Let $x, y > 0$. Then, the following properties of the natural logarithmic function hold:

1. $\ln xy = \ln x + \ln y$
2. $\ln x^a = a \ln x$
3. $\ln \frac{x}{y} = \ln x - \ln y$

EXAMPLE 3
Using Laws of Natural Logarithms

Use the properties of natural logarithms to determine the values of the following expressions.

1. $\ln e^5$
2. $(\ln e)^5$
3. $e^{5 \ln(2e)}$

Solution

1. The two operations of taking the natural logarithm of a quantity followed by expressing a quantity in base e cancel each other. So the value of the expression $\ln e^5$ is 5, by Property 4 of natural logarithms.
2. The value of $\ln e$ is 1, so:
$$(\ln e)^5 = (1)^5 = 1$$
3. By Law 2 of logarithms, we have:
$$e^{5 \ln(2e)} = \left[e^{\ln(2e)}\right]^5$$
By Property 3 of natural logarithms, $e^{\ln 2e} = 2e$, so
$$\left[e^{\ln(2e)}\right]^5 = (2e)^5 = 32e^5$$

EXAMPLE 4
Ecology

A reservoir has become polluted from an industrial waste spill. The pollution has caused a buildup of algae. The number of algae $N(t)$ present per 1,000 gallons of water t days after the spill is given by the formula:
$$N(t) = 100e^{2.1t}$$
How long will it take before the algae count reaches 10,000?

Solution

We are asked to determine the value of t that satisfies the equation
$$N(t) = 10{,}000$$
Inserting this value for $N(t)$ into the given formula, we obtain:
$$10{,}000 = 100e^{2.1t}$$
$$e^{2.1t} = 100$$
To solve this last equation, we take natural logarithms of each side and apply the properties of natural logarithms to simplify the resulting equation:
$$\ln(e^{2.1t}) = \ln 100$$
$$2.1t = \ln 100$$
$$t = \frac{\ln 100}{2.1} \approx \frac{4.6052}{2.1} \approx 2.1929$$

Thus, the number of algae per 1,000 gallons of water is 10,000 after 2.1929 days.

EXAMPLE 5
Archaeology

If $Q(t)$ denotes the amount of radioactive carbon 14 present t years after an organism dies, then $Q(t)$ is given by the formula
$$Q(t) = A_0 e^{-0.00012t}$$
where A_0 denotes the amount of carbon 14 present at the time the organism died. Suppose that archaeologists discover a parchment scroll in which they measure carbon 14 in an amount equal to 30 percent of that present in freshly manufactured parchment. How old is the parchment?

Solution

We are asked to determine the value of t for which:
$$Q(t) = 0.3 A_0$$
Using this value for $Q(t)$, we derive the equation:
$$0.3 A_0 = A_0 e^{-0.00012t}$$
We divide both sides by A_0 to remove any dependence of the equation on that number:
$$e^{-0.00012t} = 0.3$$
To solve this equation, we take natural logarithms of both sides and apply the properties of natural logarithms:
$$\ln(e^{-0.00012t}) = \ln 0.3$$
$$-0.00012t = \ln 0.3$$
$$t = -\frac{\ln 0.3}{0.00012}$$
To obtain a numerical value for t, we use a scientific calculator:
$$-\frac{\ln 0.3}{0.00012} \approx -\frac{-1.204}{0.00012} \approx 10{,}033$$
That is, the piece of parchment is 10,033 years old.

SECTION 6.4 Common Logarithms and Natural Logarithms

EXAMPLE 6
Use of CD-ROMs

A CD-ROM is a small magnetic disk used to store computer data in much the same way music data are stored on a compact audio disk. CD-ROM technology has been popularized by so-called multimedia personal computers, which combine animation, text, and sound in computer programs for education, entertainment, and other applications. According to the Wall Street Journal (Nov. 23, 1993), the number of CD-ROMs (in millions) in year x (where 1993 corresponds to 0) is projected to grow from 1993 to 1997 according to the exponential model:

$$y = 5.81e^{0.35x}$$

1. How many CD-ROMs were in use in 1993?
2. When will the number of CD-ROMs in use be 20 million?

Solution

1. The year 1993 corresponds to $x = 0$. The corresponding value of y is:
$$y = 5.81e^{0.35 \cdot 0}$$
$$= 5.81e^{0}$$
$$= 5.81 \cdot 1$$
$$= 5.81$$

That is, in 1993, 5.81 million CD-ROMs were in use.

2. If y equals 20 million, then we must solve the equation:
$$20 = 5.81e^{0.35x}$$
$$e^{0.35x} = 3.442340792$$

Take the natural logarithms of both sides to yield:
$$0.35x = \ln 3.442340792 = 1.236151703$$
$$x \approx 3.53$$

That is, by mid-1996, the number of CD-ROMs in use is projected to be 20 million.

EXAMPLE 7
Yield Curve

At any given moment, the bond market sets interest rates on bonds of various maturities. Generally speaking, the later the maturity (that is, the longer the term) of a bond, the higher the interest rate that the market will set. The curve that relates the yield to the maturity of a bond is called the yield curve. If x denotes the maturity of a bond and y denotes the corresponding interest rate, then in November 1993, the yield curve was the graph of the equation:

$$y = 0.062 - 0.031162e^{-0.2188x}$$

(The equation above is based on data reported in "Climbing the Yield Curve," *Business Week*, Dec. 6, 1993, p. 143.)

1. Sketch the graph of the yield curve.
2. What is the yield on a 10-year maturity for this yield curve?
3. What maturity will result in a 6 percent yield?

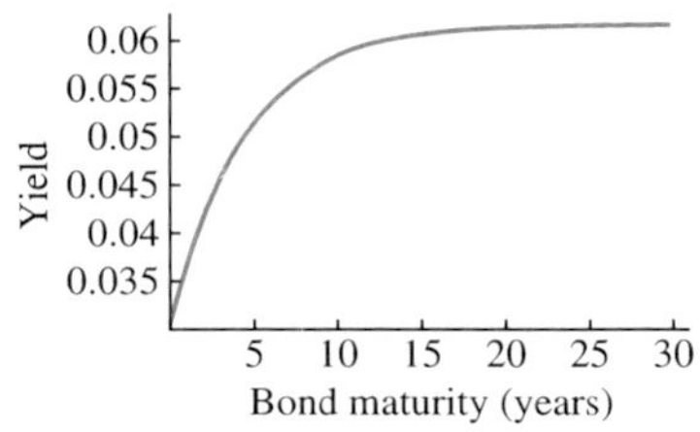

Solution

1. Use a graphing calculator to obtain the graph shown in Figure 1. Note that, as maturities get very large, the exponent $-0.2188t$ becomes a large negative number, so that the term

$$0.031162e^{-0.2188x}$$

Figure 1

is negligible and the yield is very close to 0.062. In other words, the line $y = 0.062$ is a horizontal asymptote of the graph. In financial terms, this means that bonds of very long maturities yield an interest rate of approximately 6.2 percent.

2. A 10-year maturity corresponds to $x = 10$. The corresponding yield is
$$y = 0.062 - 0.031162 e^{-0.2188 \cdot 10}$$
$$= 0.062 - 0.031162 e^{-2.188}$$
$$= 0.0585054682$$
$$\approx 5.85\%$$

That is, the yield of a 10-year bond is 5.85 percent.

3. To obtain a 6 percent yield, we set y equal to 0.06 and solve for x:
$$0.06 = 0.062 - 0.031162 e^{-0.2188x}$$
$$0.031162 e^{-0.2188x} = 0.002$$
$$e^{-0.2188x} = 0.064180732$$
$$-0.2188x = \ln 0.064180732$$
$$-0.2188x = -2.746052224$$
$$x \approx 12.5$$

That is, a bond with a maturity of about 12.5 years has the desired yield of 6 percent.

Logarithms to Various Bases

In the preceding discussion, we saw that the values of common and natural logarithms can be calculated using scientific calculators. However, these are the only special cases of the function $\log_a x$ that can be calculated directly in this way. But for any a, we can compute $\log_a x$ using the following result:

CHANGE OF BASE FORMULA

Let a and b be positive numbers. Then
$$\log_a x = \frac{\log_b x}{\log_b a}$$

Verification of the Change of Base Formula Begin with Fundamental Property 1 of logarithms:
$$x = a^{\log_a x}$$

Then,
$$\log_b x = \log_b a^{\log_a x} \qquad \text{Take } \log_b \text{ of both sides}$$
$$= \log_a x \cdot \log_b a \qquad \text{Law 3 of logarithms}$$

Hence,
$$\log_a x = \frac{\log_b x}{\log_b a}$$

The Change of Base formula asserts that any logarithmic function is a constant multiple of any other given logarithmic function. In particular, we can cal-

SECTION 6.4 Common Logarithms and Natural Logarithms

culate logarithms with base a in terms of either natural logarithms or common logarithms using the formulas:

$$\log_a x = \frac{\ln x}{\ln a}$$

or

$$\log_a x = \frac{\log x}{\log a}$$

By setting a equal to 10 in the first formula, we have the following relationship between common and natural logarithms:

$$\log x = \frac{\ln x}{\ln 10}$$

EXAMPLE 8
Using Change of Base Formula

Calculate $\log_{100} 10$ without using a calculator.

Solution

By the Change of Base formula, we have

$$\log_{100} 10 = \frac{\log_{10} 10}{\log_{10} 100}$$
$$= \frac{1}{2}$$

because $\log_{10} 10 = 1$, and $\log_{10} 100 = 2$.

EXAMPLE 9
Graphing a Function that Involves a Natural Logarithm

Sketch the graph of the function:
$$f(x) = x \ln x, \qquad x > 0$$

Solution

Use the domain $0 \leq x \leq 10$ and range $-1 \leq y \leq 20$. The graph is shown in Figure 2. Note that the calculator automatically indicates that the function $\ln x$ is not defined for $x = 0$ by not plotting a point for this value of x.

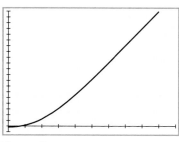
Figure 2

Exercises 6.4

Evaluate the following without using a calculator.

1. $\log_{10} 10$
2. $\log_{10} 100$
3. $\log_{10} 1,000$
4. $\log_{10} 10,000$
5. $\log_{10} 0.1$
6. $\log_{10} 0.01$
7. $\log_{10} 0.001$
8. $\log_{10} 0.0001$

9. $\ln e$
10. $\ln 1$
11. $\ln e^2$
12. $\ln e^{-7}$
13. $(\ln e)^{-7}$
14. $(\ln e)^2$
15. $e^{\ln t}$
16. $e^{\ln 7k}$
17. $e^{3\ln(4e)}$
18. $e^{-4\ln(5e^2)}$
19. $\ln \sqrt{e^8}$
20. $\ln \sqrt{e^6}$
21. $\ln e^2 + 4^3$
22. $\ln e^3 + 5^3$

Simplify the following expressions.

23. $\dfrac{1}{2\ln 10}(10)^{2x}$
24. $\dfrac{\ln x^2}{x}$
25. $\dfrac{\ln n}{\ln n^2}$
26. $\dfrac{1}{n\ln(n^3)}$
27. $e^{2\ln x}$
28. $e^{\ln x - 2\ln y}$
29. $\dfrac{1}{2}(\ln 4 - \ln 1)$
30. $\ln 8 - \ln 2$
31. $\dfrac{1}{2}e^{2(\ln 6)+1} - \dfrac{1}{2}e^{2(\ln 2)+1}$
32. $2^x(\ln 3)3^{2x} \cdot 2 + 3^{2x}(\ln 2)2^x$
33. $\ln(4^2 + 4 - 1) - \ln(1^2 + 1 - 1)$
34. $-\dfrac{1}{3}\ln\dfrac{8}{4} + \dfrac{1}{3}\ln(3 + \sqrt{10})$

Match the functions in Exercises 35–38 with their graphs below.

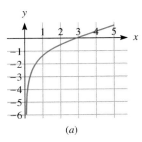
(a)

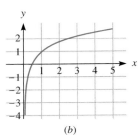
(b)

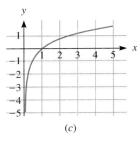
(c)

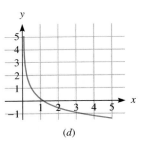
(d)

35. $f(x) = \ln x$
36. $f(x) = \ln x + 1$
37. $f(x) = \ln x - 1$
38. $f(x) = -\ln x$

Use the Change of Base formula to evaluate the following expressions.

39. $\log_7 50$
40. $\log_{12} 265$
41. $\log_{23} 0.0089$
42. $\log_8 2.347$

Convert the following to natural logarithms.

43. $\log 4{,}578$
44. $\log 2.347$
45. $\log x^2$
46. $\log \sqrt{x}$

Convert the following to common logarithms.

47. $\ln 0.987$
48. $\ln 78.56$
49. $\ln \sqrt{y}$
50. $\ln t^4$

Applications

51. **Loudness of sound.** The loudness D of a sound, measured in decibels, can be calculated in terms of the sound's intensity I using the formula
$$D(I) = 10\log\left(\dfrac{I}{I_0}\right)$$
where I is measured in watts per square centimeter, and I_0 is the minimum perceptible intensity, set by international standard at 10^{-16} watts per square centimeter. Determine the number of decibels of the sound made by an elevated train if the intensity of the sound is 10^{-3} watts per square meter.

52. **Workplace noise.** Determine the number of decibels from the sound made by a riveter if the intensity of the sound is 3.2×10^{-3} watts per square meter.

SECTION 6.4 Common Logarithms and Natural Logarithms

53. **Noise in a forest.** The noise of rustling leaves is 10 decibels. What is the intensity of a 10-decibel sound?

54. **Ear injuries from loud noises.** Sound actually hurts the ear at levels of 120 decibels or higher. What is the intensity of a 120-decibel sound?

55. **Intensity of a sound.** Two sounds have intensities I_1 and I_2, respectively. Show that the difference in their sound levels can be expressed as:
$$D(I_1) - D(I_2) = 10 \log \frac{I_1}{I_2}$$

56. **Intensity of a sound.** Find a formula for the sum of the sound levels of two sounds that have intensities I_1 and I_2.

57. **Intensity of a sound.** Solve the sound-level formula for the intensity I.

58. **Earthquake magnitude.** The magnitude R of an earthquake, as measured on the Richter scale, can be calculated in terms of an intensity I, where
$$R(I) = \log \frac{I}{I_0}$$
and I_0 is a minimum perceptible intensity. What is the magnitude of an earthquake of intensity I_0?

59. **Earthquake magnitude.** Find the magnitudes of earthquakes that have the following intensities:
   a. $10 I_0$
   b. $100 I_0$
   c. $1,000 I_0$
   d. $10,000 I_0$
   e. $100,000 I_0$
   f. $100,000,000 I_0$

60. **Earthquake magnitude.** An earthquake has a magnitude ten times as intense as another. How much higher is its magnitude on the Richter scale?

61. **Earthquake magnitude.** An earthquake has a magnitude one hundred times as intense as another. How much higher is its magnitude on the Richter scale?

62. **Earthquake magnitude.** One of the worst earthquakes ever recorded was near Lebu, Chile, on May 21–23, 1960. It had an intensity of $10^{8.9} \cdot I_0$. What was its magnitude on the Richter scale?

63. **Intensity of an earthquake.** Two earthquakes have intensities I_1 and I_2, respectively. Show that the difference in their magnitudes can be expressed as:
$$R(I_1) - R(I_2) = \log \frac{I_1}{I_2}$$

64. **Intensity of an earthquake.** Find a formula for the sum of the magnitudes of two earthquakes with intensities I_1 and I_2.

65. **Intensity of an earthquake.** Solve the Richter scale magnitude formula for I.

66. **Acidity and alkalinity.** In chemistry, the pH of a substance is defined by
$$pH = -\log_{10}(H^+)$$
where H^+ represents the concentration of hydrogen ions, in moles per liter. Find the pH of each of the following substances; their H^+ values are given.
   a. Normal blood: 3.4×10^{-8}
   b. Ammonia: 1.6×10^{-11}
   c. Wine: 4.0×10^{-4}

67. **Acidity and alkalinity.** Given the following pH values, find the corresponding hydrogen ion concentrations H^+.
   a. Eggs: pH = 7.8
   b. Tomatoes: pH = 4.2

68. **Acidity and alkalinity.** Solve the pH formula for H^+.

69. **Acidity and alkalinity.** A substance is said to be an acid if its pH is greater than 7. For what hydrogen ion concentrations is a substance an acid?

70. **Acidity and alkalinity.** A substance is said to be a base if its pH is less than 7. For what hydrogen ion concentrations is a substance a base?

71. **Ecology.** Refer to Example 4. How long will it take before the algae count reaches 20,000?

72. **Archaeology.** Use the formula in Example 5. Suppose archaeologists find a Chinese artifact that has lost 40 percent of its carbon 14. How old is the artifact?

73. **Population of the United States.** (See figure.) The population of the United States was 240 million on January 1, 1986. Think of that date as $t = 0$. The population $P(t)$, t years later, is given by
$$P(t) = P_0 e^{kt}$$
where $P_0 = 240$ million and $k = 0.009$.
   a. Estimate the population of the United States in 1995 ($t = 9$). Estimate the population in 2001.
   b. After what period of time will the population be twice that in 1986?

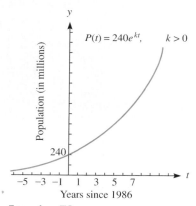

Exercise 73

74. **Exponential growth** is often modeled by the function
$$P(t) = P_0 e^{kt}$$
where P_0 is the initial population at time $t = 0$, and k is a positive constant, called the exponential growth rate.

   a. Show that the exponential growth rate k and the doubling time T are related by the formula:
$$k = \frac{\ln 2}{T}$$
   b. The growth rate of the United States population is 0.9 percent. Find the doubling time.
   c. The doubling time of the population of Texas is 23 years. Find the exponential growth rate.

75. **Population growth of California.** The population of California was 23,669,000 in 1980 and was estimated to be 27,526,000 in 1990. Let's assume this was exponential growth.
   a. Find the exponential growth rate k.
   b. Find the equation of exponential growth.
   c. Estimate the population in 2000.

76. **Interest compounded continuously.** A method of compounding interest that used to be quite prevalent was continuous compounding. (It is used less now because many people did not understand it.) A "population" of money P_0 invested at interest rate k and compounded continuously grows to an amount $P(t)$ in t years, where:
$$P(t) = P_0 e^{kt}$$
Suppose $10,000 is invested in a savings account at 8 percent interest, compounded continuously.
   a. How much is in the account after 1 year? After 2 years? 5 years? 10 years?
   b. If interest had been compounded annually, how much would be in the account after 1 year? After 2 years? 5 years? 10 years?
   c. What is the doubling time of this population of money?
   d. Suppose the money doubles in value every 9.9 years. What is the interest rate?

77. **Growth of a college fund.** A new father deposits $250 in a savings account that pays $10\frac{1}{4}$ percent interest, compounded continuously. How much will be in the account in 18 years?

78. **Growth of a college fund.** If the father in Exercise 77 wanted to have $30,000 in the savings account after 18 years, how much should he now invest?

79. **Growth of a college fund.** What would the interest rate have to be if the father wanted the $250 to grow to $40,000 in 18 years?

80. **Atmospheric pressure.** At altitude a, under standard conditions of temperature, the atmospheric pressure $P(a)$ is given by:
$$P(a) = P_0 e^{-0.0000385a}$$

At sea level $P_0 = 1{,}013$ millibars.

a. What is the pressure at 1,000 ft? at 2,000 ft? 10,000 ft? 30,000 ft?

b. A device for measuring atmospheric pressure changes from 1,013 millibars on the ground floor of the Empire State Building to 965.4038 millibars after arriving at the top of the building. How tall is the Empire State Building?

81. **Magnitude of a star.** In astronomy the magnitude M of a star is a measure of the brightness of the star. (Stars with the largest magnitudes are the faintest.) The magnitude of stars made visible through a telescope depends on the aperture (lens diameter) a of the telescope, as given by the formula:

$$M(a) = 9 + \log_{10} a$$

Calculate the magnitude of faintest star made visible by a telescope with the given apertures.

a. 3 inches

b. 30 inches

82. **Intensity of light.** A beam of light enters a medium, such as water or smoky air, with initial intensity I_0. Its intensity decreases according to the thickness of the medium. The intensity I at a water depth (or air concentration) of x units is given by:

$$I = I_0 e^{-\mu x}$$

The constant μ (pronounced "myoo"), which is the coefficient of absorption, varies with the medium. For light passing through water, $\mu = 1.4$.

a. What percentage of I_0 remains at a sea water depth of 1 m? 2 m? 3 m?

b. Plant life cannot exist below 10 meters of water. What percentage of I_0 remains at 10 meters?

83. **Salvage value.** A business estimates that the salvage value V of a piece of machinery after t years is given by $V(t) = \$80{,}000 e^{-t}$.

a. What did the machinery cost initially?

b. What is its salvage value after 4 years?

84. **Population growth of Hawaii.** The population of Hawaii in 1980 was 964,691. The total area of Hawaii is 6,425 square miles. The population grows at a rate of 1.8 percent per year. Assuming this growth is exponential, after how many years will the population be such that there is one person for every square yard of land?

Technology

Use a calculator to evaluate each of the following expressions.

85. $\log_{10} 83.47$

86. $\log_{10} 124.7$

87. $\log_{10} 3.456$

88. $\log_{10} 9.87$

89. $\log_{10} 0.458$

90. $\log_{10} 0.00778$

91. $\log_{10}(6.732 \times 10^6)$

92. $\log_{10}(4.012 \times 10^8)$

93. $\log_{10}(9.152 \times 10^{49})$

94. $\log_{10}(1.302 \times 10^{28})$

95. $\log_{10}(8.6022 \times 10^{-4})$

96. $\log_{10}(5.554432 \times 10^{-7})$

97. $\log_{10}(1.21223 \times 10^{-23})$

98. $\log_{10}(9.902 \times 10^{-19})$

Use a graphing calculator to approximate the solutions of the following equations to two significant digits.

99. $2^x - 3^x = 40$

100. $5^x = 5^{-x} - 4$

101. $3^x + x^2 = 50$

102. $2^x - 3^x + e^x = x^2 + x^3$

In your own words

103. Explain the difference between natural logarithms and common logarithms.

104. Give two mathematical models that use common logarithms.

6.5 EXPONENTIAL AND LOGARITHMIC EQUATIONS

In Section 6.4, we saw many applications of equations involving logarithms and exponential functions. The discussion in that section is only the tip of the iceberg. Equations that involve logarithms and exponential functions are common throughout mathematics, and it is important to become proficient in solving them. In this section, we solve a number of typical equations of this sort as a first step in developing that proficiency.

EXAMPLE 1
Exponential Equation

Solve the following equations.

1. $5^{x^2} = 3$
2. $e^{8x+2} = 3^x$

Solution

1. The presence of the variable in the exponent suggests that we should take logarithms of both sides. But logarithms to what base? It doesn't really matter, because logarithms to all bases are related to one another by the Change of Base formula. But it is fairly standard to perform all calculations in terms of the natural logarithmic function. Taking natural logarithms of both sides yields:

$$\ln 5^{x^2} = \ln 3$$

Therefore, we obtain:

$$x^2 \cdot \ln 5 = \ln 3$$

$$x^2 = \frac{\ln 3}{\ln 5}$$

$$x = \pm\sqrt{\frac{\ln 3}{\ln 5}} \approx \pm 0.8262$$

2. Again, we take natural logarithms of the expressions on both sides to obtain:

$$\ln e^{8x+2} = \ln 3^x$$

Therefore, we have:

$$8x + 2 = x \ln 3$$

Despite the complex appearance of this equation, it is just a linear equation in x, because $\ln 3$ is a constant. Collect terms that involve x on the left side of the equation:

$$(8 - \ln 3)x = -2$$

$$x = -\frac{2}{8 - \ln 3}$$

$$\approx -0.2898$$

EXAMPLE 2
Logarithmic Equation

Solve the equation

$$\ln \frac{x-1}{2x} = 4$$

SECTION 6.5 Exponential and Logarithmic Equations

Solution

To eliminate the natural logarithm function on the left side of the equation, we raise both sides to the power e.

$$e^{\ln\left[\frac{(x-1)}{2x}\right]} = e^4$$

$$\frac{x-1}{2x} = e^4$$

$$x - 1 = 2e^4 x$$

$$x - 2e^4 x = 1$$

$$(1 - 2e^4)x = 1$$

$$x = \frac{1}{1 - 2e^4}$$

$$\approx -0.0092425$$

EXAMPLE 3
Hyperbolic Function

The function $\sinh(x)$, called the hyperbolic sine of x, is defined by the equation:

$$\sinh(x) = \frac{e^x - e^{-x}}{2}$$

Determine all values of x for which $\sinh(x)$ is equal to $\frac{1}{4}$.

Solution

The equation $\sinh(x) = \frac{1}{4}$ is equivalent to:

$$\frac{e^x - e^{-x}}{2} = \frac{1}{4}$$

$$2e^x - 2e^{-x} = 1$$

To eliminate the negative exponent, we multiply both sides by e^x to obtain:

$$2e^x e^x - 2e^{-x} e^x = e^x$$

$$2e^{2x} - 2e^0 - e^x = 0$$

$$2e^{2x} - e^x - 2 = 0$$

To solve this last equation, we write it in the form:

$$2(e^x)^2 - e^x - 2 = 0$$

This is a quadratic equation in the variable e^x. Applying the quadratic formula gives us the solutions:

$$e^x = \frac{1 \pm \sqrt{(-1)^2 - 4(2)(-2)}}{4} = \frac{1 \pm \sqrt{17}}{4}$$

To determine the value of x, we now take natural logarithms of both sides. Recall, however, that the domain of the natural logarithmic function consists of the positive real numbers. In particular, the natural logarithmic function is undefined at the value of the right side corresponding to the minus sign, since this gives a negative value for the right side. The only possible value of x corresponds to

$$\ln e^x = x = \ln\left(\frac{1 + \sqrt{17}}{4}\right) \approx 0.2475$$

CHAPTER 6 Exponential and Logarithmic Functions

EXAMPLE 4 Solve the equation:
Another Logarithmic Equation
$$\ln(x^2 + 2x) = 0$$

Solution

Apply the natural exponential function to both sides of the equation.
$$e^{\ln(x^2+2x)} = e^0$$
$$x^2 + 2x = 1$$
$$x^2 + 2x - 1 = 0$$
$$x = \frac{-2 \pm \sqrt{2^2 - 4(1)(-1)}}{2} = \frac{-2 \pm \sqrt{8}}{2}$$
$$= \frac{-2 \pm 2\sqrt{2}}{2}$$
$$= -1 \pm \sqrt{2}$$

EXAMPLE 5 Determine the approximate values of the solutions of the equation:
Solving a Logarithmic
Equation Using
a Graphing Calculator
$$8 \ln x = \frac{11}{6}x$$

Solution

Plot the two functions
$$Y_1 = 8 \ln X$$
$$Y_2 = \frac{11}{6} X$$

on the same coordinate system. See Figure 1. We use the trace to locate the point of intersection, which is approximately (1.3684211, 2.5092605). The x-coordinate of this point is the solution to the equation. That is, the solution of the equation is approximately 1.3684211.

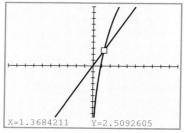

Figure 1

Exercises 6.5

Solve the following equations. Where possible, obtain exact solutions.

1. $7^{x^2} = 4$
2. $8^{x^2} = 5$
3. $e^{5x-2} = 4^x$
4. $e^{7x+4} = 5^{2x}$
5. $\ln\left(\dfrac{x+1}{3x}\right) = 6$
6. $\ln\left(\dfrac{2x-3}{5x+2}\right) = 7$
7. $e^t = 100$
8. $e^t = 60$
9. $e^{-t} = 0.01$
10. $e^{-0.02t} = 0.06$
11. $184.50 = 100e^{10k}$
12. $52 = 25e^{9k}$
13. $90 = 25e^{-10k} + 75$
14. $25e^{-0.05t} + 75 = 80$
15. $\ln(x^2 + 3x) = 0$
16. $\ln(x^2 - 5x) = 0$
17. $\log_3(6x + 5) = 2$
18. $\log_4(8 - 3x) = 3$
19. $\log x + \log(x + 9) = 1$
20. $\log(x - 9) + \log x = 1$
21. $\log(x + 3) + \log x = 1$
22. $\log(x - 3) + \log x = 1$
23. $\log_4(x + 5) + \log_4(x - 5) = 2$
24. $\log_5(x - 5) + \log_5(x + 2) = 2$
25. $\log_2 x - \log_2(x - 3) = 5$
26. $\log_3(2x - 5) - \log_3(3x + 1) = 4$
27. $(\log_2 x)^2 + \log_2 x - 6 = 0$
28. $(\log x)^2 - \log x = 6$
29. $\dfrac{\log(x + 1)}{\log x} = 2$

SECTION 6.5 Exponential and Logarithmic Equations

30. $\dfrac{\log(x+2)}{\log x} = 2$
31. $2 = 5\log x + 12(\log x)^2$
32. $\log(x-1) + \log(x+2) = \log_7 7$
33. $\log_5 x^3 + \log_5 x = 4$
34. $3\log x = \log 125$
35. $\log_3(\log_2 x) = 1$
36. $\log_5(\log_6 x) = 0$
37. $e^{2x} + 5e^x - 14 = 0$
38. $e^{4x} - e^{2x} = 12$
39. $e^x + e^{-x} - 1 = 0$
40. $5e^x - 3e^{-x} = 8$
41. $|\log_3 x| = 2$
42. $\log_2 |x| = 5$
43. $\log_2 \sqrt[3]{x-12} = 2 - \log_2 \sqrt[3]{x}$
44. $\log \sqrt{\dfrac{x+5}{x-4}} = 2$
45. $\log x^{\log x} = 9$
46. $x^{\log x} = 1{,}000 x$
47. $3^{\log_2 x} = 3x$
48. $\sqrt{\dfrac{(e^{3x} \cdot e^{-7x})^{-6}}{\dfrac{e^{-x}}{e^x}}} = e^{10}$

Solve each of the following equations for t.

49. $P = P_0 e^{kt}$
50. $N = N_0 e^{-kt}$
51. $N = \dfrac{P}{1 + Ae^{-kt}}$
52. $T = T_0 + |T_1 - T_0| e^{-kt}$
53. $\dfrac{e^t + e^{-t}}{2} = x$
54. $\dfrac{e^t - e^{-t}}{2} = x$
55. $\dfrac{e^t - e^{-t}}{e^t + e^{-t}} = x$
56. $v = v_T(1 - e^{-kt})$
57. $I = \dfrac{E}{R}\left(1 - e^{-Rt/L}\right)$
58. $N = GB^t$

Determine a function of the form $f(x) = Ce^{kx}$ that has the properties shown in the accompanying graph.

59.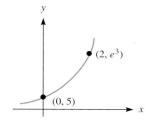

60.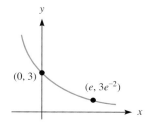

61.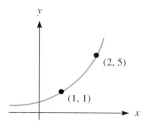

62.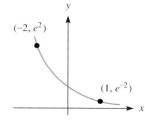

63.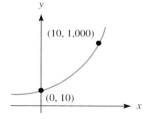

64.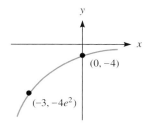

Technology

Use a graphing calculator to approximate numerically the solutions to the following equations.

65. $\log(x+1) + \log(x-3) = 4.1$
66. $(\log x)^2 - 3x = 1$
67. $\log \log x = 0.1x$
68. $x = -3\log x - 4$
69. $\log x + 10\sqrt{x} = 4$
70. $e^{3.7x} - e^{1.5x} = 8$

Mathematics and the World around Us—Population Models

In this chapter, we introduced various mathematical models for the growth of populations such as bacteria, animals, and people. We expressed these models in terms of formulas that use the exponential function. You may have the impression that the formulas used are somewhat arbitrary, as long as the graph has roughly the proper shape. But this is far from the case. The population models are derived from differential equations that are solved by using calculus. For example, consider bacterial growth. If we assume that this growth takes place with no limitations on space, food, and so on, it seems reasonable that, at any given moment, the number of bacteria is increasing at a rate that is proportional to the number of bacteria present. The rate at which the number of bacteria is increasing can be expressed using the derivative, a mathematical concept from calculus. And, using calculus, we derive the formula given earlier: The number of bacteria at time t is given by the exponential formula $N(t) = Ae^{kt}$.

Further, consider a model in which the population *is* limited by space, ecology, and food. In this case, it seems reasonable to postulate a maximum number P that the population may reach. Moreover, we can assume that at any given moment, the population is growing at a rate proportional to the product of the current population N and the remaining "room for growth" $N - P$. That is, the rate of growth at any moment is equal to

$$kN(P - N)$$

where k is a constant. Using calculus, we can show that this model leads to the formula

$$N(t) = \frac{P}{1 + Ae^{-kt}}$$

for some constant A. The graph of this function is the logistic curve we introduced in Section 6.2. This curve is S-shaped and, for large values of t, approaches the horizontal line $y = P$ as an asymptote.

In the nineteenth century, a Belgian sociologist, P. F. Verhulst, used U.S. census data accumulated every ten years since 1790 to show that the growth of the U.S. population could be modeled using a logistic curve. He predicted the U.S. population in 1950 with an error of less than 1 percent!

Chapter Review

Important Concepts—Chapter 6

- real exponent
- laws of real exponents
- exponential function with base a
- graphs of exponential functions
- graphing an increasing exponential function
- graphing a decreasing exponential function
- compound interest formula for real times
- radioactive decay

- half-life of a radioactive element
- the number e
- natural exponential function
- exponential growth
- terminal velocity
- logistic curve
- normal probability curve
- logarithmic function with base a
- fundamental properties of logarithms
- graph of $\log_a x$

- laws of logarithms
- solving exponential and logarithmic equations
- common (Briggsian) logarithms
- common logarithmic function
- natural logarithm
- natural logarithmic function
- properties of natural logarithms
- laws of natural logarithms
- Change of Base formula

Important Results and Techniques—Chapter 6

Laws of exponents	$a^x a^y = a^{x+y}$ $(a^x)^y = a^{xy}$ $(ab)^x = a^x b^x$ $\dfrac{a^x}{a^y} = a^{x-y}$	p. 296
Radioactive decay	The quantity of a radioactive substance that remains after time t is given by either of the formulas: $$Q(t) = A_0 2^{-t/H}$$ or $$Q(t) = A_0 e^{-kt}$$ where A_0 is the initial amount, H is the half-life, and k is the decay constant.	p. 326
Exponential growth	At time t, the size of a population that expands according to exponential growth is given by either of the formulas: $$N(t) = N_0 2^{t/D}$$ or $$P(t) = P_0 e^{kt}$$ where N_0 is the initial population size, D is the doubling time, and k is the growth constant.	p. 328, 332
Other exponential formulas	Velocity of a skydiver: $$v(t) = v_T(1 - e^{-kt})$$ Logistic curve, spread of a rumor, restricted growth: $$N(t) = \dfrac{P}{1 + Ae^{-kt}}$$	p. 307, 308
Properties and laws of logarithms	$\log_a x = y$ is equivalent to $a^y = x$ $\log_a a = 1$ $\log_a 1 = 0$ $a^{\log_a y} = y$ $\log_a a^x = x$ $\log_a(xy) = \log_a x + \log_a y$ $\log_a \left(\dfrac{x}{y}\right) = \log_a x - \log_a y$ $\log_a x^y = y \log_a x$ $\log x = \log_{10} x$ $\ln x = \log_e x$ $\ln xy = \ln x + \ln y$ $\ln \dfrac{x}{y} = \ln x - \ln y$ $\ln x^a = a \ln x$ $\log_a x = \dfrac{\log_b x}{\log_b a}$	p. 313, 315, 325

Review Exercises—Chapter 6

Determine the equivalent exponential equation.

1. $\log_e 2 = 0.69315$
2. $\log_{10} 2 = 0.30103$
3. $\log_{0.5} 0.25 = 2$
4. $\log_{0.25} 0.5 = \dfrac{1}{2}$

Determine the equivalent logarithmic equation.

5. $2^1 = 2$
6. $2^{-1} = \dfrac{1}{2}$

Sketch a graph of each of the following.

7. $f(x) = 2^x$
8. $f(x) = \log_2 x$
9. $f(x) = 2^{-(x-1)^2}$
10. $f(x) = |3^x - 4|$
11. $f(x) = \log_2(x + 3) - 5$
12. $f(x) = \log_3 |x|$
13. $f(x) = |\log_3 x|$
14. $f(x) = \log_3\left(\dfrac{3}{x}\right)$

Solve each of the following equations.

15. $\log_9 81 = x$
16. $\log_x \dfrac{9}{16} = 2$
17. $\ln x = 4$
18. $\log x = -3$
19. $\log_2(x^2 - 1) - \log_2(x + 1) + \log_2(x - 1) = 0$
20. $2\log_6 \sqrt{6x} - \log_6 x = 1$
21. $a^{3\log_a x} x = 16$
22. $\log_3 x = -2$
23. $27^{2x+3} - 9 = 72$
24. $\log_4(x - 2) + \log_4(x - 8) = 2$
25. $e^{-2x} = 0.01$
26. $\log_2[\log_4(\log_{10} x)] = -1$
27. $\log_{10}[\log_2(\log_x 25)] = 0$
28. $3e^x + 2 = 7e^{-x}$
29. $3(\log_8 x)^2 + \log_8 x - 2 = 0$
30. $\dfrac{e^x + e^{-x}}{2} = \dfrac{1}{3}$
31. $\log x^{\log x} = 16$
32. $21 = 60(1 - e^{-10k})$
33. $|\log_4 x| = 2$
34. $10^x = 3.6308$

Simplify each of the following expressions.

35. $6^{\log_6(5x-2)}$
36. $\log_5 25^{(3x-1)}$
37. $\log_b b$
38. $\ln 1$
39. $\log 10^3$
40. $\log_4 4x$
41. $\log_x x^4$
42. $4^{2\log_4 x + \left(\frac{1}{2}\right)\log_4 9}$
43. $2\log_a x - 3\log_a y + \log_a(x + y)$
44. $\log_a x^3 - \log_a \sqrt{x}$
45. $\dfrac{1}{3}\left[\log_a x - \dfrac{1}{2}\log_a w\right] + 3\log_a v$
46. $\log(1 - t^3) - \log(1 + t + t^2)$

Express each of the following in terms of $\log x$, $\log y$, and $\log z$.

47. $\log_3 \dfrac{x^3(y + z)}{x^2}$
48. $\log_a\left(\dfrac{x^2 y}{\sqrt{xz}}\right)$

Assume that:

$$\log_b 3 = 1.099, \quad \text{and} \quad \log_b 5 = 1.609$$

Find each of the following by applying laws of logarithms.

49. $\log_b b^2$
50. $\log_b 3b$
51. $\log_b b^b$
52. $\log_b \dfrac{b}{b + 1}$
53. $\log_b \sqrt[3]{b}$
54. $\log_b \dfrac{1}{5}$
55. $\log_b \sqrt{b}$
56. $\log_b\left(b^{4.7\sqrt{b}}\right)$

Binary Representation of Numbers. Use the function
$$B(x) = \text{INT}[\log_2(x) + 1]$$
to determine how many digits are required to represent the following integers in binary form.

57. 178,723

58. 443,876

Solve each of the following equations for t.

59. $\dfrac{e^t + e^{-t}}{e^t - e^{-t}} = y$

60. $N = e^{e^t}$

61. $\log_a y = 1 + n \log_a t$

62. $y = at^n$

63. $0.49 = 0.5(1 - e^{-0.07/t})$

64. $400 = \dfrac{4{,}800}{6 + 794e^{-800(12.2)t}}$

Applications

65. **Sales of a firm.** Sales S increase toward a limiting value of $20 million and satisfy the following function
$$S(t) = \$20(1 - e^{-kt})$$
where $k = 0.11$ and time t is in years.

    a. Find $S(t)$ for $t = 0, 1, 2, 5,$ and 10 years.

    b. Sketch a graph of $S(t)$ for $t \geq 0$.

66. **Sales of a firm.** Refer to Exercise 65. After what time will the sales be $10 million?

67. **Radioisotope power supply.** A space satellite has a radioisotope power supply. The power output $W(t)$ is given by the equation
$$W(t) = W_0 e^{-t/250}$$
where t is the time in days, $W(t)$ is measured in watts, and $W_0 = 50$ watts.

    a. How much power is available at the beginning of the flight?

    b. How much power will be available after 30 days? after 1 year?

68. **Radioisotope power supply.** Refer to the preceding problem.

    a. What is the half-life of the power supply?

    b. The equipment aboard the satellite requires 10 watts of power to operate properly. How long can the satellite operate properly?

69. **Spread of a rumor.** In a college with a population of 6,400, a group of students spread the rumor "People who study mathematics are people you can count on!" The number of people $N(t)$ who have heard the rumor after t minutes is given by
$$N(t) = \dfrac{P}{1 + Ae^{-kt}}$$
where A and k are constants. In this situation $P = 6{,}400$, $A = 8$, and $k = 0.23$.

    a. How many people start the rumor at $t = 0$?

    b. How many people have heard the rumor after 1 minute? 2 minutes? 6 minutes? 12 minutes? 30 minutes? 1 hour?

70. **Spread of a rumor.** Refer to Exercise 69.

    a. Sketch a graph of $N(t)$ for $t \geq 0$.

    b. How much time passes before half the students have heard the rumor?

71. **Dosage of medication.** A patient takes a dosage A_0 of a medication. The medication leaves the circulatory system by passage through the liver, the urinary system, the sweat glands, and by consumption by the organs it acts on. The amount of the medication present after time t, in hours, is given by
$$A(t) = A_0 e^{-kt}$$
where k is a constant that depends on the patient's metabolism and on the kind of medication. A patient takes a 2 mg tablet for which $k = 0.3$.

    a. What amount of the medication will be in the patient's system after 1 hr? 2 hrs? 4 hrs? 6 hrs?

    b. After what amount of time will the amount of medication in the patient's system be 1 mg?

72. **Dosage of medication.** For a certain medication and patient, an initial dosage of 5 mg will decrease to 3 mg after 1 hr.

    a. Find the decay constant k.

    b. Find an equation of exponential decay.

    c. After what amount of time will the amount of medication in the patient's system be 2 mg?

73. **Decay of a radioisotope.** A physicist measures the decay of a radioisotope. She finds that after 13 days, 80 percent of the original amount remains. Determine the half-life of the isotope.

74. **Population of Maryland.** The population of Maryland was 4,216,000 in 1980 and is estimated to be 4,491,000 in 1990. Find the doubling time of the population.

342 CHAPTER 6 Exponential and Logarithmic Functions

75. **Acidity and alkalinity.** Calculate the pH of a substance that has a hydrogen ion concentration H^+ of 8.2×10^{-12}.

76. **Street noise.** Determine the number of decibels in busy street traffic if the intensity of the sound I is 10^{-5} watts per square meter.

77. **Population growth of Arizona.** The population of Arizona was 2,286,000 in 1980 and was estimated to be 2,580,000 in 1990. Assume this was exponential growth.

   a. Find the exponential growth rate k.

   b. Find the equation of exponential growth.

   c. Estimate the population in 2010.

78. **Population growth of Arizona.** Refer to the previous exercise. When will the current population of Arizona increase by 50 percent over its level in 1990?

Chapter Test

1. Determine the equivalent exponential equation:
$\log_3 \sqrt{27} = \frac{3}{2}$

2. Determine the equivalent logarithmic equation: $e^2 = 7.3891$

3. Simplify the following expressions:

   a. $e^{2x} e^{x/2} e^{-5x/2}$

   b. $(e^{3x-1})^2$

   c. $\sqrt{e^{12x-10}}$

4. Sketch the graphs of the following functions.

   a. $y = 3e^{-x}$

   b. $y = -4e^x$

5. Simplify the following expressions:

   a. $\log(\sqrt{x}) - 2\log x$

   b. $\log x + 3\log y - 4\log z$

   c. $\ln \frac{5x}{y} - \ln \frac{\sqrt{x}}{y}$

6. Evaluate the following without using a calculator.

   a. $[e^3 \cdot e^{-4}]^{-1} \cdot e^{-1}$

   b. $\log_7 49$

   c. $\frac{\ln e^e}{e}$

7. Sketch the graphs of the following functions.

   a. $y = \ln(x + 1)$

   b. $y = -\ln x + 2$

8. Solve the equation $4e^{3x-2} = 12$. (Use a calculator where necessary.)

9. Solve the equation: $\log_8 x = -\frac{2}{3}$

10. Solve the equation $\ln(x - 1) + \ln(x + 1) = 1$.

11. The population of Kenya was 20 million on January 1, 1985, and it doubles every 23 years. Suppose we let $t = 0$ correspond to 1985. The population after t years is $N(t)$ where

$$N(t) = N_0 2^{t/D}$$

   and D is the doubling time.

   a. Write $N(t)$ in terms of an exponential function.

   b. What will the population of Kenya be in 23 years? in 46 years? in 1995? in 2000?

   c. How many years will it take for the population of Kenya to be 25 million?

12. **Thought question.** Describe the graph of the function $f(x) = e^{kx}$. Give examples for three positive values of k and three negative values of k.

13. **Thought question.** Define the function $\log_a x$.

CHAPTER 7
THE TRIGONOMETRIC FUNCTIONS

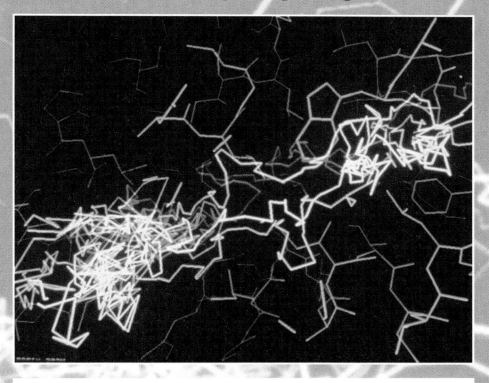

Trigonometry originated from the work of ancient Babylonian astronomers and ancient Greek geometers, who developed the mathematics of angles and triangles. The **trigonometric functions,** originally introduced to solve problems about angles and triangles, are important aids in constructing mathematical models of various physical phenomena, including electrical currents, sound waves, and predator-prey interaction. As we shall see, trigonometric functions are used in models of repetitive, or periodic, phenomena.

In Chapters 7–9, we study trigonometry from the perspectives of both geometry, to solve problems that involve angles and triangles, and of functions, to construct mathematical models.

Chapter Opening Image: *This computer visualization of an electron inside a protein illustrates the tunneling motion of electrons.*

7.1 ANGLES AND RADIAN MEASURE

Angles

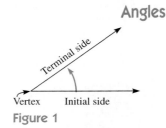

An **angle** is formed by rotating a line segment about one of its endpoints. The point around which the line segment is rotated is called the **vertex** of the angle. (See Figure 1.) The starting position of the line segment is called the **initial side** and the ending position is called the **terminal side.**

Figure 1

The angle in Figure 2 is formed by rotating the line segment OA about the point O. The initial side is OA and the terminal side is OB. We denote this angle with the notation $\angle AOB$. (The vertex is the middle letter.) This notation is especially useful in identifying angles contained in geometric figures such as triangles, rectangles, pentagons, and so forth.

Figure 2

Rotation in the counterclockwise direction is considered positive, whereas rotation in the clockwise direction is considered negative. In many instances, we specify the sign of an angle, that is, the direction of rotation, by placing an arrow pointing from the initial side to the terminal side. (See Figure 3.)

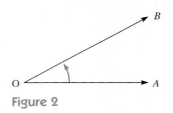

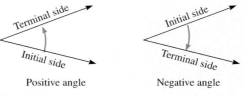

Figure 3

An angle can consist of more than one complete rotation. For example, Figure 4(a) shows an angle that consists of one and one-half rotations in the positive direction, Figure 4(b) shows an angle that consists of five-sixths of a rotation in the positive direction, and Figure 4(c) shows an angle that consists of two and one-fourth rotations in the negative direction.

Angles are often studied by placing them in a two-dimensional coordinate system, with the vertex at the origin and the initial side on the positive x-axis. Such angles are said to be in **standard position.** Figure 5 shows several angles in standard position.

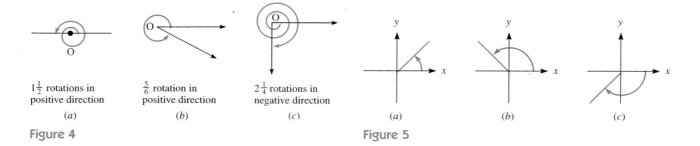

$1\frac{1}{2}$ rotations in positive direction (a)

$\frac{5}{6}$ rotation in positive direction (b)

$2\frac{1}{4}$ rotations in negative direction (c)

Figure 4

(a) (b) (c)

Figure 5

Degree Measure

The most familiar method for measuring angles is the system of degrees, minutes, and seconds. In this system of measurement, a complete rotation consists of 360 equal parts, called **degrees.** Degrees are denoted using the symbol °. An angle of 1° consists of $\frac{1}{360}$ of a complete rotation. (See Figure 6.)

SECTION 7.1 Angles and Radian Measure

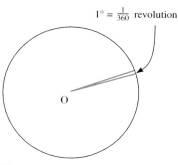

Figure 6

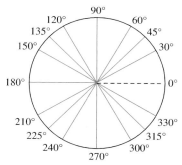

Figure 7

Here are some important angles and their measurements in degrees:

quarter revolution	90°
half revolution	180°
three-quarter revolution	270°
one revolution	360°

Figure 7 shows a circle with various angles measured in degrees. These angles are in standard position.

Each degree consists of 60 equal parts, called **minutes**. Each minute consists of 60 equal parts, called **seconds**. Minutes are denoted using the symbol ′, and seconds are denoted with the symbol ″. Using this notation, we have the following relationships: $1' = 60''$, and $1° = 60' = 3{,}600''$.

EXAMPLE 1
Drawing Angles

Draw an angle in standard position with the given measure of θ:

1. $\theta = 30°$
2. $\theta = 120°$
3. $\theta = -90°$
4. $\theta = 750°$

Solution

The specified angles are drawn in Figures 8(a)–8(d).

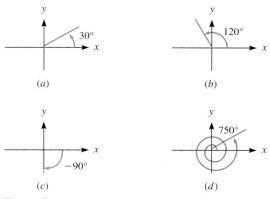

Figure 8

The most common types of angles are given special names. An angle that measures 90° is called a **right angle**. An angle that measures 180° is called a **straight angle**. A positive angle that is less than 90° is said to be **acute**, whereas an angle that is greater than 90° but less than 180° is said to be **obtuse**. Two acute angles that have measures that add up to 90° are said to be **complementary**. Two angles that have measures that add up to 180° are said to be **supplementary**. Figure 9 illustrates all of these terms.

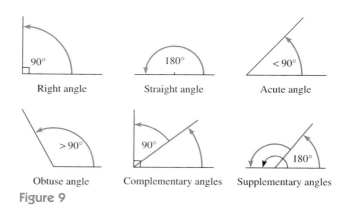

Figure 9

EXAMPLE 2
Complementary and Supplementary Angles

Suppose that $\theta = 41°28'$. Determine the measure of the following:

1. An angle that is complementary to θ.
2. An angle that is supplementary to θ.

Solution

1. The measure of an angle complementary to θ equals:
$$90° - \theta = 90° - 41°28'$$
To perform the indicated subtraction, we write 90° as 89° + 1°, or 89°60'. The difference is:
$$89°60' - 41°28' = 48°32'$$

2. The measure of an angle supplementary to θ equals:
$$180° - \theta = 180° - 41°28' = 179°60' - 41°28' = 138°32'$$

Angles that have the same sides are called **coterminal**. (See Figure 10.) In order for two angles to be coterminal, they must differ from one another by an integral number of complete rotations, that is, they must differ by an integer multiple of 360°.

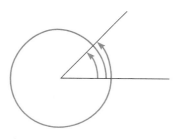

Coterminal angles
Figure 10

EXAMPLE 3
Determining Coterminal Angles

For the following angles, determine a coterminal angle with measure between 0° and 360°:

1. 530°
2. −90°
3. −400°

SECTION 7.1 Angles and Radian Measure

Solution

1. We must add or subtract a multiple of 360° so that the remainder lies between 0° and 360°. In this case, we subtract 360° from 530° to obtain the coterminal angle 170°.
2. We add 360° to −90° to obtain the coterminal angle 270°.
3. We add 2 · 360° to −400° to obtain the coterminal angle 320°.

In many applications, such as navigation and astronomy, angles are most commonly expressed in degrees, minutes, and seconds. However, for performing computations involving angles, it is easiest to express angle measure in terms of decimals, such as 42.5° rather than 42°30′. This way of expressing angle measure is called **decimal degrees.**

Radian Measure

The degree system of angle measurement dates back to the ancient Babylonians, who measured the length of a year as 360 days. One degree in their system corresponded to one day's angular motion of the earth around the sun. Of course, as we now know, the year does not contain 360 days. But the Babylonian measuring system has been retained for angle measure.

The system of **radian measure** is a much more natural method of angle measurement than the degree system. To understand this unit of measurement, consider a circle of radius 1, with its center at the origin. Such a circle is called a **unit circle.** Place an angle in standard position in the circle. As a measure of the size of the angle, let us take the distance traversed along the circumference of the circle in moving from the initial side to the terminal side of the angle. In Figure 11, the angle θ is measured by the length of this shaded arc along the circle. The units for this system of measurement are called **radians.**

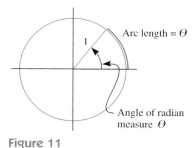
Figure 11

Definition
Radian Measure

Suppose that an angle is inscribed in the unit circle, with its vertex at the origin. The **radian measure** of the angle is the distance traveled along the circumference of the unit circle from the initial side to the terminal side.

One radian is an angle measure that corresponds to a distance of 1 unit along the unit circle. (See Figure 12(a).) Two radians is an angle measure that corresponds to a distance of 2 units along the circle. (See Figure 12(b).)

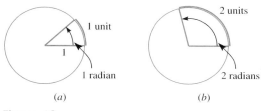
Figure 12

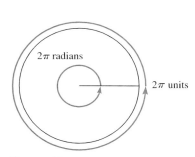
Figure 13

Because a unit circle has radius 1, its circumference is equal to:
$$C = 2\pi \cdot \text{radius} = 2\pi$$
That is, one complete rotation contains 2π radians. (See Figure 13.)

CHAPTER 7 The Trigonometric Functions

We can state the connection between the radian and degree systems of angle measurement as follows:

CONVERSION BETWEEN DEGREES AND RADIANS

$$360° = 2\pi \text{ radians}, \qquad 1° = \frac{\pi}{180} \text{ radians}$$

Here are some commonly used angles measured in degrees and their equivalent measurements in radians.

$$180° = \pi \text{ radians}$$
$$90° = \frac{\pi}{2} \text{ radians}$$
$$60° = \frac{\pi}{3} \text{ radians}$$
$$45° = \frac{\pi}{4} \text{ radians}$$
$$30° = \frac{\pi}{6} \text{ radians}$$

These angles are shown in Figure 14.

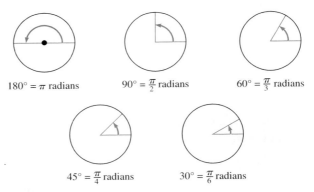

Figure 14

Using the conversion formula, we deduce the following simple procedure for converting angles from degrees to radians, and vice versa.

CONVERSION BETWEEN DEGREES AND RADIANS

1. To convert an angle measure from degrees to radians, multiply by $\pi/180$.
2. To convert an angle measure from radians to degrees, multiply by $180/\pi$.

EXAMPLE 4
Converting to Radians

Convert the following angle measures to radian measure.
1. 72°
2. −300°
3. 270°

Solution

1. To convert to radians, we multiply the degree measure by $\pi/180$:

$$72° = 72 \cdot \frac{\pi}{180} \text{ radians} = \frac{2\pi}{5} \text{ radians}$$

2. Proceeding as in part 1,

$$-300° = -300 \cdot \frac{\pi}{180} \text{ radians} = -\frac{5\pi}{3} \text{ radians}$$

3. We have:

$$270° = 270 \cdot \frac{\pi}{180} \text{ radians} = \frac{3\pi}{2} \text{ radians}$$

EXAMPLE 5 Converting to Degrees

Convert the following angle measures to degree measure (state in decimal degree form):

1. $\frac{\pi}{5}$ radians
2. $\frac{7\pi}{2}$ radians
3. 1 radian

Solution

To convert from radian measure to degree measure, multiply by $180/\pi$.

1. $\frac{\pi}{5}$ radians $= \frac{\pi}{5} \cdot \frac{180}{\pi} = 36°$
2. $\frac{7\pi}{2}$ radians $= \frac{7\pi}{2} \cdot \frac{180}{\pi} = 630°$
3. 1 radian $= 1 \cdot \frac{180}{\pi} = \frac{180}{3.14159\ldots} = 57.2958\ldots°$

Radian measure is the official unit of angle measurement in the international metric system. Moreover, radian measure is the preferred unit of angle measurement from a mathematical point of view, because the main formulas of calculus assume their simplest form in terms of radians. Throughout this book, all angles will be in radians unless explicitly stated otherwise. That is, when we speak of an angle of measure 3, we mean an angle of measure 3 radians.

Angle Calculations with a Calculator

Calculators take most of the drudgery out of doing calculations that involve angles. Exactly how much of the drudgery is automated depends on the calculator. All scientific and graphing calculators have modes that allow us to work with angle measures in either degrees or radians. We make a choice of the angle mode with a series of keystrokes, and the mode remains set, in either degrees or radians, until we change it. Most calculators remember the choice of mode even if we turn the calculator off. Please consult your calculator manual for instructions on how to set the angle mode.

We can enter angles as decimal numbers. For example, if the mode is degree, then the number 50.7 is interpreted as 50.7°. If the mode is radian, then 50.7 is interpreted as 50.7 radians.

Many calculators have methods for directly entering angles, degrees, minutes, and seconds. However, the methods of handling angles in this notation vary

EXAMPLE 6
Angle Conversions Using a Calculator

1. Convert 50.3° to radian measure.
2. Convert 3.12 radians to decimal degrees.
3. Convert 50°12′17″ to radians.

Solution

1. To convert to radians, multiply by $\pi/180$. So the radian measure of the angle is:

$$50.3 \cdot \frac{\pi}{180}$$

   The keystrokes for this calculation are

   The display reads .8779006138 radians.

2. To convert to decimal degrees, multiply by $180/\pi$:

$$3.12 \text{ radians} = 3.12 \cdot \frac{180}{\pi} \text{ degrees}$$

   The keystrokes are:

   The display reads 178.7628321°

3. Perform the calculation in two steps. First, convert to decimal degrees, and then convert to radians:

$$50°12'17'' = 50 + \frac{12}{60} + \frac{17}{3{,}600} \text{ degrees}$$
$$\approx 50.20472222 \text{ degrees}$$
$$\approx 50.20472222 \times \frac{\pi}{180} \text{ radians}$$
$$\approx .8762377028 \text{ radians}$$

Calculating Arc Length

We have defined radian measure in terms of arcs on a unit circle. However, suppose that we use a circle of radius r. Arc lengths are proportional to the radius of the circle. Thus, if we multiply the radius by r, arc lengths are also multiplied by r. That is, we have the following result:

LENGTH OF A CIRCULAR ARC

Suppose that a circle has radius r. An angle of θ radians determines an arc of length $s = r\theta$.

See Figure 15.

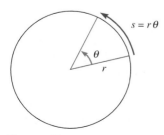

Figure 15

EXAMPLE 7
Length of a Circular Arc

Suppose that an angle inscribed within a circle of radius 4 centimeters has a measure of 2.3 radians. Determine the length of the arc defined by the angle. See Figure 16.

Solution

The angle θ is 2.3 radians. The length of the arc defined by the angle is $r\theta$, which is $4 \cdot 2.3 = 9.2$ centimeters.

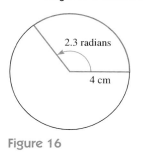

Figure 16

EXAMPLE 8
Angle Defined by a Circular Arc

Suppose that a central angle of a circle of radius 12 defines an arc of length 14. Find the radian measure of the angle that defines the arc.

Solution

Let θ denote the measure of the angle in radians. By the formula for the length of a circular arc, we see that:

$$\text{arc length} = \text{radius} \cdot \text{angle measure}$$
$$14 = 12\theta$$
$$\theta = \frac{14}{12} = \frac{7}{6} \text{ radians}$$

That is, the angle has measure $\frac{7}{6}$ radians.

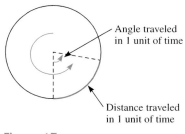

Figure 17

Suppose that a wheel rotates about an axis that is perpendicular to the wheel and passes through its center. Further, suppose that a line is drawn from the center of the wheel to a point P on its rim. The **angular speed** ω of the wheel is the angle through which P turns per unit of time. The **linear speed** V of the wheel is the distance that P travels per unit of time. (See Figure 17.) To obtain the relationship between ω and V, start with the familiar formula:

$$\text{speed} = \frac{\text{distance}}{\text{time}}$$

$$V = \frac{\text{distance } P \text{ travels in 1 sec}}{1 \text{ sec}}$$

Each second, P travels through an angle ω and covers an arc of length $r\omega$. Therefore, we can state the relationship as follows:

RELATIONSHIP BETWEEN ANGULAR AND LINEAR VELOCITY

Suppose that a point travels in a circular path of radius r with an angular velocity ω. Then its linear velocity V is given by:

$$V = r\omega$$

EXAMPLE 9
Linear and Angular Speed of a Tractor

Suppose that the wheels on a tractor have a radius of 3 feet and that the angular speed of the tires is 20 radians per second. What is the linear speed of the tractor in miles per hour?

Solution

From the preceding formula, the linear speed V is given by:
$$V = r\omega$$
$$= 3 \cdot 20 = 60 \text{ ft/sec}$$

Therefore, the tractor is moving at 60 ft/sec. We now convert this velocity to miles per hour:

$$60 \text{ ft/sec} = (60 \text{ ft/sec}) \cdot (3{,}600 \text{ sec/hr}) \cdot \left(\frac{1}{5{,}280} \text{ mi./ft}\right) = 40.91 \text{ mi./hr}$$

Exercises 7.1

Throughout this exercise set, you may use a calculator where appropriate.

Draw an angle in standard position that has the following measure.

1. 30°
2. 150°
3. −60°
4. −90°
5. −45°
6. −30°
7. 480°
8. −390°

Determine the measure of the following:
  a. An angle that is complementary to θ.
  b. An angle that is supplementary to θ.

9. $\theta = 36°24'$
10. $\theta = 78°11'$
11. $\theta = 56°34'53''$
12. $\theta = 89°42'23''$

Approximate each angle in terms of decimal degrees to the nearest ten-thousandth of a degree.

13. $\theta = 36°24'$
14. $\theta = 78°11'$
15. $\theta = 56°34'53''$
16. $\theta = 89°42'23''$
17. $\theta = -35°48'$
18. $\theta = -78°13'$
19. $\theta = 142°34'$
20. $\theta = 244°46'10''$

Approximate each angle in terms of degrees, minutes, and seconds.

21. 46.327°
22. 78.807°
23. −72.25°
24. −189.62°
25. 364.045°
26. 780.0042°
27. −1.6556°
28. −222.22222°
29. 1.90317 radians
30. 2.07312 radians
31. −2 radians
32. 0.05732 radians

Convert the following angles to radian measure.

33. 36°
34. 108°
35. 135°
36. −156.25°
37. −240°
38. 325°
39. 244.48°
40. −90°
41. 59°41'38.11''
42. 139°10''
43. −58°49'
44. 759°10'14''

Convert the following angles to decimal degrees.

45. $\dfrac{5\pi}{6}$
46. $\dfrac{17\pi}{5}$
47. $-\dfrac{7\pi}{4}$
48. $-\dfrac{3\pi}{2}$
49. 3.521
50. −1.873
51. $-\dfrac{13\pi}{6}$
52. $\dfrac{14\pi}{3}$

53. An angle has a measure of $3\pi/4$ radians. What are the degree measures of the angle and its supplement?

54. An angle has a measure of $\pi/6$ radians. What are the degree measures of the angle and its supplement?

55. An angle has a measure of 72°. What are the radian measures of the angle and its complement?

56. An angle has a measure of 15°. What are the radian measures of the angle and its complement?

Suppose that a circle has radius 5. What are the lengths of the arcs that correspond to the following central angles?

57. 20°
58. 300°
59. 1.5 radians
60. 3 radians
61. $\frac{\pi}{4}$ radians
62. $\frac{5\pi}{6}$ radians
63. a right angle
64. a straight angle

Applications

65. **Geometry.** In 35 minutes, the minute hand of a clock rotates 210°. What is this angle in radians? (See figure.)

Exercise 65

66. **Geometry.** The measure of one of the base angles of an isosceles triangle is 40°. (See figure.) What is the measure of each angle of the triangle in radians?

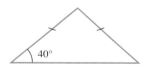

Exercise 66

67. **Geometry.** What is the angle measure in degrees and in radians of each angle of an isosceles right triangle?

68. **Physics.** A clock's pendulum is 1 m long. From one side to the other, it moves through an angle of 0.2 radians. Through what distance does the tip of the pendulum move?

69. **Sports.** A baseball player swings his bat, completely level, through an angle of 2.79 radians. If the bat is 39 inches long, through what distance does the tip of the bat move?

70. **Geometry.** For a circle with a 7-cm radius, how long is an arc that is defined by a central angle of 65°?

71. **Geometry.** What is the radius of a circle in which a central angle of 135° sweeps out an arc of length 9.42 ft?

72. **Physics.** In five hours, the tip of the needle of a circular thermometer moves 3 in. The needle is 5 in. long. What is the measure of the angle of rotation?

73. **Architecture.** A door built in the corner of a house can rotate through 2.09 radians before it is stopped by a wall. If the door is 3 ft wide, through what distance can the edge of the door move?

74. **Recreation.** A 150-ft diameter Ferris wheel makes one revolution in 65 seconds. What is the linear speed of each cab on the wheel?

75. **Recreation.** A person is seated on the end of a 5-m long see-saw. The see-saw moves up and down through a 28° angle every 3 seconds. What is the total distance the person moves in a minute?

76. **Physics.** A clock in Copenhagen, Denmark, makes one complete revolution every 25,753 years. The clock is 20 feet in diameter. How far does a point on the outside of the clock move every decade?

77. **Sports.** The Tour De France is an annual bicycle race held in France. The average speed of participants in one race is 15 mi./hr. The diameter of a bicycle wheel is 26 in. What is the average angular speed of the riders in radians/hr? (Be sure to work with the same units of length throughout the problem.)

78. **Geometry.** A building in Milwaukee has a clock with a minute hand 20 ft long. What is the linear speed of the tip of the minute hand in in./sec?

79. **Physics.** One end of a 6-inch rod is in the center of a centrifuge. The other end moves at 4,500 mph. What is the angular speed of the end of the rod in rev/sec?

80. **Fitness.** Find the linear speed of an exercise bicycle if the wheel has a radius of 10 in. and an angular speed of 0.256 radian/sec.

 In your own words

81. Explain how to calculate the measure of an angle in radians.

82. Suppose your calculator battery runs down during a test. What procedure could you use for converting degrees, minutes, and seconds to radian measurement?

83. Describe how to calculate the length of a circular arc. Give a concrete example of the calculations.

84. Draw several examples of negative angles. Explain how to distinguish between positive and negative angles.

85. Draw several examples of angles that are larger than a single revolution. Explain how to draw angles with radian measure larger than 2π.

7.2 THE TRIGONOMETRIC FUNCTIONS

In the preceding section, we introduced angle measurement using both the degree and radian systems. In this section, we define six functions that describe an angle.

Definition of Sine and Cosine

We can define the sine and cosine of an angle t as follows:[1]

Definition
Sine and Cosine of an Angle

Draw a unit circle. Let t be an angle in standard position. (See Figure 1.) Suppose that the terminal side of the angle intersects the circle at point P. The **cosine** of t, denoted $\cos t$, is the x-coordinate of P. The **sine** of t, denoted $\sin t$, is the y-coordinate of P.

---

[1] Here and in what follows, we use t or x to denote the measure of a variable angle always given in radians. We continue to use other letters, such as θ, a, b, and c, to denote measures of fixed angles, given in either radians or degrees. The reason for using t is that, in many applications involving the trigonometric functions, the variable is time.

SECTION 7.2 The Trigonometric Functions

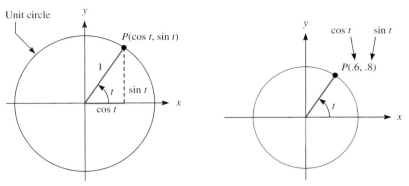

Figure 1

Figure 2

For instance, consider angle t in Figure 2. The x-coordinate gives the value of the cosine and the y-coordinate gives the value of the sine. That is:

$$\cos t = 0.6, \quad \sin t = 0.8$$

For any angle t (measured in radians or degrees), we can determine the value of the corresponding sine and cosine by drawing the angle in a unit circle and using geometry to determine the coordinates of the point P. The geometric reasoning is illustrated in the next two examples.

EXAMPLE 1
Calculating the Values of Sine and Cosine of 30°

Determine $\sin 30°$ and $\cos 30°$.

Solution

Figure 3(a) shows an angle of measure 30° in standard position. The triangle shown has angles 30°, 60°, and 90°. Recall from geometry that such a triangle has sides proportional to 1, 2, and $\sqrt{3}$, as shown in Figure 3(b). In the triangle on the left, the hypotenuse is a radius of the unit circle and is therefore equal to 1. So the length of a side of the triangle on the left is one-half the length of a side of the triangle on the right. The coordinates of the point P are therefore $(\frac{\sqrt{3}}{2}, \frac{1}{2})$. According to the definition of the sine and cosine of an angle,

$$\cos 30° = \frac{\sqrt{3}}{2}, \quad \sin 30° = \frac{1}{2}$$

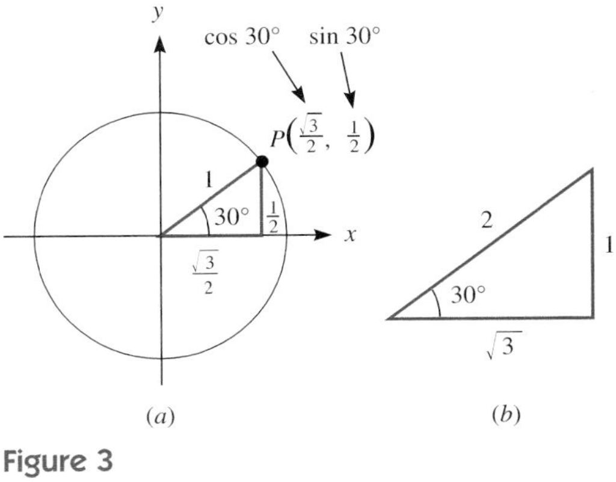

Figure 3

356 CHAPTER 7 The Trigonometric Functions

EXAMPLE 2 Determine $\sin 45°$ and $\cos 45°$.

Calculating the Values of Sine and Cosine of 45°

Solution

Figure 4(a) shows the angle 45° in standard position. Recall from geometry that a 45°-45°-90° triangle has sides proportional to 1, 1, and $\sqrt{2}$ as shown in Figure 4(b). The hypotenuse of the triangle on the left is 1, because it is a radius of the unit circle. Therefore, the length of a side of the triangle on the left is $1/\sqrt{2} = \sqrt{2}/2$ times the length of a side of the triangle on the right. So the point P has coordinates $(\sqrt{2}/2, \sqrt{2}/2)$, and, by the definition of the sign and cosine, we have:

$$\sin 45° = \frac{\sqrt{2}}{2}, \qquad \cos 45° = \frac{\sqrt{2}}{2}$$

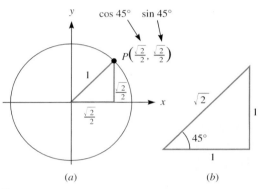

Figure 4

EXAMPLE 3 Determine $\sin 0°$ and $\cos 0°$.

Calculating Sine and Cosine of 0°

Solution

Figure 5 shows the angle 0° in standard position. We see that P has coordinates $(1, 0)$, so $\sin 0° = 0$, $\cos 0° = 1$.

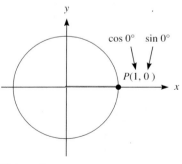

Figure 5

Here are two important points to remember about the definition of sine and cosine:

1. The circle used to define the sine and cosine is a **unit** circle (radius 1). If you use a circle of some other radius, the coordinates of the point P will not give the value of the sine and cosine of the angle t.
2. The order of the coordinates is important. The horizontal coordinate is the cosine and the vertical coordinate is the sine. Be careful not to interchange the coordinates.

The angle in a trigonometric function can be given either in degrees, as in the above examples, or in radians.

EXAMPLE 4 Determine the values of $\sin t$ and $\cos t$ for the following angles t.

Calculating Sine and Cosine of Angles in Radians

1. $\dfrac{\pi}{2}$
2. $-\dfrac{3\pi}{2}$

Solution

1. The angle of radian measure $\pi/2$ (or $90°$) is drawn in Figure 6. The coordinates of P are $(0, 1)$. Therefore, by the definition of sine and cosine,
$$\cos\frac{\pi}{2} = 0, \qquad \sin\frac{\pi}{2} = 1$$

2. The angle $-(3\pi)/2$ is obtained by making three-fourths of a revolution in the negative direction. Refer to Figure 7. From the coordinates of P, which are $(0, 1)$, we see that:
$$\cos\left(-\frac{3\pi}{2}\right) = 0, \qquad \sin\left(-\frac{3\pi}{2}\right) = 1$$

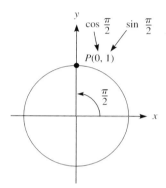
Figure 6

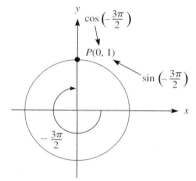
Figure 7

In Examples 1–3, we used the geometry of known triangles to determine the values of sine and cosine for various acute angles. To do similar calculations for nonacute angles, we can form a triangle and use a reference angle, as follows:

Definition
Reference Angle Let t be an angle in standard position. The corresponding **reference angle** t' is the acute angle between the terminal side of t and the horizontal axis. (See Figure 8.)

Note that the reference angle t' is always acute, no matter what the size of the angle t. (See Figure 9.)

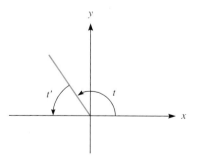
Figure 8

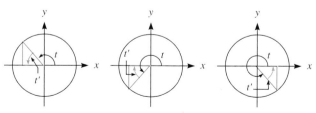
Figure 9

EXAMPLE 5
Calculating Reference Angles

Determine the reference angles for the following angles:
1. $t = -70°$
2. $t = 255°$
3. $t = \dfrac{5\pi}{3}$ radians

Solution

1. See Figure 10(a). The terminal side makes two angles with the x-axis; one is 70° and the other is 110°. The 70° angle is the reference angle, because it is an acute angle.
2. See Figure 10(b). The terminal side makes two angles with the x-axis; one is 75° and the other is 105°. The 75° angle is acute, and thus is the reference angle.
3. See Figure 10(c). The terminal side makes two angles with the x-axis; one is $\pi/3$ and the other is $2\pi/3$. The $\pi/3$ angle is acute, and thus is the reference angle.

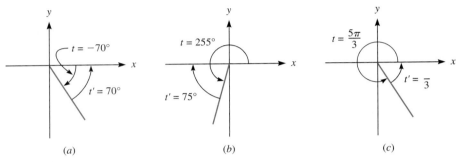

Figure 10

If the angle is embedded in a unit circle, then the reference angle defines a triangle, as shown in Figure 9. We can use this triangle to determine the values of sine and cosine, as the next example illustrates.

EXAMPLE 6
Calculating Sine and Cosine Using Reference Angles

Calculate $\sin t$ and $\cos t$ for $t = 5\pi/6$.

Solution

Draw the angle $5\pi/6$ in standard position, as shown in Figure 11. Now draw the corresponding reference angle, which is $\pi/6$. Draw the associated triangle defined by the reference angle and the unit circle. This is just a 30°-60°-90° triangle with hypotenuse 1 (because the unit circle has radius 1). Therefore, the legs are of length $\frac{1}{2}$ and $\frac{\sqrt{3}}{2}$. Because the angle $5\pi/6$ is in the second quadrant, the point P has a positive y-coordinate and a negative x-coordinate. Therefore, we see that $P = \left(-\frac{\sqrt{3}}{2}, \frac{1}{2}\right)$, so that:

$$\sin \frac{5\pi}{6} = \frac{1}{2}$$
$$\cos \frac{5\pi}{6} = -\frac{\sqrt{3}}{2}$$

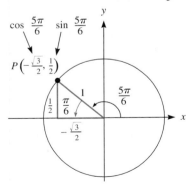

Figure 11

Determining Values of Sine and Cosine with a Calculator

The values of the trigonometric functions for particular angles t are (except for special angles) determined by referring to a table or by using a calculator. To use a table of trigonometric functions, we look up the desired value of t in the left column and read off the corresponding values of the trigonometric functions. Tables are limited in the accuracy of the angle allowed and in the number of significant digits of the trigonometric functions. In many tables, the angle t is given in hundredths of a radian and the trigonometric functions are given to five significant digits. With the advent of the scientific calculator, the significance of tables has diminished considerably and today they are rarely used.

Calculators have the keys [SIN], [COS], [TAN] for computing the values of the trigonometric functions $\sin t$, $\cos t$, and $\tan t$, usually to ten decimal places of accuracy. To calculate the value of a trigonometric function using a graphing calculator, just press the appropriate function key, followed by the angle. Then press [ENTER] or [=]. The display shows the value of the desired trigonometric function. For example, to calculate $\sin 53.1°$, first, make sure that the angle mode of the calculator is degree. Then, enter the following keystrokes: [SIN] [53.1] [ENTER]. The display will read 0.7996846585. Note that some scientific calculators require that you enter the 53.1 *before* you press [SIN].

The following example illustrates how the trigonometric function keys can be used to calculate the values of various trigonometric functions.

EXAMPLE 7
Numerical Approximations for Sine and Cosine

Use a graphing calculator to determine the values of the following trigonometric functions.

1. $\sin 42°31'$
2. $\cos 72°15'27''$

Solution

1. Set the calculator on degree angle measurement. Then enter the following key sequence:

    [SIN] [(] [42] [+] [31] [÷] [60] [)] [ENTER]

    The display reads .6758046443.

2. The angle mode is already set on degree from part 1. (There is no need to reset the angle measurement mode for each new calculation.) Enter the following key sequence:

    [COS] [(] [72] [+] [15] [÷] [60] [+] [27] [÷] [3,600] [)] [ENTER]

    The display reads .3047396281.

HISTORICAL NOTE Before the advent of scientific calculators, the values of trigonometric functions were determined from tables developed by mathematicians. To ensure accurate calculations, it was necessary to use tables that recorded the values of trigonometric functions to precise fractions of a degree and to a large number of significant figures. Some tables recorded the trigonometric function values for angle intervals of one second and to 14 decimal places. So important were such tables, that they were a necessary purchase for every student in any field of science or engineering. During the first third of the twentieth century, the center for world science was Germany. During this period, no tables of trigonometric function values were produced in English. Rather, students had to use tables imported from Germany. In order to read the explanation of how to use the tables, it was necessary for a student to learn scientific German, a standard requirement of all engineering and science students of that era.

The Other Trigonometric Functions

In addition to the trigonometric functions $\sin t$ and $\cos t$, there are four other trigonometric functions associated with the angle t. These functions are called the **tangent**, denoted $\tan t$, the **secant**, denoted $\sec t$, the **cotangent**, denoted cot *t*, and the **cosecant**, denoted csc *t*. These trigonometric functions are defined in terms of sin *t* and cos *t* using the following formulas:

$$\tan t = \frac{\sin t}{\cos t}, \qquad \cos t \neq 0$$

$$\sec t = \frac{1}{\cos t}, \qquad \cos t \neq 0$$

$$\csc t = \frac{1}{\sin t}, \qquad \sin t \neq 0$$

$$\cot t = \frac{1}{\tan t} = \frac{\cos t}{\sin t}, \qquad \sin t \neq 0$$

Note that each of the four additional trigonometric functions is defined only for certain angles *t*. We explore the domains of these functions in more detail later in the chapter.

EXAMPLE 8
Calculating Trigonometric Functions

Determine the values of tan *t*, sec *t*, csc *t*, cot *t* for $t = 5\pi/6$.

Solution

From the previous example, we have:

$$\sin \frac{5\pi}{6} = \frac{1}{2}$$

$$\cos \frac{5\pi}{6} = -\frac{\sqrt{3}}{2}$$

From these values, we apply the definitions of the tangent and secant:

$$\tan \frac{5\pi}{6} = \frac{\sin \frac{5\pi}{6}}{\cos \frac{5\pi}{6}} = \frac{\frac{1}{2}}{-\frac{\sqrt{3}}{2}} = -\frac{\sqrt{3}}{3}$$

$$\sec \frac{5\pi}{6} = \frac{1}{\cos \frac{5\pi}{6}} = \frac{1}{-\frac{\sqrt{3}}{2}} = -\frac{2\sqrt{3}}{3}$$

$$\csc \frac{5\pi}{6} = \frac{1}{\sin \frac{5\pi}{6}} = \frac{1}{\frac{1}{2}} = 2$$

$$\cot \frac{5\pi}{6} = \frac{1}{\tan \frac{5\pi}{6}} = \frac{1}{-\frac{\sqrt{3}}{3}} = -\sqrt{3}$$

EXAMPLE 9
Numerical Values of the Tangent and Secant

Use a graphing calculator to determine the values of the following trigonometric functions.

1. tan 38.523°
2. sec 63.581

Solution

1. Enter the following key sequence:

   [TAN] [38.523] [ENTER]

   The display reads .7960915412.

2. Because 63.581 does not have a degree sign, it must be an angle in radians. So set the angle measurement mode to radian. There is no secant key on the calculator. Recall, however, that the secant is the reciprocal of the cosine. So we can calculate the secant value by first calculating the cosine, then taking the reciprocal:

   [(] [COS] [63.581] [)] [x^{-1}] [ENTER]

   The display reads 1.365616341.

Signs of the Trigonometric Functions

The signs of the trigonometric functions are constant if the angle remains within a single quadrant. Let's determine these signs. Recall that the sine of an angle is given by the y-coordinate of the point at which the terminal side of the angle intersects the unit circle. Thus, $\sin t$ is positive for angles in quadrants I and II and negative for angles in quadrants III and IV. (See Figure 12(a).) The cosine of an angle is determined by the x-coordinate of the point at which the terminal side of the angle intersects the unit circle. Thus, $\cos t$ is positive for angles in quadrants I and IV and negative for angles in quadrants II and III.

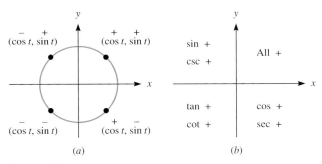

Figure 12

The signs of the other trigonometric functions can be determined directly from their respective definitions. For instance, consider the tangent function, given by the formula:

$$\tan t = \frac{\sin t}{\cos t}$$

If t is in quadrant I, the numerator and denominator are both positive, so the value of $\tan t$ is positive. If t is in quadrant II, then the numerator is positive and the denominator is negative, so that the value of $\tan t$ is negative. We use similar reasoning for the quadrants III and IV. Using similar arguments, we can derive the following scheme for determining the signs of the various trigonometric functions according to the quadrant in which the angle lies. (See Figure 12(b).)

Exercises 7.2

Determine the values of the six trigonometric functions for each value of t.

1. $t = \dfrac{5\pi}{4}$
2. $t = \dfrac{3\pi}{4}$
3. $t = \dfrac{7\pi}{6}$
4. $t = \dfrac{11\pi}{6}$
5. $t = \dfrac{4\pi}{3}$
6. $t = \dfrac{8\pi}{3}$
7. $t = -\dfrac{\pi}{4}$
8. $t = -\dfrac{\pi}{3}$
9. $t = -\dfrac{3\pi}{4}$
10. $t = -\dfrac{7\pi}{3}$
11. $t = -\dfrac{14\pi}{3}$
12. $t = -6\pi$
13. $t = 360°$
14. $t = 180°$
15. $t = 240°$
16. $t = 300°$
17. $t = 315°$
18. $t = -90°$
19. $t = -270°$
20. $t = 150°$
21. $t = -150°$
22. $t = 570°$
23. $t = -570°$
24. $t = -480°$
25. $t = 675°$
26. $t = -\dfrac{25\pi}{4}$
27. $t = \dfrac{35\pi}{6}$
28. $t = 32\pi$
29. $t = 43\pi$
30. $t = 50\pi$

Determine the exact values of the six trigonometric functions for the angle t defined by the point P on the unit circle.

31. $P = \left(\dfrac{3}{5}, \dfrac{4}{5}\right)$
32. $P = \left(\dfrac{4}{5}, -\dfrac{3}{5}\right)$
33. $P = \left(\dfrac{5}{13}, -\dfrac{12}{13}\right)$
34. $P = \left(-\dfrac{12}{13}, -\dfrac{5}{13}\right)$

Technology

Use a calculator to determine the values of the following trigonometric functions. Round your answer to five decimal places.

36. $\tan 89°$
37. $\cos 71°$
38. $\sin 34°$
39. $\tan 11°34'$
40. $\sec 23°18'$
41. $\cot 1°57'$
42. $\csc 88°59'$
43. $\cos 23°34'18''$
44. $\sin 17°51'2''$
45. $\tan 21°48'32''$
46. $\cot 75°55'25''$
47. $\csc 56.678°$
48. $\cos 76.1213°$
49. $\sin 4.34°$
50. $\tan 59.9°$
51. $\sin 0.763$
52. $\cos 1.4$
53. $\tan 1.2113$
54. $\sec 0.87$
55. $\csc 3\pi$
56. $\cot 0.8\pi$
57. $\sin 1$
58. $\cos 0.6$

In your own words

59. Give the definition of the sine and cosine of an angle.
60. Illustrate the definition of the sine and cosine for a negative angle.
61. Illustrate the definition of the sine and cosine for an angle larger than one revolution.
62. How can you calculate the tangent of an angle in terms of the sine and cosine of that angle? Illustrate with a specific angle.
63. How can you calculate the secant of an angle in terms of the sine and cosine of that angle? Illustrate with a specific angle.

7.3 TRIGONOMETRIC FUNCTIONS OF ACUTE ANGLES

In the preceding section, we introduced the trigonometric functions of an angle θ. If θ is acute, we can give an alternate definition of the trigonometric functions in terms of right triangles.[2] This alternate definition is very useful in applications of trigonometry.

The Trigonometric Functions as Ratios

Suppose that we are given a right triangle with an acute angle θ. One leg of the triangle is opposite the angle and one is adjacent, as Figure 1 shows. We can calculate the values of the trigonometric functions directly from the sides of the triangle, as follows:

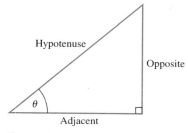

Figure 1

TRIGONOMETRIC FUNCTIONS OF ACUTE ANGLES AS RATIOS

$$\sin \theta = \frac{\text{opposite}}{\text{hypotenuse}} \quad (1)$$

$$\cos \theta = \frac{\text{adjacent}}{\text{hypotenuse}} \quad (2)$$

$$\tan \theta = \frac{\text{opposite}}{\text{adjacent}} \quad (3)$$

$$\cot \theta = \frac{\text{adjacent}}{\text{opposite}} \quad (4)$$

$$\sec \theta = \frac{\text{hypotenuse}}{\text{adjacent}} \quad (5)$$

$$\csc \theta = \frac{\text{hypotenuse}}{\text{opposite}} \quad (6)$$

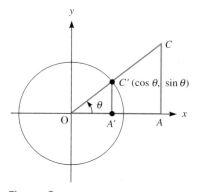

Figure 2

To prove these formulas, draw a unit circle and place the triangle so that the adjacent side lies along the positive x-axis. (See Figure 2.) Label the triangle AOC, as in the figure. Let C' be the point at which the line $\overline{OC}$ intersects the unit circle. Draw the right triangle $A'OC'$, as shown in Figure 2. Note that triangles AOC and $A'OC'$ are similar because they have at least two congruent angles. Let $d(a, b)$ denote the distance between points a and b in the plane. Because the triangles are similar, there is a positive number k such that:

$$d(A, C) = k d(A', C')$$
$$d(O, A) = k d(O, A')$$
$$d(O, C) = k d(O, C')$$

Moreover, because $\overline{OC'}$ is a radius of a unit circle, we have:

$$d(O, C') = 1$$

From the definition of the sine and cosine, we have:

$$\cos \theta = d(O, A')$$
$$= \frac{d(O, A')}{1}$$

---

[2] Throughout this section, we discuss trigonometric functions evaluated for angles of triangles. In this case, we use one of these Greek letters to denote the angle: $\theta, \alpha, \beta,$ or γ.

$$\begin{aligned}&= \frac{d(O, A')}{d(O, C')}\\&= \frac{kd(O, A')}{kd(O, C')}\\&= \frac{d(O, A)}{d(O, C)}\\&= \frac{\text{adjacent}}{\text{hypotenuse}}\end{aligned}$$

This proves formula 2. The proof of 1 is similar and is left as an exercise. To prove 3, note that $\tan\theta = \dfrac{\sin\theta}{\cos\theta}$, so, by formulas 1 and 2,

$$\begin{aligned}\tan\theta &= \frac{\left(\dfrac{\text{opposite}}{\text{hypotenuse}}\right)}{\left(\dfrac{\text{adjacent}}{\text{hypotenuse}}\right)}\\&= \frac{\text{opposite}}{\text{hypotenuse}} \cdot \frac{\text{hypotenuse}}{\text{adjacent}}\\&= \frac{\text{opposite}}{\text{adjacent}}\end{aligned}$$

This proves formula 3. The remaining formulas are proved similarly.

EXAMPLE 1 Trigonometric Functions of an Angle in a Right Triangle

A right triangle has an angle θ, as shown in Figure 3. Determine the values of $\sin\theta$, $\cos\theta$, and $\tan\theta$.

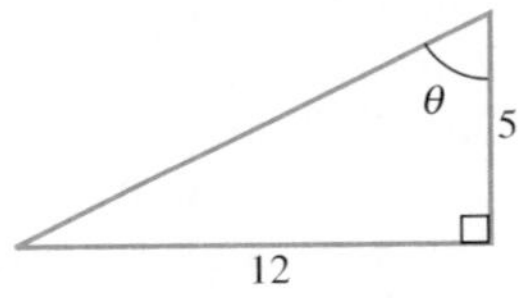

Figure 3

Solution

From the figure:

$$\text{opposite} = 12, \text{adjacent} = 5$$

Moreover, by the Pythagorean theorem,

$$\begin{aligned}\text{hypotenuse} &= \sqrt{\text{opposite}^2 + \text{adjacent}^2}\\&= \sqrt{5^2 + 12^2}\\&= \sqrt{169}\\&= 13\end{aligned}$$

From formulas 1–3, we have:

$$\begin{aligned}\sin\theta &= \frac{\text{opposite}}{\text{hypotenuse}} = \frac{12}{13}\\\cos\theta &= \frac{\text{adjacent}}{\text{hypotenuse}} = \frac{5}{13}\\\tan\theta &= \frac{\text{opposite}}{\text{adjacent}} = \frac{12}{5}\end{aligned}$$

The trigonometric functions $\cot\theta$, $\sec\theta$, and $\csc\theta$ can be calculated in terms of the other three trigonometric functions. Applying formulas 1–3, we arrive at the following formulas 4–6:

SECTION 7.3 Trigonometric Functions of Acute Angles

$$\cot \theta = \frac{1}{\tan \theta} = \frac{\text{adjacent}}{\text{opposite}}$$

$$\sec \theta = \frac{1}{\cos \theta} = \frac{\text{hypotenuse}}{\text{adjacent}}$$

$$\csc \theta = \frac{1}{\sin \theta} = \frac{\text{hypotenuse}}{\text{opposite}}$$

Notation for Labeling Triangles It is customary to adopt a labeling convention that simplifies the description of a triangle. If we label the sides a, b, c, the vertices A, B, C, and the angles α, β, γ, then each vertex and angle is always assumed to be opposite the side that has the corresponding letter. That is, vertex A and angle α are opposite side a, vertex B and angle β are opposite side b, and vertex C and angle γ are opposite side c. Moreover, in case the triangle is a right triangle, the angle γ will be the right angle. This labeling convention is used throughout the remainder of this book.

EXAMPLE 2
Computing Trigonometric Functions from a Triangle

Consider the triangle shown in Figure 4. Determine the values of the following trigonometric functions.

1. $\sin \alpha$
2. $\sin \beta$
3. $\tan \alpha$
4. $\csc \beta$

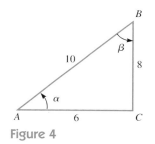
Figure 4

Solution

1. The sine of an angle equals the ratio $\frac{\text{opposite}}{\text{hypotenuse}}$. In the triangle shown, side $\overline{AB}$ is the hypotenuse and side $\overline{BC}$ is the opposite side. Therefore, we have:

$$\sin \alpha = \frac{d(B, C)}{d(A, B)} = \frac{8}{10} = 0.8$$

2. For the angle β, the hypotenuse remains unchanged, but the opposite side is $\overline{AC}$. Therefore,

$$\sin \beta = \frac{d(A, C)}{d(B, A)} = \frac{6}{10} = 0.6$$

3. We have:

$$\tan \alpha = \frac{\text{opposite}}{\text{adjacent}}$$
$$= \frac{d(B, C)}{d(A, C)}$$
$$= \frac{8}{6}$$
$$= \frac{4}{3}$$

4. We have:

$$\csc \beta = \frac{\text{hypotenuse}}{\text{opposite}}$$
$$= \frac{d(A, B)}{d(A, C)}$$
$$= \frac{10}{6}$$
$$= \frac{5}{3}$$

EXAMPLE 3
Computing Trigonometric Functions from a Known Trigonometric Function

Suppose that θ is an acute angle for which $\cos \theta = \frac{5}{7}$. Determine the values of the other five trigonometric functions.

Solution

Draw a triangle that has the angle θ, as in Figure 5. Because the cosine is the ratio of the adjacent side divided by the hypotenuse, we have labeled the adjacent side 5 and the hypotenuse 7. By the Pythagorean theorem:

$$\text{opposite} = \sqrt{7^2 - 5^2} = \sqrt{24} = 2\sqrt{6}$$

Therefore, by formulas 1–6, we have:

$$\sin \theta = \frac{\text{opposite}}{\text{hypotenuse}} = \frac{2\sqrt{6}}{7}$$

$$\tan \theta = \frac{\text{opposite}}{\text{adjacent}} = \frac{2\sqrt{6}}{5}$$

$$\sec \theta = \frac{\text{hypotenuse}}{\text{adjacent}} = \frac{7}{5}$$

$$\csc \theta = \frac{\text{hypotenuse}}{\text{opposite}} = \frac{7}{2\sqrt{6}} = \frac{7\sqrt{6}}{12}$$

$$\cot \theta = \frac{\text{adjacent}}{\text{opposite}} = \frac{5}{2\sqrt{6}} = \frac{5\sqrt{6}}{12}$$

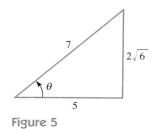
Figure 5

Determining Sides of Triangles

Once we know an angle of a triangle, we can determine the values of the trigonometric functions for the angle, using either a table or a calculator. From these trigonometric functions, we can determine the sides of the triangle, as the following examples show. Computations such as these are the crucial ingredient in solving the applied problems in Section 7.4.

EXAMPLE 4
Determining the Adjacent Side

Suppose a right triangle has a hypotenuse of length 40 and one angle of 35°. What is the length of the side adjacent to the angle?

Solution

Let x be the length of the adjacent side, as shown in Figure 6. Because the problem involves the adjacent side and the hypotenuse, we write an equation that relates these two sides to a trigonometric function, namely:

$$\frac{\text{adjacent}}{\text{hypotenuse}} = \cos 35° = 0.8191520443$$

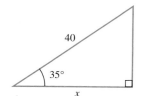
Figure 6

The last number was obtained using a calculator. From the figure, the above equation can be written:

$$\frac{x}{40} = 0.8191530443$$
$$x \approx 32.766$$

EXAMPLE 5 Determining the Opposite Side

Suppose a right triangle has one side of length 5 with an adjacent angle of 29°. What is the length of the opposite side?

Solution

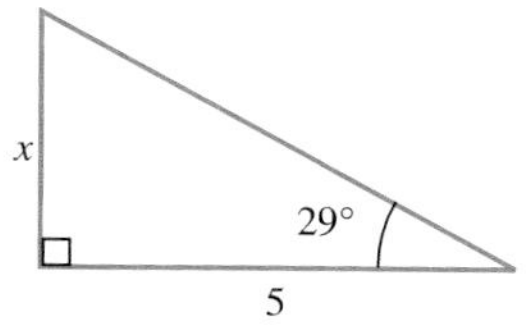

Figure 7

Let x denote the length of the opposite side, as shown in Figure 7. Because the problem involves the opposite and adjacent sides, we write an equation for the tangent of 29°:

$$\frac{\text{opposite}}{\text{adjacent}} = \tan 29°$$
$$\frac{x}{5} = 0.5543090515$$
$$x \approx 2.7715$$

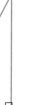

EXAMPLE 6 Determining the Hypotenuse

Suppose that a triangle has a side of length 12 and an adjacent angle of 63°. What is the length of the hypotenuse?

Solution

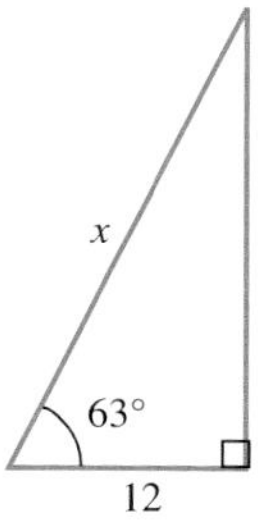

Figure 8

Let x denote the length of the hypotenuse, as shown in Figure 8. Because the problem involves the adjacent side and the hypotenuse, we write an equation for the cosine:

$$\frac{\text{adjacent}}{\text{hypotenuse}} = \cos 63°$$
$$\frac{12}{x} = 0.4539904997$$
$$x \approx 26.432$$

Determining Angles of Triangles

Suppose that we are given the value of one of the trigonometric functions $\sin\theta$, $\cos\theta$, or $\tan\theta$, for an acute angle θ. Using a calculator, we can determine the angle θ, using the $\sin^{-1}$, $\cos^{-1}$, or $\tan^{-1}$ key, respectively. For example, suppose we are given the information that:

$$\cos\theta = 0.751$$

To determine θ, we press the $\cos^{-1}$ key followed by the angle, and then we press ENTER. The display reads 41.32292462. That is,

$$\theta = 41.32292462°$$

Note that the calculator will give the angle in degrees or radians, depending on the current angle mode setting. The answer above assumes that the angle mode is set to degree.

368 CHAPTER 7 The Trigonometric Functions

EXAMPLE 7
Determining the Angles of a Right Triangle

Suppose that a right triangle has legs 4 and 7. What are the angles of the triangle?

Solution

The triangle appears in Figure 9. From the definition of the tangent, we see that:

$$\tan \alpha = \frac{\text{opposite}}{\text{adjacent}} = \frac{4}{7}$$

We determine α by keying in $\boxed{\tan^{-1}}\,\boxed{(}\,\boxed{4}\,\boxed{\div}\,\boxed{7}\,\boxed{)}$. The value displayed is:

$$\alpha \approx 29.745°$$

Because β is complementary to α and because this is a right triangle, we see that:

$$\beta = 90° - \alpha$$
$$\approx 60.255°$$

Figure 9

Exercises 7.3

For each triangle in Exercises 1–12, find the six trigonometric function values of the angle θ.

1.
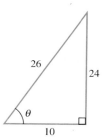

2.
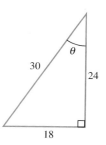

3.
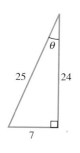

4.
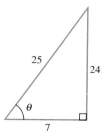

5.
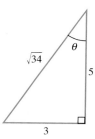

6.
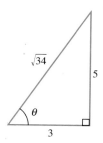

7.
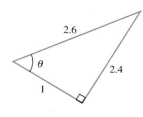

8.
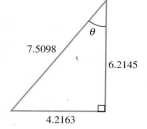

9.
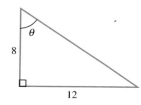

10.
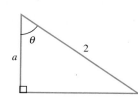

11.
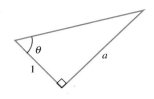

SECTION 7.3   Trigonometric Functions of Acute Angles   369

12.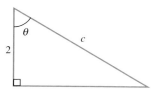

In Exercises 13–20, a value of a trigonometric function is given. Find the five other trigonometric values of the angle θ. Assume that θ is acute.

13. $\sin \theta = \dfrac{2}{3}$

14. $\cos \theta = \dfrac{1}{\sqrt{7}}$

15. $\tan \theta = \sqrt{2}$

16. $\cot \theta = 3$

17. $\sec \theta = \dfrac{25}{7}$

18. $\csc \theta = \dfrac{4}{\sqrt{5}}$

19. $\cos \theta = a$, $0 < a < 1$

20. $\tan \theta = \dfrac{1}{v}$

For each triangle, determine the side labeled x.

21.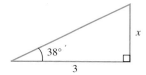

22.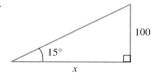

23.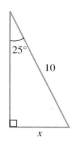

24.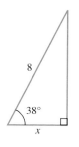

25.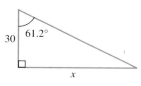

26.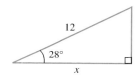

27.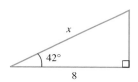

28.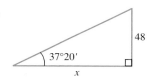

29.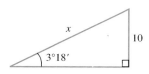

30.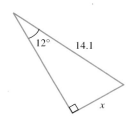

For each triangle, determine the angle α in degree measure.

31.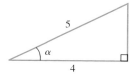

32.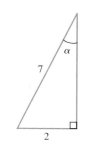

33.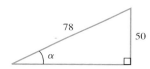

34.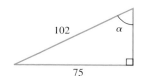

35.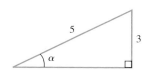

36.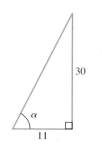

37.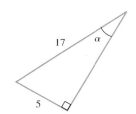

38.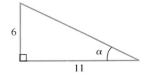
39.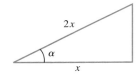
40.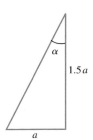

Applications

41. **Geometry.** The top of a 13-ft ladder leaning against a wall touches the wall 12 ft above the ground. Find the six trigonometric function values of the angle the ladder makes with the ground. (See figure.)

42. **Geometry.** A guy wire reaches from the top of a 20-ft pole to a point on the ground 15 ft from the bottom of the pole. (See figure.) What are the six trigonometric function values of the angle the wire makes with the ground?

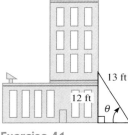
Exercise 41

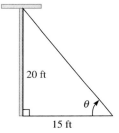
Exercise 42

Technology

Determine the angle that satisfies the given equation.

43. $\cos\theta = 0.1762$
44. $\sin\theta = 0.9974$
45. $\tan\theta = 6.7899$
46. $\sec\theta = 8$
47. The formula
$$\cos\theta \approx 1 - \frac{\theta^2}{2} + \frac{\theta^4}{24} - \frac{\theta^6}{720}$$
can be used to approximate values of the cosine function when θ is in radians. Use the formula to approximate the value of cos 0.2, and compare this value with the value produced by a calculator.

48. The formula
$$\tan\theta = \theta + \frac{\theta^3}{3} + \frac{2\theta^5}{15} + \frac{17\theta^7}{315}$$
can be used to approximate values of the tangent function when θ is in radians. Use the formula to approximate the value of tan 0.4 and compare this value with the one produced by a calculator.

In your own words

49. State the ratio formulas for each of the trigonometric functions.

50. Explain how to determine a side of a triangle given the other side and one angle. Be sure to include all possible geometric situations.

51. Explain how to determine an acute angle from the value of its sine. Use a particular example to illustrate your explanation.

7.4 APPLICATIONS OF TRIGONOMETRY

Using the definitions of the various trigonometric functions along with their numerical values, we can determine the unknown angles and sides of a right triangle from the known angles and sides. The process of determining all unknown sides and angles of a triangle is called **solving the triangle.**

Solving Triangles

The next examples illustrate how the trigonometric functions can be used to solve triangles when we start from various kinds of known information.

EXAMPLE 1
Solving a Right Triangle

The right triangle of Figure 1 has one acute angle and one side specified. Determine the measure of the other acute angle and the lengths of the other two sides.

Solution

Let's determine the various components one at a time. First, let's determine the angle α. The sum of the angles of a triangle is 180°. Therefore, because we know one acute angle and because the other angle is a right angle, we have:

$$\alpha = 180° - 90° - 38°41' = 51°19'$$

Next, let's determine the length of side a. This side is opposite the angle given as 38°41'. The side adjacent to the given angle is 55. Because we know the adjacent side and want to find the opposite side, we use a trigonometric function that involves both:

$$\tan = \frac{\text{opposite}}{\text{adjacent}}$$

$$\tan 38°41' = \frac{a}{55}$$

Using a calculator to determine the value of the tangent, we derive the equation:

$$\frac{a}{55} \approx 0.800673589$$

$$a \approx 44.037$$

Now that we know the lengths of two sides, we can determine the length of the hypotenuse c by applying the Pythagorean theorem:

$$c^2 = a^2 + 55^2$$
$$\approx (44.037)^2 + 55^2$$
$$\approx 4,964.27$$
$$c \approx 70.458$$

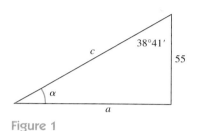

Figure 1

EXAMPLE 2
Solving a Right Triangle

The right triangle shown in Figure 2 has two sides specified. Determine the length of the unknown side and the measures of the two acute angles.

Solution

In the right triangle shown, the hypotenuse has length 190 and one side has length 100. Therefore, by the Pythagorean theorem, we have:

$$a^2 = 190^2 - 100^2$$
$$= 26,100$$
$$a = \sqrt{26,100} \approx 161.56$$

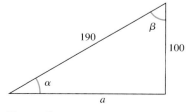

Figure 2

To find the measure of α, we can use the sine function:

$$\sin \alpha = \frac{100}{190} \approx 0.5263157895$$
$$\alpha \approx \sin^{-1} 0.5263157895$$
$$\approx 31.757°$$

Because the acute angles of a right triangle are complementary, we can determine the value of β from the value of α.

$$\alpha + \beta + 90° = 180°$$
$$\alpha + \beta = 90°$$
$$\beta = 90° - \alpha$$
$$= 58.243°$$

Applications

The process of solving a triangle can be used in many applied problems, as the next few examples illustrate.

EXAMPLE 3 *Surveying*

A surveyor wants to measure the distance from a point A on one bank of a river to a point B on the opposite bank. To do so, she stakes out a point C 100 yards from point A on the same side of the river, such that the line connecting C and A is perpendicular to the line connecting A and B. (See Figure 3.) She then uses a transit (a device for measuring angles) to determine that the angle formed by the lines AC and BC is $37°12'$. What is the distance between the two points A and B across the river from one another?

Solution

We use x to denote the distance we want to determine. (See Figure 3.) We can relate x to the known angle and known side through the tangent function:

$$\tan 37°12' = \frac{\text{opposite}}{\text{adjacent}} = \frac{x}{100}$$
$$0.75904 = \frac{x}{100}$$
$$x \approx 75.904$$

That is, the distance across the river is approximately 75.904 yards.

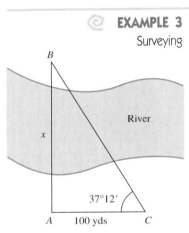

Figure 3

EXAMPLE 4 *Architecture*

A housing developer wants to create a triangular lot at the corner of two streets. The sides that form the corner are 125 feet and 160 feet. Assuming the angle at the corner is 90 degrees, determine the other two angles. (These angles are usually recorded in the land records of the local jurisdiction.) (See Figure 4.)

Solution

The angle α can be related to the two known legs of the triangle by the equation:

$$\tan \alpha = \frac{\text{opposite}}{\text{adjacent}}$$

SECTION 7.4 Applications of Trigonometry

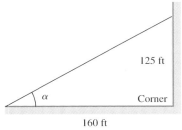

Figure 4

$$= \frac{125}{160}$$
$$\alpha = \tan^{-1} 0.78125$$
$$= 37.999°$$

The angle β is the complement of α. Therefore, we have:

$$\beta = 90° - \alpha$$
$$= 52.001°$$

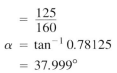

EXAMPLE 5 A surveyor wishes to determine the height of a building. He measures a distance
Surveying of 1,500 feet along the ground from the center of the base of the building, and he uses a transit to determine the angle of inclination from the end of the 1,500-ft line to the top of the building. This angle turns out to be 24°57′. How high is the building? (See Figure 5.)

Solution

Let the height of the building be denoted a. Then a is related to the given data of the problem by the formula:

$$\tan 24°57' = \frac{\text{opposite}}{\text{adjacent}}$$
$$= \frac{a}{1,500}$$
$$0.46524567 = \frac{a}{1,500}$$
$$a = 697.87 \text{ ft}$$

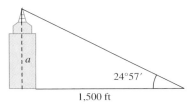

Figure 5

That is, the building is 697.87 feet tall.

HISTORICAL NOTE
The Mathematics of Surveying

The field of surveying, as it is practiced today, was founded by the German mathematician Karl Friedrich Gauss in the early nineteenth century. At that time, regular academic appointments paid very meager salaries. It was common for mathematicians to be supported by the patronage of wealthy noblemen. Gauss invented the modern field of surveying in response to his patron's request for assistance in determining the boundaries of his landholdings. The mathematical basis for modern surveying is exactly as Gauss set it out almost two centuries ago.

Gauss was one of the most important mathematicians of all time. A child prodigy, he already exhibited extraordinary mathematical talent by the age of five. When he was 16, he made a hobby of calculating the numbers of primes in various intervals of length 10,000. His tables go all the way up to 3,000,000. (Remember, Gauss had no calculator!)

Gauss' research included fundamental discoveries in number theory, astrophysics, electricity and magnetism, complex variables, and geometry.

For Gauss' two-hundredth birthday in 1977, the German government issued both a coin and a postage stamp in his honor.

Exercises 7.4

Solve the triangles in Exercises 1–4.

1.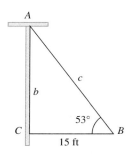

2.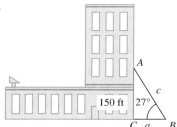

3.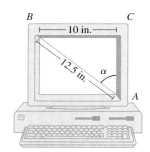

4.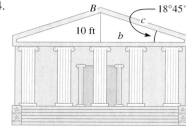

Solve these right triangles. Standard position and lettering are used.

5. $a = 3.5$, $b = 6.8$
6. $a = 11.7$, $b = 3.1$
7. $b = 23$, $c = 38$
8. $b = 2.3$, $c = 4.1$
9. $\alpha = 23°40'$, $b = 1.42$
10. $\beta = 74°10'$, $b = 21.3$
11. $\beta = 56°8'$, $a = 340$
12. $\alpha = 48°50'$, $a = 340.4$
13. $a = 32.1$, $b = 40.3$
14. $c = 103.5$, $a = 39.2$
15. $\alpha = 37°43'$, $a = 1{,}250$
16. $\beta = 58°22'$, $a = 24{,}000$
17. $a = 32.5$, $c = 63.4$
18. $b = 8.0$, $c = 14.0$

 Applications

19. **Surveying.** A surveyor wants to measure the distance from a point on one side of a river to a point directly across on the opposite side. She stakes out two points 100 ft apart so that they are on a line perpendicular to the distance she wants to determine. She then uses a transit to determine that the angle formed by the line through the two stakes and by the diagonal line that connects the second stake with the point across the river is $43°39'$. What is the distance between the two points directly across the river from each other?

20. **Architecture.** A housing developer wants to create a triangular lot at the corner, where the intersecting streets are perpendicular to each other. The sides that border the two sides of the corner are 135 ft long and 210 ft long. Determine the angles formed by the dimensions of the lot.

21. **Forestry.** The largest tree in the Sequoia National Park in California is called the General Sherman. (See figure.) To find its height, park rangers marked off a distance of 185 feet from its base. At that point, they found the angle of elevation to the top of the tree to be $55°50'$. How tall is the General Sherman?

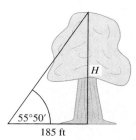

Exercise 21

22. **Pet rescue.** At a point 15 m from the base of a tree, a firefighter looks up and sees a cat stuck in the tree 35 m above ground level. The firefighter places the bottom of a ladder at his feet and extends it so it reaches the cat in the tree. How long is the extended ladder? What angle does the ladder make with the ground? What angle does the ladder make with the tree?

23. **Photography.** A photographer attaches her camera to a tripod. The camera must be 1.5 m from the floor to get the best shot, and the tripod legs are each 1.6 m long.

SECTION 7.4 Applications of Trigonometry

What is the diameter of the smallest circle on the floor that includes the three tripod legs?

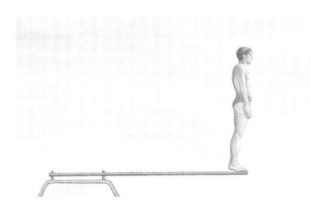

24. **Sports.** A diver 2 m tall stands on a diving board 3 m above the surface of the water and sees a spectator who is also 2 m tall standing at the other end of an Olympic-sized pool, 100 m long. How far is the diver from the spectator? What is the angle of elevation from the spectator to the diver?

25. **Navigation.** An airplane takes off from a runway at an angle of elevation of 4°20′. Two minutes later, the plane is at an altitude of 2,000 ft. How far is the airplane now from the point of takeoff?

26. **Navigation.** A 747 jet airliner takes off at an angle of elevation of 35° and flies a distance of 2,500 ft. What is its altitude?

27. **Military science.** An artillery battery spotter locates a target that is 6,480 yards west and 5,720 yards north of the present position. At what angle along the ground should the spotter fire on the target? What angle would the spotter still have to determine?

28. **Military science.** Paratroopers jump from a C-47 jet that travels at 145 mph at an altitude of 1,200 ft. In order for the paratroopers to land on target, they must jump when the angle of depression to the target is 24°10′. (See figure.) How far are they from the target horizontally when they jump? How far do they travel in the air before landing?

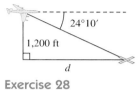

Exercise 28

29. **Measuring cloud height.** One can measure the height of clouds by pointing a bright floodlight straight up at the clouds. (See figure.) From a point 150 ft away from the floodlight, an angle of elevation is measured and found to be 72°50′. How high are the clouds?

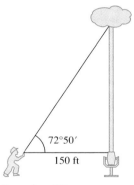

Exercise 29

30. **Meteorology.** Two tracking stations s miles apart measure the angles of elevation of a weather balloon, α and β. (See figure.) Derive a formula for the altitude h of the weather balloon in terms of s, α, and β.

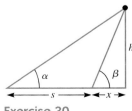

Exercise 30

31. **Geometry.** Find c and b in the following figure.

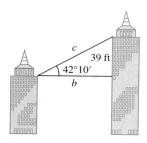

Exercise 31

32. **Geometry.** Find h in the following figure.

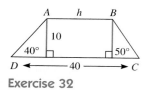

Exercise 32

33. **Civil engineering.** Engineers want to build a hanging bridge between two mountain peaks. The angle of elevation from the top of one peak to the top of the other is 18°40′. The taller peak is 2,400 m higher than the shorter one. How long will the bridge have to be?

34. **Civil engineering.** For safety reasons, civil engineers are taught that, when they build a road down a mountain, the angle of depression can be no more than 60°. (See figure.) The direct distance from point A to point B on a mountain is 4 miles, and point A is 2.5 miles vertically above point B. Can the engineers safely build a road straight down from A to B? What is the angle of depression α?

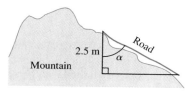

Exercise 34

35. **Home maintenance.** You have the task of putting up a new TV antenna. It should be 50 ft off the ground, and you will need three guy wires to hold it up. Each guy wire should be at an angle of 48° from level ground and should connect to the middle of the antenna. What is the total length of guy wire you will need?

36. **Geometry.** A regular octagon has sides 40.3 ft long and is inscribed within a circle. Find the radius of the circle.

37. **Geometry.** A regular hexagon is inscribed in a circle of radius 5.6 m. Find the perimeter of the hexagon.

38. **Geometry.** A regular pentagon has a perimeter of 70 yd and is inscribed in a circle. Find the radius of the circle.

39. **Military science.** The altimeter of a Navy reconnaissance helicopter records an altitude of 5,000 ft as it passes over its carrier. At that moment, it sights a submarine just under the surface. The angle of depression from the plane to the submarine is 25°40′. How far is it from the sub to the carrier?

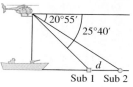

Exercises 39 and 40

40. **Military science.** Refer to Exercise 39. The helicopter spots a second submarine behind the first. The angle of depression to that submarine is 20°55′. (See figure.) How far apart are the two submarines (find d)?

41. **Aerial navigation.** Trigonometry has extensive application to aerial navigation. Pilots learn that directions are measured in degrees, clockwise from north. (See figure.) North is 0°, east is 90°, south is 180°, and west is 270°. Suppose an airplane leaves an airfield and flies 200 miles in a direction of 240°. How far south of the airfield is the plane? How far west of the airfield is the plane?

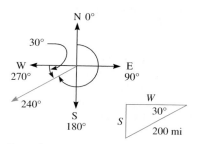

Exercise 41

42. **Aerial navigation.** From a hot air balloon that is 4,000 ft high, the angles of depression to two houses, in line with the balloon, are 46°30′ and 32°40′. How far apart are the houses?

43. **Aerial navigation.** The numbers on the end of a runway indicate the direction in which a plane takes off or lands. For example, in the movie *Airport*, the pilot is anxious to land on runway 29. This means that he would land at a direction of 290°, to the nearest 10°.

SECTION 7.4 Applications of Trigonometry

What would the direction be if the pilot were landing on the other end of this runway?

44. **Physics.** A degree can seem like a very small unit, but, at large distances, an error of one degree can be very significant. Suppose a laser beam is directed toward the visible center of the moon and that it misses its assigned target by 30 seconds. Approximate how far it is in miles from its assigned target. Assume the distance from the earth to the moon is 234,000 miles. Because the extreme distance makes variations in terrain negligible, assume the surface of the moon is flat.

45. **Surveying.** In the following figure, $\angle A_1 = \angle A_2 = 25.6°$, the distance from A_1 to A_2 is 425 ft, from A_1 to C_1 is 52 ft, and from D to C_2 is 79 ft. Find the distance from A_2 to D.

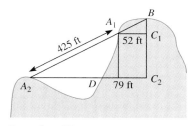

Exercise 45

46. **Navigation.** An aircraft carrier is due east of a destroyer. A lighthouse is 10 miles south of the destroyer. From the carrier, the bearing to the lighthouse is 54°28′ south of west. (This is the angle rotating from the west toward the south.) How far is the carrier from the destroyer? How far is the carrier from the lighthouse?

47. **Architecture.** The wood frame for the side of a house must be propped up on the foundation with a wood plank. (See figure.) The wood frame is 15 ft tall, the foundation is 1 ft higher than ground level, and the plank is 20 ft long. How far is the bottom end of the plank from the slab? What angle does the plank make with the ground?

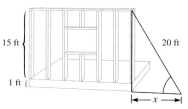

Exercise 47

48. **Carpentry.** A table is designed so that the leg is inclined at an angle of 75° to the top. (See figure.) The leg is to be 3 inches thick and 18 inches from the floor. How long a piece of wood is needed for the leg? (Find $x + y$.)

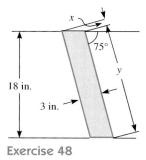

Exercise 48

In your own words

49. Suppose you are given two sides of a right triangle. How can you determine the hypotenuse?

50. Suppose you are given one acute angle of a right triangle. How can you determine the other angles?

51. Suppose you are given two sides of a right triangle. Which trigonometric function would you use to determine the angles of the triangle? Explain your answer.

52. Suppose you are given one side and the hypotenuse of a right triangle. Which trigonometric function would you use to determine the angle opposite the given side? Explain your answer.

53. Suppose you are given one side and the hypotenuse of a right triangle. Which trigonometric function would you use to determine the angle adjacent the given side? Explain your answer.

7.5 THE GRAPHS OF SINE AND COSINE

In this section, we study $\sin t$ and $\cos t$, with emphasis on their properties as functions of the variable t.

Sine and Cosine as Functions of a Real Number

For each angle measure t, we have defined the numbers $\sin t$ and $\cos t$. Let us assume that t is given in radians. Then t can be any real number. The correspondence

$$f(t) = \sin t$$

associates to the real number t the value $\sin t$, and is therefore a function of the real variable t. Similarly, the correspondence

$$g(t) = \cos t$$

associates to the real number t the value $\cos t$, and also is a function of the real variable t. Let us now study some properties of these functions.

Recall that the cosine function is defined as the x-coordinate of a certain point on the unit circle. Because x-coordinates of points on the unit circle lie between -1 and 1, we have the inequality:

$$-1 \leq \cos t \leq 1$$

Similarly, because the sine is defined as the y-coordinate of a certain point on the unit circle, we have the inequality:

$$-1 \leq \sin t \leq 1$$

These two inequalities say that the ranges of the sine and cosine functions are subsets of the interval $[-1, 1]$. In fact, because any real number in this interval can be the x-coordinate or y-coordinate of a point on the unit circle, every number in the interval $[-1, 1]$ is the value of $\cos t$ or $\sin t$ for some value of t. That is, the range of both the sine and cosine functions is precisely the interval $[-1, 1]$.

Sine and Cosine Graphs

Let's first sketch the graph of the function $\sin t$. To do this, we draw a unit circle and the coordinate system for the graph side by side. Imagine an angle t whose terminal side moves counterclockwise, beginning at the positive x-axis. For each position of the terminal side, we plot the vertical height of the point P shown in Figure 1. As the angle increases, the vertical height starts at 0 and increases to 1 when t equals $\pi/2$. The height then decreases from 1 to -1 as the angle goes from $\pi/2$ to $3\pi/2$. Finally, the height increases from -1 to 0 as the angle goes

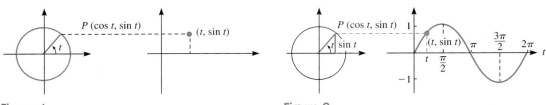

Figure 1

Figure 2

from $3\pi/2$ to 2π. The height of P equals the y-coordinate of P, which equals the value of $\sin t$. So the graph of the height of P, shown in Figure 2, is the graph of $\sin t$ for t in the interval $[0, 2\pi]$.

For angles t larger than 2π, we allow the angle in Figure 2 to continue rotating in the counterclockwise direction. We continue to plot the height of the endpoint P of the terminal side of the angle. In this way, we arrive at the graph of $\sin t$ for the remaining positive values of t. For negative values of t, we start the angle t at 0 and rotate it in the clockwise, or negative, direction and plot the height of P. This results in the portion of the graph in Figure 3 that lies to the left of the y-axis.

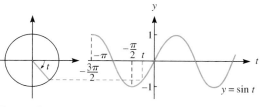

Figure 3

The horizontal position of a point on the unit circle equals the value of the cosine of the corresponding angle. So we can rotate the angle t and plot the horizontal position of the point P to sketch the graph of $\cos t$. The graph is shown in Figure 4.

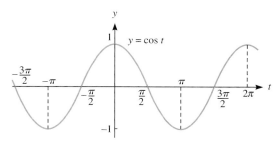

Figure 4

By examining the graphs of $\sin t$ and $\cos t$, we see that they repeat horizontally every 2π units. This is because the vertical and horizontal positions of P repeat after each revolution of 2π radians. This repetitive character of the sine and cosine functions can be expressed by the equations:

$$\sin(t + 2\pi) = \sin t$$
$$\cos(t + 2\pi) = \cos t$$

These equations represent the fact that the values of the sine and cosine functions repeat every 2π units. Mathematicians say that the sine and cosine functions are **periodic** and have the **period** 2π.

Graphing Sine and Cosine Functions on a Calculator

We can graph trigonometric functions using a graphing calculator. The procedure is exactly the same as for graphing any other function. There are three cautions, however. First, we must correctly enter the functions by using the keys [SIN], [COS], and [TAN]. Second, we must be sure that the angle mode is set on radians. Third, we must choose an appropriate x-range. For example, suppose we wish to graph $y = \sin x$ over the interval $2\pi \leq x \leq 2\pi$. To set the x-range, note that $2\pi \approx 6.28$. So we set Xmin equal to -6.28 and Xmax equal to 6.28. In order to make the graph fill the screen vertically, we set Ymin equal to -1 and Ymax equal to 1. We key the equation into the Y = menu by first pressing [SIN] and then [X]. The displayed graph is shown in Figure 5.

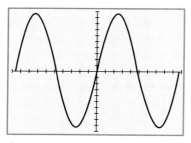

Figure 5

Periodic Phenomena

Nature provides us with many examples of repetitive phenomena. For example, the length of the day in a given location varies with the time of year, but those lengths repeat themselves each year. Sound of a particular frequency consists of vibrations repeated a certain number of times per second. The heights of the tides vary on a 28-day cycle that depends on the position of the moon relative to the earth. Blood pressure varies in a cycle determined by respiration.

These examples illustrate the importance of repetitive phenomena. We must be able to describe such phenomena mathematically and to solve problems concerning them. Repetitive phenomena can be described using functions of the form

$$f(x) = a\sin(bx + c), \qquad g(x) = a\cos(bx + c)$$

where a, b, and c are real constants.

In the rest of this section, we study such functions and their graphs, and we learn how they can be used to describe repetitive phenomena.

Throughout the rest of this section, we consider the trigonometric functions as functions of a real variable. To emphasize that the variable may or may not arise as the measure of an angle, we denote it by x, rather than t.

Periods of Sine and Cosine Functions

Let's begin by providing a mathematical description of a repetitive phenomenon.

Definition
Period of a Function

We say that the function $f(x)$ is **periodic** provided that there is a real number c such that

$$f(x + c) = f(x)$$

for all real numbers x. We say that c is a **period** of f.

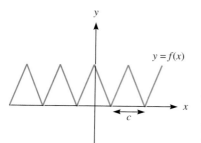

Figure 6

As an example of a periodic function, consider

$$f(x) = \sin x$$

Earlier in this section, we saw that

$$\sin(x + 2\pi) = \sin x$$

for all real numbers x. In other words, $\sin x$ is periodic, with a period of 2π.

Suppose that $f(x)$ has period c. Then the value of $f(x)$ repeats if x is increased by c units. In particular, the graph of $f(x)$ repeats every c units along the x-axis. Figure 6 shows the repetitive nature of the graph of a periodic function.

SECTION 7.5 The Graphs of Sine and Cosine

If c is a period of f, then so is $2c$, because:
$$f(x + 2c) = f((x + c) + c) = f(x + c) = f(x)$$
Similarly, any integer multiple of c is also a period of f. Among the periods of f, the least positive one is called the **fundamental period**. From the geometric way in which we have defined the sine and cosine, it is clear that these functions don't have a period less than 2π, so 2π is the fundamental period.

From our knowledge of the periodicity of the sine and cosine functions, we can deduce the following result.

PERIODS OF GENERAL SINE AND COSINE FUNCTIONS

Let b be a nonzero real number and let a and c be any real numbers. The functions
$$f(x) = a\sin(bx + c), \qquad g(x) = a\cos(bx + c)$$
have period $2\pi/b$.

Proof We prove the result for the function $f(x) = a\sin(bx + c)$. If x is any real number, then
$$f\left(x + \frac{2\pi}{b}\right) = a\sin\left(b\left(x + \frac{2\pi}{b}\right) + c\right)$$
$$= a\sin\left(bx + b \cdot \frac{2\pi}{b} + c\right)$$
$$= a\sin(bx + c + 2\pi)$$
$$= a\sin(bx + c) \qquad \text{Because } \sin x \text{ has period } 2\pi$$

That is, whenever x is increased by $2\pi/b$, the value of $f(x)$ is unchanged. So $f(x)$ has period $2\pi/b$. The proof of the theorem for the function $g(x) = a\cos(bx + c)$ is similar, so it is omitted.

EXAMPLE 1
Calculating Periods

Determine the periods of the following functions.
1. $\sin(3x)$
2. $\sin(2\pi x)$
3. $\sin\left(\frac{x}{3}\right)$
4. $\cos\left(\frac{5\pi x}{3}\right)$

Solution

In each example, the period is equal to $2\pi/b$, where b is the coefficient of x.

1. $\dfrac{2\pi}{b} = \dfrac{2\pi}{3}$
2. $\dfrac{2\pi}{b} = \dfrac{2\pi}{2\pi} = 1$
3. $\dfrac{2\pi}{b} = \dfrac{2\pi}{1/3} = 6\pi$
4. $\dfrac{2\pi}{b} = \dfrac{2\pi}{5\pi/3} = \dfrac{6}{5}$

The value of the function $f(x) = \sin bx$ oscillates between -1 and 1, with period $2\pi/b$. The next example illustrates the procedure for graphing such functions.

EXAMPLE 2
Graphing General Sine Functions

Sketch the graphs of the following functions.

1. $\sin(2x)$
2. $\sin(2\pi x)$
3. $\sin\left(\dfrac{x}{3}\right)$

Solution

1. The period of $\sin(2x)$ is $2\pi/b = 2\pi/2 = \pi$, which is half the period of $\sin x$. This means that every cycle on the graph of $\sin x$ corresponds to two cycles on the graph of $\sin(2x)$. The graphs of both functions are sketched in Figure 7. This can be easily checked by graphing both functions simultaneously using a graphing calculator.

2. The period of $\sin(2\pi x)$ is $2\pi/b = 2\pi/2\pi = 1$. Its graph is sketched in Figure 8.

3. The period of $\sin(x/3)$ equals $2\pi/b = 2\pi/(1/3) = 6\pi$. This period is three times the period of $\sin x$. Figure 9 shows the graphs of both functions.

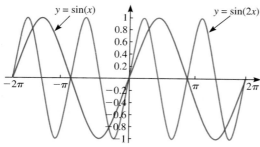

Figure 7

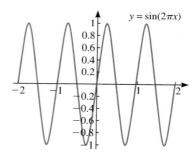

Figure 8

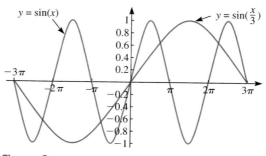

Figure 9

It takes some getting used to in order to be able to correctly identify the periods of trigonometric graphs displayed on a calculator. To become proficient at this, try to match the period in decimal notation with a rational multiple of π (such as π, 2π, $\pi/2$, and so on).

SECTION 7.5 The Graphs of Sine and Cosine

Amplitude of Sine and Cosine Functions

The graphs of the functions $a\sin(bx)$ and $a\cos(bx)$ can be obtained by scaling the graphs of $\sin(bx)$ and $\cos(bx)$, respectively, by a factor $|a|$ and, in case a is negative, reflecting the graphs in the x-axis.

Definition **Amplitude of a General Sine or Cosine Graph**	The number $|a|$ is called the **amplitude** of the graph. Geometrically, the amplitude equals one half the height of the graph or, equivalently, the amount the graph rises above (or falls below) the x-axis. Figure 10 illustrates these interpretations of the amplitude.

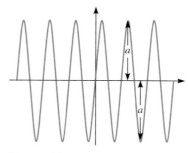

Figure 10

The next few examples illustrate the use of the amplitude in graphing.

EXAMPLE 3 Graphing a General Sine Function

Sketch the graph of the function:
$$y = 2\sin(3x)$$

Solution

The period of the function is $2\pi/b = 2\pi/3$. Moreover, the amplitude equals 2, so the function values range from -2 to 2. We sketch the graph by first drawing one cycle, in the shape of a sine curve, for all x in the interval $[0, 2\pi/3]$. The graph of this cycle corresponds to the first period. We then sketch the remainder of the graph by repeating the portion just sketched. See Figure 11.

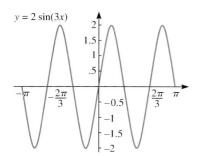

Figure 11

EXAMPLE 4 Graphing Another General Sine Function

Sketch the graph of the function:
$$y = -2\sin(6x)$$

Solution

The period of the function is $2\pi/b = 2\pi/6 = \pi/3$. The amplitude is 2, so the function values range between -2 and 2. Because the value of a is negative, the graph resembles the graph of a sine curve reflected in the x-axis. First, we sketch one period of the graph for all x in the interval $[0, \pi/3]$. Then, we sketch the remainder of the graph by repeating the portion already sketched. See Figure 12.

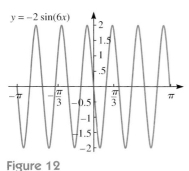

Figure 12

EXAMPLE 5
Graphing a General Cosine Function

Sketch the graph of the function:
$$y = 4\cos(2\pi x)$$

Solution

The period in this example is $2\pi/2\pi = 1$. The amplitude is 4. The graph resembles the graph of $\cos x$, except that one cycle corresponds to a distance of 1 along the x-axis, instead of a distance of 2π. We sketch one cycle of the graph for all x in the interval [0, 1]. Then, we complete the sketch by repeating, to the right and the left, the portion already sketched. See Figure 13.

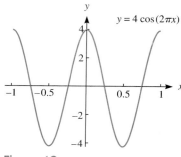

Figure 13

Applications of Sine and Cosine Functions

Many kinds of motion are periodic in nature. Such motion involves a pattern that is repeated at fixed time intervals, called **harmonic motion.** Examples of harmonic motion are pendulums and objects suspended from springs. Harmonic motion can be described using trigonometric functions, as the following example illustrates.

EXAMPLE 6
Harmonic Motion

A block is attached to a hanging spring and set in motion by stretching the spring from its resting position and releasing it. At time t seconds after the block is set in motion, the distance of the block $D(t)$ from its resting position is given by the formula
$$D(t) = 7\cos(6\pi t)$$
where $D(t)$ is measured in inches. (See Figure 14.)

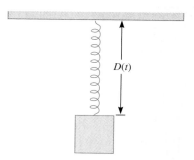

Figure 14

1. How many bounces per second does the block execute?
2. How far does the block move between the extreme ends of a bounce?
3. Sketch the graph of $D(t)$.

Solution

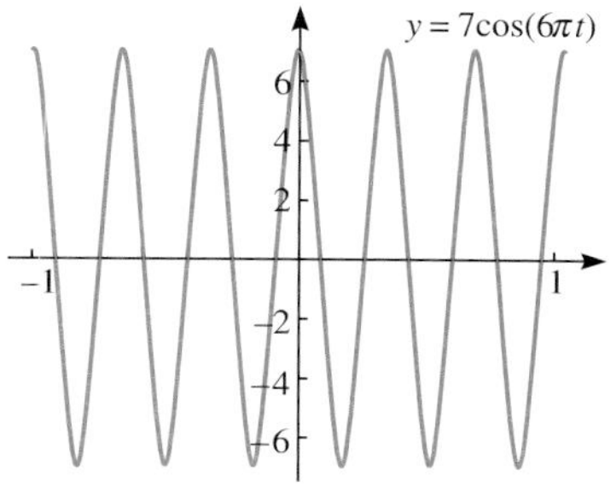

1. The period of the function $D(t)$ is $2\pi/b = 2\pi/6\pi = 1/3$. This means that each bounce of the block takes 1/3 second. Or, equivalently, there are 3 bounces per second.
2. The extreme ends of a bounce are located at a distance $|a|$ in both the positive and the negative directions from the resting position of the block, where $|a|$ is the amplitude of $D(t)$. From the given formula, the amplitude equals 7. So the distance traveled in one bounce equals the distance from 7 down to 0 plus the distance from 0 down to -7. That is, the distance equals $2 \cdot 7$, or 14 inches.
3. The graph of $D(t)$ is sketched in Figure 15.

Figure 15

EXAMPLE 7
Cyclical Animal Populations

If an animal population is in perfect ecological balance, its size is periodic, with the period depending on environmental factors, such as the size of predator populations, food supply, and so forth. Suppose that the population $P(t)$ of deer on an island at time t (in years) is given by the function:

$$P(t) = 500 + 100 \cos\left(\frac{\pi t}{3}\right)$$

1. Graph the function $P(t)$.
2. What is the maximum population of deer?
3. What is the minimum population of deer?
4. What is the time required for the deer population to go through one cycle?

Solution

1. The function $f(x) = 100 \cos(\pi t/3)$ has amplitude 100 and period:

$$\frac{2\pi}{\pi/3} = 6$$

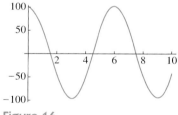

Figure 16

The graph of $f(x)$ is shown in Figure 16. The graph of $P(t)$ is obtained by translating the graph of $f(x)$ upward 500 units. The graph of $P(t)$ is shown in Figure 17.

2. The graph of $P(t)$ oscillates about the horizontal line $y = 500$. The maximum height is 100 units above the line, so the maximum deer population is $500 + 100 = 600$.
3. The minimum height of the graph is 100 units below the line, so the minimum deer population is $500 - 100 = 400$.
4. The length of a cycle is one period of $P(t)$, or 6 years. That is, the population of deer oscillates from 500 up to 600, then from 600 down to 400, then from 400 back up to 500 in a period of 6 years.

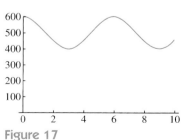

Figure 17

Exercises 7.5

Match each of the functions in Exercises 1–8 with its graph in (a)–(h).

1. $f(t) = 2 \sin t$
2. $f(t) = 3 \cos t$
3. $f(t) = \sin \dfrac{t}{2}$
4. $f(t) = 2 \cos 3t$
5. $f(t) = 3 \cos 2t$
6. $f(t) = \sin\left(t + \dfrac{\pi}{4}\right)$
7. $f(t) = 2 \cos t + 3$
8. $f(t) = -\sin 2t$

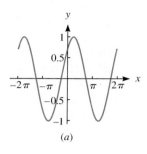
(a)

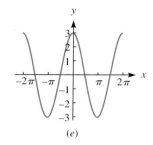
(e)

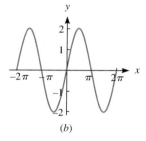
(b)

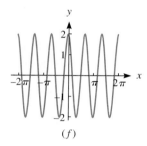
(f)

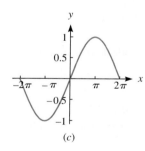
(c)

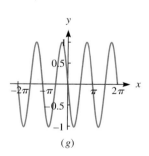
(g)

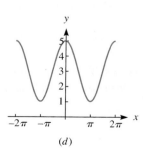
(d)

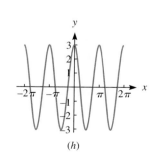
(h)

Each of the following graphs shows at least one period of a periodic function. What is the period?

9.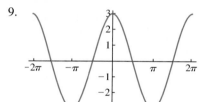

10.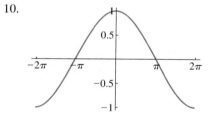

11.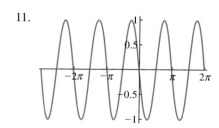

12.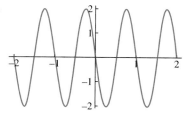

13.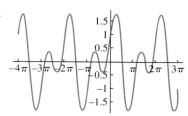

14.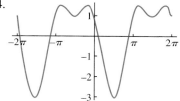

For each of the functions in Exercises 15–34, do a–c.
   a. Find the period.
   b. Find the amplitude.
   c. Sketch the graph.

15. $f(x) = \cos x$
16. $f(x) = \sin x$
17. $f(x) = \sin\left(\frac{1}{2}x\right)$
18. $f(x) = \cos(2x)$
19. $f(x) = \cos(2\pi x)$
20. $f(x) = \sin(\pi x)$
21. $f(x) = \cos(3x)$
22. $f(x) = \cos\left(\frac{1}{4}x\right)$
23. $f(x) = 2\sin x$
24. $f(x) = 3\cos x$
25. $f(x) = -\frac{1}{2}\cos x$

26. $f(x) = -2\sin x$
27. $f(x) = 2\cos(2x)$
28. $f(x) = 3\sin(3x)$
29. $f(x) = 4\sin(2\pi x)$
30. $f(x) = 2\cos(\pi x)$
31. $f(x) = -3\cos(\pi x)$
32. $f(x) = -0.25\sin(6\pi x)$
33. $f(x) = -\sin\left(\frac{2\pi x}{3}\right)$
34. $f(x) = 6\sin\left(\frac{\pi x}{4}\right)$

Which of the graphs in Exercises 35–40 represents a periodic function? For those that are periodic, determine the period.

35.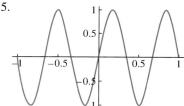

36.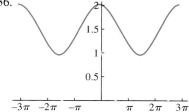

37.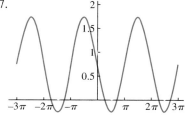

38.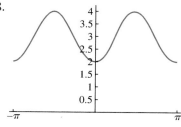

39.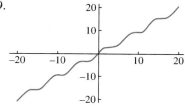

40.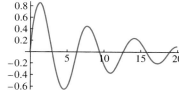

Determine a function of the form $f(x) = A \sin kx$ that has the following properties.

41. Amplitude = 2, period = π.
42. Amplitude = 4, period = $\pi/2$.
43. Amplitude = 3, $f(x) = f(x + 2\pi)$.
44. Amplitude = 1, $f(x) = f(x - 3\pi)$.
45. Amplitude = 2, period = 1.
46. Amplitude = 0.5, period = 2.

Determine a function of the form $f(x) = A \cos kx$ that has the following properties.

47. Amplitude = 1, period = π.
48. Amplitude = 1, period = 3π.
49. $f(0) = -1$, period = π.
50. $f(0) = 3\pi$, period = 1.
51. Amplitude = 2, $f(x) = f(x - 2\pi)$.
52. Amplitude = 3, $f(x) = f(x+1)$.

388 CHAPTER 7 The Trigonometric Functions

 Applications

53. **Physics.** A weight is attached to a spring and set in motion by stretching the spring from its resting position and releasing it. At time t seconds after it is set in motion, the distance of the block $D(t)$ from its resting position is given by

$$D(t) = 5\sin(4\pi t)$$

where $D(t)$ is measured in inches.

  a. How many bounces per second does the weight execute?
  b. How far does the block move between extreme ends of a bounce?
  c. What are the maximum and minimum extremes of the bounces?
  d. Sketch the graph of $D(t)$.

54. **Physics.** At time t seconds, the alternating electric current i (in amperes) in a wire is given by:

$$i(t) = 50\sin(60\pi t)$$

  a. Find the amplitude and period of this function.
  b. Sketch a graph of the function.
  c. How many cycles does the current make in one second?

 Technology

55. On the same coordinate system, plot the graphs of $y_1 = \sin x$, $y_2 = \sin 2x$, and $y_3 = \sin 3x$. Choose the x-range so that at least two periods of each function are displayed. Explain the relationships among the graphs.

56. On the same coordinate system, plot the graphs of $y_1 = \cos x$, $y_2 = \cos 2x$, and $y_3 = \cos 3x$. Choose the x-range so that at least two periods of each function are displayed. Explain the relationships among the graphs.

57. On the same coordinate system, plot the graphs of $y = \sin x$ and $y = 2\sin x$. Choose the y-range so that the graphs just fill the screen vertically. Explain the relationship between the graphs.

58. On the same coordinate system, plot the graphs of $y = \cos x$ and $y = 2\cos x$. Choose the y-range so that the graphs just fill the screen vertically. Explain the relationship between the graphs.

59. On the same coordinate system, plot the graphs of $y = \sin x$ and $y = -\sin x$. Choose the y-range so that the graphs just fill the screen vertically. Explain the relationship between the graphs.

60. On the same coordinate system, plot the graphs of $y = \sin x$, $y = -\sin x$, and $y = -2\sin x$. Choose the y-range so that the graphs just fill the screen vertically. Explain the relationships among the graphs.

 In your own words

61. Explain what is meant by a periodic function. Give examples of a function that is periodic and one that is not periodic.

62. Define the amplitude of a sine or cosine graph. Illustrate the concept with two examples.

63. Give two examples of periodic phenomena in the world around us. Explain the physical significance of the periodicity.

7.6 MORE ABOUT THE SINE AND COSINE FUNCTIONS

In this section, we use graphical analysis to further explore the sine and cosine functions.

Zeros of the Sine and Cosine Functions

By examining the graph of the sine function, we see that $\sin x$ has zeros at $x = 0$ and at $x = \pi$, within the single period $[0, 2\pi]$. Recall that the periodicity of the sine means $\sin x = \sin(x + 2\pi)$. We use this equation to deduce that the sine function has zeros for the following equations in x, where k is an integer:

$$x = 0 + 2k\pi \qquad x = \pi + 2k\pi$$
$$= 2k\pi \qquad\qquad = \pi(2k+1), \qquad k = 0, \pm 1, \pm 2, \ldots$$

SECTION 7.6 More About the Sine and Cosine Functions

The first collection of values for x consists of just the even multiples of π, and the second collection of values for x consists of the odd multiples of π. So the zeros of $\sin x$ consist of all integer multiples of π:

$$x = k\pi, \qquad k = 0, \pm 1, \pm 2, \ldots$$

Similarly, by examining the graph of $\cos x$, we see that, in the single period $[0, 2\pi]$, the function $\cos x$ has the zeros $x = \pi/2$ and $x = 3\pi/2$. Therefore, by periodicity, the zeros of the cosine function are defined by the following equations in x, where k is an integer.

$$x = \frac{\pi}{2} + 2k\pi = \frac{(4k+1)\pi}{2}$$

$$x = \frac{3\pi}{2} + 2k\pi = \frac{(4k+3)\pi}{2}$$

$$k = 0, \pm 1, \pm 2, \ldots$$

These zeros can be listed as follows:

$$x = \cdots, -\frac{7\pi}{2}, -\frac{3\pi}{2}, \frac{\pi}{2}, \frac{5\pi}{2}, \cdots$$

$$x = \cdots, -\frac{5\pi}{2}, -\frac{\pi}{2}, \frac{3\pi}{2}, \frac{7\pi}{2}, \cdots$$

Together, these consist of all odd, integer multiples of $\pi/2$:

$$x = \cdots, -\frac{7\pi}{2}, -\frac{5\pi}{2}, -\frac{3\pi}{2}, -\frac{\pi}{2}, \frac{\pi}{2}, \frac{3\pi}{2}, \frac{5\pi}{2}, \frac{7\pi}{2}, \cdots$$

We can summarize these results in the following theorem:

ZEROS OF THE GENERAL SINE AND COSINE FUNCTIONS

1. The zeros of the function $\sin x$ are the integer multiples of π.
2. The zeros of the function $\cos x$ are the odd, integer multiples of $\pi/2$.

The accompanying table summarizes the facts we have determined about the sine and cosine functions:

	$\sin t$	$\cos t$
Domain	all real numbers t	all real numbers t
Range	$[-1, 1]$	$[-1, 1]$
Zeros	integer multiples of π	odd, integer multiples of $\frac{\pi}{2}$

Phase Shift In Section 7.6, we introduced the general sine and cosine functions:

$$f(x) = a\sin(bx + c)$$
$$g(x) = a\cos(bx + c)$$

Such functions arise in many applications, and are simply translations of the functions $a \sin bx$ and $a \cos bx$, respectively. To see why, let's first consider the function $f(x)$ and write it in the form:

$$f(x) = a\sin(bx + c)$$
$$= a\sin\left(b\left[x - \left(-\frac{c}{b}\right)\right]\right)$$

We see that $f(x)$ can be obtained by replacing x with $x - (-c/b)$ in the function $f_1(x) = a\sin(bx)$. Recall our general discussion in Section 2.6, in which we showed that the graph of $f(x - h)$ can be obtained by translating the graph of $f(x)$ horizontally by h units. If h is positive, the translation is to the right; if h is negative, the translation is to the left.

In the case of the function $f(x)$, the value of h is $-c/b$, and is called the **phase shift** of $f(x)$.

PHASE SHIFT OF A GENERAL SINE OR COSINE FUNCTION

To obtain the graph of $f(x) = a \sin(bx + c)$, we shift the graph of $f_1(x) = a \sin(bx)$ horizontally by the phase shift:

$$h = -\frac{c}{b}$$

To obtain the graph of $g(x) = a \cos(bx + c)$, we shift the graph of $g_1(x) = a \cos(bx)$ horizontally by the phase shift:

$$h = -\frac{c}{b}$$

In the next two examples, we show how to use the phase shift to sketch graphs.

EXAMPLE 1
A Function with a Phase Shift

Consider the function:

$$f(x) = 3 \sin\left(2x - \frac{\pi}{3}\right)$$

1. Determine the amplitude, period, and phase shift.
2. Sketch the graph of the function.

Solution

1. In this example, we have:

$$a = 3, \qquad b = 2, \qquad c = -\frac{\pi}{3}$$

$$\text{amplitude} = a = 3$$

$$\text{period} = \frac{2\pi}{b} = \frac{2\pi}{2} = \pi$$

$$\text{phase shift} = -\frac{c}{b} = -\frac{-\pi/3}{2} = \frac{\pi}{6}$$

2. We start from the graph of $\sin x$. First, we scale vertically by the amplitude, which is 3, to obtain the graph of $3 \sin x$. The scaled graph varies vertically from -3 to 3. Next, we scale horizontally by a factor of $\frac{1}{2}$ to obtain the function $3 \sin 2x$. This compresses the graph horizontally so that a period fits in half of its original horizontal space. Finally, we translate the graph to the right by the phase shift $\pi/6$ to obtain the graph of $3 \sin 2(x - \pi/6)$. The completed sketch is shown in Figure 1.

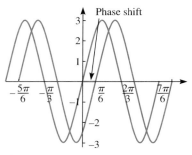

Figure 1

Vertical Translations

We now know how to graph functions of the form:

$$f(x) = a \sin(bx + c)$$
$$g(x) = a \cos(bx + c)$$

By adding constants, we arrive at even more general functions:

$$F(x) = a \sin(bx + c) + d$$

SECTION 7.6 More About the Sine and Cosine Functions

$$G(x) = a\cos(bx + c) + d$$

From our general discussion of Chapter 2, the graphs of $F(x)$ and $G(x)$ can be obtained by translating the graphs of f and g vertically by d units. The next example illustrates how this is done.

EXAMPLE 2
A Sine Function with Vertical Translation

Sketch the graph of the function:
$$F(x) = 3\sin 2x + 2$$

Solution

To obtain the graph of $F(x)$, we shift the graph of $f(x) = 3\sin 2x$ upward by 2 units. The graphs of both $f(x)$ and $F(x)$ are shown in Figure 2.

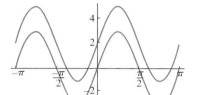

Figure 2

Applications Let's now turn to some applications of general sine and cosine functions.

EXAMPLE 3
Blood Pressure

A person's blood pressure changes in a rhythmic fashion, with a periodicity set by the beating of the heart. Suppose that a person's blood pressure at time t minutes is given by the formula:

$$P(t) = 100 + 25\sin(160\pi t)$$

1. Determine the rate at which the person's heart is beating.
2. Sketch the graph of $P(t)$.
3. What are the maximum and minimum blood pressure readings?

Solution

1. The function $P(t)$ is periodic with period equal to $2\pi/b = 2\pi/(160\pi) = \frac{1}{80}$. That is, each period of the function lasts for $\frac{1}{80}$ minute. Because each period corresponds to one heartbeat (that is, one complete blood pressure cycle), the rate at which the heart is beating is 80 beats per minute.
2. The function $P(t)$ is obtained by adding the constant 100 to the trigonometric function $25\sin(160\pi t)$. From our general principles of graphing developed in Chapter 2, the graph of $P(t)$ can be obtained by first graphing the trigonometric function and then translating the graph upward by 100 units. The graph is shown in Figure 3.

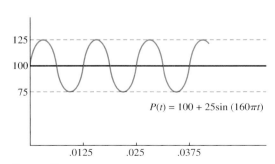

Figure 3

3. The amplitude of the trigonometric function $25\sin(160\pi t)$ is $|a| = 25$. By examining the graph in Figure 3, we see that the value of $P(t)$ oscillates between $100 - 25 = 75$ and $100 + 25 = 125$. That is, the minimum blood pressure is 75 and the maximum blood pressure is 125.

EXAMPLE 4
Number of Hours of Daylight

The number of hours $H(D)$ of daylight on a day D of the year is given by the function

$$H(D) = 12 + A\sin\left[\frac{2\pi}{365}(D - 80)\right]$$

where A depends on the latitude of the geographical location. (Note that the 80 corresponds to March 21, the vernal equinox, on which the length of day and night are equal. Leap year is ignored.) For New Orleans, the constant A is about 2.3.

1. How many hours of daylight are there on August 9?
2. How many hours of daylight are there on December 1?
3. What is the period of $H(D)$?
4. Sketch the graph of $H(D)$.
5. Which day has the most hours of daylight?
6. Which day has the fewest hours of daylight?

Solution

1. The value of D that corresponds to August 9 is found by adding the number of days in the months January through July:

$$31 + 28 + 31 + 30 + 31 + 30 + 31 + 9 = 221$$

So the number of hours of daylight in New Orleans on Aug. 9 is

$$H(221) = 12 + 2.3\sin\left[\frac{2\pi}{365}(221 - 80)\right]$$
$$= 13.507 \text{ hr}$$

2. The value of D that corresponds to December 1 is:

$$365 - 30 = 335$$

So the number of hours of daylight in New Orleans on Dec. 1 is:

$$H(221) = 12 + 2.3\sin\left[\frac{2\pi}{365}(335 - 80)\right]$$
$$= 9.8188 \text{ hr}$$

3. The period of $H(D)$ is:

$$\frac{2\pi}{2\pi/365} = 365 \text{ days}$$

That is, the number of hours of daylight repeats on a yearly basis.

4. The amplitude of $H(D)$ is 2.3, the period is 365, and the phase shift is:

$$-\frac{c}{b} = -\frac{-80\cdot(2\pi/365)}{(2\pi/365)} = -80$$

Moreover, the vertical translation is 12 units. The graph is shown in Figure 4.

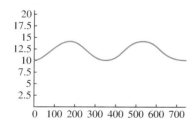

Figure 4

5. The day with the most hours of daylight corresponds to the value of D for which the sine function is 1. That is, D must satisfy:

$$\sin\left[\frac{2\pi}{365}(D-80)\right] = 1$$

The only value within the period $[0, 2\pi]$ for which $\sin x = 1$ is $x = \pi/2$. So the above equation for D implies that:

$$\frac{2\pi}{365}(D-80) = \frac{\pi}{2}$$

$$D - 80 = \frac{365}{4}$$

$$D = 171.25$$

So the day with the most hours of daylight is the 172nd day of the year, or June 21.

6. The day with the fewest hours of daylight corresponds to the value of D for which the sine function is -1. That is, D must satisfy:

$$\sin\left[\frac{2\pi}{365}(D-80)\right] = -1$$

The only value within the period $[0, 2\pi]$ for which $\sin x = -1$ is $x = 3\pi/2$. So the above equation for D implies that:

$$\frac{2\pi}{365}(D-80) = \frac{3\pi}{2}$$

$$D - 80 = \frac{365 \cdot 3}{4}$$

$$D = 353.75$$

So the day with the fewest hours of daylight is the 354th day of the year, or December 21.

Exercises 7.6

For the functions in Exercises 1–14, determine a–d.
   a. Find the period.
   b. Find the amplitude.
   c. Find the phase shift.
   d. Sketch the graph.

1. $f(x) = \sin\left(x - \frac{\pi}{2}\right)$
2. $f(x) = \sin\left(x + \frac{\pi}{4}\right)$
3. $f(x) = 3\cos(x + \pi)$
4. $f(x) = -2\sin(x - \pi)$
5. $f(x) = -2\sin(4x + \pi)$
6. $f(x) = -2\sin\left(x + \frac{3\pi}{2}\right)$
7. $f(x) = 3\cos\left(2x - \frac{\pi}{3}\right)$
8. $f(x) = 4\cos(\pi x - 2)$
9. $f(x) = \frac{3}{2}\sin(2\pi x + 1)$
10. $f(x) = \frac{5}{2}\cos(2\pi x + 2)$
11. $f(x) = -2 + \sin x$
12. $f(x) = 3 - 2\cos x$
13. $f(x) = 1 + 2\sin(\pi x)$
14. $f(x) = -3 + \frac{1}{2}\cos(2x - \pi)$

For each of the functions in Exercises 15–20, find the following:
   a. The period.
   b. The amplitude.
   c. The phase shift.

15. $f(x) = -6\cos(2x - \pi)$
16. $f(x) = \frac{1}{3}\sin\left(-3x - \frac{3\pi}{4}\right)$
17. $f(x) = 5\cos(4\pi x + 2)$
18. $f(x) = -\frac{1}{6}\sin\left(2x + \frac{2\pi}{3}\right)$
19. $f(x) = -\frac{2}{3}\sin\left(3x + \frac{\pi}{4}\right)$
20. $f(x) = 4\sin\left(3\pi x - \frac{5\pi}{4}\right)$

For Exercises 21–26, find a function of the type $f(x) = a\sin(bx + c)$ or $f(x) = a\cos(bx + c)$ (depending on which is requested) that corresponds to the given graph or that satisfies the given condition.

21.
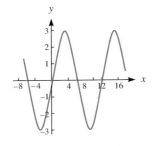

22.
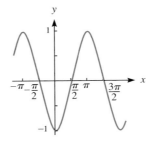

23.
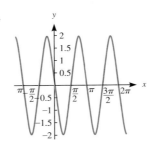

24.
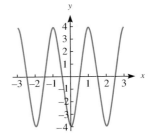

25. sin: amplitude = 4.3, period = 7, phase shift = −10.

26. cos: amplitude = 2.8, period = 1, phase shift = 6.

Applications

27. **Medicine.** A person's blood pressure changes in a rhythmic fashion, with a periodicity set by the beating of the heart. Suppose a person's blood pressure at time t minutes is given by the formula:

$$P(t) = 101 + 24\cos(160\pi t)$$

   a. Sketch the graph of $P(t)$.

   b. Determine the rate at which the person's heart is beating.

   c. What are the maximum and minimum blood pressure readings?

28. **Medicine.** The temperature of a patient during a 12-day illness is given by:

$$T(t) = 101.6° + 3\sin\left(\frac{\pi}{8}t\right)$$

   a. What are the maximum and minimum temperatures of the patient during the illness?

   b. Sketch a graph of the function over the interval [0, 12].

29. **Engineering.** See the following figure. A satellite circles the earth in such a way that it has a horizontal distance of y miles from the equator (we can ignore its height above the earth) t minutes after its launch, where:

$$y(t) = 5,000\left[\cos\frac{\pi}{45}(t - 10)\right]$$

   a. What is the period of y?

   b. What is the maximum distance the satellite can be from the equator?

   c. Sketch a graph of $y(t)$.

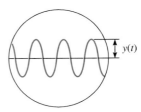

Exercise 29

30. **Sales fluctuations.** A company in a northern climate has sales of skis as given by the formula

$$S(t) = 7\left[1 - \cos\left(\frac{\pi}{6}t\right)\right]$$

where t is time, in months, and $t = 0$ corresponds to July 1, and $S(t)$ is in thousands of dollars.

a. Sketch a graph of $S(t)$ over a two-year interval [0, 24].
b. What is the minimum amount of sales and when does it occur?
c. What is the maximum amount of sales and when does it occur?
d. What is the period of the function? Does this seem reasonable? Explain.

31. **Recreation.** See the following figure. A roller coaster is constructed in such a way that its height H, in meters, x meters horizontally from the point shown in the figure, is given by:

$$H(x) = 20 \sin\left(\frac{\pi}{60}x - 2\pi\right)$$

a. Sketch a graph of $H(x)$ over the interval [0, 480].
b. What are the maximum and minimum heights of the function?
c. What is the period of the function?

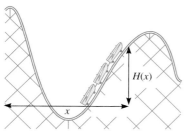

Exercise 31

32. **Recreation.** See the following figure. You are riding on a Ferris wheel. At time t, in seconds, you are at height h, in feet, where:

$$h(t) = 53 - 50 \sin\left[\frac{\pi}{5}\left(t - \frac{5}{2}\right)\right]$$

a. Find the height of the Ferris wheel when $t = 0, 2, 3, 6, 8, 9,$ and 10 seconds.
b. What is the lowest height of the wheel?
c. What is the maximum height of the wheel?
d. How long does it take for the wheel to go around once?
e. Sketch a graph of the function.

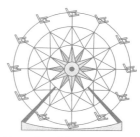

Exercise 32

33. **Ecology.** In the Hudson Bay area, the populations of lynx (a member of the cat family) and of the snowshoe hare (a member of the rabbit family) interact. When the population of prey, the hare, is high, there is lots of food for the predator, the lynx, and this causes the population of lynx to increase. As their population increases, the lynx eat more and more hare, and this causes the population of hare to decrease. Then the lynx population decreases, because there are fewer rabbits to eat. Suppose that over a certain area, during a 4-year period of time t, the populations of hare and of lynx are given by the trigonometric functions

$$H(t) = 80 + 30 \sin\left(\frac{\pi}{2}t\right)$$

and

$$L(t) = 31 + 20 \cos\left(\frac{\pi}{2}t\right)$$

where the populations are in animals per square mile and time t is in years.

a. Sketch graphs of both functions over the 4-year interval [0, 4], using the same set of axes.
b. Find the maximum and minimum values of the hare population.
c. Find the maximum and minimum values of the lynx population.

34. **Hydrology.** Water waves, sound waves, the motion of a vibrating guitar string—all of these are examples of wave motion. The general equation of wave motion is given by

$$y = A \sin\left(\frac{t}{T} - \frac{x}{\lambda}\right)$$

where y is the vertical height of a wave, x is the distance from the origin of the wave, t is the time the wave has traveled, and T is the time required for the wave to travel one wavelength λ (the Greek letter lambda).

   a. Assuming y is a function of time t, find the amplitude, period, and phase shift of a wave.

   b. Assuming y is a function of x, find the amplitude, period, and phase shift of a wave.

35. **Hydrology.** The cross-section of a water wave is given by:

$$y = 2 \sin\left(\frac{\pi}{4}x - \frac{\pi}{4}\right)$$

Sketch a graph of this function.

36. **Physics.** At time t, the electric current i in a wire located in a magnetic field is given by

$$i(t) = I \sin(\omega t + \alpha)$$

   a. Find the amplitude, period, and phase shift of the current.

   b. Sketch a graph of the function when $I = 150$, $\omega = 150\pi$, and $\alpha = -\frac{\pi}{6}$.

Technology

Use a graphing calculator to graph the functions in Exercises 37–40. Use the graph to determine the period, amplitude, and phase shift of each function.

37. $y = 4 \sin \frac{3x}{4}$

38. $y = -\cos 3x$

39. $y = 1.6 \cos(x + 1)$

40. $y = -2.2 \sin(2x - 1)$

41. Graph the function $\sin^2 x + \cos^2 x$. What conclusion can you reach from the graph?

42. Graph the functions $\sin(-x) + \sin x$. What conclusion can you reach from the graph?

43. Use a graphing calculator to determine the zeros of the function $y = x - 2 \sin x$ to two significant figures.

44. Use a graphing calculator to determine the zeros of the function $y = x^2 - \sin x$ to two significant figures.

In your own words

45. Define the concept of phase shift. Give an example, using a specific function.

46. Give an example of an application that involves a vertically translated sine function.

47. Give an example of an application that involves a function with a nonzero phase shift.

48. Give examples of two ordinary phenomena that could be modeled using general sine or cosine functions.

7.7 THE GRAPHS OF TANGENT, SECANT, COTANGENT, AND COSECANT

In the preceding section, we sketched the graphs of functions of the form $a \sin(bx + c)$. We now determine the domains for and sketch the graphs of the other trigonometric functions.

Graph of The Tangent Function

Recall that the tangent function is defined as:

$$\tan x = \frac{\sin x}{\cos x}$$

It is defined for all values of x for which the denominator, $\cos x$, is nonzero. That is, the domain of $\tan x$ is:

$$\left\{x : x \neq \cdots -\frac{3\pi}{2}, -\frac{\pi}{2}, \frac{\pi}{2}, \frac{3\pi}{2}, \cdots\right\}$$

SECTION 7.7 The Graphs of Tangent, Secant, Cotangent, and Cosecant

Because both sin x and cos x have period 2π, the same is clearly true of tan x. However, as we prove in Chapter 8, the tangent function actually has period π. For purposes of sketching the graph of the tangent function, let's accept without proof for now that the tangent function has period π.

To sketch the graph of the tangent function, we need only sketch the portion that corresponds to x in the interval $[0, \pi]$ and extend the sketch to other values of x using the periodicity of the function. To sketch the graph within this interval, we plot a set of representative points (or use a graphing calculator). The function is undefined at the midpoint $\pi/2$. The following table lists a set of representative points on the graph, including points near and on both sides of $\pi/2 = 1.57080\ldots$:

x	tan x
0	0
$\dfrac{\pi}{6}$	$\sqrt{3}/3 = 0.57735$
$\dfrac{\pi}{3}$	$\sqrt{3} = 1.73205$
1.5	14.1014
1.55	48.0785
1.6	-34.2325
1.65	-12.5993
$\dfrac{2\pi}{3}$	$-\sqrt{3} = -1.73205$
$\dfrac{5\pi}{6}$	$-\sqrt{3}/3 = -0.57735$
π	0

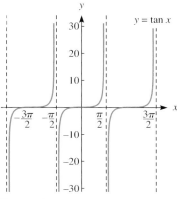

Figure 1

Figure 1 shows the plotted points and a sketch of the resulting graph. Note that the graph has as asymptotes the vertical lines:

$$x = \frac{k\pi}{2}, \qquad k = \pm 1, \pm 3, \pm 5, \ldots$$

If we use a graphing calculator, we can graph the function without making the table of numerical values.

EXAMPLE 1
Graph of a Tangent Function

Sketch the graph of the function $f(x) = \tan(3x)$.

Solution

The graph has the same general shape as the graph of tan x, except that the scale on the x-axis is changed. The asymptotes occur for:

$$3x = \frac{k\pi}{2}, \qquad k = \pm 1, \pm 3, \pm 5, \ldots$$

$$x = \frac{k\pi}{6}, \qquad k = \pm 1, \pm 3, \pm 5, \ldots$$

In particular, the first asymptotes on either side of the origin are $x = \pm \pi/6$. The distance between the asymptotes is $\pi/3$, which is the period. Figure 2 shows several periods of the graph. We can verify this graph by using a graphing calculator.

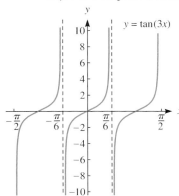

Figure 2

EXAMPLE 2
Graph of a Translated Tangent Function

Sketch the graph of the function $f(x) = \tan(3x - \pi/4)$.

Solution

Write the function in the form $f(x) = \tan 3(x - \pi/12)$. We can obtain the graph of this function from the tangent function in two steps. First, we scale horizontally by a factor of $\frac{1}{3}$. This compresses the periods of the tangent function to intervals of width $\pi/3$. Then, we translate to the right by the phase shift $\pi/12$. The resulting graph is shown in Figure 3. We can graph $f(x)$ with a graphing calculator or by hand, but the exact value of the phase shift is impossible to obtain.

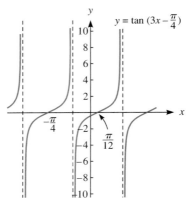

Figure 3

Graph of the Secant Function

The function sec x is defined as:

$$\sec x = \frac{1}{\cos x}$$

The domain consists of all x for which the denominator is nonzero. That is, the domain is:

$$\left\{ x : x \neq \cdots, -\frac{3\pi}{2}, -\frac{\pi}{2}, \frac{\pi}{2}, \frac{3\pi}{2}, \cdots \right\}$$

This is the same domain as for the function tan x. To graph sec x, we first sketch the graph of cos x. On the same coordinate system, we plot, for each value of x, a point that has a height equal to the reciprocal of cos x. The following table gives some points to plot.

x	cos x	sec x
0	1	1
$\frac{\pi}{4}$	-0.70711	-1.41421
$\frac{\pi}{2}$	0	undefined
$\frac{3\pi}{4}$	-0.70711	-1.41421
π	-1	-1
$\frac{5\pi}{4}$	-0.70711	-1.41421
$\frac{3\pi}{2}$	0	undefined
$\frac{7\pi}{4}$	0.70711	1.41421
2π	1	1

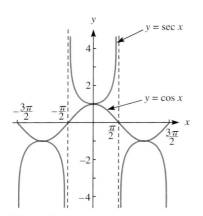

Figure 4

The resulting graph is shown in Figure 4. Note that the vertical asymptotes occur at the zeros of cos x, namely at:

$$x = \pm\frac{\pi}{2}, \pm\frac{3\pi}{2}, \pm\frac{5\pi}{2}, \cdots$$

SECTION 7.7 The Graphs of Tangent, Secant, Cotangent, and Cosecant

Note that most graphing calculators do not have a built-in secant function. However, we can enter the secant as $1/\cos x$.

EXAMPLE 3 Sketch the graph of the function $f(x) = \sec x/2$.

Graph of a Secant Function

Solution

The graph has the same shape as the graph of sec x, but the x-axis is rescaled. The first pair of vertical asymptotes occurs at:

$$\frac{x}{2} = \pm \frac{\pi}{2}$$
$$x = \pm \pi$$

Because $\sec x/2 = 1/(\cos x/2)$, the period of sec $x/2$ is the same as that of cos $x/2$, which is $2\pi/\frac{1}{2} = 4\pi$. So sec $x/2$ repeats every 4π units. In Figure 5, we show its graph.

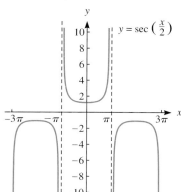

Figure 5

Graphs of the Cotangent and Cosecant Functions

The function cot x is defined by

$$\cot x = \frac{\cos x}{\sin x}$$

and can be graphed in a manner similar to tan x. The resulting graph is shown in Figure 6.

The function csc x is defined by

$$\csc x = \frac{1}{\sin x}$$

and can be graphed in a manner similar to sec x. The resulting graph is shown in Figure 7. Note that the secant and cosecant functions must be entered on a graphing calculator as the reciprocals of the sine and cosine.

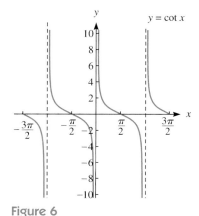

Figure 6

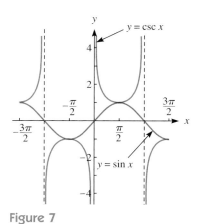

Figure 7

Graphs of Sums of Trigonometric Functions

In many applications, it is necessary to consider functions that are sums of one or more trigonometric functions. The next two examples illustrate how such functions can be graphed.

EXAMPLE 4
Graph of a Sum of Trigonometric Functions

Sketch the graph of $\sin x + 3\cos 2x$.

Solution

In Figure 8, we sketch on separate axes the graphs of the summands (terms being added) $\sin x$ and $3\cos 2x$. Because the first summand has period 2π and the second has period π, we see that the sum has period 2π. To graph the sum for x between 0 and 2π, we tabulate the function for various key values of x and plot the corresponding points. We then extend the graph to all x, using periodicity.

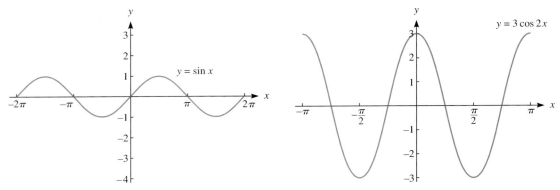

Figure 8

x	$\sin x$	$3\cos 2x$	$\sin x + 3\cos 2x$
0	0	3	3
$\frac{\pi}{4}$	$\frac{\sqrt{2}}{2}$	0	$\frac{\sqrt{2}}{2}$
$\frac{\pi}{2}$	1	-3	-2
$\frac{3\pi}{4}$	$\frac{\sqrt{2}}{2}$	0	$\frac{\sqrt{2}}{2}$
π	0	3	3
$\frac{5\pi}{4}$	$-\frac{\sqrt{2}}{2}$	0	$-\frac{\sqrt{2}}{2}$
$\frac{3\pi}{2}$	-1	-3	-4
$\frac{7\pi}{4}$	$-\frac{\sqrt{2}}{2}$	0	$-\frac{\sqrt{2}}{2}$
2π	0	3	3

Figure 9 shows the plotted points and the graph of the sum. Using a graphing calculator, you can bypass the addition of ordinates and enter the sum directly.

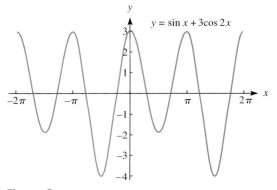

Figure 9

EXAMPLE 5
Graphing by Addition of Ordinates

Sketch the graph of the function $f(x) + g(x)$, where $f(x) = x$ and $g(x) = \sin x$.

Solution

In Figure 10, we plot the graphs of the functions $f(x) = x$ and $g(x) = \sin x$ on separate coordinate systems. The graph of the sum is obtained by finding, for each value of x, the sum of the corresponding heights on the two graphs. We can determine the shape of the sum as follows: On each successive interval of length 2π along the x-axis, the graph of $g(x) = \sin x$ oscillates between -1 and 1. The graph of $f(x) = x$ is a straight line. On each successive interval of length 2π along the x-axis, the graph of the *sum* oscillates around the line $f(x) = x$. When $\sin x$ is negative, the graph of the sum is below the line; when $\sin x$ is positive, the graph of the sum is above the line. This results in the third graph shown in Figure 10. The precise shape of the graph is obtained from a graphing calculator.

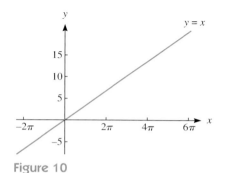
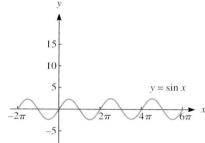
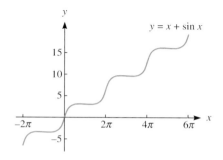

Figure 10

EXAMPLE 6
A Sum of Three Trigonometric Functions

Sketch at least two periods of the graph of:

$$f(x) = \sin x + 2 \sin 2x + 3 \sin 3x$$

Solution

This sum is sufficiently complex that we do not try to add the ordinates. Rather, we go directly to our graphing calculator. The first term has period 2π, the second has period π, and the third has period $2\pi/3$. The period of the sum is the least common multiple of all the periods of the terms. The number 2π is a common multiple of all the periods and is the least such number, and so the period of the sum is 2π. The two periods of the sum we are required to graph correspond to an interval on the x-axis of length $4\pi \approx 12.56$. So we can choose the standard range $-10 \le x \le 10$, which displays more than three periods. The graph is shown in Figure 11.

Figure 11

More Complex Graphs

Let's examine some more complex graphs involving trigonometric functions.

EXAMPLE 7
Damped Oscillation

Sketch the graph of the function $e^{-0.3x} \sin x$, $x > 0$.

Solution

The function $\sin x$ is periodic with period 2π. As x increases, the value of $e^{-0.3x}$ decreases and approaches 0 as x approaches positive infinity. On each successive interval of length 2π along the x-axis, the value of $e^{-0.3x} \sin x$ oscillates like the graph of sin x. However, the amplitude of the oscillations steadily decreases, approaching 0 as x approaches infinity. The graph, obtained from a graphing calculator, is shown in Figure 12. A graph of this sort is called a **damped sine wave**. Such graphs depict vibrations that dissipate over time, such as sound vibrations traveling in water or in air.

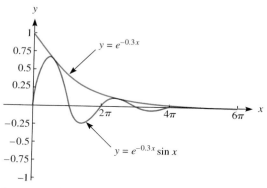

Figure 12

EXAMPLE 8
A Graph with Infinitely Many Oscillations Near $x = 0$

Use a graphing calculator to display the graph of:
$$f(x) = \sin \frac{1}{x}$$

Solution

As x approaches 0 through positive values, the quotient $1/x$ increases without bound and goes through an unlimited number of periods of the sine function. So the graph of $f(x)$ oscillates infinitely often between -1 and 1 as x approaches 0 through positive values. The graph is shown in Figure 13. Note that the poor resolution of the calculator is not able to truly capture the wild behavior of the graph near $x = 0$. For instance, the oscillations should, in each instance run from $y = -1$ to $y = +1$. However, the graph makes it appear as if the heights decrease near 0. This is due to the limited resolution of the calculator screen. If we zoom in, we get a more accurate picture of the graph near $x = 0$.

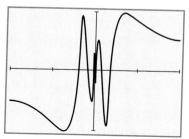

Figure 13

Exercises 7.7

Match each of the functions in Exercises 1–6 with its graph in (a)–(f).

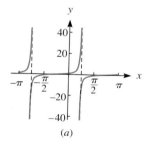

(a)

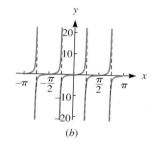

(b)

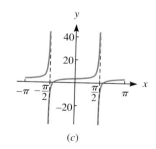

(c)

8.1 TRIGONOMETRIC IDENTITIES

In this chapter, we discuss the general nature of trigonometric identities and techniques for proving them.

Elementary Trigonometric Identities

Let's begin by reviewing the elementary trigonometric identities we have proven thus far: First are the identities used to define the trigonometric functions in terms of the sine and cosine, and the identities that are immediate consequences of these identities. Second are the Pythagorean identities. Third are the parity identities. All the identities we've seen thus far are summarized in the following table:

Elementary Trigonometric Identities

Definition Identities

$$\sin x = \frac{1}{\csc x} \qquad \csc x = \frac{1}{\sin x}$$

$$\cos x = \frac{1}{\sec x} \qquad \sec x = \frac{1}{\cos x}$$

$$\tan x = \frac{\sin x}{\cos x} = \frac{1}{\cot x} \qquad \cot x = \frac{\cos x}{\sin x} = \frac{1}{\tan x}$$

Pythagorean Identities

$$\sin^2 x + \cos^2 x = 1$$
$$\tan^2 x + 1 = \sec^2 x$$
$$\cot^2 x + 1 = \csc^2 x$$

Parity Identities

$$\sin(-x) = -\sin x \qquad \csc(-x) = -\csc x$$
$$\cos(-x) = \cos x \qquad \sec(-x) = \sec x$$
$$\tan(-x) = -\tan x \qquad \cot(-x) = -\cot x$$

Applications of the Elementary Identities

In the following examples, we discuss a number of applications of the elementary identities. Many of these applications are useful in calculus.

EXAMPLE 1 *Evaluating a Trigonometric Function*

Suppose that $\cot x = -2$ and that x is in the second quadrant. Determine the value of $\sin x$.

Solution

By the Pythagorean identity that relates $\cot x$ to $\csc x$, we have:

$$\cot^2 x + 1 = \csc^2 x$$

$$\csc x = \pm \sqrt{\cot^2 x + 1}$$

$$= \pm \sqrt{(-2)^2 + 1}$$

$$= \pm \sqrt{5}$$

Because $\sin x = 1/\csc x$, we see that:

$$\sin x = \frac{1}{\pm \sqrt{5}}$$

$$= \pm \frac{\sqrt{5}}{5} \qquad \text{\textit{Rationalizing denominator}}$$

Because x is in the second quadrant, $\sin x$ is positive, so that:

$$\sin x = \frac{\sqrt{5}}{5}$$

EXAMPLE 2
Proving a Trigonometric Identity

Prove the identity:

$$\cos(-x)\sin(-x) + \tan x = \frac{\sin^3 x}{\cos x}$$

Solution

Looking at the left side of the equation, we see that the Parity identities for sine and cosine can be applied. Moreover, by replacing the tangent with its definition in terms of the sine and cosine, we can simplify the expression further:

$$\cos(-x)\sin(-x) + \tan x = -\cos x \sin x + \tan x \quad \text{Parity identities}$$

$$= -\sin x \cos x + \frac{\sin x}{\cos x} \quad \text{Definition of } \tan x$$

$$= \frac{-\sin x \cos^2 x + \sin x}{\cos x} \quad \text{Adding fractions}$$

$$= \frac{\sin x(-\cos^2 x + 1)}{\cos x} \quad \text{Factoring}$$

$$= \frac{\sin x \cdot \sin^2 x}{\cos x} \quad \text{Pythagorean identity}$$

$$= \frac{\sin^3 x}{\cos x}$$

EXAMPLE 3
Reducing an Expression to a Single Trigonometric Function

Write the expression $\tan^3 x \cos x - \tan^2 x$ in terms of $\sin x$.

Solution

Start by factoring $\tan^2 x$ from both terms, with the hope of using the Pythagorean identity for $\tan x$:

$$\tan^3 x \cos x - \tan^2 x = \tan^2 x(\tan x \cos x - 1)$$

$$= \tan^2 x \left(\frac{\sin x}{\cos x} \cdot \cos x - 1\right) \quad \text{Definition of } \tan x$$

$$= \tan^2 x(\sin x - 1)$$

$$= (\sec^2 x - 1)(\sin x - 1) \quad \text{Pythagorean identity}$$

$$= \left(\frac{1}{\cos^2 x} - 1\right)(\sin x - 1) \quad \text{Definition of } \sec x$$

$$= \frac{1 - \cos^2 x}{\cos^2 x} \cdot (\sin x - 1)$$

$$= \frac{\sin^2 x}{1 - \sin^2 x}(\sin x - 1) \quad \text{Pythagorean identity}$$

$$= \frac{\sin^2 x}{(1 - \sin x)(1 + \sin x)}(\sin x - 1) \quad \text{Factoring}$$

$$= -\frac{\sin^2 x}{1 + \sin x}$$

EXAMPLE 4
Solving a Trigonometric Equation

Solve the equation:
$$\sec^2 x - \tan x = 1, \qquad -\frac{\pi}{2} < x < \frac{\pi}{2}$$

Solution

In order to solve an equation such as this, it is usually helpful to transform it using identities, so that the transformed equation involves a single trigonometric function. In this case, by applying the Pythagorean identity for the tangent function, we have:

$$(\tan^2 x + 1) - \tan x = 1$$
$$\tan^2 x - \tan x = 0$$
$$\tan x(\tan x - 1) = 0$$
$$\tan x = 0 \quad \text{or} \quad \tan x = 1$$
$$x = 0 \quad \text{or} \quad x = \frac{\pi}{4}$$

EXAMPLE 5
Clearing Trigonometric Fractions

Write the following expression in nonfractional form (such calculations are useful in calculus):

$$\frac{1}{\sec x - 1}$$

Solution

Multiply both numerator and denominator by $\sec x + 1$ so that we can apply the Pythagorean identity for the tangent:

$$\frac{1}{\sec x - 1} \cdot \frac{\sec x + 1}{\sec x + 1} = \frac{\sec x + 1}{\sec^2 x - 1}$$
$$= \frac{\sec x + 1}{\tan^2 x}$$
$$= \frac{1}{\tan^2 x} \cdot (\sec x + 1)$$
$$= \cot^2 x(\sec x + 1)$$

EXAMPLE 6
Calculating the Sum of Rational Trigonometric Expressions

Calculate the following sum and simplify.

$$\frac{1}{\sin y - 1} + \frac{1}{\sin y + 1}$$

Solution

Add the two fractions to obtain:

$$\frac{1}{\sin y - 1} + \frac{1}{\sin y + 1}$$
$$= \frac{1}{\sin y - 1} \cdot \frac{\sin y + 1}{\sin y + 1} + \frac{1}{\sin y + 1} \cdot \frac{\sin y - 1}{\sin y - 1} \qquad \textit{Find common denominator}$$

420   **CHAPTER 8**   **Analytic Trigonometry**

$$= \frac{\sin y + 1}{\sin^2 y - 1} + \frac{\sin y - 1}{\sin^2 y - 1} \qquad \textit{Multiply out denominators}$$

$$= \frac{(\sin y - 1) + (\sin y + 1)}{\sin^2 y - 1} \qquad \textit{Add fractions}$$

$$= \frac{2 \sin y}{-\cos^2 y} \qquad \textit{Because } \sin^2 y + \cos^2 y = 1$$

$$= -2 \sin y \cdot \frac{1}{\cos^2 y}$$

$$= -2 \sin y \sec^2 y \qquad \textit{Definition of secant}$$

There are many different ways to simplify expressions, leading to equivalent, but different, answers. In simplifying trigonometric expressions, a good rule to follow is to eliminate fractions and radicals, and to use as few different trigonometric functions as possible.

Verifying Trigonometric Identities

The identities in the preceding table are by no means the only identities for the trigonometric functions. There are literally hundreds of others! Such identities are useful for rewriting trigonometric expressions in equivalent forms, a process used repeatedly in more advanced mathematics.

Many of the other identities are based on the basic identities stated in the table at the beginning of this section. You should memorize these basic identities. (Actually, you will be using them so frequently that you probably won't need to make a special effort.)

In learning to manipulate trigonometric expressions, it is helpful to prove the validity of trigonometric identities. Here is a useful approach for proving a trigonometric identity is true:

TIPS FOR VERIFYING TRIGONOMETRIC IDENTITIES

1. Keep one side, the simpler side, unchanged throughout your proof.
2. Replace the more complex side of the equation with a series of equivalent expressions until you arrive at the simpler side.
3. Use the various identities you already know to find the equivalent expressions in step 2.

The next two examples provide concrete illustrations of how to prove identities.

EXAMPLE 7
Proving an Identity

Verify the following trigonometric identity.

$$\tan x + 1 = \sec x (\sin x + \cos x)$$

Solution

The right side involves three different trigonometric functions. Let's write the identity in terms of sines and cosines:

$$\sec x(\sin x + \cos x) = \frac{1}{\cos x}(\sin x + \cos x) \quad \text{\textit{Definition of} } \sec x$$

$$= \frac{\sin x}{\cos x} + 1 \quad \text{\textit{Distributive property}}$$

$$= \tan x + 1 \quad \text{\textit{Definition of} } \tan x$$

We have transformed the right side of the equation into the left side, using allowable algebraic operations and known identities. This proves that the identity is valid.

An identity is a mathematical statement that is valid for all values of the variable, not just for a limited set of values. This implies that if we graph both sides of an identity, we obtain the same graph. However, if an equation is not an identity, the two sides have different graphs. We can use graphing calculators to check the plausibility of an identity by graphing both sides. If we get two different graphs, then the equation is not an identity. If we get the same graph, then the equation is an identity. Of course, because the calculator only graphs a finite number of points, such a comparison of graphs cannot prove an identity is true. However, to prove that an identity is false, we need only exhibit a single value of x where the two sides differ. And this can be done using a calculator.

EXAMPLE 8
Using One Identity to Prove Another

Verify the following trigonometric identity:

$$\frac{\sin^3 x - \cos^3 x}{\sin x - \cos x} = 1 + \sin x \cos x$$

Solution

This identity is more complicated than the one in Example 7. However, we can apply the same principles we used in the preceding example. The more complex side is the left, so let's transform the left side into the right. The numerator on the left side is a difference of cubes. This suggests that we can factor it using the Difference of Cubes factorization identity to obtain:

$$\frac{(\sin x - \cos x)(\sin^2 x + \sin x \cos x + \cos^2 x)}{\sin x - \cos x}$$

$$= \sin^2 x + \sin x \cos x + \cos^2 x \quad \text{\textit{Cancel common factor}}$$

$$= 1 + \sin x \cos x \quad \text{\textit{Because} } \sin^2 x + \cos^2 x = 1$$

We have thus transformed the left side into the right. So the identity is valid.

In the preceding examples, we proved the validity of several identities. Note, however, that in order to prove that an identity is *not* valid, it is necessary to find only a single value of the variables for which the right side does not equal the left. For example, consider the alleged identity:

$$\sin x + \cos x = \tan x$$

This identity is not valid. Indeed, if we substitute x equals 0, we have an incorrect statement:

$$\sin 0 + \cos 0 = \tan 0$$
$$0 + 1 = 0$$

EXAMPLE 9
Trigonometric Substitution

In calculus, we encounter problems involving expressions of the form $\sqrt{a^2 - t^2}$. Make the substitution $t = a \sin x$, where $0 < x < \pi/2$, and write the expression in terms of the variable x. This sort of substitution is called a **trigonometric substitution**.

Solution

Substitute $a \sin x$ for t in the given expression, to obtain:

$$\sqrt{a^2 - t^2} = \sqrt{a^2 - (a \sin x)^2}$$
$$= \sqrt{a^2 - a^2 \sin^2 x}$$
$$= \sqrt{a^2(1 - \sin^2 x)}$$

Applying the Pythagorean identity, we can replace $1 - \sin^2 x$ with $\cos^2 x$ to obtain:

$$\sqrt{a^2 \cos^2 x} = \sqrt{a^2} \sqrt{\cos^2 x}$$

Because x lies in the interval $(0, \pi/2)$, the value of $\cos x$ must be positive. Therefore, the above expression is equivalent to:

$$|a| \cos x$$

This is the desired expression of $\sqrt{a^2 - t^2}$ in terms of x.

Using a Graphing Calculator to Visually Demonstrate an Identity

We can use a graphing calculator to illustrate trigonometric identities and to compare the graphs of the two sides of the identity. For instance, consider the graph in Figure 1. This is the graph of

$$f(x) = \sin^2 x + \cos^2 x$$

for x in the interval $-\pi \leq x \leq \pi$. Note that, because the periods of $\sin x$ and $\cos x$ are 2π, this interval contains a full period for each function. (In fact, it can be shown that the periods of $\sin^2 x$ and $\cos^2 x$ are both π, so that the interval contains a full period for each.) The graph shown is the straight line $y = 1$. By periodicity, the graph of $f(x)$ coincides with this line for all values of x, not just in the interval $-\pi \leq x \leq \pi$. Thus, we have visual demonstration of the identity:

$$\sin^2 x + \cos^2 x = 1$$

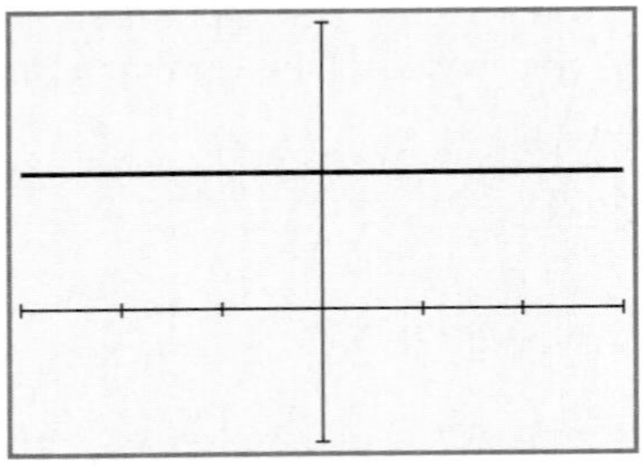

This technique of visually demonstrating identities becomes most useful when the trigonometric identities are more complicated. The exercises contain a number of identities to be demonstrated in this way.

We must emphasize that the graphical demonstration above is *not* a mathematical proof of the Pythagorean identity for sine and cosine. The graphs we examined only show certain values of the expressions on both sides. There are infinitely many other values that we didn't compare. A graphing calculator demonstration can provide us with strong evidence that an identity is correct, but cannot substitute for a logical derivation of the identity.

Figure 1

Exercises 8.1

Write the following trigonometric expressions in terms of sin x.

1. $\csc^2 x - \cot^2 x$
2. $\cot x(\cos x - 1)$
3. $\dfrac{\tan^2 x}{1 + \tan^2 x}$
4. $\dfrac{\cot^2 x + 1}{\cot^2 x}$
5. $\cos x \cot x$
6. $\sec x \tan x$
7. $\csc x \sec x \cot x$
8. $\cos x \cot x + \sin x$
9. $(1 - \sin^2 x)(\sec^2 x - 1)$
10. $\sec x[(\sin x + \cos x)^2 - 1]$

Write the following expressions in terms of cos x.

11. $\tan x \sin x$
12. $\tan x \csc x$
13. $\sec^2 x$
14. $\cot^2 x$
15. $\dfrac{1 + \tan^2 x}{\tan^2 x}$
16. $\dfrac{\cot^2 x}{1 + \cos^2 x}$
17. $\dfrac{1 - \sec^2 x}{1 - \cos^2 x}$
18. $1 + \cot^2 x$

Write the following expressions in terms of a single trigonometric function.

19. $\tan^2 x \sec x \cos x$
20. $\dfrac{\cos x}{\csc x - \sin x}$
21. $\csc x \sec x - \tan x$
22. $\dfrac{\sec^2 x}{\cos^2 x} - \dfrac{\sin^2 x}{\cos^4 x} - 1$
23. $\dfrac{\tan x(1 - \sin^2 x)}{\sin x}$
24. $\dfrac{\cos x}{1 - \sin x} - \tan x$

Make the given substitution for t, with $0 < x < \pi/2$, and write the given expressions in terms of the variable x.

25. $t = a \tan x$, for $\sqrt{a^2 + t^2}$
26. $t = a \tan x$, for $\sqrt{\dfrac{1}{t^2 + a^2}}$
27. $t = 5 \sin x$, for $\dfrac{t^2}{\sqrt{25 - t^2}}$
28. $t = 5 \sin x$, for $\sqrt{25 - t^2}$
29. $t = a \sec x$, for $\sqrt{t^2 - a^2}$
30. $t = a \sin x$, for $t \sqrt{(a \cos x)^2 + t^2}$
31. $t = \dfrac{4}{3} \tan x$, for $\sqrt{16 + 9t^2}$
32. $t = 4 \tan x$, for $t^2 \sqrt{16 + t^2}$

Prove each of the following trigonometric identities.

33. $1 = \cos^2 a + \cos^2 a \tan^2 a$
34. $\cos^2 \beta + 1 = 2 \cos^2 \beta + \sin^2 \beta$
35. $1 - 2 \sin^2 y = \cos^2 y - \sin^2 y$
36. $\tan x \sin x + \cos x = \sec x$
37. $\dfrac{\tan a - 1}{\tan a + 1} = \dfrac{1 - \cot a}{1 + \cot a}$
38. $\dfrac{1 + \sin \beta}{\cos \beta} = \dfrac{\cos \beta}{1 - \sin \beta}$
39. $\dfrac{\cos \alpha}{\cos \alpha - \sin \alpha} = \dfrac{1}{1 - \tan \alpha}$
40. $\dfrac{1 - \tan^2 a}{1 - \sec^2 a} = 1 - \cot^2 a$
41. $\dfrac{\cos \alpha + \sin \alpha}{\cos \alpha} = 1 + \tan \alpha$
42. $\dfrac{\cos \alpha}{\sec \alpha - \tan \alpha} = 1 + \sin \alpha$
43. $\dfrac{1 + \csc x}{\cot x + \cos x} = \sec x$
44. $1 + \sec^2 \alpha \tan^2 \alpha = \tan^4 \alpha + \sec^2 \alpha$
45. $1 + \sin x = \dfrac{\cos x}{\sec x - \tan x}$
46. $\dfrac{\tan^2 \alpha - 1}{\tan \alpha - \cot \alpha} = \tan \alpha$
47. $-4 \sec \alpha \tan \alpha = \dfrac{1 - \sin \alpha}{1 + \sin \alpha} - \dfrac{1 + \sin \alpha}{1 - \sin \alpha}$
48. $\sin^3 \beta \cos \beta - \sin^5 \beta \cos \beta = \sin^3 \beta \cos^3 \beta$
49. $\dfrac{\csc^2 \theta - \cot^2 \theta \csc^2 \theta}{\cot^2 \theta} = \sec^2 \theta - \csc^2 \theta$
50. $\dfrac{\sin x}{\csc x} = 1 - \dfrac{\cos x}{\sec x}$
51. $\dfrac{1 + \cos \theta}{\sin \theta} + \dfrac{\sin \theta}{1 + \cos \theta} = 2 \csc \theta$
52. $2 \sin \alpha(1 + \cos \alpha) = \dfrac{2 \sin^3 \alpha}{1 - \cos \alpha}$
53. $\dfrac{1}{\cos x} = \dfrac{\csc(-x)}{\cot(-x)}$
54. $-\cot \alpha = \cos(-\alpha) \csc(-\alpha)$
55. $(\tan \beta \sin \beta)^2 = \tan^2 \beta - \sin^2 \beta$
56. $\dfrac{\sin^3 \theta + \cos^3 \theta}{\sin \theta + \cos \theta} = 1 - \sin \theta \cos \theta$
57. $\dfrac{\sin^2 x}{1 - \cos x} = 1 + \cos x$
58. $\dfrac{\sec x}{\sec x - 1} - \dfrac{\sec x + 1}{\tan^2 x} = 1$
59. $\sec^2 x - 2 \sec x \cos x + \cos^2 x = \tan^2 x - \sin^2 x$
60. $\left(\dfrac{1 + \cos x}{\sin x}\right)^2 = \dfrac{\tan x + \sin x}{\tan x - \sin x}$
61. $\dfrac{\sin x}{1 + \cos x} = \csc x - \cot x$
62. $\dfrac{1 + \cot x}{1 - \cot x} + \dfrac{1 + \tan x}{1 - \tan x} = 0$

Technology

Use a graphing calculator to check whether the following equations could be identities. For those that appear to be identities, prove them. For those that are not, calculate a value of the variable at which the two sides are unequal.

63. $\sin x = \sqrt{1 - \cos^2 x}$
64. $(\sin x + \cos x)^2 = 1$
65. $\sqrt{\tan^2 \theta + \cot^2 \theta} = \tan \theta + \cot \theta$
66. $(\tan \theta)^2 = (\sin \theta \csc \theta)^2$
67. $\dfrac{\cos x + 1}{\sin x \cos x} - \dfrac{\sin x}{\cos x} = \dfrac{1 + \cos x}{\sin x}$
68. $\dfrac{\sec x}{\sin x} = \dfrac{\sec x + \csc x}{\sin x + \cos x}$
69. $\dfrac{\sin x}{\sin x \cot x + \cos x} = \dfrac{1}{2} \tan x$
70. $\sqrt{\tan^2 x + 1} = \sec x$
71. $\cos x = \csc x \tan x$
72. $\dfrac{1 - \sin x}{\cos x} - \dfrac{\cos x}{1 + \sin x} = 0$

Use a calculator to graph the following expressions in an appropriate range. Then use the graph to predict an identity for the expression.

73. $2 \sin x \cos x$
74. $\dfrac{1 + \cos 2x}{2}$
75. $-4 \sin^3 x - 3 \sin x$

In your own words

76. Describe a procedure for verifying a trigonometric identity. Illustrate the procedure with a specific identity.
77. Describe how to verify an identity using a graphing calculator.
78. Can a graphing calculator be used to prove an identity? Explain your answer.

8.2 THE ADDITION AND SUBTRACTION IDENTITIES

Among the most important trigonometric identities are the addition and subtraction identities, which allow us to calculate the trigonometric functions for the sum $x + y$ in terms of the trigonometric functions for x and y. In this section, we prove these identities and a number of others that follow from them. For future reference, here is a summary of the identities we prove in Section 8.2:

Addition Identities

$\sin(x + y) = \sin x \cos y + \cos x \sin y$
$\cos(x + y) = \cos x \cos y - \sin x \sin y$
$\tan(x + y) = \dfrac{\tan x + \tan y}{1 - \tan x \tan y}$

Subtraction Identities

$\sin(x - y) = \sin x \cos y - \cos x \sin y$
$\cos(x - y) = \cos x \cos y + \sin x \sin y$
$\tan(x - y) = \dfrac{\tan x - \tan y}{1 + \tan x \tan y}$

Cofunction Identities

$\cos\left(\dfrac{\pi}{2} - x\right) = \sin x$
$\sin\left(\dfrac{\pi}{2} - x\right) = \cos x$
$\tan\left(\dfrac{\pi}{2} - x\right) = \cot x$

Symmetry Identities

$\sin x = \sin(\pi - x)$
$\cos x = -\cos(\pi - x)$

Addition and Subtraction Identities for the Cosine Function

In proving the addition and subtraction identities, it is simplest to proceed in a particular order. First we prove the subtraction formula for the cosine, and then we prove the addition formula for the cosine. That is, we begin with the following identities:

ADDITION AND SUBTRACTION IDENTITIES FOR COSINE

Let x and y be any real numbers. Then we have the following identities:
$$\cos(x + y) = \cos x \cos y - \sin x \sin y$$
$$\cos(x - y) = \cos x \cos y + \sin x \sin y$$

Proof Interpret x and y as measures of angles. Draw the angles x, y, and $x - y$ in a unit circle, as shown in Figure 1. The angles AOC and BOD are both equal to $x - y$, so the arcs AC and BD are equal. By elementary geometry, this implies that the line segments AC and BD are equal.

Set $A = (1, 0)$, $B = (a_1, b_1)$, $C = (a_2, b_2)$, and $D = (a_3, b_3)$. Then, by the distance formula, we have the equation:

$$\sqrt{(a_2 - 1)^2 + (b_2 - 0)^2} = \sqrt{(a_3 - a_1)^2 + (b_3 - b_1)^2}$$

Remove the radicals by squaring both sides. Then expand the expressions using the Product of Binomials formula:

$$a_2^2 - 2a_2 + 1 + b_2^2 = a_3^2 - 2a_3 a_1 + a_1^2 + b_3^2 - 2b_3 b_1 + b_1^2$$

Because the points A, B, C, D lie on a unit circle, we have $a_1^2 + b_1^2 = 1$, $a_2^2 + b_2^2 = 1$, and $a_3^2 + b_3^2 = 1$. Substituting these equations into the last equation above, we have:

$$2 - 2a_2 = 2 - 2a_3 a_1 - 2b_3 b_1$$
$$a_2 = a_3 a_1 + b_3 b_1$$

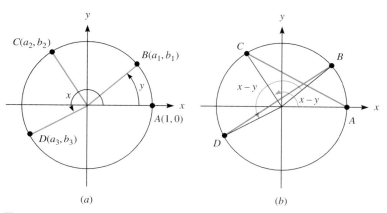

Figure 1

However, from the unit circle definition of the trigonometric functions, we have:
$$(a_1, b_1) = (\cos y, \sin y)$$
$$(a_2, b_2) = (\cos(x - y), \sin(x - y))$$
$$(a_3, b_3) = (\cos x, \sin x)$$

Substituting these values into the last displayed equation, we have:
$$a_2 = a_3 a_1 + b_3 b_1$$
$$\cos(x - y) = \cos x \cos y + \sin x \sin y$$

This completes the proof of the second identity.

To prove the first identity, replace x by $-x$ in the second identity:
$$\cos(-x - y) = \cos(-x) \cos y + \sin(-x) \cos y$$
$$\cos[-(x + y)] = \cos x \cos y + (-\sin x) \cos y \qquad \text{Parity identities}$$

Applying the parity identity again for $\cos[-(x + y)]$, we have:
$$\cos(x + y) = \cos x \cos y - \sin x \sin y$$

This completes the proof of the first identity.

EXAMPLE 1
Application of Subtraction Identity for Cosine

Determine the value of $\cos \pi/12$.

Solution

Write $\pi/12$ in the form
$$\frac{\pi}{12} = \frac{4\pi}{12} - \frac{3\pi}{12} = \frac{\pi}{3} - \frac{\pi}{4}$$

and we take the cosine of both sides of the equation:
$$\cos \frac{\pi}{12} = \cos\left(\frac{\pi}{3} - \frac{\pi}{4}\right)$$

Apply the subtraction identity for the cosine to obtain:
$$\cos \frac{\pi}{12} = \cos \frac{\pi}{3} \cos \frac{\pi}{4} + \sin \frac{\pi}{3} \sin \frac{\pi}{4}$$

Insert values for $\cos \pi/3$, $\sin \pi/3$, $\sin \pi/4$, and $\cos \pi/4$ (these values were determined in Section 7.2) to obtain:
$$\cos \frac{\pi}{12} = \frac{1}{2} \cdot \frac{\sqrt{2}}{2} + \frac{\sqrt{3}}{2} \cdot \frac{\sqrt{2}}{2}$$
$$= \frac{\sqrt{2}}{4} + \frac{\sqrt{6}}{4} = \frac{\sqrt{2} + \sqrt{6}}{4}$$

EXAMPLE 2
Application of Addition Identity for Cosine

Use the addition identity for cosine to determine an alternate expression for $\cos(t + \pi/4)$.

Solution

$$\cos\left(t + \frac{\pi}{4}\right) = \cos t \cos \frac{\pi}{4} - \sin t \sin \frac{\pi}{4}$$
$$= \cos t \cdot \frac{\sqrt{2}}{2} - \sin t \cdot \frac{\sqrt{2}}{2}$$
$$= \frac{\sqrt{2}}{2}(\cos t - \sin t)$$

SECTION 8.2 The Addition and Subtraction Identities

EXAMPLE 3
Cosine of a Sum

Suppose that angles u and v are both in the third quadrant, and suppose that $\sin u = -\sqrt{3}/2$, and $\sin v = -\frac{1}{2}$. Determine the value of $\cos(u + v)$.

Solution

By the Pythagorean identity for the sine and cosine, we have:

$$\cos u = \pm\sqrt{1 - \sin^2 u} \qquad \cos v = \pm\sqrt{1 - \sin^2 v}$$

$$= \pm\sqrt{1 - \left(-\frac{\sqrt{3}}{2}\right)^2} \qquad = \pm\sqrt{1 - \left(-\frac{1}{2}\right)^2}$$

$$= \pm\sqrt{1 - \frac{3}{4}} \qquad = \pm\sqrt{\frac{3}{4}} = \pm\frac{\sqrt{3}}{2}$$

$$= \pm\frac{1}{2}$$

Because both u and v lie in the third quadrant, the cosine of each angle is negative, so that the minus sign prevails in the last two equations. Therefore, we have:

$$\cos u = -\frac{1}{2}, \qquad \cos v = -\frac{\sqrt{3}}{2}$$

Therefore, by the addition identity for the cosine, we have:

$$\cos(u + v) = \cos u \cos v - \sin u \sin v$$

$$= \left(-\frac{1}{2}\right) \cdot \left(-\frac{\sqrt{3}}{2}\right) - \left(-\frac{\sqrt{3}}{2}\right) \cdot \left(-\frac{1}{2}\right)$$

$$= 0$$

Cofunction Identities

Our next set of identities relates each of the functions sine, cosine, and tangent to its corresponding cofunction, namely the cosine, sine, and cotangent, respectively.

COFUNCTION IDENTITIES

$$\cos\left(\frac{\pi}{2} - x\right) = \sin x$$

$$\sin\left(\frac{\pi}{2} - x\right) = \cos x$$

$$\tan\left(\frac{\pi}{2} - x\right) = \cot x$$

Proof As motivation for these identities, consider the case in which x is acute, as in the triangle in Figure 2. The second acute angle is $\pi/2 - x$, because the third angle is $\pi/2$ and the sum of the three angles is π. Using the definitions of the trigonometric functions for a right triangle, we see that

$$\sin x = \frac{\text{opposite}}{\text{hypotenuse}} = \frac{a}{c}$$

$$\cos x = \frac{\text{adjacent}}{\text{hypotenuse}} = \frac{b}{c}$$

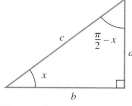

Figure 2

and that
$$\cos\left(\frac{\pi}{2} - x\right) = \frac{\text{adjacent}}{\text{hypotenuse}} = \frac{a}{c}$$
$$\sin\left(\frac{\pi}{2} - x\right) = \frac{\text{opposite}}{\text{hypotenuse}} = \frac{b}{c}$$

This verifies the first two cofunction identities. The proof of the third identity for the case in which x is acute is similar.

In the case of a nonacute angle x, we can give the following argument. To prove the first cofunction identity, apply the subtraction identity for the cosine with x replaced by $\pi/2$ and y replaced by x. This gives:
$$\cos\left(\frac{\pi}{2} - x\right) = \cos\frac{\pi}{2}\cos x + \sin\frac{\pi}{2}\sin x$$

Now use the facts that $\cos(\pi/2) = 0$ and $\sin(\pi/2) = 1$ to obtain:
$$\cos\left(\frac{\pi}{2} - x\right) = 1 \cdot \sin x$$
$$= \sin x$$

This proves the first cofunction identity.

To prove the second cofunction identity for nonacute x, start from the first cofunction identity and replace x by $\pi/2 - x$ to obtain:
$$\cos\left[\frac{\pi}{2} - \left(\frac{\pi}{2} - x\right)\right] = \sin\left(\frac{\pi}{2} - x\right)$$
$$\cos x = \sin\left(\frac{\pi}{2} - x\right)$$

This proves the second cofunction identity.

The third cofunction identity follows from the first two. Its proof is left as an exercise.

Addition and Subtraction Identities for the Sine Function

Using the cofunction identities and the addition and subtraction identities for the cosine, we can deduce addition and subtraction identities for the sine function. These identities express $\sin(x + y)$ and $\sin(x - y)$ in terms of $\sin x$, $\sin y$, $\cos x$, and $\cos y$:

ADDITION AND SUBTRACTION IDENTITIES FOR SINE

Let x and y be real numbers. Then we have the following identities:
$$\sin(x + y) = \sin x \cos y + \cos x \sin y$$
$$\sin(x - y) = \sin x \cos y - \cos x \sin y$$

Proof To prove the subtraction identity for the sine, start with the addition identity for the cosine:
$$\cos(x + y) = \cos x \cos y - \sin x \sin y$$
Replace x by $\pi/2 - x$:
$$\cos\left[\left(\frac{\pi}{2} - x\right) + y\right] = \cos\left(\frac{\pi}{2} - x\right)\cos y - \sin\left(\frac{\pi}{2} - x\right)\sin y$$

SECTION 8.2 The Addition and Subtraction Identities

$$\cos\left[\frac{\pi}{2} - (x - y)\right] = \sin x \cos y - \cos x \sin y \quad \textit{Cofunction formulas}$$

Apply the second cofunction identity to the left side:

$$\sin(x - y) = \sin x \cos y - \cos x \sin y$$

This is the subtraction identity for the sine.

To obtain the addition identity for the sine, start with the subtraction identity and replace y throughout by $-y$ to obtain:

$$\sin(x - (-y)) = \sin x \cos(-y) - \cos x \sin(-y)$$

Apply the parity identities

$$\sin(-x) = -\sin x, \qquad \cos(-x) = \cos x$$

to the previous equation:

$$\sin(x + y) = \sin x \cos y - \cos x (-\sin y)$$
$$\sin(x + y) = \sin x \cos y + \cos x \sin y$$

The Symmetry Identities Here are two additional identities that follow directly from the addition identities for the sine and cosine.

SYMMETRY IDENTITIES

$$\sin x = \sin(\pi - x)$$
$$\cos x = -\cos(\pi - x)$$

Proof By the addition identity for the sine, we have:

$$\sin(\pi - x) = \sin \pi \cos(-x) + \cos \pi \sin(-x)$$
$$= -\sin(-x) \qquad \textit{Because } \sin \pi = 0, \textit{ and } \cos \pi = -1$$
$$= \sin x \qquad \textit{Because } \sin(-x) = -\sin x$$

This proves the first symmetry identity. To prove the second, apply the addition identity for the cosine:

$$\cos(\pi - x) = \cos \pi \cos(-x) - \sin \pi \sin(-x)$$
$$= -\cos(-x) \qquad \textit{Because } \cos \pi = -1, \textit{ and } \sin \pi = 0$$
$$= -\cos x \qquad \textit{Because } \cos(-x) = \cos x$$

EXAMPLE 4
Trigonometric Functions of Special Angles

Determine the value of the following:

1. $\sin 75°$
2. $\sin 15°$

Solution

1. We can write $75° = 30° + 45°$. Therefore, by the addition identity for the sine, we have:

$$\sin 75° = \sin(30° + 45°)$$
$$= \sin 30° \cos 45° + \cos 30° \sin 45°$$

Now use the value for the trigonometric functions at 30° and 45°.

$$\sin 75° = \frac{1}{2} \cdot \frac{\sqrt{2}}{2} + \frac{\sqrt{3}}{2} \cdot \frac{\sqrt{2}}{2}$$

$$= \frac{\sqrt{2}}{4} + \frac{\sqrt{6}}{4} = \frac{\sqrt{2} + \sqrt{6}}{4}$$

2. We first note that $15° = 45° - 30°$. In this case, we apply the subtraction identity for the sine:

$$\sin 15° = \sin(45° - 30°)$$
$$= \sin 45° \cos 30° - \cos 45° \sin 30°$$
$$= \frac{\sqrt{2}}{2} \cdot \frac{\sqrt{3}}{2} - \frac{1}{2} \cdot \frac{\sqrt{2}}{2}$$
$$= \frac{\sqrt{6}}{4} - \frac{\sqrt{2}}{4} = \frac{\sqrt{6} - \sqrt{2}}{4}$$

Alternate Expressions for General Sine and Cosine Functions

The addition formula for the sine allows us to express the general sine $C \sin(x+D)$ in the form $A \sin x + B \cos x$ for suitable values of A and B. It is also possible to go backwards and express every linear combination of the form $A \sin x + B \cos x$ in the form $C \sin(x + D)$. Here's how.

The trick is to choose angle D correctly. Plot the point (A, B) on a Cartesian coordinate system. The distance of this point from the origin is $\sqrt{A^2 + B^2}$. (See Figure 3.) Draw a line from the origin to the point (A, B) and let D be the angle that has this line as its terminal side and the positive x-axis as its initial side.

From Figure 3, we see that:

$$\sin D = \frac{B}{\sqrt{A^2 + B^2}} \quad \text{and} \quad \cos D = \frac{A}{\sqrt{A^2 + B^2}}$$

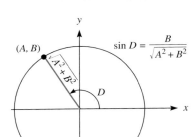

Figure 3

We then have:

$$A \sin x + B \cos x = \sqrt{A^2 + B^2} \left(\frac{A}{\sqrt{A^2 + B^2}} \cdot \sin x + \frac{B}{\sqrt{A^2 + B^2}} \cdot \cos x \right)$$

$$= \sqrt{A^2 + B^2}(\cos D \sin x + \sin D \cos x)$$

Now apply the addition identity for the sine to the expression in parentheses:

$$A \sin x + B \cos x = \sqrt{A^2 + B^2} \sin(x + D)$$

Therefore, if D is an angle as defined above, we have:

$$A \sin x + B \cos x = C \sin(x + D), \quad \text{where } C = \sqrt{A^2 + B^2}$$

EXAMPLE 5
Writing a Linear Combination of a Sine and a Cosine as a General Sine Function

Express $2 \sin x - 2 \cos x$ in the form $C \sin(x + D)$.

Solution

We apply the above formulas with $A = 2$, $B = -2$. In this case, we have:

$$C = \sqrt{A^2 + B^2} = \sqrt{2^2 + (-2)^2} = \sqrt{8} = 2\sqrt{2}$$

SECTION 8.2 The Addition and Subtraction Identities

$$\sin D = \frac{B}{C} = -\frac{2}{2\sqrt{2}} = -\frac{\sqrt{2}}{2}$$

To help us determine D from the last equation, we draw the angle D, as in Figure 4. We see that $D = 2\pi - \pi/4 = 7\pi/4$, and thus:

$$2 \sin x - 2 \cos x = C \sin(x + D) = 2\sqrt{2} \sin\left(x + \frac{7\pi}{4}\right)$$

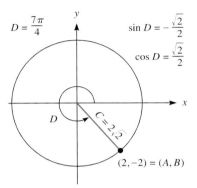

Figure 4

Addition and Subtraction Identities for the Tangent Function

From the addition and subtraction identities for the sine and cosine, we can deduce addition and subtraction identities for the tangent function. These identities express $\tan(x + y)$ and $\tan(x - y)$ in terms of $\tan x$ and $\tan y$.

ADDITION AND SUBTRACTION IDENTITIES FOR THE TANGENT

$$\tan(x + y) = \frac{\tan x + \tan y}{1 - \tan x \tan y}$$

$$\tan(x - y) = \frac{\tan x - \tan y}{1 + \tan x \tan y}$$

Proof Let's prove the first identity. By the definition of the tangent function, we have:

$$\tan(x + y) = \frac{\sin(x + y)}{\cos(x + y)}$$

Apply the addition identities for sine and cosine to the right side of the equation to obtain:

$$\tan(x + y) = \frac{\sin x \cos y + \cos x \sin y}{\cos x \cos y - \sin x \sin y}$$

Multiply the numerator and denominator on the right by $\dfrac{1}{\cos x \cos y}$ to obtain:

$$\tan(x + y) = \frac{\dfrac{\sin x \cos y + \cos x \sin y}{1}}{\dfrac{\cos x \cos y - \sin x \sin y}{1}} \cdot \frac{\dfrac{1}{\cos x \cos y}}{\dfrac{1}{\cos x \cos y}}$$

Multiplying out the rational expressions in the numerator and denominator gives:

$$\tan(x+y) = \frac{\dfrac{\sin x \cos y}{\cos x \cos y} + \dfrac{\cos x \sin y}{\cos x \cos y}}{\dfrac{\cos x \cos y}{\cos x \cos y} - \dfrac{\sin x \sin y}{\cos x \cos y}}$$

We simplify this last expression and rewrite it using the definition of the tangent:

$$\tan(x+y) = \frac{\dfrac{\sin x}{\cos x} + \dfrac{\sin y}{\cos y}}{1 - \dfrac{\sin x}{\cos x} \cdot \dfrac{\sin y}{\cos y}}$$

$$= \frac{\tan x + \tan y}{1 - \tan x \tan y}$$

This completes the proof of the addition identity for the tangent function. The proof of the subtraction identity is left as an exercise.

EXAMPLE 6
Tangent of a Special Angle

Determine the exact value of $\tan \pi/12$. Do not use a calculator.

Solution

First note that $\pi/12 = \pi/3 - \pi/4$. Therefore, we can apply the subtraction identity for the tangent to obtain:

$$\tan \frac{\pi}{12} = \frac{\tan \pi/3 - \tan \pi/4}{1 + \tan \pi/3 \tan \pi/4}$$

From calculations earlier in the chapter, we know that $\tan \pi/3 = \sqrt{3}$, and $\tan \pi/4 = 1$. Insert these values into the above identity:

$$\tan \frac{\pi}{12} = \frac{\sqrt{3} - 1}{1 + \sqrt{3} \cdot 1} = \frac{\sqrt{3} - 1}{\sqrt{3} + 1}$$

We can simplify this answer by rationalizing the denominator. We multiply both numerator and denominator by $-\sqrt{3} + 1$, which is the conjugate of the denominator. This yields:

$$\tan \frac{\pi}{12} = \frac{\sqrt{3} - 1}{\sqrt{3} + 1} \cdot \frac{-\sqrt{3} + 1}{-\sqrt{3} + 1}$$

$$= \frac{-3 + 2\sqrt{3} - 1}{-3 + 1}$$

$$= \frac{-4 + 2\sqrt{3}}{-2}$$

$$= 2 - \sqrt{3}$$

EXAMPLE 7
The Period of the Tangent

Prove that the function $\tan x$ has period π.

Solution

We must show that $\tan(x + \pi) = \tan x$ for all x. Apply the addition identity for the tangent to evaluate $\tan(x + \pi)$:

SECTION 8.2 The Addition and Subtraction Identities

$$\tan(x + \pi) = \frac{\tan x + \tan \pi}{1 - \tan x \tan \pi}$$

$$= \frac{\tan x + 0}{1 - 0 \cdot \tan x}$$

$$= \tan x$$

Angle between Two Lines

Let's now use the subtraction identity for the tangent to prove the following useful formula.

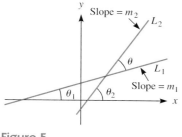

Figure 5

IDENTITY FOR THE TANGENT OF THE ANGLE BETWEEN TWO LINES

Let L_1 and L_2 be straight lines that have slopes m_1 and m_2, respectively, and let L_2 have the larger angle of inclination. (The angle of inclination is the angle the line makes with the positive x-axis.) Let θ be the angle between the two lines L_1 and L_2. Then:

$$\tan \theta = \frac{m_2 - m_1}{1 + m_1 m_2}$$

(See Figure 5.)

Proof Let θ_1 and θ_2 denote the angles of inclination of L_1 and L_2, respectively. By the definition of slope, if we start at any point on L_1 and proceed 1 unit in the positive x-direction, we must move m_1 units vertically in order to return to the line L_1. By referring to Figure 5 and using the definition of the tangent function, we see that:

$$\tan \theta_1 = \frac{\text{opposite}}{\text{adjacent}} = \frac{m_1}{1} = m_1$$

In a similar fashion, we have:

$$\tan \theta_2 = m_2$$

Referring to Figure 5, we see that the angle θ between L_1 and L_2 is $\theta_2 - \theta_1$. Therefore, we have:

$$\tan \theta = \tan(\theta_2 - \theta_1)$$

By the subtraction identity for the tangent function, we then have:

$$\tan \theta = \frac{\tan \theta_2 - \tan \theta_1}{1 + \tan \theta_1 \tan \theta_2}$$

$$= \frac{m_2 - m_1}{1 + m_1 m_2}$$

This proves the identity for the tangent of the angle between two lines.

EXAMPLE 8
Calculating the Angle between Two Lines

Determine the angle between the lines that have the equations $y = 5x + 2$ and $y = -x + 3$.

Solution

We apply the identity just proved. The slopes are $m_1 = 5$ and $m_2 = -1$. Let θ denote the angle between the lines. Then:

$$\tan\theta = \frac{m_2 - m_1}{1 + m_1 m_2}$$

$$= \frac{-1 - 5}{1 + (-5)}$$

$$= \frac{-6}{-4}$$

$$= \frac{3}{2}$$

Using a calculator, we find:

$$\theta \approx 56.310°$$

EXAMPLE 9
The Angle between Two Ships' Courses

Two ships are on intersecting courses such that, if a coordinate axis were placed with the center at their projected intersection, one ship would currently be at the point $(-10, -4)$ and the other would be at the point $(-4, 7)$. Find the angle between the courses of the two ships.

Solution

We have drawn the courses of the two ships in Figure 6. The line L_1 has slope:

$$m_1 = \frac{-4 - 0}{-10 - 0} = \frac{2}{5}$$

The line L_2 has slope:

$$m_2 = \frac{7 - 0}{-4 - 0} = -\frac{7}{4}$$

Therefore, we have:

$$\tan\theta = \frac{m_2 - m_1}{1 + m_1 m_2}$$

$$= \frac{(-\frac{7}{4}) - (\frac{2}{5})}{1 + \frac{2}{5} \cdot (-\frac{7}{4})}$$

$$= -\frac{43}{6}$$

Using a calculator, we find that:

$$\theta \approx -82.057°$$

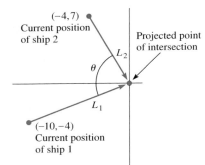

Figure 6

Exercises 8.2

Determine the exact value of the following expressions, using an addition or subtraction identity. Do not use a calculator.

1. $\cos\dfrac{7\pi}{12}$
2. $\sin\dfrac{5\pi}{12}$
3. $\tan\dfrac{17\pi}{12}$
4. $\cos\left(-\dfrac{13\pi}{12}\right)$
5. $\sin 255°$
6. $\tan 285°$
7. $\cos 15°$
8. $\sin\left(-\dfrac{11\pi}{12}\right)$
9. $\tan(-165°)$
10. $\cos 105°$
11. $\sin(-15°)$
12. $\tan\left(-\dfrac{5\pi}{12}\right)$

SECTION 8.2 The Addition and Subtraction Identities

Use the addition and subtraction identities to determine alternate expressions for each of the following.

13. $\cos\left(\dfrac{\pi}{2} + \theta\right)$

14. $\cos\left(\theta - \dfrac{\pi}{2}\right)$

15. $\sin\left(\theta + \dfrac{\pi}{3}\right)$

16. $\sin\left(\theta - \dfrac{\pi}{3}\right)$

17. $\tan\left(\theta - \dfrac{\pi}{4}\right)$

18. $\tan\left(\theta + \dfrac{\pi}{4}\right)$

19. $\cos(30° - \theta)$

20. $\cos(30° + \theta)$

21. $\sin(45° - \theta)$

22. $\sin(45° + \theta)$

23. $\tan(\theta - 135°)$

24. $\tan(\theta + 135°)$

Use the addition and subtraction identities to determine the value of each of the following expressions.

25. $\cos\left(\dfrac{\pi}{2} + x\right)$, if $\cos x = \dfrac{1}{2}$ and $0 < x < \dfrac{\pi}{2}$

26. $\cos\left(\theta - \dfrac{\pi}{2}\right)$, if $\sin\theta = \dfrac{\sqrt{3}}{2}$ and $\dfrac{\pi}{2} < \theta < \pi$

27. $\sin\left(\theta + \dfrac{\pi}{3}\right)$, if $\sin\theta = \dfrac{\sqrt{2}}{2}$

28. $\sin\left(x + \dfrac{\pi}{6}\right)$, if $\cos x = -\dfrac{1}{2}$ and $\pi < x < \dfrac{3\pi}{2}$

29. $\sin\left(\dfrac{\pi}{3} - \theta\right)$, if $\cos\theta = -\dfrac{1}{2}$ and $\pi < \theta < \dfrac{3\pi}{2}$

30. $\tan\left(\theta + \dfrac{\pi}{4}\right)$, if $\tan\theta = -1$ and $\dfrac{3\pi}{2} < \theta < 2\pi$

31. $\tan\left(\theta - \dfrac{\pi}{4}\right)$, if $\sin\theta = -\dfrac{1}{2}$ and $\dfrac{3\pi}{2} < \theta < 2\pi$

32. $\tan\left(\dfrac{\pi}{6} + x\right)$, if $\cos x = \dfrac{\sqrt{3}}{2}$ and $0 < x < \dfrac{\pi}{2}$

Determine the exact value of the indicated trigonometric expression, given the following information. Do not use a calculator.

33. $\cos(x + y)$; if $\sin x = \dfrac{\sqrt{2}}{2}$, $\dfrac{\pi}{2} < x < \pi$
    and $\sin y = \dfrac{1}{2}$, $0 < y < \dfrac{\pi}{2}$

34. $\cos(x - y)$; if $\sin x = -\dfrac{\sqrt{3}}{2}$, $\dfrac{3\pi}{2} < x < 2\pi$
    and $\sin y = -\dfrac{\sqrt{2}}{2}$, $\pi < y < \dfrac{3\pi}{2}$

35. $\sin(x + y)$; if $\cos x = -\dfrac{1}{2}$, $\dfrac{\pi}{2} < x < \dfrac{3\pi}{2}$
    and $\cos y = -\dfrac{1}{2}$, $\dfrac{\pi}{2} < y < \pi$

36. $\sin(x - y)$; if $\sin x = -\dfrac{1}{2}$, $\pi < x < \dfrac{3\pi}{2}$
    and $\sin y = -\dfrac{\sqrt{3}}{2}$, $\dfrac{3\pi}{2} < y < 2\pi$

37. $\cos 2x$; if $\sin x = \dfrac{3}{5}$, $0 < x < \dfrac{\pi}{2}$
    (**Hint:** $2x = x + x$)

38. $\sin 2x$; if $\cos x = -\dfrac{12}{13}$, $\pi < x < \dfrac{3\pi}{2}$
    (**Hint:** $2x = x + x$)

39. $\tan(x + y)$; if $\tan x = -1$, $\dfrac{3\pi}{2} < x < 2\pi$
    and $\tan y = -\sqrt{3}$, $\dfrac{3\pi}{2} < y < 2\pi$

40. $\tan(x - y)$; if $\sin x = \dfrac{1}{2}$, $0 < x < \dfrac{\pi}{2}$
    and $\cos y = -\dfrac{1}{2}$, $\dfrac{\pi}{2} < y < \pi$

41. $\cos(y + x)$; if $\cos y = \dfrac{3}{5}$, $\cos x = \dfrac{4}{5}$,
    and x, y both between 0 and $\dfrac{\pi}{2}$

42. $\cos(y - x)$; if $\cos y = -\dfrac{3}{5}$, $\dfrac{\pi}{2} < x < \pi$
    and $\cos x = \dfrac{4}{5}$, $\dfrac{3\pi}{2} < x < 2\pi$

43. $\sin(y + x)$; if $\sin y = \dfrac{4}{5}$, $0 < y < \dfrac{\pi}{2}$
    and $\cos x = -\dfrac{3}{5}$, $\pi < x < \dfrac{3\pi}{2}$

44. $\sin(y - x)$; if $\sin y = \dfrac{12}{13}$, $\dfrac{\pi}{2} < y < \pi$
    and $\cos x = -\dfrac{5}{13}$, $\dfrac{\pi}{2} < x < \pi$

45. $\tan(y + x)$; if $\cos y = -\dfrac{12}{13}$, $\dfrac{\pi}{2} < y < \pi$
    and $\sin x = -\dfrac{5}{13}$, $\dfrac{3\pi}{2} < x < 2\pi$

46. $\tan(y - x)$; if $\cos y = \dfrac{3}{5}$, $\dfrac{3\pi}{2} < y < 2\pi$
    and $\sin x = -\dfrac{4}{5}$, $\pi < x < \dfrac{3\pi}{2}$

Find the tangent of the smallest positive angle between the two given lines.

47. $2x - 2y = 6$ and $4x + 3y = 9$

48. $y = \frac{2}{3}x - 1$ and $y = -\frac{1}{2}x + 2$

49. $x = 3$ and $y = -1$

50. $y = x$ and $y = -x$

51. $y = \frac{4}{5}x + 1$ and $2x + y = 0$

52. $y = \frac{1}{5}x - 2$ and $x - 2y = 3$

53. A coordinate axis is placed with its center at the intersection of two roads. If one road goes through the point $(4, 5)$ and the other road goes through the point $(-3, -2)$, find the angle formed by the two roads at their point of intersection.

54. Two ships are on intersecting courses such that, if a coordinate axis were placed with its center at their projected intersection, one ship would be at the point $(-3, 2)$ and the other would be at the point $(-5, -3)$. Find the angle between the two ships at their point of intersection.

Prove each of the following identities.

55. $\tan\left(\frac{\pi}{2} - x\right) = \cot x$

56. $\cot\left(\frac{\pi}{2} - x\right) = \tan x$

57. $\sec\left(\frac{\pi}{2} - x\right) = \csc x$

58. $\csc\left(\frac{\pi}{2} - x\right) = \sec x$

Express each of the following in the form $C \sin(x + D)$.

59. $\sin x + \cos x$

60. $\sin x - \cos x$

61. $\sin x + \sqrt{3} \cos x$

62. $4 \sin x - 3 \cos x$

63. $5 \sin x - 13 \cos x$

64. $3 \sin x + 4 \cos x$

65. $\sin \frac{\pi}{4} t - \sqrt{3} \cos \frac{\pi}{4} t$

66. $4 \sin 2x + 3 \cos 2x$

67. If $f(x) = \sin x$, prove that:
$$\frac{f(x + h) - f(x)}{h} = \sin x \left(\frac{\cos h - 1}{h}\right) + \cos x \left(\frac{\sin h}{h}\right)$$

68. If $f(x) = \cos x$, prove that:
$$\frac{f(x + h) - f(x)}{h} = -\sin x \left(\frac{\sin h}{h}\right) + \cos x \left(\frac{\cos h - 1}{h}\right)$$

69. If $f(x) = \tan x$, prove that:
$$\frac{f(x + h) - f(x)}{h} = \frac{\sec^2 x}{\cos h - \sin h \tan x} \left(\frac{\sin h}{h}\right)$$

70. If $f(x) = \csc x$, prove that:
$$\frac{f(x + h) - f(x)}{h} = \frac{1}{h(\sin x \cos h + \cos x \sin h)} - \frac{1}{h \sin x}$$

Prove each of the following identities.

71. $2 \cos \alpha \cos \theta - \cos(\alpha - \theta) = \cos(\alpha + \theta)$

72. $\sin 2x = 2 \sin x \cos x$ (**Hint:** $2x = x + x$)

73. $\sin(x + y) - \sin(x - y) = 2 \sin y \cos x$

74. $\sin(x + y) \sec(x + y) = \frac{\tan x + \tan y}{1 - \tan x \tan y}$

75. $\cos(x + y) - \cos(x - y) = -2 \sin x \sin y$

76. $\sin(x - y) \sin(x + y) = \sin^2 x - \sin^2 y$

77. $\sin\left(\frac{\pi}{4} + x\right) - \sin\left(\frac{\pi}{4} - x\right) = \sqrt{2} \sin x$

78. $\cos\left(\frac{\pi}{6} + x\right)\cos\left(\frac{\pi}{6} - x\right) - \sin\left(\frac{\pi}{6} + x\right)\sin\left(\frac{\pi}{6} - x\right) = \frac{1}{2}$

Technology

79. Use a graphing calculator to check the plausibility of the addition identity for the sine, by testing it with five different values of y.

80. Use a graphing calculator to check the plausibility of the addition identity for the cosine, by testing it with five different values of y.

81. With a graphing calculator, demonstrate the symmetry identities.

82. With a graphing calculator, demonstrate the cofunction identities.

In your own words

83. State the addition identities for the sine and cosine. Give an illustration of each identity.

84. Compare and contrast the symmetry and cofunction identities.

8.3 MULTIPLE AND HALF-ANGLE IDENTITIES

For many applications of the trigonometric functions, especially for techniques of integration in calculus, it is necessary to use a variety of different identities to write trigonometric expressions in alternate forms. In this section, we prove a number of identities that give expressions of trigonometric functions for double angles and half angles. Here is a summary of these identities:

Double-Angle Identities

$$\sin 2u = 2 \sin u \cos u$$
$$\cos 2u = \cos^2 u - \sin^2 u$$
$$= 1 - 2\sin^2 u$$
$$= 2\cos^2 u - 1$$
$$\tan 2u = \frac{2 \tan u}{1 - \tan^2 u}$$

Square Identities

$$\sin^2 x = \frac{1 - \cos 2x}{2}$$
$$\cos^2 x = \frac{1 + \cos 2x}{2}$$
$$\tan^2 x = \frac{1 - \cos 2x}{1 + \cos 2x}$$

Half-Angle Identities

$$\sin \frac{x}{2} = \pm \sqrt{\frac{1 - \cos x}{2}}$$
$$\cos \frac{x}{2} = \pm \sqrt{\frac{1 + \cos x}{2}}$$
$$\tan \frac{x}{2} = \pm \sqrt{\frac{1 - \cos x}{1 + \cos x}} = \frac{1 - \cos x}{\sin x} = \frac{\sin x}{1 + \cos x}$$

Double-Angle Identities

Let's begin with the following identities for the trigonometric functions of double angles.

DOUBLE-ANGLE IDENTITIES

$$\sin 2u = 2 \sin u \cos u$$
$$\cos 2u = \cos^2 u - \sin^2 u$$
$$\tan 2u = \frac{2 \tan u}{1 - \tan^2 u}$$

Proof To prove the first identity, start with the addition identity for the sine and replace y by x throughout to obtain:

$$\sin(x + x) = \sin x \cos x + \cos x \sin x = 2 \sin x \cos x$$
$$\sin 2x = 2 \sin x \cos x$$

To prove the second identity, start with the addition identity for the cosine and replace y by x throughout to obtain:

$$\cos(x + x) = \cos x \cos x - \sin x \sin x = \cos^2 x - \sin^2 x$$
$$\cos 2x = \cos^2 x - \sin^2 x$$

CHAPTER 8 Analytic Trigonometry

In a similar fashion, we prove the third identity by replacing y by x in the addition identity for the tangent function. This completes the proof of the theorem.

Starting with the double-angle identity for the cosine, we replace $\sin^2 x$ by $1 - \cos^2 x$, according to the Pythagorean theorem, and we obtain the following alternate form of the double-angle identity:

$$\cos 2x = 2\cos^2 x - 1$$

Again, starting with the double-angle identity for the cosine and using the Pythagorean theorem, we replace $\cos^2 x$ with $1 - \sin^2 x$ to arrive at the second alternate form of the identity:

$$\cos 2x = 1 - 2\sin^2 x$$

EXAMPLE 1
Trigonometric Functions for Double Angles

Suppose that for a certain angle x in the first quadrant, we have $\sin x = 0.6$. Determine the value of $\sin 2x$, $\cos 2x$, and $\tan 2x$.

Solution

First, we find the values of $\cos x$ and $\tan x$.

$$\cos x = \pm\sqrt{1 - \sin^2 x} = \pm\sqrt{1 - (0.6)^2} = \pm 0.8$$

Because x is in the first quadrant, the plus sign applies:

$$\cos x = 0.8$$

$$\tan x = \frac{\sin x}{\cos x} = \frac{0.6}{0.8} = 0.75$$

Next, apply the double-angle identities and these values of the trigonometric functions at x to obtain:

$$\sin 2x = 2\sin x \cos x$$
$$= 2(0.6)(0.8)$$
$$= 0.96$$

$$\cos 2x = 2\cos^2 x - 1$$
$$= 2(0.8)^2 - 1$$
$$= 0.28$$

$$\tan 2x = \frac{2\tan x}{1 - \tan^2 x}$$
$$= \frac{2 \cdot (0.75)}{1 - (0.75)^2}$$
$$= 3.4286$$

EXAMPLE 2
Triple-Angle Identity

Express $\sin 3x$ in terms of $\sin x$.

Solution

Write

$$\sin 3x = \sin(x + 2x)$$

Now apply the addition identity for the sine function along with the double-angle identity for the sine:

$$\begin{aligned}
\sin(x + 2x) &= \sin x \cos 2x + \cos x \sin 2x \\
&= \sin x(1 - 2\sin^2 x) + \cos x \cdot 2 \sin x \cos x \\
&= -2\sin^3 x + \sin x + 2 \sin x \cos^2 x \\
&= -2\sin^3 x + \sin x + 2 \sin x(1 - \sin^2 x) \\
&= -2\sin^3 x + \sin x + 2 \sin x - 2 \sin^3 x \\
&= -4\sin^3 x + 3 \sin x
\end{aligned}$$

EXAMPLE 3
Proving an Identity Using the Double-Angle Identities

Prove the following identity:

$$\sec 2x = \frac{\sec^2 x + \sec^4 x}{2 + \sec^2 x - \sec^4 x}$$

Solution

Factor the numerator and denominator of the right side to obtain:

$$\frac{\sec^2 x + \sec^4 x}{2 + \sec^2 x - \sec^4 x} = \frac{\sec^2 x(1 + \sec^2 x)}{(2 - \sec^2 x)(1 + \sec^2 x)}$$

$$= \frac{\sec^2 x}{2 - \sec^2 x}$$

$$= \frac{\dfrac{1}{\cos^2 x}}{2 - \dfrac{1}{\cos^2 x}} \qquad \text{\textit{Definition of} } \sec x$$

$$= \frac{1}{2\cos^2 x - 1}$$

$$= \frac{1}{\cos 2x} \qquad \text{\textit{Because} } \cos 2x = 2\cos^2 x - 1$$

$$= \sec 2x \qquad \text{\textit{Definition of} } \sec x$$

This proves the identity.

EXAMPLE 4
Nautical Mile

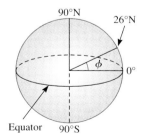

Figure 1

Latitude is used to measure north-south location on the earth between the equator and the poles. For example, the city of Hollywood, Florida, has latitude 26°N, that is, it lies 26° north of the equator. See Figure 1. In Great Britain, the nautical mile is defined as the length of a minute of arc of the earth's radius. Because the earth is flattened at the poles, a British nautical mile varies with latitude. In fact, it is given in feet, by the function $N(\phi) = 6{,}066 - 31 \cos 2\phi$, where ϕ is the latitude, in degrees.

1. What is the length of a British nautical mile of the city of Hollywood, Florida?
2. What is the length of a British nautical mile at the North Pole?
3. Express $N(\phi)$ in terms of cosine only. That is, eliminate the double angle.

Solution

1. For Hollywood, Florida, we have $\phi = 26°$, so that:
$$N(\phi) = 6{,}066 - 31 \cos 2 \cdot 26°$$
$$= 6{,}066 - 31 \cos 52°$$
$$= 6{,}046.91 \text{ ft}$$

2. At the North Pole, we have $\phi = 90°$, so that:
$$N(\phi) = 6{,}066 - 31 \cos 2\phi$$
$$= 6{,}066 - 31 \cos 2 \cdot 90°$$
$$= 6{,}066 - 31 \cos 180°$$
$$= 6{,}066 - 31 \cdot (-1)$$
$$= 6{,}097 \text{ ft}$$

3. Applying the double-angle identity for the cosine, we have:
$$N(\phi) = 6{,}066 - 31 \cos 2\phi$$
$$= 6{,}066 - 31(2\cos^2(\phi) - 1)$$
$$= 6{,}097 - 62 \cos^2 \phi$$

EXAMPLE 5
Acceleration Due to Gravity

The acceleration due to gravity is usually denoted by g. It is related to the physics of motion near the earth's surface. Usually g is considered constant; in fact, however, g is not constant, but varies slightly with latitude. If ϕ represents latitude in degrees, g is given with good approximation by the identity
$$g = 9.78049(1 + 0.005288 \sin^2 \phi - 0.000006 \sin^2 2\phi)$$
where g is measured in m/sec² at sea level.

1. Hollywood, Florida, has latitude 26°N. Find g.
2. Anchorage, Alaska, has latitude 61.8°N. Find g.
3. Express g in terms of $\sin \phi$ only. That is, eliminate the double angle.

Solution

1. In Hollywood, Florida, we have:
$$g = 9.78049(1 + 0.005288 \sin^2 \phi - 0.000006 \sin^2 2\phi)$$
$$= 9.78049(1 + 0.005288 \sin^2 26° - 0.000006 \sin^2 52°)$$
$$= 9.79039 \text{ m/sec}^2$$

2. In Anchorage, Alaska, we have:
$$g = 9.78049(1 + 0.005288 \sin^2 \phi - 0.000006 \sin^2 2\phi)$$
$$= 9.78049(1 + 0.005288 \sin^2 61.8° - 0.000006 \sin^2 123.6°)$$
$$= 9.82062 \text{ m/sec}^2$$

3. Applying the double-angle identity for the sine, we have:
$$g = 9.78049(1 + 0.005288 \sin^2 \phi - 0.000006 \sin^2 2\phi)$$
$$= 9.78049(1 + 0.005288 \sin^2 \phi - .000024 \sin^2 \phi \cos^2 \phi)$$
$$= 9.78049(1 + 0.005288 \sin^2 \phi - .000024 \sin^2 \phi(1 - \sin^2 \phi))$$
$$= 9.78049(1 + .005264 \sin^2 \phi + .000024 \sin^4 \phi)$$

SECTION 8.3 Multiple and Half-Angle Identities

Square and Half-Angle Identities

Next, let's prove a series of identities that provide alternate expressions for the squares of the sine, cosine, and tangent functions. These expressions are useful in calculus as well as in proving the half-angle identities later in this section.

SQUARE IDENTITIES

$$\sin^2 x = \frac{1 - \cos 2x}{2}$$

$$\cos^2 x = \frac{1 + \cos 2x}{2}$$

$$\tan^2 x = \frac{1 - \cos 2x}{1 + \cos 2x}$$

Proof To prove the first identity, start from the identity for $\cos 2x$:

$$\cos 2x = 1 - 2\sin^2 x$$

Solve this equation for $\sin^2 x$:

$$\cos 2x - 1 = -2\sin^2 x$$

$$-\frac{1}{2}(\cos 2x - 1) = \sin^2 x$$

$$\sin^2 x = \frac{-\cos 2x + 1}{2} = \frac{1 - \cos 2x}{2}$$

The second identity is proved in a similar fashion, starting from the double-angle identity:

$$\cos 2x = 2\cos^2 x - 1$$

To prove the third identity, start from the definition of the tangent function $\tan^2 x = (\sin x / \cos x)^2$, and apply the first two identities to obtain:

$$\tan^2 x = \left(\frac{\sin x}{\cos x}\right)^2$$

$$= \frac{\sin^2 x}{\cos^2 x}$$

$$= \frac{\frac{1 - \cos 2x}{2}}{\frac{1 + \cos 2x}{2}}$$

$$= \frac{1 - \cos 2x}{1 + \cos 2x}$$

This completes the proof of the third identity.

To close this section, let's derive identities for the sine, cosine, and tangent for half an angle, namely $x/2$.

HALF-ANGLE IDENTITIES

Let x be a real number. Then:

$$\sin \frac{x}{2} = \pm \sqrt{\frac{1 - \cos x}{2}}$$

$$\cos \frac{x}{2} = \pm \sqrt{\frac{1 + \cos x}{2}}$$

$$\tan \frac{x}{2} = \pm \sqrt{\frac{1 - \cos x}{1 + \cos x}} = \frac{1 - \cos x}{\sin x} = \frac{\sin x}{1 + \cos x}$$

where the sign depends on the quadrant in which $x/2$ lies.

Proof To prove the first identity, start from the square identity for $\sin^2 x$ and replace x by $x/2$ throughout. This gives us the identity:

$$\sin^2 \frac{x}{2} = \frac{1 - \cos(2 \cdot x/2)}{2} = \frac{1 - \cos x}{2}$$

Now take the square roots of both sides, to obtain:

$$\sin \frac{x}{2} = \pm \sqrt{\frac{1 - \cos x}{2}}$$

This identity is valid for an appropriate choice of sign on the right. Moreover, it is clear that the sign of $\sin x/2$ depends only on the quadrant of $x/2$. This establishes the first half-angle identity.

The proof of the second half-angle identity is similar to the proof of the first.

To prove the third half-angle identity, start from the first two identities to obtain:

$$\tan \frac{x}{2} = \frac{\sin \frac{x}{2}}{\cos \frac{x}{2}} = \pm \sqrt{\frac{1 - \cos x}{1 + \cos x}}$$

This is the first equality of the third half-angle identity. To prove the second equality, we write:

$$\tan \frac{x}{2} = \pm \sqrt{\frac{1 - \cos x}{1 + \cos x} \cdot \frac{1 - \cos x}{1 - \cos x}}$$

$$= \pm \sqrt{\frac{(1 - \cos x)^2}{1 - \cos^2 x}}$$

$$= \pm \sqrt{\frac{(1 - \cos x)^2}{\sin^2 x}}$$

$$= \pm \frac{1 - \cos x}{|\sin x|}$$

Note that $1 - \cos x \geq 0$ and that $\tan x/2$ and $\sin x$ have the same sign—both are positive for $0 < x < \pi$ and negative for $\pi < x < 2\pi$. Therefore the plus sign holds in the last equation:

$$\tan \frac{x}{2} = \frac{1 - \cos x}{\sin x}$$

This proves the second equality of the third half-angle identity. The final equality is proved similarly.

SECTION 8.3 Multiple and Half-Angle Identities

EXAMPLE 6
Trigonometric Functions for a Half Angle

Suppose that $\sin x = \frac{3}{5}$ and x lies in the second quadrant. Determine the values of $\sin x/2$, $\cos x/2$, and $\tan x/2$.

Solution

Draw the angle x as shown in Figure 2. From the value of $\sin x$, we can calculate that $\cos x = -\frac{4}{5}$ and $\tan x = -\frac{3}{4}$. Because $\pi/2 < x < \pi$, we have $\pi/4 < x/2 < \pi/2$, so that $x/2$ lies in the first quadrant. Therefore, by the half-angle identities, we have:

$$\sin \frac{x}{2} = +\sqrt{\frac{1 - \cos x}{2}}$$

$$= \sqrt{\frac{1 - (-\frac{4}{5})}{2}}$$

$$= \sqrt{\frac{9}{10}}$$

$$= \frac{3\sqrt{10}}{10}$$

$$\cos \frac{x}{2} = +\sqrt{\frac{1 + \cos x}{2}}$$

$$= \sqrt{\frac{1 + (-\frac{4}{5})}{2}}$$

$$= \sqrt{\frac{1}{10}}$$

$$= \frac{\sqrt{10}}{10}$$

$$\tan \frac{x}{2} = +\sqrt{\frac{1 - \cos x}{1 + \cos x}}$$

$$= \sqrt{\frac{1 - (-\frac{4}{5})}{1 + (-\frac{4}{5})}}$$

$$= 3$$

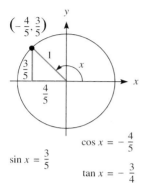

$\sin x = \frac{3}{5}$
$\cos x = -\frac{4}{5}$
$\tan x = -\frac{3}{4}$

Figure 2

EXAMPLE 7
Applying Half-Angle Identities

Use the half-angle identities to determine the exact values of the following expressions. Do not use a calculator.

1. $\sin \frac{\pi}{8}$
2. $\tan 105°$

Solution

1. The angle $\pi/8$ lies in the first quadrant. Therefore, the value of $\sin \pi/8$ is positive. By the half-angle identity, we have:

$$\sin\frac{\pi}{8} = +\sqrt{\frac{1-\cos\pi/4}{2}}$$

$$= \sqrt{\frac{1-\frac{\sqrt{2}}{2}}{2}}$$

$$= \sqrt{\frac{2-\sqrt{2}}{4}}$$

$$= \frac{\sqrt{2-\sqrt{2}}}{2}$$

2. The angle 105° lies in the second quadrant, so that the value of its tangent is negative. Twice this angle is 210°. The value of $\cos 210°$ is $-\sqrt{3}/2$. (See Figure 3.) By the half-angle identity, we have:

$$\tan 105° = -\sqrt{\frac{1-\cos 210°}{1+\cos 210°}}$$

$$= -\sqrt{\frac{1-(-\frac{\sqrt{3}}{2})}{1+(-\frac{\sqrt{3}}{2})}}$$

$$= -\sqrt{\frac{2+\sqrt{3}}{2-\sqrt{3}}}$$

$$= -\sqrt{\frac{2+\sqrt{3}}{2-\sqrt{3}} \cdot \frac{2+\sqrt{3}}{2+\sqrt{3}}}$$

$$= -\sqrt{\frac{(2+\sqrt{3})^2}{4-3}}$$

$$= -2-\sqrt{3}$$

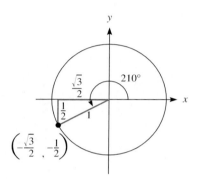

Figure 3

EXAMPLE 8
A Half-Angle Identity for Secant

Prove the identity:

$$\sec^2\frac{x}{2} = \frac{2}{1+\cos x}$$

Solution

By the half-angle identity for $\cos x$, we have:

$$\sec^2\frac{x}{2} = \frac{1}{\cos^2\frac{x}{2}}$$

$$= \frac{1}{\left(\pm\frac{\sqrt{1+\cos x}}{2}\right)^2}$$

SECTION 8.3 Multiple and Half-Angle Identities

$$= \frac{1}{\frac{1+\cos x}{2}}$$

$$= \frac{2}{1+\cos x}$$

This proves the identity.

Just because an expression involves half-angles does not necessarily mean that the easiest way to simplify the expression is to use the half-angle identity. The following example shows that sometimes the easiest method is to apply the double-angle identity.

EXAMPLE 9
Writing a Trigonometric Expression in Terms of the Sine

Write the following expression using only the sine function:

$$\sec \frac{x}{2} \csc \frac{x}{2}$$

Solution

We have:

$$\sec \frac{x}{2} \csc \frac{x}{2} = \frac{1}{\cos \frac{x}{2}} \cdot \frac{1}{\sin \frac{x}{2}}$$

$$= \frac{1}{\sin \frac{x}{2} \cos \frac{x}{2}}$$

$$= \frac{2}{2 \sin \frac{x}{2} \cos \frac{x}{2}}$$

$$= \frac{2}{\sin\left(2 \cdot \frac{x}{2}\right)} \qquad \text{Double-angle identity}$$

$$= \frac{2}{\sin x}$$

$$= 2 \cdot \frac{1}{\sin x}$$

Note that if we had used the half-angle identities, it would have been necessary to determine the sign of the resulting expression—a nasty job.

EXAMPLE 10
Proving an Identity Involving Half-Angles

Prove the identity:

$$\cos x = \frac{1 - \tan^2 \frac{x}{2}}{1 + \tan^2 \frac{x}{2}}$$

Solution

Rewrite the right-hand side using the half-angle for the tangent function:

$$\frac{1 - \tan^2 \frac{x}{2}}{1 + \tan^2 \frac{x}{2}} = \frac{1 - \left(\pm\sqrt{\frac{1 - \cos x}{1 + \cos x}}\right)^2}{1 + \left(\pm\sqrt{\frac{1 - \cos x}{1 + \cos x}}\right)^2}$$

$$= \frac{1 - \frac{1 - \cos x}{1 + \cos x}}{1 + \frac{1 - \cos x}{1 + \cos x}}$$

$$= \frac{\frac{(1 + \cos x) - (1 - \cos x)}{1 + \cos x}}{\frac{(1 + \cos x) + (1 - \cos x)}{1 + \cos x}}$$

$$= \frac{\frac{2 \cos x}{1 + \cos x}}{\frac{2}{1 + \cos x}}$$

$$= \frac{2 \cos x}{2}$$

$$= \cos x$$

This proves the identity.

Product and Factoring Identities

We can replace any product of a sine and cosine with a sum of trigonometric functions by using the so-called **factoring identities.** Or we can go backwards and replace any sum of a sine and a cosine with a product by using the **product identities.** Here are these useful identities:

Factoring Identities (Product to Sum)

$\sin x \cos y = \frac{1}{2}[\sin(x + y) + \sin(x - y)]$

$\sin x \sin y = \frac{1}{2}[\cos(x - y) - \cos(x + y)]$

$\cos x \cos y = \frac{1}{2}[\cos(x + y) + \cos(x - y)]$

Product Identities (Sum to Product)

$\sin x + \sin y = 2 \sin\left(\frac{x + y}{2}\right) \cos\left(\frac{x - y}{2}\right)$

$\sin x - \sin y = 2 \sin\left(\frac{x - y}{2}\right) \cos\left(\frac{x + y}{2}\right)$

$\cos x + \cos y = 2 \cos\left(\frac{x + y}{2}\right) \cos\left(\frac{x - y}{2}\right)$

$\cos x - \cos y = -2 \sin\left(\frac{x + y}{2}\right) \sin\left(\frac{x - y}{2}\right)$

SECTION 8.3 Multiple and Half-Angle Identities

To prove the first factoring identity, use the addition and subtraction identities for the sine:
$$\sin(x + y) = \sin x \cos y + \cos x \sin y$$
$$\sin(x - y) = \sin x \cos y - \cos x \sin y$$
Add these two equations to obtain:
$$\sin(x + y) + \sin(x - y) = 2 \sin x \cos y$$
$$\frac{1}{2}[\sin(x + y) + \sin(x - y)] = \sin x \cos y$$

To prove the second factoring identity, use the addition and subtraction identities for the cosine:
$$\cos(x + y) = \cos x \cos y - \sin x \sin y$$
$$\cos(x - y) = \cos x \cos y + \sin x \sin y$$
Subtract the first equation from the second to obtain:
$$\cos(x - y) - \cos(x + y) = 2 \sin x \sin y$$
$$\frac{1}{2}[\cos(x - y) - \cos(x + y)] = \sin x \sin y$$

Similarly, the third factoring identity follows from adding the addition and subtraction identities for the cosine. The details of the proofs of the second and third identities are left for the exercises.

To prove the first product identity, let's define new variables:
$$t = x + y, \qquad u = x - y$$
Adding these equations together, we have:
$$t + u = 2x$$
$$x = \frac{t + u}{2}$$
Similarly,
$$t - u = 2y$$
$$y = \frac{t - u}{2}$$
We write the first factoring identity in terms of the variables t and u:
$$\sin \frac{t + u}{2} \cos \frac{t - u}{2} = \frac{1}{2}[\sin t + \sin u]$$
$$2 \sin\left(\frac{t + u}{2}\right) \cos\left(\frac{t - u}{2}\right) = \sin t + \sin u$$

The other product identities are proved in a similar way.

EXAMPLE 11
Reducing to a Sum of Sines

Write $\sin 3x \cos x$ as a sum or difference of trigonometric functions.

Solution

By the first factoring identity, we have
$$\sin 3x \cos x = \frac{1}{2}[\sin(3x + x) + \sin(3x - x)]$$
$$= \frac{1}{2}[\sin 4x + \sin 2x]$$

EXAMPLE 12
Reducing to a Product of Trigonometric Functions

Express $\cos 3x - \cos 4x$ as a product of two trigonometric functions.

Solution

Use the fourth product identity with x replaced by $3x$ and y replaced by $4x$:

$$\cos 3x - \cos 4x = -2 \sin\left(\frac{3x + 4x}{2}\right) \sin\left(\frac{3x - 4x}{2}\right)$$

$$= -2 \sin \frac{7x}{2} \sin\left(-\frac{x}{2}\right)$$

$$= -2 \sin \frac{7x}{2} \cdot \left(-\sin \frac{x}{2}\right)$$

$$= 2 \sin \frac{x}{2} \sin \frac{7x}{2}$$

Exercises 8.3

Given the quadrant of angle x and one trigonometric function value for x, find the following:

a. $\sin 2x$
b. $\cos 2x$
c. $\tan 2x$

1. $0 < x < \frac{\pi}{2}$, $\sin x = \frac{1}{2}$
2. $0 < x < \frac{\pi}{2}$, $\cos x = \frac{\sqrt{2}}{2}$
3. $\frac{\pi}{2} < x < \pi$, $\cos x = -0.8$
4. $\frac{\pi}{2} < x < \pi$, $\sin x = 0.6$
5. $\pi < x < \frac{3\pi}{2}$, $\sin x = -\frac{\sqrt{3}}{2}$
6. $\pi < x < \frac{3\pi}{2}$, $\cos x = -\frac{1}{2}$
7. $\frac{3\pi}{2} < x < 2\pi$, $\tan x = -1$
8. $\frac{3\pi}{2} < x < 2\pi$, $\cot x = -\frac{\sqrt{3}}{3}$

Express the following in terms of the indicated trigonometric function or functions.

9. $\cos 3x$ in terms of $\cos x$.
10. $\tan 3x$ in terms of $\tan x$.
11. $\sin 4x$ in terms of $\sin x$ and $\cos x$.
12. $\cos 4x$ in terms of $\cos x$.

Given the quadrant of angle x and one trigonometric function value of x, find the following:

a. $\sin \frac{x}{2}$
b. $\cos \frac{x}{2}$
c. $\tan \frac{x}{2}$

13. $0 < x < \frac{\pi}{2}$, $\cos x = \frac{4}{5}$
14. $\frac{3\pi}{2} < x < 2\pi$, $\sin x = -\frac{3}{5}$
15. $\pi < x < \frac{3\pi}{2}$, $\tan x = 1$
16. $\frac{\pi}{2} < x < \pi$, $\cot x = -1$
17. $\pi < x < \frac{3\pi}{2}$, $\sec x = -4$
18. $0 < x < \frac{\pi}{2}$, $\csc x = \frac{5}{2}$
19. $\frac{3\pi}{2} < x < 2\pi$, $\sin x = -\frac{2}{3}$
20. $\pi < x < \frac{3\pi}{2}$, $\cos x = -\frac{12}{13}$

Use the given information to find the indicated trigonometric function value.

21. $\cos 2x = \frac{1}{2}$, $\pi < x < \frac{3\pi}{2}$; find $\sin x$
22. $\cos 2x = -\frac{2}{9}$, $\frac{\pi}{2} < x < \pi$; find $\sin x$.
23. $\cos 2x = -\frac{1}{8}$, $\frac{\pi}{2} < x < \pi$; find $\cos x$.
24. $\cos 2x = -\frac{4}{25}$, $0 < x < \frac{\pi}{2}$; find $\cos x$.

Write the following expressions as a single trigonometric function value.

25. $2 \cos^2 20° - 1$
26. $2 \sin 50° \cos 50°$
27. $\dfrac{2 \tan \pi/5}{1 - \tan^2 \pi/5}$
28. $\cos^2 \dfrac{\pi}{8} - \sin^2 \dfrac{\pi}{8}$
29. $\sqrt{\dfrac{1 - \cos 140°}{1 + \cos 140°}}$
30. $\dfrac{\sin \pi/5}{1 + \cos \pi/5}$
31. $-\sqrt{\dfrac{1 - \cos \pi/4}{2}}$
32. $\sqrt{\dfrac{1 + \cos 130°}{2}}$
33. $\pm\sqrt{\dfrac{1 + \cos 14x}{2}}$
34. $\dfrac{1 - \cos 40°}{\sin 40°}$

SECTION 8.3 Multiple and Half-Angle Identities

35. $\dfrac{\tan 80° + \tan 55°}{1 - \tan 80° \tan 55°}$

36. $\sin 12° \cos 42° - \sin 42° \cos 12°$

Simplify each expression.

37. $\sin 2a + (\sin a - \cos a)^2$
38. $\sin 2a - (\sin a + \cos a)^2$
39. $\cos 2x + 2 \sin^2 x$
40. $\sin x \cos^2 x - \cos 2x \sin x$
41. $\dfrac{2 \sin x}{\sin 2x}$
42. $\dfrac{2 \cos x}{\sin 2x}$

Prove each of the following identities.

43. $\tan x \sin 2x = 1 - \cos 2x$
44. $\left(\cos \dfrac{x}{2} + \sin \dfrac{x}{2}\right)^2 = 1 + \sin x$
45. $\csc x - \tan \dfrac{x}{2} = \cot x$
46. $\tan 2x = \dfrac{2 \cos x}{\csc x - 2 \sin x}$
47. $\tan x + \dfrac{\sin 3x}{\cos x} = 2 \sin 2x$
48. $\tan^2 \dfrac{x}{2} - 2 \csc x \tan \dfrac{x}{2} + 1 = 0$
49. $\dfrac{\sec^2 x}{2 - \sec^2 x} = \sec 2x$
50. $3 - 4 \sin^2 x = \dfrac{\sin 3x}{\sin x}$
51. $2 \csc 2x = \tan x + \cot x$
52. $\dfrac{2 + \sin 2x}{2} = \dfrac{\sin^3 x - \cos^3 x}{\sin x - \cos x}$
53. $\cos^2 \dfrac{x}{2} = \dfrac{\sin x + \tan x}{2 \tan x}$
54. $\csc 2x = \dfrac{1 + \tan^2 x}{2 \tan x}$
55. $\tan 2x = \dfrac{2}{\cot x - \tan x}$
56. $\cot 2x = \dfrac{\csc x - 2 \sin x}{2 \cos x}$
57. $4 \sin x \cos x \cos 2x = \sin 4x$
58. $1 + \cos 2x = \cot x \sin 2x$
59. $\tan \dfrac{x}{2} = \dfrac{\sin 2x - \sin x}{\cos 2x + \cos x}$
60. $\tan^2 \dfrac{x}{2} = \dfrac{2}{1 + \cos x} - 1$
61. $\csc x = \cot x + \tan \dfrac{x}{2}$
62. $\sin x \sec \dfrac{x}{2} = 2 \sin \dfrac{x}{2}$
63. $\dfrac{1 - \sin^2 x}{1 - \cos^2 x} = \cot^2 x$
64. $\dfrac{\sin x \cos x}{1 - 2 \sin^2 x} = \dfrac{1}{\cot x - \tan x}$
65. $\csc x \sin 4x - \cos 3x = \sin 3x \cot x$
66. $\sin^2 2x - \cos^2 2x = \dfrac{\sin x}{\csc 3x} - \dfrac{\cos 3x}{\sec x}$
67. $\dfrac{\tan 4x}{\tan x} = \dfrac{\sin 5x + \sin 3x}{\sin 5x - \sin 3x}$
68. $\cot 5x = \dfrac{\cos x + \cos 9x}{\sin x + \sin 9x}$
69. $\tan \dfrac{x}{2} = \dfrac{\sin 2x - \sin x}{\cos 2x + \cos x}$
70. $\cot \dfrac{7x}{2} = \dfrac{\cos 5x + \cos 2x}{\sin 5x + \sin 2x}$
71. $\dfrac{\sin 4x - \sin 2x}{\cos 4x + \cos 2x} = \tan x$
72. $-\cot x = \dfrac{\sin 5x + \sin 3x}{\cos 5x - \cos 3x}$
73. $\sin 2x \sin 6x = \sin^2 4x - \sin^2 2x$
74. $\cos^2 y - \sin^2 x = \cos(x + y) \cos(x - y)$
75. $\dfrac{\sin x}{\sec 2x} + \dfrac{\cos x}{\csc 2x} = \sin 3x$
76. $\dfrac{\cos 2x}{\sec x} - \dfrac{\sin x}{\csc 2x} = \cos 3x$
77. $\tan \dfrac{1}{2}(x + y) \tan \dfrac{1}{2}(x - y) = \dfrac{\cos x - \cos y}{\cos x + \cos y}$
78. $\tan \dfrac{1}{2}(x + y) = \dfrac{\sin x + \sin y}{\cos x + \cos y}$
79. $\dfrac{\cos 2x - \cos 6x}{\sin 6x - \sin 2x} = \tan 4x$
80. $\cot x = \dfrac{\sin 4x + \sin 6x}{\cos 4x - \cos 6x}$

Write the following as a sum or difference of trigonometric functions.

81. $\sin 4x \cos 3x$
82. $\sin 2a \cos a$
83. $\sin(x - y) \cos(x + y)$
84. $\cos(x + y) \cos(x - y)$

Express the given sum or difference as a product.

85. $\sin 7y + \sin 9y$
86. $\sin x + \sin 5x$
87. $\sin(x + y) - \sin(x - y)$
88. $\cos(x + y) + \cos(x - y)$
89. $\cos 2x + \cos x$
90. $\sin 8x + \sin 2x$

Technology

91. Use a graphing calculator to determine the range of x within the interval $[0, 2\pi]$ for which the half-angle identity for the sine is positive.

92. Use a graphing calculator to determine the range of x within the interval $[0, 2\pi]$ for which the half-angle identity for the cosine is positive.

In your own words

93. State the double-angle identities.

94. How can you use the double-angle identities to obtain formulas for the sine or cosine for any positive, integer multiple of x?

95. State the half-angle identities. State how you would determine the correct sign to use in a particular application of the identities.

96. State the square identities.

8.4 THE INVERSE TRIGONOMETRIC FUNCTIONS

In Chapter 5, we defined the inverse of a function and described its notation. In this section, we use that discussion to define and investigate the properties of the inverse functions of the trigonometric functions.

To start, let's summarize the discussion of Chapter 5. Suppose that $f: A \to B$ is a one-to-one function, where A and B are sets of real numbers and B is the range of f. Then we can define the inverse function $f^{-1}: B \to A$ by the identity:

$$f^{-1}(y) = x \quad \text{if and only if} \quad f(x) = y$$

where x is in A and y is in B. The following Fundamental Composition identities for inverse functions then apply:

$$(f^{-1} \circ f)(x) = x \qquad \text{(where } x \text{ is in } A\text{)}$$
$$(f \circ f^{-1})(y) = y \qquad \text{(where } y \text{ is in } B\text{)}$$

The function $\sin x$ has as its domain all real numbers and has as its range the interval $[-1, 1]$. However, it is *not* a one-to-one function, so that we cannot directly apply the composition identities above to construct an inverse function for $\sin x$. To get around this difficulty, let's restrict the domain of $\sin x$ to the interval $[-\pi/2, \pi/2]$. For this restricted domain, the function $\sin x$ is one-to-one. (See Figure 1.)

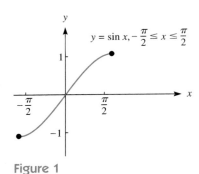

Figure 1

Definition
Inverse Sine Function

The inverse of the function $\sin x$ for x in the interval $[-\pi/2, \pi/2]$ is called the arcsin, denoted $\sin^{-1}$, and is defined by the property:

$$\sin^{-1} y = x \quad \text{if and only if} \quad \sin x = y$$
$$\left(-1 \leq y \leq 1, -\frac{\pi}{2} \leq x \leq \frac{\pi}{2}\right)$$

The domain of $\sin^{-1}$ is the interval $[-1, 1]$ and the range of $\sin^{-1}$ is the interval $[-\pi/2, \pi/2]$. The Fundamental Composition identities for the domain and range are:

$$\sin^{-1}(\sin x) = x \qquad \left(-\frac{\pi}{2} \leq x \leq \frac{\pi}{2}\right)$$
$$\sin(\sin^{-1} y) = y \qquad (-1 \leq y \leq 1)$$

The graph of the arcsine function is obtained by reflecting the graph of the portion of the sine graph $y = \sin x$, $-\pi/2 \leq x \leq \pi/2$, in the line $y = x$. The graph of the arcsine function is shown in Figure 2.

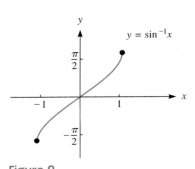

Figure 2

EXAMPLE 1
Value of an Inverse Function

Calculate the exact value of $\sin^{-1}(\sqrt{3}/2)$. Do not use a calculator.

Solution

From the definition of the arcsine function, the equation

$$\sin^{-1} \frac{\sqrt{3}}{2} = y$$

is equivalent to:
$$\sin y = \frac{\sqrt{3}}{2} \quad \left(-\frac{\pi}{2} \le y \le \frac{\pi}{2}\right)$$

There is a single value of y that satisfies this equation, namely $y = \pi/3$. Thus,
$$\sin^{-1}\frac{\sqrt{3}}{2} = \frac{\pi}{3}$$

Note that there are infinitely many values of y for which $\sin y = \sqrt{3}/2$. However, the condition $-\pi/2 \le y \le \pi/2$ restricts us to exactly one solution.

For each of the remaining trigonometric functions, it is possible to restrict the domain so that the function becomes one-to-one, just as we did for the sine. The **inverse trigonometric functions** are then defined as the inverse of the one-to-one functions. The restricted domain of the trigonometric function becomes the range of the inverse trigonometric function. The range of the trigonometric function becomes the domain of the inverse trigonometric function. The following table summarizes the domains and ranges of the various inverse trigonometric functions.

RANGES AND DOMAINS OF INVERSE TRIGONOMETRIC FUNCTIONS

Function	Range	Domain
$y = \sin^{-1} x$	$-\frac{\pi}{2} \le y \le \frac{\pi}{2}$	$-1 \le x \le 1$
$y = \cos^{-1} x$	$0 \le y \le \pi$	$-1 \le x \le 1$
$y = \tan^{-1} x$	$-\frac{\pi}{2} < y < \frac{\pi}{2}$	$-\infty < x < \infty$
$y = \sec^{-1} x$	$0 \le y < \frac{\pi}{2}$ or $\pi \le y < \frac{3\pi}{2}$	$x \ge 1$ or $x \le -1$
$y = \csc^{-1} x$	$0 < y \le \frac{\pi}{2}$ or $\pi < y \le \frac{3\pi}{2}$	$x \ge 1$ or $x \le -1$
$y = \cot^{-1} x$	$0 < y < \pi$	$-\infty < x < \infty$

Most scientific calculators have keys for the functions $\sin^{-1}$, $\cos^{-1}$, and $\tan^{-1}$. However, because the other inverse trigonometric functions are used less often, calculators usually do not have keys for them.

Figures 3 and 4 show the graphs of the functions $y = \cos^{-1} x$ and $y = \tan^{-1} x$.

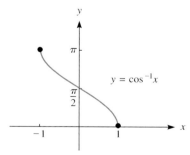

Figure 3

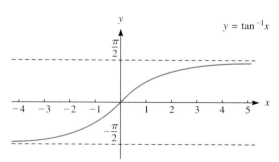

Figure 4

CHAPTER 8 Analytic Trigonometry

We can summarize the Fundamental Composition identities of these functions as follows:

FUNDAMENTAL COMPOSITION IDENTITIES OF INVERSE TRIGONOMETRIC FUNCTIONS

$$\cos^{-1}(\cos y) = y \qquad 0 \leq y \leq \pi$$
$$\cos(\cos^{-1} x) = x \qquad -1 \leq x \leq 1$$
$$\tan^{-1}(\tan y) = y \qquad -\frac{\pi}{2} < y < \frac{\pi}{2}$$
$$\tan(\tan^{-1} x) = x \qquad -\infty < x < \infty$$

EXAMPLE 2
Evaluating Inverse Trigonometric Functions

Calculate the exact values of the following. Do not use a calculator.

1. $\cos^{-1} \frac{1}{2}$
2. $\tan^{-1} 1$

Solution

1. Set $y = \cos^{-1} \frac{1}{2}$. By the definition of the inverse function, we have:

$$\cos y = \frac{1}{2}, \qquad 0 \leq y \leq \pi$$
$$y = \frac{\pi}{3}$$
$$\cos^{-1} \frac{1}{2} = \frac{\pi}{3}$$

2. Set $y = \tan^{-1} 1$. By the definition of the inverse function, we have:

$$\tan y = 1, \qquad -\frac{\pi}{2} < y < \frac{\pi}{2}$$
$$y = \frac{\pi}{4}$$
$$\tan^{-1} 1 = \frac{\pi}{4}$$

EXAMPLE 3
Composition Involving Inverse Functions

Calculate the exact value of $\cos(\tan^{-1}(-\frac{5}{4}))$. Do not use a calculator.

Solution

Let $y = \tan^{-1}(-\frac{5}{4})$. By the definition of the inverse function, we have

$$\tan y = -\frac{5}{4}$$

Because y is in the range of the inverse tangent function, it lies in the interval $(-\pi/2, \pi/2)$. Moreover, because the value of $\tan y$ is negative, y must actually lie in the interval $(-\pi/2, 0]$. Using the above equation and the Pythagorean identity, we have:

$$\sec^2 y = \tan^2 y + 1$$
$$= \left(-\frac{5}{4}\right)^2 + 1$$
$$\sec^2 y = \frac{41}{16}$$

SECTION 8.4 The Inverse Trigonometric Functions

$$\frac{1}{\cos^2 y} = \frac{41}{16}$$

$$\cos^2 y = \frac{16}{41}$$

$$\cos y = \pm\sqrt{\frac{16}{41}} = \pm\frac{4}{\sqrt{41}}$$

And because y lies in the interval $(-\pi/2, 0]$, the value of $\cos y$ is positive and the plus sign in the last equation prevails. Thus, we have:

$$\cos\left[\tan^{-1}\left(-\frac{5}{4}\right)\right] = \cos y = \frac{4}{\sqrt{41}}$$

EXAMPLE 4
Another Composition Problem

Prove the identity:

$$\sin^{-1}(\cos x) = \frac{\pi}{2} - x, \qquad 0 \le x \le \pi$$

Solution

Start from the cofunction identity:

$$\cos x = \sin\left(\frac{\pi}{2} - x\right)$$

Because $0 \le x \le \pi$, we see that $0 \ge -x \ge -\pi$, and thus:

$$\frac{\pi}{2} \ge \frac{\pi}{2} - x \ge -\frac{\pi}{2}$$

Now apply the function $\sin^{-1}$ to both sides of the last equation:

$$\sin^{-1}(\cos x) = \sin^{-1}\left[\sin\left(\frac{\pi}{2} - x\right)\right]$$

$$= \frac{\pi}{2} - x \qquad \textit{Fundamental Composition identity}$$

This proves the identity.

EXAMPLE 5
Solving a Trigonometric Equation

Solve the following equation for x in terms of y.

$$y = 4\tan^{-1}(3x - 1)$$

Solution

First divide both sides of the equation by 4 to obtain:

$$\frac{y}{4} = \tan^{-1}(3x - 1)$$

Now apply the tangent function to both sides of the equation:

$$\tan\left(\frac{y}{4}\right) = \tan(\tan^{-1}(3x - 1))$$

By the Fundamental Composition identity for inverse functions, the right side equals $(3x - 1)$. Therefore, we have:

$$3x - 1 = \tan\left(\frac{y}{4}\right)$$

$$3x = \tan\left(\frac{y}{4}\right) + 1$$

454   CHAPTER 8   Analytic Trigonometry

$$x = \frac{1}{3}\tan\left(\frac{y}{4}\right) + \frac{1}{3}$$

EXAMPLE 6
Simplification of a Trigonometric Expression

Suppose that $0 \leq x \leq 1$. Write the expression $\cos(\sin^{-1} x)$ in a form that does not involve inverse trigonometric functions.

Solution

Draw an angle t such that $\sin t = x$. We let the opposite side be x and the hypotenuse be 1. (See Figure 5.) By the Pythagorean theorem, the adjacent side is then $\sqrt{1-x^2}$, so that:

$$\cos(\sin^{-1} x) = \cos t = \frac{\text{adjacent}}{\text{hypotenuse}} = \frac{\sqrt{1-x^2}}{1} = \sqrt{1-x^2}$$

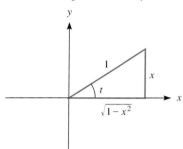

Figure 5

EXAMPLE 7
Evaluation of an Inverse Trigonometric Expression

Suppose that $y = \sin^{-1} x$, and $0 \leq x \leq 1$. Express $\sec y$ in terms of x.

Solution

From the definition of the inverse function, we have:

$$\sin y = \sin(\sin^{-1} x)$$
$$= x$$
$$\cos^2 y = 1 - \sin^2 y \quad \text{Pythagorean identity}$$
$$= 1 - x^2$$
$$\cos y = \pm\sqrt{1-x^2}$$

Because we are told that $0 \leq x \leq 1$, we know that $y = \sin^{-1} x$ lies in the interval $[0, \pi/2]$, so that $\cos y \geq 0$. Therefore, in the last equation, the plus sign prevails and we have:

$$\cos y = \sqrt{1-x^2}$$
$$\sec y = \frac{1}{\cos y}$$
$$= \frac{1}{\sqrt{1-x^2}}$$

Exercises 8.4

Calculate the value of each of the following expressions in Exercises 1–30 without using a calculator.

1. $\sin^{-1}\left(\frac{1}{2}\right)$

2. $\sin^{-1}\left(-\frac{\sqrt{3}}{2}\right)$

3. $\cos^{-1}\left(\frac{\sqrt{2}}{2}\right)$

4. $\cos^{-1}\left(\frac{1}{2}\right)$

5. $\tan^{-1}(-\sqrt{3})$

6. $\tan^{-1}(0)$

7. $\sec^{-1}(2)$

8. $\sec^{-1}(-\sqrt{2})$

9. $\csc^{-1}(\sqrt{2})$

10. $\csc^{-1}\left(\frac{2\sqrt{3}}{3}\right)$

11. $\cos^{-1}\left(-\frac{\sqrt{3}}{2}\right)$

12. $\cot^{-1}(\sqrt{3})$

13. $\cot^{-1}(-1)$

14. $\sin^{-1}\left(-\frac{\sqrt{2}}{2}\right)$

15. $\tan^{-1}\left(\frac{1}{\sqrt{3}}\right)$

16. $\sec^{-1}(\sqrt{2})$

17. $\csc\left(\csc^{-1}\frac{2}{\sqrt{3}}\right)$

18. $\sin\left(\csc^{-1}-\frac{2}{\sqrt{3}}\right)$

19. $\tan\left(\cot^{-1}\frac{1}{\sqrt{3}}\right)$

20. $\cot(\cot^{-1}(-1))$

21. $\cos\left(\cos^{-1}\frac{\sqrt{3}}{2}\right)$

22. $\csc^{-1}\left(\csc\frac{\pi}{2}\right)$

23. $\sin\left(\cos^{-1}\frac{1}{2}\right)$

24. $\cos(\sin^{-1}0)$

25. $\sec\left(\cot^{-1}\frac{3}{5}\right)$

26. $\cos\left(\sin^{-1}\frac{12}{13}\right)$

27. $\cot^{-1}\left(\cot\frac{2\pi}{3}\right)$

28. $\sin^{-1}\left(\sin\frac{5\pi}{12}\right)$

29. $\cos(\tan^{-1}(-\sqrt{3}))$

30. $\sin(\cot^{-1}0)$

Prove each identity.

31. $\tan^{-1}(-x) = -\tan^{-1}x$

32. $\cos^{-1}(-x) = \pi - \cos^{-1}x$

33. $\sin(\cos^{-1}x) = \sqrt{1-x^2}$ for $|x| \le 1$

34. $\sin(\sin^{-1}x + \cos^{-1}x) = 1$ for $|x| \le 1$

35. $\sin^{-1}x + \cos^{-1}x = \frac{\pi}{2}$ for $|x| \le 1$

36. $\sin\left(\sin^{-1}\frac{1}{2} + \cos^{-1}\frac{4}{5}\right) = \frac{4+3\sqrt{3}}{10}$

37. $\cos(\sin^{-1}(-x)) = \cos(\sin^{-1}x)$ for $0 \le x \le 1$

38. $\sin(\cos^{-1}(-x)) = \sin(\cos^{-1}x)$ for $0 \le x \le 1$

39. $\tan^{-1}x + \cot^{-1}x = \frac{\pi}{2}$ for $x \ge 0$

40. $\cos^{-1}\left[\sin\left(x + \frac{\pi}{2}\right)\right] = x$ for $0 \le x \le \pi$

41. $\sin^{-1}x = \tan^{-1}\frac{x}{\sqrt{1-x^2}}$ for $|x| < 1$

42. $\sin(2\tan^{-1}x) = \frac{2x}{1+x^2}$

43. $\cot^{-1}x = \frac{\pi}{2} - \tan^{-1}x$

44. $\cot^{-1}x = \tan^{-1}\left(\frac{1}{x}\right)$ for $x > 0$

Solve the following equations for x in terms of y.

45. $y = 2\sin^{-1}(x+1)$

46. $y = -\cos^{-1}(2x-1)$

47. $y = \frac{1}{2}\tan^{-1}(1-x)$

48. $y = 3\cot^{-1}(x+3)$

49. $y = -4\csc^{-1}(2x+1)$

50. $y = \frac{1}{3}\sec^{-1}(2-x)$

51. $y = -\frac{2}{3}\cos^{-1}(x-1)$

52. $y = -4\sin^{-1}(2x-3)$

53. $y = -\sec^{-1}(2-3x)$

54. $y = \frac{1}{5}\tan^{-1}(4x+1)$

Write the following expressions in a form that does not involve inverse trigonometric functions.

55. $\sin\left(\tan^{-1}\frac{x}{2}\right)$

56. $\sin(\cos^{-1}x)$

57. $\tan(\cos^{-1}x)$

58. $\sin\left(\sec^{-1}\frac{x}{2}\right)$

59. $\csc\left(\tan^{-1}\frac{x}{\sqrt{2}}\right)$

60. $\cos(2\tan^{-1}x)$

61. $\cos\left(\tan^{-1}\frac{x}{3}\right)$

62. $\cot\left(\cos^{-1}\frac{3}{x}\right)$

63. $\sin(\sec^{-1}x)$

64. $\sec(\tan^{-1}x)$

65. $\sin(\sin^{-1}x - \cos^{-1}x)$

66. $\sin(\cos^{-1}x - \sin^{-1}x)$

67. $\cos(\sin^{-1}x + \cos^{-1}x)$

68. $\cos(\sin^{-1}x - \cos^{-1}x)$

Find the six trigonometric functions of y in terms of x for each of the following.

69. $y = \sin^{-1}(-x)$, $\quad 0 \le x \le 1$

70. $y = \sin^{-1}(2x)$, $\quad 0 \le x \le \frac{1}{2}$

71. $y = \cos^{-1}(2x)$, $\quad 0 \le x \le \frac{1}{2}$

72. $y = \cos^{-1}\left(\frac{1}{2}x\right)$, $\quad 0 \le x \le 2$

73. $y = \csc^{-1}(x)$, $\quad x \ge 1$ or $x \le -1$

74. $y = \sec^{-1}(x)$, $\quad x \ge 1$ or $x \le -1$

75. $y = \tan^{-1}\left(\frac{1}{2}x\right)$, $\quad x \ne 0$

76. $y = \cot^{-1}(x)$, $\quad x \ne 0$

Graph each of the following.

77. $y = \cot^{-1}x$

78. $y = \sin^{-1}x$

79. $y = \sec^{-1}x$

80. $y = \csc^{-1}x$

Show that the following are *not* identities.

81. $\sin^{-1}(x) = \dfrac{1}{\sin x}$
82. $\cos^{-1}(2x) = 2\cos^{-1}(x)$
83. $[\sin^{-1}(x)]^2 + [\cos^{-1}(x)]^2 = 1$
84. $1 + [\tan^{-1}(x)]^2 = [\sec^{-1}(x)]^2$
85. $\sin^{-1}(2x) = 2\sin^{-1} x$
86. $\tan^{-1} x = \dfrac{1}{\tan x}$

Technology

Use a graphing calculator to display graphs of the following functions. From each graph, determine the domain and range of the function.

87. $y = \sin^{-1} x$
88. $y = \cos^{-1} x$
89. $y = \tan^{-1} x$
90. $y = x \tan^{-1} x$
91. $y = \sin^{-1} x + \cos^{-1} x$
92. $y = \tan(x + \tan^{-1} x)$

In your own words

93. What is the definition of the arcsine function? Be sure to give its domain, range, and relationship to the sine function.
94. What is the definition of the arctan function? Be sure to give its domain, range, and relationship to the tangent function.
95. Does the function $y = \sin x$ have an inverse function? Explain your answer.
96. Does the function $y = \tan x$ have an inverse function? Explain your answer.

8.5 SOLVING TRIGONOMETRIC EQUATIONS

Many problems can be reduced to solving equations that involve trigonometric functions. In this section, we discuss some techniques for solving many of the most common of these equations. The simplest equations involving the trigonometric functions are those that assert that a trigonometric function equals a constant. Some examples of such equations are:

$$\sin x = 0.5$$
$$\tan x = 0.573$$
$$\sec x = -2$$

The following example illustrates how to solve this kind of equation:

EXAMPLE 1
A Trigonometric Equation

Determine all solutions to the equation $\sin x = 0.5$.

Solution

By examining the unit circle, we see there are two points that have y-coordinates equal to 0.5. Corresponding to these points, the equation has two solutions:

$$x = \frac{\pi}{6}, \frac{5\pi}{6}$$

By periodicity, we can add any integer multiple of 2π to these solutions in order to obtain other solutions. So the equation has two infinite families of solutions.

$$x = \frac{\pi}{6} + 2k\pi \quad (k = 0, \pm 1, \pm 2, \ldots)$$

$$x = \frac{5\pi}{6} + 2k\pi \quad (k = 0, \pm 1, \pm 2, \ldots)$$

SECTION 8.5 Solving Trigonometric Equations

EXAMPLE 2
Another Trigonometric Equation

Determine all solutions of the equation:
$$\sin x = 0.452 \qquad (0 \leq x \leq 2\pi)$$

Solution

There are two points on the unit circle that have y-coordinate 0.452 (or any other y-coordinate in the interval $(-1, 1)$, for that matter). So the equation has two solutions. One solution is obtained by taking the arcsine of both sides of the equation:
$$x = \sin^{-1}(\sin x) = \sin^{-1}(0.452) = 0.46901$$

To find the second solution, we use the identity:
$$\sin x = \sin(\pi - x)$$
$$\sin(\pi - x) = 0.452$$
$$\pi - x = 0.46901$$
$$x = 2.67258$$

Problems such as this are easily solved using the trace function on a graphing calculator: First, graph the two equations $y = \sin x$ and $y = 0.452$. Then, use the trace function to locate the x-coordinate of the point of intersection.

EXAMPLE 3
Using Periodicity in Solving an Equation

Determine all solutions of the equation:
$$2\tan\left(3x - \frac{\pi}{4}\right) = 2, \qquad -\frac{\pi}{2} < x < \frac{\pi}{2}$$

Solution

Divide both sides of the equation by 2 to obtain
$$\tan\left(3x - \frac{\pi}{4}\right) = 1$$

An angle that has tangent equal to 1 is $\pi/4$. Moreover, because the graph of the tangent function is increasing in the interval $(-\pi/2, \pi/2)$, we see that $\pi/4$ is the only angle in this interval for which the tangent has the value 1. Because the tangent function has period π, the following angles all have tangent 1:
$$\frac{\pi}{4} + n\pi, \qquad n = 0, \pm 1, \pm 2, \ldots$$

Therefore, the solutions of the equation are given by:
$$3x - \frac{\pi}{4} = \frac{\pi}{4} + n\pi$$
$$3x = \frac{\pi}{2} + n\pi$$
$$x = \frac{\pi}{6} + \frac{n\pi}{3}, \qquad n = 0, \pm 1, \pm 2, \ldots$$

The condition $-\pi/2 < x < \pi/2$ imposes the inequality:
$$-\frac{\pi}{2} < \frac{\pi}{6} + \frac{n\pi}{3} < \frac{\pi}{2}$$
$$-\frac{\pi}{6} - \frac{\pi}{2} < \frac{n\pi}{3} < \frac{\pi}{2} - \frac{\pi}{6}$$

$$-\frac{2\pi}{3} < \frac{n\pi}{3} < \frac{\pi}{3}$$

$$-2 < n < 1$$

The only possibilities for n are thus $n = -1, 0$, and the only solutions of the equation are:

$$x = \frac{\pi}{6}, -\frac{\pi}{6}$$

Many trigonometric equations can be solved by regarding them as an algebraic equation in which the variable is a particular trigonometric function, say, $\sin x$ or $\cos x$. The next two examples provide illustrations of such equations that can be regarded as quadratic equations with a trigonometric function as the variable.

EXAMPLE 4
A Quadratic Equation in sin x

Determine all solutions to the equation:

$$\sin^2 x - \sin x = 0$$

Solution

The given equation can be regarded as a quadratic equation with variable $\sin x$. Factoring the quadratic expression on the left yields the equation:

$$\sin x(\sin x - 1) = 0$$

Equating each of the factors to 0 yields:

$\sin x = 0$ $\qquad\qquad\qquad\qquad$ $\sin x = 1$

$x = n\pi, \qquad n = 0, \pm 1, \pm 2, \ldots$ $\qquad$ $x = \frac{\pi}{2} \pm 2n\pi, \qquad n = 0, 1, 2, \ldots$

$x = \ldots, -\pi, 0, \pi, \ldots$ $\qquad\qquad\qquad$ $x = \ldots, -\frac{3\pi}{2}, \frac{\pi}{2}, \frac{5\pi}{2}, \ldots$

These two infinite families of numbers are the solutions of the given equation.

EXAMPLE 5
Another Quadratic Equation in sin x

Determine the solutions of the equation:

$$2\sin^2 x + \sin x - 1 = 0, \qquad 0 \leq x < 2\pi$$

Solution

The given equation is a quadratic in the variable $\sin x$. Factor the equation:

$$2\sin^2 x + \sin x - 1 = 0$$

$$(2\sin x - 1)(\sin x + 1) = 0$$

Equate the factors to 0:

$2\sin x - 1 = 0 \qquad\qquad \sin x + 1 = 0$

$\sin x = \frac{1}{2} \qquad\qquad\qquad \sin x = -1$

There are an infinite number of values of x for which each of these equations holds. However, the only ones lying within the interval $[0, 2\pi]$ are $x = \pi/6, 5\pi/6$ (solutions of the first equation) and $x = 3\pi/2$ (solution of the second equation).

SECTION 8.5 Solving Trigonometric Equations

In solving trigonometric equations, it is often useful to simplify the equation using one of the various identities we have proved. The next two examples provide illustrations of some typical equations that can be solved in this way.

EXAMPLE 6
Equation Involving Two Trigonometric Functions

Find the solutions of the equation:
$$\sin x \cos x = \frac{\sqrt{3}}{4}, \quad 0 \le x < 2\pi$$

Solution

Let's apply the double-angle identity for the sine to rewrite the left side of the equation. To do so, multiply both sides by 2 to obtain:
$$2 \sin x \cos x = \frac{\sqrt{3}}{2}$$
$$\sin 2x = \frac{\sqrt{3}}{2}$$
$$2x = \frac{\pi}{3}, \frac{2\pi}{3}, \frac{7\pi}{3}, \frac{8\pi}{3} \quad \text{Because } 0 \le 2x \le 4\pi$$
$$x = \frac{\pi}{6}, \frac{\pi}{3}, \frac{7\pi}{6}, \frac{4\pi}{3}$$

EXAMPLE 7
Equation Involving Trigonometric Functions of Several Angles

Solve the equation:
$$\cos 2x - 2\cos^2 x = \cos x, \quad 0 \le x < 2\pi$$

Solution

This equation involves one trigonometric function, cosine, but it is evaluated at two different values. As a first step in solving the equation, let's use the double-angle identity for the cosine to get an equivalent form of the equation in which all the trigonometric functions are evaluated at the same value. We use the double-angle identity: $\cos 2x = 2\cos^2 x - 1$. This yields:
$$(2\cos^2 x - 1) - 2\cos^2 x = \cos x$$
$$\cos x = -1$$
$$x = \pi$$

Exercises 8.5

Determine all solutions of the following equations.

1. $\sin x = 0$
2. $\cos x = 0$
3. $\tan x = 0$
4. $\cot x = 0$
5. $\sec x = 0$
6. $\csc x = 0$
7. $\sin x = -1$
8. $\cos x = -1$
9. $\tan x = 1$
10. $\cot x = -1$
11. $\sec x = 2$
12. $\csc x = -\frac{2\sqrt{3}}{3}$
13. $\tan x = -\frac{\sqrt{3}}{3}$
14. $\sin x = \frac{\sqrt{3}}{2}$
15. $\cos x = -\frac{1}{2}$
16. $\sec x = -2$
17. $\tan x = 1.621$
18. $\cos x = 0.9224$

19. $\cot x = -7.770$

20. $\sin x = 0.4318$

21. $\sin x = -3.2$

22. $\tan x = 4.449$

23. $2\cot\left(3x - \dfrac{\pi}{4}\right) = 2$,
    $-\dfrac{\pi}{2} < x < \dfrac{\pi}{2}$

24. $2\cot\left(3x - \dfrac{\pi}{4}\right) = -2$,
    $-\dfrac{\pi}{2} < x < \dfrac{\pi}{2}$

25. $2\sin\left(2x + \dfrac{\pi}{3}\right) = 1$,
    $-\dfrac{\pi}{2} < x < \dfrac{\pi}{2}$

26. $-2\cos(2x + \pi) = 1$,
    $-\dfrac{\pi}{2} < x < \dfrac{\pi}{2}$

27. $-\dfrac{1}{2}\sin\left(x - \dfrac{\pi}{4}\right) = \dfrac{\sqrt{3}}{4}$,
    $-\dfrac{\pi}{2} < x < \dfrac{\pi}{2}$

28. $-\sqrt{3}\tan\left(2x + \dfrac{\pi}{4}\right) = 1$,
    $-\dfrac{\pi}{2} < x < \dfrac{\pi}{2}$

29. $2\cos\left(\dfrac{\pi}{4} - x\right) = -\sqrt{2}$,
    $-\dfrac{\pi}{2} < x < \dfrac{\pi}{2}$

30. $3\sin\left(\dfrac{\pi}{3} - x\right) = 3$,
    $-\dfrac{\pi}{2} < x < \dfrac{\pi}{2}$

Determine all solutions to the following equations over the interval $0 \leq x < 2\pi$, without a calculator or table.

31. $2\cos(x - \pi) = -\sqrt{3}$

32. $-\sin\left(x + \dfrac{\pi}{4}\right) = \dfrac{1}{2}$

33. $\tan^2 x - \tan x = 0$

34. $2\sin^2 x + 3\sin x + 1 = 0$

35. $\cos^3 x = \cos x$

36. $\tan^2 x - \sqrt{3}\tan x = 0$

37. $2\sec^2 x = 2 - 3\sec x$

38. $2\cos^2 x - \sqrt{3}\cot x = 0$

39. $2\cos^2 x - 3\cos x = -1$

40. $1 - 3\tan^2 x = 0$

41. $\csc^2 x + \csc x - 2 = 0$

42. $\sec^2 x + 2 = 3\sec x$

43. $\cos^2 x + \sin x + 1 = 0$

44. $\sqrt{3}\cot x + 1 = \csc^2 x$

45. $\sin^2 3x = 0$

46. $\cos^2 \dfrac{x}{2} = \dfrac{1}{4}$

47. $\sec x \csc x = 2\csc x$

48. $2\sin x + \csc x = 0$

49. $4\sin x \cos x = 1$

50. $\sin 2x \sin x - \cos x = 0$

Technology

Determine all solutions to the following equations over the interval $0 \leq x < 2\pi$. Use a calculator where necessary.

51. $\sec x + \tan x = 1$

52. $\cot x + 1 = \csc x$

53. $2\tan^2 x + 7\tan x = 15$

54. $4\cot^2 x - 4\cot x = 3$

55. $\sec^2 x = 6\sec x + 2$

56. $\sin^2 x = 20 - \sin x$

57. $\cos 3x = \cos x$

58. $\sin^2 3x - \sin^2 x = 0$

59. $3\sin x = \sqrt{3}\cos x$

60. $3\sin^2 x - \cos^2 x = 1$

61. $\cot x + \tan x = \sec x \csc x$

62. $\cos x = \sin 2x$

63. $\cos 2x + 1 = \cos x$

64. $\cos 2x = 2\sin^2 x - 2$

65. $\sin 3x + \sin x = 0$

66. $\cos 3x \cos x - \sin x \sin 3x = \dfrac{1}{2}$

67. $2\cot x \sin x - \cot x = 0$

68. $\cos x + 2\cos^2 \dfrac{x}{2} = 2$

69. $\cos 3x = 1 - \sin 3x$

70. $1 + \sin x = \cos x$

71. $\tan\left(x + \dfrac{\pi}{4}\right) - \tan\left(x - \dfrac{\pi}{4}\right) = 4$

72. $\tan\left(x + \dfrac{\pi}{4}\right) + \tan\left(x - \dfrac{\pi}{4}\right) = 2$

73. $2\sin x \tan x + \tan x - 2\sin x - 1 = 0$

74. $2\tan x \csc x + 2\csc x + \tan x + 1 = 0$

75. $\sec^2 x = 5 - 3\cot^2 x$

76. $\sin x = 3\cos x$

 In your own words

77. Explain how to solve an equation of the form $\sin x = A$. Illustrate your explanation with a particular equation.

78. Why do some trigonometric equations have an infinite number of solutions while others have only a finite number?

Mathematics and the World around Us—Chaos Theory

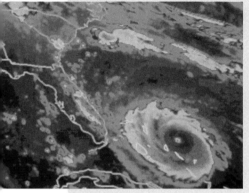

In designing a mathematical model, we attempt to mirror as closely as possible the behavior of the system we are modeling. Some systems, such as the solar system, obey physical laws that can be used as the basis for models. Other systems, such as traffic on a highway, exhibit a seeming random character, so that statistical techniques can be used as the basis for models. Other systems show sudden changes of state that occur as a result of natural forces that are not clearly understood. For example, a weather system suddenly forms a tornado, or the stock market can suddenly change from orderly trading to a panic, or a normal cell can suddenly change its division behavior and turn cancerous. Modeling such phenomena is difficult, because most of the traditional mathematical tools are more easily adapted to building models of continuous, gradual movements, rather than abrupt transitions. It is only in the last 30 years or so that mathematicians have turned their attention to modeling such discontinuous behavior. The branch of mathematics that has resulted is called chaos theory.

First introduced by the Massachusetts Institute of Technology meteorologist Edward Lorenz in 1960 to model weather patterns, chaos theory has been successfully used to build models of biological systems, stock market trading, and many other phenomena that are difficult to model using older modeling techniques. Chaos theory employs high-powered mathematical ideas drawn from a number of different branches of mathematics, including topology, differential equations, and dynamics. Both mathematical and scientific fields of research had evolved into isolated disciplines in the first two-thirds of the twentieth century. But chaos theory has served as a bridge between research fields. For an excellent introduction to chaos theory in lay terms, see *Chaos—Making a New Science* by James Gleick (New York: Penguin, 1987).

Chapter Review

Important Concepts—Chapter 8

- definition identities
- Pythagorean identities
- parity identities
- tips for verifying trigonometric identities
- trigonometric substitution
- addition identities
- subtraction identities
- cofunction identities
- symmetry identities

- identity for the tangent of the angle between two lines
- double-angle identities
- square identities
- half-angle identities
- factoring identities (product to sum)
- product identities (sum to product)

- inverse sine function; arcsine
- inverse trigonometric functions
- ranges and domains of inverse trigonometric functions
- Fundamental Composition identities of inverse trigonometric functions
- solving trigonometric equations

Important Results and Techniques—Chapter 8

Definition identities	$\sin x = \dfrac{1}{\csc x}$ $\qquad$ $\csc x = \dfrac{1}{\sin x}$ $\cos x = \dfrac{1}{\sec x}$ $\qquad$ $\sec x = \dfrac{1}{\cos x}$ $\tan x = \dfrac{\sin x}{\cos x} = \dfrac{1}{\cot x}$ $\qquad$ $\cot x = \dfrac{\cos x}{\sin x} = \dfrac{1}{\tan x}$	p. 417
Pythagorean identities	$\sin^2 x + \cos^2 x = 1$ $\tan^2 x + 1 = \sec^2 x$ $\cot^2 x + 1 = \csc^2 x$	p. 417
Parity identities	$\sin(-x) = -\sin x$ $\qquad$ $\csc(-x) = -\csc x$ $\cos(-x) = \cos x$ $\qquad$ $\sec(-x) = \sec x$ $\tan(-x) = -\tan x$ $\qquad$ $\cot(-x) = -\cot x$	p. 417
Addition and subtraction identities	$\cos(x + y) = \cos x \cos y - \sin x \sin y$ $\cos(x - y) = \cos x \cos y + \sin x \sin y$	p. 425
	$\sin(x + y) = \sin x \cos y + \cos x \sin y$ $\sin(x - y) = \sin x \cos y - \cos x \sin y$	p. 428
	$\tan(x + y) = \dfrac{\tan x + \tan y}{1 - \tan x \tan y}$ $\tan(x - y) = \dfrac{\tan x - \tan y}{1 + \tan x \tan y}$	p. 431
Cofunction identities	$\cos\left(\dfrac{\pi}{2} - x\right) = \sin x$ $\sin\left(\dfrac{\pi}{2} - x\right) = \cos x$ $\tan\left(\dfrac{\pi}{2} - x\right) = \cot x$	p. 427
Symmetry identities	$\sin x = \sin(\pi - x)$ $\cos x = -\cos(\pi - x)$	p. 429
Identity for the tangent of an angle between two lines	Tangent of the smallest angle formed by two intersecting lines: $\tan \theta = \dfrac{m_2 - m_1}{1 + m_1 m_2}$	p. 433

Double-angle identities	$\sin 2x = 2\sin x \cos x$ $\cos 2x = \cos^2 x - \sin^2 x$ $ = 1 - 2\sin^2 x$ $ = 2\cos^2 x - 1$ $\tan 2x = \dfrac{2\tan x}{1 - \tan^2 x}$	p. 437
Square identities	$\sin^2 x = \dfrac{1 - \cos 2x}{2}$ $\cos^2 x = \dfrac{1 + \cos 2x}{2}$ $\tan^2 x = \dfrac{1 - \cos 2x}{1 + \cos 2x}$	p. 441
Half-angle identities	$\sin \dfrac{x}{2} = \pm\sqrt{\dfrac{1 - \cos x}{2}}$ $\cos \dfrac{x}{2} = \pm\sqrt{\dfrac{1 + \cos x}{2}}$ $\tan \dfrac{x}{2} = \pm\sqrt{\dfrac{1 - \cos x}{1 + \cos x}}$ $\phantom{\tan \dfrac{x}{2}} = \dfrac{1 - \cos x}{\sin x}$ $\phantom{\tan \dfrac{x}{2}} = \dfrac{\sin x}{1 + \cos x}$	p. 441–442
Factoring identities	$\sin x \cos y = \dfrac{1}{2}[\sin(x + y) + \sin(x - y)]$ $\sin x \sin y = \dfrac{1}{2}[\cos(x - y) - \cos(x + y)]$ $\cos x \cos y = \dfrac{1}{2}[\cos(x + y) + \cos(x - y)]$	p. 446
Product identities	$\sin x + \sin y = 2\sin\left(\dfrac{x + y}{2}\right)\cos\left(\dfrac{x - y}{2}\right)$ $\sin x - \sin y = 2\sin\left(\dfrac{x - y}{2}\right)\cos\left(\dfrac{x + y}{2}\right)$ $\cos x + \cos y = 2\cos\left(\dfrac{x + y}{2}\right)\cos\left(\dfrac{x - y}{2}\right)$ $\cos x - \cos y = -2\sin\left(\dfrac{x + y}{2}\right)\sin\left(\dfrac{x - y}{2}\right)$	p. 446

464 CHAPTER 8 Analytic Trigonometry

Ranges and domains of inverse trigonometric functions	Function	Range	Domain	p.451
	$y = \sin^{-1}$	$-\frac{\pi}{2} \leq y \leq \frac{\pi}{2}$	$-1 \leq x \leq 1$	
	$y = \cos^{-1}$	$0 \leq y \leq \pi$	$-1 \leq x \leq 1$	
	$y = \tan^{-1}$	$-\frac{\pi}{2} < y < \frac{\pi}{2}$	$-\infty < x < \infty$	
	$y = \sec^{-1}$	$0 \leq y \leq \frac{\pi}{2}$ or $\pi \leq y \leq \frac{3\pi}{2}$	$x \geq 1$ or $x \leq -1$	
	$y = \csc^{-1}$	$0 < y \leq \frac{\pi}{2}$ or $\pi < y \leq \frac{3\pi}{2}$	$x \geq 1$ or $x \leq -1$	
	$y = \cot^{-1}$	$0 < y < \pi$	$-\infty < x < \infty$	

Review Exercises—Chapter 8

Write the following expressions in terms of a single trigonometric function.

1. $\cos x \csc x$
2. $\sec x \csc x - \tan x$
3. $\dfrac{\cot x + \tan x}{\csc x}$
4. $\sin x \sec x$
5. $\dfrac{\sec x}{\cot x + \tan x}$
6. $\dfrac{1 + \csc x}{\sec x} - \cot x$

Given the quadrant for angle x and one trigonometric function value, find the other five trigonometric function values.

7. x in the fourth quadrant, $\sin x = -\frac{1}{2}$.
8. x in the second quadrant, $\tan x = -\sqrt{3}$.
9. x in the first quadrant, $\sec x = \frac{5}{3}$.
10. x in the third quadrant, $\cos x = -\frac{12}{13}$.

Make the given substitution for t, with $0 < x < \dfrac{\pi}{2}$, and write the given expression in terms of x.

11. $\sqrt{9 - t^2}$, $\quad t = 3 \sin x$
12. $\sqrt{t^2 - 1}$, $\quad t = \sec x$

Prove each of the following identities.

13. $\dfrac{\tan^2 x + 1}{\cot^2 x + 1} = \tan^2 x$
14. $1 + \sin x = \dfrac{\cos x}{\sec x - \tan x}$
15. $\dfrac{1}{\sec x + \tan x} = \dfrac{1 - \sin x}{\cos x}$
16. $\sec^4 x - \tan^4 x = \dfrac{1 + \sin^2 x}{\cos^2 x}$

Determine the value of each of the following expressions without the use of a calculator or trigonometric tables.

17. $\tan \dfrac{5\pi}{12}$
18. $\sin -\dfrac{7\pi}{12}$
19. $\cos -15°$
20. $\tan 255°$

Write each of the following expressions in a form that involves only trigonometric functions evaluated at x.

21. $\sin\left(\dfrac{\pi}{3} + x\right)$
22. $\tan\left(x + \dfrac{\pi}{4}\right)$
23. $\tan(135° - x)$
24. $\cos(45° - x)$

Determine the value of the following expressions without using a calculator or trigonometric tables.

25. $\cos\left(x + \dfrac{\pi}{6}\right)$ if $\cos x = -\dfrac{1}{2}$ and $\pi < x < \dfrac{3\pi}{2}$
26. $\sin\left(\dfrac{\pi}{3} - x\right)$ if $\sin x = -\dfrac{3}{5}$ and $\dfrac{3\pi}{2} < x < 2\pi$
27. $\tan(45° - x)$ if $\sin x = -\dfrac{12}{13}$ and $\pi < x < \dfrac{3\pi}{2}$
28. $\cos(60° + x)$ if $\sec x = 2$ and $0 < x < \dfrac{\pi}{2}$

29. $\cos(x+y)$ if $\sin x = \frac{3}{5}$, $0 < x < \frac{\pi}{2}$ and $\sin y = -\frac{4}{5}$, $\pi < y < \frac{3\pi}{2}$

30. $\sin(x-y)$ if $\sin x = \frac{12}{13}$, $\frac{\pi}{2} < x < \pi$ and $\cos y = -\frac{5}{13}$, $\pi < y < \frac{3\pi}{2}$

31. $\tan(x+y)$ if $\tan x = 1$, $\pi < x < \frac{3\pi}{2}$ and $\tan y = -\sqrt{3}$, $\frac{\pi}{2} < y < \pi$

32. $\tan(y-x)$ if $\cos y = \frac{3}{5}$, $\frac{3\pi}{2} < y < 2\pi$ and $\sin x = \frac{4}{5}$, $0 < x < \frac{\pi}{2}$

Find the value of the smallest positive angle between the two intersecting lines.

33. $x + 2y = 6$ and $2x - 3y = 4$

34. $y = \frac{1}{5}x + 2$ and $y = \frac{2}{3}x - 5$

35. If $f(x) = \cos x$, prove that:
$$\frac{f(x+y) - f(x)}{y} = \cos x \left(\frac{\cos y - 1}{y}\right) - \sin x \left(\frac{\sin y}{y}\right)$$

Prove the following identities.

36. $\cos(\pi + x)\cos(\pi - x) + \sin(\pi + x)\sin(\pi - x) = \cos 2x$

37. $\cot\left(\frac{\pi}{4} + x\right) = \frac{1 - \tan x}{\tan x + 1}$

38. $\cot(x+y) = \frac{\cot x \cot y - 1}{\cot x + \cot y}$

Exercises 39–42 give the quadrant where angle x is located and a trigonometric function value of x. Find a–f.

a. $\sin 2x$
b. $\cos 2x$
c. $\tan 2x$
d. $\sin \frac{x}{2}$
e. $\cos \frac{x}{2}$
f. $\tan \frac{x}{2}$

39. $0 < x < \frac{\pi}{2}$, $\cos x = \frac{3}{5}$

40. $\frac{\pi}{2} < x < \pi$, $\sin x = \frac{\sqrt{3}}{2}$

41. $\pi < x < \frac{3\pi}{2}$, $\tan x = \sqrt{3}$

42. $\frac{3\pi}{2} < x < 2\pi$, $\cot x = -\frac{3}{4}$

Express the following in terms of the indicated trigonometric function or functions.

43. $\sin 3x$ in terms of $\sin x$
44. $\tan 14x$ in terms of $\tan 7x$
45. $\cos 10x$ in terms of $\sin 5x$
46. $\cos 3x$ in terms of $\cos x$

Prove each of the following identities.

47. $\frac{1 - \cos 2x}{\sin 2x} = \tan x$

48. $\sec 2x = \frac{1}{1 - 2\sin^2 x}$

49. $\tan 3x = \frac{3\tan x - \tan^3 x}{1 - 3\tan^2 x}$

50. $1 + \cos x = 2\cos^2 \frac{x}{2}$

51. $\tan \frac{x}{2} = \frac{1 - \cos x}{\sin x}$

52. $\sec \frac{x}{2} = \frac{\sqrt{2 + 2\cos x}}{1 + \cos x}$

Find the indicated trigonometric function value without using a calculator or tables.

53. Find $\cos x$ if $\cos 2x = \frac{1}{2}$ and $\pi < x < \frac{3\pi}{2}$

54. Find $\tan x$ if $\cos 2x = -\frac{2}{9}$ and $\frac{\pi}{2} < x < \pi$

55. Find $\cot x$ if $\cos 2x = -\frac{1}{8}$ and $\frac{\pi}{2} < x < \pi$

56. Find $\sin x$ if $\cos 2x = -\frac{4}{25}$ and $0 < x < \frac{\pi}{2}$

Without the use of a calculator or a trigonometric table, find the value of the following.

57. $\sin 67.5°$
58. $\tan 22.5°$
59. $\cos -\frac{3\pi}{8}$
60. $\sec -\frac{\pi}{8}$

Write as a single trigonometric function value or give the value.

61. $2\sin 15° \cos 15°$
62. $2\cos^2 22.5 - 1$
63. $\cos^2 \frac{\pi}{5} - \sin^2 \frac{\pi}{5}$
64. $\sqrt{\frac{1 + \cos \pi/4}{2}}$

Simplify each expression.

65. $\frac{\sin 2x}{2 \sin x}$
66. $\cos^2 x - \cos 2x$
67. $\sin 2x \cos x + \cos 2x \sin x$
68. $(\sin x + \cos x)^2 - \sin 2x$

Write each expression as a sum or a difference of trigonometric functions.

69. $2 \sin x \sin 9x$
70. $2 \cos 3x \cos 5x$
71. $2 \sin 3x \cos 5x$
72. $2 \sin 3x \cos x$

Write each expression as a product.

73. $\sin 60° - \sin 20°$
74. $\sin 7y + \sin 9y$

75. $\cos 3x - \cos x$

76. $\cos 40° - \cos 80°$

Prove each of the following identities.

77. $\dfrac{\cos 4x + \cos 2x}{\sin 4x - \sin 2x} = \cot x$

78. $\dfrac{\cos x + \cos 3x}{\sin 3x - \sin x} = \cot x$

79. $\dfrac{\cos 2x - \cos 4x}{2 \sin 3x} = \sin x$

80. $\dfrac{\sin x + \sin y}{\cos x + \cos y} = \tan\left(\dfrac{x+y}{2}\right)$

Calculate the exact values of the following quantities. Do not use a calculator.

81. $\cos^{-1}\left(\dfrac{\sqrt{3}}{2}\right)$

82. $\tan^{-1}(-\sqrt{3})$

83. $\sin(\csc^{-1} 2)$

84. $\cos\left(\tan^{-1}\dfrac{3}{5}\right)$

Solve each equation for x in terms of y.

85. $y = 3\sin^{-1}(x+1)$

86. $y = \dfrac{1}{2}\tan^{-1}(4x - 3)$

87. $y = -\dfrac{3}{2}\cos^{-1}(1 - x)$

88. $y = -2\csc^{-1}(2x + 1)$

Write each expression in a form that does not involve inverse trigonometric functions.

89. $\sin\left(\tan^{-1}\dfrac{x}{3}\right)$

90. $\cos\left(\tan^{-1}\dfrac{2}{x}\right)$

91. $\cos\left(\sin^{-1}\dfrac{x}{5}\right)$

92. $\tan(\cos^{-1} x)$

Find the six trigonometric functions of y in terms of x.

93. $y = \sin^{-1}(x)$, $\quad 0 \le x \le 1$

94. $y = \cos^{-1}\left(\dfrac{x}{2}\right)$, $\quad 0 \le x \le 2$

95. $y = \csc^{-1}(-x)$, $\quad x \ge 1$ or $x \le 1$

96. $y = \cot^{-1}(-x)$, $\quad x \ne 0$

Find $\sin(x+y)$, $\cos(x+y)$, $\sin 2x$, $\cos 2x$, and $\tan 2x$.

97. $x = \sin^{-1}\left(\dfrac{\sqrt{3}}{2}\right)$, $y = \cos^{-1}\left(\dfrac{1}{2}\right)$

98. $x = \tan^{-1}\left(-\dfrac{3}{5}\right)$, $y = \tan^{-1}\dfrac{4}{3}$

Determine all solutions to the following equations. Do not use a calculator or tables.

99. $2\sin x = 1$

100. $-2\cos x = \sqrt{3}$

101. $\tan x = -1.621$

102. $\cot x = 7.770$

103. $3\cot\left(3x + \dfrac{\pi}{4}\right) = -3$, $-\dfrac{\pi}{2} < x < \dfrac{\pi}{2}$

104. $\sqrt{3}\tan\left(2x + \dfrac{\pi}{4}\right) = -1$, $-\dfrac{\pi}{2} < x < \dfrac{\pi}{2}$

Determine all solutions to the following equations over the interval $0 \le x < 2\pi$.

105. $4\sin^2 x - 3 = 0$

106. $\sin^2 x = 3\cos^2 x$

107. $2\sin x + \csc x = 0$

108. $(\cos 2x + \sin 2x)^2 = 1$

109. $2\sin^2 x + 7\sin x - 15 = 0$

110. $4\cos^2 x = 4\cos x - 1$

111. $20\cos^2 x - \cos x = 1$

112. $\sec^2 x + \sec x + 1 = 0$

Chapter Test

Solve the following problems. You may use a calculator where appropriate.

1. Write the following trigonometric expression in terms of $\cos x$: $[4 - (2\sin x)^2](\sec^2 x - 1)$.

2. Prove the identity $\dfrac{1 - \tan^2 x}{1 - \cot^2 x} = 1 - \sec^2 x$.

3. Determine the exact value of $\sin\left(-\dfrac{13\pi}{12}\right)$ without using a calculator.

4. Use the addition identities to determine an alternate expression for $\sin\left(y + \dfrac{\pi}{3}\right)$.

5. Find the tangent of the smallest positive angle between the lines that have the equations $y = 2x - 1$ and $y = -\dfrac{2}{3}x + 3$.

6. Express $\sin 3x$ in terms of $\sin x$.

7. Suppose that $\sin x = -\dfrac{3}{5}$ and that x lies in the fourth quadrant. Determine $\cos x/2$.

8. Simplify the expression $\sin 2a + (\sin a - \cos a)^2$.

9. Prove the identity:

$$\dfrac{\sin x + \tan x}{\cos^2 \dfrac{x}{2}} = 2\tan x$$

10. Write $\cos 5x \sin 3x$ in terms of the trigonometric functions $\sin 8x$ and $\sin 2x$.

11. Determine $\cos(\tan^{-1} x)$ as an algebraic expression in x.

12. Determine $\cos(2\tan^{-1} x)$ as an algebraic expression in x.

13. Solve the equation $y = 5 \sin^{-1}(x + 1)$ for x in terms of y.

14. Determine all solutions of the equation $5 \sin(3x - 1) = 2.5$.

15. **Thought question.** Define the functions $\sin^{-1} x$, $\cos^{-1} x$, and $\tan^{-1} x$, and state the domains and ranges of each.

16. **Thought question.** True or false: Because the trigonometric functions are periodic, each equation involving a trigonometric function has an infinite number of solutions. Explain your answer.

CHAPTER 9

APPLICATIONS OF TRIGONOMETRY

Trigonometry has many applications to other fields as well as in mathematics itself. In this chapter, we discuss a number of these applications.

We started our discussion of trigonometry by solving right triangles. In this chapter, first we extend that discussion to solving oblique (nonright) triangles, using the Laws of Sines and Cosines. Second, we use trigonometry to introduce polar coordinate systems and to discuss their relationship to rectangular coordinate systems. Third, we explore the connection between trigonometry and complex numbers. We use De Moivre's theorem to calculate the nth roots of complex numbers.

Finally, we use trigonometry to discuss some elementary properties of vectors, which are useful in engineering and physics.

Chapter Opening Image: *A mathematical model of a severe storm.*

9.1 THE LAW OF SINES

In the preceding chapter, we saw how solving right triangles could be used in many applied problems. In this section, we take up the problem of solving oblique triangles—triangles that do not contain a right angle. The fundamental result we introduce in this section to handle such triangles is the **Law of Sines**.

LAW OF SINES

Suppose that a triangle has sides of length a, b, and c. Suppose that the angle opposite a is α, the angle opposite b is β, and the angle opposite c is γ. (See Figure 1.) Then the following equations hold:

$$\frac{\sin \alpha}{a} = \frac{\sin \beta}{b} = \frac{\sin \gamma}{c}$$

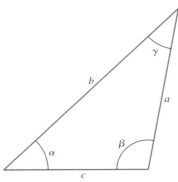

Figure 1

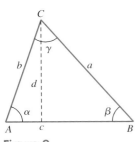

Figure 2

Proof Let's prove the first equality. The proof of the second is similar and will be left for the exercises. Drop a perpendicular from the vertex opposite side c, as shown in Figure 2. Denote by d the length of the perpendicular. Then by the geometric definition of the sine we have:

$$\sin \alpha = \frac{d}{b} \quad \text{so that} \quad d = b \sin \alpha$$

$$\sin \beta = \frac{d}{a} \quad \text{so that} \quad d = a \sin \beta$$

Equating the two expressions for d, we have:

$$b \sin \alpha = a \sin \beta$$

The last equation can be rewritten in the form:

$$\frac{\sin \alpha}{a} = \frac{\sin \beta}{b}$$

This proves the first equality of the Law of Sines.

Solving Triangles Using the Law of Sines

One application of the Law of Sines is to determine the sides and angles of a triangle given two angles and a side opposite one of them, as illustrated in the following example.

EXAMPLE 1
Solving a Triangle

Suppose a triangle has its sides and angles labeled as in Figure 1 and that $a = 112$, $\alpha = 31°45'$, and $\beta = 72°37'$. Determine the remaining sides and angles.

Solution

The third angle is easily deduced from the geometric fact that the angles of a triangle add up to 180°:

$$\gamma = 180° - 31°45' - 72°37' = 75°38'$$

From the Law of Sines, we have:

$$\frac{\sin \alpha}{a} = \frac{\sin \beta}{b}$$

$$\frac{\sin 31°45'}{112} = \frac{\sin 72°37'}{b}$$

$$b = \frac{112 \sin 72°37'}{\sin 31°45'} = 203.12$$

We can calculate the length of the third side c by applying the Law of Sines again:

$$\frac{\sin \alpha}{a} = \frac{\sin \gamma}{c}$$

$$c = \frac{a \sin \gamma}{\sin \alpha}$$

$$= \frac{112 \sin 75°38'}{\sin 31°45'}$$

$$= 206.19$$

The triangle with all its parts labeled is shown in Figure 3.

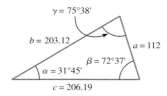

Figure 3

($\gamma = 75°38'$, $b = 203.12$, $a = 112$, $\beta = 72°37'$, $\alpha = 31°45'$, $c = 206.19$)

In the preceding example, we solved a triangle given two angles and a side opposite one of them. Another possibility is to be given two sides, say, a and b, and an angle opposite one of them, say, α. In such examples, we use the Law of Sines to solve the equation

$$\frac{\sin \alpha}{a} = \frac{\sin \beta}{b}$$

$$\sin \beta = \frac{b \sin \alpha}{a}$$

for β. There may be no, one, or two solutions. (See Figure 4.) The various cases for α acute may be distinguished geometrically as follows: Let d be the length of the perpendicular to side c. As we have seen, $d = b \sin \alpha$. The various cases are summarized in the following table.

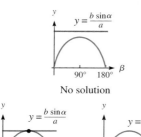

Figure 4

α Acute	
If $d = a$, there is one solution with the given data.	
If $d > a$, there is no solution with the given data.	
If $a > b$, there is one solution with the given data.	
If $d < a < b$, there are two solutions with the given data.	

For α obtuse, there are two cases, as described in the following table.

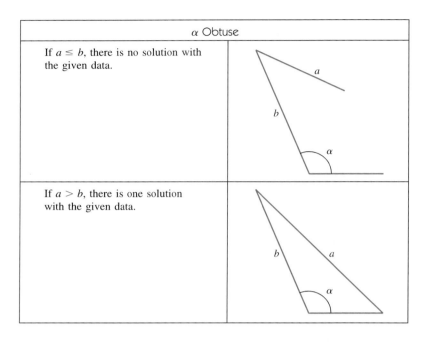

α Obtuse	
If $a \leq b$, there is no solution with the given data.	
If $a > b$, there is one solution with the given data.	

If there are two distinct triangles with the given data, we call this an **ambiguous case of the Law of Sines**. The following examples show how to recognize the various possibilities as they occur in practice.

EXAMPLE 2
Data with No Solution

Determine all triangles for which $a = 50$, $b = 150$, and $\alpha = 60°$.

Solution

Applying the Law of Sines yields:

$$\frac{\sin \alpha}{a} = \frac{\sin \beta}{b}$$

$$\frac{\sin 60°}{50} = \frac{\sin \beta}{150}$$

$$\sin \beta = \frac{150 \sin 60°}{50}$$

$$= \frac{3\sqrt{3}}{2} > 1$$

Because the sine of an angle must lie in the interval $[-1, 1]$, the above equation has no solution β. Thus, there are no triangles satisfying the given data.

EXAMPLE 3
Data with Two Solutions

Determine all triangles for which $b = 520$, $c = 952$, and $\beta = 13°$.

Solution

By the Law of Sines, we have:

$$\frac{\sin \beta}{b} = \frac{\sin \gamma}{c}$$

$$\frac{\sin 13°}{520} = \frac{\sin \gamma}{952}$$

$$\sin \gamma = \frac{952 \sin 13°}{520} = 0.41183$$

Solving the last equation using the techniques of Section 7.3, we determine that there are two solutions: $\gamma = 24.32°, 155.68°$.

Let's consider the first solution. In this case, the value of α is given by:

$$\alpha = 180° - \beta - \gamma$$
$$= 180° - 13° - 24.32°$$
$$= 142.68°$$

The value of a can then be determined from the Law of Sines:

$$\frac{\sin \alpha}{a} = \frac{\sin \beta}{b}$$

$$a = \frac{b \sin \alpha}{\sin \beta}$$

$$= \frac{520 \cdot \sin 142.68°}{\sin 13°}$$

$$= 1{,}401.45$$

The triangle corresponding to this set of data is shown in Figure 5(a). In a similar fashion, we can determine the values of α and a corresponding to the second value of γ:

$$\alpha = 180° - \beta - \gamma$$
$$= 180° - 13° - 155.68°$$
$$= 11.32°$$

$$\frac{\sin \alpha}{a} = \frac{\sin \beta}{b}$$

$$a = \frac{b \sin \alpha}{\sin \beta}$$

$$= \frac{520 \cdot \sin 11.32°}{\sin 13°}$$

$$= 453.74$$

The triangle corresponding to the second set of data is shown in Figure 5(b).

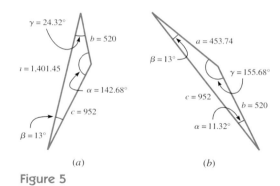

Figure 5

EXAMPLE 4
Data with One Solution

Determine all triangles for which $a = 10$, $c = 20$, and $\alpha = 30°$.

Solution

By the Law of Sines, we have:

$$\frac{\sin \alpha}{a} = \frac{\sin \gamma}{c}$$

$$\sin \gamma = \frac{20 \sin 30°}{10} = 1$$

$$\gamma = 90°$$

In this case, there is a single value for γ and the resulting triangle is a right triangle. The third angle is given by:

$$\beta = 180° - \alpha - \gamma$$
$$= 180° - 30° - 90°$$
$$= 60°$$

To determine the third side, we can apply the Pythagorean theorem:

$$a^2 + b^2 = c^2$$
$$10^2 + b^2 = 20^2$$
$$b^2 = 300$$
$$b = 10\sqrt{3}$$

The triangle corresponding to the given data is shown in Figure 6.

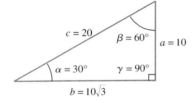

Figure 6

Applications of the Law of Sines

Oblique triangles arise quite frequently in applications. The next example shows how the Law of Sines can be used to solve a problem in marine navigation.

 EXAMPLE 5
Navigation

A ship receives notification of a storm to its east. To avoid the storm, it first proceeds at a heading 48° north of east for 160 nautical miles. It then changes heading to 54° south of east. (See Figure 7.)

1. How far does the ship go until it intersects its original course?
2. How far out of its way did the ship go to avoid the storm?

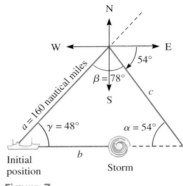

Figure 7

SECTION 9.1 The Law of Sines

Solution

1. Let angle α and side a be as shown in Figure 7. From geometry, $\alpha = 54°$ (alternate interior angles). Therefore, by the Law of Sines, we have:

$$\frac{\sin 48°}{c} = \frac{\sin 54°}{160}$$

$$c = \frac{160 \sin 48°}{\sin 54°}$$

$$= 146.97 \text{ nautical miles}$$

Therefore, the total distance the boat travels to avoid the storm is $146.97 + 160 = 306.97$ nautical miles.

2. If the ship did not avoid the storm, it would have traversed the third side of the triangle. By the Law of Sines, this side has length given by:

$$\frac{\sin \beta}{b} = \frac{\sin \alpha}{a}$$

$$\frac{\sin 78°}{b} = \frac{\sin 54°}{160}$$

$$b = 193.45 \text{ nautical miles}$$

The distance the ship went out of its way equals $306.97 - 193.45 = 113.52$ nautical miles.

Exercises 9.1

Determine the remaining sides and angles in Exercises 1–16.

1.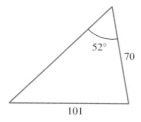

2.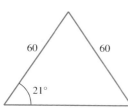

3.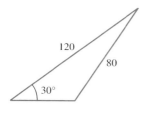

4.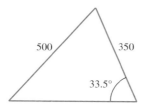

5.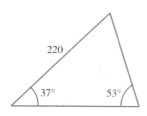

6.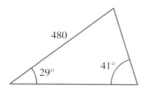

7. $\angle A = 115.5°, \angle C = 30°, a = 36$

8. $\angle A = 30°, \angle C = 52°, b = 20$

9. $b = 5, \alpha = 75°, \beta = 30°$

10. $c = 3, \beta = 37°, \gamma = 30°$

11. $a = 200, \alpha = 32°21', \gamma = 21°39'$

12. $a = 50, \alpha = 37°30', \beta = 71°10'$

13. $\angle A = 100°15', a = 48, b = 16$

14. $\beta = 2°45', b = 6.2, c = 5.8$

15. $\angle B = 15°30', a = 4.5, b = 6.8$

16. $\angle A = 145°, b = 4, a = 14$

In Exercises 17–24, determine the number of solutions for the given triangle and find them.

17. $b = 125, c = 100, \beta = 110°$

18. $\alpha = 64°9', \beta = 13°, a = 12.3$

19. $b = 9, a = 25.6, \beta = 58°$

20. $c = 84.8, a = 50, \angle C = 58°$

21. $a = 10, b = 30, \alpha = 60°$

22. $b = 260, c = 476, \beta = 13°$

23. $a = 20, c = 40, \alpha = 30°$

24. $a = 15, b = 24, \beta = 36°$

25. By using the Law of Sines and the formula Area = $\frac{1}{2}bh$ for finding the area of any triangle, develop the new formula Area = $\frac{1}{2}bc \sin A$ for finding the area of any triangle. (**Hint:** solve for h and use substitution.) (See figure.)

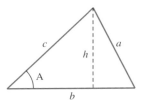

Exercise 25

Find the area for each of the following triangles.

26.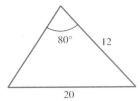

27.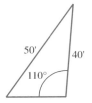

28. $\angle A = 25°, a = 10, b = 25$

29. $\angle C = 42°20', \angle A = 60°, c = 30$

30.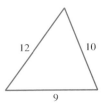

Prove that the following are true for any triangle ABC.

31. $\dfrac{a+b}{b} = \dfrac{\sin A + \sin B}{\sin B}$

32. $\dfrac{a-b}{b} = \dfrac{\sin A - \sin B}{\sin B}$

33. $\dfrac{a-b}{a+b} = \dfrac{\sin A - \sin B}{\sin A + \sin B}$

34. $\dfrac{a+b}{a-b} = \dfrac{\sin A + \sin B}{\sin A - \sin B}$

 Applications

35. **Surveying.** Two lighthouses at points X and Y are 40 kilometers apart. Each has a visual contact with a fishing boat at point Z. If $\angle ZXY = 20°30'$ and $\angle ZYX = 115°$, how far is the fishing boat from lighthouse Y?

36. **Forestry.** Two fire towers located 10 km apart sight a forest fire. Electronic equipment allows them to determine that the fire is at an angle of 71° from one tower and 100° from the other tower. Which tower is closer to the fire, and how far is it?

37. **Agriculture.** A tree grows vertically on the side of a hill that slopes upward from the horizontal by 8°. When the angle of elevation of the sun measures 20° from the side of the hill, the shadow of the tree falls 42 m down the hill. How tall is the tree?

38. **Navigation.** Two Coast Guard cutters on an east-west line and 4.6 km apart receive a distress call from a freighter that has a bearing of 47°40′ from the cutter farther west and a bearing of 302°30′ from the easternmost cutter. Which cutter has the shortest distance to travel to reach the freighter, and how far is it?

39. **Surveying.** To find the distance across a swamp, LM, a distance MN meters is measured off on one side of the swamp. In the triangle LMN, the measure of angle LMN is 112°10′ and the measure of angle LNM is 15°20′. Find the distance across the swamp.

40. **Surveying.** A tree stands at point Z across a river from point X. A base line XY is measured on one side of the river. The measure of XY is 80 meters. The measure of ∠YXZ is 54°20′ and that of ∠ZYX is 74°10′. The angle of elevation of the top of the tree from X measures 10°20′. How tall is the tree?

41. **Geometry.** Two angles of a triangle are 55° and 30°, and the longest side measures 34 feet. Find the length of the shortest side.

42. **Surveying.** From two points 400 meters apart on the bank of a river flowing due south, the bearings of a point on the opposite shore are 28°20′ and 148°20′, respectively. How wide is the river?

43. **Surveying.** A railroad surveyor measures the angle from point A to point B as 63° west of south. She also measures the angle from point A to point C as 38° west of south. If the distance from A to B is 239 meters and the distance from C to B is 374 meters, find the distance, across a river, from C to A.

44. **Navigation.** An airplane was flying a course at 6,500 ft altitude. When the plane was at point A, the angle of depression to a lighthouse at point C was 75°. When the plane reached point B, 10 seconds later, the angle of depression to the lighthouse was 50°30′. Find the ground speed of the plane in miles per hour.

45. **Surveying.** A tree stands on a hill that has a 15° incline from the horizontal. The tree leans at a 20° angle down the hill from the vertical. The angle of elevation to the sun from the side of the hill over the top of the tree is 49° from the end of the tree's 50-foot shadow down the hill. How tall is the tree?

46. **Transportation.** At 1 P.M. a train leaves Detroit at 30 miles per hour in a direction due east, and a second train leaves Detroit, traveling south. At 3 P.M. the first train is 49.18° west of north from the second train. How far apart are the trains at 3 P.M.?

47. **Military science.** A battery commander B is ordered to shoot at a target T from a position G, from which T is visible. To check on the range GT as found by a range finder, B locates an observation point H from which T is visible. The bearing of T from G is 12°48′ east of north, the bearing of T from H is 6°23′ west of north, and the bearing of H from G is 82°53′ east of south. From a map, it is found that GH = 3,250 meters. Find the range of the target from G.

48. **Navigation.** A lighthouse stands at the top of a cliff. From a buoy at sea, out from the base of the cliff, the angle of elevation to the top of the lighthouse is 40° and the angle of elevation to the bottom of the lighthouse is 25°. From a ship 150 meters from the buoy, on line with the lighthouse, the angle of elevation to the top of the lighthouse is 32°. Find the height of the lighthouse.

49. **Forestry.** Two fire towers are located 7 miles apart on a mountain ridge. If tower A is directly north of tower B, and a fire is spotted bearing 265° from tower B and 250° from tower A, how far is the fire from tower B?

50. **Navigation.** A ship is sighted at point C from two observation posts, A and B, on shore. If points A and B are 24 kilometers apart, the measure of angle CAB is 41°40′, and the measure of angle CBA is 36°10′, find the distance from observation post A to the ship.

In your own words

51. State the Law of Sines.
52. Describe the three possible cases for solving a triangle using the Law of Sines. Give an example of a triangle corresponding to each case.
53. Give an example of the ambiguous case of the Law of Sines. What is it that is ambiguous?

9.2 THE LAW OF COSINES

In this section, we discuss another set of identities useful in solving triangles, the so-called **Law of Cosines**. In addition, we derive several formulas for the area of a triangle that follow as consequences of the Law of Cosines.

LAW OF COSINES

In any triangle, the square of the length of any side equals the sum of the squares of the other two sides minus twice the product of the lengths of the other two sides times the cosine of the angle between them. That is, suppose that a triangle has sides of length a, b, and c, respectively. Suppose that the angle opposite a is α, the angle opposite b is β, and the angle opposite c is γ. Then the following equations hold:

$$a^2 = b^2 + c^2 - 2bc\cos\alpha$$
$$b^2 = a^2 + c^2 - 2ac\cos\beta$$
$$c^2 = a^2 + b^2 - 2ab\cos\gamma$$

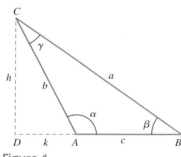

Figure 1

Proof Let's prove the first formula. The proofs of the second and third are similar. Orient the triangle so that side c is horizontal. Draw a perpendicular to side c (or its extension), as in Figure 1.

Let's assume that α is obtuse, as in the picture. At the end of the proof, we will indicate the modifications necessary if α is acute. Suppose that the length of $\overline{DA}$ is k and that the length of $\overline{DC}$ is h. From the definition of the sine, we have:

$$\sin\alpha = \frac{h}{b}$$
$$h = b\sin\alpha$$

Because α is obtuse, $\cos\alpha$ is negative. Moreover, from the definition of the cosine, we have:

$$\cos\alpha = -\frac{k}{b}$$
$$k = -b\cos\alpha$$

By the Pythagorean theorem applied to triangle BCD, we have:

$$\begin{aligned} a^2 &= h^2 + (c+k)^2 \\ &= h^2 + c^2 + k^2 + 2ck \\ &= (b\sin\alpha)^2 + c^2 + (-b\cos\alpha)^2 + 2c(-b\cos\alpha) \\ &= b^2(\sin^2\alpha + \cos^2\alpha) + c^2 - 2bc\cos\alpha \\ &= b^2 + c^2 - 2bc\cos\alpha \end{aligned}$$

SECTION 9.2 The Law of Cosines

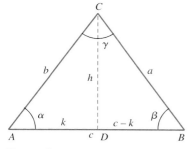

Figure 2

This proves the first equation of the Law of Cosines in case α is obtuse. In case α is acute, the reasoning is the same, except the first two equations in the above argument read:

$$\sin \alpha = \frac{h}{b}, \qquad \cos \alpha = \frac{k}{b}$$

$$h = b \sin \alpha, \qquad k = b \cos \alpha$$

(See Figure 2.) Moreover, in this case, the Pythagorean theorem yields

$$a^2 = h^2 + (c - k)^2$$

Working through the algebra gives the same result as in the first case.

Solving Triangles Using the Law of Cosines

The Law of Cosines can be used to solve a triangle given two sides and the angle between them. (In geometry, this data is abbreviated SAS—side-angle-side.) The following example illustrates how this is done.

EXAMPLE 1
Solving a Triangle Given Two Sides and an Angle

Suppose that a triangle has sides $a = 300$, $b = 225$, and an angle $\gamma = 51°$. Determine the remaining parts of the triangle.

Solution

From the third equation in the Law of Cosines, we have:

$$\begin{aligned} c^2 &= a^2 + b^2 - 2ab \cos \gamma \\ &= 300^2 + 225^2 - 2 \cdot 300 \cdot 225 \cos 51° \\ &= 90{,}000 + 50{,}625 - 84{,}958 \\ &= 55{,}667 \\ c &= 235.94 \end{aligned}$$

We can determine the two unknown angles using the other two equations in the Law of Cosines. Applying the first equation, we have:

$$a^2 = b^2 + c^2 - 2bc \cos \alpha$$

$$300^2 = 225^2 + (235.94)^2 - 2(225)(235.94) \cos \alpha$$

$$\cos \alpha = \frac{-16{,}292}{-106{,}172} = 0.15345$$

$$\alpha = 81.17°$$

We could determine the third angle using the second equation in the Law of Cosines. However, at this point, it is simplest to use the fact that the sum of the angles of a triangle is 180°, so that:

$$\begin{aligned} \beta &= 180° - \alpha - \gamma \\ &= 180° - 81.17° - 51° \\ &= 47.83° \end{aligned}$$

(The two missing angles could also have been found using the Law of Sines.)

The Law of Cosines can also be used to determine the angles of a triangle when the lengths of the sides are given.

EXAMPLE 2
Solving a Triangle Given Three Sides

Suppose that a triangle has sides of length 45, 75, and 90 feet. Determine the angles of the triangle. (See Figure 3.)

Solution

From each of the three equations in the Law of Cosines, we can determine one of the angles. As in Example 1, it is simplest to determine two of the angles in this way and then determine the remaining angle using the fact that the sum of the three angles is 180°. Here are the calculations:

$$a^2 = b^2 + c^2 - 2bc \cos \alpha$$
$$45^2 = 75^2 + 90^2 - 2 \cdot 75 \cdot 90 \cos \alpha$$
$$\cos \alpha = 0.86667$$
$$\alpha = 29.926°$$

$$b^2 = a^2 + c^2 - 2ac \cos \beta$$
$$75^2 = 45^2 + 90^2 - 2 \cdot 45 \cdot 90 \cos \beta$$
$$\cos \beta = 0.55556$$
$$\beta = 56.251°$$

$$\gamma = 180° - \alpha - \beta$$
$$= 180° - 29.926° - 56.251°$$
$$= 93.823°$$

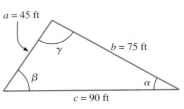

Figure 3

The Area of a Triangle

Many applications require us to calculate the area of a triangle. In elementary geometry, we learned the following formula for the area:

$$\text{Area of a triangle} = \frac{1}{2} \cdot \text{length of base} \cdot \text{height}$$

This formula requires that we know (or can compute) the height and the length of the base. As an alternative, here is a formula that allows us to calculate the area from the lengths of two sides and the angle opposite the third side.

Suppose that a triangle has sides a and b and that the included angle is γ. (See Figure 4.) Then the area of the triangle is given by the formula:

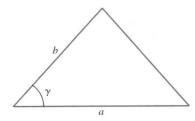

Figure 4

AREA OF A TRIANGLE

$$\text{Area of a triangle} = \frac{ab}{2} \sin \gamma$$

SECTION 9.2 The Law of Cosines

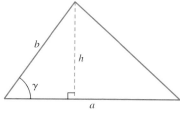

Figure 5

Proof The proof is a simple application of trigonometry. Draw a perpendicular from the vertex opposite side a. (See Figure 5.) From elementary geometry, we have:

$$\text{Area of triangle} = \frac{1}{2}ah$$

$$\sin \gamma = \frac{h}{b}$$

$$h = b \sin \gamma$$

$$\text{Area of triangle} = \frac{1}{2}ah = \frac{1}{2}ab \sin \gamma$$

This completes the proof.

EXAMPLE 3
Area of a Triangle Determined from Two Sides and an Angle

Suppose that a triangle has sides of length 22 feet and 31 feet and that the angle between these sides is 37°. What is the area of the triangle?

Solution
By the formula for the area of a triangle,

$$\text{Area} = \frac{ab}{2} \sin \gamma$$

$$= \frac{1}{2}(22 \cdot 31 \sin 37°)$$

$$= 205.22 \text{ ft}^2$$

The area of a triangle can be calculated from the lengths of the three sides using a particularly elegant result known as **Heron's Formula.**

Heron's Formula Here is a very useful formula for finding the area of a triangle.

HERON'S FORMULA

Suppose that a triangle has sides of length a, b, and c. Then the area of the triangle is given by the formula:

$$\text{Area} = \sqrt{s(s-a)(s-b)(s-c)}$$

where

$$s = \frac{1}{2}(a + b + c)$$

Proof Let's rewrite the expression under the radical solely in terms of a, b, and c. We have:

$$s(s-a)(s-b)(s-c) = \frac{a+b+c}{2} \cdot \frac{-a+b+c}{2} \cdot \frac{a-b+c}{2} \cdot \frac{a+b-c}{2}$$

$$= \left[\frac{(b+c)+a}{2} \cdot \frac{(b+c)-a}{2}\right]\left[\frac{a+(b-c)}{2} \cdot \frac{a-(b-c)}{2}\right]$$

$$= \left[\frac{(b+c)^2 - a^2}{4}\right]\left[\frac{a^2 - (b-c)^2}{4}\right]$$

$$= \left[\frac{b^2 + c^2 + 2bc - a^2}{4}\right]\left[\frac{a^2 - b^2 - c^2 + 2bc}{4}\right]$$

Let's now apply the Law of Cosines to replace a^2 in each of the expressions in the numerator:

$$\frac{b^2 + c^2 + 2bc - a^2}{4} = \frac{-(b^2 + c^2 - 2bc\cos\alpha) + 2bc + c^2 + b^2}{4}$$

$$= \frac{bc + bc\cos\alpha}{2}$$

$$= \frac{bc(1 + \cos\alpha)}{2}$$

$$\frac{a^2 - b^2 - c^2 + 2bc}{4} = \frac{(b^2 + c^2 - 2bc\cos\alpha) - b^2 - c^2 + 2bc}{4}$$

$$= \frac{-bc\cos\alpha + bc}{2}$$

$$= \frac{bc(1 - \cos\alpha)}{2}$$

This gives us

$$s(s-a)(s-b)(s-c) = \left[\frac{bc(1 + \cos\alpha)}{2}\right]\left[\frac{bc(1 - \cos\alpha)}{2}\right]$$

$$= b^2c^2\frac{(1 - \cos^2\alpha)}{4}$$

$$= \frac{b^2c^2\sin^2\alpha}{4}$$

$$\sqrt{s(s-a)(s-b)(s-c)} = \frac{|bc\sin\alpha|}{2}$$

However, by the formula for the area of a triangle, the right side of the last equation equals the area of the triangle. This completes the proof of Heron's formula.

EXAMPLE 4

Area of a Triangle Found by Using Heron's Formula

Suppose that a triangle has sides of length 12 cm, 15 cm, and 11 cm. Use Heron's formula to determine the area of the triangle.

Solution

In this case, we have

$$s = \frac{1}{2}(12 + 15 + 11)$$

$$= 19$$

Therefore, by Heron's formula, we have

$$\text{Area} = \sqrt{s(s-a)(s-b)(s-c)}$$

$$= \sqrt{19 \cdot (19 - 12) \cdot (19 - 15) \cdot (19 - 11)}$$

$$= \sqrt{4256}$$

$$\approx 65.24 \text{ cm}^2$$

Exercises 9.2

Determine the remaining sides and angles. (Determine sides to the nearest tenth and angles to the nearest tenth of a degree.)

1. $a = 4, b = 5, \angle C = 30°$
2. $a = 1, b = 4, \angle C = 30°$
3. $c = 3, b = 2, \angle A = 60°$
4. $c = 4, b = \sqrt{3}, \angle A = 30°$
5. $b = 7, c = \sqrt{2}, \angle A = 135°$
6. $b = 9, c = \sqrt{5}, \angle A = 145°$
7. $a = 6, b = 9, \angle C = 70°$
8. $a = 4.5, c = 30.2, \angle B = 60°$
9. $b = 8, c = 12, \angle A = 53°$
10. $x = 27, z = 21, \angle Y = 112°$
11. $m = 10, n = 40, \angle N = 60°$
12. $a = 3.6, b = 7.2, \angle A = 32°20'$
13. $a = 3.2, b = 7.5, \angle C = 15°25'$
14. $b = 10.2, c = 15.5, \angle A = 72°50'$
15. $a = 14, b = 18, c = 12$
16. $a = 17, b = 25, c = 17$
17. $a = 2, b = 3, c = 4$
18. $a = 70, b = 240, c = 250$
19. $a = 16, b = 20, c = 28$
20. $a = 8.3, b = 16.4, c = 11.8$
21. $a = 7.2, b = 5.1, c = 11.4$
22. $a = 5, b = 12, c = 13$
23. $b = 13, c = 23, \angle A = 69°40'$
24. $a = 0.94, c = 0.35, \angle B = 72°$

Find the area of the following triangles.

25. $a = 15, b = 8, \angle C = 30°$
26. $b = 6, c = 14, \angle A = 150°$
27. $a = 18, b = 25, \angle B = 148°40'$
28. $b = 8, \angle B = 16°40', \angle C = 145°$
29. $a = 26, b = 30, c = 13$
30. $a = 22, b = 10, c = 12\sqrt{2}$
31. Prove the following formula for the area of triangle ABC.

$$\text{Area} = \frac{1}{2}a^2 \frac{\sin B \sin C}{\sin A}$$

A triangle has sides of length a, b, and c with angle α opposite side a. Prove the following.

32. $A = \frac{1}{2}ab\sqrt{1 - \left(\frac{a^2 + b^2 - c^2}{2ab}\right)^2}$

33. $1 + \cos \alpha = \frac{(b + c + a)(b + c - a)}{2bc}$

34. $1 - \cos \alpha = \frac{(a - b + c)(a + b - c)}{2bc}$

Applications

35. **Surveying.** To measure the length of a lake, a surveyor measures the distances from point A, on the side of the lake, to points B and C, at the opposite ends of the lake, and finds them to be 52 meters and 47 meters, respectively. He then determines the measure of the angle BAC to be 55°. How long is the lake?

36. **Geography.** Pontiac is located approximately 14 miles due north of Southfield. Highland is located approximately 17 miles from Pontiac on a line that is 13° north of west from Pontiac. Find the distance from Southfield to Highland.

37. **Navigation.** The distances from a sailboat, at point C, to two points A and B on the shore are known to be 100 meters and 80 meters, respectively. If the angle ACB measures 55°, find the distance between points A and B on the shore.

38. **Geometry.** Two sides and a diagonal of a parallelogram are 7, 9, and 15 yards, respectively. Find the measures of the angles of the parallelogram.

39. **Navigation.** A pilot is flying from Jackson, Michigan, to Chicago, Illinois, a distance of approximately 200 miles. As she leaves Jackson, she flies 20° off course for 50 miles. How far is she then from Chicago? By how much must she change her course to correct her error?

40. **Mining.** A gold mine is dug into the face of a hill that slopes upward from the horizontal by 12°. If a 50-foot shaft is dug into the hill so that the shaft makes an angle 12° down from the horizontal, how far is the end of the shaft vertically from the surface?

41. **Navigation.** Two planes, one flying at 300 mph and the other flying at 400 mph, leave Cleveland at the same time. If their courses diverge by 40°, how far apart are the planes after 2 hours of flying?

42. **Navigation.** In a storm cloud, an airplane meets an air current flowing vertically upward at a rate of 100 mph. The pilot aims the plane 58° from the horizontal downward. Airspeed then reads 250 mph. If these conditions hold steady for 6 minutes, how far (the horizontal distance) will the plane have traveled? What would then be the ground speed of the plane?

43. **Geography.** Points X and Y are on opposite sides of Lake Fenton. From a third point Z, the angle between the sight lines to X and Y is 46°20'. If XZ is 350 meters and YZ is 286 meters, how far apart are X and Y?

44. **Geometry.** An isosceles triangle has a base of 24 cm. If the vertex angle measures 54°, what is the perimeter of the triangle?

45. **Navigation.** Two ships start from Pearl Harbor at the same time. If one cruises at 30 km per hour, the other cruises at 40 km per hour, and their courses diverge by 40°, how far apart are they after 2 hours?

46. **Navigation.** Two planes leave St. Louis at 6 A.M. One plane travels 300 mph due east and the other plane travels 400 mph 20° west of north. How far apart are the two planes at 8 A.M.?

47. **Navigation.** Two lighthouses are 12 miles apart at points D and E. A freighter is observed between the two at point F, and an observer at lighthouse D notes that $\angle FDE$ measures 70°30'. An observer at lighthouse E notes that $\angle FED$ measures 23°50'. How far is the freighter from each lighthouse?

48. **Navigation.** A navigator locates his position, C, on a map. He finds that he is 52 miles due east of lighthouse A, 43 miles from Port B, and the distance between the lighthouse and the port is 22 miles. Find the heading he must take from the north to reach Port B.

In your own words

49. State the Law of Cosines.
50. Describe how to apply the Law of Cosines to solve a triangle given three sides.
51. Describe how to apply the Law of Cosines to solve a triangle given two sides and the included angle.

9.3 POLAR COORDINATES

Throughout the book, we have graphed points and curves using rectangular coordinate systems. In many applications, especially in calculus and physics, polar coordinate systems provide a useful alternative.

Definition of a Polar Coordinate System

To define a polar coordinate system, we proceed as follows. Select a point O in the plane and a ray extending outward from the point along the positive x-axis. (See Figure 1.) The point is called the **pole** and the ray is called the **polar axis**. The **polar coordinates** of a point P are a pair of numbers (r, θ), where r is the distance of P from the pole and θ is the angle that the line segment $\overline{OP}$ makes with the polar axis. (See Figure 2.) The pole is assigned the coordinates $(0, \theta)$, where θ is any angle.

Pole •────────────▶
   O Polar axis

Figure 1

SECTION 9.3 Polar Coordinates

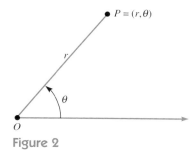

Figure 2

The polar coordinates of a point P are a pair of real numbers (r, θ), with $r \geq 0$. We also assign a point to a pair (r, θ) with $r < 0$, namely, the point obtained by going $|r|$ units in the negative direction on the ray with angle θ to the polar axis.

EXAMPLE 1
Plotting Points in Polar Coordinates

Plot the following points given in polar coordinates:

1. $(2, 30°)$
2. $(3, 135°)$
3. $\left(-1, -\dfrac{\pi}{3}\right)$

Solution

1. To plot the point with polar coordinates $(2, 30°)$, we first draw an angle of $30°$ with the polar axis as the initial side and the vertex at the pole. We then start at the pole and proceed 2 units along the terminal side of the angle. The point we come to is the one with the given polar coordinates. (See Figure 3.)

2. To plot the point with polar coordinates $(3, 135°)$, we first draw an angle of $135°$ with the polar axis as the initial side and the vertex at the pole. We then start at the pole and proceed 3 units along the terminal side of the angle. The point we come to is the one with the given polar coordinates. (See Figure 4.)

3. To plot the point with polar coordinates $(-1, -\pi/3)$, we first draw an angle of $-\pi/3$ with the polar axis as the initial side and the vertex at the pole. We then start at the pole and proceed 1 unit in the negative direction along the terminal side of the angle. The point we come to is the one with the given polar coordinates. (See Figure 5.)

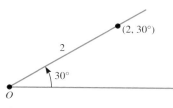

Figure 3

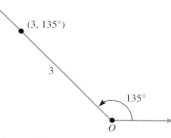

Figure 4

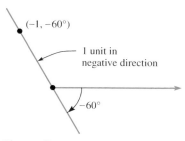

Figure 5

Polar coordinates are usually plotted on special coordinate paper, which shows various rays from the origin corresponding to the most common angles θ and various concentric circles centered at the origin corresponding to various values of r. (See Figure 6.)

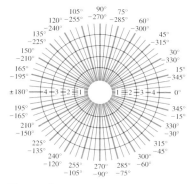

Figure 6

The polar coordinates (2, 30°) and (2, 390°) have angles that differ from one another by 360°, but have the same value of r. These polar coordinates correspond to the same point. More generally, two polar coordinates (r, θ), (r, θ') correspond to the same point if θ and θ' differ from one another by a multiple of 360°. In particular, any point P has an infinite number of different polar coordinate representations. Among this infinite collection of representations, it is most common to use the one for which $0 \leq \theta < 360°$.

Converting between Rectangular and Polar Coordinates

Suppose that we have an x–y coordinate system in the plane. We can place a polar coordinate system in the same plane by putting the pole at the origin and the polar axis along the positive x-axis. Then every point in the plane has two sets of coordinates: rectangular and polar. We can relate the two sets of coordinates as follows:

CONVERTING BETWEEN RECTANGULAR AND POLAR COORDINATES

Suppose that a polar coordinate system is superimposed on a rectangular coordinate system with the pole at the origin and the polar axis along the positive x-axis. Let P be a point with rectangular coordinates (x, y) and polar coordinates (r, θ). Then the coordinates are related by the equations:

$$x = r \cos \theta$$
$$y = r \sin \theta$$
$$r = \sqrt{x^2 + y^2}$$
$$\tan \theta = \frac{y}{x}, \qquad x \neq 0$$

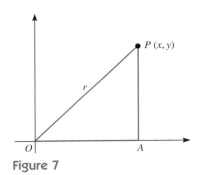

Figure 7

Proof Draw the coordinate systems and the point P as in Figure 7. From the definitions of the sine and cosine, we have:

$$\cos \theta = \frac{d(O, A)}{d(O, P)} = \frac{x}{r}$$
$$x = r \cos \theta$$
$$\sin \theta = \frac{d(A, P)}{d(O, P)} = \frac{y}{r}$$
$$y = r \sin \theta$$

This proves the first two equations of the result. To obtain the third, add the squares of the first two equations:

$$x^2 + y^2 = (r \cos \theta)^2 + (r \sin \theta)^2$$
$$= r^2 \cos^2 \theta + r^2 \sin^2 \theta$$
$$= r^2 (\sin^2 \theta + \cos^2 \theta)$$
$$= r^2 \cdot 1$$
$$= r^2$$
$$r = \sqrt{x^2 + y^2}$$

This proves the third equation. Finally, to prove the last equation, assume that $x \neq 0$ and divide the second equation by the first:

$$\frac{y}{x} = \frac{r \sin \theta}{r \cos \theta}$$

$$= \frac{\sin \theta}{\cos \theta}$$

$$= \tan \theta$$

This proves the fourth equation.

The next example provides some practice in converting rectangular coordinates to polar coordinates and polar coordinates to rectangular coordinates.

EXAMPLE 2
Rectangular to Polar Coordinates

1. A point P has rectangular coordinates $(3, 6)$. Determine its polar coordinates.
2. Determine the rectangular coordinates of the point with polar coordinates $(2, 135°)$.

Solution

1. Let (r, θ) be the polar coordinates of P. We have:

$$\tan \theta = \frac{y}{x} = \frac{6}{3} = 2$$

$$\theta = \tan^{-1} 2 = 63.435°$$

$$r = \sqrt{x^2 + y^2}$$

$$= \sqrt{3^2 + 6^2}$$

$$= \sqrt{45}$$

$$= 3\sqrt{5}$$

Therefore, the polar coordinates of the point are $(3\sqrt{5}, 63.435°)$.

2. We have:

$$x = r \cos \theta$$

$$= 2 \cos 135°$$

$$= 2 \cdot \left(-\frac{\sqrt{2}}{2}\right)$$

$$= -\sqrt{2}$$

$$y = r \sin \theta$$

$$= 2 \sin 135°$$

$$= 2 \cdot \frac{\sqrt{2}}{2}$$

$$= \sqrt{2}$$

So the rectangular coordinates of the point are $(-\sqrt{2}, \sqrt{2})$.

488 CHAPTER 9 Applications of Trigonometry

Note: Most scientific calculators have automated the process of converting between rectangular and polar coordinates. However, the keystrokes vary considerably among the various brands of calculators. Consult your calculator manual in this regard.

The next two examples provide some practice in converting equations in rectangular coordinates into polar equations and vice versa.

EXAMPLE 3
Converting an Equation to Polar Coordinates

Write the equation $x - 2y = 3$ in terms of polar coordinates.

Solution

Apply the first two conversion equations to obtain alternate expressions for x and y. Substitute these expressions into the given equation to obtain:

$$x - 2y = 3$$
$$r\cos\theta - 2r\sin\theta = 3$$
$$r(\cos\theta - 2\sin\theta) = 3$$
$$r = \frac{3}{\cos\theta - 2\sin\theta}$$

The last equation is the form of the original equation in polar coordinates.

EXAMPLE 4
Converting a Polar Equation to Rectangular Coordinates

Write the polar equation

$$r = \frac{1}{1 + 2\sin\theta}$$

in terms of rectangular coordinates.

Solution

Multiply both sides of the equation by the denominator of the right side to obtain:

$$r(1 + 2\sin\theta) = 1$$
$$r + 2r\sin\theta = 1$$
$$r + 2y = 1 \qquad y = r\sin\theta$$
$$\sqrt{x^2 + y^2} + 2y = 1 \qquad r = \sqrt{x^2 + y^2}$$
$$\sqrt{x^2 + y^2} = 1 - 2y$$
$$x^2 + y^2 = (1 - 2y)^2 \qquad \text{Squaring}$$
$$x^2 + y^2 = 1 - 4y + 4y^2$$
$$x^2 - 3y^2 + 4y = 1$$

Graphs of Polar Equations

Just as we graphed equations in rectangular (x–y) coordinates, we can graph an equation in polar coordinates by determining and plotting all points (r, θ) that satisfy the equation. This can be done manually or using a graphing calculator. The next three examples illustrate the manual procedure.

EXAMPLE 5
A Polar Graph

Graph the polar equation $r = 2\cos\theta$.

Solution

Because the right side is periodic in θ with period $360°$, it suffices to consider angles θ such that $0 \le \theta < 360°$. We take a representative sample of angles in this range and compute the corresponding values of r. These are contained in the following table. Because $\cos\theta = \cos(360° - \theta)$ (the cosine function is even), the points corresponding to values of θ between 180° and 360° lead to repetition of the points obtained for values between $\theta = 0°$ and $\theta = 180°$. We plot these points and draw a smooth curve through them. (See Figure 8.) Note that the graph appears to be a circle, which can be verified by converting the equation to rectangular form.

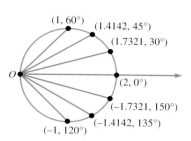

Figure 8

θ	$r = 2\cos\theta$
0°	2
30°	1.7321
45°	1.4142
60°	1
90°	0
120°	-1
135°	-1.4142
150°	-1.7321
180°	-2

EXAMPLE 6
Three-Leafed Rose

Graph the polar equation
$$r = 4\sin 3\theta$$

Solution

To sketch this graph, first recall that the sine function has period 2π. Therefore, the function $\sin 3\theta$ has period $2\pi/3$. As θ ranges over the interval $[0, 2\pi/6]$ (the first half period), the sine function first increases from 0 to 1 and then decreases from 1 to 0. So in this interval for θ, the value of r first increases from 0 to 4 and then decreases from 4 to 0. In the next half period, the value of r decreases from 0 to -4 and then increases from -4 to 0. Each half of the first period yields a portion of the graph that has the shape of a rose petal. Similarly, the portions of the graph corresponding to θ in the intervals $[2\pi/3, 4\pi/3]$ and $[4\pi/3, 2\pi]$ each correspond to two rose petals. You might think that because there are six half periods, there will be six rose petals. However, upon careful examination, you will see that each petal is traced out twice. For instance, the petal corresponding to the first half period is repeated as the petal for the fourth half period. The completed sketch is shown in Figure 9.

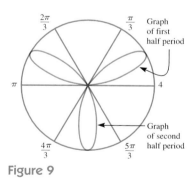

Figure 9

EXAMPLE 7
Spiral of Archimedes

Graph the polar equation
$$r = \theta, \qquad \theta > 0, \qquad \theta \text{ in radians}$$

Solution

As θ increases, so does the value of r. Note, however, that the right side is not periodic, so we must consider all positive values of θ. With rotation through 360°, the value of r increases by 2π. The graph in this case is a spiral, as shown in Figure 10.

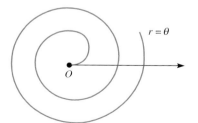

Figure 10

Polar Coordinates and Graphing Calculators

Most graphing calculators have a polar mode for displaying graphs of equations in polar coordinates. On the TI-82, you set the mode using the menu obtained from the keystroke (MODE). Select **Pol** on the third line and exit the menu.

When you now press the key (Y=), the equations will all be in the form r equals a function of θ. Enter the equation just as you would a rectangular equation, using the key labeled (X, T, θ) to enter the variable. The (WINDOW) key allows you to set the maximum and minimum values for θ (in degrees or radians, depending on the current angle measurement setting), as well as the other parameters used in specifying a rectangular plot (X_{min}, X_{max}, Y_{min}, Y_{max}, and so on). Pressing (GRAPH) displays the graph of the currently defined equations.

Polar graphing on other calculators (or other models of Texas Instruments calculators) requires different keystrokes. For these, refer to your calculator manual.

Once you have figured out polar graphing on your calculator, you should rework the preceding examples using your calculator to verify the graphs we have drawn.

Exercises 9.3

Plot the following points given in polar coordinates.

1. $A = (4, 30°)$
2. $A = (6, 45°)$
3. $A = \left(5, \dfrac{3\pi}{2}\right)$
4. $A = \left(4, -\dfrac{\pi}{2}\right)$
5. $A = \left(6, \dfrac{4\pi}{3}\right)$
6. $A = \left(5, \dfrac{2\pi}{3}\right)$
7. $A = (10, -120°)$
8. $A = (2, -135°)$

Graph the given polar equation. Use a graphing calculator if available.

9. $r = \sin\theta$
10. $r = 1 - \sin\theta$
11. $r = \cos 2\theta$ (Four-leafed rose)
12. $r = \cos 4\theta$ (Eight-leafed rose)
13. $r = 1 + 2\sin\theta$
14. $r = \sec\theta$
15. $r = \sin\dfrac{\theta}{2}$
16. $r = \sin 5\theta$ (Lemniscate)
17. $r = 2\theta, \theta > 0, \theta$ in radians
18. $r = \sin 3\theta$ (Three-leafed rose)
19. $r = 2 + 4\cos\theta$
20. $r = 2 - \cos\theta$
21. $r = 4 - 4\sin\theta$
22. $r = 5$
23. $r = 3 + \sin\theta$ (Limacon)
24. $r = 5 + \cos\theta$ (Limacon)
25. $r = e^{\theta}$ (Logarithmic spiral)
26. $\theta = \dfrac{\log r}{2}$ (Logarithmic spiral)
27. $r = \cos 2\theta \sec\theta$ (Strophoid)
28. $r = \sin\theta \tan\theta$ (Cissoid)
29. $r = 2\cos 2\theta - 1$ (Bow tie)
30. $r = \cos 2\theta - 2$ (Peanut)
31. $r = \dfrac{\tan^2\theta \sec\theta}{4}$ (Semicubical parabola)
32. $r = 2\theta + \cos\theta$ (Twisted sister)

Determine polar coordinates for each of the given points. Give θ in degrees to one decimal place. Use your calculator.

33. $(4, 4)$
34. $(-9, 9)$
35. $\left(-\dfrac{1}{2}, \dfrac{\sqrt{3}}{2}\right)$
36. $\left(\dfrac{\sqrt{2}}{2}, -\dfrac{\sqrt{2}}{2}\right)$
37. $(\sqrt{3}, 1)$
38. $(2\sqrt{3}, -2)$
39. $(3, 4)$
40. $(-4, 3)$
41. $(-2, -2)$
42. $(-\sqrt{3}, -1)$
43. $(3, 2)$
44. $(-2, 3)$

Write the given equations in terms of polar coordinates.

45. $3x + 2y = 4$

46. $y = 2$
47. $x^2 + y^2 = 16$
48. $y^2 = 25x$
49. $y = 6$
50. $x + y = 4$
51. $x^2 - 4y^2 = 4$
52. $x = 5$
53. $x^2 + 9y^2 = 36$
54. $3x + 4y = 5$

Write the given polar equation in rectangular form.

55. $r = 5$
56. $r = 3\sin\theta + 2\cos\theta$
57. $r = -\dfrac{2}{\sin\theta}$
58. $r = 2\sec\theta$
59. $r - 6\sin\theta = 0$
60. $r(1 - \sin\theta) = 2$
61. $\theta = \dfrac{\pi}{4}$
62. $r(\cos\theta + \sin\theta) = 2$
63. $r = 4\cos\theta$
64. $r + 2\sin\theta = -2\cos\theta$

Determine coordinates of the points of intersection of the polar graphs for each pair of equations. You can use a calculator.

65. $r = 2 + 4\sin\theta, r = 2$
66. $r = 1 + \sin\theta, r = 2\sin\theta$

In your own words

67. Describe how to plot a point in polar coordinates.
68. Does a point in the plane have one and only one set of corresponding polar coordinates? Explain your answer.
69. Explain how to convert an equation in rectangular coordinates to polar coordinates. Illustrate the procedure with a specific equation.
70. Explain how to convert an equation in polar coordinates to rectangular coordinates. Illustrate the procedure with a specific equation.
71. Give an example of an application in which polar coordinates might be useful.

9.4 TRIGONOMETRIC FORM OF A COMPLEX NUMBER

In Section 4.4, we introduced the complex number system and explored its elementary algebraic properties. Let's now discuss the connections between the complex numbers and the trigonometric functions. As we will see, each complex number can be represented in a trigonometric form. Moreover, by using this form, we can give simple interpretations of various operations among complex numbers, including multiplication, division, and extraction of nth roots.

Absolute Value and Argument of a Complex Number

To introduce the trigonometric form of a complex number, let's first provide a geometric representation of a complex number. Recall that a complex number z has a unique representation in the form $z = a + bi$, where i is the imaginary unit and a and b are real numbers. For $z = a + bi$, let (a, b) be the corresponding point in the plane. (See Figure 1.) Then there is a one-to-one correspondence between points of the plane and complex numbers. When the plane is so identified, it is called the **complex plane**.

The points on the horizontal axis correspond to complex numbers for which $b = 0$, that is, real numbers. For this reason, the horizontal axis is also called the **real axis**. Similarly, the points on the vertical axis correspond to complex numbers for which $a = 0$, that is, the pure imaginary numbers. For this reason, the vertical axis is also called the **imaginary axis**.

Recall that we previously introduced the absolute value $|x|$ of a real number x. This number is equal to the distance of the point x from the origin of the number line. In a similar fashion, let's define the absolute value of a complex number.

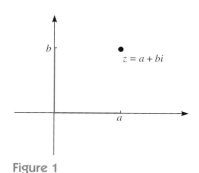

Figure 1

492 CHAPTER 9 Applications of Trigonometry

Definition
Absolute Value of a Complex Number

Let $z = a + bi$. Then the **absolute value of a complex number** z, denoted $|z|$, is defined to be the distance of the point z from the origin of the complex plane. (See Figure 2.) $|z|$ is also referred to as the **modulus** of z.

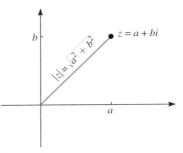

Figure 2

Referring to Figure 2, we see that if $z = a + bi$, $|z|$ is equal to the length of the hypotenuse of a right triangle with legs of length a and b. So by the Pythagorean theorem, we have the formula:

$$|z| = \sqrt{a^2 + b^2}$$

EXAMPLE 1
Calculating Absolute Values

Evaluate the following:
1. $|3 + 4i|$
2. $|-2i|$
3. $|-\sqrt{2}|$

Solution

1. $|3 + 4i| = \sqrt{3^2 + 4^2} = \sqrt{25} = 5$
2. $|-2i| = |0 - 2i| = \sqrt{0^2 + (-2)^2} = \sqrt{4} = 2$
3. $|-\sqrt{2}| = |-\sqrt{2} + 0i| = \sqrt{(-\sqrt{2})^2 + 0^2} = \sqrt{2}$

EXAMPLE 2
Absolute Value Identity

Let z be a complex number. Prove the identity:

$$z\bar{z} = |z|^2$$

Solution

Suppose that $z = a + bi$. Then:

$$\bar{z} = \overline{a + bi} = a - bi$$
$$z\bar{z} = (a + bi)(a - bi)$$
$$= a^2 + b^2$$
$$= \left(\sqrt{a^2 + b^2}\right)^2$$
$$= |z|^2$$

This proves the given identity.

EXAMPLE 3
Absolute Value of a Product

Let z and w be complex numbers. Prove the following identity:

$$|zw| = |z||w|$$

Solution

Let $z = a + bi$ and $w = c + di$. Then:

$$|z| = \sqrt{a^2 + b^2}$$

SECTION 9.4 Trigonometric Form of a Complex Number

$$|w| = \sqrt{c^2 + d^2}$$
$$|z||w| = \sqrt{a^2 + b^2}\sqrt{c^2 + d^2}$$
$$= \sqrt{(a^2 + b^2)(c^2 + d^2)} \quad (1)$$

On the other hand, we have:

$$zw = (a + bi)(c + di)$$
$$= (ac - bd) + i(ad + bc)$$

Therefore,

$$|zw| = \sqrt{(ad + bc)^2 + (ac - bd)^2}$$
$$= \sqrt{(a^2d^2 + b^2c^2 + 2adbc) + a^2c^2 + b^2d^2 - 2acbd}$$
$$= \sqrt{a^2d^2 + b^2c^2 + a^2c^2 + b^2d^2}$$
$$= \sqrt{(a^2 + b^2)(c^2 + d^2)}$$
$$= |z||w| \qquad \text{By equation (1)}$$

This completes the proof of the identity.

Suppose that $z = a + bi$ is a nonzero complex number. Represent z as a point in the complex plane. Draw a line $\overline{OA}$ from the origin to z, as shown in Figure 3.

Let $r = |z|$ denote the modulus of z. Let θ be the angle that $\overline{OA}$ makes with the positive real axis. The angle θ is called the **argument** of z.

The reason for assuming that z is nonzero is that for $z = 0$, the line $\overline{OA}$ reduces to a point and the angle θ is undefined. That is, the argument of the complex number 0 is undefined.

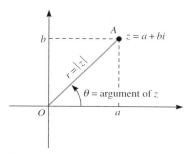

Figure 3

Polar Form of a Complex Number

Referring to Figure 3, the definitions of sine and cosine yield the relationships:

$$\frac{a}{r} = \cos\theta, \qquad \frac{b}{r} = \sin\theta$$
$$a = r\cos\theta, \qquad b = r\sin\theta$$
$$z = a + bi$$
$$= r\cos\theta + ir\sin\theta$$
$$= r(\cos\theta + i\sin\theta)$$

That is, we have the following result:

TRIGONOMETRIC FORM OF A COMPLEX NUMBER

Let z be a nonzero complex number with modulus r and argument θ. Then z can be written in the following trigonometric form (also called **polar form**):

$$z = r(\cos\theta + i\sin\theta)$$

Again referring to Figure 3, we see that

$$a = r\cos\theta, \qquad b = r\sin\theta$$
$$\frac{b}{a} = \frac{r\sin\theta}{r\cos\theta} = \frac{\sin\theta}{\cos\theta} = \tan\theta$$

EXAMPLE 4
Determining Polar Form

Determine the polar form of the following complex numbers. For each, determine the modulus and argument.

1. $z = 1 - i$
2. $z = -3i$

Solution

1. Draw the point corresponding to z, as shown in Figure 4. The radius from the origin to z makes a $-45°$ angle with the positive x-axis. So the argument is $315°$, or:

$$\theta = \frac{7\pi}{4}$$

The modulus r is given by:

$$r = |z|$$
$$= \sqrt{1^2 + (-1)^2}$$
$$= \sqrt{2}$$

Therefore, the polar form of z is:

$$z = \sqrt{2}\left(\cos\frac{7\pi}{4} + i\sin\frac{7\pi}{4}\right)$$

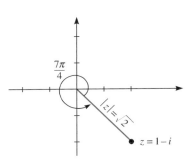

Figure 4

2. Draw the point corresponding to z, as shown in Figure 5. From the figure, we see that the modulus is $\theta = 3\pi/2$. The amplitude is:

$$r = |z|$$
$$= \sqrt{0^2 + (-3)^2}$$
$$= 3$$

Therefore, the polar form is given by:

$$z = r(\cos\theta + i\sin\theta)$$
$$= 3\left(\cos\frac{3\pi}{2} + i\sin\frac{3\pi}{2}\right)$$

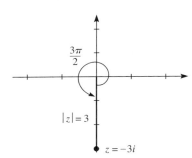

Figure 5

EXAMPLE 5
Plotting from Polar Form

Describe the points corresponding to the following complex numbers in terms of their argument and modulus.

1. $2\left(\cos\frac{5\pi}{6} + i\sin\frac{5\pi}{6}\right)$
2. $3(\cos(-\pi) + i\sin(-\pi))$

Solution

1. The argument is $5\pi/6$ radians and the modulus is 2. Draw an angle of $5\pi/6$ radians in standard position. Start at the origin and move two units along the terminal side of the angle. (See Figure 6.) The point in rectangular form is $-\sqrt{3} + i$.

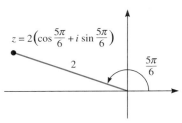

Figure 6

SECTION 9.4 Trigonometric Form of a Complex Number

2. The argument is $-\pi$ radians and the amplitude is 3. Draw an angle of $-\pi$ radians in standard position. Start at the origin and move three units along the terminal side of the angle. The resulting point corresponds to the given polar form. (See Figure 7.) The point in rectangular form is $z = -3 + 0i$.

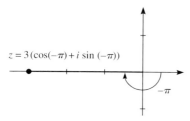

Figure 7

Suppose that the complex number z is written in the polar form:

$$z = r(\cos\theta + i\sin\theta)$$

The argument θ is not uniquely determined by z. Indeed, because the sine and cosine functions have period 2π, you can add any integer multiple of 2π to θ and obtain another polar form for z:

$$z = r(\cos(\theta + 2n\pi) + i\sin(\theta + 2n\pi)), \qquad n = 0, \pm 1, \pm 2, \ldots$$

Another way of saying this is that z has an infinite number of arguments, differing from one another by integer multiples of 2π. A complex number has an infinite number of polar forms, one corresponding to each possible argument.

Among the arguments for a complex number z, there is precisely one lying in the interval $[0, 2\pi)$. This argument is called the **principal argument of a complex number** z. In the polar forms determined in the last example, argument 1 is principal, whereas argument 2 is not.

Multiplication and division of complex numbers have simple interpretations in terms of their respective polar forms. Namely, we have the following result.

MULTIPLICATION AND DIVISION IN TERMS OF POLAR FORMS

Let the complex numbers z_1 and z_2 have polar forms:

$$z_1 = r(\cos\theta_1 + i\sin\theta_1)$$
$$z_2 = r(\cos\theta_2 + i\sin\theta_2)$$

Then,

$$z_1 z_2 = r_1 r_2(\cos(\theta_1 + \theta_2) + i\sin(\theta_1 + \theta_2))$$
$$\frac{z_1}{z_2} = \frac{r_1}{r_2}(\cos(\theta_1 - \theta_2) + i\sin(\theta_1 - \theta_2))$$

Proof Let's prove the first formula. The proof of the second is similar. The key to the proof is to apply the addition formula for the sine and cosine functions:

$$z_1 z_2 = [r_1(\cos\theta_1 + i\sin\theta_1)] \cdot [r_2(\cos\theta_2 + i\sin\theta_2)]$$
$$= r_1 r_2(\cos\theta_1\cos\theta_2 - \sin\theta_1\sin\theta_2 + i(\cos\theta_1\sin\theta_2 + \cos\theta_2\sin\theta_1))$$
$$= r_1 r_2(\cos(\theta_1 + \theta_2) + i\sin(\theta_1 + \theta_2)) \qquad \textit{Addition formulas}$$

This completes the proof.

EXAMPLE 6
Multiplication and Division from Polar Form

Suppose that z and w are given by the respective polar forms:

$$z = 5\left(\cos\frac{3\pi}{5} + i\sin\frac{3\pi}{5}\right)$$

$$w = 7\left(\cos\frac{\pi}{3} + i\sin\frac{\pi}{3}\right)$$

Calculate a polar form of zw and z/w.

Solution

In this example, we have $r_1 = 5, r_2 = 7$ and $\theta_1 = 3\pi/5, \theta_2 = \pi/3$. Therefore,

$$\theta_1 + \theta_2 = \frac{3\pi}{5} + \frac{\pi}{3}$$
$$= \frac{14\pi}{15}$$
$$\theta_1 - \theta_2 = \frac{3\pi}{5} - \frac{\pi}{3}$$
$$= \frac{4\pi}{15}$$

Thus, by the preceding theorem, we have:

$$zw = r_1 r_2 (\cos\theta_1 + \theta_2) + i\sin(\theta_1 + \theta_2)$$
$$= 5 \cdot 7\left(\cos\frac{14\pi}{15} + i\sin\frac{14\pi}{15}\right)$$
$$= 35\left(\cos\frac{14\pi}{15} + i\sin\frac{14\pi}{15}\right)$$
$$\frac{z}{w} = \frac{r_1}{r_2}(\cos(\theta_1 - \theta_2) + i\sin(\theta_1 - \theta_2))$$
$$= \frac{5}{7}\left(\cos\frac{4\pi}{15} + i\sin\frac{4\pi}{15}\right)$$

Exercises 9.4

Evaluate the following.

1. $|2 - 3i|$
2. $|5 + 2i|$
3. $|-\sqrt{3}|$
4. $|5|$
5. $|-3i|$
6. $|i\sqrt{5}|$
7. $|5 + i\sqrt{3}|$
8. $|-3 - 5i|$

Determine the polar form of the following complex numbers; also determine the amplitude and argument of each. (Assume that the argument is in $[0, 2\pi)$.)

9. $z = 2 + 2i$
10. $z = 3 + 3i$
11. $z = 4i$
12. $z = 5i$
13. $z = 3$
14. $z = -10$
15. $z = 3 - 3i$
16. $z = 2 - 2i$
17. $z = 4 + 4i$
18. $z = -6 + 6i$
19. $z = \sqrt{3} - i$
20. $z = -1 - i\sqrt{3}$
21. $z = 1 + i\sqrt{3}$
22. $z = -\sqrt{3} + i$
23. $z = -2\sqrt{3} - 2i$
24. $z = 4 + 4i\sqrt{3}$

SECTION 9.4 Trigonometric Form of a Complex Number

25. $z = -2$
26. $z = 4$
27. $z = 5i$
28. $z = -2i$

Graph the points corresponding to the following complex numbers given in polar form.

29. $-2(\cos \pi + i \sin \pi)$
30. $3\left(\cos \frac{\pi}{2} + i \sin \frac{\pi}{2}\right)$
31. $\frac{1}{2}\left(\cos \frac{\pi}{2} + i \sin \frac{\pi}{2}\right)$
32. $4(\cos \pi + i \sin \pi)$
33. $(\cos(-45°) + i \sin(-45°))$
34. $\frac{1}{3}(\cos(-60°) + i \sin(-60°))$
35. $4\left[\cos\left(-\frac{5\pi}{6}\right) + i \sin\left(-\frac{5\pi}{6}\right)\right]$
36. $2\left[\cos\left(-\frac{3\pi}{2}\right) + i \sin\left(-\frac{3\pi}{2}\right)\right]$
37. $3(\cos 60° + i \sin 60°)$
38. $(\cos 240° + i \sin 240°)$
39. $5(\cos 30° + i \sin 30°)$
40. $3(\cos 150° + i \sin 150°)$
41. $2\left[\cos\left(-\frac{2\pi}{3}\right) + i \sin\left(-\frac{2\pi}{3}\right)\right]$
42. $2(\cos(-\pi) + i \sin(-\pi))$
43. $3(\cos 270° + i \sin 270°)$
44. $-5(\cos 225° + i \sin 225°)$

For the following complex numbers in polar form, z and w, calculate a polar form of zw and z/w.

45. $z = 5(\cos 180° + i \sin 180°)$
    $w = 2(\cos 45° + i \sin 45°)$
46. $z = 2(\cos 360° + i \sin 360°)$
    $w = 3(\cos 135° + i \sin 135°)$
47. $z = \left[\cos\left(-\frac{\pi}{2}\right) + i \sin\left(-\frac{\pi}{2}\right)\right]$
    $w = 4\left[\cos \frac{3\pi}{2} + i \sin \frac{3\pi}{2}\right]$
48. $z = 4\left[\cos\left(-\frac{\pi}{4}\right) + i \sin\left(-\frac{\pi}{4}\right)\right]$
    $w = 3\left(\cos \frac{3\pi}{2} + i \sin \frac{3\pi}{2}\right)$

49. $z = 2\left[\cos\left(-\frac{5\pi}{6}\right) + i \sin\left(-\frac{5\pi}{6}\right)\right]$
    $w = 3\left[\cos\left(-\frac{\pi}{2}\right) + i \sin\left(-\frac{\pi}{2}\right)\right]$
50. $z = 3\left[\cos\left(-\frac{5\pi}{3}\right) + i \sin\left(-\frac{5\pi}{3}\right)\right]$
    $w = 2\left[\cos\left(-\frac{4\pi}{3}\right) + i \sin\left(-\frac{4\pi}{3}\right)\right]$
51. $z = 5\left(\cos \frac{5\pi}{6} + i \sin \frac{5\pi}{6}\right)$
    $w = 2\left[\cos\left(-\frac{3\pi}{4}\right) + i \sin\left(-\frac{3\pi}{4}\right)\right]$
52. $z = 8(\cos 2\pi + i \sin 2\pi)$
    $w = 4\left[\cos\left(-\frac{\pi}{4}\right) + i \sin\left(-\frac{\pi}{4}\right)\right]$
53. $z = 6(\cos 60° + i \sin 60°)$
    $w = 2(\cos 90° + i \sin 90°)$
54. $z = 2(\cos 60° + i \sin 60°)$
    $w = 6(\cos 135° + i \sin 135°)$
55. $z = 9[\cos(-225°) + i \sin(-225°)]$
    $w = 3[\cos(-60°) + i \sin(-60°)]$
56. $z = 10[\cos(-90°) + i \sin(-90°)]$
    $w = 2[\cos(-120°) + i \sin(-120°)]$
57. $z = 4\left(\cos \frac{5\pi}{6} + i \sin \frac{5\pi}{6}\right)$
    $w = 8\left[\cos\left(-\frac{\pi}{2}\right) + i \sin\left(-\frac{\pi}{2}\right)\right]$
58. $z = \left[\cos\left(-\frac{\pi}{4}\right) + i \sin\left(-\frac{\pi}{4}\right)\right]$
    $w = 6\left(\cos \frac{4\pi}{3} + i \sin \frac{4\pi}{3}\right)$

Write the following complex numbers in the form $a + bi$.

59. $2\left(\cos \frac{3\pi}{4} + i \sin \frac{3\pi}{4}\right)$
60. $5[\cos(-\pi) + i \sin(-\pi)]$
61. $6\left[\cos\left(-\frac{\pi}{2}\right) + i \sin\left(-\frac{\pi}{2}\right)\right]$
62. $3\left(\cos \frac{2\pi}{3} + i \sin \frac{2\pi}{3}\right)$
63. $\cos(-150°) + i \sin(-150°)$
64. $2[\cos(-60°) + i \sin(-60°)]$

65. $\sqrt{3}[\cos(-30°) + i\sin(-30°)]$

66. $\sqrt{2}(\cos 135° + i\sin 135°)$

Find the following complex numbers and write the results in $a + bi$ form.

67. $[2(\cos 90° + i\sin 90°)][5(\cos 45° + i\sin 45°)]$

68. $[3(\cos 120° + i\sin 120°)][2(\cos 30° + i\sin 30°)]$

69. $[5(\cos \pi + i\sin \pi)]\left[3\left(\cos\left(-\frac{\pi}{2}\right) + i\sin\left(-\frac{\pi}{2}\right)\right)\right]$

70. $\left[2\left(\cos \frac{3\pi}{2} + i\sin \frac{3\pi}{2}\right)\right][\cos(2\pi) + i\sin(2\pi)]$

71. $[4(\cos 45° + i\sin 45°)] \div [3(\cos(-30°) + i\sin(-30°))]$

72. $[3(\cos 60° + i\sin 60°)] \div [6(\cos(-45°) + i\sin(-45°))]$

73. $\left[2\left(\cos\left(-\frac{3\pi}{2}\right) + i\sin\left(-\frac{3\pi}{2}\right)\right)\right] \div [4(\cos(-2\pi) + i\sin(-2\pi))]$

74. $\left[12\left(\cos \frac{5\pi}{4} + i\sin \frac{5\pi}{4}\right)\right] \div [3(\cos \pi + i\sin \pi)]$

Prove the following.

75. $\left|\dfrac{z}{w}\right| = \dfrac{|z|}{|w|}$

76. $|z||\bar{z}| = |z \cdot \bar{z}|$

Technology

Use a scientific calculator to find zw and z/w.

77. $z = 3(\cos 40° + i\sin 40°)$
    $w = 2(\cos 130° + i\sin 130°)$

78. $z = 2[\cos(-20°) + i\sin(-20°)]$
    $w = 4[\cos(305°) + i\sin(305°)]$

79. $z = 5\left[\cos\left(-\frac{\pi}{5}\right) + i\sin\left(-\frac{\pi}{5}\right)\right]$
    $w = 2\left(\cos \frac{5\pi}{9} + i\sin \frac{5\pi}{9}\right)$

80. $z = 10\left(\cos \frac{\pi}{10} + i\sin \frac{\pi}{10}\right)$
    $w = 2\left(\cos \frac{17\pi}{15} + i\sin \frac{17\pi}{15}\right)$

In your own words

81. Describe how to compute the absolute value of a complex number.

82. What is the geometric significance of the absolute value of a complex number?

83. Describe how to compute the principal argument of a complex number.

84. What is the geometric significance of the principal argument of a complex number?

9.5 THE nTH ROOT OF A COMPLEX NUMBER

Let z be a complex number and n a positive integer. An nth root of z is a number w whose nth power is z. That is, w is an nth root of z provided that:

$$w^n = z$$

In this section, we will use the trigonometric form of a complex number to determine its nth roots.

De Moivre's Theorem

As a first step toward determining the nth roots of a complex number, let's prove the following result for determining a polar form for the nth power of a complex number.

SECTION 9.5 The nth Root of a Complex Number

DE MOIVRE'S THEOREM

Let n be an integer and $z = r(\cos\theta + i\sin\theta)$ be a nonzero complex number in polar form. Then:
$$z^n = [r(\cos\theta + i\sin\theta)]^n = r^n(\cos n\theta + i\sin n\theta)$$
That is, the modulus of z^n equals the nth power of the modulus of z and an amplitude of z^n equals n times the amplitude of z.

Proof From the formula for multiplying complex numbers in polar form, we see that:
$$\begin{aligned} z^2 &= z \cdot z \\ &= r \cdot r(\cos(\theta + \theta) + i\sin(\theta + \theta)) \\ &= r^2(\cos 2\theta + i\sin 2\theta) \end{aligned}$$
This is the desired result for $n = 2$. For $n = 3$, note that $z^3 = z \cdot z^2$. So applying the result for $n = 2$, we have
$$\begin{aligned} z^3 &= z \cdot z^2 \\ &= [r(\cos\theta + i\sin\theta)][r^2(\cos 2\theta + i\sin 2\theta)] \\ &= r^3(\cos(\theta + 2\theta) + i\sin(\theta + 2\theta)) \\ &= r^3(\cos 3\theta + i\sin 3\theta) \end{aligned}$$
This is the desired result for $n = 3$. Following the same reasoning, the result for each value of n can be deduced from the result for the value preceding it. We will fill in the details of the proof once we introduce the method of mathematical induction.

EXAMPLE 1
Powers of a Complex Number

Use De Moivre's theorem to calculate the following powers:
1. $z^3, z = 2\left(\cos\dfrac{\pi}{5} + i\sin\dfrac{\pi}{5}\right)$
2. $(1 - i)^{10}$

Solution
1. By De Moivre's theorem, we have:
$$\begin{aligned} z^3 &= 2^3\left[\cos\left(3 \cdot \frac{\pi}{5}\right) + i\sin\left(3 \cdot \frac{\pi}{5}\right)\right] \\ &= 8\left(\cos\frac{3\pi}{5} + i\sin\frac{3\pi}{5}\right) \end{aligned}$$
2. We have previously determined the polar form of $1 - i$, namely:
$$1 - i = \sqrt{2}\left(\cos\frac{7\pi}{4} + i\sin\frac{7\pi}{4}\right)$$
Therefore, by De Moivre's theorem, we have:
$$\begin{aligned} (1 - i)^{10} &= \left(\sqrt{2}\right)^{10}\left(\cos\left(10 \cdot \frac{7\pi}{4}\right) + i\sin\left(10 \cdot \frac{7\pi}{4}\right)\right) \\ &= 32\left(\cos\frac{35\pi}{2} + i\sin\frac{35\pi}{2}\right) \end{aligned}$$

Because the sine and cosine functions have period 2π, we can rewrite the last expression as follows:

$$(1-i)^{10} = 32\left(\cos\left(\frac{6\pi}{4} + \frac{64\pi}{4}\right) + i\sin\left(\frac{6\pi}{4} + \frac{64\pi}{4}\right)\right)$$

$$= 32\left(\cos\left(\frac{3\pi}{2} + 16\pi\right) + i\sin\left(\frac{3\pi}{2} + 16\pi\right)\right)$$

$$= 32\left(\cos\frac{3\pi}{2} + i\sin\frac{3\pi}{2}\right)$$

$$= 32(0 + i(-1))$$

$$= -32i$$

nth Roots of a Complex Number

Suppose that we are given a nonzero complex number z. Let's determine a formula for its nth roots. Let a polar form of z be:

$$z = r(\cos\theta + i\sin\theta)$$

For each of the integers $j = 0, 1, 2, \ldots, n-1$, set:

$$t_j = r^{1/n}\left(\cos\frac{\theta + 2\pi j}{n} + i\sin\frac{\theta + 2\pi j}{n}\right)$$

By De Moivre's theorem, the nth power of t_j is equal to:

$$t_j^n = \left(r^{1/n}\right)^n\left(\cos\left(n\cdot\frac{\theta + 2\pi j}{n}\right) + i\sin\left(n\cdot\frac{\theta + 2\pi j}{n}\right)\right)$$

$$= r(\cos(\theta + 2\pi j) + i\sin(\theta + 2\pi j))$$

$$= r(\cos\theta + i\sin\theta) \qquad \textit{Periodicity}$$

$$= z$$

That is, each of the numbers t_j is an nth root of z. Moreover, any two consecutive nth roots have arguments differing by $2\pi/n$. In particular, the t_j are all different. Thus, z has n nth roots, and we have proved the following result:

---

nTH ROOTS OF A COMPLEX NUMBER

Let $z = r(\cos\theta + i\sin\theta)$ be a nonzero complex number and n a positive integer. Then there are n distinct complex numbers that are nth roots of z. These numbers are given by the formula:

$$t_j = r^{1/n}\left[\cos\left(\frac{\theta + 2\pi j}{n}\right) + i\sin\left(\frac{\theta + 2\pi j}{n}\right)\right], \qquad (j = 0, 1, \ldots, n-1)$$

EXAMPLE 2 Calculate the cube roots of 2.

Complex Cube Roots

Solution

The polar form of 2 is:

$$2(\cos 0 + i\sin 0)$$

Therefore, the cube roots of 2 are given by:

$$t_j = 2^{1/3}\left(\cos\frac{0+2\pi j}{3} + i\sin\frac{0+2\pi j}{3}\right)$$

$$= 2^{1/3}\left(\cos\frac{2\pi j}{3} + i\sin\frac{2\pi j}{3}\right) \quad (j = 0, 1, 2)$$

More explicitly, we have:

$$t_0 = 2^{1/3}(\cos 0 + i\sin 0) = 2^{1/3}(1 - 0i) = 2^{1/3}$$

$$t_1 = 2^{1/3}\left(\cos\frac{2\pi}{3} + i\sin\frac{2\pi}{3}\right)$$

$$= 2^{1/3}\left(-\frac{1}{2} + i\frac{\sqrt{3}}{2}\right)$$

$$t_2 = 2^{1/3}\left(\cos\frac{4\pi}{3} + i\sin\frac{4\pi}{3}\right)$$

$$= 2^{1/3}\left(-\frac{1}{2} - i\frac{\sqrt{3}}{2}\right)$$

(See Figure 1.)

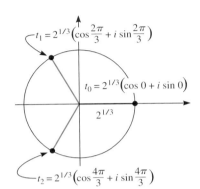

Figure 1

EXAMPLE 3
Square Roots of a Complex Number

Determine the square roots of i.

Solution

A polar form of i is given by:

$$i = 1 \cdot \left(\cos\frac{\pi}{2} + i\sin\frac{\pi}{2}\right)$$

(See Figure 2.) Therefore, the square roots of i are equal to t_0, t_1, where:

$$t_j = 1 \cdot \left(\cos\frac{\frac{\pi}{2}+2\pi j}{2} + i\sin\frac{\frac{\pi}{2}+2\pi j}{2}\right), \quad (j = 0, 1)$$

$$t_0 = \cos\frac{\pi}{4} + i\sin\frac{\pi}{4} = \frac{\sqrt{2}}{2} + \frac{\sqrt{2}}{2}i$$

$$t_1 = \cos\frac{5\pi}{4} + i\sin\frac{5\pi}{4} = -\frac{\sqrt{2}}{2} - \frac{\sqrt{2}}{2}i$$

(See Figure 3.)

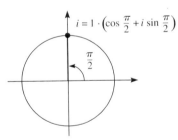

Figure 2

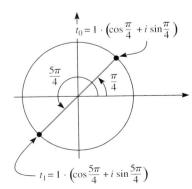

Figure 3

CHAPTER 9 Applications of Trigonometry

EXAMPLE 4
Complex Roots of an Equation

Determine the complex solutions of the equation $x^4 - 16 = 0$.

Solution

The solutions of the equation satisfy:
$$x^4 = 16$$
That is, they are fourth roots of 16. The polar form of 16 is:
$$16 = 16(\cos 0 + i \sin 0)$$
Therefore, the formula for the nth roots of a complex number implies that its fourth roots are t_0, t_1, t_2, t_3, where:
$$t_j = 16^{1/4}\left(\cos \frac{2\pi j}{4} + i \sin \frac{2\pi j}{4}\right), \qquad (j = 0, 1, 2, 3)$$

$$t_0 = 2\left(\cos \frac{2\pi \cdot 0}{4} + i \sin \frac{2\pi \cdot 0}{4}\right)$$
$$= 2(\cos 0 + i \sin 0)$$
$$= 2$$

$$t_1 = 2\left(\cos \frac{2\pi \cdot 1}{4} + i \sin \frac{2\pi \cdot 1}{4}\right)$$
$$= 2\left(\cos \frac{\pi}{2} + i \sin \frac{\pi}{2}\right)$$
$$= 2i$$

$$t_2 = 2\left(\cos \frac{2\pi \cdot 2}{4} + i \sin \frac{2\pi \cdot 2}{4}\right)$$
$$= 2(\cos \pi + i \sin \pi)$$
$$= -2$$

$$t_3 = 2\left(\cos \frac{2\pi \cdot 3}{4} + i \sin \frac{2\pi \cdot 3}{4}\right)$$
$$= 2\left(\cos \frac{3\pi}{2} + i \sin \frac{3\pi}{2}\right)$$
$$= -2i$$

EXAMPLE 5
Complex Roots of an Equation

Determine the roots of the equation:
$$x^4 - 2x^2 + 2 = 0$$

Solution

Only even powers of x appear, so let's make the substitution
$$y = x^2$$

to obtain the new equation:
$$y^2 - 2y + 2 = 0$$
By the quadratic formula, we have:
$$y = \frac{-(-2) \pm \sqrt{(-2)^2 - 4(1)(2)}}{2}$$
$$= 1 \pm i$$
$$x^2 = 1 \pm i$$

So to determine x, we must find the square roots of $1 + i$ and $1 - i$. The moduli of these complex numbers are:
$$|1 + i| = \sqrt{1^2 + 1^2} = \sqrt{2}$$
$$|1 - i| = \sqrt{1^2 + (-1)^2} = \sqrt{2}$$

The arguments are $\pi/4$ and $7\pi/4$, respectively. So we have the polar forms:
$$2 + i = \sqrt{2}\left[\cos\frac{\pi}{4} + i\sin\frac{\pi}{4}\right]$$
$$2 - i = \sqrt{2}\left[\cos\left(\frac{7\pi}{4}\right) + i\sin\left(\frac{7\pi}{4}\right)\right]$$

So we have the square roots:
$$\sqrt{2+i} = \sqrt[4]{2}\left[\cos\frac{\pi}{8} + i\sin\frac{\pi}{8}\right], \quad \sqrt[4]{2}\left[\cos\frac{9\pi}{8} + i\sin\frac{9\pi}{8}\right]$$
$$\sqrt{2-i} = \sqrt[4]{2}\left[\cos\left(\frac{7\pi}{8}\right) + i\sin\left(\frac{7\pi}{8}\right)\right], \quad \sqrt[4]{2}\left[\cos\left(\frac{15\pi}{8}\right) + i\sin\left(\frac{15\pi}{8}\right)\right]$$

These last four complex numbers are the solutions to the equation.

Exercises 9.5

Use De Moivre's theorem to calculate the following. Express the results in $a + bi$ form.

1. z^{20}, $z = \cos 210° + i\sin 210°$
2. z^{12}, $z = \cos 180° + i\sin 180°$
3. z^{10}, $z = 2\left(\cos\frac{\pi}{4} + i\sin\frac{\pi}{4}\right)$
4. z^{30}, $z = 2\left(\cos\frac{5\pi}{4} + i\sin\frac{5\pi}{4}\right)$
5. z^3, $z = 5[\cos(-30°) + i\sin(-30°)]$
6. z^4, $z = 6[\cos(-60°) + i\sin(-60°)]$
7. $\left(-\frac{\sqrt{3}}{2} - \frac{1}{2}i\right)^6$
8. $\left(\frac{\sqrt{3}}{2} + \frac{1}{2}i\right)^4$
9. $(5 + 5i)^{14}$
10. $\left(-\frac{\sqrt{2}}{2} - \frac{\sqrt{2}}{2}i\right)^{13}$
11. $(-2\sqrt{3} + 2i)^8$
12. $(-1 + \sqrt{3}i)^{10}$
13. $[3(\cos 15° + i\sin 15°)]^4$
14. $[2(\cos 10° + i\sin 10°)]^{18}$
15. $[2(\cos 20° + i\sin 20°)]^6$
16. $[3(\cos 50° + i\sin 50°)]^6$

17. $\left[2\left(\cos\left(-\dfrac{\pi}{18}\right) + i\sin\left(-\dfrac{\pi}{18}\right)\right)\right]^3$

18. $\left[2\left(\cos\left(-\dfrac{\pi}{12}\right) + i\sin\left(-\dfrac{\pi}{12}\right)\right)\right]^4$

19. $\left[2\left(\cos\dfrac{\pi}{8} + i\sin\dfrac{\pi}{8}\right)\right]^{10}$

20. $\left[3\left(\cos\dfrac{\pi}{10} + i\sin\dfrac{\pi}{10}\right)\right]^{20}$

Find the indicated roots. Where possible, express the roots in $a + bi$ form.

21. cube roots of -8
22. cube roots of 8
23. tenth roots of $-\sqrt{3} - i$
24. sixth roots of $-\sqrt{3} + i$
25. cube roots of i
26. fourth roots of -1
27. square roots of $9i$
28. fifth roots of $1 - i$
29. fifth roots of $1 + i$
30. square roots of $-16i$
31. fourth roots of $8 - 8i\sqrt{3}$
32. square roots of $-2 + 2\sqrt{3}i$
33. cube roots of $27(\cos 180° + i \sin 180°)$
34. fifth roots of $32(\cos 150° + i \sin 150°)$
35. fourth roots of $81(\cos 120° + i \sin 120°)$
36. cube roots of $125(\cos 135° + i \sin 135°)$
37. square roots of $16\left[\cos\left(\dfrac{\pi}{2}\right) + i\sin\left(\dfrac{\pi}{2}\right)\right]$
38. square roots of $100\left[\cos\left(-\dfrac{\pi}{3}\right) + i\sin\left(-\dfrac{\pi}{3}\right)\right]$

Determine the complex solutions for the following.

39. $a^5 = 243$
40. $m^6 - 64 = 0$
41. $m^4 + 81i = 0$
42. $x^5 = -243$
43. $z^4 = -8\sqrt{3} + 8i$
44. $x^3 + 4\sqrt{3} - 4i = 0$
45. $x^5 - 1 = 0$
46. $z^5 - i = 0$
47. $x^3 + 27 = 0$
48. $x^4 = -16$
49. $x^2 + 25 = 0$
50. $x^6 - 1 = 0$
51. $x^3 = 1 - \sqrt{3}i$
52. $x^5 = \sqrt{2} - \sqrt{2}i$
53. $y^5 + \sqrt{2} - \sqrt{2}i = 0$
54. $a^3 - 1 - \sqrt{3}i = 0$

Compute the following.

55. $\left(-\dfrac{1}{3} - \dfrac{1}{3}i\right)^{-3}$
56. $(2\sqrt{3} - 2i)^{-4}$
57. $\dfrac{(1 + i)^3}{(\sqrt{3} + i)^4}$
58. $\dfrac{(2 - 2i)^4}{(1 + i)^3}$
59. $(\sqrt{3} + i)^4(1 - i)^3$
60. $(2 - 2i)^3(1 + i)^4$
61. $\left[2\left(\cos\dfrac{\pi}{2} + i\sin\dfrac{\pi}{2}\right)\right]^{-2}\left[3\left(\cos\dfrac{\pi}{2} + i\sin\dfrac{\pi}{2}\right)\right]^2$
62. $\left[3\left(\cos\dfrac{2\pi}{3} + i\sin\dfrac{2\pi}{3}\right)\right]^3\left[2\left(\cos\dfrac{\pi}{3} + i\sin\dfrac{\pi}{3}\right)\right]^2$
63. $\dfrac{\left[4\left(\cos\left(-\dfrac{\pi}{4}\right) + i\sin\left(-\dfrac{\pi}{4}\right)\right)\right]^3}{[2(\cos 2\pi + i \sin 2\pi)]^4}$
64. $\dfrac{\left[3\left(\cos\dfrac{3\pi}{4} + i\sin\dfrac{3\pi}{4}\right)\right]^4}{\left[9\left(\cos\left(-\dfrac{\pi}{6}\right) + i\sin\left(-\dfrac{\pi}{6}\right)\right)\right]^2}$

In your own words

65. State De Moivre's theorem.
66. Describe how to calculate the nth root of a complex number.

9.6 VECTORS

Another application of trigonometry is to elementary vector analysis, which is of critical importance in physics and engineering. In this section, we will provide a brief introduction to vectors in the plane.

Scalar and Vector Quantities

In modeling the world around us, some quantities are described by their magnitudes alone. Such quantities are called **scalar quantities.** Examples of such quantities are length, weight, area, and volume. Scalar quantities are represented by real numbers. Models also require using quantities that are described by both a **magnitude** and a **direction.** Such quantities are called **vector quantities.** Examples of vector quantities abound in physics and include velocity, acceleration, and force.

It is customary to represent a vector geometrically as an arrow. The length of the arrow represents the magnitude of the vector, and the direction of the arrow represents its direction. For instance, the vector in Figure 1 shows a velocity of 50 miles per hour in a direction 30° north of east.

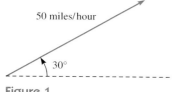
Figure 1

The starting point of a vector is called its **initial point** and the ending point its **terminal point.**

Two vectors are equal to one another if and only if they have the same magnitude and direction. For example, the vectors **A** and **B** in Figure 2 are equal to one another. However, the vectors **C** and **D** are not equal because they have the same direction but different magnitudes. Similarly, the vectors **E** and **F** are not equal to one another because they have the same magnitude but different directions. The location of the vector in the plane is immaterial. Two vectors with the same magnitude and direction are equal even though they are in different locations.

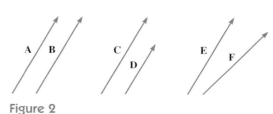
Figure 2

If **A** is a vector, its magnitude is denoted $|\mathbf{A}|$. The **zero vector,** denoted **0**, is the vector with magnitude 0. By convention, the direction of the zero vector is undefined.

To distinguish between scalars and vectors, it is customary to write scalars in lowercase, lightfaced italic letters, such as a, b, x, y. Vectors are denoted using boldfaced letters, either uppercase or lowercase, such as **A, B, i, j**.

A vector does not depend on its initial point. For many purposes, it is convenient to move the vector so that its initial point is at the origin. In this representation, the vector can be completely described by the coordinates of its terminal point. In Figure 3, the vector **v** has initial point at the origin and terminal point (x, y). We denote this vector with the notation $\langle x, y \rangle$, called the **coordinate representation** of **v**.

By the Pythagorean theorem, the magnitude of the vector $\mathbf{v} = \langle x, y \rangle$ is given by the formula:

$$|\mathbf{v}| = \sqrt{x^2 + y^2}$$

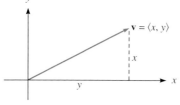
Figure 3

EXAMPLE 1
The Magnitude of a Vector

Draw the vector $\mathbf{v} = \langle 3, -4 \rangle$. Calculate its magnitude.

Solution

See Figure 4. The magnitude of $\mathbf{v}$ is given by:
$$|\mathbf{v}| = \sqrt{x^2 + y^2} = \sqrt{3^2 + (-4)^2} = 5$$

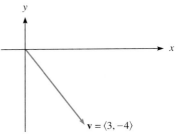

Figure 4

Multiplication of a Vector by a Scalar

Let $\mathbf{A}$ be a vector. The vector $-\mathbf{A}$ is defined to be the vector with the same length as $\mathbf{A}$ but with the opposite direction. (See Figure 5.)

If k is a scalar, the product $k\mathbf{A}$ is defined to be the vector with magnitude $|k|$ times the magnitude of $\mathbf{A}$, pointing in the same direction as $\mathbf{A}$ if $k > 0$ and in the opposite direction if $k < 0$. (See Figure 6.)

In terms of the coordinate representation, scalar multiplication is accomplished through coordinates. That is, if $\mathbf{v} = \langle x, y \rangle$, then:
$$k\mathbf{v} = \langle kx, ky \rangle$$

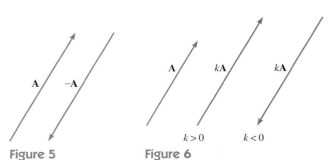

Figure 5   Figure 6

EXAMPLE 2
Scalar Multiplication

Suppose that $\mathbf{w} = \langle 4, -7 \rangle$. Calculate $2\mathbf{w}$.

Solution

We have:
$$2\mathbf{w} = 2\langle 4, -7 \rangle = \langle 2 \cdot 4, 2 \cdot (-7) \rangle = \langle 8, -14 \rangle$$

Vector Addition

Suppose that $\mathbf{A}$ and $\mathbf{B}$ are given vectors. We can form the sum $\mathbf{A} + \mathbf{B}$ as follows: Place the initial point of $\mathbf{B}$ on the terminal point of $\mathbf{A}$, as shown in Figure 7. The sum is the vector that connects the initial point of $\mathbf{A}$ to the terminal point of $\mathbf{B}$.

Another way of interpreting the sum is to draw a copy of $\mathbf{B}$ from the same initial point as $\mathbf{A}$ and draw a parallelogram with the two vectors as sides. Then the sum $\mathbf{A} + \mathbf{B}$ is just the diagonal of the parallelogram. (See Figure 8.) For this reason, vector addition is often called the **parallelogram rule.**

Vector addition can be accomplished by adding coordinates. For example, if $\mathbf{A} = \langle x_1, y_1 \rangle$ and $\mathbf{B} = \langle x_2, y_2 \rangle$, then:

$$\mathbf{A} + \mathbf{B} = \langle x_1 + x_2, y_1 + y_2 \rangle$$

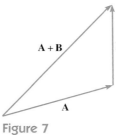

Figure 7

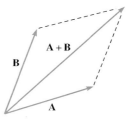

Figure 8

EXAMPLE 3
Vector Arithmetic

Let $\mathbf{A}$ and $\mathbf{B}$ be as drawn in Figure 9. Use the parallelogram rule to describe the vectors $\mathbf{A} + (-\mathbf{B})$ and $-\tfrac{1}{2}\mathbf{A}$.

Solution

See Figure 10.

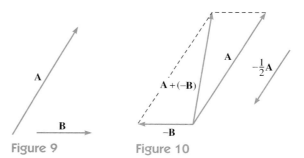

Figure 9

Figure 10

EXAMPLE 4
Adding Vectors

Suppose that $\mathbf{A} = \langle -1, 3 \rangle$, $\mathbf{B} = \langle -6, 9 \rangle$. Calculate $\mathbf{A} + \mathbf{B}$.

Solution

Just add the respective coordinates:

$$\begin{aligned}\mathbf{A} + \mathbf{B} &= \langle -1, 3 \rangle + \langle -6, 9 \rangle \\ &= \langle -1 + (-6), 3 + 9 \rangle \\ &= \langle -7, 12 \rangle\end{aligned}$$

EXAMPLE 5
Vector Arithmetic

Suppose that $\mathbf{A} = \langle 3, 4 \rangle$, $\mathbf{B} = \langle -1, 10 \rangle$. Determine the vector $3\mathbf{A} - 2\mathbf{B}$.

Solution

We have:

$$\begin{aligned}3\mathbf{A} &= \langle 9, 12 \rangle \\ -2\mathbf{B} &= \langle 2, -20 \rangle \\ 3\mathbf{A} - 2\mathbf{B} &= \langle 9, 12 \rangle + \langle 2, -20 \rangle \\ &= \langle 11, -8 \rangle\end{aligned}$$

EXAMPLE 6
Calculating the Heading of an Airplane

Suppose that an airplane is heading east at 450 mph. Suddenly a 40-mph northerly wind is encountered. In what direction will the aircraft go, and what will be its speed in that direction?

Solution

Represent the original velocity of the plane as a vector **v** and the wind as the vector **w**, as shown in Figure 11. The new velocity of the plane is represented by the vector **v** + **w**. In terms of coordinates, we have:

$$\mathbf{v} + \mathbf{w} = \langle 450, 0 \rangle + \langle 0, 40 \rangle$$
$$= \langle 450, 40 \rangle$$

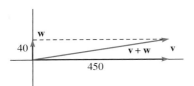

Figure 11

The new speed is the magnitude of this vector:

$$|\mathbf{v} + \mathbf{w}| = \sqrt{450^2 + 40^2} \approx 451.77 \text{ mph}$$

Let θ denote the angle this vector makes with the positive x-axis. (See Figure 12.) Then:

$$\tan \theta = \frac{40}{450} = 0.088889$$

$$\theta = \tan^{-1} 0.088889 \approx 5.0796°$$

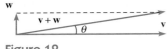

Figure 12

That is, the airplane goes 451.77 mph at a heading 5.0796° north of east.

Resolution of Vectors into Components

Suppose we are given a vector **v** and two distinct directions, as shown in Figure 13. We can write **v** as a sum of vectors $\mathbf{v}_1$ and $\mathbf{v}_2$, one in each of the directions, as shown in the figure. The vectors $\mathbf{v}_1$ and $\mathbf{v}_2$ are called the **components** of **v** in the given directions. The process of writing **v** as a sum is called **resolution into components**. This procedure is of fundamental importance in physics and engineering.

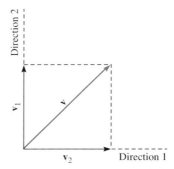

Figure 13

EXAMPLE 7
Resolution of a Vector into Horizontal and Vertical Components

Suppose that a force of 100 lb acts along a rope strung in the direction shown in Figure 14. Resolve the force into horizontal and vertical components.

Solution

Draw the components, as shown in Figure 15. We determine the magnitudes of the components using trigonometry:

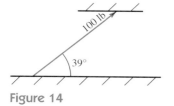

Figure 14

$$|\mathbf{F}_{\text{horizontal}}| = |\mathbf{F}| \cos 39° = 100 \cdot 0.79864 = 79.864 \text{ lb}$$

$$|\mathbf{F}_{\text{vertical}}| = |\mathbf{F}| \sin 39° = 100 \cdot 0.62932 = 69.932 \text{ lb}$$

Let $\mathbf{i}$ denote a vector of length 1 pointed in the positive x-direction and $\mathbf{j}$ a vector of length 1 pointed in the positive y-direction. Then we can write the horizontal and vertical components as:

$$\mathbf{F}_{\text{horizontal}} = 79.864\mathbf{i}$$

$$\mathbf{F}_{\text{vertical}} = 69.932\mathbf{j}$$

$$\mathbf{F} = \mathbf{F}_{\text{horizontal}} + \mathbf{F}_{\text{vertical}} = 79.864\mathbf{i} + 69.932\mathbf{j}$$

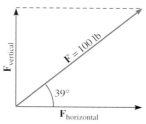

Figure 15

Applications of Vectors

One of the fundamental laws of physics is Newton's first law, which implies that if a body is at rest, then the vector sum of all forces acting on the body is zero. This law is used extensively in analyzing engineering structures, as the next example shows.

EXAMPLE 8
Forces on a Structure

A 50-kg weight is supported by two ropes, as shown in Figure 16. Determine the forces acting on the two ropes.

Solution

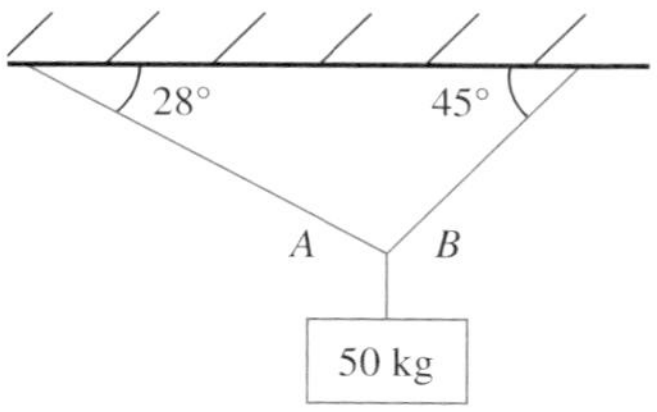

Consider the point where the ropes are joined. This point is at rest, so the vector sum of all forces acting on it is 0. The forces in each rope act in the direction of the rope, as shown in Figure 17. Resolve each force into vertical and horizontal components. The sum of the vertical components must add to 0, so we obtain the equation:

$$|\mathbf{F}_A| \sin 28° + |\mathbf{F}_B| \sin 45° - 50 = 0$$

The sum of the horizontal components must add to 0, so we obtain the equation:

$$-|\mathbf{F}_A| \cos 28° + |\mathbf{F}_B| \cos 45° = 0$$

From the second equation, we obtain:

$$|\mathbf{F}_B| = \frac{\cos 28°}{\cos 45°} |\mathbf{F}_A|$$

$$\approx 1.2487 |\mathbf{F}_A|$$

Substituting this value for $|\mathbf{F}_B|$ into the first equation, we obtain:

$$|\mathbf{F}_A| \sin 28° + 1.2487 |\mathbf{F}_A| \cdot \sin 45° = 50$$

$$|\mathbf{F}_A| \approx 36.970$$

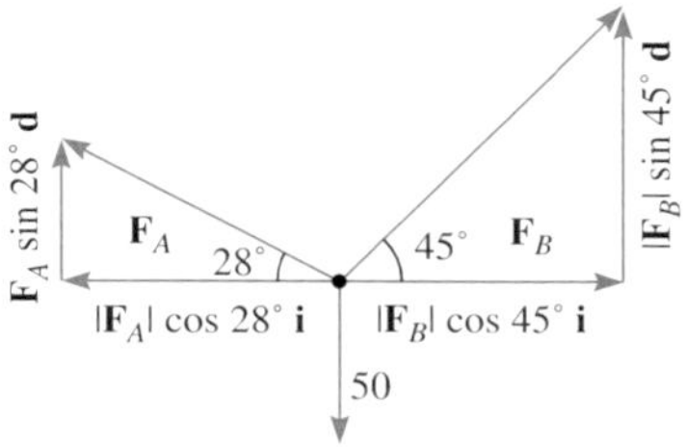

Figure 16

Figure 17

$$|\mathbf{F}_B| \approx 1.2487 \quad |\mathbf{F}_A| \approx 46.164$$

Thus, the force in rope A is 36.97 lb and the force in rope B is 46.164 lb.

EXAMPLE 9
Circular Motion

A race car is traveling at 180 miles per hour on a circular track. The car hits a patch of oil and changes direction by 30°. (See Figure 18.) We can write the new velocity as a sum of two vectors, one directed along the radius of the circle (the radial component) and one along the tangent line to the circle (the tangential component). What are the magnitudes of the radial and tangential components of the car's velocity?

Solution

Referring to Figure 18, we see that the magnitude of the radial component of the car's velocity is:

$$180 \sin 30° = 180 \cdot \tfrac{1}{2}$$
$$= 90 \text{ mph}$$

The magnitude of the tangential component of the velocity vector is:

$$180 \cos 30° = 180 \cdot \frac{\sqrt{3}}{2}$$
$$\approx 155.88 \text{ mph}$$

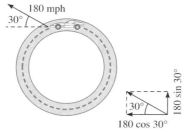

Figure 18

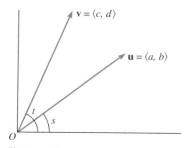

Scalar Product of Two Vectors

Let $\mathbf{u} = \langle a, b \rangle$, $\mathbf{v} = \langle c, d \rangle$ be two nonzero vectors based at the origin, and let s and t be, respectively, the angles these vectors make with the positive x-axis. (See Figure 19.) From the definitions of the sine and cosine, we have the identities:

$$\sin s = \frac{b}{\sqrt{a^2 + b^2}} = \frac{b}{|\mathbf{u}|}, \qquad \cos s = \frac{a}{\sqrt{a^2 + b^2}} = \frac{a}{|\mathbf{u}|}$$

$$\sin t = \frac{d}{\sqrt{c^2 + d^2}} = \frac{d}{|\mathbf{v}|}, \qquad \cos t = \frac{c}{\sqrt{c^2 + d^2}} = \frac{c}{|\mathbf{v}|}$$

Multiplying the first and third identities and the second and fourth, and then adding the result, gives:

$$\frac{ac + bd}{|\mathbf{u}||\mathbf{v}|} = \cos s \cos t + \sin s \sin t = \cos(s - t) \qquad (1)$$

Figure 19

However, $s - t$ is just the angle between the two vectors, which we will denote by θ. So the last identity can be written in the form:

$$ac + bd = |\mathbf{u}||\mathbf{v}| \cos \theta \qquad (2)$$

This identity was proven under the assumption that the vectors were nonzero. However, it clearly holds as well if one or both vectors are zero. Let's assign a name to the expression on the left side of the equation:

Definition
Dot Product

Let $\mathbf{u} = \langle a, b \rangle$, $\mathbf{v} = \langle c, d \rangle$ be two vectors. Then their **dot product** (also called the **scalar product**), denoted $\mathbf{u} \cdot \mathbf{v}$, is the real number defined by:

$$\mathbf{u} \cdot \mathbf{v} = ac + bd \qquad (3)$$

Warning: Don't confuse the dot product of two vectors with the scalar product of a real number and a vector. The result of a dot product is a real number, whereas the result of a scalar product is a vector.

We can restate equation (2) as follows:

GEOMETRIC INTERPRETATION OF THE DOT PRODUCT

Let $\mathbf{u} = \langle a, b \rangle$ and $\mathbf{v} = \langle c, d \rangle$ be two vectors, and let θ denote the positive angle from $\mathbf{u}$ to $\mathbf{v}$. Then:

$$\mathbf{u} \cdot \mathbf{v} = |\mathbf{u}||\mathbf{v}|\cos\theta$$

If $\mathbf{u} \neq \mathbf{0}$, $\mathbf{v} \neq \mathbf{0}$, then $|\mathbf{u}| \neq 0$, $|\mathbf{v}| \neq 0$, so the preceding relation can be written in the form:

$$\cos\theta = \frac{\mathbf{u} \cdot \mathbf{v}}{|\mathbf{u}| \cdot |\mathbf{v}|}, \qquad \mathbf{u} \neq \mathbf{0}, \quad \mathbf{v} \neq \mathbf{0}$$

EXAMPLE 10
Determining the Angle between Two Vectors

Determine the angle between the vectors $\mathbf{u} = \langle 1, -1 \rangle$, $\mathbf{v} = \langle 3, 0 \rangle$.

Solution

Let θ be the positive angle from $\mathbf{u}$ to $\mathbf{v}$. We have:

$$\cos\theta = \frac{\mathbf{u} \cdot \mathbf{v}}{|\mathbf{u}||\mathbf{v}|}$$

$$= \frac{3}{\sqrt{1^2 + (-1)^2} \cdot \sqrt{3^2 + 0^2}}$$

$$= \frac{\sqrt{2}}{2}$$

$$\theta = \frac{\pi}{4}$$

Two vectors are perpendicular if and only if $\cos\theta = 0$, so we have:

DOT PRODUCT CRITERION FOR PERPENDICULARITY

The vectors $\mathbf{u}$ and $\mathbf{v}$ are perpendicular if and only if $\mathbf{u} \cdot \mathbf{v} = 0$.

EXAMPLE 11
Testing Vectors for Perpendicularity

Are the vectors

$$\mathbf{v} = \left\langle \sqrt{2}, -\sqrt{2} \right\rangle, \qquad \mathbf{w} = \langle 5, 5 \rangle$$

perpendicular to one another?

Exercises 9.6

Given vectors **A**, **B**, and **C**, pictured in the figure below, draw the following.

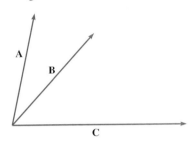

1. $-\mathbf{A}$
2. $-\mathbf{B}$
3. $\mathbf{A} + \mathbf{B}$
4. $\mathbf{A} - \mathbf{B}$
5. $\mathbf{A} - \mathbf{C}$
6. $\mathbf{B} + \mathbf{C}$
7. $2\mathbf{A}$
8. $-2\mathbf{C}$
9. $\mathbf{A} + 2\mathbf{C}$
10. $-\mathbf{A} - \mathbf{C}$

For the following vectors (**A**) with initial point at the origin and the given terminal point, first determine $|\mathbf{A}|$, and then determine the horizontal and vertical components of **A**.

11. terminal point $(1, -1)$
12. terminal point $(-2, -2)$
13. terminal point $(3, 3)$
14. terminal point $(-5, 5)$
15. terminal point $(-1, \sqrt{3})$
16. terminal point $(4\sqrt{3}, 4)$

Given vectors $\mathbf{A} = \langle 3, 2 \rangle$, $\mathbf{B} = \langle -2, 4 \rangle$, $\mathbf{C} = \langle -1, -2 \rangle$, and $\mathbf{D} = \langle 4, -3 \rangle$, determine the following vectors.

17. $2\mathbf{A}$
18. $-3\mathbf{B}$
19. $\mathbf{A} + \mathbf{B}$
20. $\mathbf{B} - 2\mathbf{D}$
21. $3\mathbf{C} + 2\mathbf{A}$
22. $3\mathbf{A} - 5\mathbf{B}$
23. $2\mathbf{A} - 3\mathbf{D}$
24. $4\mathbf{C} + 2\mathbf{B}$
25. $\mathbf{A} - 2\mathbf{C}$
26. $2\mathbf{B} + 3\mathbf{A}$

Find the dot product of each pair of vectors.

27. $\mathbf{u} = \langle 1, \sqrt{2} \rangle$, $\mathbf{v} = \langle -3, -5 \rangle$
28. $\mathbf{u} = \langle -23, 25 \rangle$, $\mathbf{v} = \langle 12, -50 \rangle$

Prove the following for dot products.

29. $c_1(\mathbf{v}_1 \cdot \mathbf{v}_2) = (c_1 \mathbf{v}_1) \cdot \mathbf{v}_2$
30. $(c_1 \mathbf{v}_1) \cdot (c_2 \mathbf{v}_2) = (c_1 c_2)(\mathbf{v}_1 \cdot \mathbf{v}_2)$
31. $\mathbf{v}_1 \cdot (\mathbf{v}_2 + \mathbf{v}_3) = \mathbf{v}_1 \cdot \mathbf{v}_2 + \mathbf{v}_1 \cdot \mathbf{v}_3$
32. $\mathbf{v}_2 \cdot \mathbf{v}_1 = \mathbf{v}_1 \cdot \mathbf{v}_2$

Determine which of the following pairs of vectors are perpendicular to one another.

33. $\langle 1, 1 \rangle$, $\langle 1, -1 \rangle$
34. $\langle 3, 4 \rangle$, $\langle 4, -3 \rangle$
35. $\langle 1, -2 \rangle$, $\langle 2, 5 \rangle$
36. $\langle 2, 3 \rangle$, $\langle -4, 1 \rangle$

Determine the angle between the following vectors.

37. $\mathbf{u} = \langle 1, 2 \rangle$, $\mathbf{v} = \langle 1, 1 \rangle$
38. $\mathbf{u} = \langle -1, 1 \rangle$, $\mathbf{v} = \langle 4, -4 \rangle$
39. $\mathbf{u} = \langle 0, 3 \rangle$, $\mathbf{v} = \langle -2, 1 \rangle$
40. $\mathbf{u} = \langle 5, -\frac{5}{2} \rangle$, $\mathbf{v} = \langle -1, 0 \rangle$

 Applications

41. **Navigation.** Neglecting the current, a ship would head through water on a compass heading of 30° at a speed of 20 knots. It is traveling in a current that causes the ship to move on a path with a heading of 45° at 15 knots. Find the speed of the current if it is flowing directly from the north.

42. **Navigation.** A plane that was flying due south at 250 mph is blown off course by the wind, which is blowing at 40 mph from the east. What is the angle that the plane is blown off course, and what is the actual ground speed of the plane?

43. **Navigation.** A steamer sails 100 miles east, and then 40 miles on a heading of 120°. How far is the steamer, and what is its bearing from its starting point?

44. **Navigation.** A sailboat is headed east at 18 mph relative to the water. An ocean current is carrying the water south at 3 mph. Find the course of the ship and its speed with respect to the ocean floor.

45. **Navigation.** An airplane must fly at a ground speed of 450 mph on a course 170° to be on schedule. The wind velocity is 25 mph in the direction 40°. Find the necessary heading to the nearest degree and the necessary air speed to the nearest unit that the plane must fly to be on schedule.

46. **Navigation.** A plane flies 650 mph on a bearing of 175°. A 25-mph wind, from a direction of 266°, blows against the plane. Find the resulting bearing of the plane.

47. **Navigation.** At what bearing and speed should a pilot head to fly due south at 520 mph when a 40-mph east wind is blowing?

48. **Navigation.** An airplane is headed 230° at 320 knots when a 15-knot wind is blowing from 50°. Find the ground speed of the plane and the wind correction angle.

49. **Navigation.** A river is flowing at 2.4 mph when a girl rows across it. If the girl rows at a still-water speed of 3.1 mph and heads the boat perpendicular to the direction of the current, find the ground speed of the boat.

50. **Navigation.** The velocity produced by the engines of a ship on a heading of 157° has a magnitude of 35.2 km/hr. The water current has a velocity of 8 km/hr in the direction of 213°. Find the magnitude of the actual velocity (ground speed) of the ship.

51. **Navigation.** Two ships leave a harbor, one traveling at 20 knots on a course of 80° and the other at 24 knots on a course of 140°. How far apart are the ships after two hours? What is the bearing from the first ship to the second at that time?

52. **Navigation.** An airplane flies on a compass heading of 90° at 200 mph. The wind affecting the plane is blowing from 300° at 30 mph. What is the true course and ground speed of the airplane?

53. **Physics.** A sled on an inclined plane weighs 500 lb, and the plane makes an angle of 50° with the horizontal. What force, perpendicular to the plane, is exerted on the sled by the plane?

54. **Physics.** A 150-lb box is dragged up a runway inclined 42° to the horizontal. Find the pressure of the box against the runway and the force required to drag the box.

55. **Physics.** If a force of 25 lb is required to move an 80-lb sled up a hill, what angle does the hill make with the horizontal?

56. **Physics.** What would be the force required to push a 100-lb object up a ramp that is inclined 10° with the horizontal?

57. **Physics.** A weight of 15 lb is placed on a smooth plane inclined at an angle of 24° with the horizontal. What force pushing along the plane will just prevent the weight from slipping?

58. **Physics.** A painting weighing 25 lb is supported at the top corners by a taut wire that hangs from a nail embedded in the wall. If the wire forms an angle of 120° at the nail, what is the total pull on the wire?

59. **Physics.** A truck weighing 6,875 lb moves up a bridge inclined 7°32′ from the horizontal. Find the pressure of the truck against the bridge.

60. **Physics.** What would be the largest weight a person could drag up a slope inclined 35° from the horizontal, if that person is able to pull with a force of 125 lb?

61. **Physics.** When a girl pulls her sled with a rope, the rope makes an angle of 35° with the horizontal. If a pull of 16 lb on the rope is needed to move the sled, what is the horizontal component force?

62. **Physics.** If a force of 52.1 lb is needed to keep a 75-lb block from sliding down an incline, what is the angle that the incline makes with the horizontal?

In your own words

63. Compare and contrast scalars and vectors.
64. List three vector quantities in the world around you.
65. List three scalar quantities in the world around you.
66. Describe how to add vectors geometrically. Give a concrete example.

Mathematics and the World around Us—Simulation as a Modeling Tool

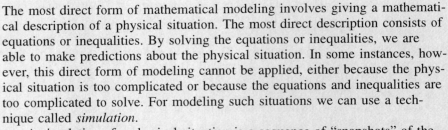

The most direct form of mathematical modeling involves giving a mathematical description of a physical situation. The most direct description consists of equations or inequalities. By solving the equations or inequalities, we are able to make predictions about the physical situation. In some instances, however, this direct form of modeling cannot be applied, either because the physical situation is too complicated or because the equations and inequalities are too complicated to solve. For modeling such situations we can use a technique called *simulation*.

A simulation of a physical situation is a sequence of "snapshots" of the situation at successive time intervals. Each snapshot is related to the preceding one using the mathematics and physical details of the situation. By observing the sequence of snapshots, you can, in effect, observe the situation in slow motion. Moreover, by varying physical details, you can analyze their effect on the sequence of events. Simulations typically involve massive amounts of calculations that often require supercomputers to carry out.

Simulations are used to determine the effects of wind on car and airplane designs. Analysis by simulation is much cheaper and less time-consuming than wind tunnel experiments, because it doesn't require a physical model of the car or plane.

Simulations have recently been used to analyze the effects of car design on occupant safety during various types of crashes. Simulations have also been used to analyze the effects of hurricanes and tidal waves on buildings and the effects of volcanic eruptions on the atmosphere and weather.

Calculating the various snapshots in a simulation often requires sophisticated mathematical approximations belonging to the field of finite element analysis, which is part of the field of partial differential equations.

Chapter Review

Important Concepts—Chapter 9

- Law of Sines
- ambiguous case of the Law of Sines
- Law of Cosines
- Heron's formula
- pole
- polar axis
- polar coordinates
- conversion between rectangular and polar coordinates
- graphs of polar equations
- complex plane
- real axis
- imaginary axis

- absolute value of a complex number; modulus
- argument of a complex number
- trigonometric form of a complex number; polar form
- principal argument of a complex number
- multiplication and division in polar form
- nth root of a complex number
- De Moivre's theorem
- scalar quantities
- vector quantities
- magnitude of a vector

- direction of a vector
- initial point of a vector
- terminal point of a vector
- zero vector
- coordinate representation
- multiplication of a vector by a scalar
- parallelogram rule for adding vectors
- components of a vector
- resolution into components
- dot product; scalar product
- dot product criterion for perpendicularity

Important Results and Techniques—Chapter 9

Law of Sines	$\dfrac{a}{\sin \alpha} = \dfrac{b}{\sin \beta} = \dfrac{c}{\sin \gamma}$	p. 469
Law of Cosines	$a^2 = b^2 + c^2 - 2bc \cos \alpha$ $b^2 = a^2 + c^2 - 2ac \cos \beta$ $c^2 = a^2 + b^2 - 2ab \cos \delta$	p. 478
Area of triangle	$A = \tfrac{1}{2} ab \sin \gamma$ $A = \tfrac{1}{2} bc \sin \alpha$ $A = \tfrac{1}{2} ac \sin \beta$ $A = \sqrt{s(s-a)(s-b)(s-c)}$, where $s = \dfrac{a+b+c}{2}$	p. 480–481
Conversion between rectangular and polar coordinates	$x = r \cos \theta$ $y = r \sin \theta$ $r = \sqrt{x^2 + y^2}$ $\tan \theta = \dfrac{y}{x}, \quad x \neq 0$	p. 486
Polar form	$z = r(\cos \theta + i \sin \theta)$, where $r = \sqrt{a^2 + b^2}, \quad \tan \theta = b/a$	p. 493
Polar product	$z_1 z_2 = r_1 r_2 [\cos(\theta_1 + \theta_2) + i \sin(\theta_1 + \theta_2)]$	p. 495
Polar division	$\dfrac{z_1}{z_2} = \dfrac{r_1}{r_2} [\cos(\theta_1 - \theta_2) + i \sin(\theta_1 - \theta_2)]$	p. 495
De Moivre's theorem	$z^n = r^n (\cos n\theta + i \sin n\theta)$	p. 499
nth roots	$t_j = r^{1/n} \left[\cos \dfrac{\theta + 2\pi j}{n} + i \sin \dfrac{\theta + 2\pi j}{n} \right]$, $(j = 0, 1, \ldots, n-1)$	p. 500
Vectors	$|\mathbf{A}| = \sqrt{a^2 + b^2}$	p. 505
	Scalar product of k and $\mathbf{A}$: $k\mathbf{A} = \langle ka_1, kb_1 \rangle$	p. 506
	Sum of $\mathbf{A} = \langle a_1, b_1 \rangle$ and $\mathbf{B} = \langle a_2, b_2 \rangle$ $\mathbf{A} + \mathbf{B} = \langle a_1 + a_2, b_1 + b_2 \rangle$	p. 507
	Dot product: $\langle a, b \rangle \cdot \langle c, d \rangle = ac + bd$ $\mathbf{u} \cdot \mathbf{v} = |\mathbf{u}| |\mathbf{v}| \cos \theta$, $\theta =$ the angle between $\mathbf{u}$ and $\mathbf{v}$	p. 511

Review Exercises—Chapter 9

Determine the remaining sides and angles.

1. $a = 15, b = 3, \angle A = 30°$
2. $c = 10, \angle C = 45°, \angle B = 30°$

Determine the number of solutions and find them for the given triangle.

3. $a = 260, b = 476, \angle A = 13°$
4. $b = 20, a = 40, \angle C = 30°$

Determine the remaining sides and angles of the following triangles.

5. $a = 10, b = 12, \angle C = 45°$
6. $c = 20, a = 30, \angle C = \dfrac{\pi}{6}$

Find the area of the following triangles.

7. $a = 6, b = 14, \angle C = \dfrac{5\pi}{6}$
8. $a = 26, b = 30, c = 8$
9. $a = 18, b = 25, \angle A = 120°$
10. $a = 10, b = 15, \angle C = \dfrac{\pi}{3}$

Evaluate the following.

11. $|1 - 2i|$
12. $|3 + 3i|$
13. $|-1 - \sqrt{3}i|$
14. $|-\sqrt{2} + \sqrt{2}i|$

Determine (a) the polar form, (b) the amplitude, and (c) the argument of each of the following complex numbers.

15. $z = 2i$
16. $z = 3 - 3i$
17. $z = -4 - \sqrt{3}i$
18. $z = 5i$

Graph the point corresponding to the following.

19. $3(\cos 30° + i \sin 30°)$
20. $4(\cos(-45°) + i \sin(-45°))$
21. $2\left[\cos\left(-\dfrac{\pi}{3}\right) + i \sin\left(-\dfrac{\pi}{3}\right)\right]$
22. $5\left[\cos\dfrac{2\pi}{3} + i \sin\dfrac{2\pi}{3}\right]$

Compute zw and z/w for the following complex numbers in polar form.

23. $z = 5(\cos 90° + i \sin 90°)$
    $w = 2(\cos 45° + i \sin 45°)$
24. $z = 2(\cos(-30°) + i \sin(-30°))$
    $w = 3(\cos 60° + i \sin 60°)$
25. $z = 3\left[\cos\left(-\dfrac{2\pi}{3}\right) + i \sin\left(-\dfrac{2\pi}{3}\right)\right]$
    $w = 2(\cos 2\pi + i \sin 2\pi)$
26. $z = 4\left[\cos\left(-\dfrac{\pi}{3}\right) + i \sin\left(-\dfrac{\pi}{3}\right)\right]$
    $w = 2\left[\cos\left(-\dfrac{5\pi}{6}\right) + i \sin\left(-\dfrac{5\pi}{6}\right)\right]$

Change from polar form to complex (standard) form.

27. $\sqrt{2}(\cos 45° + i \sin 45°)$
28. $\sqrt{3}[\cos(-60°) + i \sin(-60°)]$
29. $5[\cos(-2\pi) + i \sin(-2\pi)]$
30. $2\left(\cos\dfrac{\pi}{6} + i \sin\dfrac{\pi}{6}\right)$

Calculate the following and express results in $a + bi$ form.

31. $z^{10}, \quad z = \left(\cos\dfrac{7\pi}{6} + i \sin\dfrac{7\pi}{6}\right)$
32. $z^{12}, \quad z = \cos \pi + i \sin \pi$
33. $z^{30}, \quad z = 2(\cos 225° + i \sin 225°)$
34. $z^3, \quad z = 3\left[\cos\left(-\dfrac{\pi}{6}\right) + i \sin\left(-\dfrac{\pi}{6}\right)\right]$
35. $(2\sqrt{3} + 2i)^8$
36. $(3\sqrt{3} - 3i)^4$

Find the indicated roots and express in $a + bi$ form if exact; otherwise, leave in polar form.

37. cube roots of 27
38. square roots of $16i$
39. fifth roots of $2 + 2i$
40. fourth roots of $8 + 8i\sqrt{3}$
41. square roots of $100[\cos(-60°) + i \sin(-60°)]$
42. cube roots of $125\left[\cos\left(\dfrac{3\pi}{4}\right) + i \sin\left(\dfrac{3\pi}{4}\right)\right]$

Determine the complex solutions for the following.

43. $x^5 + 1 = 0$
44. $a^3 = 8$
45. $a^3 = 1 - \sqrt{3}i$
46. $x^4 + 8\sqrt{3} - 8i = 0$

Given vectors **A** and **B** in the figure, draw the following vectors.

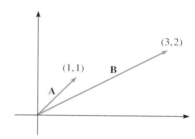

47. $\mathbf{A} + \mathbf{B}$
48. $2\mathbf{A} - \mathbf{B}$
49. $3\mathbf{B} - \mathbf{A}$
50. $\mathbf{A} + 2\mathbf{B}$

For the given vector **A**, with initial point at the origin and the given terminal point, determine $|\mathbf{A}|$, and determine the horizontal and vertical components of **A**.

51. vector **A** terminal point $(2, 2)$
52. vector **A** terminal point $(-1, -\sqrt{3})$
53. vector **A** terminal point $(-\sqrt{2}, +\sqrt{2})$
54. vector **A** terminal point $(3, -3)$

Determine the following vectors, given $\mathbf{A} = \langle 2, 5 \rangle$, $\mathbf{B} = \langle -2, -3 \rangle$, and $\mathbf{C} = \langle -1, -2 \rangle$.

55. $3\mathbf{A}$
56. $\mathbf{A} + \mathbf{B} + \mathbf{C}$
57. $2\mathbf{B} - \mathbf{C}$
58. $2\mathbf{C} - 3\mathbf{B}$

Plot the following points given in polar coordinates.

59. $(-1, 180°)$
60. $(3, -30°)$
61. $\left(-3, \dfrac{-3\pi}{4}\right)$
62. $\left(3, -\dfrac{7\pi}{4}\right)$

Graph the given polar equation.

63. $r = 5$
64. $r = 4\cos\theta$
65. $r = -6 - 6\cos\theta$
66. $r = 2 + 4\sin\theta$

Determine the polar coordinates for each of the following points.

67. $(-2, 2)$
68. $(1, -\sqrt{3})$
69. $(3, -3\sqrt{3})$
70. $(-4, -4)$

Write the given equations in terms of polar coordinates.

71. $x = -3$
72. $y = 6$
73. $x^2 + y^2 = 4$
74. $x^2 + y^2 + 2y = 0$

Write the given polar equations in rectangular form.

75. $r = 2$
76. $r = 2\sec\theta$
77. $r\cos\theta = 5$
78. $r = \tan\theta$
79. $r^2 = \dfrac{1}{\cos 2\theta}$
80. $r^2(4\sin^2\theta - 9\cos^2\theta) = 36$

Applications

81. **Geometry.** How long is a pole if, when it leans away from the sun at an angle of 7° to the vertical, the angle of elevation of the sun is 51° and the pole casts a shadow 47 ft long on level ground?

82. **Geometry.** A tree grows vertically on the side of a hill that slopes upward from the horizontal by 8°. When the angle of elevation of the sun measures 20°, the shadow of the tree falls 42 ft down the hill. How tall is the tree?

518   CHAPTER 9   Applications of Trigonometry

83. **Geometry.** One end of a 20-ft ladder is placed on the ground at a point 8 ft from the start of a 41° incline. If the other end rests on the incline, how far up the incline is this other end?

84. **Geometry.** Two angles of a triangle are 30° and 55° and the longest side is 34 m. Find the length of the shortest side.

85. **Geometry.** Points A and B are on opposite sides of Lake Tyrone. From a third point C, the angle between the line of sight to A and B is 46.3°. If AC is 350 ft long and BC is 286 ft long, find AB.

86. **Geometry.** A parallelogram has adjacent sides of length 8 and 10 inches, and the measure of one included angle is 35°. Find the length of each diagonal of the parallelogram.

87. **Navigation.** At what compass heading and air speed should an aircraft fly if a wind of 40 mph is blowing from the north and the pilot wants to maintain a speed of 200 mph on a true course of 90°?

88. **Navigation.** Two planes leave an airport together, traveling on courses that have an angle of 135°40′ between them. If they each travel 402 miles, how far apart are they?

89. **Navigation.** For a boat traveling at 30 mph to travel directly north across a river, it must aim at a point that has a bearing of 15°. If the current is flowing directly west, find the rate of the current.

90. **Navigation.** Two submarines, one cruising at 25 knots and the other at 20 knots, left a naval base at the same

time. Three hours later they were 100 nautical miles apart. What was the measure of the angle between their courses?

91. **Physics.** A cable is attached to the top of a pole, and a 100-lb ball at the other end of the cable swings with a constant speed in a circle parallel to the ground. As it swings around, the cable makes an angle of 30° with its support. What are the centrifugal force on the ball and the force exerted on the ball by the cable?

92. **Physics.** A loading ramp makes an angle of 22° with the horizontal. A box weighing 130 lb slides down the ramp at a constant velocity. What is the force of friction acting on the box?

Chapter Test

Solve the following problems. You may use a calculator where appropriate.

1. Suppose that two angles and a side of a triangle are given by $b = 6$, $\alpha = 70°$, $\beta = 28°$. Determine the other angle and the other sides.

2. A ship receives notification of a storm to its north. To avoid the storm it first proceeds at a heading 31.5° north of east for 200 nautical miles. It then changes heading to 40° north of west.

   a. How far does the ship go until it intersects its original course?

   b. How far out of its way did the ship go to avoid the storm?

3. Prove the following identity for a triangle with sides a, b, c and angles α, β, γ:
$$\frac{a}{a+b} = \frac{\sin \alpha}{\sin \alpha + \sin \beta}$$

4. Determine the remaining sides and angles of a triangle with $b = 10$, $c = 13$, $\alpha = 49.6°$.

5. The distances from a sailboat at point C to two points A and B on the shore are known to be 350 m and 200 m, respectively. If the angle ACB measures 58°, find the distance between points A and B on the shore.

6. Plot the point $(5, 5\pi/4)$ in polar coordinates.

7. a. Convert the polar coordinates $(3, 5\pi/6)$ into rectangular coordinates.

    b. Convert the rectangular coordinates $(-3, 2)$ into polar coordinates.

8. Sketch the graph of the polar equation $r = 2(1 - \sin\theta)$.
9. Write the complex number $5 + 5i$ in polar form.
10. Determine the cube roots of i.
11. Let $\mathbf{A} = \langle 5, 7 \rangle$, $\mathbf{B} = \langle 3, -2 \rangle$.

    a. Determine the horizontal and vertical components of $\mathbf{A}$.

    b. Determine the vector $\mathbf{A} - 3\mathbf{B}$.

    c. Determine the angle between the vectors $\mathbf{A}$ and $\mathbf{B}$.

12. **Thought question.** Describe how to calculate the *n*th roots of a complex number.
13. **Thought question.** How would you go about converting a polar equation to rectangular coordinates?

CHAPTER 10
LINEAR AND NONLINEAR SYSTEMS

As we have seen in the preceding chapters, applications give rise to relationships between variables. These relationships may be expressed as equations or as inequalities. By solving the equations or inequalities arising in an application, we can answer the applied question. So far, our equations and inequalities have involved only a single variable. However, in many applications it is necessary to consider equations and inequalities in several variables. In fact, a single problem involving several variables can lead to a group, or system, of equations or inequalities. In this chapter, we learn to solve such systems.

Chapter Opening Image: *A fractal image.*

10.1 SYSTEMS OF EQUATIONS IN TWO VARIABLES

In our applications thus far the problems we have considered have boiled down to solving a single equation in one unknown. However, in many applied problems it is necessary to simultaneously satisfy several equations in several unknowns. For instance, consider the following problem. Suppose a pension fund seeks to invest $1 million. In order to satisfy its current and future obligations, the fund must earn $90,000 per year. Suppose the fund has two sorts of investments open to it: stocks paying 8 percent per year and corporate bonds paying 12 percent per year. How much should the pension fund invest in each in order to achieve exactly the desired income?

To solve the problem, let x be the amount invested in stocks and y be the amount invested in bonds. Denote money amounts in multiples of $1 million. For instance, the required annual income from the portfolio is $0.09 million. The requirement that $1 million be invested can be expressed through the equation:

$$x + y = 1$$

The requirement that the income of the portfolio be 0.09 can be expressed as:

$$0.08x + 0.12y = 0.09$$

To solve the problem, we must determine x and y so that the two equations are *simultaneously* satisfied. The pair of equations is an example of a **system of equations** in the variables x and y.

In general, a system of equations is a list of equations in two or more variables. In this section we restrict ourselves to systems of two equations in two variables. For such systems let us make the following definition.

Definition
Solution of a System of Equations

A **solution of a system in two variables** x and y is an ordered pair of numbers (a, b) such that all equations of the system are satisfied for $x = a$ and $y = b$.

For example, consider the system:

$$\begin{cases} x - y = -1 \\ 5x - 4y = 3 \end{cases}$$

The ordered pair $(x, y) = (7, 8)$ is a solution to the system, because when we substitute $x = 7$ and $y = 8$, both equations are satisfied:

$$\begin{cases} 7 - 8 = -1 \\ 5(7) - 4(8) = 3 \end{cases}$$

As we shall soon see, this is the only solution of the system. However, as we illustrate in this chapter, some systems have a single solution, some a finite number of solutions, others an infinite number of solutions, and yet others no solution at all.

Our goal in this chapter is to develop techniques for determining the solutions of systems of equations (if there are any). This is quite an ambitious goal, because as the number of equations and number of variables increase, it becomes much more difficult to determine the solutions. However, we will go part way toward achieving our goal by developing techniques that can be used to solve a wide variety of systems.

The Method of Substitution

Very often, one of the equations of a system can be solved for one variable in terms of another. When this is possible, we can often determine the solutions of the system by using the **method of substitution**, described as follows:

METHOD OF SUBSTITUTION

To solve a system of two equations in two variables:

1. Solve an equation for one of the variables in terms of the other.
2. Substitute the expression found in step 1 into the other equation.
3. Solve the resulting equation for the possible values of one variable.
4. Substitute each value found in step 3 into the expression of step 1 to determine the corresponding value of the other variable.

The method of substitution is illustrated in the following example.

EXAMPLE 1
Applying Substitution to a System of Two Linear Equations in Two Variables

Determine all solutions of the following system:
$$\begin{cases} x - y = 1 \\ 3x + 2y = 7 \end{cases}$$

Solution

Let's begin by solving the first equation for y in terms of x:
$$x - y = 1$$
$$y = x - 1$$

Next we substitute the expression for y just obtained into the second equation:
$$3x + 2y = 7$$
$$3x + 2(x - 1) = 7$$
$$3x + 2x - 2 = 7$$
$$5x = 9$$
$$x = \frac{9}{5}$$

To obtain the value of y, we substitute the value for x into the above equation that gives y in terms of x.
$$y = x - 1$$
$$= \frac{9}{5} - 1$$
$$= \frac{4}{5}$$

Thus $x = \frac{9}{5}$ and $y = \frac{4}{5}$, so the system has a single solution, namely $\left(\frac{9}{5}, \frac{4}{5}\right)$. We can check this solution by substituting the values for x and y into each equation:

$$x - y = 1 \qquad\qquad 3x + 2y = 7$$
$$\frac{9}{5} - \frac{4}{5} = 1 \qquad\qquad 3\left(\frac{9}{5}\right) + 2\left(\frac{4}{5}\right) = 7$$
$$\frac{5}{5} = 1 \qquad\qquad \frac{27}{5} + \frac{8}{5} = 7$$

SECTION 10.1 Systems of Equations in Two Variables

$$1 = 1 \qquad \frac{35}{5} = 7$$
$$7 = 7$$

Let's now apply the method of substitution to solve the pension fund problem stated at the beginning of this section.

EXAMPLE 2
Solving the Pension Fund Problem

Recall the pension fund problem. We set $x =$ the amount of money invested in stocks (in millions) and $y =$ the amount of money invested in bonds (in millions). We wish to determine all solutions x and y of the system:

$$\begin{cases} x + y = 1 \\ 0.08x + 0.12y = 0.09 \end{cases}$$

Solution

Solve the first equation for y in terms of x:

$$y = 1 - x$$

Now substitute this expression into the second equation:

$$0.08x + 0.12(1 - x) = 0.09$$
$$-0.04x + 0.12 = 0.09$$
$$0.04x = 0.03$$
$$x = 0.75$$

Now substitute the value of x into the expression for y:

$$y = 1 - x = 1 - 0.75 = 0.25$$

That is, the pension fund should invest $750,000 in stocks and $250,000 in bonds.

Examples 1 and 2 applied the method of substitution to systems whose equations were linear. Here is an example of the method applied to solve a nonlinear system.

EXAMPLE 3
Applying Substitution to a Nonlinear System

Solve the following system of equations in two variables:

$$\begin{cases} x^2 + 3y^2 = 28 \\ x + y = 4 \end{cases}$$

Solution

We apply the method of substitution. The second equation can be simply solved for y in terms of x:

$$x + y = 4$$
$$y = 4 - x$$

Substitute this last expression for y into the first equation:

$$x^2 + 3y^2 = 28$$
$$x^2 + 3(4 - x)^2 = 28$$
$$x^2 + 3(16 - 8x + x^2) = 28$$

$$x^2 + 3x^2 - 24x + 48 = 28$$
$$4x^2 - 24x + 20 = 0$$
$$4(x^2 - 6x + 5) = 0$$
$$4(x - 5)(x - 1) = 0$$
$$x = 5, 1$$

In this example there are two values of x. For each of these values we go back to the equation expressing y in terms of x and obtain a corresponding value of y. Two pairs are solutions of the system:

$$y = 4 - x \qquad\qquad y = 4 - x$$
$$y = 4 - 5 = -1 \qquad y = 4 - 1 = 3$$
$$(x, y) = (5, -1), (1, 3)$$

We check each solution by substituting it into the original equations of the system.

EXAMPLE 4
A System with No Solutions

Solve the following system of linear equations:
$$\begin{cases} 3x - y = 1 \\ -6x + 2y = 4 \end{cases}$$

Solution

Solve the first equation for y in terms of x:
$$y = 3x - 1$$

Substitute this expression into the second equation:
$$-6x + 2y = 4$$
$$-6x + 2(3x - 1) = 4$$
$$-6x + 6x - 2 = 4$$
$$-2 = 4$$

The last equation is inconsistent and so has no solution. Therefore, the system has no solution.

Graphical Analysis of Systems in Two Variables

As we have seen in preceding chapters, an equation in two variables has a graph that is a curve in the x-y plane. The points on the curve correspond to the ordered pairs (x, y) that satisfy the equation.

We can graph a system of equations by graphing each equation of the system separately. The solutions of the system then correspond to the points where the graphs intersect. For example, Figure 1 shows the graph of the system of Example 1. The graph of each equation is a straight line. The intersection point of the two lines has coordinates that correspond to the solution of the system. Figure 2 shows the graph of the system in the pension fund problem. Again the graph of each equation is a straight line, and the intersection of the lines corresponds to the solution of the system. Figure 3 shows the graph of the nonlinear system of Example 3. In this case there are several solutions, one for each intersection point of the two graphs. Finally, Figure 4 shows the graph of the system of Example 4. The fact that the two lines don't intersect corresponds to the result of Example 4 that the system has no solutions.

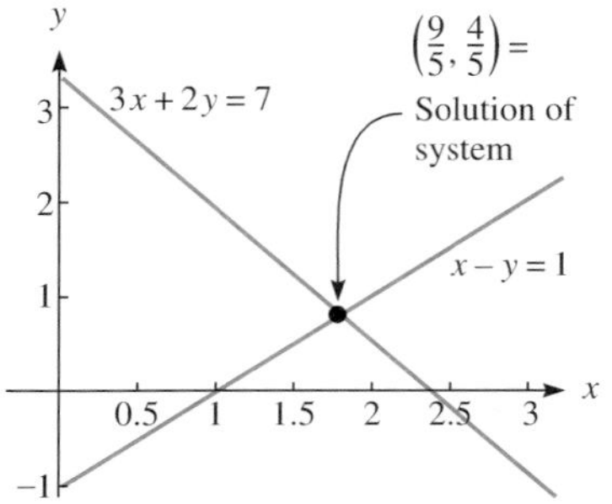

Figure 1

SECTION 10.1 Systems of Equations in Two Variables

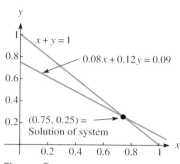

Figure 2

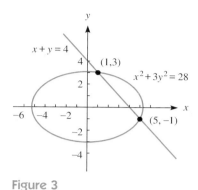

Figure 3

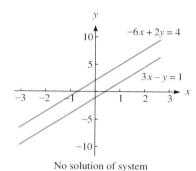

Figure 4

Solving Systems Using a Graphing Calculator

Sometimes it is difficult to solve a system using algebraic methods. In such cases the solutions can usually be approximated numerically using a graphing calculator, as shown in the following example.

EXAMPLE 5 Solving a System Using a Graphing Calculator

Use a graphing calculator to determine all solutions of the system:
$$\begin{cases} y = 2x^3 - 5x - 1 \\ y = 2x \end{cases}$$

Solution

To solve this system by substitution would require that we solve a cubic equation of the form

$$2x^3 - 5x - 1 = 2x$$

which is difficult to do. However, we can approximate the solutions of the system numerically by drawing the graph, as shown in Figure 5. It is clear from the graph that the system has three solutions, corresponding to values of x approximately equal to -1.5, -0.5, and 1.5. By enlarging the portion of the graph near each of these values of x, as shown in Figures 6–8, and using the trace function, we arrive at the approximate solutions of the system:

$$(x, y) = (1.938537191, 3.877074382),$$
$$(-1.794832142, -3.589664284),$$
$$(-0.1437050495, -0.2874100990)$$

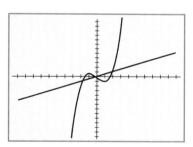

Figure 5

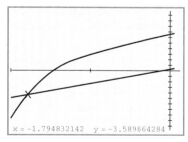

Figure 6

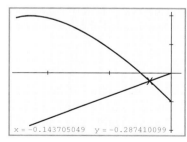

Figure 7

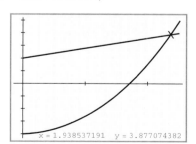

Figure 8

Applications of Systems in Two Variables

Here are some applied problems that give rise to systems of two equations in two variables.

EXAMPLE 6
Farming

A rectangular cow pen is fenced on three sides, with the fourth side a barn. The fence opposite the barn costs $4 per foot. The other sides cost $3 per foot. The perimeter of the pasture is 400 feet, and the cost of the fencing is $1,400. (See Figure 9.) What are the dimensions of the pen?

Solution

In the figure we have labeled the length of the side opposite the barn as x and the length of the other sides as y. The perimeter of the pen is $x + 2y$, so we have the equation:

$$x + 2y = 400$$

The cost of the side opposite the barn is $4x$, and the cost of the other sides is $3y + 3y = 6y$. The total cost of the fencing is therefore $4x + 6y$, so we have the equation:

$$4x + 6y = 1,400$$

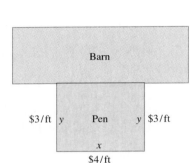

Figure 9

Thus we must solve the system of linear equations:

$$\begin{cases} x + 2y = 400 \\ 4x + 6y = 1,400 \end{cases}$$

To do so, we use the method of substitution:

$$x = 400 - 2y$$

$$4x + 6y = 1,400$$
$$4(400 - 2y) + 6y = 1,400 \quad \text{Subsitution in second equation}$$
$$-2y = -200$$
$$y = 100$$
$$x = 400 - 2 \cdot 100$$
$$= 200$$

That is, the side opposite the barn is 200 ft and the other two sides are each 100 ft. A simple check shows that these lengths satisfy the conditions set forth.

EXAMPLE 7
Architectural Design

A cylindrical water cistern is to contain 10,000 cubic feet of water, and its sides are to be constructed of 5,000 square feet of aluminum. What are the dimensions of the cistern?

Solution

Let r denote the radius of the cistern and h its height. See Figure 10. By the formula for the surface area of a cylinder, the area of the side of the cistern is equal to $2\pi rh$. Therefore we have the equation:

$$2\pi rh = 5,000$$

SECTION 10.1 Systems of Equations in Two Variables

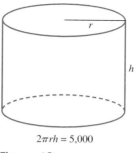

Figure 10

By the formula for the volume of a cylinder, the volume of the cistern is equal to $\pi r^2 h$. Therefore we have the second equation:

$$\pi r^2 h = 10{,}000$$

So to solve the problem, we must solve the system of equations:

$$\begin{cases} 2\pi r h = 5{,}000 \\ \pi r^2 h = 10{,}000 \end{cases}$$

From the first equation we have:

$$h = \frac{5{,}000}{2\pi r} = \frac{2{,}500}{\pi r}$$

Substituting this expression into the second equation, we have:

$$\pi r^2 \left(\frac{2{,}500}{\pi r} \right) = 10{,}000$$

$$2{,}500 r = 10{,}000$$

$$r = 4 \text{ ft}$$

$$h = \frac{2{,}500}{\pi r} = \frac{2{,}500}{4\pi} \approx 198.94 \text{ ft}$$

Exercises 10.1

Determine all solutions of the following systems.

1. $\begin{cases} 3x - 6y = -5 \\ 2x + 9y = 1 \end{cases}$

2. $\begin{cases} 3x + 8y = 4 \\ 12y + 6x = -1 \end{cases}$

3. $\begin{cases} x - 3y = 17 \\ 2x + y = 6 \end{cases}$

4. $\begin{cases} x - 3y = 10 \\ 3x + 2y = 2 \end{cases}$

5. $\begin{cases} \frac{3}{2}x + y = 2 \\ \frac{5}{4}x + \frac{3}{4}y = \frac{1}{2} \end{cases}$

6. $\begin{cases} \frac{3}{2}x = 2 + \frac{5}{4}y \\ \frac{1}{2}x = \frac{3}{2} - \frac{5}{3}y \end{cases}$

7. $\begin{cases} x^2 + y^2 = 16 \\ y - 2x = 3 \end{cases}$

8. $\begin{cases} x + y = 1 \\ x^2 - y^2 = 4 \end{cases}$

9. $\begin{cases} x + 2y = 3 \\ y^2 = 9 - x^2 \end{cases}$

10. $\begin{cases} x + 3 = y \\ 4y^2 - 16 = x^2 \end{cases}$

11. $\begin{cases} 4y = 1 + 3x \\ 2x^2 - 8y^2 + 3x + 2y = 2 \end{cases}$

12. $\begin{cases} 5x - 2y = 6 \\ 4x^2 + 4x - y^2 - 4y = 12 \end{cases}$

13. $\begin{cases} xy + x^2 + 2y = 3 \\ 3x - 2y + 2 = 0 \end{cases}$

14. $\begin{cases} 5xy = 4x + y \\ x = 5 - 4y \end{cases}$

15. $\begin{cases} 3x^2 - 2y^2 = 9 \\ x^2 - y^2 = -1 \end{cases}$

16. $\begin{cases} x^2 + y^2 = 4 \\ x^2 - 2y^2 = 1 \end{cases}$

17. $\begin{cases} xy = 8 \\ y = -\frac{1}{2}x^2 + 2x + 2 \end{cases}$

18. $\begin{cases} xy = 8 \\ \frac{1}{x} + \frac{1}{y} = \frac{1}{4} \end{cases}$

Applications

19. **Chemistry.** The atomic number of antimony is one less than four times the atomic number of aluminum. If twice the atomic number of aluminum is added to that of antimony, the result is 77. What are the atomic numbers of aluminum and antimony?

20. **Temperature conversion.** Two temperature scales are established. In the x scale water freezes at $10°$ and boils at $300°$, and in the y scale water freezes at $30°$ and boils at $1{,}430°$. If x and y are linearly related, find an expression for any temperature y in terms of a temperature x.

21. **Recreation.** Tickets for a pre-Broadway showing of a play sold at $40.00 for the main floor and $27.50 for the balcony. If the receipts from the sale of 1,600 tickets totaled $55,250.00, how many tickets were sold at each price?

22. **Recreation.** Tickets to a class play were $2.50 and $5.00. In all, 275 tickets were sold, and the total receipts were $1,187.50. Find the number of tickets sold at each price.

23. **Stock market investment.** You have $5,000 to invest for one year. Part of your investment went into a risky stock that paid 12 percent annual interest, and the rest went into a savings account that paid 6 percent annual interest. If your annual return was $540 from both investments, how much was invested at each rate?

24. **Investment analysis.** Suppose you invested a certain sum of money at 8.5 percent and another sum at 9.5 percent. If your total investment was $17,000 and your total interest for one year of both investments was $1,535, how much did you invest at each rate?

25. **Home projects.** Marvin wants to make a window box from a rectangular sheet of metal whose area is 680 cm². He cuts a 6-cm square from each corner and folds the sides to make a box with base area of 176 cm². Find the dimensions of the original metal sheet.

26. **Catering costs.** A caterer finds that the expenses for a breakfast consist of fixed costs and costs per guest. A meal function for 40 guests costs $150 and for 100 guests costs $225. Find the fixed costs and the cost per guest.

27. **Geometry.** Find the value of C for which the graph of $y = 2x + C$ is tangent to the graph of $y = x^2 + 1$. (**Hint:** There is only 1 point of intersection.)

28. **Geometry.** Find the value of C for which the graph of $y = 2x + C$ is tangent to $x^2 + y^2 = 25$.

29. **Travel.** Airline schedules allow 3 hrs, 31 min (terminal to terminal) to fly from Chicago to Phoenix and 3 hrs, 9 min to fly from Phoenix to Chicago. Of this, about $\frac{1}{2}$ hr might be used in getting off the ground and getting back down, so the actual flying time at cruising speed might be 3 hrs in one direction and 2.7 hrs in the other. The difference is in the jet stream, a wind that usually blows west to east.

   a. If there were no wind, or jet stream, how much time would you expect a flight to take?
   b. The airline distance between Chicago and Phoenix is 1,453 mi. Estimate the average speed of the jet stream and the average speed of a plane.

30. **Travel.** Dave made a round trip between two cities 270 miles apart. On his return trip, heavy snow reduced his average rate by 5 miles per hour and increased his traveling time by $\frac{3}{4}$ hours. What was his rate going?

31. **Puzzle.** The sum of the tens digit and the ones digit of a three-digit number is 5. The product of the three digits is 24, and the hundreds digit is 2 less than twice the ones digit. Find the number.

32. **Puzzle.** The sum of the digits of a three-digit number is 8, the product of the digits is 10, and if the hundreds digit is subtracted from the ones digit, the result is 4. Find the number.

Technology

Use a graphing calculator to solve the following systems.

33. $\begin{cases} y = 25x^2 - 3x + 2 \\ y = x^2 \end{cases}$

34. $\begin{cases} y = 10x^2 + x \\ y = 5x - x^3 \end{cases}$

35. $\begin{cases} y = x^2 + 1 \\ y = \dfrac{5}{x} \end{cases}$

36. $\begin{cases} y = e^x \\ y = 3x^2 \end{cases}$

37. $\begin{cases} y = \sqrt{1 - x^2} \\ y = 0.5x \end{cases}$

38. $\begin{cases} y = 1 - \ln x \\ y = x \end{cases}$

In your own words

39. Describe the method of substitution. Illustrate the various steps of the method using a specific example.

40. What systems of equations would the method of substitution *not* be appropriate for?

10.2 LINEAR SYSTEMS IN TWO VARIABLES AND ELIMINATION

In this section we concentrate on a second method for solving systems of two linear equations in two unknowns. This method, called **elimination**, is to eliminate one of the variables from the system to obtain a single equation in one unknown. This procedure is illustrated in the following example.

 **EXAMPLE 1**
Solving a Linear System Using Elimination

Use elimination to solve the following system:
$$\begin{cases} 2x + 3y = 10 \\ x - y = 5 \end{cases}$$

Solution

Let's eliminate x by creating equations with coefficients of x that are negatives of each other. Then by adding we can eliminate the variable x.

$$\begin{cases} -2x - 3y = -10 & \text{Multiply first equation by } -1 \\ 2x - 2y = 10 & \text{Multiply second equation by 2} \end{cases}$$

$$-5y = 0 \qquad \text{Add second equation to first}$$
$$y = 0$$
$$2x + 3(0) = 10 \qquad \text{Substitute into first equation}$$
$$2x = 10$$
$$x = 5 \qquad \text{Solve for } x$$

We can summarize the process of elimination as follows:

METHOD OF ELIMINATION

To solve a system of two linear equations in two variables:

1. Multiply one or both equations of the system by suitable numbers to obtain a new system in which the coefficients of one of the variables are equal to or negatives of each other.
2. Add or subtract the equations to obtain a linear equation in one variable.
3. Solve the equation obtained in step 2.
4. Substitute the value obtained into either of the original system equations and solve for the second variable.

530 CHAPTER 10 Linear and Nonlinear Systems

Graphical Analysis of the Solutions

We can graph the equations of a linear system in two variables. The graph of each equation is a straight line. The solutions of the system correspond to points where the lines intersect. There are three possible configurations:

1. The graphs of the equations are two distinct lines. In this case the system has a single solution, corresponding to the point of intersection of the lines. A linear system of this type is called **independent**. See Figure 1.
2. The two straight lines coincide. In this case the system has infinitely many solutions, corresponding to all points on the line. A linear system of this type is called **dependent**. See Figure 2.
3. The two straight lines are different but parallel. In this case the system has no solution. A linear system of this type is called **inconsistent**. See Figure 3.

Example 1 provides an illustration of a system with a single solution. The graphs of the two equations and the solution (intersection point) are shown in Figure 4. The next example illustrates the second case.

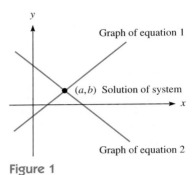

Figure 1

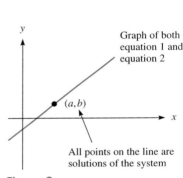

Figure 2

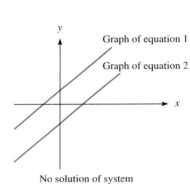

Figure 3

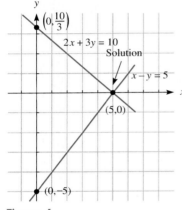

Figure 4

EXAMPLE 2
A System in Two Variables with Infinitely Many Solutions

Solve the following system of linear equations:

$$\begin{cases} 3x + 8y = 12 \\ \dfrac{3}{4}x + 2y = 3 \end{cases}$$

Solution

Eliminate x:

$$\begin{cases} 3x + 8y = 12 \\ 3x + 8y = 12 \quad \text{Multiply equation 2 by 4} \\ 0 = 0 \quad \text{Subtract equation 2 from equation 1} \end{cases}$$

This equation is satisfied by any value of y. For a particular value of y, the corresponding value of x is obtained by solving one of the equations for x in terms of y:

$$3x = 12 - 8y$$

$$x = 4 - \frac{8}{3}y$$

Figure 5

SECTION 10.2 Linear Systems in Two Variables and Elimination

Figure 5 shows the graphs of the two equations of the system. Both equations have the same graph because one is 4 times the other. Any point on the line is a solution to the system.

EXAMPLE 3 Determine all solutions of the system:

A System in Two Variables with No Solutions

$$\begin{cases} 2x - 10y = 1 \\ 3x - 15y = 7 \end{cases}$$

Solution

Eliminate x:

$$\begin{cases} 6x - 30y = 3 & \text{Multiply equation 1 by 3} \\ 6x - 30y = 14 & \text{Multiply equation 2 by 2} \\ 0 = -11 & \text{Subtract equation 2 from equation 1} \end{cases}$$

Clearly, the last equation has no solution, so the same can be said of the system. In this case the graphs of the two equations are parallel, as shown in Figure 6.

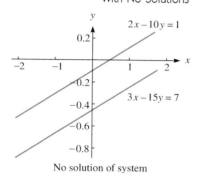

Figure 6

Solving Linear Systems in Two Variables Using a Graphing Calculator

We can solve a linear system in two variables using a graphing calculator. Just graph the solutions and use the trace function to locate the point of intersection of the graphs.

EXAMPLE 4 Use a graphing calculator to solve the system:

Solving a Linear System with a Graphing Calculator

$$\begin{cases} 5.1x - 3.2y = 10 \\ -4.7x + 6.9y = 15.1 \end{cases}$$

Solution

Solve each equation for y in terms of x and graph the resulting equations on a single screen, as shown in Figure 7. It takes some experimentation to determine that the x-range from 0 to 10 clearly shows the point of intersection. Using the trace function, we can estimate the point of intersection to obtain the solution of the system:

$$(x, y) = (5.8244681, 6.157746)$$

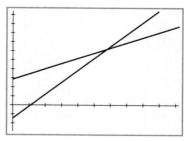

Figure 7

Application Let's now apply the method of elimination to solve a linear system that arises in an application.

EXAMPLE 5
Aviation

A plane can fly against the wind and cover a 500-mile distance in 2 hours. The same plane can fly with the wind and cover the same route in the opposite direction in 1.5 hours. What is the speed of the plane in still air and what is the speed of the wind?

Solution

Let v denote the speed of the plane in still air and let w denote the speed of the wind. When flying against the wind, the speed of the plane is $v - w$. Because the plane covers 500 miles in 2 hours, we have:

$$[\text{distance}] = [\text{rate}][\text{time}]$$
$$500 = (v - w) \cdot 2$$
$$2v - 2w = 500$$

When flying with the wind, the speed of the plane is $v + w$. Because it takes 1.5 hours to cover the same 500 miles in this case, we have:

$$500 = 1.5(v + w)$$
$$1.5v + 1.5w = 500$$

To solve the problem, we must therefore solve the following system:

$$\begin{cases} 2v - 2w = 500 \\ 1.5v + 1.5w = 500 \end{cases}$$

$$\begin{cases} 6v - 6w = 1{,}500 & \text{\textit{Multiply equation 1 by 3}} \\ 6v + 6w = 2{,}000 & \text{\textit{Multiply equation 2 by 4}} \end{cases}$$

$$12v = 3{,}500 \quad \text{\textit{Add the equations}}$$
$$v = 291.67 \text{ mph} \quad \text{\textit{Solve}}$$
$$2 \cdot 291.67 - 2w = 500 \quad \text{\textit{Substitute in equation 1}}$$
$$w = 41.67 \text{ mph}$$

Thus, the airplane goes 291.67 mph in still air, and the wind speed is 41.67 mph.

Exercises 10.2

Use elimination to determine all solutions of the following systems of equations.

1. $\begin{cases} 2x - y = 0 \\ 3x + y = 5 \end{cases}$

2. $\begin{cases} 3x + y = 7 \\ x - y = 1 \end{cases}$

3. $\begin{cases} 3x + 4y = -3 \\ 2x + 2y = 2 \end{cases}$

4. $\begin{cases} 3x - y = 3 \\ 2x + 2y = -6 \end{cases}$

5. $\begin{cases} 2x + 4y = 0 \\ 3x + 6y = 6 \end{cases}$

6. $\begin{cases} 3x - 6y = -5 \\ 9x - 18y = 1 \end{cases}$

7. $\begin{cases} 3x + 8y = 4 \\ 12y + 6x = -1 \end{cases}$

8. $\begin{cases} x - 3y = 17 \\ 2x + y = 6 \end{cases}$

9. $\begin{cases} x - 3y = 10 \\ 3x + 2y = 2 \end{cases}$

10. $\begin{cases} 3x + 2y = 4 \\ 5x + 3y = 2 \end{cases}$

11. $\begin{cases} \frac{3}{2}x = 2 + \frac{5}{4}y \\ \frac{1}{2}x = \frac{3}{2} - \frac{5}{3}y \end{cases}$

12. $\begin{cases} \frac{9}{2}x - 4y = -3 \\ \frac{4}{3}x - \frac{1}{2}y = \frac{7}{6} \end{cases}$

Solve the following systems for x and y in terms of a and b.

13. $\begin{cases} x + y = 2a \\ 2x + y = 3a + b \end{cases}$

14. $\begin{cases} 2x + y = \frac{a}{b} \\ x + y = 0 \end{cases}$

Applications

15. **Recreation.** The perimeter of a rectangular pool is 84 meters. The width of the pool is $\frac{3}{4}$ of the length. What are the length and the width of the pool?

16. **Recreation.** The perimeter of a rectangular pool is 70 meters. The length is $\frac{4}{3}$ the width of the pool. What are the width and the length of the pool?

17. **Small business.** A lawn service company has two mowers, a new one and an old one. With both mowers working together, a certain job is done in 2 hours and 24 minutes. On another job of the same kind, the old mower worked alone for 3 hours, and was then joined by the new mower, and the two mowers finished the job in an additional 1 hour and 12 minutes. How long would it take each mower operating alone to do the second job?

18. **Bricklaying.** One bricklayer can build the wall of a standard basement in 6 hours less than a second bricklayer. If the faster bricklayer works for 2 hours alone on a standard basement and is then joined by the second bricklayer, they complete the basement in an additional $1\frac{3}{7}$ hours working together. How long would it take each bricklayer working alone to build a standard basement?

19. **Boating.** Traveling downstream, the riverboat *New Baltimore* can travel 12 miles in 2 hours. Going up the same river, it can travel $\frac{2}{3}$ of this distance in twice the time. How fast does the river flow?

534 CHAPTER 10 Linear and Nonlinear Systems

20. **Savings.** A bank contains $37 in quarters and dimes. The number of dimes is 10 more than twice the number of quarters. How many dimes and how many quarters are there in the bank?

21. **Recreation.** Tickets to a high school play cost $4.50 and $6.00. If 450 tickets were sold and the receipts were $2,550, how many of each ticket were sold?

22. **Flying.** Flying with the wind, a SAC bomber can cover a distance of 3,360 miles to its target in 5 hours and 15 minutes. On the bomber's return against the same wind, the trip would require 6 hours. Find the wind speed.

Technology

Solve the following linear systems using a graphing calculator.

23. $\begin{cases} 2x - y = 0 \\ 3x + y = 5 \end{cases}$

24. $\begin{cases} 3x + y = 7 \\ x - y = 1 \end{cases}$

25. $\begin{cases} 3x + 4y = -3 \\ 2x + y = 2 \end{cases}$

26. $\begin{cases} 3x - y = 3 \\ 2x + 2y = -6 \end{cases}$

In your own words

27. Describe the method of elimination. Illustrate the steps of the method with a specific example.

28. Discuss the different types of solutions that can occur in a linear system in two variables. Illustrate the various cases graphically.

10.3 LINEAR SYSTEMS IN ANY NUMBER OF VARIABLES

In the preceding section we learned to solve systems of two linear equations in two variables using elimination. Let's now consider systems of linear equations in any number of variables. We develop a method, called **Gauss elimination**, for determining all solutions of such a system.

Systems in Triangular Form

Let's begin by introducing a type of linear system that is extremely easy to solve. Consider a linear system in the variables $x, y, z, w, \ldots$, where the variables appear in that order. We say that the system is in **triangular form** (also called **echelon form**) provided that:

1. The second equation does not involve x, the third equation does not involve x and y, the fourth equation does not involve x, y, and z, and so forth.
2. The first nonzero coefficient of each equation is 1.

Here are two systems in triangular form:

$$\begin{cases} x + y + z = -1 \\ \phantom{x + {}} y - 2z = 4 \\ \phantom{x + y + {}} z = 4 \end{cases}$$

$$\begin{cases} x - 5y + z = -1 \\ \phantom{x - 5y + {}} z = 6 \end{cases}$$

In each system the second equation does not involve x and the third equation does not involve x and y. Note that in the second system, the coefficients of the second

SECTION 10.3 Linear Systems in Any Number of Variables

equation are all 0 and that equation is omitted. However, in order to consider it in triangular form, we consider it as part of the system.

A triangular system is simple to solve. Just start from the last equation and determine the value of the last variable, then use the next-to-last equation to determine the value of the next-to-last variable, and so forth. The next example illustrates how this is done.

EXAMPLE 1
Solving a Triangular System

Solve the following system of equations in triangular form:

$$\begin{cases} x + 2y - 5z = -1 \\ y - 4z = 7 \\ z = -2 \end{cases}$$

Solution

Start from the bottom equation and work upward. The bottom equation gives the value of z, namely $z = -2$. Proceed upward to the next equation and substitute the value of z just obtained:

$$y - 4z = 7$$
$$y - 4(-2) = 7$$
$$y = -1$$

Now we have determined the values of y and z. Move up to the first equation and substitute these values:

$$x + 2y - 5z = -1$$
$$x + 2(-1) - 5(-2) = -1$$
$$x = -9$$

Thus, the system has the single solution:

$$(x, y, z) = (-9, -1, -2)$$

The method just illustrated can be used to solve any system in triangular form. It is called **back substitution**.

EXAMPLE 2
Triangular System with Infinitely Many Solutions

Solve the triangular system:

$$\begin{cases} x - y + z = 5 \\ y - z = 4 \end{cases}$$

Solution

The equation that would determine the value of z is missing. So let z be any real number. We can find y from the second equation in terms of this choice of z:

$$y = 4 + z$$

And we can find x from the first equation:

$$x = y - z + 5$$
$$= (4 + z) - z + 5$$
$$= 9$$

The solution of the system is:
$$z = \text{any value}, \quad y = 4 + z, \quad x = 9$$
This system has infinitely many solutions.

EXAMPLE 3
Another System with Infinitely Many Solutions

Determine all solutions of the triangular system:
$$\begin{cases} x + 3y - 2z = 1 \\ \\ z = 4 \end{cases}$$

Solution

The equation that would determine y is missing. So let y have any value. In terms of this value, we obtain the solution:
$$z = 4$$
$$x = -3y + 2z + 1$$
$$= -3y + 2 \cdot 4 + 1$$
$$= -3y + 9$$

So the solutions of the system are:
$$y = \text{any value}, \quad x = -3y + 9, \quad z = 4$$

Elementary Operations on a Linear System

We can transform a linear system into triangular form using a sequence of the following operations, which do not change the solutions of the system:

ELEMENTARY OPERATIONS ON LINEAR SYSTEMS

1. Interchange two equations.
2. Multiply an equation by a nonzero real number.
3. Add to an equation a multiple of another equation.

The next example illustrates how these operations can be used to transform a system into triangular form.

EXAMPLE 4
Reducing a System in Two Variables to Triangular Form

Consider the following system of linear equations:
$$\begin{cases} 3x - y = -5 \\ x + y = 2 \end{cases}$$

1. Use a sequence of elementary row operations to transform the system into echelon form.
2. Determine all solutions of the system.

Solution

1. Start with interchanging the first equation with the second (to avoid fractions where possible):

SECTION 10.3 Linear Systems in Any Number of Variables

$$\begin{cases} x + y = 2 \\ 3x - y = -5 \end{cases}$$

Multiply the first equation by -3 and add it to the second equation:

$$\begin{cases} x + y = 2 \\ -4y = -11 \end{cases}$$

Multiply the second equation by $-\frac{1}{4}$:

$$\begin{cases} x + y = 2 \\ y = \frac{11}{4} \end{cases}$$

2. Back substituting $y = \frac{11}{4}$ into the first equation, we have:

$$x + \frac{11}{4} = 2$$

$$x = -\frac{3}{4}$$

Therefore, the system has the single solution:

$$(x, y) = \left(-\frac{3}{4}, \frac{11}{4}\right)$$

The main advantages of using the echelon form in solving a linear system is that the method works for any number of equations in any number of variables and that although it is tedious, the method is simple to embody in a computer program. It is the preferred method for solving large systems. In the next section we discuss in detail how it works for solving linear systems in three or more variables.

EXAMPLE 5
Reducing a System with Three Variables to Triangular Form

Use a sequence of elementary operations to transform the following system into triangular form. Determine all solutions of the system.

$$\begin{cases} 3y + 4z = 15 \\ x - y - z = -4 \\ 2x + 5y + 9z = 39 \end{cases}$$

Solution

Because the first equation has no x term, we interchange the first and second equations to obtain the system:

$$\begin{cases} x - y - z = -4 \\ 3y + 4z = 15 \\ 2x + 5y + 9z = 39 \end{cases}$$

Next, multiply the second equation by $\frac{1}{3}$ to obtain:

$$\begin{cases} x - y - z = -4 \\ y + \frac{4}{3}z = 5 \\ 2x + 5y + 9z = 39 \end{cases}$$

Next, add (-2) times the first equation to the third:
$$\begin{cases} x - y - z = -4 \\ y + \frac{4}{3}z = 5 \\ 7y + 11z = 47 \end{cases}$$
Next, add (-7) times the second equation to the third equation:
$$\begin{cases} x - y - z = -4 \\ y + \frac{4}{3}z = 5 \\ \frac{5}{3}z = 12 \end{cases}$$
Finally, multiply the third equation by $\frac{3}{5}$ to obtain the triangular form:
$$\begin{cases} x - y - z = -4 \\ y + \frac{4}{3}z = 5 \\ z = \frac{36}{5} \end{cases}$$
We now use back substitution to solve the triangular system:
$$z = \frac{36}{5}$$
$$y = 5 - \frac{4}{3}z = 5 - \frac{48}{5} = -\frac{23}{5}$$
$$x = y + z - 4 = -\frac{23}{5} + \frac{36}{5} - 4 = -\frac{7}{5}$$

Matrices and Gauss Elimination

The elementary row operations on a system of linear equations involve only the numbers appearing in the equations, and do not involve the variable names in any way. Therefore, as a shorthand, it is convenient to write the system
$$\begin{cases} 3x - 2y = 4 \\ x + 5y = 0 \end{cases}$$
as the rectangular array of numbers:
$$\begin{bmatrix} 3 & -2 & 4 \\ 1 & 5 & 0 \end{bmatrix}$$
Rectangular arrays such as this one occur often in mathematics and are called **matrices**. An extensive theory of matrices has been developed over the last century. Matrices have been used to describe diverse phenomena, including elementary particles in physics and the interactions of the various segments of an economy.

Definition
Matrix, Entries of a Matrix

A **matrix** is a rectangular array:
$$\begin{bmatrix} a_{11} & a_{12} & \cdots & a_{1n} \\ a_{21} & a_{22} & \cdots & a_{2n} \\ \vdots & \vdots & & \vdots \\ a_{m1} & a_{m2} & \cdots & a_{mn} \end{bmatrix}$$
where $a_{11}, a_{21}, \ldots$ are numbers, called the **elements,** or **entries,** of the matrix.

SECTION 10.3 Linear Systems in Any Number of Variables

The size of a matrix is measured by the numbers of rows and columns it contains. If a matrix contains m rows and n columns, it is called an $m \times n$ (read m by n) **matrix**. For example, the matrix

$$A = \begin{bmatrix} 1 & 2 \\ 3 & 4 \end{bmatrix}$$

is 2×2, whereas the matrix

$$B = \begin{bmatrix} -2 & 0 & 3 \\ 1 & 5 & 7 \end{bmatrix}$$

is 2×3. Note that in specifying the size of a matrix, the number of rows always comes first.

The rows and columns are numbered beginning from 1, with the top row numbered 1 and the left-hand column numbered 1. The elements of a matrix are often denoted, as above, using the subscript notation a_{ij}. In this notation the first subscript, i, refers to the row number; the second subscript, j, refers to the column number. Thus, for example, in the matrix B above a_{23} refers to the element in row 2, column 3, and therefore equals 7.

The elements whose row and column subscripts are equal, $a_{11}, a_{22}, \ldots,$ are said to form the **main diagonal** of the matrix. A matrix with an equal number of rows and columns is called a **square matrix**.

Suppose we are given a system of linear equations:

$$\begin{cases} a_{11}x_1 + a_{12}x_2 + \cdots + a_{1n}x_n = b_1 \\ a_{21}x_1 + a_{22}x_2 + \cdots + a_{2n}x_n = b_2 \\ \quad \vdots \\ a_{m1}x_1 + a_{m2}x_2 + \cdots + a_{mn}x_n = b_m \end{cases}$$

The matrix

$$\begin{bmatrix} a_{11} & a_{12} & \cdots & a_{1n} \\ a_{21} & a_{22} & \cdots & a_{2n} \\ \vdots & \vdots & & \vdots \\ a_{m1} & a_{m2} & \cdots & a_{mn} \end{bmatrix}$$

is called the **coefficient matrix** of the system. The matrix

$$\begin{bmatrix} a_{11} & a_{12} & \cdots & a_{1n} & b_1 \\ a_{21} & a_{22} & \cdots & a_{2n} & b_2 \\ \vdots & \vdots & & \vdots & \vdots \\ a_{m1} & a_{m2} & \cdots & a_{mn} & b_m \end{bmatrix}$$

is called the **augmented matrix** of the system. For example, consider the system:

$$\begin{cases} 5x - 2y = 13 \\ 3x + 9y = 1 \end{cases}$$

The augmented matrix is:

$$\begin{bmatrix} 5 & -2 & 13 \\ 3 & 9 & 1 \end{bmatrix}$$

In order to solve a system of linear equations, we can perform elementary row operations on the augmented matrix rather than equation operations on the system of equations. Here is a summary of the elementary row operations in matrix form:

ELEMENTARY ROW OPERATIONS ON MATRICES

E_{ij}. Interchange rows i and j.
kE_i. Multiply row i by a nonzero real number k.
$kE_i + E_j$. Add k times row i to row j.

The notations introduced in the above list are a compact way of describing elementary row operations without using long phrases. The next example provides some practice in using this notation.

EXAMPLE 6
Elementary Row Operations

Consider the matrix:

$$\begin{bmatrix} 3 & 0 & 1 \\ -1 & 5 & 8 \\ 0 & -4 & 7 \end{bmatrix}$$

Perform the following elementary row operations:

1. E_{13}
2. $\frac{1}{2}E_2$
3. $3E_1 + E_2$

Solution

1. The elementary row operation specified requires us to interchange rows 1 and 3. The result is:

$$\begin{bmatrix} 0 & -4 & 7 \\ -1 & 5 & 8 \\ 3 & 0 & 1 \end{bmatrix}$$

2. The elementary row operation specified requires us to multiply row 2 by $\frac{1}{2}$. The result is:

$$\begin{bmatrix} 3 & 0 & 1 \\ -\frac{1}{2} & \frac{5}{2} & 4 \\ 0 & -4 & 7 \end{bmatrix}$$

3. The elementary row operation specified requires us to add three times row 1 to row 2. The result is:

$$\begin{bmatrix} 3 & 0 & 1 \\ 8 & 5 & 11 \\ 0 & -4 & 7 \end{bmatrix}$$

SECTION 10.3 Linear Systems in Any Number of Variables

Corresponding to the echelon form of a linear system, we make the following definition:

Definition
Echelon Form

Let A be a matrix. We say that A is in **echelon form** provided that:

1. In each nonzero row, the first nonzero element is a 1.
2. The position of the first 1 in each row is to the right of the first 1 in the row above it.
3. Any all-zero rows are at the bottom of the matrix.

Here are two matrices in echelon form:

$$\begin{bmatrix} 1 & 1 & 2 & -5 \\ 0 & 1 & 3 & 2 \\ 0 & 0 & 1 & 4 \end{bmatrix}$$

$$\begin{bmatrix} 1 & -1 & 5 & 4 \\ 0 & 0 & 1 & 9 \\ 0 & 0 & 0 & 0 \end{bmatrix}$$

By using a sequence of elementary operations, we can transform a matrix into echelon form. The next example illustrates a typical calculation.

EXAMPLE 7
Transforming a Matrix into Echelon Form

Use a sequence of elementary operations to transform the following matrix into echelon form:

$$\begin{bmatrix} 4 & 1 & 0 & -1 \\ 1 & 0 & 1 & 0 \\ 2 & -1 & 3 & 0 \end{bmatrix}$$

Solution

We transform one row at a time, proceeding from the left of the row to the right, taking subsequent rows in left-to-right order.

$$\begin{bmatrix} 4 & 1 & 0 & -1 \\ 1 & 0 & 1 & 0 \\ 2 & -1 & 3 & 0 \end{bmatrix} \xrightarrow{\frac{1}{4}E_1} \begin{bmatrix} 1 & \frac{1}{4} & 0 & -\frac{1}{4} \\ 1 & 0 & 1 & 0 \\ 2 & -1 & 3 & 0 \end{bmatrix}$$

$$\xrightarrow{(-1)E_1 + E_2} \begin{bmatrix} 1 & \frac{1}{4} & 0 & -\frac{1}{4} \\ 0 & -\frac{1}{4} & 1 & \frac{1}{4} \\ 2 & -1 & 3 & 0 \end{bmatrix}$$

$$\xrightarrow{(-2)E_1 + E_3} \begin{bmatrix} 1 & \frac{1}{4} & 0 & -\frac{1}{4} \\ 0 & -\frac{1}{4} & 1 & \frac{1}{4} \\ 0 & -\frac{3}{2} & 3 & \frac{1}{2} \end{bmatrix}$$

$$\xrightarrow{(-4)E_2} \begin{bmatrix} 1 & \frac{1}{4} & 0 & -\frac{1}{4} \\ 0 & 1 & -4 & -1 \\ 0 & -\frac{3}{2} & 3 & \frac{1}{2} \end{bmatrix}$$

$$\xrightarrow{(\frac{3}{2})E_2 + E_3} \begin{bmatrix} 1 & \frac{1}{4} & 0 & -\frac{1}{4} \\ 0 & 1 & -4 & -1 \\ 0 & 0 & -3 & -1 \end{bmatrix}$$

$$\xrightarrow{(-\frac{1}{3})E_3} \begin{bmatrix} 1 & \frac{1}{4} & 0 & -\frac{1}{4} \\ 0 & 1 & -4 & -1 \\ 0 & 0 & 1 & \frac{1}{3} \end{bmatrix}$$

The last matrix is the echelon form for the given matrix.

Once we have the echelon form of the matrix of a linear system, we can determine the solutions of the system using back substitution. This method of solving a linear system is called **Gauss elimination**, named for the nineteenth-century mathematician Karl Friedrich Gauss.

THE METHOD OF GAUSS ELIMINATION

To solve a system of linear equations in any number of variables:
1. Form the augmented matrix of the system.
2. Use elementary row operations to convert the augmented matrix to echelon form.
3. Write the system of linear equations corresponding to the echelon form. This system is triangular.
4. Use back substitution to solve the triangular system.

The following examples illustrate the computational procedures of Gauss elimination.

EXAMPLE 8
Solving a System in Three Variables Using Gauss Elimination

Solve the following system of linear equations:
$$\begin{cases} 4x + y = -1 \\ x + z = 0 \\ 2x - y + 3z = 0 \end{cases}$$

Solution

The augmented matrix of the system is precisely the matrix considered in the preceding example. We found that the echelon form of the matrix is:

$$\begin{bmatrix} 1 & \frac{1}{4} & 0 & -\frac{1}{4} \\ 0 & 1 & -4 & -1 \\ 0 & 0 & 1 & \frac{1}{3} \end{bmatrix}$$

SECTION 10.3 Linear Systems in Any Number of Variables

The corresponding linear system is:

$$x + \frac{1}{4}y = -\frac{1}{4}$$
$$y - 4z = -1$$
$$z = \frac{1}{3}$$

Using back substitution, we find that the solution is $\left(-\frac{1}{3}, \frac{1}{3}, \frac{1}{3}\right)$.

The following example illustrates how to handle a system in which the echelon form has a row of zeros.

EXAMPLE 9
An Echelon Form with a Row of Zeros

Solve the linear system:
$$\begin{cases} 3x - 15y = 36 \\ x - 5y = 12 \end{cases}$$

Solution

The augmented matrix is:
$$\begin{bmatrix} 3 & -15 & 36 \\ 1 & -5 & 12 \end{bmatrix}$$

We now transform the matrix into echelon form:

$$\begin{bmatrix} 3 & -15 & 36 \\ 1 & -5 & 12 \end{bmatrix} \xrightarrow{\frac{1}{3}E_1} \begin{bmatrix} 1 & -5 & 12 \\ 1 & -5 & 12 \end{bmatrix}$$

$$\xrightarrow{-E_1 + E_2} \begin{bmatrix} 1 & -5 & 12 \\ 0 & 0 & 0 \end{bmatrix}$$

The echelon form has all zeros in the final row, so in the triangular form there is no corresponding equation. Here is the triangular form:

$$x - 5y = 12$$

There is no equation corresponding to y, so the value of y can be anything. The value of x can be determined in terms of y by solving the single equation for x in terms of y. So the solution of the linear system is:

$$\begin{cases} y = \text{any value} \\ x = 12 + 5y \end{cases}$$

Nonsquare Systems

In the preceding examples we considered only **square linear systems,** that is, systems in which the number of equations equals the number of variables. However, Gauss elimination works just as well on **nonsquare systems,** as the next example shows.

EXAMPLE 10
A Nonsquare System

Solve the following system of linear equations:
$$\begin{cases} 2x - 3y + z - w = 4 \\ 2x - 9z - 3w = -2 \end{cases}$$

Solution

The augmented matrix of the system is:
$$\begin{bmatrix} 2 & -3 & 1 & -1 & 4 \\ 2 & 0 & -9 & -3 & -2 \end{bmatrix}$$

We transform the first column of the matrix, proceeding as in the preceding example:

$$\begin{bmatrix} 2 & -3 & 1 & -1 & 4 \\ 2 & 0 & -9 & -3 & -2 \end{bmatrix} \xrightarrow{\frac{1}{2}E_1} \begin{bmatrix} 1 & -\frac{3}{2} & \frac{1}{2} & -\frac{1}{2} & 2 \\ 2 & 0 & -9 & -3 & -2 \end{bmatrix}$$

$$\xrightarrow{(-2)E_1 + E_2} \begin{bmatrix} 1 & -\frac{3}{2} & \frac{1}{2} & -\frac{1}{2} & 2 \\ 0 & 3 & -10 & -2 & -6 \end{bmatrix}$$

$$\xrightarrow{\frac{1}{3}E_2} \begin{bmatrix} 1 & -\frac{3}{2} & \frac{1}{2} & -\frac{1}{2} & 2 \\ 0 & 1 & -\frac{10}{3} & -\frac{2}{3} & -2 \end{bmatrix}$$

The two rows are both in the proper form, so the last matrix is the echelon form. The corresponding system is:

$$x - \frac{3}{2}y + \frac{1}{2}z - \frac{1}{2}w = 2$$

$$y - \frac{10}{3}z - \frac{2}{3}w = -2$$

The equations from which the values of z and w would be determined are not present in this echelon form. Accordingly, the values of z and w can be any real numbers. The echelon form can be used to determine the values of x and y in terms of the values chosen for z and w. From the second equation, we have:

$$y = \frac{10}{3}z + \frac{2}{3}w - 2$$

Therefore, from the first equation we have:

$$x = \frac{3}{2}y - \frac{1}{2}z + \frac{1}{2}w + 2$$

$$= \frac{3}{2}\left(\frac{10}{3}z + \frac{2}{3}w - 2\right) - \frac{1}{2}z + \frac{1}{2}w + 2$$

$$= \frac{9}{2}z + \frac{3}{2}w - 1$$

Therefore, the solution of the system is:

$$z = \text{any value}$$
$$w = \text{any value}$$
$$x = \frac{9}{2}z + \frac{3}{2}w - 1$$
$$y = \frac{10}{3}z + \frac{2}{3}w - 2$$

We see that the system has an infinite number of solutions. Here are three particular solutions of the system:

$$\left(\frac{7}{2}, \frac{4}{3}, 1, 0\right), \quad \left(\frac{1}{2}, -\frac{4}{3}, 0, 1\right), \quad (-1, -2, 0, 0)$$

SECTION 10.3 Linear Systems in Any Number of Variables

EXAMPLE 11
A System with No Solution

Determine all solutions of the following system of linear equations.
$$\begin{cases} x + y - 2z = 1 \\ y - 3z = 4 \\ 2x + 3y - 7z = 5 \end{cases}$$

Solution

The first two rows are already in the proper echelon form. We use row operations on the system matrix to transform the third row, proceeding from left to right:

$$\begin{bmatrix} 1 & 1 & -2 & 1 \\ 0 & 1 & -3 & 4 \\ 2 & 3 & -7 & 5 \end{bmatrix} \xrightarrow{(-2)E_1 + E_3} \begin{bmatrix} 1 & 1 & -2 & 1 \\ 0 & 1 & -3 & 4 \\ 0 & 1 & -3 & 3 \end{bmatrix}$$

$$\xrightarrow{(-1)E_2 + E_3} \begin{bmatrix} 1 & 1 & -2 & 1 \\ 0 & 1 & -3 & 4 \\ 0 & 0 & 0 & -1 \end{bmatrix}$$

Note that the last row of the final matrix corresponds to the equation:
$$0 = -1$$

This equation is inconsistent. Therefore, the system has no solutions. Whenever the echelon form of a system contains a row consisting of zeros in all columns except the rightmost one, the system has no solutions. Such systems are called **inconsistent**.

Row Operations Using a Calculator

Matrix arithmetic can be quite tedious to do manually. However, much of the tedium can be reduced by appropriate application of technology. In this section we indicate how to use a calculator to perform various matrix calculations. Up to this point our calculator discussions have applied to most graphing calculators. However, from here on in the book, we must be more specific because there is great variation in how different calculators handle matrices, binomial coefficients, and other topics that we discuss. In order to be specific, we will use the TI-82 calculator as a model. If you have a different calculator, consult your calculator manual for the corresponding keystrokes in the following discussions.

A TI-82 calculator can store three matrices, denoted $[A]$, $[B]$, and $[C]$. You will find these indicated on keys at the bottom of the calculator.

Let's begin our discussion by learning to enter matrices into the calculator. To enter $[A]$, press (MATRX). You will see the menu shown in Figure 1. Select *Edit* from the first row and press (ENTER). You will see the display shown in Figure 2.

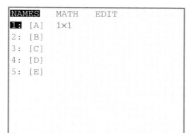

Figure 1

Figure 2

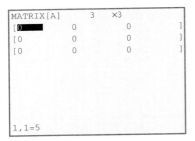

Figure 3

First enter the size of the matrix in the first row by moving the cursor to the blank to the left of ×, entering 3, and then moving the cursor to the right of × and entering 3 again. Now press ENTER. The cursor moves to the first matrix entry in column 1. (See Figure 3.) Enter the value for this entry and press ENTER. The cursor moves to the next entry in row 1. Continue in this fashion. When you have finished keying in the entries, press 2nd QUIT. This takes you back to the main screen of the calculator. You can now enter the values of other matrices or perform calculations.

After a matrix has been entered into the calculator, it remains stored until you change it. You can perform elementary row operations on the matrices stored in the calculator, as illustrated in the next example.

EXAMPLE 12
Doing Row Operations Using a Calculator

Perform the following consecutive row operations on the matrix

$$A = \begin{bmatrix} 5 & -1 \\ 3 & 4 \end{bmatrix}$$

using a TI-82 calculator.

1. E_{12}
2. $3E_2 + E_1$
3. $\left(\frac{1}{2}\right)E_1$
4. $E_1 + E_2$

Solution

Begin by storing the matrix as A using the matrix EDIT menu, as described above. When done, press ENTER to return to the main screen.

1. To swap the first and second rows, press MATRX MATH and select the option RowSwap(from the menu. (Just position the cursor on the option and press ENTER.) On the screen you will see the line:

    RowSwap(

    Fill in this line with the matrix and the lines to be swapped as follows:

    RowSwap([A],1,2)

    (To enter the matrix [A], press MATH and select [A] from the NAMES menu.) Now press ENTER. The calculator displays the new matrix:

    $$\begin{bmatrix} 3 & 4 \\ 5 & -1 \end{bmatrix}$$

2. Select the *Row+(operation from the MATRX menu. On the screen you will see the line:

    *Row+(

    Fill in this line with the matrix and rows to be used, as follows:

    *Row+(3,[A],2,1)

    Now press ENTER. The calculator displays the new matrix:

    $$\begin{bmatrix} 14 & 11 \\ 3 & 4 \end{bmatrix}$$

SECTION 10.3 Linear Systems in Any Number of Variables

3. Select the *Row(operation from the [MATRX] menu. On the screen you will see the line:

    *Row(

   Fill in this line with the matrix and rows to be used, as follows:

    *Row(.5,[A],1

   Now press [ENTER]. The calculator displays the new matrix:

   $$\begin{bmatrix} 2.5 & -0.5 \\ 3 & 4 \end{bmatrix}$$

4. Select the Row+(operation from the [MATRX] menu. On the screen you will see the line:

    Row+(

   Fill in this line with the matrix and rows to be used, as follows:

    Row+([A],1,2

   Now press [ENTER]. The calculator displays the new matrix:

   $$\begin{bmatrix} 5 & -1 \\ 8 & 3 \end{bmatrix}$$

Exercises 10.3

Solve the following systems of equations in echelon form.

1. $\begin{cases} x + 2y - z = 4 \\ 5y - 2z = 8 \\ z = 1 \end{cases}$

2. $\begin{cases} 2x + y + 2z = 12 \\ 4y - 3z = 17 \\ z = 15 \end{cases}$

3. $\begin{cases} 3x + 2y - z = 15 \\ y - 3z = 5 \\ z = -2 \end{cases}$

4. $\begin{cases} x + 3y - z = 11 \\ 4y - 7z = 14 \\ z = 2 \end{cases}$

5. $\begin{cases} x + y + 5z = 13 \\ y - 4z = -10 \\ z = 3 \end{cases}$

6. $\begin{cases} 3x - 5y - z = 7 \\ 6y - 2z = 8 \\ z = -4 \end{cases}$

Consider the matrix:

$$\begin{bmatrix} 4 & 1 & 0 \\ -1 & 2 & 3 \\ 0 & -1 & 5 \end{bmatrix}$$

Perform the following elementary row operations.

7. E_{12}
8. E_{13}
9. $\frac{1}{2}E_2$
10. $-2E_1$
11. $2E_1 - E_2$
12. $4E_3 + 2E_2$
13. $-1E_1 + 2E_2 + E_3$
14. $E_2 - E_3 + 2E_1$

Transform the following matrices into echelon form.

15. $\begin{bmatrix} 1 & 2 & 0 & 3 \\ 4 & 5 & 1 & 0 \\ -2 & 0 & 1 & -1 \end{bmatrix}$

16. $\begin{bmatrix} 1 & -1 & 5 & -6 \\ 3 & 3 & -1 & 10 \\ 1 & 3 & 2 & 5 \end{bmatrix}$

17. $\begin{bmatrix} 1 & 1 & 2 \\ 2 & 2 & 5 \end{bmatrix}$

18. $\begin{bmatrix} 2 & 6 & 28 \\ 4 & -3 & -19 \end{bmatrix}$

19. $\begin{bmatrix} 1 & 3 & 2 & 1 \\ 2 & 1 & -1 & 2 \\ 1 & 1 & 1 & 2 \end{bmatrix}$

20. $\begin{bmatrix} 2 & 1 & 1 & 3 \\ 3 & -4 & 2 & -7 \\ 1 & 1 & 1 & 2 \end{bmatrix}$

21. $\begin{bmatrix} 1 & 5 & 6 \\ 0 & 1 & 1 \end{bmatrix}$

22. $\begin{bmatrix} 2 & 7 & 1 \\ 5 & 0 & -15 \end{bmatrix}$

23. $\begin{bmatrix} 3 & 2 & 1 & 1 \\ 0 & 2 & 4 & 22 \\ -1 & -2 & 3 & 15 \end{bmatrix}$

24. $\begin{bmatrix} 4 & 2 & 6 & 24 \\ 4 & -3 & 0 & 10 \\ 5 & 0 & -4 & -11 \end{bmatrix}$

25. $\begin{bmatrix} 1 & 2 & 3 & 0 \\ 1 & 0 & -2 & 3 \\ 0 & 1 & 1 & 1 \end{bmatrix}$

26. $\begin{bmatrix} 2 & 1 & 2 & 1 \\ 1 & 2 & -3 & 4 \\ 3 & -1 & 1 & 0 \end{bmatrix}$

Solve the following systems of linear equations.

27. $\begin{cases} x + 2y + 3z = 0 \\ x - 2z = 3 \\ y + z = 1 \end{cases}$

28. $\begin{cases} x + y + 4z = 1 \\ -2x - y + z = 2 \\ 3x - 2y + 3z = 5 \end{cases}$

29. $\begin{cases} 3x + 2y - z = 4 \\ 2x - y + 3z = 5 \\ x + 3y + 2z = -1 \end{cases}$

30. $\begin{cases} 4x - 2y + 3z = 4 \\ 5x - y + 4z = 7 \\ 3x + 5y + z = 7 \end{cases}$

31. $\begin{cases} 2x + 3y + z = 1 \\ x - y + 2z = 3 \end{cases}$

32. $\begin{cases} -5x + 3y - z = 1 \\ 2x + y - z = 4 \end{cases}$

33. $\begin{cases} 2x - z = 0 \\ y + 2z = -2 \\ x + y = 1 \end{cases}$

34. $\begin{cases} x + z = 1 \\ x + y = -1 \\ y + z = 4 \end{cases}$

35. $\begin{cases} x + y + z + w = -2 \\ -2x + 3y + 2z - 3w = 10 \\ 3x + 2y - z + 2w = -12 \\ 4x - y + z + 2w = 1 \end{cases}$

36. $\begin{cases} x - y - 3z - 2w = 2 \\ 4x + y + z + 2w = 2 \\ x + 3y - 2z - w = 9 \\ 3x + y - z + w = 5 \end{cases}$

37. $\begin{cases} x + 2y + 3z = 0 \\ x + z = 1 \\ y + z = 1 \end{cases}$

38. $\begin{cases} 3x - 5y = 6 \\ -6x + 10y = 7 \end{cases}$

39. $\begin{cases} 3x - 6y = 12 \\ -2x + 4y = -8 \end{cases}$

40. $\begin{cases} x - y + z + w = 0 \\ 5x + 3z + w = 1 \end{cases}$

41. $\begin{cases} -5x + 3y - z = 1 \\ 2x + y - z = 4 \end{cases}$

42. $\begin{cases} x - y = 5 \\ y - z = 6 \\ z - w = 7 \\ x + w = 8 \end{cases}$

43. $\begin{cases} x + y - z + w = 4 \\ 2x + 3y + z - w = -1 \\ -x - 3y + 2z + 3w = 3 \\ 3x + 2y - 3z + w = 8 \end{cases}$

44. $\begin{cases} x + 2y + 3z - w = 7 \\ x - 4y + z = 3 \\ 2y - 3z + w = 4 \end{cases}$

Applications

45. **Recreation.** Flying with the wind, an airplane can travel 1,080 miles in six hours, but flying against the same wind, it goes only $\frac{1}{3}$ this distance in half the time. Find the speed of the plane in still air and the wind speed.

46. **Corporate finance.** To start a record company, Barry G. borrowed $50,000 from three different banks. He borrowed some money at 8 percent, and he borrowed $3,000 more than that amount at 6.9 percent. The rest was borrowed at 12 percent. If his annual interest totals $4,755 from all three loans, how much was borrowed at each rate?

47. **Manufacturing.** A furniture manufacturer makes chairs and sofas. Each requires both carpentry and upholstering. A sofa requires 1 hour of carpentry and 2 hours of upholstering. A chair requires 3 hours of carpentry and 1 hour of upholstering. Both the carpentry and upholstery departments operate 15 hours a day. How many of each furniture item can be produced in a five-day work week under these conditions?

48. **Retailing.** A famous national appliance store sells both Sony and Sanyo stereos. The Sony sells for $280, and the Sanyo sells for $315. During a one-day sale, a total of 85 Sonys and Sanyos was sold for a total of $23,975. How many of each brand were sold during this one-day sale?

Technology

In Exercises 49–54 let:

$$A = \begin{bmatrix} 1 & 2 & 0 & 3 \\ 4 & 5 & 1 & 0 \\ -2 & 0 & 1 & -1 \end{bmatrix}$$

Use a calculator to perform the following row operations.

49. E_{13}

50. E_{23}

51. $(-2)E_2$

52. $3E_1$

53. $4E_3 - E_2$

54. $E_1 + E_2$

In your own words

55. Describe the method of Gauss elimination. Use a specific linear system in your description.

56. Explain the difference between the method of substitution and Gauss elimination.

57. What are the advantages of Gauss elimination over substitution? Its disadvantages?

10.4 SYSTEMS OF INEQUALITIES

In Chapter 2 we introduced inequalities involving a single variable and showed how to graph such inequalities on the number line. Let's now turn our attention to inequalities in two variables x and y. Here are some typical inequalities of this sort:

$$3x + 5y < 4$$
$$x^2 + 3xy \leq 10$$
$$x + \frac{1}{y} > 1$$

Suppose we are given an inequality in two variables. An ordered pair (a, b) is said to **satisfy the inequality** (or **is a solution of the inequality**) provided that the inequality is true when x is replaced by a and y is replaced by b. For instance, the ordered pair $(-1, 2)$ is not a solution of the first inequality above because when x is replaced by -1 and y is replaced by 2, we have:

$$3x + 5y = 3(-1) + 5(2)$$
$$= -3 + 10$$
$$= 7$$

Because 7 is not less than 4, we see that $(-1, 2)$ is not a solution. However, $(-4, 1)$ is a solution of the inequality, because when x is replaced by -4 and y is replaced by 1, we have:

$$3x + 5y = 3(-4) + 5(1)$$
$$= -7$$
$$< 4 \quad \text{True}$$

The set of ordered pairs that are solutions of an inequality is called the **solution set** of the inequality. The solution set can be represented graphically as a region in the x–y plane. The next few examples determine the solution sets of some typical inequalities in two variables and the corresponding plane regions.

EXAMPLE 1
Determining the Solution Set

Graph the inequality $x^2 + y^2 < 4$.

Solution

Note that $x^2 + y^2$ equals the square of the distance of the point (x, y) from the origin. The inequality specifies that the square of the distance of (x, y) be less than 4; that is, the distance is less than 2. These are the points in the interior of

550   CHAPTER 10   Linear and Nonlinear Systems

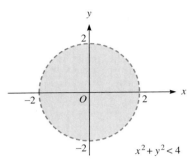

Figure 1

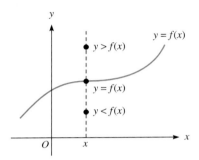

Figure 2

the circle shown in Figure 1. The dotted line indicates that the points on the circle are excluded from the shaded region.

In Figure 2 appears the graph of an equation of the form $y = f(x)$, where f is a function of the variable x. A point (x, y) for which $y = f(x)$ lies on the graph. A point (x, y) for which $y > f(x)$ lies above the graph. A point (x, y) for which $y < f(x)$ lies below the graph. That is, we have the following result.

GRAPHING SOLUTION SETS

The graph of the inequality $y > f(x)$ consists of the region above the graph of the equation $y = f(x)$. (See Figure 3(a).) The graph of the inequality $y < f(x)$ consists of the region below the graph of the equation $y = f(x)$. (See Figure 3(b).)

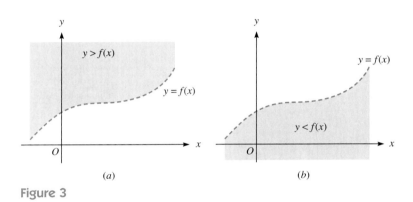

Figure 3

The next example shows how to apply the above result.

EXAMPLE 2 Graph the inequality $y \geq x^2 - 3$.

Graphing a Solution Set

Solution

In Figure 4 the curve of the equation $y = x^2 - 3$ is drawn. The graph of the given inequality is the region above and on the curve. This is the shaded region in Figure 4.

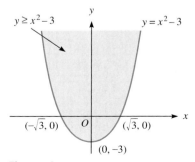

Figure 4

EXAMPLE 3
Solution Set of a Linear Inequality

Graph the inequality $5x - 3y < 12$.

Solution

Let's transform the inequality so that y is isolated on the left side. First subtract $5x$ from both sides to obtain:

$$-3y < 12 - 5x$$

Next, multiply by $-\frac{1}{3}$. Because this quantity is negative, we must reverse the direction of the inequality:

$$y > -\frac{1}{3}(12 - 5x)$$

$$y > -4 + \frac{5}{3}x$$

We graph the equation $y = -4 + \frac{5}{3}x$, as shown in Figure 5. The graph of the given inequality is the region above the line in the figure.

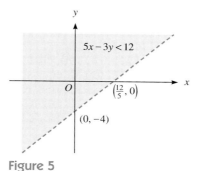

Figure 5

EXAMPLE 4
Solution Set of a Nonlinear Inequality

Graph the solution set of the inequality:

$$xy \geq 1$$

Solution

As in the preceding example, we apply the laws of inequalities to isolate y on the left side. Basically, we wish to multiply the inequality by $1/x$. However, this quantity is positive if $x > 0$ and negative if $x < 0$. In the first case the inequality is equivalent to:

$$y \geq \frac{1}{x}, \qquad x > 0$$

In the second case the inequality is equivalent to:

$$y \leq \frac{1}{x}, \qquad x < 0$$

For the first case, draw the graph

$$y = \frac{1}{x}, \qquad x > 0$$

as shown in Figure 6(a). The points corresponding to solutions of the inequality are those lying on or above the graph. These are the shaded points in the figure. For the second case draw the graph

$$y = \frac{1}{x}, \quad x < 0$$

as shown in Figure 6(b). The points corresponding to solutions of the inequality are those lying on or below the graph. These are the shaded points in the figure. The solution set of the given inequality consists of the points belonging to the solution sets of both the cases. This set is the shaded region shown in Figure 6(c).

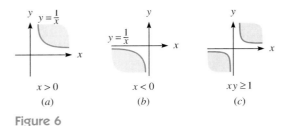

Figure 6

Another method of graphing an inequality is to first replace the inequality sign by an equals sign and graph the resulting equation. The graph of the inequality consists of all points lying to one side of the graph. To determine which side, choose a point (a, b) not on the graph of the equation and substitute it into the inequality. If the result is a true inequality, the point lies on the same side as the graph of the inequality. Otherwise the point lies on the opposite side from the graph of the inequality. The point (a, b) is called a **test point**.

For example, consider the inequality $x^2 + y^2 < 4$ of Example 1 and the test point (2, 2). The graph of the equation $x^2 + y^2 = 4$ is a circle, and the test point lies outside the circle. At the test point we have:

$$x^2 + y^2 = 2^2 + 2^2$$
$$= 8$$

That is, the inequality is false at the test point because $8 < 4$ is false. This means that the region containing the test point, the exterior of the circle, does not belong to the graph of the inequality. The graph consists of the points within the circle. And because the inequality is $<$ rather than $\leq$, the circle itself is not included in the graph.

In the previous sections of this chapter, we studied systems of equations and their solutions. By analogy, let's now turn to systems of *inequalities* and their solutions.

A **system of inequalities** is a collection of one or more inequalities. Here we concentrate on systems of inequalities in two variables. A **solution** of a system of inequalities is an ordered pair that is a solution of each of the inequalities of the system. The set of the solutions of a system of inequalities is called the **feasible set** (or **graph**) of the system. One method for determining the feasible set is as follows.

DETERMINING THE FEASIBLE SET

1. Solve each inequality of the system separately.
2. Graph on the same coordinate system the solution set of each inequality. Shade each solution set with a different sort of shading (hatching).
3. The solution set of the system consists of the region that is shaded with all the shading types.

The following example illustrates this method.

EXAMPLE 5
Graphing a Feasible Set

Determine the feasible set of the following system of inequalities:
$$\begin{cases} y \geq x^2 - 3 \\ 5x + 3y < 12 \end{cases}$$

Solution
In Figure 7(a) is sketched the solution of the first inequality of the system. In Figure 7(b) appears the solution set of the second inequality. Figure 7(c) shows the two solution sets superimposed on a single coordinate system. The solution set of the system consists of the region that is shaded twice. This is the shaded region in Figure 7(d).

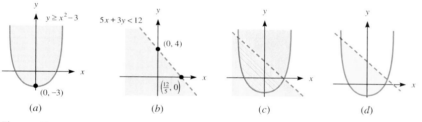

Figure 7

EXAMPLE 6
Feasible Set of a Linear System

Sketch the graph of the following system of linear inequalities:
$$\begin{cases} x \leq 8 \\ y \leq 3x - 7 \\ y > x + 4 \end{cases}$$

Solution
In this case, there are three solution sets to determine, one for each inequality. Figure 8(a) shows the three solution sets, each shaded with a different pattern. The region that is shaded three times is the solution set of the system. See Figure 8(b).

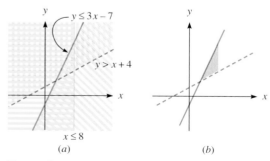

Figure 8

EXAMPLE 7 Manufacturing

A pharmaceutical company manufactures two drugs using the same ingredients. Each case of drug 1 requires 300 grams of ingredient A and 600 grams of ingredient B. Each case of drug 2 requires 500 grams of ingredient A and 100 grams of ingredient B. Market conditions dictate that at most 100,000 grams of ingredient A and 50,000 grams of ingredient B are available each week.

1. Determine a system of inequalities describing the production limitations of the two drugs.
2. Determine the feasible set of the system of inequalities.

Solution

Let

$$x = \text{number of cases of drug 1}$$
$$y = \text{number of cases of drug 2}$$

First, the values of x and y must be nonnegative. That is:

$$x \geq 0, \quad y \geq 0$$

The condition on ingredient A is:

$$\text{total ingredient A} \leq 100{,}000$$
$$\text{ingredient A in drug 1} + \text{ingredient A in drug 2} \leq 100{,}000$$
$$300x + 500y \leq 100{,}000$$

Similarly, the condition on ingredient B is:

$$\text{total of ingredient B} \leq 50{,}000$$
$$\text{ingredient B in drug 1} + \text{ingredient B in drug 2} \leq 50{,}000$$
$$600x + 100y \leq 50{,}000$$

Thus, the production limitations are described by the following system of four inequalities:

$$\begin{cases} x \geq 0, \quad y \geq 0 \\ 3x + 5y \leq 1{,}000 \\ 6x + 1y \leq 500 \end{cases}$$

In Figure 9(a) appears the system of inequalities. The inequalities $x \geq 0$ and $y \geq 0$ imply that we are limited to the first quadrant. The region in the first quadrant that is shaded twice is the feasible set of the system. This is the shaded region in Figure 9(b).

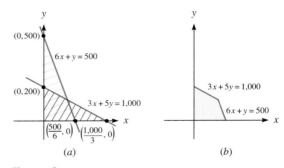

Figure 9

SECTION 10.4 Systems of Inequalities

Exercises 10.4

Graph the inequalities.

1. $x + 2y \leq 4$
2. $3x - y > 10$
3. $x^2 + y^2 < 4$
4. $x^2 + 9y^2 \leq 36$
5. $x < -y^2$
6. $x + y^2 \geq -4$
7. $xy > 4$
8. $x \leq \dfrac{16}{y}$
9. $x^2 + 4y^2 - 10x + 24y + 57 > 0$
10. $x^2 - y^2 + 2x - 2y \geq 0$
11. $3x - 2y > 6$
12. $x < y$
13. $y \leq x^2 - 4x + 2$
14. $y \geq -1 - x^2$

Determine the feasible set of each the following systems of inequalities.

15. $\begin{cases} x + 2y > 6 \\ x^2 > 2y \end{cases}$

16. $\begin{cases} x^2 + y^2 \leq 9 \\ \dfrac{x^2}{16} + \dfrac{y^2}{4} \geq 1 \end{cases}$

17. $\begin{cases} 4x + 3y < 12 \\ 4x + y > -4 \end{cases}$

18. $\begin{cases} x^2 + y^2 \leq 36 \\ -4 \leq x \leq 4 \end{cases}$

19. $\begin{cases} 9x^2 - y^2 \geq 0 \\ x^2 + y^2 \leq 25 \end{cases}$

20. $\begin{cases} x^2 + 4y^2 < 16 \\ x^2 - 2xy \geq 0 \end{cases}$

21. $\begin{cases} y \geq (x - 2)^2 + 3 \\ 3x - 4y \leq -16 \end{cases}$

22. $\begin{cases} y \leq -(x - 3)^2 + 2 \\ 3x + 5y \geq -20 \end{cases}$

23. $\begin{cases} |x| \geq 6 \\ |y| \leq 1 \end{cases}$

24. $\begin{cases} y > |x + 2| \\ 3 \geq |x| \end{cases}$

25. $\begin{cases} x + y > 6 \\ 2x - y \leq 0 \\ y \geq -2 \end{cases}$

26. $\begin{cases} 2x - y \leq 6 \\ x + y > 6 \\ x \geq -1 \end{cases}$

27. $\begin{cases} x - y \leq 2 \\ x + 2y \geq 8 \\ y \leq 4 \end{cases}$

28. $\begin{cases} 2x - y \geq -1 \\ 2x + y \geq 1 \\ x \leq 2 \end{cases}$

29. $\begin{cases} x \geq 0 \\ y \geq 0 \\ 2x + 3y \leq 12 \\ 3x + y \leq 6 \end{cases}$

30. $\begin{cases} x \geq 0 \\ y \geq 0 \\ 2x + y \leq 4 \\ 2x - 3y \leq 6 \end{cases}$

31. $\begin{cases} x \geq 2 \\ y \geq 5 \\ 3x - y \geq 12 \\ x + y \leq 15 \end{cases}$

32. $\begin{cases} x \geq 10 \\ y \geq 20 \\ 2x + 3y \leq 100 \\ 5x + 4y \leq 200 \end{cases}$

 Applications

For the following problems:
a. Determine a system of inequalities describing the given problems.
b. Determine the feasible set of the systems of inequalities.

33. **Manufacturing.** A manufacturer of skis can produce as many as 60 pairs of recreational skis and as many as 45 pairs of racing skis per day. It takes 3 hours of labor to produce a pair of recreational skis and 4 hours of labor to produce a pair of racing skis. The company has up to 240 hours of labor available for ski production each day.

34. **Product design.** Wheat contains 80 g of protein and 40 mg of iron per kilogram, and oats contain 100 g of protein and 30 mg of iron per kilogram. General Mills wants to make a cereal from a mixture of oats and wheat that will contain at least 88 g of protein and at least 36 mg of iron per kilogram of cereal.

35. **Education.** You are taking a test comprising 16 questions in which items of type x are worth 10 points and items of type y are worth 15 points. It takes 3 minutes to answer each item of type x and 6 minutes to answer each item of type y. Total time allowed is 60 minutes.

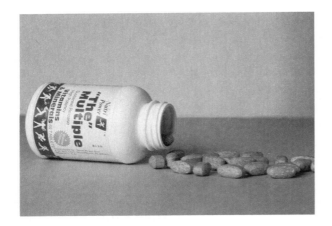

38. **Elevators.** Many elevators have a capacity of 1 metric ton (1,000 kg). Suppose c children, each weighing 35 kg, and a adults, each weighing 75 kg, are on an elevator.

36. **Nutrition.** Every day Marvin B. needs a dietary supplement of 4 mg of vitamin A, 11 mg of vitamin B, and 100 mg of vitamin C. Either of two brands of vitamin pills can be used: brand x at 6 cents a pill or brand y at 8 cents a pill. Brand x supplies 2 mg of vitamin A, 3 mg of vitamin B, and 25 mg of vitamin C per pill. Likewise a brand y pill supplies 1, 4, and 50 mg of vitamins A, B, and C, respectively.

37. **Area of a basketball floor.** Floor areas of basketball courts vary considerably because of building sizes and other constraints, such as cost. The length L of a planned basketball court is to be at most 74 ft and the width w is to be at most 50 ft.

39. **Hockey.** A hockey team figures that it needs at least 60 points for the season to make the play-offs. A win is worth 2 points in the standings and a tie is worth 1 point. Let w be the number of wins and t the number of ties.

In your own words

40. Explain how to determine whether the boundary of the feasible set is included in the graph of an inequality in two variables.

41. Explain how to graph a system of inequalities in two variables.

42. What sort of geometric figure can be the feasible set of a system of linear inequalities?

10.5 LINEAR PROGRAMMING

Linear programming is an important application involving systems of linear inequalities. To introduce this important subject, let's return to the pharmaceutical company example (Example 6) at the end of the preceding section. Suppose the company earns $50 for each case of drug 1 and $75 for each case of drug 2. How many cases of each drug should it produce in order to maximize its profit?

Let's state the problem in algebraic form: Suppose the drug company produces x cases of drug 1 and y cases of drug 2. The profit P it earns is given by:

$$P = 50x + 75y$$

At first you might ask, Why not just manufacture a huge amount of each type of drug and thereby make a huge profit? The answer is that production has a number of constraints imposed on it, as stated in Example 6. Namely, production is subject to the inequalities:

$$300x + 500y \leq 100{,}000$$

$$600x + 100y \leq 50{,}000$$

$$x \geq 0, \qquad y \geq 0$$

The four inequalities are called **constraints**, because they constrain the choice of x and y. The set of ordered pairs (x, y) that satisfies the system of inequalities is the feasible set of the system and consists of the points in the shaded region in Figure 9(b) of Section 10.4. Our problem is to choose the point (x, y) of the feasible set at which P is a maximum.

A problem of the sort just described is called a **linear programming problem**. More precisely, suppose we are given a function

$$P = ax + by$$

of two variables x and y and a system of linear inequalities in the same variables. Consider the following:

Definition
Linear Programming Problem

Determine the ordered pair (x, y) that is a solution of the system of inequalities and for which the value of P is maximized (minimized).

The function P is called the **objective function**, and the inequalities are called **constraints**. The corners of the feasible set are called **vertices**. We have the following theorem:

FUNDAMENTAL THEOREM OF LINEAR PROGRAMMING

Suppose the feasible set is a polygon.[1] The objective function assumes its maximum (minimum) value at a vertex of the feasible set.

We omit the proof of this theorem.

[1] In more advanced treatments of linear programming, the case of feasible sets that are "unbounded" (that is, sets that extend to infinity in either the horizontal or vertical direction) is also treated. For simplicity, we will not discuss problems with such feasible sets.

558   CHAPTER 10   Linear and Nonlinear Systems

EXAMPLE 1
Manufacturing

Determine the number of cases of each type of drug that the pharmaceutical company must produce in order to maximize its profit.

Solution

According to the Fundamental Theorem of Linear Programming, the maximum profit occurs at a vertex of the feasible set. Let's determine the vertices. In Figure 1, we draw the feasible set and label each of the vertices. Also, each of the bounding lines of the feasible set is labeled with its equation. Vertex A is the origin $(0, 0)$. Vertex B is the intersection of the lines with equations:

$$y = 0$$
$$600x + 100y = 50{,}000$$

The solution of this system is obtained by substituting $y = 0$ into the second equation:

$$600x + 100(0) = 50{,}000$$
$$x = \frac{250}{3}$$
$$B = \left(\frac{250}{3}, 0\right)$$

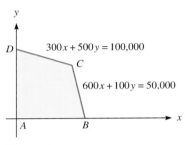

Figure 1

Vertex C is the intersection of the lines:

$$300x + 500y = 100{,}000$$
$$600x + 100y = 50{,}000$$

To solve this system, first divide both equations by 100 (to keep the numbers small).

$$3x + 5y = 1{,}000$$
$$6x + y = 500$$

Now multiply the second equation by -5 and add it to the first to obtain:

$$\begin{cases} 3x + 5y = 1{,}000 \\ -30x - 5y = -2{,}500 \end{cases}$$
$$-27x = -1{,}500$$
$$x = \frac{1{,}500}{27} = \frac{500}{9}$$
$$y = 500 - 6x$$
$$= 500 - 6\left(\frac{500}{9}\right)$$
$$= \frac{1{,}500}{9}$$
$$= \frac{500}{3}$$
$$C = \left(\frac{500}{9}, \frac{500}{3}\right)$$

The final vertex, D, is the intersection of the lines with the next two equations:

$$x = 0$$
$$300x + 500y = 100{,}000$$
$$300 \cdot 0 + 500y = 100{,}000$$

$$y = \frac{1{,}000}{5} = 200$$

$$D = (0, 200)$$

Let's now make a table of the vertices and the value of the function P at each vertex.

Vertex	$P = 50x + 75y$
$A: (0, 0)$	0
$B: \left(\frac{250}{3}, 0\right)$	$\frac{12{,}500}{3} = 4{,}166.67$
$C: \left(\frac{500}{9}, \frac{500}{3}\right)$	$15{,}277.78 \leftarrow$ Maximum
$D: (0, 200)$	$15{,}000$

From the table, we see that the maximum profit occurs for vertex C. By the Fundamental Theorem of Linear Programming, the maximum value of the function P for (x, y) in the feasible set is 15,277.78, and this value occurs for $x = \frac{500}{9}$, $y = \frac{500}{3}$. That is, the company should produce $\frac{500}{9}$ cases of drug 1 and $\frac{500}{3}$ cases of drug 2 per week.

EXAMPLE 2
Minimization of a Linear Function

Determine the minimum value of the function $P = 3x - 5y$ if x and y are constrained to lie in the triangle having vertices $(1, 1)$, $(3, 2)$, $(1, 5)$. See Figure 2.

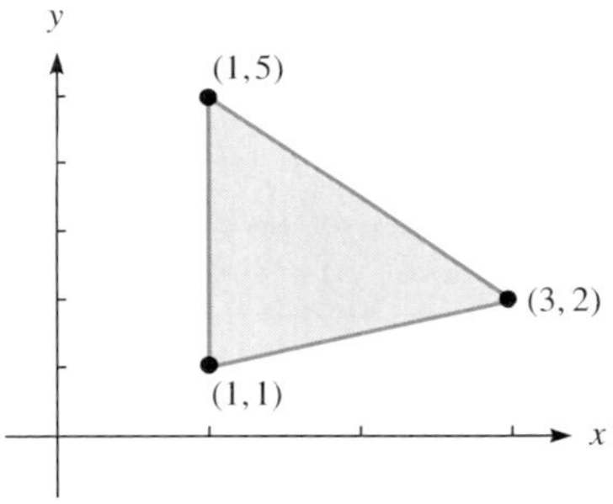

Solution
By the Fundamental Theorem of Linear Programming, the minimum point must occur at one of the vertices of the triangle. So let's tabulate the values of the function at the vertices.

Vertex	$P = 3x - 5y$
$(1, 1)$	-2
$(3, 2)$	-1
$(1, 5)$	-22

In the table we see that the minimum value of $3x - 5y$ is -22, which occurs for $x = 1$, $y = 5$.

Figure 2

Exercises 10.5

For every function and feasible set that follow, determine the maximum and minimum values of the function and the values of x and y where they occur.

1. $f(x, y) = 3x + 5y$; vertices at $(4, 8), (2, 4), (1, 1), (5, 2)$

2. $f(x, y) = x + 4y$; vertices at $(0, 7), (0, 0), (6, 2), (5, 4)$

3. $Q = x + 3y$ subject to
$$\begin{cases} x \geq 0 \\ y \geq 0 \\ 5x + 2y \leq 20 \\ 2y \geq x \end{cases}$$

4. $F = 2x - 3y$ subject to
$$\begin{cases} x \geq 0 \\ y \geq 0 \\ 3x + 2y \leq 15 \\ 5x + 3y \geq 15 \end{cases}$$

5. $f(x) = 10x + 12y$ subject to
$$\begin{cases} 2x + 5y \geq 22 \\ 4x + 3y \geq 28 \\ 2x + 2y \leq 17 \\ x \geq 0 \\ y \geq 0 \end{cases}$$

6. $f(x) = -3x + y$ subject to
$$\begin{cases} x + 2y \leq 10 \\ 2x + y \geq 12 \\ x + y \leq 8 \\ x \geq 0 \\ y \geq 0 \end{cases}$$

560 CHAPTER 10 Linear and Nonlinear Systems

Find the maximum and minimum values of the given functions in the indicated regions.

7. $f(x, y) = 2x - 3y$

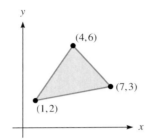

9. $f(x, y) = 5x + y$

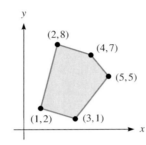

11. $f(x, y) = 8x + 2y$

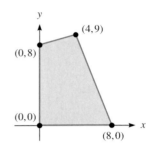

8. $f(x, y) = x + 5y$

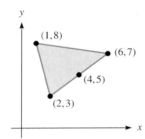

10. $f(x, y) = 3x + 8y$

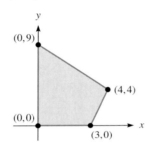

12. $f(x, y) = 5x + 3y$

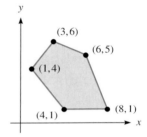

Applications

Solve each of the following linear programming problems.

13. **Manufacturing.** RCA Manufacturing Company in Ft. Wayne, Indiana, makes a $60 profit on each 19-inch TV it produces and a $40 profit on each 13-inch TV. A 19-inch TV requires 1 hour on machine X, 1 hour on machine Y, and 4 hours on machine Z. The 13-inch TV requires 2 hours on X, 1 hour on Y, and 1 hour on Z. In a given day, machines X, Y, and Z can work a maximum of 16, 9, and 24 hours, respectively. How many 19-inch TVs and how many 13-inch TVs should be produced per day to maximize the profit?

14. **Manufacturing.** The Buick manufacturing plant in Flint, Michigan, must fill orders for Park Avenues from two dealers. The first dealer, Farmington Buick, has ordered 20 Park Avenues, and the second dealer, Brighton Buick, has ordered 30. The manufacturer has the cars stored in two different areas, southeast Flint and southwest Flint. There are 40 cars in southeast Flint and only 15 cars in southwest Flint. The shipping costs, per car, are $15 from southeast Flint to Brighton, $13 from southeast Flint to Farmington, $14 from southwest Flint to Brighton, and $16 from southwest Flint to Farmington.

With these conditions find the number of cars to be shipped from each area to each dealer if the total shipping cost is to be a minimum. What is the cost?

15. **Product design.** Kellogg's of Battle Creek, Michigan, is going to produce a new cereal from a mixture of bran and rice that contains at least 88 g of protein and at least 36 mg of iron. Knowing that bran contains 80 g of protein and 40 mg of iron per kilogram and that rice contains 100 g of protein and 30 mg of iron per kilogram, find the minimum cost of producing this new cereal, "Rice Bran," if bran costs 50 cents per kilogram and rice costs 40 cents per kilogram.

16. **Manufacturing.** Sanyo makes stereo receivers. It produces a 30-watt receiver that it sells for $100 profit and a 50-watt receiver that it sells for a $150 profit. The 30-watt receiver requires 3 hours to manufacture, and the 50-watt receiver takes 5 hours. The cabinet shop spends one hour on a 30-watt receiver and 3 hours on a 50-watt receiver. Packing takes 2 hours for both types of receivers. Per week, Sanyo has available 3,900 work hours for manufacturing, 2,100 hours for cabinet making, and 2,200 hours for packing. How many receivers of each type should Sanyo produce per week to maximize its profit, and what is the maximum profit per week?

17. **Manufacturing.** Almosttexas makes two types of calculators. Deluxe sells for $12 and Top of the Line sells for $10. It costs Almosttexas $9 to produce a Deluxe and $8 to produce a Top of the Line calculator. In one week, Almosttexas can produce from 200 to 300 Deluxe calculators and from 100 to 250 Top of the Line calculators, but no more than 500 total calculators. How many of each type should be produced per week to maximize the profits for Almosttexas?

18. **Portfolio analysis.** You have $40,000 to invest in stocks and bonds. The least you are allowed to invest in stocks is $6,000, and you cannot invest more than $22,000 in stocks. You can also invest no more than $30,000 in bonds. The interest on stocks is 8 percent tax-free, and the interest on bonds is $7\frac{1}{2}$ percent tax-free. How much should you invest in each type to maximize your income? What is your income from the $40,000 invested?

19. **Agriculture.** The Glodfeltys' 312-acre farm grows corn and beans. The work of growing and picking the corn takes 35 work hours per acre, the work of growing and picking the beans is 27 work hours per acre. Only 9,500 work hours are available for these crops. The profit per acre of corn is $173, and the profit per acre of beans is $152. With this in mind, how many acres should be planted in corn to maximize the Glodfeltys' profits? What are the maximum profits?

20. **Construction.** Fenton Cement Company has been constructed to haul 360 tons of cement per day for a highway construction job on US 23. Fenton has 7 six-ton trucks, 4 ten-ton trucks, and 9 drivers to haul cement a day. The six-ton trucks can make 8 trips per day, and the ten-ton trucks can make only 6 trips per day; the costs are $15 and $24, respectively, per day. If Fenton uses all 9 drivers, how many trucks of each type should be used to minimize the cost per day? What is that cost?

21. **Building code.** The North Manchester building code requires that the area of the windows must be at least $\frac{1}{8}$ the area of the walls and roof of all new buildings. The annual heating cost of a new building is $3 per square foot of window area and $1 per square foot of wall and roof area. To the nearest square foot, what is the largest surface area a new building can have if its annual heating cost cannot exceed $1,000?

22. **Order fulfillment.** Gary Smelting Company receives a monthly order for at least 40 tons of iron, 60 tons of copper, and 40 tons of lead. It can fill this order by smelting either alloy A or alloy B. Each railroad carload of A will produce 1 ton of iron, 3 tons of copper, and 4 tons of lead after smelting. Each railroad carload of B will produce 2 tons of iron, 2 tons of copper and 1 ton of lead after smelting. If the cost of smelting one carload of alloy A is $350 and the cost of smelting one carload of alloy B is $200, how many carloads of each should be used to fill the order at a minimum cost to Gary Smelting? What is that minimum cost?

23. **Production planning.** A lumber company can convert logs into either lumber or plywood. In a given week the mill can turn out 400 units of production, of that 100 units of lumber and 150 units of plywood are required by regular customers. The profit on a unit of lumber is $20, and the profit on a unit of plywood is $30. How many units of each type should the mill produce per week in order to maximize profit?

24. **Agriculture.** A farm consists of 240 acres of cropland. The farmer wishes to plant part or all of the acreage in corn or oats. The profit per acre in corn production is $40, and that in oats is $30. An additional restriction is that the total hours of labor during the production is no more than 320. Each acre of land in corn production uses 2 hours of labor during the production period, but production of oats requires only 1 hour per acre. How many acres of land should be planted in corn and how many in oats in order to maximize profit?

Mathematics and the World around Us—Compression of Data

You might think that the most important mathematical ideas are necessarily complicated. However, this is often not true. Some of most significant mathematical ideas are really quite simple. An illustration is provided by the various algorithms for data compression that have been developed in the last 35 years.

Computer storage devices enable us to store and access huge volumes of data. The typical personal computer is capable of storing tens of thousands of pages of information on its hard disk. Optical storage systems enable storage of hundreds of thousands or even millions of pages.

In order to improve the efficiency of such storage, you can use so-called **data compression algorithms,** which encode data so that they occupy less space. At first, this might seem like magic. (That's the way it seemed to me, when I first learned about it!) However, the algorithms are based on a fairly simple mathematical idea.

All data are stored in binary form, that is, as a sequence of 0's and 1's. Text is stored by means of a coding system in which letters and symbols are represented by sequences of 0's and 1's of fixed length. For example, the ASCII system, used in all personal computers, uses a sequence of eight 0's and 1's to encode a single letter or symbol. Such a sequence is called a **byte**. Each digit of a byte can be chosen in two ways, so the number of different bytes is $2^8 = 256$. This allows for 256 different symbols to be encoded.

The idea of a data compression algorithm is to use fewer than eight digits for some symbols and possibly more than eight digits for others. This can result in compression if you encode the most commonly used symbols using the fewest digits. So to compress a piece of text, you would first determine the relative frequency of each of the symbols in the text. Encode the two most frequently used symbols with 0 and 1. Code the next four most frequently used symbols with 00, 01, 10, and 10. Code the next eight most frequently used symbols with 000, 001, 010, 011, 100, 101, 110, and 111. And so forth. Of course, in addition to the encoded data, you must store a dictionary for translating from the compressed coding to the original eight-digit coding. The dictionary is then stored with the data. Depending on the text, compression of 50 percent to 90 percent can result.

This idea also works for compressing pictures. A black and white picture can be represented by an array of dots, which in turn can be represented numerically, with 0 representing a white dot and 1 representing a black dot. You can divide the numerical version of the picture into eight-digit groups, just as if it were text. Using the same compression idea as used for text can lead to dramatic condensation of data, especially for pictures with large areas of pure white or pure black.

Many data compression algorithms are available, but the basic idea behind them all is similar to what we have described.

 In your own words

25. Explain the meaning of the terms *objective function* and *constraint*. Illustrate them using a specific linear programming problem.

26. Give an example of a feasible set for a system of linear inequalities that is not a polygon.

27. Can any triangle be the feasible set of a system of linear inequalities? Explain your answer.

28. Can any four-sided figure be the feasible set of a system of linear inequalities? Explain your answer.

Chapter Review

Important Concepts—Chapter 10

- system of equations
- solution of a system of equations
- method of substitution for solving a system of equations
- linear system of equations
- nonlinear system of equations
- graphical analysis of systems in two variables
- method of elimination for solving systems of linear equations
- independent
- dependent
- inconsistent
- linear system in triangular form (echelon form)
- back substitution

- elementary operations on a linear system
- method of Gauss elimination
- matrix
- elements (entries) of a matrix
- main diagonal of a matrix
- square matrix
- coefficient matrix of a linear system
- augmented matrix
- elementary row operations on matrices
- matrix in echelon form
- square linear system
- nonsquare linear system

- system of inequalities in two variables
- satisfying an inequality
- solution of a system of inequalities
- solution set of an inequality
- test point
- system of inequalities
- feasible set; graph
- determining the feasible set
- linear programming problem
- constraint
- vertex of a feasible set
- objective function
- fundamental theorem of linear programming

Important Results and Techniques—Chapter 10

Method of substitution for systems of two equations in two variables	1. Solve an equation for one of the variables in terms of the other. 2. Substitute the expression found in step 1 into the other equation. 3. Solve the resulting equation for the possible values of one variable. 4. For each value found in step 3, substitute into the expression of step 1 to determine the corresponding value of the other variable.	p. 522
Method of elimination for systems of two linear equations in two variables	1. Multiply one or both equations of the system by suitable numbers to obtain a new system in which the coefficients of one of the variables are equal to or negatives of each other. 2. Subtract the equations to obtain a linear equation in one variable. 3. Solve the equation obtained in step 2. 4. Substitute the value obtained into either of the original system equations and solve for the second variable.	p. 529
Graphical analysis of solutions of two linear equations in two variables	1. The graphs of the equations correspond to two distinct lines. In this case the system has a single solution, corresponding with the point of intersection of the lines.	p. 530

		2. The two straight lines coincide. In this case the system has infinitely many solutions, corresponding to all points on the line. 3. The two straight lines are different but parallel. In this case the system has no solution.	
	Elementary operations on linear systems	1. Interchange two equations. 2. Multiply an equation by a nonzero real number. 3. Add to an equation a multiple of another equation.	p. 536
	Determining the feasible set of a system of inequalities	1. Solve each inequality of the system separately. 2. Graph on the same coordinate system the solution sets of all the inequalities. Shade each solution set with a different sort of shading (hatching). 3. The solution set of the system consists of the region that is shaded with all shading types.	p. 553
	Fundamental Theorem of Linear Programming	The objective function assumes its maximum (minimum) value at a vertex of the feasible set.	p. 557

Review Exercises—Chapter 10

Determine all solutions of the following systems of equations.

1. $\begin{cases} 3x + y = 11 \\ 2x - 5y = 13 \end{cases}$

2. $\begin{cases} 4x + 3y = 7 \\ 2x + 4y = 16 \end{cases}$

3. $\begin{cases} x + y = 1 \\ x^2 + y^2 = 25 \end{cases}$

4. $\begin{cases} x - y - z = 0 \\ xy = 6 \\ 3x + 4y = 5 \end{cases}$

Solve the following linear systems.

5. $\begin{cases} x + y + z = 9 \\ 3y + 2z = 7 \\ z = -1 \end{cases}$

6. $\begin{cases} 3x - 2y = 1 \\ 4x + y = 5 \end{cases}$

7. $\begin{cases} x + 2y = 3 \\ 2x + 3y = 4 \end{cases}$

8. $\begin{cases} 3x - 4y = -1 \\ x + 6y = -4 \\ -2x + 8y = -2 \end{cases}$

Graph the inequalities.

9. $x + 3y \geq 12$
10. $x - y < 4$
11. $x^2 + y^2 < 25$
12. $x^2 + 4y^2 \leq 36$
13. $xy > 16$
14. $x + y^2 < -4$
15. $y > x^2 + 16$
16. $xy \leq 25$

Determine the feasible sets of the following systems of inequalities.

17. $\begin{cases} x \geq 0 \\ y \geq 0 \\ x + y \leq 8 \end{cases}$

18. $\begin{cases} x \geq 1 \\ y < 4 \\ x + y > 8 \end{cases}$

19. $\begin{cases} x^2 + y^2 \leq 36 \\ -2 \leq x \leq 2 \end{cases}$

20. $\begin{cases} 9x^2 - y^2 \geq 0 \\ x^2 + y^2 \leq 25 \end{cases}$

Sketch the graphs of the following systems of linear inequalities.

21. $\begin{cases} 4x + 7y \leq 28 \\ 2x - 3y \geq -6 \\ y \geq -2 \end{cases}$

22. $\begin{cases} 3x + 9y \leq 18 \\ 6x - 4y \geq 8 \\ 3y \geq 0 \end{cases}$

23. $\begin{cases} x + y \leq 10 \\ x - 3y \geq -18 \\ x \geq 0 \\ y \geq 0 \end{cases}$

24. $\begin{cases} 2x + 3y \geq -120 \\ 6x - 3y \geq -150 \\ x \leq 0 \\ y \leq 0 \end{cases}$

For every function and feasible set that follow, determine the maximum and minimum values of the functions and the values of x and y where they occur.

25. $f(x, y) = 2x + 3y$; vertices at $(3, 1), (5, 2), (1, 4), (8, 1)$

26. $f(x, y) = 4x + y$; vertices at $(2, 1), (1, 9), (3, 5), (2, 7)$

27. $Q = 2x + 3y$ subject to:
$\begin{cases} x \geq 0 \\ y \geq 0 \\ x + 2y \leq 7 \\ 5x - 8y \leq -3 \end{cases}$

28. $F = \frac{3}{2}x + y$ subject to:
$\begin{cases} x \leq 6 \\ y \geq 0 \\ x + 2y \geq 4 \\ x - 2y \geq 2 \end{cases}$

Find the maximum and the minimum values of the given functions in the indicated regions.

29. $f(x, y) = x + 4y$

30. $f(x, y) = 3x + y$

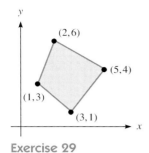

Exercise 29

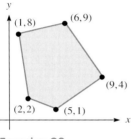

Exercise 30

Applications

31. **Investment decision.** Suppose you invest a certain sum of money at 12.5 percent and another sum at 14 percent. If your total investment was $60,000 and your total yearly interest for one year on both investments was $8,190, how much did you invest at each rate?

32. **Geometry.** The perimeter of a rectangular pool is 68 meters. If the width of the pool is 6 meters less than the length, what are the dimensions of the pool?

33. **Recreation.** Tickets to the Fisher Theater cost $20.00 for the main floor and $16.00 for the balcony. If the receipts from the sale of 1,420 tickets were $26,060, how many tickets were sold at each price?

34. **Brain teaser.** The sum of the digits of a three-digit number is 11, the product of the digits is 40, and if the hundreds digit is subtracted from the ones digit, the result is 1. Find the number.

For each of the following linear programming problems:
   a. Determine a system of inequalities describing the given problem.
   b. Determine the feasible set of the systems of inequalities.
   c. Solve the problem.

35. **Manufacturing.** A cabinet company makes two types of cabinet drawers, one plain and one fancy. Each plain drawer takes 2 hours of work to assemble and 1 hour to sand. Each fancy drawer takes 1 hour of work to assemble and 4 hours to sand. The 4 assembly workers and 6 sanding workers are each used 12 hours per day. Each plain drawer nets a $3 profit, and each fancy drawer nets a $5 profit. If the company can sell all the drawers it makes, how many of each kind should be produced each day in order to maximize profit?

36. **Manufacturing.** Seiberling Tire Company of Akron, Ohio, has 1,000 units of raw rubber to use in producing radial tires for cars and tractors. Each car tire requires 5 units of rubber, and each tractor tire requires 20 units of rubber. Labor costs are $8 for a car tire and $12 for a tractor tire. If Seiberling does not want to pay more than $1,500 in labor costs and wishes to make a profit of $10 per car tire and $25 per tractor tire, how many of each type of tire should be made in order to maximize profits?

Chapter Test

1. Determine all solutions of the following system of equations:
$$\begin{cases} x = 3 - 2y \\ x^2 = 9 - y^2 \end{cases}$$

2. Determine all solutions of the following system of equations:
$$\begin{cases} x + y + z = 0 \\ xy = 6 \\ 2x - y = 0 \end{cases}$$

3. Solve the following linear system:
$$\begin{cases} x - y + 5z = 13 \\ y - 4z = -10 \\ z = 3 \end{cases}$$

4. Solve the following linear system:
$$\begin{cases} 2x - y + 5z = 5 \\ 4x - y + 6z = -1 \\ x - z = 1 \end{cases}$$

5. Solve the following linear system:
$$\begin{cases} 2x - 3y = -12 \\ -10x + 15y = 60 \\ 6x - 9y = -36 \end{cases}$$

6. Graph the inequalities:
   a. $3x - 2y < 4$
   b. $x \geq 9$
   c. $x^2 + y^2 \leq 9$

7. Determine the feasible set of the following system of inequalities:
$$\begin{cases} x + y \leq 5 \\ y \geq 3 + x \\ x \geq 0 \\ y \geq 0 \end{cases}$$

8. Given a feasible set with vertices below, determine the maximum and minimum values of the function $f(x, y)$ and the values of x and y where they occur:
$$f(x, y) = 2x + 7y; \quad (8, 0), (0, 9), (5, 5), (2, 7)$$

9. The perimeter of a rectangular pool is 140 meters. If the width of the pool is 12 meters less than the length, what are the dimensions of the pool?

10. A furniture company makes two types of desks, one plain and one fancy. Each plain desk takes 3 hours of work to assemble and 1 hour to finish. Each fancy desk takes 2 hours of work to assemble and 4 hours to finish. The 4 assembly workers and 6 sanding workers are each used 12 hours per day. Each plain desk nets a $80 profit, and each fancy desk nets a $200 profit. If the company can sell all the desks it makes, how many of each kind should be produced each day in order to maximize profit?

11. **Thought question.** Suppose a system consists of three linear equations in five variables. What can you say about the solutions?

12. **Thought question.** Give an example of a system of inequalities whose feasible set is *not* a polygon. (The Fundamental Theorem of Linear Programming does not necessarily apply to such feasible sets.)

CHAPTER 11

MATRICES AND APPLICATIONS OF LINEAR SYSTEMS

In the preceding chapter we developed the mathematics of systems of linear equations and inequalities, and as part of that development, we introduced matrices. In this chapter, we pursue the subjects of linear systems and matrices in greater detail: we develop the algebra of matrices and learn to calculate determinants and inverses; we use determinants in Cramer's Rule for solving linear systems; we use matrix inverses in input-output analysis, which describes the interrelationships between sectors of an economy; and finally, we use linear systems to discuss partial fractions, which are useful in calculus.

Chapter Opening Image: *Computer art that explores variations of color over time.*

11.1 MATRICES

In Chapter 10 we introduced the notion of a matrix and showed how elementary row operations on a matrix can be used to solve systems of linear equations. Matrices are an important mathematical notion in their own right, and their applications extend throughout pure and applied mathematics. Matrices are used by physicists extensively, for instance in describing mechanics problems and in representing elementary particles. Economists use matrices to describe the interactions of the various sectors of an economy. Matrices are used by computer scientists in describing computer graphics calculations. Mathematicians have developed an extensive algebraic theory of matrices, which parallels (and draws its motivation from) the elementary algebraic notions for real numbers. In this section we provide an introduction to the algebra of matrices.

Recall that a matrix is a rectangular array of real numbers. Matrices are usually denoted by capital letters, such as A, B, C, X, Y, Z. The entries of a matrix are typically denoted using the lowercase version of the matrix name, with subscripts to indicate position. For instance, the entries of the matrix A are denoted a_{ij}, where i denotes the row number and j denotes the column number. We write

$$A = [a_{ij}]$$

to indicate that a typical entry of A is a_{ij}.

Suppose $A = [a_{ij}]$ and $B = [b_{ij}]$ are matrices. We say that the two matrices are equal, denoted $A = B$, if the two matrices are of the same size and their corresponding entries are equal; that is, $a_{ij} = b_{ij}$ for all possible i and j.

A matrix in which all entries are 0 is called a **zero matrix** and is denoted **0**. There is a zero matrix corresponding to each possible matrix size.

Definition
Sum of Two Matrices; Product of a Matrix and a Real Number

Suppose A and B are matrices of the same size. The **sum of A and B**, denoted $A + B$, is the matrix obtained by adding corresponding entries of A and B. Suppose k is a real number. The **product kA** is the matrix for which the entries are the entries of A each multiplied by k.

EXAMPLE 1
Matrix Arithmetic

Let:

$$A = \begin{bmatrix} 5 & 0 \\ 3 & -1 \end{bmatrix}, \quad B = \begin{bmatrix} -2 & 4 \\ 1 & 3 \end{bmatrix}$$

Calculate the following:
1. $A + B$
2. $(-2)A$
3. $3A + 4B$

Solution

1.
$$A + B = \begin{bmatrix} 5 & 0 \\ 3 & -1 \end{bmatrix} + \begin{bmatrix} -2 & 4 \\ 1 & 3 \end{bmatrix}$$
$$= \begin{bmatrix} 5-2 & 0+4 \\ 3+1 & -1+3 \end{bmatrix}$$
$$= \begin{bmatrix} 3 & 4 \\ 4 & 2 \end{bmatrix}$$

2.
$$(-2)A = (-2)\begin{bmatrix} 5 & 0 \\ 3 & -1 \end{bmatrix}$$
$$= \begin{bmatrix} (-2)\cdot 5 & (-2)\cdot 0 \\ (-2)\cdot 3 & (-2)\cdot(-1) \end{bmatrix}$$
$$= \begin{bmatrix} -10 & 0 \\ -6 & 2 \end{bmatrix}$$

3.
$$3A + 4B = 3\begin{bmatrix} 5 & 0 \\ 3 & -1 \end{bmatrix} + 4\begin{bmatrix} -2 & 4 \\ 1 & 3 \end{bmatrix}$$
$$= \begin{bmatrix} 3\cdot 5 + 4\cdot(-2) & 3\cdot 0 + 4\cdot 4 \\ 3\cdot 3 + 4\cdot 1 & 3\cdot(-1) + 4\cdot 3 \end{bmatrix}$$
$$= \begin{bmatrix} 7 & 16 \\ 13 & 9 \end{bmatrix}$$

Addition of matrices obeys both the commutative and associative laws. That is, the following formulas hold for matrices of the same size:

$$A + B = B + A \qquad \textit{Commutative law of addition}$$
$$A + (B + C) = (A + B) + C \qquad \textit{Associative law of addition}$$

Moreover, the zero matrix (of appropriate size) acts as the identity for the operation of addition. That is, we have:

$$A + \mathbf{0} = \mathbf{0} + A = A \qquad \textit{Identity}$$

The **additive inverse of the matrix A,** denoted $-A$, is the matrix for which entries are the negatives of the corresponding entries of A. The additive inverse has the property:

$$A + (-A) = (-A) + A = \mathbf{0}$$

The difference $A - B$ of two matrices is defined via the formula:

$$A - B = A + (-B)$$

All the above formulas should remind you of the corresponding properties of real numbers with respect to the operations of addition and subtraction. In fact, from the above definitions we can deduce analogues of many of the elementary algebraic properties of real numbers. The exercises contain a number of these for you to deduce.

A matrix consisting of a single row is called a **row matrix;** a matrix consisting of a single column is called a **column matrix.** Suppose we are given a row matrix and a column matrix, each having n entries. The product of the row times the column is the 1×1 matrix formed by multiplying corresponding entries of the two matrices and adding. That is, we have:

$$[a_1\ a_2\ \cdots\ a_n]\begin{bmatrix} b_1 \\ b_2 \\ \vdots \\ b_n \end{bmatrix} = [a_1 b_1 + a_2 b_2 + \cdots + a_n b_n]$$

Thus, for example, we have:

$$[2\ 1\ -3]\begin{bmatrix} 5 \\ 2 \\ -4 \end{bmatrix} = [2\cdot 5 + 1\cdot 2 + (-3)\cdot(-4)]$$
$$= [24]$$

We have just considered multiplication of a row matrix by a column matrix. We compute more general matrix products AB by multiplying each row of A by each column of B. Let's see how this is done in the case of 2×2 matrices.

EXAMPLE 2
Product of 2×2 Matrices

Compute the following matrix product.

$$\begin{bmatrix} 5 & -1 \\ 2 & 0 \end{bmatrix} \begin{bmatrix} 3 & 1 \\ 0 & 9 \end{bmatrix}$$

Solution

The number of columns of the left matrix equals the number of rows of the right matrix, so the product is defined. Moreover, the number of rows in the product equals the number of rows (2) in the left matrix; the number of columns in the product equals the number of rows (2) in the right matrix. That is, the product is a 2×2 matrix. Let's now work out the entries in the product. The entry in row 1, column 1 is obtained by multiplying the first row on the left by the first column on the right. The product is:

$$5 \cdot 3 + (-1) \cdot 0 = 15$$

This gives us :

$$\begin{bmatrix} 5 & -1 \\ 2 & 0 \end{bmatrix} \begin{bmatrix} 3 & 1 \\ 0 & 9 \end{bmatrix} = \begin{bmatrix} 15 & * \\ * & * \end{bmatrix}$$

The entry in row 1, column 2 of the product is obtained as the product of the first row on the left times the second column on the right. This product is:

$$5 \cdot 1 + (-1) \cdot 9 = -4$$

That is,

$$\begin{bmatrix} 5 & -1 \\ 2 & 0 \end{bmatrix} \begin{bmatrix} 3 & 1 \\ 0 & 9 \end{bmatrix} = \begin{bmatrix} 15 & -4 \\ * & * \end{bmatrix}$$

Next, we compute the entry in row 2, column 1 as the product of the second row times the first column:

$$\begin{bmatrix} 5 & -1 \\ 2 & 0 \end{bmatrix} \begin{bmatrix} 3 & 1 \\ 0 & 9 \end{bmatrix} = \begin{bmatrix} 15 & -4 \\ 6 & * \end{bmatrix}$$

Finally, we compute the entry in row 2, column 2 as the product of the second row times the second column:

$$\begin{bmatrix} 5 & -1 \\ 2 & 0 \end{bmatrix} \begin{bmatrix} 3 & 1 \\ 0 & 9 \end{bmatrix} = \begin{bmatrix} 15 & -4 \\ 6 & 2 \end{bmatrix}$$

Now that we know how to multiply 2×2 matrices, let's consider multiplication of matrices in general.

Definition
Matrix Multiplication

Suppose A is an $m \times n$ matrix and B is an $n \times k$ matrix. The product AB is the $m \times k$ matrix with entries as follows: The entry in the ith row and the jth column of AB is obtained by multiplying the ith row of A by the jth column of B.

Suppose the ith row of A is equal to
$$[a_{i1}\ a_{i2}\ \cdots\ a_{in}]$$
and the jth row of B is equal to
$$\begin{bmatrix} b_{1j} \\ b_{2j} \\ \vdots \\ b_{nj} \end{bmatrix}$$
as shown below:

$$\begin{bmatrix} a_{11} & a_{12} & \cdots & a_{1n} \\ \vdots & \vdots & \vdots & \vdots \\ a_{i1} & a_{i2} & \cdots & a_{in} \\ \vdots & \vdots & \vdots & \vdots \\ a_{m1} & a_{m2} & \cdots & a_{mn} \end{bmatrix} \begin{bmatrix} b_{11} & \cdots & b_{1j} & \cdots & b_{1k} \\ b_{21} & \cdots & b_{2j} & \cdots & b_{2k} \\ \vdots & \vdots & \vdots & \vdots & \vdots \\ b_{n1} & \cdots & b_{nj} & \cdots & b_{nk} \end{bmatrix}$$

Then the entry c_{ij} in the ith row and jth column of the product is given by:
$$c_{ij} = a_{i1}b_{1j} + a_{i2}b_{2j} + \cdots + a_{in}b_{nj}$$

Note that the definition of matrix multiplication is based on the fact that we can form the products of the rows of the first matrix times the columns of the second matrix. In order for this to be possible, the number of columns of the first matrix must equal the number of rows of the second. If this is not the case, the product is not defined.

EXAMPLE 3
Which Products Are Defined?

Determine which of the following products are defined and the size of those products.

1. A is 2×5; B is 5×7.
2. A is 3×4; B is 2×4.
3. A is 4×4; B is 4×3.

Solution

1. The product AB is defined because the *inner dimensions* of A and B match. That is, the number of columns of A (5) equals the number of rows of B (5). The size of the product is obtained from the *outer dimensions* and is 2×7.
2. The product is undefined because the number of columns of A (4) is unequal to the number of rows of B (2).
3. The product AB is defined and is of size 4×3.

EXAMPLE 4
Multiplying 3×3 Matrices

Determine the matrix product AB, where:
$$A = \begin{bmatrix} 1 & 0 & -2 \\ 3 & 1 & 0 \\ 4 & 0 & -1 \end{bmatrix}, \quad B = \begin{bmatrix} 0 & 6 & 0 \\ -5 & 1 & 1 \\ 0 & 3 & -2 \end{bmatrix}$$

Solution

The product is formed by multiplying each row of A by each column of B. Here is the calculation:

$$\begin{bmatrix} 1 & 0 & -2 \\ 3 & 1 & 0 \\ 4 & 0 & -1 \end{bmatrix} \begin{bmatrix} 0 & 6 & 0 \\ -5 & 1 & 1 \\ 0 & 3 & -2 \end{bmatrix}$$

$$= \begin{bmatrix} 1 \cdot 0 + 0 \cdot (-5) + (-2) \cdot 0 & 1 \cdot 6 + 0 \cdot 1 + (-2) \cdot 3 & 1 \cdot 0 + 0 \cdot 1 + (-2) \cdot (-2) \\ 3 \cdot 0 + 1 \cdot (-5) + 0 \cdot 0 & 3 \cdot 6 + 1 \cdot 1 + 0 \cdot 3 & 3 \cdot 0 + 1 \cdot 1 + 0 \cdot (-2) \\ 4 \cdot 0 + 0 \cdot (-5) + (-1) \cdot 0 & 4 \cdot 6 + 0 \cdot 1 + (-1) \cdot 3 & 4 \cdot 0 + 0 \cdot 1 + (-1) \cdot (-2) \end{bmatrix}$$

$$= \begin{bmatrix} 0 & 0 & 4 \\ -5 & 19 & 1 \\ 0 & 21 & 2 \end{bmatrix}$$

Matrix multiplication obeys the distributive and associative laws. That is, the following formulas hold:

$$A(BC) = (AB)C$$
$$A(B + C) = AB + AC$$

We omit the proofs of these facts.

Unlike multiplication of real numbers, matrix multiplication does not necessarily obey the commutative law, as the following example illustrates.

EXAMPLE 5 Matrix Multiplication Is Noncommutative

Suppose

$$A = \begin{bmatrix} 5 & 1 \\ 2 & 3 \end{bmatrix}, \quad B = \begin{bmatrix} 1 & 4 \\ 3 & 0 \end{bmatrix}$$

Show that $AB \neq BA$.

Solution

We determine the values of AB and BA by direct calculation:

$$AB = \begin{bmatrix} 5 & 1 \\ 2 & 3 \end{bmatrix} \begin{bmatrix} 1 & 4 \\ 3 & 0 \end{bmatrix} = \begin{bmatrix} 8 & 20 \\ 11 & 8 \end{bmatrix}$$

$$BA = \begin{bmatrix} 1 & 4 \\ 3 & 0 \end{bmatrix} \begin{bmatrix} 5 & 1 \\ 2 & 3 \end{bmatrix} = \begin{bmatrix} 13 & 13 \\ 15 & 3 \end{bmatrix}$$

Two matrices are equal only when their corresponding entries are the same. By examining these entries, we see that they are certainly not the same, so $AB \neq BA$.

For each positive integer n, let I_n denote the $n \times n$ matrix with 1's down the main diagonal and 0's everywhere else. For instance,

$$I_1 = [1]$$

$$I_2 = \begin{bmatrix} 1 & 0 \\ 0 & 1 \end{bmatrix}$$

$$I_3 = \begin{bmatrix} 1 & 0 & 0 \\ 0 & 1 & 0 \\ 0 & 0 & 1 \end{bmatrix}$$

Then I_n is called the **identity matrix of size** n. Indeed, I_n acts as the identity for multiplication with square matrices of size $n \times n$. That is, if A is an $n \times n$ matrix, one can prove that:
$$I_n A = A I_n = A \tag{1}$$
In the case $n = 2$, this statement follows from the equations:
$$I_2 \cdot \begin{bmatrix} a & b \\ c & d \end{bmatrix} = \begin{bmatrix} 1 & 0 \\ 0 & 1 \end{bmatrix} \begin{bmatrix} a & b \\ c & d \end{bmatrix}$$
$$= \begin{bmatrix} 1 \cdot a + 0 \cdot c & 1 \cdot b + 0 \cdot d \\ 0 \cdot a + 1 \cdot c & 0 \cdot b + 1 \cdot d \end{bmatrix} = \begin{bmatrix} a & b \\ c & d \end{bmatrix}$$
Similarly,
$$\begin{bmatrix} a & b \\ c & d \end{bmatrix} \cdot I_2 = \begin{bmatrix} a & b \\ c & d \end{bmatrix}$$
The proof of equation (1) for general n is omitted here.

Matrix Addition and Multiplication Using a Calculator

In the preceding chapter, we learned to store matrices in a TI-82 calculator and to perform elementary row operations using the calculator. The following example shows how to use the calculator to perform matrix addition and multiplication using these stored matrices.

EXAMPLE 6
Matrix Addition and Multiplication Using a Calculator

Let
$$A = \begin{bmatrix} 5 & -1 & 4 \\ 3 & 0 & 9 \\ 2 & -5 & 8 \end{bmatrix}, \qquad B = \begin{bmatrix} 0.1 & 0.15 & 0.3 \\ 1.1 & 2.7 & 4.0 \\ 0.3 & 1.0 & 0.9 \end{bmatrix}$$

Use a calculator to determine:

1. $A + B$ 2. AB 3. BA

Solution

First enter the matrices into the calculator using the procedure described in Section 10.3.

1. Enter the expression for the sum $A + B$ as [MATRX] [A] [+] [MATRX] [B] [ENTER]. You will see the sum as shown in Figure 1.*
2. Enter AB using the keystrokes [MATRX] [A] [×] [MATRX] [B] [ENTER]. You will see the product as shown in Figure 2.*
3. See Figure 3.

```
[A]+[B]
[[5.1  -.85  4.3]
 [4.1  2.7   13 ]
 [2.3  -4    8.9]]
```

Figure 1

```
[A]*[B]
[[.6   2.05  1.1  ]
 [3    9.45  9    ]
 [-2.9 -5.2  -12.2]]
```

Figure 2

```
[B]*[A]
[[1.55  -1.6   4.15]
 [21.6  -21.1  60.7]
 [6.3   -4.8   17.4]]
```

Figure 3

*Note that in the actual calculator display you need to use the right arrow key to see the rightmost edge of the matrix

Exercises 11.1

Given

$$A = \begin{bmatrix} 2 & 0 \\ 1 & 3 \end{bmatrix}, \quad B = \begin{bmatrix} -1 & 3 \\ 0 & -2 \end{bmatrix}, \quad C = \begin{bmatrix} 4 & 2 \\ -1 & 0 \end{bmatrix},$$

$$D = \begin{bmatrix} 1 & 0 \\ 0 & 1 \end{bmatrix}, \quad E = \begin{bmatrix} 2 & -1 \\ 0 & 0 \end{bmatrix}, \quad F = \begin{bmatrix} 1 & 3 \\ 2 & 6 \end{bmatrix}$$

calculate the following.

1. $A + B$
2. $A + E$
3. $2F$
4. $3C$
5. $B - C$
6. $C - F$
7. $2A + 3D$
8. $4F - 2B$
9. $B + D$
10. $E + D$
11. $F - 2A$
12. $E + 2B$
13. $-5B$
14. $-3E$
15. AB
16. EF
17. CD
18. BC
19. $(AB)C$
20. $A(BF)$

Given

$$A = \begin{bmatrix} 1 & 2 \\ 3 & 4 \end{bmatrix}, \quad B = \begin{bmatrix} -1 \\ 7 \end{bmatrix}, \quad C = \begin{bmatrix} 3 & -4 & 1 \\ 5 & 0 & 2 \end{bmatrix},$$

$$D = \begin{bmatrix} -2 & -3 & -4 \\ 2 & -1 & 0 \\ 4 & -2 & 3 \end{bmatrix}, \quad E = \begin{bmatrix} -1 & 2 \\ 3 & -2 \end{bmatrix}, \quad F = \begin{bmatrix} 5 \\ -2 \end{bmatrix},$$

$$G = \begin{bmatrix} -1 & 2 & 3 \\ 0 & 1 & 0 \end{bmatrix}, \quad H = \begin{bmatrix} 0 & 1 & 4 \\ 1 & 2 & -1 \\ 3 & 2 & -2 \end{bmatrix},$$

$$I = \begin{bmatrix} 4 & 0 & -1 \end{bmatrix}$$

find each of the following matrices, whenever possible.

21. AB
22. DH
23. CD
24. EF
25. FG
26. GH
27. CH
28. BI
29. DG
30. EA
31. IH
32. IF
33. FE
34. AF

Determine which of the following products are defined and determine the size of those products.

35. A is 3×2; B is 2×5.
36. A is 4×5; B is 5×2.
37. A is 2×2; B is 1×2.
38. A is 5×7; B is 5×7.
39. A is 1×4; B is 4×3.
40. A is 5×1; B is 1×3.

Find the values of w, x, y, and z in the following matrix equations.

41. $\begin{bmatrix} 3 & 1 \\ 4 & 5 \end{bmatrix} = \begin{bmatrix} x & y \\ z & 5 \end{bmatrix}$

42. $\begin{bmatrix} 3 & 5 & x \\ 2 & y & 3 \end{bmatrix} = \begin{bmatrix} z & 5 & 2 \\ 2 & 7 & w \end{bmatrix}$

43. $\begin{bmatrix} x-7 & 4y & 8z \\ 6w & 2 & 5 \end{bmatrix} + \begin{bmatrix} -9 & 8y & 3 \\ 2 & 5 & 4 \end{bmatrix} = \begin{bmatrix} 2 & 36 & 27 \\ 20 & 7 & 9 \end{bmatrix}$

44. $\begin{bmatrix} x+2 & 3y+1 & 5w \\ 4z & 0 & 18 \end{bmatrix} + \begin{bmatrix} 3x & 2y & 5w \\ 2z & 7 & -6 \end{bmatrix} = \begin{bmatrix} 10 & -14 & 80 \\ 10 & 7 & 12 \end{bmatrix}$

Let

$$A = \begin{bmatrix} a_{11} & a_{12} \\ a_{21} & a_{22} \end{bmatrix}, \quad B = \begin{bmatrix} b_{11} & b_{12} \\ b_{21} & b_{22} \end{bmatrix},$$

$$C = \begin{bmatrix} c_{11} & c_{12} \\ c_{21} & c_{22} \end{bmatrix}, \quad I_2 = \begin{bmatrix} 1 & 0 \\ 0 & 1 \end{bmatrix}$$

Prove the following.

45. $I_2 A = AI_2 = A$
46. $A + B = B + A$
47. $k(A + B) = kA + kB$
48. $(A + B) + C = A + (B + C)$

Technology

Let

$$A = \begin{bmatrix} 0.17 & 0.51 & 0.38 \\ 1.15 & 0.98 & 1.22 \\ 3.75 & 0.87 & 0.96 \end{bmatrix}, \quad B = \begin{bmatrix} 5.7 & 3.8 & 4.9 \\ -1.8 & 4 & 9.8 \\ -4.7 & -3.1 & 4.5 \end{bmatrix}$$

Use your calculator to perform the following matrix operations.

49. $A + B$
50. $A - B$
51. AB
52. BA
53. A^2 (that is, $A \cdot A$)
54. B^2

In your own words

55. Explain how to add matrices.

56. Explain how to multiply a row matrix by a column matrix.

57. Explain how to multiply matrices in general.

58. Explain how to determine whether a matrix product is defined.

11.2 DETERMINANTS

Let A be a square matrix. Associated with A is a real number called its **determinant**. The determinant plays a significant role in matrix theory, as we see in the next section, where we discuss Cramer's rule for calculating the solution of a linear system.

The determinant is defined only for square matrices. Accordingly, throughout this section all matrices are assumed to be square.

The determinant of a square matrix A is denoted $|A|$ and is defined in terms of determinants of matrices of smaller sizes. So we begin by defining the determinants of 1×1 and 2×2 matrices.

The determinant of a 1×1 matrix is defined as follows: Suppose $A = [a]$ is a 1×1 matrix. Then $|A|$ is defined as a.

Definition
Determinant of a 2 × 2 Matrix

Suppose A is the 2×2 matrix

$$A = \begin{bmatrix} a & b \\ c & d \end{bmatrix}$$

The determinant $|A|$ is defined by the formula:

$$|A| = ad - cb$$

That is, $|A|$ is the product of the elements along the diagonal from top left to bottom right minus the product of the elements along the diagonal from bottom left to top right.

EXAMPLE 1
Determinant of a 2 × 2 Matrix

Calculate $|A|$, where A is the matrix:

$$A = \begin{bmatrix} 2 & -3 \\ 4 & 1 \end{bmatrix}$$

Solution

From the definition given above, the value of the determinant equals:

$$|A| = 2 \cdot 1 - 4 \cdot (-3)$$
$$= 14$$

Definition
Minor of a 3 × 3 Matrix

By eliminating a row and a column of the matrix A, we can arrive at a 2×2 matrix. The determinant of this matrix is called the **minor** of A associated with the eliminated row and column.

For instance, suppose:

$$A = \begin{bmatrix} -1 & 0 & 3 \\ 2 & 1 & 0 \\ 3 & 4 & -1 \end{bmatrix}$$

The minor obtained by eliminating the second row and third column is denoted M_{23} and is given by:

$$M_{23} = \begin{vmatrix} -1 & 0 \\ 3 & 4 \end{vmatrix} = (-1) \cdot 4 - 3 \cdot 0 = -4$$

The subscript in the notation M_{23} indicates that the minor was formed from A by eliminating the second row and the third column. Note that the row comes first, then the column. More generally, we denote by M_{ij} the minor obtained by eliminating the ith row and jth column.

Definition
Cofactor of a Minor

Suppose A is a 3×3 matrix. The **cofactor** A_{ij} corresponding to the minor M_{ij} is defined through the formula:

$$A_{ij} = (-1)^{i+j} M_{ij}$$

For instance, let A be the matrix defined above. The cofactor A_{23} is equal to

$$A_{23} = (-1)^{2+3} M_{23}$$
$$= -(-4)$$
$$= 4$$

The cofactor of any entry of A is the minor of that entry multiplied by $+1$ or -1 according to the scheme:

$$\begin{bmatrix} + & - & + \\ - & + & - \\ + & - & + \end{bmatrix}$$

The determinant of a 3×3 matrix can be defined in terms of cofactors as follows:

Definition
Determinant of a 3 × 3 Matrix

Let A be a 3×3 matrix. Its determinant $|A|$ is defined by the formula:

$$|A| = \begin{vmatrix} a_{11} & a_{12} & a_{13} \\ a_{21} & a_{22} & a_{23} \\ a_{31} & a_{32} & a_{33} \end{vmatrix} = a_{11}A_{11} - a_{12}A_{12} + a_{13}A_{13}$$

That is, to calculate the determinant of a 3×3 matrix, form the sum of every element of the first row times its corresponding cofactor, that is, the cofactor formed by eliminating the row and column containing the element.

We can derive an alternative formula for the determinant of a 3×3 matrix by using the definitions of the cofactors. We have:

$$A_{11} = (-1)^{1+1} M_{11} = \begin{vmatrix} a_{22} & a_{23} \\ a_{32} & a_{33} \end{vmatrix}$$
$$= a_{22}a_{33} - a_{32}a_{23}$$

$$A_{12} = (-1)^{1+2} M_{12} = (-1) \begin{vmatrix} a_{21} & a_{23} \\ a_{31} & a_{33} \end{vmatrix}$$

$$= -(a_{21}a_{33} - a_{31}a_{23})$$

$$A_{13} = (-1)^{1+3}M_{13} = \begin{vmatrix} a_{21} & a_{22} \\ a_{31} & a_{32} \end{vmatrix}$$

$$= a_{21}a_{32} - a_{31}a_{22}$$

Therefore, by the definition of a determinant for a 3×3 matrix, this gives us

$$|A| = a_{11}A_{11} + a_{12}A_{12} + a_{13}A_{13}$$
$$= a_{11}(a_{22}a_{33} - a_{32}a_{23}) - a_{12}(a_{21}a_{33} - a_{31}a_{23})$$
$$+ a_{13}(a_{21}a_{32} - a_{31}a_{22})$$
$$= a_{11}a_{22}a_{33} - a_{11}a_{32}a_{23} - a_{12}a_{21}a_{33} + a_{12}a_{31}a_{23}$$
$$+ a_{13}a_{21}a_{32} - a_{13}a_{31}a_{22}$$

When calculating determinants, it is usually most convenient to go back to the definition rather than use this last formula.

EXAMPLE 2
Determinant of a 3×3 Matrix

Calculate the determinant of the matrix:

$$A = \begin{bmatrix} 4 & -1 & 5 \\ 0 & -1 & 1 \\ 1 & -1 & 1 \end{bmatrix}$$

Solution

We form the sum of the products of the elements of the first row times their corresponding cofactors:

$$|A| = 4 \cdot (-1)^{1+1} \begin{vmatrix} -1 & 1 \\ -1 & 1 \end{vmatrix} + (-1) \cdot (-1)^{1+2} \begin{vmatrix} 0 & 1 \\ 1 & 1 \end{vmatrix}$$
$$+ 5 \cdot (-1)^{1+3} \begin{vmatrix} 0 & -1 \\ 1 & -1 \end{vmatrix}$$
$$= 4 \cdot ((-1) \cdot 1 - (-1) \cdot 1) + 1 \cdot (0 \cdot 1 - 1 \cdot 1)$$
$$+ 5 \cdot (0 \cdot (-1) - 1 \cdot (-1))$$
$$= 4 \cdot 0 + 1 \cdot (-1) + 5 \cdot 1$$
$$= 4$$

The above definition of a determinant involves products of elements of the first row times the corresponding cofactors. The determinant can also be computed in analogous fashion using any row or column. We describe this computation as **expanding the determinant about a given row or column.** For example, here is the calculation of the determinant of Example 2 expanding about the second column.

$$|A| = (-1) \cdot (-1)^{1+2} \begin{vmatrix} 0 & 1 \\ 1 & 1 \end{vmatrix} + (-1) \cdot (-1)^{2+2} \begin{vmatrix} 4 & 5 \\ 1 & 1 \end{vmatrix}$$
$$+ (-1) \cdot (-1)^{3+2} \begin{vmatrix} 4 & 5 \\ 0 & 1 \end{vmatrix}$$
$$= 1 \cdot (0 \cdot 1 - 1 \cdot 1) - 1 \cdot (4 \cdot 1 - 1 \cdot 5) + 1 \cdot (4 \cdot 1 - 0 \cdot 5)$$
$$= -1 + 1 + 4$$
$$= 4$$

This agrees with the result derived in Example 2. You can check that we arrive at the same result in this case no matter what row or column we use. This observation is a special case of the following theorem.

CALCULATING A DETERMINANT BY COFACTORS

Let A be a 3×3 matrix. The determinant of A can be computed as the sum of the products of the elements in a particular row or column times the corresponding cofactors.

We omit the proof of this result, which properly belongs to a course in linear algebra.

In our discussion so far, we have defined the determinants of square matrices of size 1, 2, and 3. Let's now define the determinants of square matrices of any size. The general idea is to define the determinant of a matrix of a given size in terms of determinants of matrices of smaller size. The definition is exactly the one given above.

Definition
Determinant of an $n \times n$ Matrix

Let A be an $n \times n$ matrix. The determinant $|A|$ is defined by the formula:
$$|A| = a_{11}A_{11} + a_{12}A_{12} + \cdots + a_{1n}A_{1n}$$
where A_{ij} denotes the cofactor determined by eliminating the ith row and jth column.

Note that the size of the determinants involved in the cofactors is $n - 1$. This means, for example, that the above formula defines the determinant of a 4×4 matrix in terms of determinants of 3×3 matrices, the determinant of a 5×5 matrix in terms of determinants of 4×4 matrices, and so forth.

As with 3×3 matrices, the definition given involves expansion about the first row. However, as with 3×3 matrices the determinant can be calculated by expanding about any row or column.

The next example illustrates how this definition can be applied to calculate the determinant of a 4×4 matrix.

EXAMPLE 3
Determinant of a 4×4 Matrix

Calculate the determinant of the matrix:
$$A = \begin{bmatrix} 2 & 0 & 1 & 0 \\ 0 & 4 & 2 & 0 \\ 0 & 1 & 0 & 1 \\ 0 & -1 & 1 & 0 \end{bmatrix}$$

Solution

Note that the first column has a single nonzero element. So the computation is simplified if we expand the determinant about the first column, because in this case all but the first products are 0. We have:
$$|A| = 2 \cdot (-1)^{1+1} \begin{vmatrix} 4 & 2 & 0 \\ 1 & 0 & 1 \\ -1 & 1 & 0 \end{vmatrix}$$

The simplest calculation for expanding the matrix on the right is about the third column, because two of the resulting products are 0. The result is:

$$|A| = 2 \cdot (-1)^{1+1} \cdot 1 \cdot (-1)^{2+3} \begin{vmatrix} 4 & 2 \\ -1 & 1 \end{vmatrix}$$
$$= -2 \cdot (4 \cdot 1 - (-1) \cdot 2) = -2(4+2) = -2(6)$$
$$= -12$$

Cramer's Rule

Cramer's rule gives a general formula for the solution of a system of linear equations in terms of determinants formed from the coefficients of the system. To motivate the general statement of Cramer's rule, let's consider the specialized case of a linear system in two variables:

$$\begin{cases} ax + by = S \\ cx + dy = T \end{cases}$$

Use the elimination method. Multiply the first equation by c, multiply the second equation by a, and subtract:

$$cax + cby = cS$$
$$cax + ady = aT$$
$$\overline{cby - ady = cS - aT}$$
$$(cb - ad)y = cS - aT$$
$$y = \frac{aT - cS}{ad - bc}, \qquad ad - bc \neq 0$$

Substituting this value for y into the first equation of the system, we have:

$$ax + by = S$$
$$ax + b\left(\frac{aT - cS}{ad - bc}\right) = S$$
$$ax = S - \frac{baT - bcS}{ad - bc}$$
$$= \frac{adS - baT}{ad - bc}$$
$$x = \frac{Sd - Tb}{ad - bc}, \qquad ad - bc \neq 0$$

The numerators and denominators of x and y can each be expressed as a determinant, namely:

$$x = \frac{\begin{vmatrix} S & b \\ T & d \end{vmatrix}}{\begin{vmatrix} a & b \\ c & d \end{vmatrix}}, \qquad y = \frac{\begin{vmatrix} a & S \\ c & T \end{vmatrix}}{\begin{vmatrix} a & b \\ c & d \end{vmatrix}}$$

We can describe these formulas as follows:

1. The denominator in each is the determinant of the matrix of coefficients A of the system. We have assumed that this determinant is nonzero, and we have shown that with this assumption the system has a unique solution.

2. In the formula for x, the column of coefficients a and c of x is replaced by the column of numbers S and T on the left side of the system. The matrix appearing in the numerator is denoted A_1, so that:

$$x = \frac{|A_1|}{|A|}$$

3. In the formula for y, the column of coefficients b and d of y is replaced by the column of numbers S and T on the right side of the system. The matrix appearing in the numerator is denoted A_2, so that:

$$y = \frac{|A_2|}{|A|}$$

The formulas

$$x = \frac{|A_1|}{|A|}, \qquad y = \frac{|A_2|}{|A|}$$

are a special case of **Cramer's rule.**

EXAMPLE 4
Application of Cramer's Rule

Use Cramer's rule to solve the system:

$$\begin{cases} 5x - 2y = 8 \\ 3x + 4y = -1 \end{cases}$$

Solution

The determinant of the coefficients is:

$$|A| = \begin{vmatrix} 5 & -2 \\ 3 & 4 \end{vmatrix} = 5 \cdot 4 - 3 \cdot (-2) = 26$$

This value is nonzero, so the system has a unique solution. Applying Cramer's rule, we have:

$$x = \frac{|A_1|}{|A|} = \frac{\begin{vmatrix} 8 & -2 \\ -1 & 4 \end{vmatrix}}{26} = \frac{30}{26} = \frac{15}{13}$$

$$y = \frac{|A_2|}{|A|} = \frac{\begin{vmatrix} 5 & 8 \\ 3 & -1 \end{vmatrix}}{26} = \frac{-29}{26} = -\frac{29}{26}$$

Therefore the system has the solution $(x, y) = (\frac{15}{13}, -\frac{29}{26})$.

Cramer's rule can be generalized to systems of linear equations in n variables. Suppose we are given a system with coefficient matrix A. Let A_n denote the matrix obtained by replacing the nth column of A with the column of numbers on the right side of the system. Then we have the following result.

CRAMER'S RULE FOR $n \times n$ LINEAR SYSTEMS

If a linear system of n equations in n variables $x, y, z, \ldots$ has a nonzero determinant $|A|$, the system has a unique solution given by the formulas:

$$x = \frac{|A_1|}{|A|}, \qquad y = \frac{|A_2|}{|A|}, \qquad z = \frac{|A_3|}{|A|}, \ldots$$

Note that Cramer's rule only applies to systems in which the number of equations equals the number of variables. Contrast this with Gauss elimination, which can be used to solve any linear system. The proof of Cramer's rule is beyond the scope of this book and will be omitted.

EXAMPLE 5
Application of Cramer's Rule

Use Cramer's rule to solve the system:
$$\begin{cases} x + + 2z = 1 \\ y - 3z = 0 \\ 2x - y = 4 \end{cases}$$

Solution

In this case the matrix of coefficients is:
$$A = \begin{bmatrix} 1 & 0 & 2 \\ 0 & 1 & -3 \\ 2 & -1 & 0 \end{bmatrix}$$

Moreover, we have:
$$A_1 = \begin{bmatrix} 1 & 0 & 2 \\ 0 & 1 & -3 \\ 4 & -1 & 0 \end{bmatrix}$$

$$A_2 = \begin{bmatrix} 1 & 1 & 2 \\ 0 & 0 & -3 \\ 2 & 4 & 0 \end{bmatrix}$$

$$A_3 = \begin{bmatrix} 1 & 0 & 1 \\ 0 & 1 & 0 \\ 2 & -1 & 4 \end{bmatrix}$$

By simple computations we compute the determinants of these matrices. Then we apply Cramer's rule.

$$|A| = -7, \quad |A_1| = -11, \quad |A_2| = 6, \quad |A_3| = 2$$

$$x = \frac{|A_1|}{|A|} = \frac{-11}{-7} = \frac{11}{7}$$

$$y = \frac{|A_2|}{|A|} = \frac{6}{-7} = -\frac{6}{7}$$

$$z = \frac{|A_3|}{|A|} = \frac{2}{-7} = -\frac{2}{7}$$

Equations of Lines and Areas of Triangles

Determinants can be used to write many of the formulas in analytic geometry and physics in a simple and elegant form. For example, consider the equation of a line passing through the points (x_1, y_1) and (x_2, y_2). We have seen that this equation can be written in the form:

$$y - y_1 = \frac{y_2 - y_1}{x_2 - x_1}(x - x_1)$$

Clear the fraction and write the equation in the form:
$$(y - y_1)(x_2 - x_1) - (y_2 - y_1)(x - x_1) = 0$$
$$x_2y - x_1y - x_2y_1 + x_1y_1 - xy_2 + x_1y_2 + xy_1 - x_1y_1 = 0$$
$$(xy_1 - x_1y) - (xy_2 - x_2y) + (x_1y_2 - x_2y_1) = 0$$

Each of the differences is a determinant:
$$\begin{vmatrix} x & y \\ x_1 & y_1 \end{vmatrix} - \begin{vmatrix} x & y \\ x_2 & y_2 \end{vmatrix} + \begin{vmatrix} x_1 & y_1 \\ x_2 & y_2 \end{vmatrix} = 0$$

It is now easy to recognize this expression as the expansion of the 3×3 determinant
$$\begin{vmatrix} x & y & 1 \\ x_1 & y_1 & 1 \\ x_2 & y_2 & 1 \end{vmatrix}$$
in minors about the third column. Thus, we have proven the following result.

DETERMINANT FORM OF THE EQUATION OF A LINE

The equation of a line passing through the points (x_1, y_1) and (x_2, y_2) is given in determinant form:
$$\begin{vmatrix} x & y & 1 \\ x_1 & y_1 & 1 \\ x_2 & y_2 & 1 \end{vmatrix} = 0$$

Suppose we are given three points (x_1, y_1), (x_2, y_2), and (x_3, y_3). They lie on a common line provided that the third point lies on the line determined by the first two. Therefore, by the preceding result, we have the following:

DETERMINANT CONDITION FOR COLLINEARITY

Three points (x_1, y_1), (x_2, y_2), and (x_3, y_3) are collinear provided that:
$$\begin{vmatrix} x_1 & y_1 & 1 \\ x_2 & y_2 & 1 \\ x_3 & y_3 & 1 \end{vmatrix} = 0$$

One can calculate the area of a triangle in terms of a determinant according to the following result:

DETERMINANT FORMULA FOR THE AREA OF A TRIANGLE

The area of a triangle with vertices (x_1, y_1), (x_2, y_2), and (x_3, y_3) is given by the formula:
$$\text{Area} = \pm\frac{1}{2} \begin{vmatrix} x_1 & y_1 & 1 \\ x_2 & y_2 & 1 \\ x_3 & y_3 & 1 \end{vmatrix}$$

EXAMPLE 6
Area of a Triangle

Verify the last result in the case that the triangle is a right triangle with the right angle at the origin.

Solution

In this case we can set $(x_1, y_1) = (0, 0)$, $(x_2, y_2) = (x_2, 0)$, and $(x_3, y_3) = (0, y_3)$. The determinant formula then gives:

$$\text{Area} = \pm \frac{1}{2} \begin{vmatrix} 0 & 0 & 1 \\ x_2 & 0 & 1 \\ 0 & y_3 & 1 \end{vmatrix}$$

$$= \pm \frac{1}{2} \left[0 \cdot \begin{vmatrix} 0 & 1 \\ y_3 & 1 \end{vmatrix} - 0 \cdot \begin{vmatrix} x_2 & 1 \\ 0 & 1 \end{vmatrix} + 1 \cdot \begin{vmatrix} x_2 & 0 \\ 0 & x_3 \end{vmatrix} \right]$$

$$= \pm \frac{1}{2} x_2 y_3$$

$$= \pm \frac{1}{2} \cdot \text{base} \cdot \text{altitude}$$

Graphing Calculators and Determinants

Many graphing calculators have built-in determinant functions. By using such a function you can avoid the considerable calculation involved in evaluating determinants, especially large ones.

EXAMPLE 7
Using a Calculator to Evaluate Determinants

Let:

$$B = \begin{bmatrix} 0.1 & 0.15 & 0.3 \\ 1.1 & 2.7 & 4.0 \\ 0.3 & 1.0 & 0.9 \end{bmatrix}$$

Use a calculator to determine $|B|$.

Solution

First enter the matrix B. (If you haven't changed matrix $[B]$ from Example 6 of Section 11.1, it is still stored in your calculator.) Now press [MATRX] and select the option MATH from the menu displayed. Choose the option det from the menu shown. (See Figure 1.) Indicate the matrix B by pressing [MATRX] and selecting [B] from the menu displayed. The display is as shown in Figure 2. Now press [ENTER]. The determinant, -0.0385, is then displayed.

Figure 1

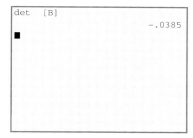

Figure 2

Exercises 11.2

Calculate $|A|$, where A is the matrix given. Do not use a calculator.

1. $\begin{bmatrix} 1 & 2 \\ 3 & -1 \end{bmatrix}$

2. $\begin{bmatrix} 4 & -2 \\ 1 & 3 \end{bmatrix}$

3. $\begin{bmatrix} -5 & 0 \\ -2 & 1 \end{bmatrix}$

4. $\begin{bmatrix} 0 & -2 \\ 1 & 3 \end{bmatrix}$

5. $\begin{bmatrix} 3 & -1 \\ -2 & -4 \end{bmatrix}$

6. $\begin{bmatrix} -1 & -3 \\ -2 & -5 \end{bmatrix}$

Let

$$A = \begin{bmatrix} 4 & -1 & -6 \\ 3 & 0 & 7 \\ 1 & 2 & -1 \end{bmatrix} \text{ and }$$

$$B = \begin{bmatrix} 2 & -1 & 0 & 3 \\ -1 & 4 & 1 & 3 \\ 0 & 1 & 0 & 2 \\ -1 & 3 & -2 & 1 \end{bmatrix}$$

Calculate the following. You may use a calculator.

7. Find A_{21}.
8. Find A_{12}.
9. Find B_{43}.
10. Find B_{31}.
11. Find A_{11}.
12. Find A_{23}.
13. Find B_{22}.
14. Find B_{44}.

Calculate $|A|$, where A is the given matrix. You may use a calculator.

15. $\begin{bmatrix} 1 & 2 & -1 \\ 2 & -1 & 1 \\ 4 & 0 & 2 \end{bmatrix}$

16. $\begin{bmatrix} 1 & 2 & -3 \\ 4 & 5 & -9 \\ 0 & 0 & 1 \end{bmatrix}$

17. $\begin{bmatrix} 3 & 3 & -1 \\ 2 & 6 & 0 \\ -6 & -6 & 2 \end{bmatrix}$

18. $\begin{bmatrix} 1 & 2 & -1 \\ 3 & 2 & 1 \\ 1 & 0 & -2 \end{bmatrix}$

19. $\begin{bmatrix} -2 & 0 & 4 & 2 \\ 3 & 6 & 0 & 4 \\ 0 & 0 & 0 & 3 \\ 9 & 0 & 2 & -1 \end{bmatrix}$

20. $\begin{bmatrix} 7 & -8 & 1 & 2 \\ 21 & 4 & 3 & -1 \\ -35 & 8 & 3 & -2 \\ 14 & 16 & 0 & 1 \end{bmatrix}$

21. $\begin{bmatrix} 4 & 2 & 1 & 0 \\ -2 & 4 & -1 & 7 \\ -5 & 2 & 3 & 1 \\ 6 & 4 & -3 & 2 \end{bmatrix}$

22. $\begin{bmatrix} 1 & -1 & 0 & 2 \\ 0 & 1 & -1 & 0 \\ 2 & 1 & 0 & -1 \\ -2 & 2 & 1 & 1 \end{bmatrix}$

23. $\begin{bmatrix} 11 & -15 & 20 \\ 16 & 24 & -8 \\ 6 & 9 & 15 \end{bmatrix}$

24. $\begin{bmatrix} 2 & 1 & 1 \\ 2 & -3 & -1 \\ -4 & 5 & 2 \end{bmatrix}$

25. $\begin{bmatrix} -3 & 0 & 2 & 6 \\ 2 & 4 & 0 & -1 \\ -1 & 0 & -5 & 2 \\ 0 & -1 & -2 & -3 \end{bmatrix}$

26. $\begin{bmatrix} 4 & 0 & 0 & 2 \\ -1 & 0 & 3 & 0 \\ 2 & 4 & 0 & 1 \\ 0 & 0 & 1 & 2 \end{bmatrix}$

Use Cramer's rule to solve the following linear systems.

27. $\begin{cases} 4x - y = 10 \\ 3x + 5y = 19 \end{cases}$

28. $\begin{cases} 2x + 5y = 18 \\ 3x + 4y = 7 \end{cases}$

29. $\begin{cases} 2x + 3y = -9 \\ 3x + 5y = -13 \end{cases}$

30. $\begin{cases} 4x - y = 3 \\ 2x - 3y = -1 \end{cases}$

31. $\begin{cases} x + 2y = 5 \\ 2x - 5y = -8 \end{cases}$

32. $\begin{cases} 2x + 3y + z = 1 \\ x - 3y + 2z = 3 \end{cases}$

33. $\begin{cases} 3x - 5y = 6 \\ -6x + 10y = 7 \end{cases}$

34. $\begin{cases} 3x - 6y = 12 \\ -2x + 4y = -8 \end{cases}$

35. $\begin{cases} x + 2y + 3z = 0 \\ x - 2z = 3 \\ y + z = 1 \end{cases}$

36. $\begin{cases} x + y + 4z = 1 \\ -2x - y + z = 2 \\ 3x - 2y + 3z = 5 \end{cases}$

37. $\begin{cases} 3x + 2y - z = 4 \\ 2x - y + 3z = 5 \\ x + 3y + 2z = -1 \end{cases}$

38. $\begin{cases} 4x - 2y + 3z = 4 \\ 5x - y + 4z = 7 \\ 3x + 5y + z = 7 \end{cases}$

Solve the following equations.

39. $\begin{vmatrix} -20 & 2 & -3 \\ 6 & -4 & 1 \\ -1 & -1 & 2 \end{vmatrix} = 12x$

40. $\begin{vmatrix} 2 & -20 & -3 \\ 1 & 6 & 1 \\ 4 & -1 & 2 \end{vmatrix} = 2x + 1$

41. $\begin{vmatrix} a - 4 & 0 & 0 \\ 0 & a + 4 & 0 \\ 0 & 0 & a + 1 \end{vmatrix} = 0$

42. $\begin{vmatrix} 1 & m & m^2 \\ 1 & 1 & 1 \\ 4 & 5 & 0 \end{vmatrix} = 0$

Calculate $|A|$, where A is the given matrix and x, y, and z are real numbers.

43. $\begin{bmatrix} x & y & z \\ 0 & -4 & 2 \\ -1 & 3 & 1 \end{bmatrix}$

44. $\begin{bmatrix} x - 1 & 4x \\ y & 2 & 4y \\ z & -3 & 4z \end{bmatrix}$

Prove the following.

45. $\begin{vmatrix} x_{11} & x_{12} & x_{13} & x_{14} \\ x_{21} & x_{22} & x_{23} & x_{24} \\ 0 & 0 & x_{33} & x_{34} \\ 0 & 0 & x_{43} & x_{44} \end{vmatrix} = \begin{vmatrix} x_{11} & x_{12} \\ x_{21} & x_{22} \end{vmatrix} \cdot \begin{vmatrix} x_{33} & x_{34} \\ x_{43} & x_{44} \end{vmatrix}$

46. $\begin{vmatrix} x_1 & y_1 & z_1 \\ x_2 & y_2 & z_2 \\ x_3 & y_3 & z_3 \end{vmatrix} = -\begin{vmatrix} x_2 & y_2 & z_2 \\ x_1 & y_1 & z_1 \\ x_3 & y_3 & z_3 \end{vmatrix}$

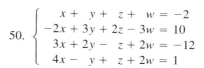

Technology

Use a calculator to compute the determinants of the following matrices.

47. $\begin{bmatrix} -10 & 21 & -3 \\ 6 & -5 & 1 \\ -1 & -1 & -1 \end{bmatrix}$

48. $\begin{bmatrix} -5 & 0 & 2 & 6 \\ 2 & 4 & 0 & -1 \\ -4 & 0 & -6 & 9 \\ 1 & -3 & -2 & -7 \end{bmatrix}$

Use Cramer's rule and a calculator to solve the following linear systems.

49. $\begin{cases} x + 2y + 3z - w = 7 \\ x - 4y + z = 3 \\ 2y - 3z + w = 4 \\ 2x - 5y + 11w = -2 \end{cases}$

50. $\begin{cases} x + y + z + w = -2 \\ -2x + 3y + 2z - 3w = 10 \\ 3x + 2y - z + 2w = -12 \\ 4x - y + z + 2w = 1 \end{cases}$

51. $\begin{cases} x - y - 3z - 2w = 2 \\ 4x + y + z + 2w = 2 \\ x + 3y - 2z - w = 9 \\ 3x + y - z + w = 5 \end{cases}$

52. $\begin{cases} x - y = 5 \\ y - z = 6 \\ z - w = 7 \\ x + w = 8 \end{cases}$

In your own words

53. Describe how to calculate the determinant of a 2×2 matrix.

54. What is the cofactor of an element in a matrix? Use a specific example.

55. Describe how to calculate the determinant of a matrix using cofactors. Use a 3×3 matrix as an example.

11.3 INVERSE OF A SQUARE MATRIX

In this section we introduce the inverse of a matrix and provide a technique for calculating the inverse. As motivation for the inverse of a matrix, consider a real number a. We say that a^{-1} is the multiplicative inverse of a provided that a^{-1} satisfies:

$$aa^{-1} = a^{-1}a = 1$$

Recall that the number 1 is the identity for multiplication.

Definition
Inverse of a Square Matrix

Suppose A is an $n \times n$ matrix. By analogy with real numbers, we say that A^{-1} is an **inverse** of A if A^{-1} satisfies the equations:

$$AA^{-1} = A^{-1}A = I_n$$

Here I_n is the identity matrix of size n, which is the identity for multiplication for square matrices of size $n \times n$.

For instance, suppose:
$$A = \begin{bmatrix} 5 & 7 \\ 2 & 3 \end{bmatrix}$$

An inverse of A is given by:
$$A^{-1} = \begin{bmatrix} 3 & -7 \\ -2 & 5 \end{bmatrix}$$

Indeed, a simple computation shows that
$$\begin{bmatrix} 5 & 7 \\ 2 & 3 \end{bmatrix} \begin{bmatrix} 3 & -7 \\ -2 & 5 \end{bmatrix} = \begin{bmatrix} 1 & 0 \\ 0 & 1 \end{bmatrix}$$
$$AA^{-1} = I_2$$

and similarly that
$$A^{-1}A = I_2$$

This shows that A^{-1} is an inverse of A.

The inverse of a 2×2 matrix can be determined from the following result:

INVERSE OF A 2×2 MATRIX

Let
$$A = \begin{bmatrix} a & b \\ c & d \end{bmatrix}$$
be a 2×2 matrix for which $|A| \neq 0$. Then A has an inverse given by the formula:
$$A^{-1} = \begin{bmatrix} \dfrac{d}{\Delta} & -\dfrac{b}{\Delta} \\ -\dfrac{c}{\Delta} & \dfrac{a}{\Delta} \end{bmatrix}, \quad \Delta = |A| = ad - cb$$

This formula is easy to verify. Let's calculate the product AA^{-1} and show that it equals I_2.

$$AA^{-1} = \begin{bmatrix} a & b \\ c & d \end{bmatrix} \begin{bmatrix} \dfrac{d}{\Delta} & -\dfrac{b}{\Delta} \\ -\dfrac{c}{\Delta} & \dfrac{a}{\Delta} \end{bmatrix}$$

$$= \begin{bmatrix} a \cdot \dfrac{d}{\Delta} + b \cdot -\dfrac{c}{\Delta} & a \cdot -\dfrac{b}{\Delta} + b \cdot \dfrac{a}{\Delta} \\ c \cdot \dfrac{d}{\Delta} + d \cdot -\dfrac{c}{\Delta} & c \cdot -\dfrac{b}{\Delta} + d \cdot \dfrac{a}{\Delta} \end{bmatrix}$$

$$= \begin{bmatrix} \dfrac{ad - bc}{\Delta} & 0 \\ 0 & \dfrac{ad - bc}{\Delta} \end{bmatrix}$$

$$= \begin{bmatrix} 1 & 0 \\ 0 & 1 \end{bmatrix}$$

$$= I_2$$

because $ad - bc = \Delta$. Similarly, $A^{-1}A = I_2$.

It is possible to prove that a matrix has at most one inverse. So instead of referring to "an inverse," we can refer to "the inverse," provided, of course, that the matrix has an inverse.

EXAMPLE 1
Matrix Inverse

Calculate the inverse of the matrix:
$$A = \begin{bmatrix} 5 & -1 \\ 4 & 3 \end{bmatrix}$$

Solution

In this case we have:
$$\Delta = |A| = 5 \cdot 3 - 4 \cdot (-1) = 19$$

Therefore, by the above result, we have
$$A^{-1} = \begin{bmatrix} \frac{3}{19} & -\frac{-1}{19} \\ -\frac{4}{19} & \frac{5}{19} \end{bmatrix} = \begin{bmatrix} \frac{3}{19} & \frac{1}{19} \\ -\frac{4}{19} & \frac{5}{19} \end{bmatrix}$$

The above result provides a formula for the inverse of a 2×2 matrix in case Δ is nonzero. It is possible to show that if $\Delta = 0$, the matrix has no inverse. We prove this in particular cases in the exercises.

There is a straightforward computational procedure for determining the inverse of any square matrix A (provided that the inverse exists). The procedure relies on performing elementary row operations. It is described as follows:

1. Form the matrix
$$\begin{bmatrix} a_{11} & a_{12} & \cdots & a_{1n} & 1 & 0 & \cdots & 0 \\ a_{21} & a_{22} & \cdots & a_{2n} & 0 & 1 & \cdots & 0 \\ \vdots & & & & \vdots & & & \\ a_{n1} & a_{n2} & \cdots & a_{nn} & 0 & 0 & \cdots & 1 \end{bmatrix}$$
where the matrix on the right is the identity matrix I_n.
2. If possible, perform elementary row operations on this matrix to transform the left half into the identity matrix I_n.
3. After the transformation, the matrix in the right half is A^{-1}.

The proof that this method works belongs to a more advanced book and won't be included here. The next example illustrates this technique for calculating the inverse.

EXAMPLE 2
Inverse of a 3×3 Matrix

Determine the inverse of the matrix:
$$\begin{bmatrix} 4 & 8 & 0 \\ 3 & 1 & 4 \\ 0 & 0 & 2 \end{bmatrix}$$

Solution

We first form the matrix:
$$\begin{bmatrix} 4 & 8 & 0 & 1 & 0 & 0 \\ 3 & 1 & 4 & 0 & 1 & 0 \\ 0 & 0 & 2 & 0 & 0 & 1 \end{bmatrix}$$

Now we perform a sequence of elementary row operations on this matrix to transform the left half into the identity matrix I_3:

$$\begin{bmatrix} 4 & 8 & 0 & | & 1 & 0 & 0 \\ 3 & 1 & 4 & | & 0 & 1 & 0 \\ 0 & 0 & 2 & | & 0 & 0 & 1 \end{bmatrix} \xrightarrow{-2E_3+E_2} \begin{bmatrix} 4 & 8 & 0 & | & 1 & 0 & 0 \\ 3 & 1 & 0 & | & 0 & 1 & -2 \\ 0 & 0 & 2 & | & 0 & 0 & 1 \end{bmatrix}$$

$$\xrightarrow{-8E_2+E_1} \begin{bmatrix} -20 & 0 & 0 & | & 1 & -8 & 16 \\ 3 & 1 & 0 & | & 0 & 1 & -2 \\ 0 & 0 & 2 & | & 0 & 0 & 1 \end{bmatrix}$$

$$\xrightarrow{-\frac{1}{20}E_1} \begin{bmatrix} 1 & 0 & 0 & | & -\frac{1}{20} & \frac{8}{20} & -\frac{16}{20} \\ 3 & 1 & 0 & | & 0 & 1 & -2 \\ 0 & 0 & 2 & | & 0 & 0 & 1 \end{bmatrix}$$

$$\xrightarrow{-3E_1+E_2} \begin{bmatrix} 1 & 0 & 0 & | & -\frac{1}{20} & \frac{8}{20} & -\frac{16}{20} \\ 0 & 1 & 0 & | & \frac{3}{20} & -\frac{4}{20} & \frac{8}{20} \\ 0 & 0 & 2 & | & 0 & 0 & 1 \end{bmatrix}$$

$$\xrightarrow{\frac{1}{2}E_3} \begin{bmatrix} 1 & 0 & 0 & | & -\frac{1}{20} & \frac{8}{20} & -\frac{16}{20} \\ 0 & 1 & 0 & | & \frac{3}{20} & -\frac{4}{20} & \frac{8}{20} \\ 0 & 0 & 1 & | & 0 & 0 & \frac{1}{2} \end{bmatrix}$$

The left side of the matrix is now the identity matrix I_3. The inverse can then be read off the right side:

$$A^{-1} = \begin{bmatrix} -\frac{1}{20} & \frac{2}{5} & -\frac{4}{5} \\ \frac{3}{20} & -\frac{1}{5} & \frac{2}{5} \\ 0 & 0 & \frac{1}{2} \end{bmatrix}$$

This result can be checked by verifying that A^{-1} satisfies the equations:

$$AA^{-1} = A^{-1}A = I_3$$

We leave the calculations as an exercise.

Matrix Solutions of Linear Systems

Our original motivation for introducing matrices was to have a convenient way of performing row operations on a system of linear equations. Actually, the connection between matrices and systems of linear equations is much more fundamental than our previous discussion would indicate. Let's now explore this connection in greater depth. Consider the following linear system:

$$\begin{cases} 3x - 8y = 7 \\ -x + 4y = 6 \end{cases}$$

Let A denote the matrix of coefficients, B the matrix of numbers on the right side of the system, and X the matrix of variables, as follows:

$$A = \begin{bmatrix} 3 & -8 \\ -1 & 4 \end{bmatrix}, \quad B = \begin{bmatrix} 7 \\ 6 \end{bmatrix}, \quad X = \begin{bmatrix} x \\ y \end{bmatrix}$$

The matrix product AX is given by
$$AX = \begin{bmatrix} 3x - 8y \\ -x + 4y \end{bmatrix}$$
The system can therefore be written as the single matrix equation:
$$AX = B$$
The advantage of writing the system in this form is that it can be simply solved using inverses. If we multiply both sides by the inverse matrix A^{-1}, we have:
$$A^{-1} \cdot AX = A^{-1} \cdot B$$
But by the property of inverse matrices, we have:
$$A^{-1} \cdot A = I_2$$
where I_2 is the identity matrix of order 2. Therefore, because $I_2 X = X$, the above matrix equation can be written in the form:
$$I_2 X = A^{-1} B$$
$$X = A^{-1} B$$

This is a solution for the matrix of unknowns in terms of B and the inverse A^{-1}. We can determine the solution numerically as follows: We first calculate A^{-1} using the formula determined earlier:
$$\Delta = 3 \cdot 4 - (-1) \cdot (-8) = 4$$
$$A^{-1} = \begin{bmatrix} \frac{4}{4} & -\frac{-8}{4} \\ -\frac{-1}{4} & \frac{3}{4} \end{bmatrix}$$
$$= \begin{bmatrix} 1 & 2 \\ \frac{1}{4} & \frac{3}{4} \end{bmatrix}$$

Thus,
$$X = A^{-1} B$$
$$= \begin{bmatrix} 1 & 2 \\ \frac{1}{4} & \frac{3}{4} \end{bmatrix} \begin{bmatrix} 7 \\ 6 \end{bmatrix}$$
$$= \begin{bmatrix} 1 \cdot 7 + 2 \cdot 6 \\ \frac{1}{4} \cdot 7 + \frac{3}{4} \cdot 6 \end{bmatrix}$$
$$= \begin{bmatrix} 19 \\ \frac{25}{4} \end{bmatrix}$$

Because $X = \begin{bmatrix} x \\ y \end{bmatrix}$, we see that the solution of the system is:
$$x = 19, \qquad y = \frac{25}{4}$$

The above computation can be generalized into a method for solving square systems of linear equations. Suppose we are given a system of n linear equations in n variables. Let A denote the matrix of coefficients, B the matrix of numbers on the right side of the system, and X the matrix of variables. We write the system of linear equations in the form:
$$AX = B$$

Let's assume that A has an inverse. Multiplying both sides of the equation by A^{-1}, we have:

$$AX = B$$
$$A^{-1}(AX) = A^{-1}B$$
$$(A^{-1}A)X = A^{-1}B \quad \text{Associative law}$$
$$I_n X = A^{-1}B \quad \text{Definition of inverse}$$
$$X = A^{-1}B \quad I_n \text{ is the identity for multiplication}$$

Thus we have the following solution of the system in matrix form:

$$X = A^{-1}B$$

The matrix X is just the column matrix of variables. The right side is a matrix whose value can be calculated in terms of the constants of the system. By equating corresponding entries on both sides of the equation, we arrive at a solution of the system. We have now derived a new method for solving a square system of linear equations:

SOLUTION OF A SQUARE LINEAR SYSTEM IN TERMS OF MATRICES

1. Calculate A^{-1}.
2. Form the product $A^{-1}B$.
3. Use the equation $X = A^{-1}B$ to determine the values of the variables.

The next example illustrates how to solve a system using this method.

EXAMPLE 3
Matrix Solution of a Linear System

Solve the following system using inverse matrices.

$$\begin{cases} 4x + 8y = -1 \\ 3x + y + 4z = 2 \\ 2z = 5 \end{cases}$$

Solution

Let

$$A = \begin{bmatrix} 4 & 8 & 0 \\ 3 & 1 & 4 \\ 0 & 0 & 2 \end{bmatrix}, \quad B = \begin{bmatrix} -1 \\ 2 \\ 5 \end{bmatrix}, \quad X = \begin{bmatrix} x \\ y \\ z \end{bmatrix}$$

Then the system can be written in the matrix form:

$$AX = B$$
$$X = A^{-1}B$$

We calculated A^{-1} in Example 2, where we found that:

$$A^{-1} = \begin{bmatrix} -\frac{1}{20} & \frac{2}{5} & -\frac{4}{5} \\ \frac{3}{20} & -\frac{1}{5} & \frac{2}{5} \\ 0 & 0 & \frac{1}{2} \end{bmatrix}$$

SECTION 11.3   Inverse of a Square Matrix

We therefore have:

$$X = A^{-1}B$$

$$= \begin{bmatrix} -\frac{1}{20} & \frac{2}{5} & -\frac{4}{5} \\ \frac{3}{20} & -\frac{1}{5} & \frac{2}{5} \\ 0 & 0 & \frac{1}{2} \end{bmatrix} \begin{bmatrix} -1 \\ 2 \\ 5 \end{bmatrix}$$

$$= \begin{bmatrix} -\frac{63}{20} \\ \frac{29}{20} \\ \frac{5}{2} \end{bmatrix}$$

Thus, the solution of the system is $x = -\frac{63}{20}$, $y = \frac{29}{20}$, $z = \frac{5}{2}$.

Note that the method employed in the preceding example makes use of the inverse matrix. We should emphasize that not every matrix has an inverse, so the above method is not applicable to solving all systems of linear equations but only those for which the matrix of coefficients has an inverse. It can be proved that these are precisely the systems with a single solution. As we have previously seen, there are, in addition, systems that have no solution and systems with an infinite number of solutions. For such systems, we cannot use the method of the preceding example. Instead, we must rely on the method of Gauss elimination, discussed earlier in Section 10.3.

Graphing Calculators and Matrix Inverses

Calculating the inverse of a matrix is usually quite tedious. Let us now show how to use a TI-82 calculator to determine the inverse of a matrix. (For other calculators, refer to your calculator manual.)

EXAMPLE 4
Using a Calculator to Determine the Inverse of a Matrix

Let:

$$A = \begin{bmatrix} 5 & -1 & 4 \\ 3 & 0 & 9 \\ 2 & -5 & 8 \end{bmatrix}$$

Use a calculator to evaluate A^{-1}.

Solution

Enter the matrix A. (If you haven't changed matrix [A] from Example 6 in Section 11.1, it is still stored in your calculator.) Indicate the inverse of A with the keystrokes MATRX [A] ENTER x^{-1}. On the display you will see $[A]^{-1}$. Now press ENTER. The inverse will be displayed as shown in Figure 1. The rows of dots on the right of the display indicate that the entire inverse can't be displayed at one time. To see the remaining columns, use the arrow keys to shift the display to the right.

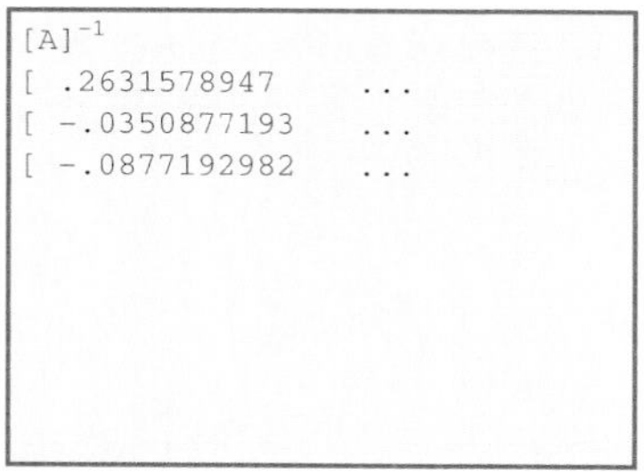

```
[A]⁻¹
[ .2631578947   ...
[ -.0350877193  ...
[ -.0877192982  ...
```

Figure 1

Exercises 11.3

If a graphing calculator is available, it may be used to solve the exercises in this set.

Calculate the inverse of the given matrix if it exists.

1. $\begin{bmatrix} 3 & 5 \\ -2 & -14 \end{bmatrix}$

2. $\begin{bmatrix} 3 & 8 \\ 2 & 5 \end{bmatrix}$

3. $\begin{bmatrix} -3 & 4 \\ 1 & -2 \end{bmatrix}$

4. $\begin{bmatrix} 2 & 4 \\ 1 & -1 \end{bmatrix}$

5. $\begin{bmatrix} 2 & -4 \\ 1 & -2 \end{bmatrix}$

6. $\begin{bmatrix} 2 & -3 \\ -3 & 5 \end{bmatrix}$

7. $\begin{bmatrix} 2 & -1 & 1 \\ 1 & -2 & 3 \\ 4 & 1 & 2 \end{bmatrix}$

8. $\begin{bmatrix} 1 & -4 & 8 \\ 1 & -3 & 2 \\ 2 & -7 & 10 \end{bmatrix}$

9. $\begin{bmatrix} -1 & -1 & -1 \\ 4 & 5 & 0 \\ 0 & 1 & -3 \end{bmatrix}$

10. $\begin{bmatrix} 2 & 4 & 6 \\ -1 & -4 & -3 \\ 0 & 1 & -1 \end{bmatrix}$

Determine whether the given matrices are inverses of each other.

11. $\begin{bmatrix} 2 & -1 & 1 \\ 1 & -2 & 3 \\ 4 & 1 & 2 \end{bmatrix}$ and $\begin{bmatrix} \frac{7}{15} & -\frac{1}{5} & \frac{1}{15} \\ -\frac{2}{3} & 0 & \frac{1}{3} \\ -\frac{3}{5} & \frac{2}{5} & \frac{1}{5} \end{bmatrix}$

12. $\begin{bmatrix} 1 & -1 & 2 \\ 0 & 1 & 3 \\ 2 & 1 & -2 \end{bmatrix}$ and $\begin{bmatrix} \frac{1}{3} & 0 & \frac{1}{3} \\ -\frac{2}{5} & \frac{2}{5} & \frac{1}{5} \\ \frac{2}{15} & \frac{1}{5} & -\frac{1}{15} \end{bmatrix}$

13. $\begin{bmatrix} 11 & 3 \\ 7 & 2 \end{bmatrix}$ and $\begin{bmatrix} 2 & -3 \\ -7 & 11 \end{bmatrix}$

14. $\begin{bmatrix} 3 & 2 \\ 5 & 3 \end{bmatrix}$ and $\begin{bmatrix} -3 & -2 \\ 5 & -3 \end{bmatrix}$

15. $\begin{bmatrix} 1 & 3 & 3 \\ 1 & 4 & 3 \\ 1 & 3 & 4 \end{bmatrix}$ and $\begin{bmatrix} 7 & -3 & -3 \\ -1 & 1 & 0 \\ -1 & 0 & 1 \end{bmatrix}$

16. $\begin{bmatrix} -1 & 0 & 2 \\ 3 & 1 & 0 \\ 0 & 2 & -3 \end{bmatrix}$ and $\begin{bmatrix} -\frac{1}{5} & \frac{4}{15} & -\frac{2}{15} \\ \frac{3}{5} & \frac{1}{5} & \frac{2}{5} \\ \frac{2}{5} & \frac{2}{15} & -\frac{1}{15} \end{bmatrix}$

Solve the following systems of equations using inverse matrices.

17. $\begin{cases} x + 2y = 6 \\ 2x + y = 9 \end{cases}$

18. $\begin{cases} x + 3y = 8 \\ 3x + 4y = 9 \end{cases}$

19. $\begin{cases} 3x - 2y = 9 \\ 2x + 5y = 8 \end{cases}$

20. $\begin{cases} 2x + 4y = 6 \\ x - 3y = 3 \end{cases}$

21. $\begin{cases} 4x - y = 3 \\ 6x + 4y = -1 \end{cases}$

22. $\begin{cases} 3x - 6y = 1 \\ -5x + 9y = -1 \end{cases}$

23. $\begin{cases} 4x - 3y = -23 \\ -3x + 2y = 16 \end{cases}$

24. $\begin{cases} 2x + 3y = 13 \\ x + 2y = 8 \end{cases}$

State the conditions under which A^{-1} exists, and then find a formula for A^{-1}.

25. $A = \begin{bmatrix} a & 0 \\ 0 & b \end{bmatrix}$

26. $\begin{bmatrix} x & 0 & 0 \\ 0 & y & 0 \\ 0 & 0 & z \end{bmatrix}$

27. Suppose A is a 2×2 matrix. Prove that if $\Delta = 0$, the inverse of A does not exist.

28. Suppose A is a 3×3 matrix. Prove that if $\Delta = 0$, the inverse of A does not exist.

Technology

Use a calculator to determine the inverses of the following matrices.

29. $\begin{bmatrix} 3 & 1 & 0 \\ 1 & 1 & 1 \\ 1 & -1 & 2 \end{bmatrix}$

30. $\begin{bmatrix} 2 & -1 & 0 \\ 3 & 0 & 1 \\ -2 & 4 & 0 \end{bmatrix}$

31. $\begin{bmatrix} 1 & -2 & 3 & 0 \\ 0 & 1 & -1 & 1 \\ -2 & 2 & -2 & 4 \\ 0 & 2 & -3 & 1 \end{bmatrix}$

32. $\begin{bmatrix} 1 & 2 & 3 & 4 \\ 0 & 1 & 3 & -5 \\ 0 & 0 & 1 & -2 \\ 0 & 0 & 0 & 1 \end{bmatrix}$

34. $\begin{cases} x + 2y + 3z = -1 \\ 2x - 3y + 4z = 2 \\ -3x + 5y - 6z = 4 \end{cases}$

35. $\begin{cases} x + y - 3z = 4 \\ 2x + 4y - 4z = 8 \\ -x + y + 4z = -3 \end{cases}$

36. $\begin{cases} 2x - 2y = 5 \\ 4y + 8z = 7 \\ 2z = 1 \end{cases}$

Use a calculator to solve the following systems using a matrix equation.

33. $\begin{cases} 4x + z = 1 \\ 2x + 2y = 3 \\ x - y + z = 4 \end{cases}$

In your own words

37. Describe the concept of an inverse matrix. Use a specific example.

38. What is the analogy between an inverse matrix and the multiplicative inverse of a real number?

39. Describe Cramer's rule. Use a specific example.

40. What are the advantages of Cramer's rule over Gauss elimination? The disadvantages?

11.4 INPUT–OUTPUT ANALYSIS

Matrix arithmetic is used extensively in economics. One of the most significant applications is called **input–output analysis**, which is used to analyze the interactions of various segments of an economy in trying to satisfy given consumption and export conditions. Input–output analysis was developed by the mathematician/economist Vasilly Leontieff, whose work on this technique won him the 1973 Nobel prize in economics.

Consider an economic entity, which may be a country, a company, or a department of a company. Divide the entity into sectors, each of which both produces output and consumes output of other sectors. For example, if the entity is a country, there are many sectors, including steel, oil, computers, automobiles, services, and so forth. The number of sectors used depends on how detailed a model you wish to consider. Some econometric models for the United States include hundreds of sectors. A simple model of a single company may include only a few sectors.

To describe the interdependence of the various sectors, we use an **input–output matrix,** constructed as follows. Each sector has a corresponding column of the matrix. The column for a particular sector describes the amount of output from each sector (in dollars) required to produce $1 of output from the particular sector. For example, to produce $1 of output from the automobile sector requires certain amounts of input from the steel, service, oil, and other sectors. Here is a typical input–output matrix for an economic entity with three sectors:

$$\text{From} \begin{cases} \text{Sector 1} \\ \text{Sector 2} \\ \text{Sector 3} \end{cases} \begin{matrix} & \text{Input requirements of} \\ & \text{Sector 1} \quad \text{Sector 2} \quad \text{Sector 3} \end{matrix} \begin{bmatrix} \cdots & \cdots & \cdots \\ \cdots & \cdots & \cdots \\ \cdots & \cdots & \cdots \end{bmatrix}$$

The input–output matrix expresses the amount of output demanded from each sector by every other sector. However, some of the output is used by end consumers other than the various sectors. The amount of output demanded by these end consumers is expressed as a column matrix representing the dollar amount of output demanded from each sector. This matrix is called the **final demand matrix**:

$$[\text{final demand}] = \begin{bmatrix} \text{Demand for sector 1} \\ \text{Demand for sector 2} \\ \cdots \end{bmatrix}$$

The following example provides some practice in setting up demand and input–output matrices.

EXAMPLE 1
Multinational Production

A computer company has three divisions—semiconductors, computers, and printers. Each $1 of output of semiconductors requires $.02 of semiconductors, $.15 of computers, and $.01 of printers. Each $1 output of computers requires $.23 of semiconductors, $.02 of computers, and $.01 of printers. Each $1 output of printers requires $.08 of semiconductors, $.05 of computers, and $.01 of printers. Suppose that in the most recent fiscal year, the company sold to consumers (outside the company) a total of $130 million of semiconductors, $175 million of computers, and $58 million of printers. Determine the input–output matrix A and the final demand matrix D of the company.

Solution

The input–output matrix has one column for each division. The first column corresponds to output from the semiconductor division and equals:

$$\begin{bmatrix} 0.02 \\ 0.15 \\ 0.01 \end{bmatrix}$$

The second column corresponds to output from the computer division and equals:

$$\begin{bmatrix} 0.23 \\ 0.02 \\ 0.01 \end{bmatrix}$$

The third column corresponds to output from the printer division and equals:

$$\begin{bmatrix} 0.08 \\ 0.05 \\ 0.01 \end{bmatrix}$$

Putting the three columns together, we arrive at the input–output matrix:

$$A = \begin{bmatrix} 0.02 & 0.23 & 0.08 \\ 0.15 & 0.02 & 0.05 \\ 0.01 & 0.01 & 0.01 \end{bmatrix}$$

The final demand matrix, D, is a column displaying the end consumer demand for each of three divisions, so:

$$D = \begin{bmatrix} 130 \\ 175 \\ 58 \end{bmatrix}$$

SECTION 11.4 Input–Output Analysis

EXAMPLE 2
Multinational Production, Continued

Suppose the three divisions of the company of Example 1 manufactured $150 million of semiconductors, $200 million of computers, and $70 million of printers. Determine the amount of output from each division that is used in the manufacturing process.

Solution

We must subtract from the output of each division the amount of the output used by the other divisions. Consider, for example, the semiconductor division. It manufactures $150 million of output. Referring to the first column of the input–output matrix, we see that of each dollar of this $150 million, the amount used by the semiconductor division itself is $.02, or $0.02 \cdot 150$ million. Similarly, consulting the second column of the input–output matrix, the computer division uses $.23 of semiconductors for each dollar of output it produces, or $0.23 \cdot 200$ million. Similarly, the printer division uses $0.08 \cdot 70$ million of semiconductors in its manufacturing. We see that the total amount of semiconductors used in the manufacturing process is

$$0.02 \cdot 150 + 0.23 \cdot 200 + 0.08 \cdot 70$$

This quantity is just the first entry in the matrix product

$$\begin{bmatrix} 0.02 & 0.23 & 0.08 \\ 0.15 & 0.02 & 0.05 \\ 0.01 & 0.01 & 0.01 \end{bmatrix} \begin{bmatrix} 150 \\ 200 \\ 70 \end{bmatrix} = \begin{bmatrix} 54.6 \\ 30 \\ 4.2 \end{bmatrix}$$

Similarly, the second and third entries in the product give the amounts of computers and printers used in the manufacturing process, respectively.

We can generalize the method used in the preceding example as follows. Suppose an economic entity has an input–output matrix A and the matrix X is a column giving the total amount produced by each division. Then the matrix product AX is a column giving the amount of output from each division that is used in the manufacturing process.

Consider the following problem: Suppose we are given a final demand matrix and an input–output matrix. What quantity should each sector produce in order to satisfy the given final demand as well as the demand required from the other sectors?

This problem is simple to solve using the preceding observation that the total amount of output used in the manufacturing process is given by the matrix AX. The amount left over for end-consumer demand is therefore equal to the matrix difference

$$AX - X$$

However, this matrix is just equal to the final demand D:

$$AX - X = D$$

We can solve this equation for the matrix X as follows: Let I be an identity matrix of the same size as A. Then:

$$X - AX = D$$
$$IX - AX = D$$
$$(I - A)X = D$$
$$X = (I - A)^{-1}D$$

That is, we have the following result.

DETERMINING THE OUTPUT OF AN ECONOMIC ENTITY

Suppose an economic entity with n sectors has an input–output matrix A and a final demand matrix D. Let X be the output matrix corresponding to final demand D. Then
$$X = (I - A)^{-1} D$$
where I is the identity matrix of order n.

EXAMPLE 3 Consider the company of Example 1. Determine the amount each division should produce in order to satisfy the stated final demand.

Determining Final Demand

Solution

According to the result just stated, we must determine the matrix X given by the equation
$$X = (I - A)^{-1} D$$
where
$$A = \begin{bmatrix} 0.02 & 0.23 & 0.08 \\ 0.15 & 0.02 & 0.05 \\ 0.01 & 0.01 & 0.01 \end{bmatrix}$$
and
$$D = \begin{bmatrix} 130 \\ 175 \\ 58 \end{bmatrix}$$

The identity matrix I is 3×3, so
$$(I - A)^{-1} = \left(\begin{bmatrix} 1 & 0 & 0 \\ 0 & 1 & 0 \\ 0 & 0 & 1 \end{bmatrix} - \begin{bmatrix} 0.02 & 0.23 & 0.08 \\ 0.15 & 0.02 & 0.05 \\ 0.01 & 0.01 & 0.01 \end{bmatrix} \right)^{-1}$$

$$= \begin{bmatrix} 0.98 & -0.23 & -0.08 \\ -0.15 & 0.98 & -0.05 \\ -0.01 & -0.01 & 0.99 \end{bmatrix}^{-1}$$

$$= \begin{bmatrix} 1.06 & 0.25 & 0.10 \\ 0.16 & 1.06 & 0.07 \\ 0.01 & 0.01 & 1.01 \end{bmatrix}$$

where we have computed the entries in the inverse to two decimal places (using Gauss elimination and a calculator). We therefore have:

$$X = (I - A)^{-1} D$$

$$= \begin{bmatrix} 1.06 & 0.25 & 0.10 \\ 0.16 & 1.06 & 0.07 \\ 0.01 & 0.01 & 1.01 \end{bmatrix} \begin{bmatrix} 130 \\ 175 \\ 58 \end{bmatrix}$$

$$= \begin{bmatrix} 187.35 \\ 213.36 \\ 61.63 \end{bmatrix}$$

SECTION 11.4 Input–Output Analysis

That is, to achieve the given final demand, the semiconductor division should output $187.35 million, the computer division $213.36 million, and the printer division $61.63 million.

Exercises 11.4

Calculate $(I - A)^{-1}$ with entries accurate to two decimal places.

1. I = identity matrix of order 2;
$$A = \begin{bmatrix} 0.10 & 0.20 \\ 0.20 & 0.10 \end{bmatrix}$$

2. I = identity matrix of order 2;
$$A = \begin{bmatrix} 0.20 & 0.10 \\ 0.00 & 0.30 \end{bmatrix}$$

3. I = identity matrix of order 3;
$$A = \begin{bmatrix} 0.10 & 0.10 & 0.00 \\ 0.20 & 0.00 & 0.00 \\ 0.00 & 0.10 & 0.10 \end{bmatrix}$$

4. I = identity matrix of order 3;
$$A = \begin{bmatrix} 0.10 & 0.00 & 0.00 \\ 0.00 & 0.20 & 0.25 \\ 0.00 & 0.10 & 0.10 \end{bmatrix}$$

Applications

5. **Determining output.** Consider the company of Example 1. Suppose the semiconductor division outputs $100 million worth of product, the computer division $50 million, and the printer division $100 million. Determine the amount of output from each division used in the production process.

6. **Satisfying final demand.** Consider the company of Example 1. Suppose the final demand is $350 million of semiconductors, $100 million of computers, and $200 million of printers. Determine the amount each division should produce to satisfy the final demand.

A multinational company has branches in the United States, Canada, and Mexico. The branches in each country purchase goods from each other. For each $1 of output of the U.S. branch, $.03 is used within that branch, $.15 is purchased by the Canadian branch, and $.08 is purchased by the Mexican branch. For each $1 of output of the Canadian branch, $.10 is used by the branch itself, $.15 is purchased by the U.S. branch, and $.18 is purchased by the Mexican branch.

For each $1 of output of the Mexican branch, $.02 is used by the branch itself, $.20 by the U.S. branch, and $.09 by the Canadian branch.

7. **Determining input–output matrix for a multinational.** Determine the input–output matrix of the company.

8. **Determining output.** Determine the amount of output of each branch that is used within the company if the U.S. branch produces $80 million, the Canadian branch $50 million, and the Mexican branch $40 million.

9. **Determining final demand.** Suppose production is as described in Exercise 8. What is the final demand?

10. **Satisfying given final demand.** Suppose final demand is $100 million for the U.S. branch, $75 million for the Canadian branch, and $50 million for the Mexican branch. What should the production levels of the branches be to satisfy the final demand?

Technology

It can be shown that if A is a square matrix all of whose entries are less than 1 in absolute value, the inverse matrix $(I - A)^{-1}$, where I is an identity matrix the same size as A, can be approximated by the matrix sum:
$$I + A + A^2$$
where $A^2 = A \cdot A$. Use a calculator to determine the difference
$$(I - A)^{-1} - (I + A + A^2)$$

where A is given by the matrix in:

11. Exercise 1.
12. Exercise 2.
13. Exercise 3.
14. Exercise 4.

11.5 PARTIAL FRACTIONS

In Section 1.4, we learned to add rational expressions in a single variable. We found that the sum of rational expressions is another rational expression. Moreover, the denominator of the sum is the product of the denominators of the summands (before any reduction is carried out). For example:

$$\frac{1}{x-2} + \frac{2}{x-1} = \frac{3x-5}{(x-2)(x-1)}$$

In many contexts (most notably in integration in calculus) an expression such as the one on the left is far easier to work with than one such as the expression on the right. The method of **partial fractions** is a procedure for writing a rational expression (such as the one on the right) in terms of a sum of simple rational expressions (such as the ones on the left). Explicitly, here is what the method of partial fractions accomplishes.

Suppose you are given a rational expression $f(x)/g(x)$, where $f(x)$ and $g(x)$ are polynomials with real coefficients. By the division algorithm for polynomials, we can divide $f(x)$ by $g(x)$ to obtain

$$\frac{f(x)}{g(x)} = q(x) + \frac{h(x)}{g(x)}$$

where $q(x)$ is the quotient and $h(x)$ is the remainder and has degree less than the degree of $g(x)$. We then write the rational expression $h(x)/g(x)$ as a sum of simple expressions as follows: As we discussed in Section 5.3, the polynomial $g(x)$ can be factored into a product of linear and irreducible quadratic factors, all with real coefficients. A linear factor is of the form $(ax+b)^n$. Corresponding to each such factor, the expression for $h(x)/g(x)$ has a sum of simple expressions of the form

$$\frac{A_1}{ax+b} + \frac{A_2}{(ax+b)^2} + \cdots + \frac{A_n}{(ax+b)^n}$$

for suitable real numbers $A_1, A_2, \ldots, A_n$. A quadratic factor is of the form $(cx^2 + dx + e)^n$. Corresponding to each such factor, the expression for $h(x)/g(x)$ has a sum of simple expressions of the form

$$\frac{A_1 + B_1 x}{cx^2 + dx + e} + \frac{A_2 + B_2 x}{(cx^2 + dx + e)^2} + \cdots + \frac{A_n + B_n x}{(cx^2 + dx + e)^n}$$

where $A_1, B_1, A_2, B_2, \ldots, A_n, B_n$ are suitable real numbers.

The expression of $f(x)/g(x)$ as a sum of a polynomial and rational expressions of the above form is called a **partial fraction expansion**. We can summarize the above discussion as follows:

DETERMINING THE PARTIAL FRACTION EXPANSION OF $f(x)/g(x)$

1. Use long division to write $f(x)/g(x)$ in the form

$$\frac{f(x)}{g(x)} = q(x) + \frac{h(x)}{g(x)}, \qquad \deg(h) < \deg(g)$$

2. Factor $g(x)$ into a product of distinct real factors, some linear and some quadratic.

3. For a linear factor $(ax+b)^n$, the partial fraction expansion has a component:

$$\frac{A_1}{ax+b} + \frac{A_2}{(ax+b)^2} + \cdots + \frac{A_n}{(ax+b)^n}$$

4. For an irreducible quadratic factor $(cx^2 + dx + e)^n$, the partial fraction expansion has a component:

$$\frac{A_1 + B_1 x}{cx^2 + dx + e} + \frac{A_2 + B_2 x}{(cx^2 + dx + e)^2} + \cdots + \frac{A_n + B_n x}{(cx^2 + dx + e)^n}$$

5. The partial fraction expansion is the sum of $q(x)$ in step 1 and all the components in steps 3 and 4.

The principal difficulty in determining the partial fraction expansion of a given rational expression is in determining the values of the various coefficients listed above. The following example shows how this can be done.

EXAMPLE 1
Simple Partial Fraction Expansion

Determine the partial fraction expansion of the rational expression

$$\frac{3x - 5}{(x - 2)(x - 1)}$$

Solution

Because the degree of the numerator is less than the degree of the denominator, it is unnecessary to carry out the division indicated in the discussion above. (The result would indicate that $q(x)$ is the zero polynomial and that $h(x)$ equals $f(x)$.) The denominator is already factored. According to the above discussion, there is a sum corresponding to each factor in the denominator. The sum corresponding to the factor $x - 1$ is of the form:

$$\frac{A}{x - 1}$$

The sum corresponding to the factor $x - 2$ is of the form:

$$\frac{B}{x - 2}$$

That is, we must have:

$$\frac{3x - 5}{(x - 2)(x - 1)} = \frac{A}{x - 1} + \frac{B}{x - 2}$$

Multiply both sides by $(x - 2)(x - 1)$ to obtain:

$$3x - 5 = A(x - 2) + B(x - 1)$$

This last equation must hold for all values of x. Let's choose two particular values of x to use in determining the values of A and B. The most convenient values are $x = 1$ and $x = 2$. Substituting these values into the last equation, we obtain:

For $x = 1$:

$$3(1) - 5 = A(1 - 2) + B(0)$$
$$-A = -2$$
$$A = 2$$

For $x = 2$:

$$3(2) - 5 = A(0) + B(2 - 1)$$
$$B = 1$$

Thus, the partial fraction expansion of the given rational expression is

$$\frac{3x - 5}{(x - 2)(x - 1)} = \frac{2}{x - 1} + \frac{1}{x - 2}$$

The calculation we just performed reversed the addition of rational functions we performed at the beginning of the section.

EXAMPLE 2
Partial Fractions with Three Factors

Determine the partial fraction expansion of the following rational expression:

$$\frac{2x^4 - 1}{x^3 - x}$$

Solution

Begin by dividing the numerator by the denominator: The quotient is $q(x) = 2x$, and the remainder is $r(x) = 2x^2 - 1$. That is, we have:

$$\frac{2x^4 - 1}{x^3 - x} = 2x + \frac{2x^2 - 1}{x^3 - x}$$

Next, factor the denominator:

$$x^3 - x = x(x^2 - 1)$$
$$= x(x - 1)(x + 1)$$

The partial fraction expansion has the form:

$$\frac{2x^2 - 1}{x^3 - x} = \frac{A}{x} + \frac{B}{x - 1} + \frac{C}{x + 1}$$

Multiply both sides by $x(x - 1)(x + 1)$ to obtain the identity:

$$2x^2 - 1 = A(x - 1)(x + 1) + Bx(x + 1) + Cx(x - 1)$$

To obtain the values of A, B, and C, we substitute three different values of x into the above identity. The three most convenient values are $x = 0$, $x = 1$, and $x = -1$, because with these values some of the terms equal 0:

For $x = 0$:

$$2(0)^2 - 1 = A(0 - 1)(0 + 1) + B \cdot 0(0 + 1) + C \cdot 0(0 - 1)$$
$$-1 = -A$$
$$A = 1$$

For $x = 1$:

$$2(1)^2 - 1 = A(1 - 1)(1 + 1) + B \cdot 1(1 + 1) + C \cdot 1(1 - 1)$$
$$1 = 2B$$
$$B = \frac{1}{2}$$

For $x = -1$:

$$2(-1)^2 - 1 = A(-1 - 1)(-1 + 1) + B \cdot (-1)(-1 + 1) + C \cdot (-1)(-1 - 1)$$
$$1 = 2C$$
$$C = \frac{1}{2}$$

Thus, the partial fraction expansion is given by:

$$\frac{2x^4 - 1}{x^3 - x} = 2x + \frac{1}{x} + \frac{1/2}{x - 1} + \frac{1/2}{x + 1}$$

$$= 2x + \frac{1}{x} + \frac{1}{2(x - 1)} + \frac{1}{2(x + 1)}$$

In the preceding two examples we were able to determine the unknown coefficients in the partial fraction expansion one at a time. However, in some instances, the coefficients must be determined by solving a system of linear equations. This phenomenon is illustrated in the next example.

EXAMPLE 3
Partial Fractions with a Multiple Factor

Determine the partial fraction expansion of the following rational expression:

$$\frac{3x^3 - 5x + 1}{x^4 + x^2}$$

Solution

The degree of the numerator is less than the degree of the denominator, so the initial division step is unnecessary. Factoring the denominator gives us:

$$x^4 + x^2 = x^2(x^2 + 1)$$

Note that the quadratic factor

$$x^2 + 1$$

cannot be factored further because the factorization is assumed to be factored into polynomials with real coefficients. The general form of the partial fraction expansion is then:

$$\frac{3x^3 - 5x + 1}{x^4 + x^2} = \frac{A}{x} + \frac{B}{x^2} + \frac{Cx + D}{x^2 + 1}$$

Multiplying both sides by $x^4 + x^2$ yields the identity:

$$3x^3 - 5x + 1 = Ax(x^2 + 1) + B(x^2 + 1) + (Cx + D)x^2$$

Substituting $x = 0$ yields the value of B:

$$1 = 0 + B(0^2 + 1) + 0$$
$$B = 1$$

Inserting this value back into the identity gives us:

$$3x^3 - 5x + 1 = Ax(x^2 + 1) + (x^2 + 1) + (Cx + D)x^2$$
$$3x^3 - 5x + 1 = (Ax^3 + Ax) + (x^2 + 1) + (Cx^3 + Dx^2)$$
$$3x^3 - 5x + 1 = (A + C)x^3 + (D + 1)x^2 + Ax + 1$$

We now have an identity between two polynomials in x. For them to be equal, the coefficients of their powers of x must be equal. That is, we have the system of equations:

$$A + C = 3 \quad \text{Coefficient of } x^3$$
$$D + 1 = 0 \quad \text{Coefficient of } x^2$$
$$A = -5 \quad \text{Coefficient of } x$$

Solving this system gives us:
$$D = -1$$
$$A = -5$$
$$-5 + C = 3$$
$$C = 8$$

Therefore, the desired partial fraction expansion is:
$$\frac{3x^3 - 5x + 1}{x^4 + x^2} = \frac{-5}{x} + \frac{1}{x^2} + \frac{8x - 1}{x^2 + 1}$$

Note that in our solution we compared like coefficients on both sides of the identity rather than substituting assorted values of x to arrive at a system of equations. Both approaches are valid. For instance, we could just as well have arrived at a (different) system by substituting for x the values $x = 0, 1, 2, 3$. Why not carry out the calculations and verify that the resulting system leads to the same partial fraction expansion?

Exercises 11.5

Determine the partial fractional expansions of the given rational expressions.

1. $\dfrac{6x - 2}{(x + 1)(x - 1)}$

2. $\dfrac{4x + 1}{(x - 2)(x + 1)}$

3. $\dfrac{4x + 2}{x(x + 1)(x + 2)}$

4. $\dfrac{2x^2 - 6x - 2}{x(x + 2)(x - 1)}$

5. $\dfrac{7x - 10}{(2x - 1)(x - 2)}$

6. $\dfrac{2x + 8}{(x - 3)(3x - 2)}$

7. $\dfrac{x^2 - 17x + 35}{(x^2 + 1)(x - 4)}$

8. $\dfrac{7x^2 - 29x + 24}{(2x - 1)(x - 2)^2}$

9. $\dfrac{13x + 5}{3x^2 - 7x - 6}$

10. $\dfrac{4x - 13}{2x^2 + x - 6}$

11. $\dfrac{3x^3 + 4x^2 - x}{x^2 + x - 2}$

12. $\dfrac{2x^3 - x^2 - 13x + 6}{x^2 - x - 6}$

13. $\dfrac{3x^3 - 8x^2 + 9x - 6}{x^2 - 3x + 2}$

14. $\dfrac{4x^3 - 16x^2 + 9x - 15}{2x^2 - 9x + 4}$

15. $\dfrac{5x + 5}{x^2 - x - 6}$

16. $\dfrac{11x + 2}{2x^2 + x - 1}$

17. $\dfrac{-2x^2 + 7x + 2}{x^3 - 2x^2 + x}$

18. $\dfrac{-3x^2 - 10x - 4}{x(x + 1)^2}$

19. $\dfrac{-x^2 + 26x + 6}{(2x - 1)(x + 2)^2}$

20. $\dfrac{5x^2 + 9x - 56}{(x - 4)(x - 2)(x + 1)}$

21. $\dfrac{6x^3 + 29x^2 - 6x - 5}{2x^2 + 9x - 5}$

22. $\dfrac{-9x^2 + 7x - 4}{x^3 - 3x^2 - 4x}$

23. $\dfrac{x^3 + x^2 + 2}{(x^2 + 2)^2}$

24. $\dfrac{2x^3 + 8x^2 + 2x + 4}{(x + 1)^2(x^2 + 3)}$

25. $\dfrac{6x^3 + 5x^2 - 7}{3x^2 - 2x - 1}$

26. $\dfrac{3x^2 - 11x - 26}{(x^2 - 4)(x + 1)}$

27. $\dfrac{3x^2 + 10x + 9}{(x + 2)^3}$

28. $\dfrac{x^2 - x - 4}{(x - 2)^3}$

29. $\dfrac{5}{(x + 2)^2(x + 1)}$

30. $\dfrac{-8x + 23}{2x^2 + 5x - 12}$

Expand into partial fractions.

31. $\dfrac{x}{x^4 - a^4}$

32. $\dfrac{2x^2 + 2ax + 3x + 2a^2 - 3a}{x^3 - a^3}$

Expand into partial fractions and graph by addition of ordinates.

33. $y = \dfrac{x - 11}{x^2 - 1}$

34. $y = \dfrac{5x - 1}{x^2 - x - 2}$

35. $y = \dfrac{2x^2 + 4x - 1}{x^2 + x}$

36. $y = \dfrac{9}{x^2 + x - 2}$

Find constants x, y, and z such that the following statements are true.

37. $\dfrac{2a + 1}{(a + 1)(a^2 - 3a + 5)} = $

$\dfrac{x}{(a + 1)} + \dfrac{ya + z}{a^2 - 3a + 5}$

38. $\dfrac{a + 2}{a^3 + 2a^2 + a} = $

$\dfrac{x}{a} + \dfrac{ya + z}{a^2 + 2a + 1}$

Mathematics and the World around Us—Error-Correcting Codes

Our modern society depends on the accurate transmission of data from one place to another. For instance, when a computer accesses information in a database at a remote location or an automatic teller machine checks your personal identification code and your bank balance, data are transferred over telephone lines. When a spacecraft transmits a picture of the surface of Venus, data are transmitted using radio waves. Guaranteeing the accuracy of transmitted data is of paramount importance. Unfortunately, that is not very simple. Electronic malfunctions and static over telephone lines introduce errors into data before or during transmission. Because such errors are random and largely unavoidable, it is crucial that the receiving party be able to detect that a piece of data contains an error. Mathematicians have developed very ingenious methods for doing this using so-called **error-correcting codes**.

All data in computers are stored in binary form, that is, as a sequence of 0's and 1's. Exactly how to translate letters and numbers into this form is not important for this discussion. Let's just assume that it has been accomplished somehow and that the data are stored in binary form in the sender's computer. The idea of an error-correcting code is to add to the data, at regular intervals, extra data that can be used to detect changes in the data during transmission.

The simplest such addition is the use of a **parity bit**, which works as follows. Data are typically sent in groups of eight binary digits (0's or 1's). At the end of each such group, add an extra digit, the parity bit, determined as follows: For each group of eight digits, count the number of 1's. If this number is even, the parity bit is 0; if this number is odd, the parity bit is 1. The parity bit is then transmitted with the group of eight digits. The receiver checks whether the parity bit is correct. The most likely error is for a single digit to be altered. If this happens, the receiver will detect it. Note, however, that if two digits are altered, the parity bit will appear correct and the error will go undetected. However, an error in two digits is much less likely than a single error. So at the expense of increasing the data size by one digit for every eight, we give ourselves a reasonable chance at detecting errors in transmission.

Use of a parity bit is common in computers of all sizes. However, it is not a perfect safeguard of data. It is only a probabilistic check. Moreover, if the parity bit indicates an error, there is no way of determining what the error is. The only way of correcting the error is to ask for a retransmission. In some circumstances this is not practical. For instance, in transmitting space photography, the satellite can go out of range before it is able to retransmit its current pictures. In this case, the spacecraft's computer must embed enough information in the photo data transmitted to enable errors to be corrected by the receiver. This is done using error-correcting codes.

Just as with a parity bit, an error-correcting code uses extra digits embedded in data to indicate errors. However, the extra digits are constructed in such a way that not only are the most common errors detected, but the correction can be determined.

Error-detecting codes are studied in more advanced algebra courses. They were first invented by the MIT computer scientist and engineer R. W. Hamming in the 1940s. Today they are in common use in guaranteeing the integrity of data in our computers and other electronic equipment.

Chapter Review

Important Concepts and Results—Chapter 11

- subscript notation for matrix elements
- zero matrix
- sum of two matrices
- product of a matrix and a real number
- additive inverse of a matrix
- row matrix
- column matrix
- matrix multiplication
- when products are defined
- matrix multiplication is noncommutative
- identity matrix of size n
- determinant of a square matrix
- determinant
- determinant of a 2×2 matrix
- minor of a row and a column in a matrix
- minor of a 3×3 matrix
- cofactor of a minor
- determinant of a 3×3 matrix
- expanding a determinant about a given row or column
- calculating a determinant by cofactors
- determinant of an $n \times n$ matrix
- Cramer's rule
- determinant form of the equation of a line
- condition for collinearity
- determinant formula for the area of a triangle
- inverse of a square matrix
- inverse of a 2×2 matrix
- matrix solution of a square linear system
- input–output analysis
- input–output matrix
- final demand matrix
- partial fraction
- partial fraction expansion of a rational function

Important Formulas and Techniques—Chapter 11

Matrices	Subscript notation for matrix entries.	p. 568
	Zero matrix: Matrix of all zeros.	
	Sum of two matrices: Add corresponding entries.	
	Row matrix: Only one row. Column matrix: Only one column	p. 569
	Matrix multiplication: Suppose A is an $m \times n$ matrix and B is an $n \times k$ matrix. The product AB is the $m \times k$ matrix with entries as follows: The entry in the ith row and the jth column of AB is obtained by multiplying the ith row of A by the jth column of B.	p. 570
	Identity matrix I_n of size n.	p. 573
Determinant of an $n \times n$ matrix	Let A be an $n \times n$ matrix. The determinant $|A|$ is defined by the formula: $$|A| = a_{11}A_{11} + a_{12}A_{12} + \cdots + a_{1n}A_{1n}$$ where A_{ij} denotes the cofactor determined by eliminating the ith row and jth column.	p. 578

Cramer's rule for $n \times n$ linear systems	If a linear system of n equations in n variables $x, y, z, \ldots$ has a nonzero determinant $|A|$, the system has a unique solution given by the formulas: $$x = \frac{|A_1|}{|A|}, \quad y = \frac{|A_2|}{|A|}, \quad z = \frac{|A_3|}{|A|}, \quad \ldots$$	p. 580
Determinant form of the equation of a line	The equation of a line passing through the points (x_1, y_1) and (x_2, y_2) is given in determinant form: $$\begin{vmatrix} x & y & 1 \\ x_1 & y_1 & 1 \\ x_2 & y_2 & 1 \end{vmatrix} = 0$$	p. 582
Determinant condition for collinearity	Three points (x_1, y_1), (x_2, y_2), and (x_3, y_3) are collinear provided that: $$\begin{vmatrix} x_1 & y_1 & 1 \\ x_2 & y_2 & 1 \\ x_3 & y_3 & 1 \end{vmatrix} = 0$$	p. 582
Determinant formula for the area of a triangle	The area of the triangle with vertices (x_1, y_1), (x_2, y_2), and (x_3, y_3) is given by the formula: $$\text{Area} = \pm \frac{1}{2} \begin{vmatrix} x_1 & y_1 & 1 \\ x_2 & y_2 & 1 \\ x_3 & y_3 & 1 \end{vmatrix}$$	p. 582
Inverse of a square matrix	A^{-1} is an inverse of A provided that A^{-1} satisfies the equations: $$AA^{-1} = A^{-1}A = I_n$$	p. 585
Computing the inverse of a square matrix	1. Form the matrix $$\begin{bmatrix} a_{11} & a_{12} & \ldots & a_{1n} & 1 & 0 & \ldots & 0 \\ a_{21} & a_{22} & \ldots & a_{2n} & 0 & 1 & \ldots & 0 \\ \vdots & & & & \vdots & & & \\ a_{n1} & a_{n2} & \ldots & a_{nn} & 0 & 0 & \ldots & 1 \end{bmatrix}$$ where the matrix on the right is the identity matrix I_n. 2. If possible, perform elementary row operations on this matrix to transform the left half into the identity matrix I_n. 3. After the transformation, the matrix in the right half is A^{-1}.	p. 587
Solution of a square linear system in terms of matrices	The solution of the matrix equation $AX = B$ is $X = A^{-1}B$.	p. 590

Determining the partial fraction expansion of $f(x)/g(x)$	1. Use long division to write $f(x)/g(x)$ in the form: $$\frac{f(x)}{g(x)} = q(x) + \frac{h(x)}{g(x)}, \quad \deg(h) < \deg(g)$$ 2. Factor $g(x)$ into a product of distinct real factors, some linear and some quadratic. 3. For a linear factor $(ax+b)^n$, the partial fraction expansion has a component: $$\frac{A_1}{ax+b} + \frac{A_2}{(ax+b)^2} + \cdots + \frac{A_n}{(ax+b)^n}$$ 4. For an irreducible quadratic factor $(cx^2+dx+e)^n$, the partial fraction expansion has a component: $$\frac{A_1+B_1x}{cx^2+dx+e} + \frac{A_2+B_2x}{(cx^2+dx+e)^2} + \cdots + \frac{A_n+B_nx}{(cx^2+dx+e)^n}$$ 5. The partial fraction expansion is the sum of $q(x)$ in step 1 and all the components in steps 3 and 4.	p. 598

Review Exercises—Chapter 11

Given

$$A = \begin{bmatrix} 4 & 2 \\ 1 & -3 \end{bmatrix}, \quad B = \begin{bmatrix} 0 \\ 3 \end{bmatrix}, \quad C = \begin{bmatrix} 2 & -1 & 4 \\ 3 & 0 & 4 \end{bmatrix},$$

$$D = \begin{bmatrix} -2 & 4 & 5 \\ 1 & -1 & 2 \\ 3 & 2 & -3 \end{bmatrix}, \quad E = \begin{bmatrix} 3 & 4 \\ -1 & 5 \end{bmatrix},$$

$$F = \begin{bmatrix} 5 \\ -2 \end{bmatrix}, \quad G = \begin{bmatrix} -1 & 3 & 4 \\ 0 & 2 & -1 \end{bmatrix},$$

$$H = \begin{bmatrix} 0 & 4 & -2 \\ 2 & -1 & 3 \\ 3 & 5 & 2 \end{bmatrix}, \quad I = \begin{bmatrix} 4 & 1 & 0 \end{bmatrix}$$

find the following matrices, if they are defined.

1. DH
2. EF
3. $4A + 3E$
4. $2E - 3A$
5. GH
6. CH
7. AB
8. CD
9. BI
10. FG

Determine which of the following products are defined and the size of those products.

11. A is 3×5; B is 4×5.
12. A is 3×4; B is 4×3.
13. A is 1×3; B is 3×1.
14. A is 2×2; B is 4×2.

Find the values of $w, x, y,$ and z in the following matrix equations.

15. $\begin{bmatrix} x & 5 & 2 \\ 2 & 7 & y \end{bmatrix} = \begin{bmatrix} 3 & 5 & z \\ 2 & w & 3 \end{bmatrix}$

16. $\begin{bmatrix} 4 & 1 \\ 2 & 4 \end{bmatrix} = \begin{bmatrix} x & w \\ y & 4 \end{bmatrix}$

Calculate $|A|$, where A is the matrix given.

17. $\begin{bmatrix} 1 & -1 \\ 2 & 3 \end{bmatrix}$

18. $\begin{bmatrix} 4 & 3 \\ -1 & -2 \end{bmatrix}$

19. $\begin{bmatrix} -1 & -2 \\ -4 & 5 \end{bmatrix}$

20. $\begin{bmatrix} 0 & 3 \\ 2 & -5 \end{bmatrix}$

Use $A = \begin{bmatrix} 1 & 3 & -1 \\ 4 & 5 & 1 \\ 3 & -1 & 2 \end{bmatrix}$ and $B = \begin{bmatrix} 7 & 2 & 0 & 4 \\ 2 & -1 & 2 & 2 \\ 1 & 4 & 3 & -1 \\ -3 & 5 & 1 & 5 \end{bmatrix}$ for the following exercises.

21. Find A_{12}.
22. Find A_{11}.
23. Find B_{44}.
24. Find B_{22}.

Calculate $|A|$, where A is the given matrix.

25. $\begin{bmatrix} 1 & 3 & -1 \\ 2 & 4 & 5 \\ -1 & 2 & 0 \end{bmatrix}$

26. $\begin{bmatrix} 0 & 3 & 1 \\ 4 & -2 & -1 \\ -1 & 4 & 2 \end{bmatrix}$

27. $\begin{bmatrix} 2 & 0 & 1 & 4 \\ 1 & 2 & 3 & 0 \\ 4 & -1 & 0 & 2 \\ 0 & -2 & 3 & 1 \end{bmatrix}$

28. $\begin{bmatrix} -3 & 0 & 2 & 1 \\ 2 & 4 & 0 & -1 \\ 3 & 0 & 1 & 2 \\ 0 & 2 & -1 & 4 \end{bmatrix}$

Solve the following for x.

29. $\begin{vmatrix} 3 & -1 \\ 2 & 5 \end{vmatrix} = 2x - 1$

30. $\begin{vmatrix} 2 & -1 & 1 \\ 3 & 4 & 0 \\ 1 & 0 & 2 \end{vmatrix} = x - 5$

Calculate the inverse of the given matrix, if it exists.

31. $\begin{bmatrix} 3 & 5 \\ 1 & 2 \end{bmatrix}$

32. $\begin{bmatrix} 4 & -3 \\ 1 & 2 \end{bmatrix}$

33. $\begin{bmatrix} 1 & -4 & 8 \\ 1 & -3 & 2 \\ 2 & -7 & 10 \end{bmatrix}$

34. $\begin{bmatrix} 1 & 2 & 3 & 4 \\ 0 & 1 & 3 & -5 \\ 0 & 0 & 1 & -2 \\ 0 & 0 & 0 & -1 \end{bmatrix}$

Determine whether the given matrices are inverses of each other.

35. $\begin{bmatrix} 1 & 2 & -1 \\ 3 & 4 & 2 \\ 1 & 1 & 0 \end{bmatrix}$ and $\begin{bmatrix} 1 & 1 & 0 \\ 3 & 4 & 2 \\ 1 & 2 & -1 \end{bmatrix}$

36. $\begin{bmatrix} 2 & 3 \\ 1 & 5 \end{bmatrix}$ and $\begin{bmatrix} \frac{1}{2} & \frac{1}{3} \\ 1 & \frac{1}{5} \end{bmatrix}$

Solve the following systems of equations using inverse matrices.

37. $\begin{cases} 2x + y = 7 \\ 3x + 2y = 9 \end{cases}$

38. $\begin{cases} 6x + y = 0 \\ 3x + 5y = 9 \end{cases}$

39. $\begin{cases} x + y = 4 \\ y + z = 2 \\ 2x - z = 1 \end{cases}$

40. $\begin{cases} x + z = 1 \\ 2x + y = 3 \\ x - y + z = 4 \end{cases}$

Determine the partial fraction expansion of the given rational expressions.

41. $\dfrac{-3a^2 - 10a - 4}{a(a+1)^2}$

42. $\dfrac{13y + 5}{2y^2 - 7y + 6}$

43. $\dfrac{7x - 1}{6x^2 - 5x + 1}$

44. $\dfrac{5x^2 + 9x - 56}{(x-4)(x-2)(x+1)}$

Chapter Test

1. In which cases is the matrix product AB defined? What is the size of the product?

    a. A is 3×5 and B is 5×7.

    b. A is 5×3 and B is 5×1.

    c. A is 1×3 and B is 3×3.

    d. A is 2×2 and B is 2×2.

Suppose
$$A = \begin{bmatrix} 5 & 0 \\ -1 & 2 \end{bmatrix}, \quad B = \begin{bmatrix} 3 & -3 \\ 2 & 1 \end{bmatrix}$$

Calculate the following matrices.

2. $A + B$
3. AB
4. BA
5. $|A|$

Suppose
$$A = \begin{bmatrix} 1 & 4 & -1 \\ 2 & 3 & 1 \\ 0 & -1 & 2 \end{bmatrix}$$

Calculate the following:

6. A_{31}
7. A_{13}
8. $|A|$

Calculate the inverse of the following matrices if the inverse exists:

9. $\begin{bmatrix} 1 & -1 & 2 \\ 0 & 1 & 3 \\ 2 & 1 & -2 \end{bmatrix}$

10. $\begin{bmatrix} -2 & -3 & 4 & 1 \\ 0 & 1 & 1 & 0 \\ 0 & 4 & -6 & 1 \\ -2 & -2 & 5 & 1 \end{bmatrix}$

11. Solve the following system using inverse matrices:
$$\begin{cases} 5x + 2y = 7 \\ -7x - 3y = 9 \end{cases}$$

12. A multinational company has branches in the United States, Canada, and Mexico. The branches in each country purchase goods from each other. Suppose that for each $1 of output of the U.S. branch, $.05 is used within that branch, $.18 is purchased by the Canadian branch, and $.11 is purchased by the Mexican branch. Suppose that for each $1 of output of the Canadian branch, $.10 is used by the branch itself, $.09 is purchased by the U.S. branch, and $.22 is purchased by the Mexican branch. Suppose that for each $1 of output of the Mexican branch, $.02 is used by the branch itself, $.40 by the U.S. branch, and $.20 by the Canadian branch.

    a. Determine the input–output matrix of the company.

    b. Determine the amount of output of each branch that is used within the company if the U.S. branch produces $500 million, the Canadian branch $120 million, and the Mexican branch $400 million.

13. **Thought question.** Describe how to use Cramer's rule to calculate the solutions to a system of linear equations. Are there systems to which Cramer's rule does not apply? Explain your answer.

14. **Thought question.** Describe how to calculate the determinant of a 5×5 matrix.

CHAPTER 12

SEQUENCES, SERIES, AND PROBABILITY

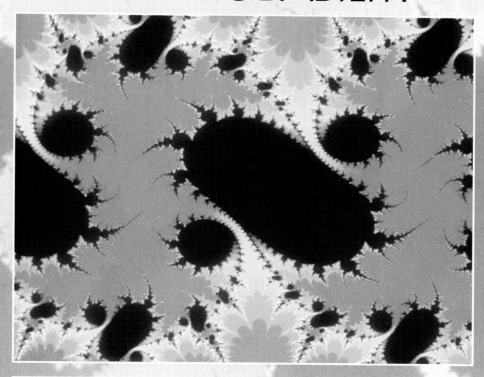

In this chapter, we discuss a number of topics that are preludes to more advanced mathematics courses. We begin by discussing the binomial theorem, used to raise binomials to positive, integer powers. To prove the binomial theorem (and many other results as well), we introduce the method of mathematical induction. We then discuss the algebra of sequences, both arithmetic and geometric, useful in developing models showing the state of systems at discrete time intervals. We use the results about geometric sequences to solve problems from the mathematics of finance. Finally, we use the formulas from the binomial theorem to discuss elementary counting and probability problems.

Chapter Opening Image: *A fractal image constructed using a geometrical object called the Julia set.*

12.1 THE BINOMIAL THEOREM

A **binomial** is an algebraic expression that is a sum of two terms. Here are some examples:

$$a + b, \quad 2x^2 - y, \quad \frac{5}{t} - 2t, \quad \sqrt{x} + 3y$$

Note that we have been working with such expressions throughout the book. In many applications, it is necessary to raise a binomial to a positive, integer power, such as:

$$(a + b)^2$$
$$(a + b)^3$$
$$(2x^2 - y)^5$$
$$(4\sqrt{x} + 3y)^{18}$$

It is possible to compute such powers by repeated multiplication. For example, if we do just that in the case of the first two powers above, we come up with the following formulas from Chapter 1:

$$(a + b)^2 = a^2 + 2ab + b^2$$
$$(a + b)^3 = a^3 + 3a^2b + 3ab^2 + b^3$$

For higher powers, such results are tedious to obtain by direct multiplication. However, the **binomial theorem** provides a shortcut.

To motivate the binomial theorem, let's examine some particular expansions:

$$(a + b)^0 = 1$$
$$(a + b)^1 = a + b$$
$$(a + b)^2 = a^2 + 2ab + b^2$$
$$(a + b)^3 = a^3 + 3a^2b + 3ab^2 + b^3$$
$$(a + b)^4 = a^4 + 4a^3b + 6a^2b^2 + 4ab^3 + b^4$$
$$(a + b)^5 = a^5 + 5a^4b + 10a^3b^2 + 10a^2b^3 + 5ab^4 + b^5$$

Based on these expansions, we can guess at the general pattern for:

$$(a + b)^n$$

1. The first term is:
$$a^n b^0 = a^n \cdot 1 = a^n$$

2. Each successive term is formed by decreasing the exponent of a by 1 and increasing the exponent of b by 1.

3. The sum of the exponents of a and b in any term is n.

4. The last term is:
$$a^0 b^n = 1 \cdot b^n = b^n$$

5. Each term has a numerical coefficient that must be determined.

If we denote the coefficient by $c_0, c_1, \ldots, c_n$, then we have the following result:

SECTION 12.1 The Binomial Theorem

BINOMIAL THEOREM (FIRST FORM)

The expansion of $(a + b)^n$ can be written in the form:
$$(a + b)^n = c_0 a^n + c_1 a^{n-1} b + c_2 a^{n-2} b^2 + \cdots + c_{n-1} a b^{n-1} + c_n b^n$$
where the coefficients on the right are calculated as follows:
1. The first coefficient c_0 equals 1.
2. To compute c_i for $i \geq 1$, first multiply the preceding coefficient by the exponent of a in the preceding term.
3. Then divide the result by the exponent of b namely i.

We will prove the binomial theorem in the next section.

EXAMPLE 1
Cube of a Binomial

Use the binomial theorem to verify the Cube of a Binomial formula:
$$(a + b)^3 = a^3 + 3a^2 b + 3ab^2 + b^3$$

Solution

We calculate the coefficients beginning with that for the term 0 coefficient:
$$c_0 = 1$$
Next, we calculate the term 1 coefficient:
$$c_1 = c_0 \cdot \frac{3}{1} = 1 \cdot \frac{3}{1} = 3$$
Next, we find the term 2 coefficient: The preceding term has exponent 2 on a. This term has exponent 1 on b.
$$c_2 = c_1 \cdot \frac{2}{2} = 3 \cdot \frac{2}{2} = 3$$
Finally, we find the term 3 coefficient: The preceding term has exponent 1 on a. This term has exponent 3 on b.
$$c_3 = c_2 \cdot \frac{1}{3} = 3 \cdot \frac{1}{3} = 1$$
These coefficients agree with those cited in the expansion:
$$(a + b)^3 = a^3 + 3a^2 b + 3ab^2 + b^3$$

EXAMPLE 2
Sixth Power of a Binomial

a. Calculate the coefficients in the expansion of $(a + b)^6$.
b. Write out the expansion of $(a + b)^6$.

Solution

1. We follow the calculational procedure described in Example 1.
$$c_0 = 1$$
$$c_1 = c_0 \cdot \frac{6}{1} = 6$$
$$c_2 = c_1 \cdot \frac{5}{2} = 6 \cdot \frac{5}{2} = 15$$

$$c_3 = c_2 \cdot \frac{4}{3} = 15 \cdot \frac{4}{3} = 20$$

$$c_4 = c_3 \cdot \frac{3}{4} = 20 \cdot \frac{3}{4} = 15$$

$$c_5 = c_4 \cdot \frac{2}{5} = 15 \cdot \frac{2}{5} = 6$$

$$c_6 = c_5 \cdot \frac{1}{6} = 6 \cdot \frac{1}{6} = 1$$

2. From the above calculation of the coefficients, the expansion of
$$(a + b)^6$$
is equal to:
$$a^6 + 6a^5b + 15a^4b^2 + 20a^3b^3 + 15a^2b^4 + 6ab^5 + b^6$$

Factorials, Binomial Coefficients, and Applications

The method just described allows us to calculate the coefficients of any binomial expansion. However, the method suffers from a serious difficulty: It requires that we compute the coefficients in order. To determine the tenth coefficient, say, we must compute the first nine coefficients. In many applications, such as probability and statistics, it is necessary to calculate single terms in binomial expansions. For such applications, the method we have given is very cumbersome. Let's now present another method that can be used to compute any specified term of a binomial expansion.

The method we described uses the notion of a factorial.

Definition
Factorial

Let n be a positive integer. Then **n factorial**, denoted $n!$, is the product of all positive integers from 1 to n. That is,
$$n! = n(n-1)(n-2) \cdots 2 \cdot 1$$

Here are some calculations of factorials:

$$1! = 1$$
$$2! = 2 \cdot 1 = 2$$
$$3! = 3 \cdot 2 \cdot 1 = 6$$
$$4! = 4 \cdot 3 \cdot 2 \cdot 1 = 24$$
$$5! = 5 \cdot 4 \cdot 3 \cdot 2 \cdot 1 = 120$$

EXAMPLE 3
Factorials

Calculate the value of $11!/(6!5!)$.

Solution

This type of quotient arises in calculating the coefficients in binomial expansions. In computing such a quotient, it is best to write out the definitions of all the factorials and do whatever simplification is possible.

SECTION 12.1 The Binomial Theorem

$$\frac{11!}{6!5!} = \frac{11 \cdot 10 \cdot 9 \cdot 8 \cdot 7 \cdot \cancel{6} \cdot \cancel{5} \cdot \cancel{4} \cdot \cancel{3} \cdot \cancel{2} \cdot \cancel{1}}{(\cancel{6} \cdot \cancel{5} \cdot \cancel{4} \cdot \cancel{3} \cdot \cancel{2} \cdot \cancel{1})(5 \cdot 4 \cdot 3 \cdot 2 \cdot 1)}$$

$$= \frac{11 \cdot \cancel{10}(2) \cdot \cancel{9}(3) \cdot \cancel{8} \cdot 7}{\cancel{5} \cdot \cancel{4} \cdot \cancel{3} \cdot \cancel{2} \cdot 1}$$

$$= 11 \cdot 2 \cdot 3 \cdot 7$$

$$= 462$$

The above definition of factorials defines $n!$ for a positive integer n. For use in connection with the binomial theorem, it is convenient to define $0!$ according to the formula:

$$0! = 1$$

Let's now introduce a new notation. The coefficient of $a^{n-k}b^k$ in the binomial expansion of $(a + b)^n$ is customarily denoted with the symbol:

$$\binom{n}{k}$$

This symbol is called a **binomial coefficient** and is read "binomial n over k." Note, however, that "n over k" does not mean the quotient of n divided by k. Rather, the term is just a convenient way of reminding us of the notation for the binomial coefficient.

From the binomial expansion for $(a + b)^3$, for example, we can read off the values of the following binomial coefficients:

$$\binom{3}{0} = 1, \quad \binom{3}{1} = 3, \quad \binom{3}{2} = 3, \quad \binom{3}{3} = 1$$

Note that $\binom{3}{0}$ is the first coefficient in the expansion, $\binom{3}{1}$ the second, and so forth. In general, $\binom{n}{k}$ is the $(k + 1)$st coefficient in the binomial expansion of $(a + b)^n$.

We can calculate binomial coefficients using the following result:

FACTORIAL FORMULA FOR BINOMIAL COEFFICIENTS

The binomial coefficients can be computed using the formula:

$$\binom{n}{k} = \frac{n!}{k!(n - k)!}$$

We delay proving this result until the next section. However, in the meantime, we can use it to evaluate binomial coefficients, as illustrated in the next example.

EXAMPLE 4
Calculating Binomial Coefficients

Use the factorial formula to compute the following binomial coefficients.

1. $\binom{3}{2}$ 2. $\binom{4}{3}$ 3. $\binom{7}{2}$

Solution

1. $\binom{3}{2} = \dfrac{3!}{2!(3-2)!} = \dfrac{3!}{2! \cdot 1!} = \dfrac{3 \cdot \cancel{2} \cdot 1}{\cancel{2} \cdot 1 \cdot 1} = 3$

   Note that this result agrees with the value of the coefficient stated earlier.

2. $\binom{4}{3} = \dfrac{4!}{3!(4-3)!} = \dfrac{4!}{3! \cdot 1!} = \dfrac{4 \cdot \cancel{3} \cdot \cancel{2} \cdot 1}{\cancel{3} \cdot \cancel{2} \cdot 1 \cdot 1} = 4$

3. $\binom{7}{2} = \dfrac{7!}{2!(7-2)!}$

   $= \dfrac{7!}{2! \cdot 5!}$

   $= \dfrac{7 \cdot 6 \cdot \cancel{5} \cdot \cancel{4} \cdot \cancel{3} \cdot \cancel{2} \cdot \cancel{1}}{(2 \cdot 1)(\cancel{5} \cdot \cancel{4} \cdot \cancel{3} \cdot \cancel{2} \cdot \cancel{1})}$

   $= \dfrac{7 \cdot \cancel{6}(3)}{\cancel{2} \cdot 1}$

   $= 7 \cdot 3 = 21$

Using the binomial coefficient notation, we can compute the $(k+1)$st term in the expansion of the binomial $(a+b)^n$ according to the formula:

$$(k+1)\text{st term} = \binom{n}{k} a^{n-k} b^k$$

Using this last formula, we can restate the expansion of a binomial in a slightly different form.

BINOMIAL THEOREM (SECOND FORM)

Let n be a nonnegative integer. Then:

$$(a+b)^n = \binom{n}{0} a^n + \binom{n}{1} a^{n-1} b + \binom{n}{2} a^{n-2} b^2 + \cdots + \binom{n}{n} b^n$$

Using the factorial formula for the binomial coefficients, we can easily calculate any particular term of a binomial expansion, as illustrated in the following example.

EXAMPLE 5
Determining a Single Term in a Binomial Expansion

Determine the fifth term in the expansion of

$$(x+y)^9$$

Solution

We have:

$$n = 9$$
$$k + 1 = 5$$
$$k = 4$$

The fifth term of the expansion is:

$$\binom{9}{4}x^{9-4}y^4 = \frac{9!}{4!(9-4)!}x^5y^4$$
$$= \frac{9!}{4!5!}x^5y^4$$
$$= \frac{\cancel{9}(3)\cdot\cancel{8}\cdot 7\cdot 6}{\cancel{4}\cdot\cancel{3}\cdot\cancel{2}\cdot 1}x^5y^4$$

Simplifying the fractional coefficient gives the equivalent expression:

$$3\cdot 7\cdot 6x^5y^4 = 126x^5y^4$$

EXAMPLE 6
Medicine

The recovery rate for a certain serious viral infection is 80 percent. A group of 100 people is infected with the virus. The probability that exactly k will recover is given by the $(k+1)$st term of the binomial expansion $(0.8 + 0.2)^{100}$. Determine the probability that exactly three people will recover from the infection.

Solution

The described probability is given by the fourth term in the expansion, which is equal to:

$$\binom{100}{3}(0.8)^{100-3}(0.2)^3 = \frac{100!}{3!(100-3)!}(0.8)^{97}(0.2)^3$$
$$= \frac{100!}{3!97!}(0.8)^{97}(0.2)^3$$
$$= \frac{100\cdot 99\cdot 98}{3\cdot 2\cdot 1}(0.8)^{97}(0.2)^3$$

To compute a numerical value of this last expression, we use a calculator. (Computing a 97th power by hand is not fun!) We find that the value of the above expression is approximately $5.1467E - 7 = .00000051467$.

Using a Calculator to Compute Binomial Coefficients

For large values of n, calculation of binomial coefficients can be very tedious. The TI-82 calculator allows us to calculate the binomial coefficient $\binom{n}{r}$ using the notation $_nC_r$ (which is introduced later in this chapter). The next example shows how this feature works.

EXAMPLE 7
Using a Calculator to Determine a Binomial Coefficient

Use a calculator to determine $\binom{20}{9}$.

Solution

Key the number 20 into the calculator. Then press [MATH] to get the menu shown in Figure 1. Select the option PRB in the first row. You will see the display shown in Figure 2. Use the arrow keys to select $_nC_r$ from the menu. Now press [ENTER]. You will be returned to the main screen. Then key in the number 9. See Figure 3. Press [ENTER]. The value of the binomial coefficient will be displayed. See Figure 4.

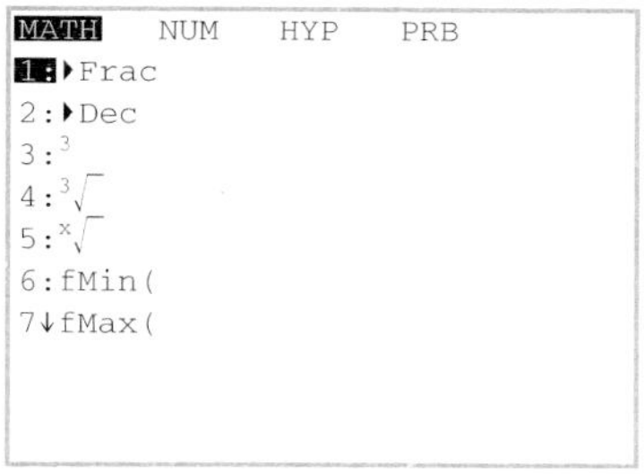

```
MATH   NUM   HYP   PRB
1:▶Frac
2:▶Dec
3:³
4:³√
5:ˣ√
6:fMin(
7↓fMax(
```

Figure 1

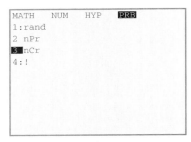

Figure 2

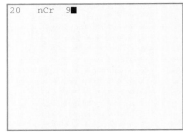

Figure 3

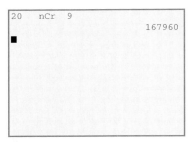

Figure 4

Expansions Using the Binomial Theorem

The binomial theorem gives the expansion of a binomial $(a + b)^n$. By replacing a and b with algebraic expressions, we can use the binomial theorem to compute the expansion of any binomial to a nonnegative, integer power. The following examples illustrate some typical expansions.

EXAMPLE 8
Expansion of a More Complex Binomial

Determine the expansion of the following binomial:
$$(2x^2 - y)^3$$

Solution

We apply the binomial theorem with a replaced by $2x^2$ and b replaced by $-y$. The result is:

$$(2x^2 - y)^3 = \binom{3}{0}(2x^2)^3(-y)^0 + \binom{3}{1}(2x^2)^2(-y)^1$$
$$+ \binom{3}{2}(2x^2)^1(-y)^2 + \binom{3}{3}(2x^2)^0(-y)^3$$

Now we replace the binomial coefficients by their respective values and simplify the various terms to obtain the expansion:

$$(2x^2)^3 + 3(2x^2)^2(-y)^1 + 3(2x^2)^1(-y)^2 + (-y)^3$$
$$= 8x^6 - 12x^4y + 6x^2y^2 - y^3$$

EXAMPLE 9
A Binomial Involving a Rational Expression

Determine the expansion of the following binomial:
$$\left(\frac{5}{t} - 2t\right)^4$$

Solution

We use the binomial theorem with a replaced by $5/t$ and b replaced by $-2t$:

$$\left(\frac{5}{t} - 2t\right)^4 = \binom{4}{0}\left(\frac{5}{t}\right)^4 + \binom{4}{1}\left(\frac{5}{t}\right)^3(-2t) + \binom{4}{2}\left(\frac{5}{t}\right)^2(-2t)^2$$
$$+ \binom{4}{3}\left(\frac{5}{t}\right)^1(-2t)^3 + \binom{4}{4}(-2t)^4$$

Inserting the values of the binomial coefficients, we have the following expansion:

$$\left(\frac{5}{t}\right)^4 + 4\left(\frac{5}{t}\right)^3(-2t) + 6\left(\frac{5}{t}\right)^2(-2t)^2 + 4\left(\frac{5}{t}\right)(-2t)^3 + (-2t)^4$$

SECTION 12.1 The Binomial Theorem

Simplifying each of the terms, we obtain the following expansion:
$$\frac{625}{t^4} - \frac{1{,}000}{t^2} + 600 - 160t^2 + 16t^4$$

EXAMPLE 10
A Binomial Involving a Radical

Determine the fifth term in the expansion of the following binomial:
$$\left(\sqrt{x} + 3y\right)^{18}$$

Solution

The desired term is given by the expression:
$$\binom{18}{4}\left(\sqrt{x}\right)^{18-4}(3y)^4 = \frac{18!}{4!(18-4)!}\left(\sqrt{x}\right)^{14}(3y)^4$$
$$= \frac{18 \cdot 17 \cdot 16 \cdot 15}{4 \cdot 3 \cdot 2 \cdot 1} x^7 \cdot 81 y^4$$
$$= 247{,}860 x^7 y^4$$

Pascal's Triangle

The seventeenth-century mathematician and philosopher Blaise Pascal rediscovered a simple method for computing binomial coefficients using only addition. (Up to this point, the methods we've used involve multiplying numbers that tend to be rather large.)

To understand Pascal's method, list the binomial coefficients that correspond to $n = 1, 2, 3, \ldots$, with the coefficients corresponding to a given value of n written across a row, as follows:

```
             1
           1   1
         1   2   1
       1   3   3   1
     1   4   6   4   1
   1   5  10  10   5   1
```

Each row of the table can be computed from the preceding row, as follows: The first element of the row is always 1. Each subsequent element is obtained by adding the two elements in the previous row that are directly above it: For example, consider row 3. The first element is 1. The next element, 2, is obtained by adding the two elements immediately above it in the preceding row, namely 1 + 1. The next element, 1, is obtained by adding a 1 and a blank space (which has a value of zero).

Now consider row 4. The first element is 1. The next is obtained by adding 1 + 2 from the row above. The next element is obtained by adding 2 + 1 from the row above. The final element is obtained by adding 1 and a blank from the row above.

In this manner, we can calculate the rows of the table one after another. Each row gives the set of binomial coefficients corresponding to one particular value of n. This method of computing the binomial coefficients is called **Pascal's triangle**.

The advantage of Pascal's triangle is that it allows us to calculate the binomial coefficients using addition rather than multiplication. The disadvantage, however, is that to apply Pascal's triangle, we must calculate all rows preceding the one we are interested in. If we need a row corresponding to a large value of n, then a considerable amount of calculation is involved.

Exercises 12.1

Evaluate the following expressions.
1. $0!$
2. $1!$
3. $8!$
4. $5!$
5. $\dfrac{6!}{4!2!}$
6. $\dfrac{8!}{5!3!}$
7. $\dfrac{7!}{2!5!}$
8. $\dfrac{10!}{6!4!}$
9. $\binom{6}{1}$
10. $\binom{8}{0}$
11. $\binom{9}{5}$
12. $\binom{12}{3}$

Determine the indicated term of the expansion.

13. fourth term, $(x+y)^7$
14. sixth term, $(a+b)^8$
15. eleventh term, $(t-3)^{15}$
16. tenth term, $(y-4)^{13}$
17. fifth term, $\left(\dfrac{a}{5} - \dfrac{1}{a^2}\right)^{10}$
18. fourth term, $\left(\sqrt{x} + 4t^3\right)^9$
19. middle two terms, $\left(\sqrt{t} - \sqrt{2}\right)^9$
20. middle term, $(5a^2 + 7b^3)^8$

Expand the following expressions using the binomial theorem.

21. $(a+b)^4$
22. $(p-q)^5$
23. $(4a-b)^6$
24. $(a^2+3b)^5$
25. $\left(\sqrt{2} + 2\sqrt{3}\right)^4$
26. $\left(t - \dfrac{t}{2}\right)^8$
27. $\left(\dfrac{a^2}{b} - b^3\right)^6$
28. $\left(2 + \dfrac{t}{2}\right)^8$
29. $(3x^2 + 2xy^3)^5$
30. $\left(\dfrac{3a^2}{2b} - \dfrac{2b^2}{3a}\right)^4$
31. $\left(\dfrac{1}{\sqrt{x}} + \sqrt{x}\right)^6$
32. $(a^3 - a^{-3})^5$
33. $(2-2)^n$
34. $(1+1)^n$
35. $\left(1 + \sqrt{2}\right)^5 - \left(1 + \sqrt{2}\right)^5$
36. $\left(1 - \sqrt{3}\right)^4 + \left(1 - \sqrt{3}\right)^4$
37. $\left(a + \sqrt{5}\right)^6$
38. $\left(a - \sqrt{3}\right)^4$
39. $(2a-b)^{10}$ to four terms.
40. $(2p+3q)^9$ to four terms.
41. $\left(\dfrac{1}{\sqrt{x}} - 3x^2\right)^8$ to three terms.
42. $\left(\sqrt{a} + \sqrt{b}\right)^{10}$ to five terms.
43. $\left(3x - \dfrac{1}{y}\right)^{12}$, the last four terms.
44. $\left(\dfrac{1}{a} + \dfrac{1}{3}a\sqrt{b}\right)^{10}$, the last three terms.
45. Find the term that does not contain a in the expansion of $[(3a^{-2}/2) + \tfrac{1}{3}a]^{12}$.
46. Find the term containing $1/t^3$ in $(4 - \tfrac{5}{2}t)^9$.

Use Pascal's triangle to find the coefficients of the binomial expansion of $(a+b)^n$. Then find each expansion.

47. $n=7$
48. $n=8$
49. $n=9$
50. $n=10$
51. Divide the fourth term of $(x^2 + 3yx^{1/3})^5$ by the third term and simplify.
52. Expand and simplify: $[(x+h)^5 - x^5]/h$.
53. Expand and simplify: $[(x+h)^n - x^n]/h$.
54. Find a formula for $(a+b+c)^n$.

In calculus, it can be shown that the formula for the binomial expansion can be altered in such a way that includes powers other than nonnegative integers. The formula is as follows:

$$(1+x)^n = 1 + nx + \dfrac{n(n-1)}{2!}x^2 + \dfrac{n(n-1)(n-2)}{3!}x^3 + \cdots,$$

where n is any real number, x is any real number such that $|x| < 1$, and the ellipsis ... means that the expansion continues indefinitely. Find each of the following expansions to eight terms, assuming $|x| < 1$.

55. $(1+x)^{-1}$
56. $(1-x)^{-2}$
57. $(1+x)^{1/2}$
58. $(1+x)^{1/3}$

Simplify the following expressions.

59. $\dfrac{\dfrac{(n+2)^2 2^{n+1}}{(n+1)!}}{\dfrac{n^2 2^n}{n!}}$

60. $\dfrac{\dfrac{(3x+6)^{n+1}}{(n+1)!}}{\dfrac{(3x+6)^n}{n!}}$

Applications

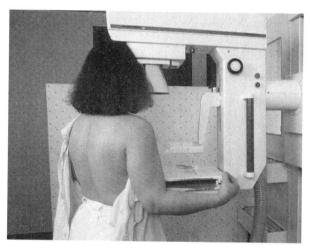

61. **Medicine.** About 9 percent of women will contract breast cancer sometime in their lifetimes. There are 100 women at a sales meeting. The probability that exactly 4 of them will contract breast cancer in their lifetimes is given by the fifth term of the binomial expansion of $(0.09 + 0.91)^{100}$. Find this term and use your calculator to approximate the probability.

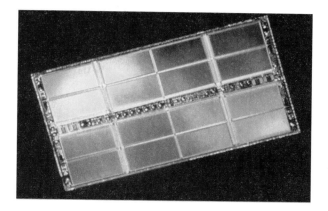

62. **Quality control.** An electrical company manufactures computer chips. About 2 percent of the chips are defective. A quality control inspector pulls 50 chips at random off an assembly line. The probability that exactly 10 of them are defective is given by the eleventh term of the binomial expansion of $(0.02 + 0.98)^{50}$. Find this term and use your calculator to approximate the probability.

Technology

Calculate the first five terms of the following expansions, using a calculator to determine the coefficients.

63. $(x + y)^{500}$
64. $(a - b)^{37}$

Use a calculator to determine the following binomial coefficients.

65. $\binom{50}{13}$
66. $\binom{30}{10}$

67. Before the advent of modern calculators and computers, mathematicians were always looking for ways to approximate the results of difficult or lengthy computations. For example, consider 1.01^{17}. An approximation might have been found by considering this as $1.01^{17} = (1.00 + 0.01)^{17}$. Use the sum of the first four terms of this binomial expansion to approximate 1.01^{17} to six decimal places. Approximate 1.01^{17} to six decimal places on your calculator. Compare both approximations.

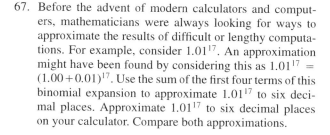

In your own words

68. Describe the binomial theorem in your own words. Illustrate it with two examples.
69. What is $n!$?
70. What is a binomial coefficient? Show how to calculate it using a specific example.

12.2 MATHEMATICAL INDUCTION

Let n be a positive integer. Consider the following formula, which we prove later in this section:

$$1 + 2 + \cdots + n = \frac{n(n+1)}{2}$$

It can be viewed as a series of statements, one corresponding to each value of n. Let's denote the above statement by P_n. Then P_1 is the statement:

$$1 = \frac{1 \cdot (1+1)}{2}$$

P_2 is the statement

$$1 + 2 = \frac{2 \cdot (2+1)}{2}$$

and so forth. The following result can be used to prove a series of such statements:

PRINCIPLE OF MATHEMATICAL INDUCTION

Suppose that for each positive integer k there is given a statement P_k. Furthermore, suppose that the following two conditions are fulfilled:

1. P_1 is true.
2. By assuming that P_k is true, we can deduce the truth of P_{k+1}. (Here, k is any positive integer.)

Then P_k is true for every positive integer k.

The operation of the Principle of Mathematical Induction is analogous to a row of dominoes. The truth of statement P_n is represented by the falling of the nth domino. Condition 1 is analogous to the first domino falling. Condition 2 is analogous to domino n knocking down domino $n+1$. Conditions 1 and 2 together are analogous to all the dominoes falling. That is, the statement P_n is true for every value of n.

In more advanced courses, the Principle of Mathematical Induction (in a slightly different form) is used as the fundamental tool for proving many of the most important properties of the natural numbers. In this text, we assume its truth. The next three examples show how the Principle of Mathematical Induction can be used as a technique of proof, called **proof by mathematical induction**.

EXAMPLE 1
Sum of the First n Integers

Let n be a positive integer. Use the mathematical induction to prove the formula:

$$1 + 2 + \cdots + n = \frac{n(n+1)}{2}$$

Solution

According to the Principle of Mathematical Induction, we can prove the formula by verifying Conditions 1 and 2. Condition 1 is simple, because P_1 is the statement:

$$1 = \frac{1 \cdot (1+1)}{2}$$

And this is a true statement. As for Condition 2, let's assume that statement P_n is true. That is, we assume that:

$$1 + 2 + \cdots + n = \frac{n(n+1)}{2}$$

Let's deduce from this statement the truth of statement P_{n+1}. To do so, start from the last equation and add $n+1$ to each side to obtain:

$$1 + 2 + \cdots + n + (n+1) = \frac{n(n+1)}{2} + (n+1)$$

Let's simplify the right side by combining the two expressions found there:

$$1 + 2 + \cdots + n + (n+1) = \frac{n(n+1) + 2(n+1)}{2}$$

Factoring $n+1$ from each term in the numerator, we have the equation:

$$1 + 2 + \cdots + n + (n+1) = \frac{(n+1)(n+2)}{2}$$

But this is precisely the statement P_{n+1}. Indeed, if we take the statement P_n and replace n with $n+1$ everywhere, we arrive at the last equation. Thus, starting from the truth of P_n, we have deduced the truth of P_{n+1}. So Condition 2 is verified. Because Conditions 1 and 2 hold, the Principle of Mathematical Induction says that P_n holds for all n. That is, the given formula holds for every positive integer n. This completes the proof.

Condition 2 requires that you prove that P_n implies P_{n+1}. This part of a proof by induction is often called the **induction step**.

The next example provides further practice in constructing proofs by induction.

EXAMPLE 2
Sum of the First n Odd Integers

Use mathematical induction to prove that for each positive integer n the following formula holds:

$$1 + 3 + 5 + \cdots + (2n - 1) = n^2$$

Solution

In this example, the statement P_1 is the formula:

$$1 = 1^2$$

which is clearly true. So Condition 1 holds. To prove the inductive step, assume that P_n holds. That is, assume that

$$1 + 3 + 5 + \cdots + (2n - 1) = n^2$$

We must deduce P_{n+1} from this formula. That is, we must deduce from the last formula the following result:

$$1 + 3 + 5 + \cdots + [2(n+1) - 1] = (n+1)^2$$

To see how to proceed, let's rewrite what we wish to deduce. Explicitly write out the next-to-last term to obtain:

$$1 + 3 + 5 + \cdots + (2n - 1) + [2(n + 1) - 1] = (n + 1)^2$$

Note that all the terms but the last form the expression on the left side of the statement P_n. This suggests that we start from P_n and add $2(n + 1) - 1 = 2n + 1$ to both sides of the equation to obtain:

$$1 + 3 + \cdots + (2n - 1) + (2[n + 1] - 1) = n^2 + [2(n + 1) - 1]$$
$$= n^2 + 2n + 1$$

Factoring the right side yields:

$$1 + 3 + \cdots + (2[n + 1] - 1) = (n + 1)^2$$

This is the statement P_{n+1}. That is, we have deduced P_{n+1} from P_n. This is the inductive step, which completes the proof of the formula.

Recall that, in the preceding section, we stated and applied the binomial theorem but did not supply a proof. Let's now remedy that omission.

EXAMPLE 3
Binomial Theorem

Let n be a positive integer. Use mathematical induction to prove the binomial theorem:

$$(a + b)^n = a^n + \binom{n}{1}a^{n-1}b + \binom{n}{2}a^{n-2}b^2 + \cdots + \binom{n}{n-1}ab^{n-1} + b^n$$

Solution

Here P_1 is the statement

$$(a + b)^1 = a^1 + b^1$$

This is clearly true. So let's prove the inductive step. Assume that P_n is true. Namely, assume that:

$$(a + b)^n = a^n + \binom{n}{1}a^{n-1}b + \binom{n}{2}a^{n-2}b^2 + \cdots + \binom{n}{n-1}ab^{n-1} + b^n$$

Let's multiply both sides of the equation by $a + b$. On the left, we obtain $(a+b)^{n+1}$, so that the equation reads:

$$(a + b)^{n+1} = (a + b)\left[a^n + \binom{n}{1}a^{n-1}b + \binom{n}{2}a^{n-2}b^2 + \cdots + \binom{n}{n-1}ab^{n-1} + b^n\right]$$

Apply the distributive law to the right side to obtain:
$(a + b)^{n+1} =$

$$\left[\binom{n}{0}a^{n+1} + \binom{n}{1}a^n b + \binom{n}{2}a^{n-1}b^2 + \cdots + \binom{n}{n-1}a^2 b^{n-1} + \binom{n}{n}ab^n\right]$$

$$+ \left[\binom{n}{0}a^n b + \binom{n}{1}a^{n-1}b^2 + \binom{n}{2}a^{n-2}b^3 + \cdots + \binom{n}{n-1}ab^n + \binom{n}{n}b^{n+1}\right]$$

Collecting like terms, we then have:

$$(a+b)^{n+1} =$$

$$\binom{n}{0}a^{n+1} + \left[\binom{n}{0}+\binom{n}{1}\right]a^n b + \left[\binom{n}{1}+\binom{n}{2}\right]a^{n-1}b^2$$

$$+ \left[\binom{n}{2}+\binom{n}{3}\right]a^{n-2}b^3 + \cdots + \left[\binom{n}{n-1}+\binom{n}{n}\right]ab^n + \binom{n}{n}b^{n+1}$$

$$= a^{n+1} + \left[\binom{n}{0}+\binom{n}{1}\right]a^n b + \left[\binom{n}{1}+\binom{n}{2}\right]a^{n-1}b^2$$

$$+ \left[\binom{n}{2}+\binom{n}{3}\right]a^{n-2}b^3 + \cdots + \left[\binom{n}{n-1}+\binom{n}{n}\right]ab^n + b^{n+1}$$

To complete simplification of the right side, let's obtain an alternate formula for the sums of binomial coefficients that appear. Let j be one of the integers $0, 1, \ldots, n-1$. Then we have:

$$\binom{n}{j}+\binom{n}{j+1} = \frac{n!}{j!(n-j)!} + \frac{n!}{(j+1)!(n-(j+1))!}$$

$$= n!\left(\frac{1}{j!(n-j)!} + \frac{1}{j!\cdot(j+1)\cdot(n-j-1)!}\right)$$

$$= \frac{n!}{j!}\left(\frac{1}{(n-j-1)!\cdot(n-j)} + \frac{1}{(j+1)\cdot(n-j-1)!}\right)$$

$$= \frac{n!}{j!(n-j-1)!}\left(\frac{1}{n-j} + \frac{1}{j+1}\right)$$

$$= \frac{n!}{j!(n-j-1)!} \cdot \frac{n+1}{(n-j)(j+1)}$$

$$= \frac{(n+1)!}{(j+1)!(n-j)!}$$

$$= \binom{n+1}{j+1}$$

Using this formula, we replace the sums of the binomial coefficients in the preceding equation to obtain:

$$(a+b)^{n+1} = a^{n+1} + \binom{n+1}{1}a^n b + \binom{n+1}{2}a^{n-1}b^2 + \binom{n+1}{3}a^{n-2}b^3 + \cdots$$

$$+ \binom{n+1}{n}ab^n + b^{n+1}$$

But this equation is precisely statement P_{n+1}. This completes the inductive step and, with it, the proof of the binomial theorem.

Note that, in the first two examples, we added a term to each side of the formula to obtain the inductive step. In the third example, the inductive step was more complicated and involved first multiplying both sides of the equation by

624   CHAPTER 12   Sequences, Series, and Probability

$a + b$, then performing considerable simplification, and then proving an identity about binomial coefficients. This is fairly typical of the ingenuity that is sometimes required to prove the inductive step.

The preceding examples used induction to prove formulas. Induction can also be used to prove inequalities, as the next example shows.

EXAMPLE 4
Inductive Proof of an Inequality

Let n be a positive integer. Prove that $2n + 1 \leq 3^n$.

Solution

In this case, the statement P_1 is the true inequality:
$$2 \cdot 1 + 1 \leq 3^1$$
$$3 \leq 3$$

Let's now prove the inductive step. We assume that P_n holds. That is, we assume that:
$$2n + 1 \leq 3^n$$

We must prove that:
$$2(n + 1) + 1 \leq 3^{n+1}$$

To prove this, let's start from the assumed inequality and add 2 to both sides to obtain:
$$(2n + 1) + 2 \leq 3^n + 2$$
$$2(n + 1) + 1 \leq 3^n + 2$$

Because n is a positive integer, we have:
$$3^n \geq 3 > 2$$

Therefore, applying this fact to the preceding inequality, we have:
$$2(n + 1) + 1 \leq 3^n + 2$$
$$< 3^n + 3^n = 2 \cdot 3^n \qquad \textit{Because } 3^n > 2$$
$$\leq 3 \cdot 3^n = 3^{n+1}$$
$$2(n + 1) + 1 \leq 3^{n+1}$$

This last inequality is the assertion for $n + 1$. This proves the inductive step and completes the proof of the inequality.

Exercises 12.2

Use mathematical induction to prove that the following formulas hold for all positive integers n.

1. $1 + 5 + 9 + \cdots + (4n - 3) = 2n^2 - n$

2. $1 + 4 + 7 + \cdots + (3n - 2) = \dfrac{n(3n - 1)}{2}$

3. $1 \cdot 2 + 3 \cdot 4 + 5 \cdot 6 + \cdots + (2n - 1)2n = \dfrac{n(n + 1)(4n - 1)}{3}$

4. $1^2 + 3^2 + 5^2 + \cdots + (2n - 1)^2 = \dfrac{n(4n^2 - 1)}{3}$

5. $\dfrac{1}{2} + \dfrac{1}{4} + \dfrac{1}{8} + \cdots + \dfrac{1}{2^n} = \dfrac{2^n - 1}{2^n}$

6. $3 + 3^2 + 3^3 + \cdots + 3^n = \dfrac{3^{n+1} - 3}{2}$

7. $1^3 + 2^3 + 3^3 + \cdots + n^3 = \left[\dfrac{n(n + 1)}{2}\right]^2$

8. $\dfrac{1}{1 \cdot 2} + \dfrac{1}{2 \cdot 3} + \dfrac{1}{3 \cdot 4} + \cdots + \dfrac{1}{n(n + 1)} = \dfrac{n}{n + 1}$

9. $1^2 + 2^2 + 3^2 + \cdots + n^2 = \dfrac{n(n+1)(2n+1)}{6}$

10. $2 + 2^2 + 2^3 + \cdots + 2^n = 2(2^n - 1)$

11. $2 + 4 + 6 + \cdots + 2n = n^2 + n$

12. $1 + 3 + 5 + \cdots + (2n - 1) = n^2$

13. $1 + 2 + 3 + \cdots + n = \dfrac{n(n+1)}{2}$

14. $1 + 2 + 3 + \cdots + n < \dfrac{(2n+1)^2}{8}$

15. $n(n^2 + 5)$ is a multiple of 6.

16. $3^{2n} - 1$ is a multiple of 8.

17. Three is a factor of $n^3 - n + 3$.

18. Four is a factor of $5^n - 1$.

19. $n^2 + n$ is divisible by 2.

20. $n^3 - n$ is divisible by 6.

21. Seven is a factor of $11^n - 4^n$.

22. Three is a factor of $n^3 + 2n$.

23. $4^n > n^4$ for $n \geq 5$ (**Hint:** Start by proving P_5 is true).

24. $3^n < 3^{n+1}$

25. $n^2 < 2^n$ for $n > 4$.

26. $2n < 2^n$ for $n \geq 3$.

For the following functions $f(n)$ in Exercises 27–30, do a–c.
a. Complete the table.
b. On the basis of the results in the table, what would you believe to be the value of $f(7)$? Compute $f(7)$.
c. Make a conjecture about the value of $f(n)$.

27. $f(n) = 1 + 2 + 4 + \cdots + 2^{n-1}$

n	1	2	3	4	5	6
$f(n)$						

28. $f(n) = 3 + 7 + 11 + \cdots + (4n - 1)$

n	1	2	3	4	5	6
$f(n)$						

29. $f(n) = 2 + 9 + 16 + \cdots + (7n - 5)$

n	1	2	3	4	5	6
$f(n)$						

30. $f(n) = 1 + 3 + 5 + \cdots + (2n - 1)$

n	1	2	3	4	5	6
$f(n)$						

31. Prove that it is possible to pay any debt of \$4, \$5, \$6, \$7, ..., \$n and so on, by using only \$2 and \$5 bills.

32. Prove that it is possible to pay any debt of \$8, \$9, \$10, ..., \$n and so on, by using only \$3 and \$5 bills.

33. Prove that the number of diagonals of an n-sided convex polygon (a polygon for which all diagonals lie in the interior of the polygon) is $\dfrac{n(n-3)}{2}$ for $n \geq 3$.

34. Prove that the sum of the measures of the angles in an n-sided convex polygon is $180°(n - 2)$ for $n \geq 3$.

Find the smallest positive integer n for which the given statement is true. Then prove that the statement is true for all integers greater than or equal to that smallest value.

35. $n + 5 < 2^n$

36. $\log n < n$

37. $3^n > 2^n + 20$

38. $(1 + x)^n \geq 1 + nx$, if $x \geq -1$

What can you conclude from the given information about the sequence of statements? For example, given that P_4 is true and that P_k implies P_{k+1} is true for any k, you can conclude that P_n is true for every integer $n \geq 4$.

39. P_{18} is true and P_k implies P_{k+1}.

40. P_{18} is not true and P_k implies P_{k+1}.

41. P_1 is true but P_k does not imply P_{k+1}.

42. P_1 is true and P_k implies P_{k+2}.

43. P_1 and P_2 are true and P_k and P_{k+1} together imply P_{k+2}.

44. P_1 is true and P_k implies P_{4k}.

In your own words

45. Describe the Principle of Mathematical Induction. Illustrate it with a particular example.

46. Give an example of mathematical induction in everyday life.

12.3 SEQUENCES AND SERIES

Many applications involve quantities that are reported at discrete intervals. For instance, the balance on a mortgage is computed monthly, the unemployment rate is reported monthly, and the temperature is reported hourly by the weather bureau. To record and manipulate such data, mathematicians use sequences.

626 CHAPTER 12 Sequences, Series, and Probability

Definition Sequence A **sequence** is a list of real numbers organized in a particular order. The numbers contained in a sequence are called its **terms**.

Here are some examples of sequences:

$$1, \frac{1}{2}, \frac{1}{4}, \frac{1}{8}, \ldots$$

$$1, -1, 1, -1, \ldots$$

$$5, 4, 3, 2, 1$$

The first two sequences above contain infinitely many terms and are called **infinite sequences**. The third sequence above contains a finite number of terms and is therefore called a **finite sequence**. The terms of a sequence are customarily labeled using positive integer subscripts to indicate the term's position. For example, here is how a typical sequence is written using subscript notation:

$$a_1, a_2, a_3, \ldots, a_n, \ldots$$

That is, a_1 denotes the first term of the sequence, a_2 denotes the second term, and a_n denotes the nth term.

One method for defining a finite or infinite sequence is to specify a formula for computing a_n in terms of n. In the next example, we calculate terms of sequences specified in this fashion.

EXAMPLE 1
Determining Terms of Sequences

Determine terms 1, 2, 3, and 7 of the sequences defined by:

1. $a_n = \dfrac{1}{n}$

2. $a_n = 1 + \dfrac{(-1)^n}{n}$

3. $a_n = \dfrac{n^n}{n!}$

Solution

In each case, we calculate a_n by substituting the value of n in the given formula for a_n.

1. $a_1 = \dfrac{1}{1} = 1$

   $a_2 = \dfrac{1}{2}$

   $a_3 = \dfrac{1}{3}$

   $a_7 = \dfrac{1}{7}$

2. $a_1 = 1 + \dfrac{(-1)^1}{1} = 1 + \dfrac{-1}{1} = 0$

   $a_2 = 1 + \dfrac{(-1)^2}{2} = 1 + \dfrac{1}{2} = \dfrac{3}{2}$

   $a_3 = 1 + \dfrac{(-1)^3}{3} = 1 + \dfrac{-1}{3} = \dfrac{2}{3}$

   $a_7 = 1 + \dfrac{(-1)^7}{7} = 1 + \dfrac{-1}{7} = \dfrac{6}{7}$

SECTION 12.3 Sequences and Series

3. $a_1 = \dfrac{1^1}{1!} = \dfrac{1}{1} = 1$

$a_2 = \dfrac{2^2}{2!} = \dfrac{2 \cdot 2}{1 \cdot 2} = 2$

$a_3 = \dfrac{3^3}{3!} = \dfrac{3 \cdot 3 \cdot 3}{1 \cdot 2 \cdot 3} = \dfrac{9}{2}$

$a_7 = \dfrac{7^7}{7!} = \dfrac{7^6}{1 \cdot 2 \cdot 3 \cdot 4 \cdot 5 \cdot 6} = \dfrac{117{,}649}{720}$

All the sequences in the preceding example were defined by formulas for calculating a_n in terms of n. In many applications, it is most convenient to define a sequence by specifying how to calculate a_n in terms of preceding terms of the sequence $a_1, a_2, \ldots, a_{n-1}$. Such sequences are said to be defined **recursively**.

For instance, let a_n denote the amount of money in a bank account at the end of n months, where the account earns 0.5 percent interest per month. Then the amount a_n is related to the amount at the end of the preceding month, a_{n-1}, by the formula:

$$a_n = 1.005 a_{n-1}$$

If the amount in the account at the end of month 1 is \$1,000, then $a_1 = 1{,}000$. The two equations $a_n = 1.005 a_{n-1}$ and $a_1 = 1{,}000$ suffice to define the sequence of bank balances recursively.

EXAMPLE 2
Calculating Terms from One Another

Determine the first five terms of the bank balance sequence defined above.

Solution

The value of a_1 gives the first term. Each successive term can be computed from its predecessor using the given formula. Here are the calculations for the first five terms:

$a_1 = 1{,}000$

$a_2 = 1.005 \cdot a_1 = 1.005 \cdot 1{,}000 = 1{,}005$

$a_3 = 1.005 \cdot a_2 = 1.005 \cdot 1{,}005 = 1{,}010.025$

$a_4 = 1.005 \cdot a_3 = 1.005 \cdot 1{,}010.025 = 1{,}015.075125$

$a_5 = 1.005 \cdot a_4 = 1.005 \cdot 1{,}015.0751 = 1{,}020.150501$

The sum of the first n terms of a sequence is called the **nth partial sum** of the sequence and is denoted S_n. That is,

$$S_n = a_1 + a_2 + \cdots + a_n$$

Thus, for example, if we consider the sequence $a_n = 2^{-n}$, then:

$$S_1 = a_1 = 2^{-1} = \dfrac{1}{2}$$

$$S_2 = a_1 + a_2$$
$$= 2^{-1} + 2^{-2}$$
$$= \dfrac{1}{2} + \dfrac{1}{4} = \dfrac{3}{4}$$

628 CHAPTER 12 Sequences, Series, and Probability

$$S_3 = a_1 + a_2 + a_3$$
$$= 2^{-1} + 2^{-2} + 2^{-3}$$
$$= \frac{1}{2} + \frac{1}{4} + \frac{1}{8} = \frac{7}{8}$$

A sequence is a function of an integer variable, namely its subscript. For each value n of the subscript, there is a value a_n that is the corresponding term of the sequence. In dealing with sums of consecutive terms of a sequence, it is often convenient to use a special notation:

Definition
Summation Notation

Suppose that A and B are integers with $A < B$. The symbol

$$\sum_{n=A}^{B} a_n$$

is another way of denoting the sum:

$$a_A + a_{A+1} + \cdots + a_B$$

The notation

$$\sum_{n=A}^{B} a_n$$

tells us to add up the terms a_n, where the variable n begins with A and ends with B. The variable n is called the **summation index**. The statement $n = A$ is written under the summation symbol $\sum$, and B is written above the summation symbol.

The next example provides some practice in using the summation notation.

EXAMPLE 3
Evaluating Sums Given by Summation Notation

Evaluate the following expressions:

1. $\sum_{n=1}^{5} n$

2. $\sum_{n=2}^{5} n^2$

3. $\sum_{n=0}^{3} \binom{3}{n}$

Solution

1. We add the terms of the summation corresponding to the values $n = 1, 2, 3, 4, 5$:

$$\sum_{n=1}^{5} n = 1 + 2 + 3 + 4 + 5 = 15$$

2. The terms of the summation are obtained by evaluating n^2 in succession, for $n = 2, 3, 4, 5$.

$$\sum_{n=2}^{5} n^2 = 2^2 + 3^2 + 4^2 + 5^2 = 54$$

3. The terms of the summation are obtained by substituting, in succession, $n = 0, 1, 2, 3$.

$$\sum_{n=0}^{3} \binom{3}{n} = \binom{3}{0} + \binom{3}{1} + \binom{3}{2} + \binom{3}{3}$$
$$= 1 + 3 + 3 + 1$$
$$= 8$$

Summation notation provides a convenient way of writing many mathematical facts in a shorthand form. For instance, the definition of the nth partial sum of a sequence $a_1, a_2, \ldots, a_n, \ldots$ can be written in summation notation as:

$$S_n = \sum_{j=1}^{n} a_j$$

Note that in this formula, we used j as the summation index. The right side of the above equation tells us to add up all the terms a_j for j beginning with 1 and ending with n. Note that we couldn't use n to denote the summation index, because this letter is already in use, namely as the index of the last term to be added. You can use any variable name you wish for the summation index, so long as it won't be confused with a variable in use for some other purpose. For instance, we could have just as easily denoted the summation index k and written the last formula in the form:

$$\sum_{k=1}^{n} a_k$$

Because the summation index can be replaced by any unused variable, it is often referred to as a **dummy variable**. The most common are $i, j,$ and k.

Here are some further applications of summation notation. A formula we proved by induction in Section 12.2 is:

$$1 + 2 + 3 + \cdots + N = \frac{N(N+1)}{2}$$

In summation notation, this formula can be written:

$$\sum_{n=1}^{N} n = \frac{N(N+1)}{2}$$

The binomial theorem states that for a positive integer N, we have:

$$(a+b)^N = a^N + \binom{N}{1}a^{N-1}b + \binom{N}{2}a^{N-2}b^2 + \cdots + \binom{N}{N-1}ab^{N-1} + b^N$$

In terms of summation notation, this theorem can be written:

$$(a+b)^N = \sum_{n=0}^{N} \binom{N}{n} a^{N-n} b^n$$

Indeed, the first term in the above summation corresponds to $n = 0$:

$$\binom{N}{0} a^{N-0} b^0 = 1 \cdot a^N \cdot 1 = a^N$$

The second term corresponds to $n = 1$:

$$\binom{N}{1} a^{N-1} b^1 = \binom{N}{1} a^{N-1} b$$

The final term corresponds to $n = N$:

$$\binom{N}{N} a^{N-N} b^N = 1 \cdot a^0 b^N = b^N$$

As you can see, the terms in the summation correspond exactly to the terms on the right side of the binomial theorem. As another example of an application of the summation notation, consider the polynomial:

$$a_0 + a_1 X + a_2 X^2 + \cdots + a_N X^N$$

In terms of the summation notation, this polynomial can be written:

$$\sum_{n=0}^{N} a_n X^n$$

Exercises 12.3

Determine terms 1, 2, 3, and 8 of the sequences defined by the following formulas.

1. $a_n = n + 2$
2. $a_n = n - 4$
3. $a_n = \dfrac{n+1}{n-2}$
4. $a_n = \dfrac{n-1}{n+1}$
5. $a_n = 2^{n-3}$
6. $a_n = 3^{n-4}$
7. $a_n = n^{-2}$
8. $a_n = n^3$
9. $a_n = (-2)\sqrt[n]{n}$
10. $a_n = (-1)^{n-1} \sqrt{2n}$
11. $a_n = 3^{n-1} - 3^n$
12. $a_n = 2^{n+1} - 2^{n-1}$
13. $a_n = \dfrac{n^2-1}{n^2-2}$
14. $a_n = \dfrac{n^3+1}{n^3-2}$

Find the first five terms for the following sequences.

15. $a_1 = 2$, $a_n = a_{n-1} + 5$, $n > 1$
16. $a_1 = -3$, $a_n = 2a_{n-1}$, $n > 1$
17. $a_1 = 9$, $a_n = \dfrac{1}{3} a_{n-1}$, $n > 1$
18. $a_1 = 1$, $a_n = \left(-\dfrac{1}{3}\right)^n a_{n-1}$, $n > 1$
19. $a_n = n + \dfrac{1}{n}$
20. $a_n = n^3 + 1$
21. $a_n = \left(-\dfrac{1}{2}\right)^n (n^{-1})$
22. $a_n = \dfrac{1}{(n+1)!}$
23. $a_n = \dfrac{1}{2n} \log 1{,}000^n$
24. $a_n = n \log 10^{n-1}$
25. $a_1 = 3$, $a_{n+1} = \dfrac{1}{a_n}$
26. $a_1 = -2$, $a_{n+1} = -a_n^{-1}$

Evaluate the following summations.

27. $\sum_{n=1}^{4} (n - 2)$
28. $\sum_{n=1}^{3} (2n + 1)$
29. $\sum_{n=1}^{3} (n^2 + 1)$
30. $\sum_{n=1}^{5} (-2)^n$
31. $\sum_{n=2}^{4} n^{-1}$
32. $\sum_{n=0}^{3} (4n - 1)$
33. $\sum_{n=0}^{7} \log 100^n$
34. $\sum_{n=1}^{6} \log 1{,}000^{n-1}$
35. $\sum_{n=1}^{50} 3$
36. $\sum_{n=2}^{200} -1$
37. $\sum_{n=1}^{10} [1 + (-1)^n]$
38. $\sum_{n=1}^{20} (-2n)$

Determine the nth term of the sequence.

39. $1, 3, 5, 7, 9, \ldots$
40. $3, 9, 27, 81, 243, \ldots$
41. $2, -4, 8, -16, 32, -64, \ldots$
42. $1, -\dfrac{1}{2}, \dfrac{1}{3}, -\dfrac{1}{4}, \dfrac{1}{5}, -\dfrac{1}{6}, \ldots$
43. $\dfrac{1}{2}, \dfrac{1}{8}, \dfrac{1}{32}, \dfrac{1}{128}, \ldots$
44. $17, 21, 25, 29, \ldots$

Rewrite each of the following summations using sigma notation.

45. $1 + \frac{1}{2} + \frac{1}{3} + \frac{1}{4} + \frac{1}{5}$

46. $3 + 6 + 12 + 24 + 48 + 96$

47. $1+2+3+4+5+6+7+8+9+10$

48. $-2+4-8+16-32+64-128+256$

49. $-3 + 9 - 27 + 81 - 243 + 729$

50. $1 + 5 + 9 + 13 + 17$

Solve the following equations for x.

51. $\sum_{n=4}^{6} nx = 30$

52. $\sum_{n=1}^{12} (nx - 5) = 64$

53. $\sum_{n=3}^{5} (-1)^n (x + 1) = 45$

54. $\sum_{n=1}^{7} (-1)^n x = -2$

55. $\sum_{n=0}^{5} n(n - x) = 25$

56. $\sum_{n=1}^{6} (x - 3n) = -3$

Technology

Use a calculator to determine the tenth term of the following sequences:

57. $a_n = (-1)^n 5^{n+1}$

58. $a_n = \dfrac{3}{\sqrt{2n + 1}}$

59. $a_n = 2a_{n-1}, \quad a_1 = 10$

60. $a_n = 1.05a_{n-1} - 10, \quad a_1 = 100$

Use a calculator to evaluate the following summations.

61. $\sum_{n=1}^{10} (2n + 1)^2$

62. $\sum_{n=1}^{15} n^2$

63. $\sum_{n=1}^{10} \frac{1}{n}$

64. $\sum_{j=1}^{10} \frac{1}{2j + 1}$

In your own words

65. Explain the concept of a sequence. Use a specific example.

66. Give two examples of sequences you have encountered in magazines or newspapers.

67. Explain how summation notation works.

68. Explain the advantages of using summation notation.

12.4 ARITHMETIC SEQUENCES

In this section and the next, we study two particular types of sequences that play an important role in both pure and applied problems, namely, arithmetic and geometric sequences.

Definition
Arithmetic Sequence An **arithmetic sequence** is a sequence of real numbers in which the differences between consecutive terms are constant.

EXAMPLE 1
Proving a Sequence Is Arithmetic

Prove that the following sequence is arithmetic:

$$6, 11, 16, \ldots, 5n + 1, \ldots$$

Solution

Let's compute the difference between two typical consecutive terms, namely a_{n+1} and a_n. We have:

$$a_{n+1} - a_n = [5(n + 1) + 1] - (5n + 1)$$
$$= (5n + 6) - (5n + 1)$$
$$= 5$$

This computation shows that the differences between consecutive terms are constant and, in fact, all equal to 5. Thus the sequence is arithmetic.

Suppose that an arithmetic sequence has first term a_1 and that the difference between consecutive terms is d. Then the terms of the sequence are:

$$a_1$$
$$a_2 = a_1 + d$$
$$a_3 = a_2 + d$$
$$= (a_1 + d) + d$$
$$= a_1 + 2d$$
$$a_4 = a_3 + d$$
$$= (a_1 + 2d) + d$$
$$= a_1 + 3d$$

These formulas generalize to the following result.

---

nTH TERM OF AN ARITHMETIC SEQUENCE

Suppose that an arithmetic sequence has first term a_1 and has difference between consecutive terms d. Then the nth term of the sequence is given by the formula:

$$a_n = a_1 + (n - 1)d$$

We can verify the formula for the sum of an arithmetic progression using induction. We have shown that the formula holds for $n = 1, 2, 3, 4$. Assume now that the formula holds for n. That is, assume that:

$$a_n = a_1 + (n - 1)d$$

We want to prove that:

$$a_{n+1} = a_1 + [(n + 1) - 1]d$$

Because the sequence is arithmetic with difference d, we see that:

$$a_{n+1} - a_n = d$$
$$a_{n+1} = a_n + d$$
$$a_{n+1} = [a_1 + (n - 1)d] + d \quad \text{\textit{Because we assume} } a_n = a_1 + (n - 1)d$$
$$= a_1 + nd$$
$$= a_1 + [(n + 1) - 1]d$$

This last equation is the formula for a_{n+1}. This completes the inductive step and the proof of the desired formula.

EXAMPLE 2
Evaluation of the Term of an Arithmetic Sequence

Suppose that the first term of an arithmetic sequence is 5 and the difference between consecutive terms is 7. What is the value of the thirtieth term of the sequence?

Solution

Use the formula for the nth term of an arithmetic progression with $a_1 = 5$ and $d = 7$. We have:
$$a_{30} = a_1 + (30 - 1)d$$
$$= 5 + 29 \cdot 7$$
$$= 208$$

EXAMPLE 3
Calculating the First Term and Difference of an Arithmetic Sequence

Suppose that $a_1, a_2, \ldots$ is an arithmetic sequence. Suppose also that $a_{15} = 40$ and $a_{20} = 50$. Determine the first term of the sequence and the difference d between consecutive terms of the sequence.

Solution

Use the formula for the nth term of an arithmetic progression. We have:
$$a_{15} = 40$$
$$a_1 + (15 - 1)d = 40$$
$$a_1 + 14d = 40 \qquad (1)$$
$$a_{20} = 50$$
$$a_1 + (20 - 1)d = 50$$
$$a_1 + 19d = 50 \qquad (2)$$

Subtracting equation (1) from equation (2), we have:
$$19d - 14d = 50 - 40$$
$$5d = 10$$
$$d = 2$$

Substituting this value of d into equation (1) gives us:
$$a_1 + 14d = 40$$
$$a_1 + 14 \cdot 2 = 40$$
$$a_1 = 12$$

Thus, the sequence has first term 12 and difference 2.

The following result provides a simple formula for calculating the sum of an arithmetic sequence:

PARTIAL SUM OF AN ARITHMETIC SEQUENCE

Let $a_1, a_2, a_3, \ldots$ be an arithmetic sequence with difference d, and let S_n be its nth partial sum. Then:
$$S_n = \frac{n}{2}(a_1 + a_n)$$
$$= na_1 + \frac{n(n-1)}{2}d$$

Let's verify these formulas using a variety of facts. From the definition of the nth partial sum, we have:
$$S_n = a_1 + a_2 + \cdots + a_n$$
Let's now use the formula for the nth term of an arithmetic progression in each term on the right side:
$$\begin{aligned} S_n &= a_1 + (a_1 + d) + (a_1 + 2d) + \cdots + [a_1 + (n-1)d] \\ &= (a_1 + a_1 + \cdots + a_1) + [d + 2d + \cdots + (n-1)d] \\ &= na_1 + [1 + 2 + \cdots + (n-1)]d \end{aligned} \qquad (1)$$
In Section 12.2, we proved the formula:
$$1 + 2 + \cdots + n = \frac{n(n+1)}{2}$$
Replacing n by $n - 1$ in this formula, we see that:
$$1 + 2 + \cdots + (n-1) = \frac{(n-1)(n-1+1)}{2} = \frac{n(n-1)}{2}$$
Substituting this last formula into equation (1), we have:
$$S_n = na_1 + \frac{n(n-1)}{2}d$$
This is the second formula for the sum of an arithmetic sequence. To get the first formula, rewrite the last formula as follows:
$$\begin{aligned} S_n &= \frac{2na_1}{2} + \frac{n(n-1)d}{2} \\ &= \frac{n}{2}[2a_1 + (n-1)d] \\ &= \frac{n}{2}[a_1 + a_1 + (n-1)d] \\ &= \frac{n}{2}(a_1 + a_n) \end{aligned}$$
This is the desired statement.

EXAMPLE 4
Calculating the Sum of an Arithmetic Sequence

Compute the sum of the first fifty terms of the arithmetic sequence:
$$1, 4, 7, 10, \ldots, 3n - 2, \ldots$$

Solution

The sequence has $a_1 = 1, d = 3$. Therefore, by the formula for the sum of an arithmetic progression, we have:
$$S_n = na_1 + \frac{n(n-1)}{2}d$$
$$S_{50} = 50 \cdot 1 + \frac{50 \cdot (50-1)}{2} \cdot 3 = 3{,}725$$

EXAMPLE 5
Seats in an Auditorium

A concert auditorium has 30 rows of seats. The first row contains 50 seats. As you move to the rear of the auditorium, each row has two more seats than the previous one. How many seats are in the auditorium?

SECTION 12.4 Arithmetic Sequences

Solution

Let a_n denote the number of seats in the nth row. Then the sequence $a_1, a_2, \ldots, a_{30}$ is an arithmetic sequence with first term 50 and difference 2. The number of seats in the auditorium is the sum of the first 30 terms of the sequence, or S_{30}. By the formula for the sum of an arithmetic sequence, this quantity equals:

$$S_n = na_1 + \frac{n(n-1)}{2}d$$

$$S_{30} = 30 \cdot 50 + \frac{30 \cdot 29}{2} \cdot 2 = 2,370$$

That is, the auditorium has 2,370 seats.

EXAMPLE 6 Evaluate the sum:
Evaluating the Sum of Terms
in an Arithmetic Sequence

$$\sum_{n=1}^{10}(5n + 3)$$

Solution

The numbers being summed are:

$$5 \cdot 1 + 3, 5 \cdot 2 + 3, \ldots, 5 \cdot 10 + 3$$

This arithmetic sequence has first term 8 and last term 53. By the formula for the sum of an arithmetic progression, the sum of these 10 terms is equal to:

$$S_{10} = \frac{10}{2}(a_1 + a_{10})$$
$$= 5 \cdot (8 + 53)$$
$$= 305$$

Exercises 12.4

Prove that the following sequences are arithmetic.

1. $1, 3, 5, 7, \ldots, 2n - 1, \ldots$
2. $4, 5, 6, \ldots, n + 3, \ldots$
3. $4, 7, 10, \ldots, 1 + 3n, \ldots$
4. $0, 2, 4, \ldots, 2n - 2, \ldots$
5. $7, 8, 9, \ldots, n + 6, \ldots$
6. $9, 14, 19, \ldots, 4 + 5n, \ldots$
7. $1, -1, -3, \ldots, 3 - 2n, \ldots$
8. $3, -1, -5, \ldots, 7 - 4n, \ldots$

Find the indicated term for the given arithmetic sequences.

9. If $a_1 = 2$ and $d = -3$, find a_{10}.
10. If $a_1 = -5$ and $d = 7$, find a_{12}.
11. If $a_1 = 0$ and $d = 9$, find a_{30}.
12. If $a_1 = 4$ and $d = -5$, find a_{40}.
13. If $a_1 = 9$ and $d = -4$, find a_8.
14. If $a_1 = -7$ and $d = -4$, find a_{20}.
15. If $a_1 = -3$ and $d = -9$, find a_{19}.
16. If $a_1 = 0$ and $d = \frac{1}{2}$, find a_{35}.
17. If $a_1 = 3$ and $d = -\frac{1}{2}$, find a_{40}.
18. If $a_1 = -\frac{1}{2}$ and $d = 2$, find a_8.

For each of the following arithmetic sequences, find the first term a_1 and the difference d between consecutive terms of the sequence.

19. $a_{10} = -26, a_{14} = -7$
20. $a_{18} = 49, a_{20} = 28$
21. $a_{15} = 16, a_{19} = -12$
22. $a_{10} = 3, a_{16} = 21$
23. $a_5 = 3, a_{50} = 30$
24. $a_5 = \frac{1}{2}, a_{20} = \frac{7}{8}$
25. $a_{17} = -40, a_{28} = -73$
26. $a_{10} = -11, a_{40} = -71$
27. $a_{20} = 66, a_{59} = 222$
28. $a_{16} = 60\frac{1}{2}, a_{41} = 160\frac{1}{2}$

The following exercises refer to arithmetic sequences. Find the indicated value.

29. If $d = 4$ and $a_8 = 33$, what is a_1?
30. If $a_1 = 8$ and $a_{11} = 26$, what is d?
31. If $a_7 = \frac{7}{3}$ and $d = -\frac{2}{3}$, what is a_{15}?

32. If $d = -7$ and $a_3 = 37$, what is a_{17}?
33. If $a_1 = -2$, $d = 7$, and $a_n = 138$, what is n?
34. If $a_3 = -7$, $d = -5$, and $a_n = -142$, what is n?
35. If $a_2 = 5$ and $a_4 = 1$, what is a_{10}?
36. If $a_4 = 5$ and $d = -2$, what is a_1?
37. If $a_3 = 5$, $d = -3$, and $a_n = -76$, what is n?
38. If $a_3 = 25$, $d = -14$, and $a_n = -507$, what is n?

Compute the indicated sums for the following arithmetic sequences.

39. $3, 5, 7, \ldots, 2n + 1, \ldots$ Find S_{20}.
40. $2, 5, 8, \ldots, 3n - 1, \ldots$ Find S_{30}.
41. $-1, -4, -7, \ldots, 2 - 3n, \ldots$ Find S_{19}.
42. $7, 9, 11, \ldots, 5 + 2n, \ldots$ Find S_{35}.
43. $2, 2\frac{1}{2}, 3, \ldots, \dfrac{3 + n}{2}, \ldots$ Find S_{44}.
44. $1\frac{1}{2}, \frac{1}{2}, -\frac{1}{2}, \ldots, \dfrac{5 - 2n}{2}, \ldots$ Find S_{50}.
45. $-4, -1, 2, \ldots, 3n - 7, \ldots$ Find S_7.
46. $3, 9, 15, \ldots, 6n - 3, \ldots$ Find S_{15}.
47. If $a_1 = 10$ and $d = -3$, find S_{30}.
48. If $a_1 = 15$ and $d = 9$, find S_{28}.
49. If $a_5 = -15$ and $d = 6$, find S_{45}.
50. If $a_3 = -12$ and $d = -2$, find S_{89}.

Applications

51. **Oil Exploration.** A well-drilling firm charges $2.50 to drill the first foot, $2.65 for the second foot, and so on, in arithmetic sequence. At this rate, what would be the cost to drill the last foot of a well that is 350 feet deep?

52. **Word Processing.** A student doing some typing finds that she can type 5 words per minute faster each half hour that she types. If she starts typing at 35 words per minute at 6:30 P.M., how fast is she typing at 10:00 P.M.?

53. **Construction.** A log pyramid has 16 logs on the bottom row, 15 on the next row, and so on, until there is just one log on the top row. How many logs are in the pyramid?

54. **Real Estate.** The developer of a housing development finds that the sale of the first house gives him a $200 profit. His profit on the second sale is $325, and he finds that each additional house sold increases his profit per house by $125. How many houses did he sell if his profit on the last house sold was $6,450?

55. Consider the sequence for which $a_n = n^2$, starting with $n = 0$. If $b_n = a_n - a_{n-1}$, show that $b_1, b_2, b_3, \ldots$ is an arithmetic sequence.

56. Prove that an expression for the nth term of an arithmetic sequence defines a linear function.

Technology

Use a calculator to determine the following.

57. The twentieth term of the arithmetic sequence $a_n = 2.1n + 3.55$.
58. The thirtieth term of the arithmetic sequence $a_n = 75.9n - 23.4$.
59. The sum of the first 100 terms of the arithmetic sequence that has first term 5 and difference 12.
60. The sum of the first 50 terms of the arithmetic sequence that has first term 1,000 and difference -5.77.

In your own words

61. Explain the concept of an arithmetic sequence.
62. Give two examples of arithmetic sequences from magazines or newspapers.
63. Explain how to determine the nth term of an arithmetic sequence. Illustrate with a specific example.
64. Explain how to determine the sum of n terms of an arithmetic sequence. Illustrate with a specific example.

12.5 GEOMETRIC SEQUENCES AND SERIES

Let's now proceed to the second special type of sequence that arises most commonly in applications, namely, the geometric sequence. As we saw in the preceding section, consecutive terms of an arithmetic sequence arise by adding a constant. For a geometric sequence, we multiply rather than add.

Definition
Geometric Sequence

A sequence $a_1, a_2, \ldots, a_n$ is a **geometric sequence** provided that each term is a fixed multiple of its immediate predecessor. That is, there is a fixed real number r, called the **ratio**, such that, for each positive integer n, we have:

$$a_{n+1} = ra_n$$

An example of a geometric sequence is:

$$\frac{1}{2}, \frac{1}{2^2}, \frac{1}{2^3}, \ldots$$

Here, $r = \frac{1}{2}$. Indeed, we have $a_n = 1/2^n$, so that:

$$a_{n+1} = \frac{1}{2^{n+1}}$$
$$= \frac{1}{2 \cdot 2^n}$$
$$= \frac{1}{2} \cdot \frac{1}{2^n}$$
$$= \frac{1}{2} \cdot a_n$$

The next example provides another illustration of a geometric sequence and offers practice in proving that a sequence is geometric.

EXAMPLE 1
Proving a Sequence is Geometric

Consider the sequence given by:

$$a_n = (-1)^{n-1}\left(\frac{4}{3}\right)^n$$

Prove that this sequence is a geometric sequence.

Solution

In looking for a possible value of r, we need to compute the ratio:

$$r = \frac{a_{n+1}}{a_n} = \frac{(-1)^{(n+1)-1}\left(\frac{4}{3}\right)^{n+1}}{(-1)^{n-1}\left(\frac{4}{3}\right)^n}$$
$$= (-1)\left(\frac{4}{3}\right)^{(n+1)-n}$$
$$= -\left(\frac{4}{3}\right)^1$$
$$= -\frac{4}{3}$$

Because this ratio is a constant (it does not depend on the value of n), the sequence is geometric with ratio $-\frac{4}{3}$.

638 CHAPTER 12 Sequences, Series, and Probability

Our development of geometric sequences parallels our discussion of arithmetic sequences in Section 12.4, where we derived a formula for the nth term of an arithmetic sequence in terms of the first term, the value of n, and the difference. Here is the analogous formula for the nth term of a geometric sequence in terms of the first term, the index, and the ratio.

---

nTH TERM OF A GEOMETRIC SEQUENCE

Suppose that $a_1, a_2, \ldots, a_n, \ldots$ is a geometric sequence with ratio r. Then:
$$a_n = a_1 r^{n-1}$$

Let's verify this formula by induction. The initial case corresponds to $n = 1$, for which case the formula asserts that:
$$a_1 = a_1 r^{1-1}$$
$$= a_1 r^0$$
$$= a_1$$

So the result is true in case $n = 1$. Let's now assume that the result holds for a particular value of n. That is, we assume that:
$$a_n = a_1 r^{n-1}$$

We must deduce from this the truth of the result for case $n + 1$. That is, we must prove that:
$$a_{n+1} = a_1 r^{[(n+1)-1]} = a_1 r^n$$

To do so, let's first recall that the sequence is geometric, so that:
$$a_{n+1} = r a_n$$

Into this last equation, substitute the assumed formula for a_n. This gives us:
$$a_{n+1} = r a_n$$
$$= r \cdot a_1 r^{n-1}$$
$$= a_1 r^n$$
$$= a_1 r^{(n+1)-1}$$

This is the statement of the result for $n + 1$. So the inductive step is proved and with it the formula we wished to verify.

The next two examples provide applications of the formula for the nth term of a geometric progression.

EXAMPLE 2
Calculating a Term of a Geometric Sequence

Suppose that a geometric sequence has first term 3 and ratio $\frac{2}{3}$. Determine the fifth term of the sequence.

Solution

By the formula for the nth term of a geometric progression, the nth term a_n of the sequence is given by:

SECTION 12.5 Geometric Sequences and Series

$$a_n = a_1 r^{n-1}$$
$$= 3 \cdot \left(\frac{2}{3}\right)^{n-1}$$

In particular, for $n = 5$, this equation yields:

$$a_5 = 3 \cdot \left(\frac{2}{3}\right)^{5-1}$$
$$= 3 \cdot \left(\frac{2}{3}\right)^{4}$$
$$= 3 \cdot \frac{16}{81}$$
$$= \frac{16}{27}$$

EXAMPLE 3
Formula for the *n*th Term

Suppose that $a_1, a_2, \ldots$ is a geometric sequence. Further, suppose that $a_2 = \frac{5}{4}$ and $a_5 = \frac{5}{32}$. Determine a formula for a_n.

Solution

Let r denote the ratio of the sequence. Then, we have:

$$a_2 = \frac{5}{4} = a_1 r^{2-1}$$
$$\frac{5}{4} = a_1 r$$

and

$$a_5 = \frac{5}{32} = a_1 r^{5-1}$$
$$\frac{5}{32} = a_1 r^4$$

Dividing the second equation by the corresponding terms of the first equation, we have:

$$\frac{\frac{5}{32}}{\frac{5}{4}} = \frac{a_1 r^4}{a_1 r}$$
$$\frac{1}{8} = r^3$$
$$r = \frac{1}{2}$$

Inserting this value of r into the first equation gives:

$$\frac{5}{4} = a_1 r$$

$$\frac{5}{4} = a_1 \cdot \left(\frac{1}{2}\right)$$

$$a_1 = \frac{5}{2}$$

Inserting the values of a_1 and r into the formula for the nth term of a geometric progression, we find the desired formula for a_n, namely:

$$a_n = a_1 r^{n-1}$$
$$= \frac{5}{2} \cdot \left(\frac{1}{2}\right)^{n-1}$$

EXAMPLE 4
Compound Interest

Suppose that a bank account earns compound interest at a rate of i per period. (The period can be any unit of time.) Further, suppose that at the start of period 1, the account has balance P. Then the balance a_n in the account at the end of period n is given by the formula:

$$a_n = P(1 + i)^n$$

1. Show that the sequence $a_0, a_1, a_2, \ldots, a_n$ is a geometric sequence with ratio $1 + i$.
2. Suppose that $400 is deposited in a bank account that pays 8 percent interest compounded monthly. Calculate the amount in the account at the end of 5 years.

Solution

1. Note that the first index of this sequence is $n = 0$. In dealing with financial calculations, it is usually most convenient to begin sequences with the 0th term rather than the first term. This allows the sequence to start its description with time 0 (the beginning of the first time period). Note that if n is any nonnegative integer, then:

$$a_{n+1} = P(1 + i)^{n+1}$$
$$= P(1 + i)^n \cdot (1 + i)$$
$$= (1 + i)a_n$$

Thus, the sequence is geometric with ratio $1 + i$.

2. The 8 percent interest refers to an annual rate. Because the interest is compounded monthly, the length of a single financial period is a month and the amount of interest for one month is $i = \frac{8}{12} = \frac{2}{3}$ percent $= \frac{1}{150}$. Five years corresponds to 60 months, that is, 60 financial periods. So the balance in the account at the end of five years is:

$$a_{60} = 400\left(1 + \frac{1}{150}\right)^{60}$$
$$= \$595.94$$

(We use a scientific calculator to calculate the numerical value of the expression.)

Let's now turn to the problem of calculating the partial sum S_n of a geometric sequence. We have the following formula:

SECTION 12.5 Geometric Sequences and Series

PARTIAL SUM OF A GEOMETRIC SEQUENCE

Suppose that $a_1, a_2, \ldots, a_n, \ldots$ is a geometric sequence and that S_n is its nth partial sum. Then:

$$S_n = a_1 \frac{1 - r^n}{1 - r} \qquad (r \neq 1)$$

$$ = na_1 \qquad (r = 1)$$

Let's verify these formulas. We have:

$$S_n = a_1 + a_2 + a_3 + \cdots + a_n$$
$$= a_1 + a_1 r + a_1 r^2 + \cdots + a_1 r^{n-1}$$
$$= a_1(1 + r + r^2 + \cdots + r^{n-1})$$

In the case $r = 1$, each of the terms within the parentheses on the right is equal to 1 and the formula for S_n reads:

$$S_n = a_1(1 + 1 + 1 + \cdots + 1) = na_1$$

Let's now assume that $r \neq 1$. We can write the right side in the form:

$$S_n = a_1(1 + r + r^2 + \cdots + r^{n-1})$$
$$= a_1(1 + r + r^2 + \cdots + r^{n-1}) \frac{1 - r}{1 - r}$$
$$= a_1 \frac{1 - r^n}{1 - r}$$

This completes the verification of the formulas for the sum of a geometric sequence.

EXAMPLE 5 Determine the value of the following sum:
Sum of a Geometric Sequence

$$3 + 3\left(\frac{3}{4}\right) + 3\left(\frac{3}{4}\right)^2 + \cdots + 3\left(\frac{3}{4}\right)^9$$

Solution

The sum is the partial sum S_{10} of a geometric sequence with $a_1 = 3$ and $r = \frac{3}{4}$. Therefore, by the first formula for the partial sum of a geometric sequence, we have:

$$S_{10} = a_1 \frac{1 - r^{10}}{1 - r}$$

$$= 3 \cdot \frac{1 - \left(\frac{3}{4}\right)^{10}}{1 - \frac{3}{4}}$$

$$= 12 \cdot \left(1 - \left(\frac{3}{4}\right)^{10}\right) = 11.324$$

EXAMPLE 6
Value of a Pension Fund

Each year, for 20 years, a woman deposits $5,000 into a pension fund. The money earns 9 percent interest compounded annually. How much money is in the pension fund after 20 years?

Solution

There are 20 deposits to the pension fund, each earning interest for a different length of time. The last deposit earns interest for 0 years, the next-to-last deposit earns interest for 1 year, and the first deposit earns interest for 19 years. Let a_1 denote the amount of interest generated by the last deposit, a_2 the amount of interest generated by the next-to-last deposit, and a_{20} the amount generated by the first deposit. The amount of money represented by a_n earns interest for $n-1$ years at 9 percent compounded annually. So by our previous discussion, we have:

$$a_n = 5{,}000(1 + .09)^{n-1}$$
$$= 5{,}000 \cdot 1.09^{n-1}$$

The amount in the pension fund at the end of 20 years equals the sum of the amounts attributable to each of the deposits. This equals:

$$a_1 + a_2 + a_3 + \cdots + a_{20}$$

This sum is the partial sum S_{20} of a geometric series with $a_1 = 5{,}000$ and $r = 1.09$. By the first formula for the partial sum of a geometric sequence, this sum equals:

$$a_1 \frac{1-r^n}{1-r} = 5{,}000 \cdot \frac{1 - 1.09^{20}}{1 - 1.09}$$
$$= -\frac{5{,}000}{0.09}(1 - 1.09^{20})$$
$$= \$255{,}800.60$$

That is, the pension fund balance after 20 years equals $255,800.60.

Now let us consider the infinite geometric sequence:

$$1, \frac{1}{2}, \frac{1}{2^2}, \frac{1}{2^3}, \ldots$$

Here are its first few partial sums:

$$S_1 = 1$$
$$S_2 = \frac{3}{2}$$
$$S_3 = \frac{7}{4}$$
$$S_4 = \frac{15}{8}$$
$$S_5 = \frac{31}{16}$$

As more and more terms of the sequence are included, the partial sums approach ever closer to the value 2. In this circumstance, we say that the sum of the infinite geometric sequence equals 2, and we write:

SECTION 12.5 Geometric Sequences and Series

$$\sum_{n=0}^{\infty} \frac{1}{2^n} = 2$$

In a similar fashion, we can consider the sums of other infinite geometric sequences. It is possible to prove that if $|r| < 1$, then the partial sums of the sequence $a_n = r^n$ approach the limiting value $1/(1-r)$ (2 in the above example). The limit of the partial sums can be thought of as an "infinite sum"

$$1 + r + r^2 + \cdots$$

and the limiting value of the partial sums can be thought of as the value of the infinite sum.

There are other circumstances in which the partial sums of an infinite geometric sequence approach a limiting value, as described in the following theorem:

SUM OF AN INFINITE GEOMETRIC SEQUENCE

Suppose that $|r| < 1$. The partial sums of the infinite geometric sequence $a, ar, ar^2, \ldots, ar^n$ approach the limiting value $a/(1-r)$. This fact is expressed in summation notation as:

$$\sum_{n=0}^{\infty} ar^n = \frac{a}{1-r}$$

EXAMPLE 7
Calculating the Sum of an Infinite Geometric Sequence

Calculate the sum of the infinite geometric sequence:

$$3, -3 \cdot \frac{1}{5}, 3 \cdot \left(\frac{1}{5}\right)^2, -3 \cdot \left(\frac{1}{5}\right)^3, \ldots$$

Solution

The geometric sequence has $a_1 = 3$, $r = -\frac{1}{5}$. Therefore, by the formula for the sum of a geometric sequence, the partial sums of the sequence approach a limiting value and this value equals:

$$\frac{a}{1-r} = \frac{3}{1-(-\frac{1}{5})}$$

$$= \frac{3}{\frac{6}{5}}$$

$$= \frac{5}{2}$$

Note that if $|r| \geq 1$, then the partial sums of the geometric series $a_n = ar^n$ do not approach a limiting value.

Infinite sums, such as those we have just introduced, are discussed in detail in the study of infinite series, a part of calculus.

Exercises 12.5

Prove that the following are geometric sequences.

1. $a_n = 3\left(-\dfrac{1}{3}\right)^{n-1}$
2. $a_n = 24\left(\dfrac{1}{2}\right)^{n-1}$
3. $a_n = 5(4)^{n-1}$
4. $a_n = 2^{n-1}$
5. $a_n = 3^{-n}$
6. $a_n = \left(-\dfrac{1}{2}\right)^{n+1}$

Find the indicated term for each of the following geometric sequences.

7. If $a_1 = \tfrac{1}{2}$ and $r = \tfrac{1}{2}$, find a_{10}.
8. If $a_1 = \tfrac{2}{3}$ and $r = \tfrac{2}{3}$, find a_9.
9. If $a_1 = 1$ and $r = -\sqrt{2}$, find a_7.
10. If $a_1 = 81$ and $r = 1/\sqrt{3}$, find a_{12}.
11. Given $54, 36, 24, \ldots$, find a_8.
12. Given $32, 16, 8, \ldots$, find a_{15}.
13. If $a_2 = 3$ and $r = 2$, find a_{13}.
14. If $a_4 = 64$ and $r = -4$, find a_{10}.
15. If $a_4 = 6$ and $a_5 = 12$, find a_{14}.
16. If $a_6 = 9$ and $a_7 = 3$, find a_2.
17. Given $300, -30, 3, \ldots$, find a_4.
18. Given $-\sqrt{5}, 5, -5\sqrt{5}, \ldots$, find a_{20}.
19. Given $2, \tfrac{2}{3}, \tfrac{2}{9}, \ldots$, find a_5.
20. Given $-162, 54, -18, 6, \ldots$, find a_8.
21. Which term of the geometric sequence $2, -6, 18, \ldots$ is 162?
22. If $a_7 = 256$ and $a_1 = 4$, find a_5.
23. If $r = 3$ and $a_5 = 324$, find a_1.
24. Which term of the geometric sequence $-243, 81, -27, \ldots$ is $\tfrac{1}{9}$?
25. Fill in the blanks for this geometric sequence: ___, -5, ___, ___, ___, -405.
26. Fill in the blanks for this geometric sequence: ___, ___, 1, ___, 729.

In the following geometric sequences, three of the five real numbers $a_1, a_n, r, n,$ and S_n are given. Find the other two.

27. $a_1 = 5, r = -2, S_n = -25$
28. $a_1 = 5, a_n = 320, r = 2$
29. $a_1 = 4, a_n = 324, r = 3$
30. $n = 9, r = 2, S_n = 1{,}022$
31. $n = 5, r = -3, S_n = 244$
32. $a_1 = 10, r = 3, n = 5$
33. $a_n = 384, n = 7, r = 2$
34. $a_n = 729, n = 7, r = \dfrac{3}{2}$
35. $a_1 = 162, r = -\dfrac{1}{3}, n = 8$
36. $a_1 = 4, r = -\dfrac{3}{2}, n = 6$
37. $a_n = 9, r = -\sqrt{3}, S_n = 12 - 3\sqrt{3}$
38. $a_n = -625\sqrt{5}, r = -\sqrt{5}, S_n = 780 - 781\sqrt{5}$

Find a formula for the following geometric sequences.

39. $a_2 = -4$ and $a_5 = \dfrac{4}{27}$
40. $a_3 = 6$ and $a_6 = 18\sqrt{3}$
41. $a_3 = 50$ and $a_6 = -6{,}550$
42. $a_2 = 3$ and $a_5 = 54$

Find the sums of the following infinite geometric sequences, if the sums exist.

43. $1 - \dfrac{1}{2} + \dfrac{1}{4} - \dfrac{1}{8} + \dfrac{1}{16} - \cdots$
44. $1 - \dfrac{1}{5} + \dfrac{1}{25} - \dfrac{1}{125} + \cdots$
45. $-\dfrac{5}{3} - \dfrac{10}{9} - \dfrac{20}{27} - \dfrac{40}{81} \cdots$
46. $2 + \dfrac{2}{3} + \dfrac{2}{9} + \dfrac{2}{27} + \dfrac{2}{81} + \cdots$
47. $\dfrac{9}{10} + \dfrac{9}{100} + \dfrac{9}{1{,}000} + \dfrac{9}{10{,}000} + \cdots$
48. $\dfrac{1}{2} + \dfrac{1}{3} + \dfrac{2}{9} + \dfrac{4}{27} + \cdots$
49. $1 + 2 + 4 + 8 + \cdots$
50. $-1 + 1 - 1 + 1 - 1 + \cdots$

In the following problems, S denotes the sum of an infinite geometric sequence with ratio r.

51. If $S = 14$ and $a_1 = 21$, then find r.
52. If $r = \tfrac{2}{3}$ and $S = 18$, then find a_1.
53. If $r = -\tfrac{1}{2}$ and $S = 20$, then find a_1.
54. If $S = 35$ and $a_1 = 7$, then find r.
55. If $a_1 = 4$ and $r = \tfrac{1}{2}$, then find S.
56. If $a_1 = \tfrac{1}{3}$ and $S = \tfrac{2}{3}$, then find r.

Applications

57. **Merit salary plan.** On a merit salary plan, yearly raises are determined as a percent of the present salary. If a worker were to begin with a $15,000 yearly salary and receive merit increases of 8.5 percent for 9 years, then what would be her yearly salary during her ninth year?

58. **Sports salaries.** In place of a salary of $1 million, a baseball pitcher agrees to accept $10 for his first win, $20 for his second, $40 for his third, and so on. How many games must he win in order for his earnings to exceed $1 million? What would be his salary if he were to win one more game?

59. Depreciation. If a new printing press costs $50,000 and it has a yearly depreciation of 25 percent of its value at the beginning of each year, then what is its value after 5 years?

60. Installment buying. Highway Appliance Store has CD players that can be purchased on a daily installment plan. Customers pay 1 cent the first day, 3 cents the second day, 9 cents the third day, and so on, for 10 days. How much will the CD player cost?

61. Genealogy. A child states that one of her tenth-generation ancestors was on the Mayflower. Assume that each of her ancestors had a single child. How many relatives does she have in the tenth generation?

62. Compound interest. If the interest on a ten-year certificate of deposit is set at 9 percent compounded semiannually, what would be the value of an $8,000 investment after ten years?

63. Population change. The city of Akron, Ohio, lost $\frac{1}{100}$ of its population each year from 1972 through 1980. If the population in 1972 was 350,000, what was the population at the end of 1980?

64. Multiplier effect. Imagine the federal government puts an extra billion dollars into the economy such that each business and individual saves 25 percent of its income and spends the rest. Of the one billion dollars, 75 percent is spent by individuals and businesses, then 75 percent of the remaining amount is spent, and so forth. What is the total increase in spending due to the government action?

65. Business loan. To begin a business, an entrepreneur must borrow $80,000 at 12 percent interest, compounded monthly. If the loan is to be paid off in one lump sum at the end of 2 years, what will be the amount of the payment?

66. Retirement planning. For retirement, you elect to deposit $1,000 a year in a savings account that pays 11 percent, compounded annually. How much will be in your account at the end of 40 years?

67. Car loan. To buy a used car, you borrow $1,000 at 10 percent interest, compounded monthly. You also elect to pay off this loan in one lump-sum payment at the end of 4 years. What will be the amount of your payment?

68. Saving for college. A couple with a 1-year-old son decides to put $500 per year into a savings account paying 12 percent interest, compounded annually, for their son's college money. At the end of 18 years, what will be the amount in this account?

69. Physics. A superball is dropped 40 feet and rebounds on each bounce $\frac{2}{5}$ of the height from which it fell. How far will it travel before coming to rest?

70. Physics. If a ball is dropped from a height of 12 meters and rebounds $\frac{1}{3}$ of the distance it fell on each bounce, then how far will it travel before coming to rest?

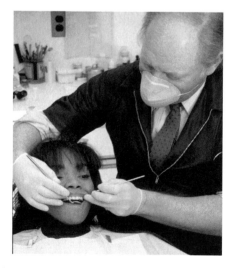

71. Salary growth. If a dentist earns $40,000 during his first year of practice, and if each succeeding year his salary increases 10 percent, then what would be the total of the dentist's salary over the first 9 years?

72. **Farming.** Below is a diagram illustrating the path taken by some grapes that travel from the grower to you.

grower → trucker → regional market → trucker

→ wholesaler → trucker → retailer → you

Assume that each person or organization in this chain makes a 25 percent profit, that all grapes are sold, and that the grower sells her grapes for $0.20 per pound. How much would you have to pay per pound?

73. **Farming.** If the grower in Exercise 72 receives $10,000 for the grapes she sold to the first trucker, then what is the sum of all the profits made when you buy these grapes?

74. **Salary growth.** How many years would the dentist in Exercise 71 have to work for the salary total to exceed one million dollars?

Technology

Use a calculator to determine the following.

75. The tenth term of a geometric sequence with first term 50 and ratio $\frac{2}{3}$.

76. The eighth term of a geometric sequence with first term 1,275 and ratio 0.6.

77. The sum of the first 20 terms of the geometric sequence in Exercise 75.

78. The sum of the first 30 terms of the geometric sequence in Exercise 76.

In your own words

79. Explain the concept of a geometric sequence. Use a specific example.

80. Give two examples of geometric sequences found in newspapers or magazines.

81. Explain how to calculate the nth term of a geometric sequence.

82. Explain how to calculate the sum of a geometric sequence.

83. Explain what is meant by the sum of an infinite geometric sequence.

84. Explain the differences between arithmetic and geometric sequences.

85. Suppose that you proceed to the door of the room by walking $\frac{1}{2}$ the distance to the door, then stopping, then moving half the remaining distance to the door, then stopping, and so forth. You never cover the total distance to the door, and, hence, you can never leave the room. Is this statement true or false? Explain your answer.

12.6 MATHEMATICS OF FINANCE

Throughout our lives, we must solve problems about financial matters: What are the monthly payments on a mortgage? How large a car loan will fit within the budget. How much must be saved to yield a given monthly retirement income? In this section, we learn to solve such problems using techniques based on the formula for the sum of a geometric series.

The Fundamental Difference Equation

Suppose that a financial transaction has an associated balance at periodic intervals. For example, a mortgage has a balance at the end of each month, a retirement fund has a balance at the end of each quarter, and a savings account has a balance at the end of each year. Let y_n denote the balance at the end of the nth period. Then the balances at the ends of the various periods form a sequence:

$$y_0, y_1, y_2, \ldots$$

Here y_0 denotes the initial balance, that is, the balance at the start of the transaction.

SECTION 12.6 Mathematics of Finance

For many of the most common financial transactions, the terms y_n of the sequence are related by an equation of the form

$$y_{n+1} = ay_n + b \quad (1)$$

where a and b are constants independent of n. Such an equation is called a **difference equation** for the sequence.

EXAMPLE 1
The Difference Equation of a Mortgage

Suppose that a mortgage has a 6 percent annual rate and a $700 monthly payment. Let y_n denote the balance at the end of the nth month. Find a difference equation satisfied by the sequence.

Solution

The interest is a 6 percent annual rate. Each month, $\frac{1}{12}$ of the interest is charged on the balance for the month. That is, the interest charge added during the $(n+1)$st month is:

$$\frac{0.06}{12} \times [\text{balance during } n\text{th month}] = 0.005 y_n$$

Therefore:

$$\begin{bmatrix} \text{balance at end of} \\ \text{month } n+1 \end{bmatrix} = \begin{bmatrix} \text{balance at end of} \\ \text{month } n \end{bmatrix} - [\text{payment}] + [\text{interest}]$$

$$y_{n+1} = y_n - 700 + 0.005 y_n$$
$$= 1.005 y_n - 700$$

Assume that equation (1) holds and that $a \neq 1$. Then, by repeatedly applying the equation, we can determine a formula for y_n as follows:

$$y_n = a y_{n-1} + b$$
$$= a(a y_{n-2} + b) + b$$
$$= a^2 y_{n-2} + ab + b$$
$$= a^2(a y_{n-3} + b) + ab + b$$
$$= a^3 y_{n-3} + a^2 b + ab + b$$
$$= \cdots$$
$$= a^n y_0 + a^{n-1} b + a^{n-2} b + \cdots + ab + b$$

Factoring out b, we obtain:

$$y_n = a^n y_0 + b(a^{n-1} + a^{n-2} + \cdots + a + 1)$$
$$y_n = a^n y_0 + b \frac{a^n - 1}{a - 1} \quad \text{Sum of a geometric series}$$

Writing this formula in a slightly different form, we have:

SOLUTION TO A LINEAR DIFFERENCE EQUATION

$$y_n = a^n \left(y_0 + \frac{b}{a - 1} \right) - \frac{b}{a - 1} \quad (2)$$

This fundamental equation can be used to solve a wide variety of financial problems, as we demonstrate in Example 2.

EXAMPLE 2
Solving a Difference Equation

Consider the mortgage in Example 1. Suppose that the initial balance is $50,000.
1. Determine a formula for y_n, the balance owed at the end of month n.
2. Determine the balance owed after three years.

Solution

1. As we determined in Example 1, the sequence of balances satisfies the difference equation:
$$y_{n+1} = 1.005 y_n - 700$$
In this example, $a = 1.005$, $b = -700$, and $y_0 = 50,000$. By equation (2), we therefore have:
$$y_n = 1.005^n \left(50,000 + \frac{-700}{1.005 - 1} \right) - \frac{-700}{1.005 - 1}$$
$$= 140,000 - 90,000 \cdot 1.005^n$$

2. Three years corresponds to 36 months. The balance owed after 36 months is y_{36}, which equals:
$$y_{36} = 140,000 - 90,000 \cdot 1.005^{36}$$
$$\approx \$32,298.75$$

Mortgages and Consumer Loans

Most consumer loans (mortgages, car loans, and personal loans) quote interest on an annual rate, but the payments and interest charges are calculated monthly. Each month, $\frac{1}{12}$ of the annual interest is charged on the balance remaining. Let's now solve some typical problems in personal finance using the method of difference equations introduced above.

EXAMPLE 3
Determining the Monthly Payment of a Mortgage

What is the monthly payment on a $200,000, 30-year mortgage with an annual interest rate of 6.75 percent?

Solution

Let y_n denote the balance owed at the end of month n. As in Example 1, we can determine a difference equation for the sequence of payments:

$$\begin{bmatrix} \text{balance at end of} \\ \text{month } n+1 \end{bmatrix} = \begin{bmatrix} \text{balance at end of} \\ \text{month } n \end{bmatrix} - [\text{payment}] + [\text{interest}]$$

Let P denote the monthly payment. Then the above word equation yields:
$$y_{n+1} = y_n - P + \frac{0.0675}{12} y_n$$
$$y_{n+1} = 1.005625 y_n - P$$

Applying equation (2) gives us:
$$y_n = \left(200,000 + \frac{-P}{1.005625 - 1} \right) \cdot 1.005625^n - \frac{-P}{1.005625 - 1}$$
$$= 177.78P + (200,000 - 177.78P) \cdot 1.005625^n$$

SECTION 12.6 Mathematics of Finance

Because the mortgage is paid off after 30 years, or 360 months, we know that $y_{360} = 0$, so that:

$$177.7777777P + (200{,}000 - 177.777777P) \cdot 1.005625^{360} = 0$$
$$177.7777777P + (200{,}000 - 177.777777P) \cdot 7.533245477 = 0$$
$$1{,}161.465862P = 1{,}506{,}649.095$$
$$P = 1{,}297.20$$

That is, the monthly payment is $1,297.20.

EXAMPLE 4
Determining the Maximum Affordable Car Loan

Suppose the current annual rate of interest for car loans is 9 percent and suppose that you can afford payments of $250 per month for 48 months. How large a loan can you afford?

Solution

Let y_n denote the balance at the end of month n. Then the difference equation for the sequence of balances is:

$$y_{n+1} = y_n - 250 + \frac{0.09}{12} y_n$$
$$y_{n+1} = 1.0075 y_n - 250$$

The solution of this difference equation is:

$$y_n = \left(y_0 + \frac{-250}{1.0075 - 1}\right) \cdot 1.0075^n - \frac{-250}{1.0075 - 1}$$
$$= \left(y_0 - \frac{100{,}000}{3}\right) \cdot 1.0075^n + \frac{100{,}000}{3}$$

Because the loan must be paid off in 48 months, we have:

$$y_{48} = 0$$

$$\left(y_0 - \frac{100{,}000}{3}\right) \cdot 1.0075^{48} + \frac{100{,}000}{3} = 0$$

$$y_0 \approx \$10{,}046.20$$

That is, the maximum loan you can afford is $10,046.20.

Other Personal Financial Transactions

Let's now turn to other common personal financial transactions—saving for college expenses and drawing payments from a retirement fund. The next examples show how to solve some typical problems that arise.

EXAMPLE 5
Accumulating a College Fund

At the present rate of increase, the cost of four years of attendance at Princeton for a freshman who begins in 2005 will be $225,000. How much must be deposited each month, for ten years, into a college fund that earns 6 percent interest, compounded monthly, in order for the fund to have enough to pay for a four-year Princeton education in the year 2005?

Solution

Let D denote the monthly deposit into the fund. The difference equation satisfied by the balances is:

$$\begin{bmatrix} \text{balance at end of} \\ \text{month } n+1 \end{bmatrix} = \begin{bmatrix} \text{balance at end of} \\ \text{month } n \end{bmatrix} + [\text{deposit}] + [\text{interest}]$$

$$y_{n+1} = y_n + D + \frac{0.06}{12}y_n$$

$$y_{n+1} = 1.005y_n + D$$

The solution of this difference equation is:

$$y_n = \left(y_0 + \frac{D}{1.005 - 1}\right)1.005^n - \frac{D}{1.005 - 1}$$

$$= (y_0 + 200D)1.005^n - 200D$$

Because the fund starts out with a balance of 0, we have $y_0 = 0$. To accumulate the desired amount of money, we must have $y_{120} = 225{,}000$. This gives the equation:

$$225{,}000 = 200D \cdot 1.005^{120} - 200D$$

Solving this equation for D gives:

$$D = \$1{,}372.96$$

EXAMPLE 6
Payout of an Annuity

Suppose that a retirement fund earns 6 percent interest, compounded monthly. How large must the fund be to pay a monthly annuity of \$3,000 for 20 years?

Solution

Let y_n be the balance in the retirement fund in month n after payments begin. The sequence of payments satisfies the model:

$$\begin{bmatrix} \text{balance at end of} \\ \text{month } n+1 \end{bmatrix} = \begin{bmatrix} \text{balance at end of} \\ \text{month } n \end{bmatrix} - [\text{payout}] + [\text{interest}]$$

$$y_{n+1} = y_n - 3{,}000 + \frac{0.06}{12}y_n$$

$$y_{n+1} = 1.005y_n - 3{,}000$$

The solution of the difference equation is:

$$y_n = \left(y_0 + \frac{-3{,}000}{1.005 - 1}\right)1.005^n - \frac{-3{,}000}{1.005 - 1}$$

$$= (y_0 - 600{,}000)1.005^n + 600{,}000$$

Because the fund must last for 20 years, that is, 240 months, we must determine y_0 so that $y_{240} = 0$:

$$0 = (y_0 - 600{,}000)1.005^{240} + 600{,}000$$

$$y_0 = \$418{,}742.32$$

That is, if the retirement fund starts with \$418,742.32, it will provide a payout of \$3,000 per month for 20 years and will run out of money after exactly 20 years.

Exercises 12.6

Solve the following difference equations:

1. $y_{n+1} = 1.1y_n$, $y_0 = 1$
2. $y_{n+1} = 2y_n$, $y_0 = 1$
3. $y_{n+1} = 2y_n - 1$, $y_0 = 10$
4. $y_{n+1} = 3y_n + 2$, $y_0 = 4$
5. $y_{n+1} = 1.001y_n - 500$, $y_0 = 10{,}000$
6. $y_{n+1} = 1.05y_n + 20$, $y_0 = 50{,}000$
7. $y_{n+1} = 1.01y_n + 40$, $y_0 = 500$
8. $y_{n+1} = 1.015y_n - 700$, $y_0 = 100{,}000$

In Exercises 9–12, consider the difference equation:

$$y_{n+1} = 3y_n - 4, \qquad y_0 = 2$$

Determine the following terms:

9. y_1
10. y_2
11. y_{10}
12. y_{20}

Applications

13. **Mortgage.** What is the monthly payment on a $100,000, 30-year mortgage with an annual interest rate of 7 percent?

14. **Mortgage.** What is the monthly payment on a $750,000, 20-year mortgage with an annual interest rate of 7.5 percent?

15. **Refinancing.** Suppose that the Joneses currently have a mortgage on which the payments are $900 per month. The balance owed is $55,000. If they refinance the mortgage with a 15-year loan at 7 percent annual interest, by how much will their monthly payment decrease?

16. **Home equity loan.** The Mings currently have a home equity loan with a balance of $30,000 and monthly payments of $315, and a mortgage with a balance of $150,000 and monthly payments of $1,550. How much will their monthly payments decrease if they replace the two loans with a single 30-year mortgage at a 6.5 percent annual rate?

17. **Mortgage.** What is the monthly payment on a $100,000, 30-year mortgage with an annual interest rate of 6.5 percent?

18. **Mortgage.** What is the monthly payment on a $350,000, 15-year mortgage with an annual interest rate of 8 percent?

19. **Car loan.** Suppose the current annual rate of interest for car loans is 9 percent and that you can afford payments of $350 per month for 36 months. How large a loan can you afford?

20. **Car loan.** Suppose the current annual rate of interest for car loans is 10 percent and that you can afford payments of $175 per month for 48 months. How large a loan can you afford?

21. **Cost of college.** At the present rate of increase, the cost of four years of attendance at Harvard for a freshman who begins in 2000 will be $175,000. If monthly deposits are made for 7 years, how much must be deposited each month into a bond fund earning 8 percent interest, compounded annually, in order for the fund to have enough to pay for a Harvard education in the year 2000?

22. **Saving for college.** How much must the Santeros deposit each month into a savings account earning 7 percent interest, compounded quarterly (every three months), in order to accumulate $50,000 for a college fund when their newborn daughter turns 18?

23. **Saving for retirement.** Suppose that a retirement fund earns 10 percent interest, compounded monthly. How large must the fund be to pay a monthly annuity of $5,000 for 10 years?

24. **Saving for retirement.** Suppose that a retirement fund earns 8 percent interest, compounded monthly. How large must the fund be to pay a monthly annuity of $2,000 for 15 years?

Technology

25. Suppose a $50,000 mortgage has an annual interest rate of 7 percent and a monthly payment of $332.65. Make a table that shows the amount of interest paid and the balance owed at the end of each of the first 12 months.

26. Suppose that you deposit $50 each month into a bond fund paying 6 percent annual interest, compounded monthly. Make a table that shows the interest earned and the balance in the account at the end of each of the first 12 months.

In your own words

27. Explain how to construct the difference equation satisfied by a mortgage. Give an example.

28. Explain how to construct the difference equation satisfied by an annuity. Give an example.

12.7 PERMUTATIONS AND COMBINATIONS

Many applied problems require you to count the number of possible occurrences of a particular type. Here are some examples:

1. A communications network consists of 500 cables from point A to point B and 300 cables from point B to point C. How many ways are there to choose a transmission path from point A to point C?
2. A state lottery requires you to pick a sequence of five digits. How many different outcomes of the lottery drawing are possible?
3. A certain genetic combination consists of eight traits, for each of which there are 4 possibilities. How many genetic combinations are possible?

The field of mathematics that deals with counting problems is called combinatorics. In this section, we provide a brief introduction to combinatorics by considering a few of the simplest types of counting problems.

Elementary Counting Problems

One of the basic techniques for solving counting problems is to break the problem down into a sequence of independent choices, such that each possibility can be counted. In counting such sequences, we use the following basic result:

FUNDAMENTAL PRINCIPLE OF COUNTING

Suppose that tasks $A_1, A_2, \ldots, A_n$ are independent of one another. Further, suppose that it is possible to perform task A_1 in m_1 ways, task A_2 in m_2 ways, $\ldots$, and task A_n in m_n ways. Then the number of possible ways to perform the sequence of tasks $A_1, A_2, \ldots, A_n$ is equal to $m_1 m_2 \ldots m_n$.

We assume the truth of the Fundamental Principle of Counting in what follows. The next examples illustrate how it can be applied in solving counting problems.

EXAMPLE 1
Counting License Plates

Car license plates in a certain state consist of three letters followed by three digits. How many different license plates are possible?

Solution

The process of selecting a license plate can be seen as a sequence of six tasks, corresponding to the selection of the six letters and digits. Each of the first three tasks can be accomplished in 26 different ways. Each of the last three tasks can be accomplished in 10 different ways. By the Fundamental Principle of Counting, the sequence of six tasks can then be accomplished in

$$26 \cdot 26 \cdot 26 \cdot 10 \cdot 10 \cdot 10 = 17{,}576{,}000$$

different ways. That is, there are 17,576,000 different possible license plates.

EXAMPLE 2
Hexadecimal Numbers

In the hexadecimal number system, used extensively in computer work, there are 16 digits, namely 0–9 and A–F. A typical hexadecimal number consists of a sequence of hexadecimal digits. How many 4-digit hexadecimal numbers are possible if the first digit is neither 0 nor 1?

Solution

The specified 4-digit hexadecimal number can be understood as a sequence of four independent tasks, each corresponding to choosing one of the 16 digits. The first task can be chosen in only 14 different ways because the hexadecimal digit can't be 0 or 1. Each of the remaining tasks can be accomplished in 16 different ways. So the sequence of tasks can be accomplished in

$$14 \cdot 16 \cdot 16 \cdot 16 = 57{,}344$$

ways. That is, there are 57,344 such hexadecimal numbers.

Permutations

One of the most common types of counting problems involves permutations, which are defined as follows:

Definition
Permutation

Suppose that we are given n different objects. A selection of r of them arranged in a particular order is called a **permutation of n objects taken r at a time** and the number of permutations of n objects taken r at time is denoted $_nP_r$.

For instance, suppose that $n = 3$ and that the objects are the three letters A, B, C. Then, if $r = 2$, there are six permutations of these objects taken two at a time, namely, AB, BA, AC, CA, BC, CB. We see that $_3P_2 = 6$. We can compute $_nP_r$ from n and r using the following result:

FUNDAMENTAL THEOREM OF PERMUTATIONS

Suppose that n and r are positive integers with $r \leq n$. Then

$$_nP_r = n(n-1)(n-2)\cdots(n-r+1)$$

for a total of r factors. That is, to calculate $_nP_r$, start with n and multiply r consecutive, decreasing integers.

To verify this formula, consider the task of choosing a permutation of n objects taken r at a time. This can be viewed as a sequence of r independent tasks. Select the first object in the permutation; this can be done in n ways. Next, select the second object in the permutation; because one object has already been selected, this second task can be done in $n - 1$ ways. Next, select the third object in the permutation; because two objects have already been selected, the second task can be done in $n - 2$ ways. And so on. The final task is selecting the rth object. The preceding tasks have used up $r - 1$ objects, so the final task can be done in $n - (r - 1) = n - r + 1$ ways. By the Fundamental Principle of Counting, the sequence of r tasks can be accomplished in

$$n(n-1)(n-2)\cdots(n-r+1)$$

ways. This proves the Fundamental Theorem of Permutations.

EXAMPLE 3
Calculating Numbers of Permutations

Determine the values of the following:

1. $_5P_3$
2. $_{10}P_4$
3. $_nP_3$

Solution

1. $_5P_3 = 5 \cdot 4 \cdot 3 = 60$
2. $_{10}P_4 = 10 \cdot 9 \cdot 8 \cdot 7 = 5{,}040$
3. The second subscript, 3, determines the number of factors. So we have:

$$_nP_3 = n(n-1)(n-2)$$

EXAMPLE 4
Assignment of Jobs

There are 30 students in a certain class. In how many ways can the instructor assign students to perform four jobs labeled A through D?

Solution

Each assignment of students to jobs can be understood as a permutation of 30 objects (students) taken four at a time. The ordering of the permutation creates the assignment to the jobs: The first student is assigned to job A, the second student to job B, and so forth. The number of ways of assigning the students is $_{30}P_4$. The numerical value of this symbol is:

$$_{30}P_4 = 30 \cdot 29 \cdot 28 \cdot 27 = 657{,}720$$

That is, there are 657,720 ways to assign 30 students to four jobs.

EXAMPLE 5
Communications Engineering

In a communications network, 13 cables connect two points, and five messages, labeled A through E, are to be sent from one point to the other point. In how many ways can the cables for the transmission be selected if each cable carries a single message?

Solution

The assignment of a message to each cable is a permutation of 13 objects (the cables) taken five at a time. The number of such permutations is equal to:

$$_{13}P_5 = 13 \cdot 12 \cdot 11 \cdot 10 \cdot 9 = 154{,}440$$

So the messages can be sent in 154,440 different ways.

Combinations

Now let's consider a second common type of counting problem, namely one involving combinations, which are defined as follows:

Definition
Combination

Suppose that we are given n distinct objects. A selection of r of them, without regard to the order of arrangement of the objects, is called a **combination of n objects taken r at a time**. The number of combinations of n objects taken r at a time is denoted $_nC_r$.

Note that a combination differs from a permutation in that a combination does not take into account the order of arrangement. For instance, we saw that the permutations of the three letters A, B, C taken two at a time are:

AB, BA, AC, CA, BC, CB

However, when we form combinations of these same three letters taken two at a time, the order of the letters does not count. For instance, the two permutations AB and BA are regarded as the same combination. There are only three combinations in this example, namely:

$$AB, AC, BC$$

The number $_nC_r$ can be calculated using the following result:

FUNDAMENTAL THEOREM OF COMBINATIONS

The number of combinations of n objects taken r at a time is equal to:

$$_nC_r = \frac{n!}{r!(n-r)!}$$

To verify this formula, let's begin by forming all permutations of the n objects taken r at a time. The number of such permutations is $_nP_r$. Each permutation corresponds to a combination. However, permutations that are obtained by rearranging the same objects in a different order represent the same combination. How many such rearrangements are there? We can choose the first object in a rearrangement in r ways, the second in $r-1$ ways, the third in $r-2$ ways, and so forth. By the Fundamental Principle of Counting, we see that each permutation can be rearranged in:

$$r(r-1)(r-2)\cdots(2)(1) = r!$$

ways. So each combination is counted $r!$ times. That is, we have:

$$_nP_r = r! \cdot {_nC_r}$$

$$_nC_r = \frac{_nP_r}{r!}$$

$$= \frac{n(n-1)\cdots(n-r+1)}{r!}$$

$$= \frac{n(n-1)\cdots(n-r+1)(n-r)(n-r-1)\cdots(1)}{r!(n-r)(n-r-1)\cdots(1)}$$

$$= \frac{n!}{r!(n-r)!}$$

This proves the Fundamental Theorem of Combinations.

Note that the right side of the formula in the Fundamental Theorem of Combinations is just a binomial coefficient. That is, another way of stating the theorem is:

$$_nC_r = \binom{n}{r}$$

Because $_nP_r = r! \cdot {_nC_r}$, we also conclude that:

$$_nP_r = \frac{n!}{(n-r)!}$$

Let's now use the Fundamental Theorem of Combinations to solve some applied counting problems.

EXAMPLE 6
Calculating Number of Combinations

Calculate the number of combinations of 10 objects taken four at a time.

Solution

The specified number equals $_{10}C_4$. And by the Fundamental Theorem of Combinations, this symbol has the value:

$$_{10}C_4 = \frac{10!}{4!(10-4)!}$$

$$= \frac{10 \cdot 9 \cdot 8 \cdot 7 \cdot \cancel{6} \cdot \cancel{5} \cdot \cancel{4} \cdot \cancel{3} \cdot \cancel{2} \cdot \cancel{1}}{(4 \cdot 3 \cdot 2 \cdot 1)(\cancel{6} \cdot \cancel{5} \cdot \cancel{4} \cdot \cancel{3} \cdot \cancel{2} \cdot \cancel{1})}$$

$$= \frac{10 \cdot \cancel{9}(3) \cdot \cancel{8} \cdot 7}{\cancel{4} \cdot \cancel{3} \cdot \cancel{2} \cdot 1}$$

$$= 210$$

Note that in the computation of Example 6, we canceled factors shared by the denominator and numerator in order to ease the burden of working with large numbers. Actually, we can always perform such cancellation in working with $_nC_r$, as the following result shows:

FORMULA FOR COMBINATIONS

Let n and r be positive numbers, with $r \leq n$. Then:

$$_nC_r = \frac{n(n-1)\cdots(n-r+1)}{r!}$$

Proof Start from the preceding formula for $_nC_r$:

$$_nC_r = \frac{n!}{r!(n-r)!}$$

$$= \frac{n(n-1)(n-2)\cdots(n-r+1)\cancel{(n-r)(n-r-1)\cdots(2)(1)}}{r!\cancel{(n-r)(n-r-1)\cdots(2)(1)}}$$

$$= \frac{n(n-1)(n-2)\cdots(n-r+1)}{r!}$$

This proves the Formula for Combinations.

EXAMPLE 7
Communications Engineering

In a communications network, 13 cables connect two points. In how many ways can four cables be selected to send messages from one point to the other?

Solution

Each selection of cables is a combination of 13 objects (the cables) taken four at a time. By the Formula for Combinations,

$$_{13}C_4 = \frac{13 \cdot 12 \cdot 11 \cdot 10}{4!}$$

$$= 715$$

SECTION 12.7 Permutations and Combinations

EXAMPLE 8
Baseball

A baseball team has 14 players. How many different nine-player teams (disregarding the order of players) are possible?

Solution

The number of such teams equals the number of combinations of 14 objects taken nine at a time. This number equals:

$$_{14}C_9 = \frac{14 \cdot 13 \cdot 12 \cdot 11 \cdot 10 \cdot \cancel{9} \cdot \cancel{8} \cdot \cancel{7} \cdot \cancel{6}}{\cancel{9} \cdot \cancel{8} \cdot \cancel{7} \cdot \cancel{6} \cdot 5 \cdot 4 \cdot 3 \cdot 2 \cdot 1}$$

$$= \frac{14 \cdot 13 \cdot \cancel{12}(2) \cdot 11 \cdot \cancel{10}}{\cancel{5} \cdot \cancel{4} \cdot \cancel{3} \cdot \cancel{2} \cdot 1}$$

$$= 2{,}002$$

That is, the team can be selected in 2,002 ways.

EXAMPLE 9
Poker

A standard deck of playing cards contains 52 cards. A poker hand contains 5 cards. How many different poker hands are possible?

Solution

A poker hand is a combination of 52 objects taken five at a time. The number of such objects equals:

$$_{52}C_5 = \frac{52 \cdot 51 \cdot 50 \cdot 49 \cdot 48}{5!} = 2{,}598{,}960$$

That is, there are 2,598,960 possible poker hands.

Using a Calculator in Permutation and Combination Problems

Most scientific calculators have built-in keystroke sequences for calculating $_nP_r$ and $_nC_r$. In Section 12.1, we used the TI-82 to calculate $_nC_r$, the coefficient of a binomial expansion. Let's now give the procedure for calculating $_nP_r$. The following example illustrates the procedure for a particular case.

EXAMPLE 10
Calculating $_nP_r$ Using a Calculator

Use the TI-82 to determine the value of $_{30}P_{19}$.

Solution

Use the following procedure.

1. Key the number 30 into the calculator.
2. Press (MATH)
3. Use the arrow keys to select the option **PRB** in the first row of the menu displayed.
4. Now select the second option, namely $_n\mathbf{P_r}$, from the menu displayed.
5. Press (ENTER). You are returned to the main screen, which now displays:

    30 $_nP_r$

6. Key in the number 19.
7. Press (ENTER)
8. The value of the binomial coefficient is displayed.

The answer to this problem is 6.645143394E24.

Exercises 12.7

Determine the following permutations and combinations.

1. $_4P_4$
2. $_7P_4$
3. $_6P_2$
4. $_9P_P$
5. $_3P_1$
6. $_9P_6$
7. $_nP_5$
8. $_nP_4$
9. $_7P_3$
10. $_6P_1$
11. $_8C_3$
12. $_9C_2$
13. $_5C_4$
14. $_6C_1$
15. $_{10}C_2$
16. $_{12}C_8$
17. $_nC_{n-1}$
18. $_nC_4$
19. $_4C_0$
20. $_nC_n$

Solve the following equations for n.

21. $_nC_2 = {}_{100}C_{98}$
22. $_nC_5 = {}_nC_3$
23. $_nP_4 = 4(_nP_3)$
24. $_nP_3 = (_{n-1}P_2)$

Prove the following.

25. $_nP_4 - {}_nP_3 = (n-4)(_nP_3)$
26. $_5P_r = 5(_4P_{r-1})$

Applications

27. How many numbers consisting of 1, 2, or 3 digits, without repeating digits, can you form using the digits 1, 2, 3, 4, 5, 6?

28. How many truck license plates can be made if each plate will have 3 numbers followed by 2 letters followed by 1 number?

29. How many 5-digit numbers begin with an even digit and end with an even digit?

30. In how many different orders can you arrange 4 books on a shelf?

31. Marv's Pizza Palace offers pepperoni, onion, sausage, mushrooms, and anchovies as toppings for its plain cheese base of pizza. How many different pizzas can be made?

32. If someone has 6 different skirts and 10 different blouses, how many different skirt and blouse outfits are possible?

33. If 6 people run a race, in how many different orders can they all finish if there are no ties?

34. For a 9-player baseball team, how many different batting orders are possible?

35. If you have 6 different colored shirts and 8 different ties, how many different combinations of shirt and tie can you make?

36. A club consists of 30 men and 70 women. In how many ways can a president, vice-president, secretary, and sergeant at arms be chosen if the president must be a woman, the sergeant at arms must be a man, and one person cannot hold more than one office?

37. How many arrangements can be formed from the letters of the word *FRAGMENTS*, taken 4 at a time?

38. A Navy signaler has 6 different flags. How many different signals can be sent by flying 3 flags at a time, one above the other, on a flag pole?

39. On leaving a train, a commuter finds that he has a nickel, a dime, a quarter, and a half-dollar in his pocket. In how many ways can he tip the porter?

40. In how many ways can 3 books be chosen from 7 different books and arranged in 3 spaces on a bookshelf?

41. How many permutations are there of 5 cards, taken from a deck of 52 cards? (The order of the cards is significant.)

42. How many arrangements can be formed from the letters of the word *HYPERBOLA*, using all the letters in each arrangement?

SECTION 13.1 Conic Sections and Parabolas

$$y + 3 = 3(x - 1)^2$$
$$(x - 1)^2 = \tfrac{1}{3}(y + 3)$$
$$(x - 1)^2 = 4 \cdot \tfrac{1}{12}(y + 3)$$

From the last form of the equation, we see that the graph is a parabola with vertex $(h, k) = (1, -3)$ and $p = \tfrac{1}{12}$. The last equation is in the second standard form, so the directrix is horizontal and the axis is vertical. The x-coordinate of the focus is the same as that of the vertex, namely 1. The y-coordinate of the focus is obtained by adding p to the y-coordinate of the vertex: $-3 + p = -3 + \tfrac{1}{12} = -\tfrac{35}{12}$. So the focus is at $(1, -\tfrac{35}{12})$. The directrix is:

$$y = -3 - p = -3 - \tfrac{1}{12} = -\tfrac{37}{12}$$

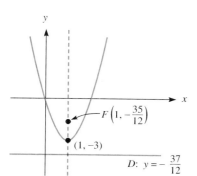

Figure 6

The graph of the equation is sketched in Figure 6.

EXAMPLE 3
Reducing to Standard Form

Determine the focus and the directrix of the parabola with equation:

$$y^2 + 2y = -x - 5$$

Sketch the parabola.

Solution

In this example, y appears in a term of the second degree, so we complete the square on the left:

$$y^2 + 2y + 1 = (-x - 5) + 1$$
$$(y + 1)^2 = -1 \cdot (x + 4)$$
$$[y - (-1)]^2 = 4 \cdot \left(-\tfrac{1}{4}\right) \cdot [x - (-4)]$$

From the equation, we read off the vertex: $(-4, -1)$. Also from the equation, we see that $p = -\tfrac{1}{4}$. Because the y-term is of degree 2, the directrix is vertical and the axis is horizontal. Because the value of p is negative, the focus is to the left of the vertex and is at:

$$\left(-4 - \tfrac{1}{4}, -1\right) = \left(-\tfrac{17}{4}, -1\right)$$

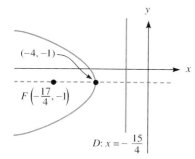

Figure 7

The directrix lies a directed distance p to the right of the vertex and is therefore the line $x = -4 - p = -4 + \tfrac{1}{4} = -\tfrac{15}{4}$. Figure 7 shows the graph of the equation.

Graphing Parabolas with a Graphing Calculator

Parabolas with horizontal directrixes have equations that can be written in the form $y = $ a function of x. These parabolas can be graphed just like any of the equations we have considered earlier, as the following example illustrates.

EXAMPLE 4
Graphing a Parabola on a Calculator—Horizontal Directrix

Use a graphing calculator to sketch the graph of the parabola:

$$(x - 1)^2 = 12(y - 3)$$

Solution

Solve the equation for y in terms of x:

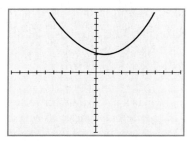

Figure 8

$$12y - 36 = (x - 1)^2$$
$$y = \frac{1}{12}[(x-1)^2 + 36]$$

Input the equation in this form into the calculator. The graph is shown in Figure 8.

EXAMPLE 5 Sketch the parabola $y^2 = 2x$ using a graphing calculator.

Graphing a Parabola on a Calculator—Vertical Directrix

Solution

Solve for y in terms of x. Taking the square root gives us two equations:

$$y = \pm\sqrt{2x}$$

We enter the expressions for y into the calculator as two equations:

$$y = \sqrt{2x}$$
$$y = -\sqrt{2x}$$

We graph both equations on the same display, as shown in Figure 9.

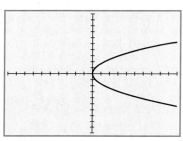

Figure 9

The Reflection Property of the Parabola

When a light or sound wave hits a surface, it is reflected in such a way that the angle made by the incoming wave equals the angle made by the reflected wave (see Figure 10). This is a physical law that was first observed in the seventeenth century. The parabola has a geometric property that, when combined with the law of reflection of light and sound waves, allows for many interesting and useful applications. Figure 11 shows a parabola with a vertical axis. A series of light

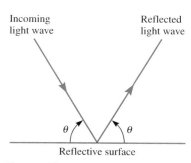

Figure 10

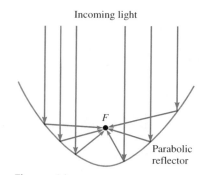

Figure 11

(or sound) waves moves parallel to the axis and is reflected off the surface of the parabola. It can be proved that the reflected waves all pass through the focus. This result is known as the **reflection property** of the parabola.

The reflection property can be used to construct a telescope. The observed light waves travel the length of the telescope tube and are reflected off a parabolic mirror. The reflected light rays all pass through the focus of the parabola, where the eyepiece of the telescope is located. In this case, the reflection property of the parabola states that the observed image is displayed at the eyepiece.

Another application using a similar design principle is the parabolic microphone often used in sporting events. This device consists of a parabolic dish that reflects sound to a microphone located at the focus. By concentrating diverse sound waves at the focus, the microphone can pick up and amplify sounds that originate from a long distance away.

Parabolas occur in many other applications in physics and engineering. For instance, the cable of a suspension bridge forms a parabola. Moreover, if you rotate a bucket of water around a vertical axis through the center of the bucket, then a vertical cross section of the surface of the water forms a parabola.

To prove the reflection property of the parabola requires calculus. By assuming a simple result from calculus, we outline a proof of the reflection property in the exercises at the end of this section.

EXAMPLE 6
Geometry of a Reflecting Telescope

The vertical cross section of the parabolic lens of a 6-in.-diameter reflecting telescope has the equation

$$x^2 = 240y$$

where the origin is at the center of the lens.

1. What is the focal length of the lens?
2. What is the depth of the lens at the center?

Solution

1. In standard form, the equation is:

$$x^2 = 4 \cdot 60y$$

This shows that the value of p is 60, so that the focal length is 60 in. (This is a typical value. A 6-in.-diameter telescope has the eyepiece located 60 in. from the center of the lens.)

2. The depth is the value of y at the outer edge of the lens. Because the diameter is 6 in., we have:

$$6^2 = 240y$$
$$y = 0.15 \text{ in.}$$

Exercises 13.1

Match each equation in Exercises 1–6 with its graph in (a)–(f).

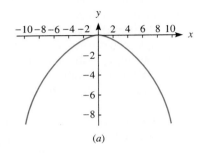
(a)

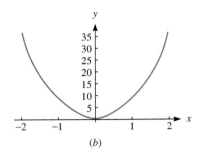
(b)

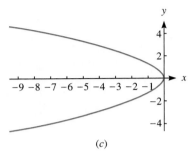
(c)

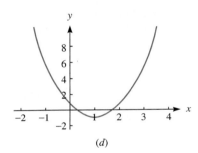
(d)

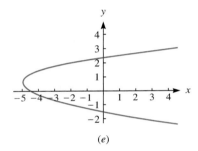
(e)

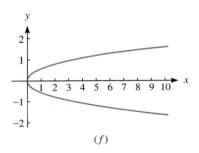
(f)

1. $y = 9x^2$
2. $x = 4y^2$
3. $y^2 = -2x$
4. $x^2 = -12y$
5. $y + 1 = 2(x - 1)^2$
6. $(2y - 1)^2 = 3(x + 5) - 1$

Find the vertex, focus, and directrix of each of the following parabolas.

7. $y = 9x^2$
8. $x = 4y^2$
9. $y^2 = -2x$
10. $x^2 = -12y$
11. $y + 1 = 2(x - 1)^2$
12. $3y - 4 = 6(x + 2)^2$
13. $y^2 - (y + 1)^2 = x^2$
14. $y = 4x^2 - 4x + 1$
15. $(3x + 1)^2 = 4y + 12$
16. $(2y - 1)^2 = 3(x + 5) - 1$
17. $y^2 + 3y - 6x = 8$
18. $y^2 + 3y - 6x = 0$
19. $25x^2 + 10x - y + 5 = 0$
20. $2x^2 + 6x + 5y - 20 = 0$

Determine the equation of the parabola that has the following properties.

21. vertex: (0, 0)
    directrix: $x = 2$
22. vertex: (0, 0)
    directrix: $x = -3$
23. vertex: (2, 1)
    directrix: $y = -1$
24. vertex: (1, 1)
    directrix: $y = 2$
25. vertex: (0, 0)
    focus: (1, 0)
26. vertex: (0, 0)
    focus: (-2, 0)
27. vertex: (3, -2)
    focus: (3, 4)
28. vertex: (4, 2)
    focus: (4, 0)
29. focus: (1, 2)
    directrix: $y = -2$
30. focus: (1, 2)
    directrix: $y = 5$
31. focus: (-1, -4)
    directrix: $x = 3$
32. focus: (1, 0)
    directrix: $x = 0$

SECTION 13.1 Conic Sections and Parabolas

33. The parabola passes through the three points $(0, 0)$, $(-2, 4)$, and $(3, 6)$, and its axis is vertical.

34. The parabola passes through the point $(3, 10)$, its axis is vertical, and its vertex is at the origin.

35. The parabola passes through the three points $(0, 0)$, $(2, 5)$, and $(3, 12)$, and its axis is horizontal.

36. The parabola passes through the origin, its axis is horizontal, and its vertex is at $(2, 1)$.

In calculus, it is proved that the slope m of the tangent line to the parabola that has the equation $4p(y - k) = (x - h)^2$ is given by the formula

$$m = \frac{1}{2p}(a - h)$$

at the point where $x = a$. Determine the equations of the following lines.

37. The line tangent to $y = x^2$ at $(2, 4)$.

38. The line tangent to $y = 2(x - 3)^2$ at the point where $x = 1$.

39. The line that passes through the point of tangency and is perpendicular to the line that is tangent to $y = 2x^2 + 3$ at the point where $x = -2$.

40. The line that passes through the origin and is parallel to the line that is tangent to $y = x^2$ at $x = 3$.

Applications

41. **Engineering.** The headlight of a car has a cross section that is a parabola. The light source is located at the focus. Suppose that the light source is 2 inches from the back of the headlight and that the radius of the headlight is 6 inches. Determine the equation of the parabola. Explain what the reflection property of the parabola has to do with the operation of the headlight.

42. **Engineering.** The main span of a suspension bridge is 5,200 feet long. Suppose that the towers hold the ends of the cable 300 feet above the roadway. The cables form a parabola. Assuming there is a coordinate system with horizontal axis along the roadway and vertical axis along the left tower, determine the equation of the parabola.

43. **Astronomy.** The parabolic mirror of a telescope has a diameter of 6 inches and a depth of .01 inches at the center. Determine the distance from the center of the mirror to the focus.

44. **Music.** The wall and ceiling in back of the orchestra of a symphony hall have a cross section that is a parabola. Suppose that the conductor stands at the focus, which is 50 feet from the wall. The height of the wall above the conductor is 60 feet. Determine the equation of the parabola. Explain what the reflection property of the parabola has to do with the design of the wall.

45. **Engineering.** The main span of a suspension bridge is 4,000 feet long. Suppose that the towers are 500 feet high and the roadway is 300 feet above the surface of the water. The cables form a parabola. Assuming there is a coordinate system with horizontal axis along the roadway and vertical axis along the left tower, determine the equation of the parabola.

Technology

46. Use a graphing calculator to draw a parabola with its vertex at the origin, its vertical directrix, and its focus at $(p, 0)$, for $p = 0.5$. Then, on the same graph, draw the same parabola for $p = 0.75, 1, 2,$ and 3. Describe what happens to the graph as the focal length p increases.

Use a graphing utility to graph the following parabolas.

47. $3y^2 = 4x + 1$
48. $-\dfrac{y^2}{2} = \dfrac{4x}{7}$
49. $(y - 3)^2 = 2(x + 5) + 1$
50. $(3y + 2)^2 = x - 3$

In your own words

51. Give the definition of a parabola.
52. Give three examples of parabolas that occur in everyday life.
53. What is the reflection property of the parabola?
54. Why does the reflection property of the parabola make parabolic shapes useful?

13.2 ELLIPSES

Let us now consider two additional conic sections—circles and ellipses. Circles are special cases of ellipses, so let's start the section by concentrating on ellipses.

Definition and Standard Equation of an Ellipse

We can define an ellipse in geometric terms, as follows. Suppose we are given two fixed points F_1 and F_2 that are foci of the ellipse and a positive number $2a$. (*Foci* is the plural of *focus*.)

Definition
Ellipse
An **ellipse** with foci F_1 and F_2 is the set of all points P such that the sum of the distances $d(P, F_1)$ and $d(P, F_2)$ is $2a$.

In Figure 1, the two foci are F_1 and F_2. A typical point on the ellipse is $P(x, y)$. The distance from P to F_1 is denoted d_1, and the distance from P to F_2 is denoted

SECTION 13.2 Ellipses

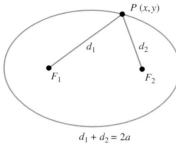

Figure 1

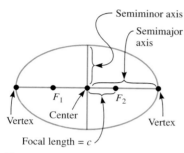

Figure 2

d_2. The definition of an ellipse states that:

$$d_1 + d_2 = 2a \qquad (1)$$

(The reason for writing $2a$ rather than a is explained shortly.)

In Figure 2, we draw a line through the foci that intersects the ellipse in two points called **vertices**. The line segment connecting the vertices is called the **major axis**. The midpoint of the major axis is called the **center** of the ellipse. Half the length of the major axis, from the center to a vertex, is called the **semimajor axis** of the ellipse. It is clear that the center of the ellipse is midway between the two foci and the two vertices. Now we draw a line segment perpendicular to the major axis and passing through the center, with both endpoints on the ellipse. This line segment is called the **minor axis**. Half the length of the minor axis, from the center to the endpoint on the ellipse, is called the **semiminor axis** of the ellipse.

Assume that the center of the ellipse is at the origin and that the distance from the center to each focus is c. The distance c is called the **focal length** of the ellipse. (See Figure 2.) Let us assume that the major axis is horizontal. Then the foci are at $(\pm c, 0)$. From the definition of the ellipse, if the right vertex is at $(x, 0)$, and if d_1 and d_2 are the distances from the right vertex to F_1 and F_2, respectively, then:

$$d_1 + d_2 = 2a$$
$$(x + c) + (x - c) = 2a$$
$$2x = 2a$$
$$x = a$$

So the right vertex is at $(a, 0)$. Similarly, the left vertex is at $(-a, 0)$. Moreover, from these coordinates, we see that $2a$ is the length of the major axis of the ellipse.

Let's now write equation (1) in terms of the coordinates of the foci and the point $P(x, y)$ on the ellipse:

$$d_1 + d_2 = 2a$$
$$d(F_1, P) + d(F_2, P) = 2a$$
$$\sqrt{[x - (-c)]^2 + y^2} = 2a - \sqrt{(x - c)^2 + y^2}$$
$$[x - (-c)]^2 + y^2 = 4a^2 + (x - c)^2 + y^2 - 4a\sqrt{(x - c)^2 + y^2}$$
$$-4a\sqrt{(x - c)^2 + y^2} = 4a^2 - 4cx$$
$$a\sqrt{(x - c)^2 + y^2} = a^2 - cx$$
$$a^2(x - c)^2 + a^2 y^2 = (a^2 - cx)^2$$
$$a^2 x^2 + a^2 c^2 - 2a^2 cx + a^2 y^2 = a^4 + c^2 x^2 - 2a^2 cx$$
$$a^2 x^2 - c^2 x^2 + a^2 y^2 = a^4 - a^2 c^2$$
$$(a^2 - c^2) x^2 + a^2 y^2 = a^2(a^2 - c^2)$$
$$\frac{x^2}{a^2} + \frac{y^2}{a^2 - c^2} = 1 \qquad (2)$$

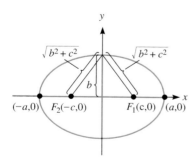

Figure 3

From Figure 3, we see that, if b denotes half the length of the minor axis, then, by equation (1), we have:

$$2\sqrt{b^2 + c^2} = 2a$$
$$a^2 = b^2 + c^2$$
$$a^2 - c^2 = b^2$$

Therefore, we can rewrite equation (2) in the form:

$$\frac{x^2}{a^2} + \frac{y^2}{b^2} = 1$$

This derivation assumes that the center of the ellipse is at the origin. However, if the center is at (h, k), we can obtain the equation of the ellipse by translating the equation we just derived. We obtain the result:

STANDARD FORM OF THE EQUATION OF AN ELLIPSE (MAJOR AXIS IS HORIZONTAL)

$$\frac{(x - h)^2}{a^2} + \frac{(y - k)^2}{b^2} = 1 \tag{3}$$

This is the standard form of the equation of an ellipse for which the major axis is horizontal.

If we assume that the major axis is vertical, we arrive at the equation:

STANDARD FORM OF THE EQUATION OF AN ELLIPSE (MAJOR AXIS IS VERTICAL)

$$\frac{(x - h)^2}{b^2} + \frac{(y - k)^2}{a^2} = 1 \tag{4}$$

In either case, the foci are a distance c from the center along the major axis, where:

$$a^2 = c^2 + b^2 \tag{5}$$

A circle is an ellipse in which the semimajor and semiminor axes are equal (to the radius). In this case, the above standard equations of an ellipse reduce to the familiar equation of a circle. Indeed, if an ellipse has $a = b = r$ and its center is (h, k), then the equation of the ellipse is:

$$\frac{(x - h)^2}{r^2} + \frac{(y - k)^2}{r^2} = 1$$
$$(x - h)^2 + (y - k)^2 = r^2 \tag{6}$$

This is the standard form of a circle with center at (h, k) and radius r.

Ellipses and Their Equations

In the next several examples, we provide practice in determining equations of ellipses from geometric data and determining geometric data from equations of ellipses.

SECTION 13.2 Ellipses

EXAMPLE 1
Determining the Equation of an Ellipse from Foci and Minor Axis

Determine the equation of the ellipse that has foci at (3, 5) and (9, 5) and minor axis of length 10.

Solution

Because the foci have the same y-coordinate, the major axis is horizontal. The center of the ellipse is midway between the two foci and is therefore (6, 5), and the distance from the center to each focus is given by $c = 3$. The length of the semiminor axis b is given by:

$$2b = 10$$
$$b = 5$$

The length of the semimajor axis a is:

$$a^2 = c^2 + b^2$$
$$a^2 = 9 + 25$$
$$a^2 = 34$$
$$a = \sqrt{34}$$

Therefore, the equation of the ellipse is:

$$\frac{(x-6)^2}{(\sqrt{34})^2} + \frac{(y-5)^2}{5^2} = 1$$

$$\frac{(x-6)^2}{34} + \frac{(y-5)^2}{25} = 1$$

The graph is sketched in Figure 4.

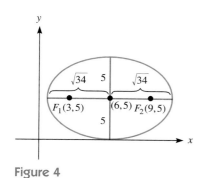

Figure 4

EXAMPLE 2
Determining the Geometry of an Ellipse from Its Equation

Sketch the graph of the ellipse with equation:

$$4x^2 + 25y^2 = 100$$

Determine the foci, vertices, semimajor axis, and semiminor axis.

Solution

Because both x and y appear to the second power and the coefficients of both have the same sign, the graph of the equation is an ellipse. In the standard form of the equation of an ellipse, the constant term is 1. To reduce the equation to this form, we divide both sides by 100:

$$\frac{4x^2}{100} + \frac{25y^2}{100} = 1$$

$$\frac{x^2}{25} + \frac{y^2}{4} = 1$$

$$\frac{x^2}{5^2} + \frac{y^2}{2^2} = 1$$

From the equation, we see that the ellipse is centered at the origin and has a semimajor axis of length 5 along the x-axis and a semiminor axis of length 2 along the y-axis. The vertices are ($\pm 5, 0$). The foci are at a distance c from the center, where:

$$c^2 + b^2 = a^2$$
$$c^2 = 5^2 - 2^2$$

$$= 21$$
$$c = \sqrt{21}$$

Therefore, the foci are $(\pm\sqrt{21}, 0)$. The graph is sketched in Figure 5.

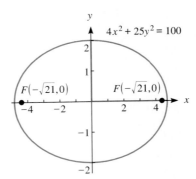

Figure 5

EXAMPLE 3

Reducing the Equation of an Ellipse to Standard Form

Sketch the graph of the ellipse that has the equation:

$$5x^2 + 10x + 4y^2 + 8y - 5 = 0$$

Determine the foci, vertices, semimajor axis, and semiminor axis.

Solution

Note that the equation involves both x and y raised to the second power and that the coefficients of the squared terms have the same sign. This means that the graph of the equation is an ellipse. To transform the equation into standard form, we complete the square in both x and y:

$$5x^2 + 10x + 4y^2 + 8y - 5 = 0$$
$$5(x^2 + 2x) + 4(y^2 + 2y) = 5$$
$$5(x^2 + 2x + 1) + 4(y^2 + 2y + 1) = 5 + 5 + 4$$
$$5(x + 1)^2 + 4(y + 1)^2 = 14$$
$$\frac{5(x+1)^2}{14} + \frac{4(y+1)^2}{14} = 1$$
$$\frac{(x+1)^2}{\frac{14}{5}} + \frac{(y+1)^2}{\frac{7}{2}} = 1$$
$$\frac{(x+1)^2}{\left(\sqrt{\frac{14}{5}}\right)^2} + \frac{(y+1)^2}{\left(\sqrt{\frac{7}{2}}\right)^2} = 1$$

This last equation has a graph that is an ellipse with center $(-1, -1)$. Because $\frac{7}{2} = 3.5$ is greater than $\frac{14}{5} = 2.8$, we see that the semimajor axis is vertical and the semiminor axis is horizontal. The lengths of the axes are:

$$\text{semimajor axis} = \sqrt{\frac{7}{2}} = \frac{\sqrt{14}}{2}, \quad \text{vertical}$$

$$\text{semiminor axis} = \sqrt{\frac{14}{5}} = \frac{\sqrt{70}}{5}, \quad \text{horizontal}$$

The foci are at a distance c from the center, where:

$$\begin{aligned} c^2 &= a^2 - b^2 \\ &= \left(\sqrt{\frac{7}{2}}\right)^2 - \left(\sqrt{\frac{14}{5}}\right)^2 \\ &= \frac{7}{2} - \frac{14}{5} \\ &= \frac{7}{10} \end{aligned}$$

$$c = \sqrt{\frac{7}{10}} = \frac{\sqrt{70}}{10}$$

Therefore, the foci are $(-1, -1 \pm \sqrt{70}/10)$. The graph is sketched in Figure 6.

Figure 6

Graphing Ellipses with a Graphing Calculator

We can use a graphing calculator to display ellipses. To do so, just solve the equation of the ellipse for y in terms of x. Because a square root results, there are two expressions for y. We enter these expressions as separate equations and graph both equations on the same display. The following example supplies the details.

EXAMPLE 4 Using a Graphing Calculator to Graph an Ellipse

Display the graph of the ellipse

$$\frac{x^2}{3^2} + \frac{y^2}{2^2} = 1$$

using a graphing calculator. Determine the points on the ellipse for which x has the value 2.3.

Solution

Solve the equation for y in terms of x:

$$\begin{aligned} y^2 &= 4\left(1 - \frac{x^2}{9}\right) \\ &= \frac{4}{9}(9 - x^2) \\ y &= \pm \frac{2}{3}\sqrt{9 - x^2} \end{aligned}$$

Enter the two equations into the calculator:

$$y = \frac{2}{3}\sqrt{9 - x^2}$$

$$y = -\frac{2}{3}\sqrt{9 - x^2}$$

686   CHAPTER 13   Topics in Analytic Geometry

From the equation, we see that the ellipse is centered at the origin and has horizontal semimajor axis 3 and vertical semiminor axis 2. In order to maximize the size of the ellipse on the display, we choose the following display parameters:

Xmin = −3
Xmax = 3
Ymin = −2
Ymax = 2

The graph is shown in Figure 7. To determine the points on the ellipse for which $x = 2.3$, we use the trace function. First, we trace along the upper-half of the ellipse, which corresponds to the first equation, and we obtain the point shown in Figure 8. We then switch to the lower curve by pressing the down arrow key, and we trace along the bottom curve to obtain the point shown in Figure 9.

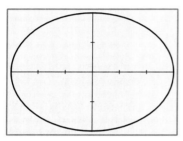

Figure 7

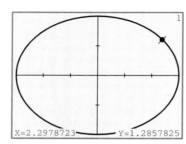

Figure 8

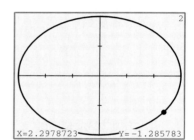

Figure 9

Eccentricity of an Ellipse

The eccentricity of an ellipse, denoted e, measures the "ovalness" of an ellipse. The eccentricity always lies between 0 and 1. The closer the eccentricity is to 0, the closer the ellipse is to a circle. The closer the eccentricity is to 1, the more elongated the ellipse.

Definition
Eccentricity

Suppose that an ellipse has semimajor axis a and focal length c. Then the **eccentricity** e of the ellipse is defined as:

$$e = \frac{c}{a}$$

Note that, because the focal length is less than the semimajor axis, the eccentricity as defined by the above formula lies between 0 and 1.

EXAMPLE 5
Calculating Eccentricity

Determine the eccentricity of the ellipse in Example 3.

Solution

In Example 3, we showed that $a = \sqrt{14}/2$ and $c = \sqrt{70}/10$. Therefore, we have:

$$e = \frac{c}{a} = \frac{\sqrt{70}/10}{\sqrt{14}/2} = \frac{\sqrt{5}}{5}$$

EXAMPLE 6
Eccentricity of a Circle

Show that an ellipse of eccentricity 0 is a circle.

Solution

If $e = 0$, then $c = 0$, so that:

$$0 = a^2 - b^2$$
$$a^2 = b^2$$
$$a = b \quad \text{Because } a \text{ and } b \text{ are positive}$$

That is, the semimajor and semiminor axes are equal and the ellipse is a circle.

One of the major accomplishments of physics is the discovery of Kepler's laws, which state that the planets travel in elliptical orbits around the sun, with the sun located at a focus of the ellipse. Using Newton's law of gravitation, it is possible to deduce the precise equations of the orbits in terms of the masses of the planets, the mass of the sun, and the observed distances and velocities of the planets at particular times.

EXAMPLE 7
Orbit of the Earth

The eccentricity of the earth's orbit about the sun is approximately 0.0167. The closest distance between the earth and the sun is approximately 93 million miles. What is the furthest distance between the earth and the sun? (See Figure 10.)

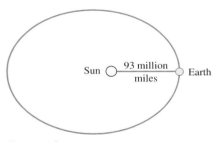

Figure 10

Solution

Let a be the semimajor axis of the orbit. Assume that the center of the ellipse is $(0, 0)$. Suppose that the sun is at the focus $(c, 0)$. Then:

$$a - c = 93$$
$$a = c + 93$$

Therefore:

$$e = \frac{c}{a} = \frac{c}{c + 93} = 0.0167$$
$$c = 0.0167c + 1.5531$$
$$0.9833c = 1.5531$$
$$c = 1.5795$$

The furthest distance of the earth from the sun is $a + c = 2c + 93$. Therefore,

$$a + c = 2c + 93 = 96.159 \text{ million miles}$$

Exercises 13.2

Match each equation in Exercises 1–6 with its graph in (a)–(f).

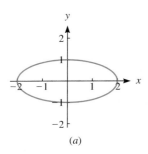
(a)

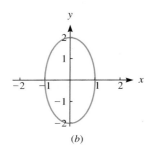
(b)

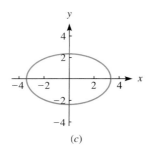
(c)

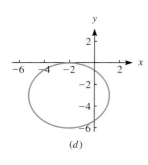
(d)

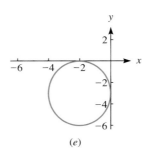
(e)

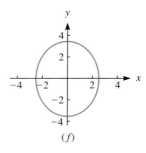
(f)

1. $x^2 + \dfrac{y^2}{4} = 1$

2. $\dfrac{x^2}{4} + y^2 = 1$

3. $\dfrac{x^2}{12.25} + \dfrac{y^2}{6.25} = 1$

4. $\dfrac{x^2}{6.25} + \dfrac{y^2}{12.25} = 1$

5. $\dfrac{(x+2)^2}{4} + \dfrac{(y+3)^2}{9} = 1$

6. $\dfrac{(x-2)^2}{9} + \dfrac{(y+2)^2}{9} = 1$

Determine the equation of the ellipse that has the following properties.

7. vertices: $(0, \pm 3)$
   foci: $(0, \pm 2)$

8. vertices: $(5, 1)$, $(-5, 1)$
   foci: $(3, 1)$, $(-3, 1)$

9. vertices: $(-1, 4)$, $(-1, 10)$
   foci: $(-1, 5)$, $(-1, 9)$

10. vertices: $(-3, 2)$, $(5, 2)$
    foci: $(-1, 2)$, $(3, 2)$

11. vertices: $(0, \pm 7)$
    semiminor axis: 3

12. vertices: $(\pm 5, 0)$
    semiminor axis: 2

13. vertices: $(-1, 2)$, $(3, 2)$
    semiminor axis: 1

14. vertices: $(5, -3)$, $(5, 7)$
    semiminor axis: 3

15. vertices: $(5, -3)$, $(5, 7)$
    eccentricity: $\dfrac{2}{5}$

16. vertices: $(\pm 3, 0)$
    eccentricity: $\dfrac{3}{4}$

17. vertices: $(0, \pm 1)$
    foci: $(0, \pm \dfrac{1}{3})$

18. vertices: $(\pm 5, 0)$
    foci: $(\pm 1, 0)$

19. vertices: $(5, 1)$, $(15, 1)$
    foci: $(7, 1)$, $(13, 1)$

20. vertices: $(3, 7)$, $(3, -5)$
    foci: $(3, 5)$, $(3, -3)$

21. vertices: $(0, \pm 4)$
    passing through: $(2, 0)$

22. vertices: $(\pm 3, 0)$
    passing through: $(0, 2)$

23. vertices: $(\pm 3, 0)$
    passing through: $(2, 1)$

24. vertices: $(0, \pm 5)$
    passing through: $(-2, 3)$

Graph each of the following ellipses. Determine the semimajor and semiminor axes, the foci, and the eccentricity.

25. $\dfrac{x^2}{9} + \dfrac{y^2}{4} = 1$

26. $\dfrac{x^2}{4} + \dfrac{y^2}{9} = 1$

27. $2x^2 + 4y^2 = 16$

28. $4x^2 + 2y^2 = 16$

29. $\dfrac{(x+2)^2}{4} + \dfrac{(y-2)^2}{16} = 1$

30. $\dfrac{(x+2)^2}{16} + \dfrac{(y-2)^2}{4} = 1$

31. $3(x-1)^2 + 6(y-2)^2 = 108$

32. $6(x-1)^2 + 3(y-2)^2 = 108$

33. $x^2 - 4x + 5y^2 + 30y + 24 = 0$

34. $4x^2 + 8x + 6y^2 + 24y + 4 = 0$

35. $36x^2 + 4y^2 - 144 = 0$

36. $x^2 + 9y^2 - 9 = 0$

37. $9x^2 + y^2 - 54x - 4y + 49 = 0$

38. $4x^2 + 25y^2 + 8x + 150y + 129 = 0$

39. $\dfrac{(x+1)^2}{16} + \dfrac{(y+1)^2}{25} = 1$

40. $16(x+1)^2 + 25(y+1)^2 = 400$

41. $x^2 + y^2 + 6y = 0$

42. $3x^2 + 3y^2 + 6x = 1$

43. The latera recta of an ellipse are the line segments that are perpendicular to the major axis, drawn through the foci, and have endpoints on the ellipse. Determine the endpoints of the latera recta of the ellipse:
$$\dfrac{x^2}{4} + \dfrac{y^2}{9} = 1$$

44. Show that the length of the latera recta are $(2b)^2/a$.

45. Suppose that an ellipse has semimajor axis a, semiminor axis b, and eccentricity e. Prove that $b^2 = a^2(1 - e^2)$.

46. Suppose that the semimajor and semiminor axes of an ellipse are both multiplied by the same factor k. Show that the eccentricity is unchanged.

Applications

47. A communications satellite is launched into orbit around the earth, with the low point of the orbit at 22,000 miles and the high point of the orbit 24,000 miles above the earth's surface. Determine the equation of the orbit.

48. Refer to Exercise 47. What is the eccentricity of the orbit?

Technology

Use a graphing calculator to graph the following ellipses.

49. $\dfrac{x^2}{2.25} + \dfrac{y^2}{6.25} = 1$

50. $3x^2 + 5y^2 = 13$

51. $\dfrac{(x-2)^2}{9} + \dfrac{(y+2)^2}{20.25} = 1$

52. $7(x-2)^2 + 6(y-2)^2 = 17$

53. Graph ellipses with horizontal major axis and eccentricities 0.25, 0.5, 0.75, 0.9. What happens as the eccentricities increase?

54. Graph ellipses with vertical major axis and eccentricities 0.25, 0.5, 0.75, 0.9. What happens as the eccentricities increase?

In your own words

55. Give the definition of an ellipse.

56. Describe how you can use a pencil, string, and paper to draw an ellipse with given foci.

57. Refer to Exercise 56. What is the relationship of the length of the string to the geometry of the ellipse?

58. Give an example of an ellipse in the world around us.

13.3 HYPERBOLAS

Definition and Standard Equation of a Hyperbola

Let's now turn our attention to the final conic section, the hyperbola.

Suppose we are given two foci F_1 and F_2 and a positive number $2a$.

Definition
Hyperbola

A **hyperbola** with foci F_1 and F_2 is the set of all points P satisfying the equation
$d(P, F_1) - d(P, F_2) = 2a$

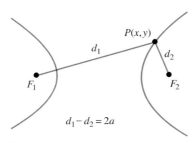

Figure 1

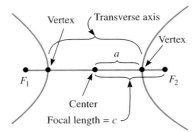

Figure 2

Note the similarity of this definition to the definition of an ellipse in Section 13.2. An ellipse is the set of points P such that the *sum* of the distances from P to the foci is a constant. To get the definition of a hyperbola, just replace "sum" with "difference."

The two foci of the hyperbola in Figure 1 are F_1 and F_2. A typical point on the hyperbola is $P(x, y)$. The distance from P to F_1 is denoted d_1 and the distance from P to F_2 is denoted d_2. The definition of a hyperbola states that there is a constant $2a$, independent of the point P, such that:

$$d_1 - d_2 = 2a \tag{1}$$

We draw a line connecting the foci, as in Figure 2. This line intersects the hyperbola in two points called the **vertices**. The line connecting the vertices is called the **transverse axis** of the hyperbola. The midpoint of the transverse axis is called the **center** of the hyperbola. The distance from the center to either of the vertices is a. (The proof of this fact is similar to that for the case of an ellipse, in Section 13.2.) Let c be the distance from each focus to the center. Then c is called the **focal length**. We define b by the equation:

$$b^2 = c^2 - a^2$$

Then, by reasoning as in the case of the ellipse, we can deduce the following equations, where (h, k) is the center.

STANDARD FORMS OF THE EQUATION OF A HYPERBOLA		
$\dfrac{(x-h)^2}{a^2} - \dfrac{(y-k)^2}{b^2} = 1$	*Transverse axis is horizontal*	(2)
$\dfrac{(y-k)^2}{a^2} - \dfrac{(x-h)^2}{b^2} = 1$	*Transverse axis is vertical*	(3)

We leave the derivation of these equations as an exercise.

EXAMPLE 1

Determining the Equation of a Hyperbola from Foci and Transverse Axis

Determine the equation of a hyperbola that has foci at $(-3, 1)$ and $(5, 1)$ and transverse axis of length 4.

Solution

The transverse axis is horizontal because the two foci lie on a horizontal line. The center, which is halfway between the foci, is the point $(h, k) = (1, 1)$. Because the length of the transverse axis is 4, the vertices are $(3, 1)$ and $(-1, 1)$, and $a = 2$. The distance c of each focus from the center is 4. Thus, the value of b is given by:

$$b^2 = c^2 - a^2$$
$$= 4^2 - 2^2$$
$$= 12$$
$$b = \sqrt{12} = 2\sqrt{3}$$

Thus, the equation of the hyperbola is:

$$\frac{(x-1)^2}{2^2} - \frac{(y-1)^2}{(2\sqrt{3})^2} = 1$$

$$\frac{(x-1)^2}{4} - \frac{(y-1)^2}{12} = 1$$

Graphing Hyperbolas

Let's now examine the graph of a hyperbola. First, we consider the standard form of a hyperbola with a horizontal transverse axis:

$$\frac{(x-h)^2}{a^2} - \frac{(y-k)^2}{b^2} = 1$$

Solve the equation for $y - k$:

$$\frac{(y-k)^2}{b^2} = \frac{(x-h)^2}{a^2} - 1$$

$$\frac{y-k}{b} = \pm\sqrt{\left(\frac{x-h}{a}\right)^2 - 1}$$

$$y - k = \pm\frac{b}{a}\sqrt{(x-h)^2 - a^2}$$

The square root on the right yields a real number only if $|x - h| \geq a$. Therefore, the x-coordinates of the points on the hyperbola must satisfy the inequality:

$$|x - h| \geq a$$

This is equivalent to:

$$x \leq h - a \quad \text{or} \quad x \geq h + a$$

That is, a point on the hyperbola is either to the right of the right vertex or to the left of the left vertex. As the absolute value of x increases, so does the value of:

$$\frac{b}{a}\sqrt{(x-h)^2 - a^2}$$

Moreover, the $\pm$ indicates that, for each value of x, there are two points on the graph, symmetrically placed, above and below the line $y = k$. The graph is shown in Figure 3. We can write the equation of the hyperbola in the form:

$$y - k = \pm\frac{b}{a}\sqrt{(x-h)^2\left(1 - \frac{a^2}{(x-h)^2}\right)}$$

$$= \pm\frac{b}{a}(x-h)\sqrt{1 - \frac{a^2}{(x-h)^2}}$$

As x approaches either infinity or minus infinity, the expression under the square root approaches 1, so that the value of y approaches:

$$k \pm \frac{b}{a}(x-h)$$

This means that, as x approaches either infinity or minus infinity, the graph approaches the lines with equations:

$$y = k + \frac{b}{a}(x - h)$$

$$y = k - \frac{b}{a}(x - h)$$

These lines are called **asymptotes** of the hyperbola. The lines are drawn in Figure 3. Note that, if we draw a rectangle centered at (h, k) of width $2a$ and height $2b$, then the asymptotes are the diagonals of this rectangle. (See Figure 3.) The vertical line that passes through the center of the hyperbola and extends b units above and b units below the center is called the **conjugate axis** of the hyperbola.

We repeat the same reasoning for a hyperbola that has a vertical transverse axis to obtain the graph shown in Figure 4. In this case, the asymptotes are the equations:

$$y = k + \frac{a}{b}(x - h)$$

$$y = k - \frac{a}{b}(x - h)$$

The conjugate axis in this case is horizontal.

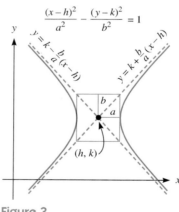

Figure 3

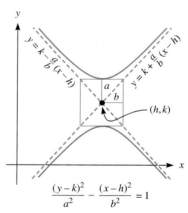

Figure 4

EXAMPLE 2
Graphing Hyperbolas

Sketch the graphs of the following equations.

1. $\dfrac{x^2}{4} - \dfrac{y^2}{9} = 1$

2. $(x - 3)^2 - \dfrac{(y - 1)^2}{4} = 1$

3. $y^2 - \dfrac{x^2}{4} = 1$

Solution

1. We first write the equation in one of the two standard forms.

$$\frac{x^2}{2^2} - \frac{y^2}{3^2} = 1$$

We see that $a = 2$ and $b = 3$. Moreover, because the positive sign is on the x-term, the axis is horizontal. The center, (h, k), is $(0, 0)$. The vertices are $(\pm a, 0) = (\pm 2, 0)$. We draw the transverse axis of length $2a = 4$ and the conjugate axis of length $2b = 6$, centered at $(0, 0)$. Around the axes, we draw the associated rectangle. Next, we draw the diagonals of the rectangle, which are the asymptotes of the hyperbola. Finally, we draw the hyperbola, symmetric with respect to the transverse and conjugate axes, and asymptotic to the diagonal lines. See Figure 5.

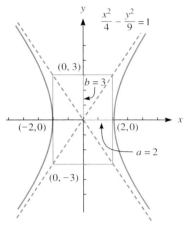

Figure 5

2. We write the equation in the standard form:

$$\frac{(x-3)^2}{1^2} - \frac{(y-1)^2}{2^2} = 1$$

From this form, we determine that $a = 1$ and $b = 2$. Because the positive sign is on the x-term, the transverse axis of the hyperbola is horizontal. Moreover, the center of the hyperbola is the point $(3, 1)$. We plot the center and then draw the transverse axis of length $2a = 2$ and the conjugate axis of length $2b = 4$. We use the axes to draw a rectangle. The diagonals of the rectangle are the asymptotes. Next, we draw the hyperbola, symmetric with respect to both axes and asymptotic to the diagonal lines. See Figure 6.

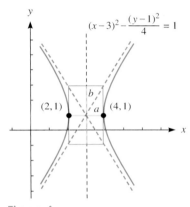

Figure 6

3. We write the equation in standard form:

$$\frac{y^2}{1^2} - \frac{x^2}{2^2} = 1$$

In this case, $a = 1$ and $b = 2$. Because $h = 0$ and $k = 0$, the center of the hyperbola is the point $(0, 0)$. Moreover, because the positive sign is on the y-term, the transverse axis is vertical. The transverse axis has length $2a = 2$. The conjugate axis has length $2b = 4$ and is horizontal. To sketch the graph, we plot the center and draw the axes. We use the axes to draw a rectangle and the diagonals of the rectangle, which are the asymptotes. Next, we draw the hyperbola, symmetric with respect to both axes and asymptotic to the diagonal lines. See Figure 7.

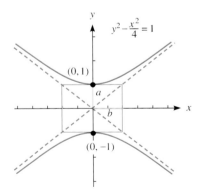

Figure 7

EXAMPLE 3
Reducing the Equation of a Hyperbola to Standard Form

Sketch the graphs of the following equations.

1. $x^2 - 2y^2 = 1$
2. $x^2 - 2x - y^2 + 6y - 7 = 0$

Solution

1. We write the equation in standard form:

$$x^2 - \frac{y^2}{\frac{1}{2}} = 1$$

$$\frac{x^2}{1^2} - \frac{y^2}{\left(1/\sqrt{2}\right)^2} = 1$$

From the last equation, we see that the hyperbola has a horizontal transverse axis of length $2a = 2$ and a vertical conjugate axis of length $2b = 2/\sqrt{2} = \sqrt{2}$. To sketch the graph, we first draw in the rectangle determined by the vertices $(\pm 1, 0)$ and the points $(0, \pm \sqrt{2}/2)$. Next, we draw the diagonals of the rectangle. These are the asymptotes of the hyperbola. Finally, we draw the hyperbola. (See Figure 8.)

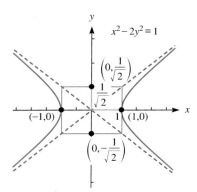

Figure 8

2. We begin by putting the equation into standard form. To do this, we complete the square with respect to both x and y:

$$(x^2 - 2x) - (y^2 - 6y) = 7$$
$$(x^2 - 2x + 1) - (y^2 - 6y + 9) = 7 + 1 - 9$$
$$(x - 1)^2 - (y - 3)^2 = -1$$
$$(y - 3)^2 - (x - 1)^2 = 1$$
$$\frac{(y - 3)^2}{1^2} - \frac{(x - 1)^2}{1^2} = 1$$

From this equation, we see that the hyperbola is centered at $(1, 3)$. Because the y-term appears with a positive sign, the transverse axis of the hyperbola is vertical. Moreover, the lengths of the two axes are both 2. This provides us with the data to sketch the graph. (See Figure 9.)

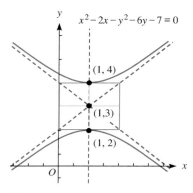

Figure 9

Graphing Hyperbolas with a Graphing Calculator

We can use a graphing calculator to graph hyperbolas. Just solve the equation of the hyperbola for y in terms of x. Because of the square root that results, we must graph two equations on the same display.

EXAMPLE 4
Using a Graphing Calculator to Graph a Hyperbola

Consider the hyperbola that has the equation:

$$\frac{x^2}{2^2} - \frac{y^2}{5^2} = 1$$

1. Use a graphing calculator to display the hyperbola.
2. Determine all points on the graph for which $x = 4.7$.

Solution

1. Solve the equation for y in terms of x:

$$y^2 = 5^2\left(\frac{x^2}{2^2} - 1\right)$$

$$= \frac{5^2}{2^2}(x^2 - 4)$$

$$y = \pm\frac{5}{2}\sqrt{x^2 - 4}$$

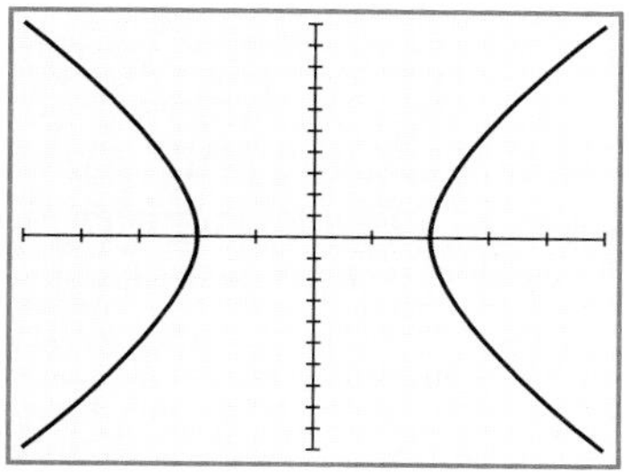

We enter the following two equations in the calculator:

$$y = \frac{5}{2}\sqrt{x^2 - 4}$$

$$y = -\frac{5}{2}\sqrt{x^2 - 4}$$

and display their graphs in **dots only** mode. The display shows the graph in Figure 10. We used the display parameters:

Xmin = −5
Xmax = 5
Ymin = −10
Ymax = 10

Figure 10

2. Use the trace to determine the points where $x = 4.7$. There are two points, one on the upper half of the graph and one on the lower half. These two halves come from different equations. So we trace the graph of the first equation and obtain the point shown in Figure 11. Then we use the down arrow key to trace the lower curve to obtain the point shown in Figure 12.

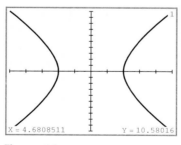

Figure 11

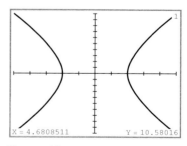

Figure 12

Applications of Hyperbolas

Hyperbolas are used in navigation, as illustrated in the next example.

EXAMPLE 5
Hyperbolic Navigation

Suppose that two rescue teams are searching a wooded area for a trapped camper. Both teams hear the camper's shouts. Suppose that the two rescue teams are 8 miles apart, and Team 2 hears the shouts of the camper 5 seconds after Team 1 hears them. Use a coordinate system that has its origin at Team 1, and assume the positive x-axis extends from the first rescue team to the second. Write an equation describing the possible location of the camper. Assume that sound travels at 1,100 feet per second.

Solution

In Figure 13, we draw the coordinate system and label the distance of the camper from the two rescue teams with d_1 and d_2. Because the distance between the rescue teams is in miles, we must convert the speed of sound into miles per second:

$$1{,}100 \text{ ft/sec} = 1{,}100 \cdot \frac{1}{5{,}280} \text{ ft/sec} \cdot \text{mi./ft}$$
$$= 0.20833 \text{ mi./sec}$$
$$= a$$

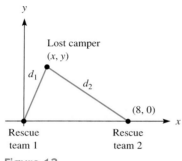

Figure 13

By the formula
$$\text{distance} = \text{rate} \times \text{time}$$
we see that the times for sound to travel from the camper to each rescue team are d_1/a and d_2/a. Moreover, because the times are 5 seconds apart, we have

$$\frac{d_2}{a} - \frac{d_1}{a} = 5$$
$$d_2 - d_1 = 5a$$

However, referring to the figure and applying the distance formula, this last equation yields:

$$\sqrt{(x-8)^2 + y^2} - \sqrt{x^2 + y^2} = 5a$$
$$\sqrt{(x-8)^2 + y^2} = 5a + \sqrt{x^2 + y^2}$$

Square both sides of this last equation and simplify the result:

$$(x-8)^2 + y^2 = 25a^2 + 10a\sqrt{x^2 + y^2} + (x^2 + y^2)$$
$$x^2 - 16x + 64 + y^2 = 25a^2 + 10a\sqrt{x^2 + y^2} + (x^2 + y^2)$$
$$(25a^2 - 64) + 16x = -10a\sqrt{x^2 + y^2}$$
$$[(25a^2 - 64) + 16x]^2 = 100a^2(x^2 + y^2)$$
$$(25a^2 - 64)^2 + 256x^2 + 32x(25a^2 - 64) = 100a^2x^2 + 100a^2y^2$$
$$-(256 - 100a^2)x^2 + 32(25a^2 - 64)x + 100a^2y^2 = (25a^2 - 64)^2$$

Approximating the coefficients gives us the equation:

$$-251.66x^2 - 2013.3x + 4.0334y^2 = 3958.3$$

This is clearly the equation of a hyperbola because the coefficients of x^2 and y^2 have opposite signs.

Eccentricity of a Hyperbola

The definition of the eccentricity of a hyperbola is parallel to the definition for the eccentricity of an ellipse. We define the eccentricity of a hyperbola as the focal length c divided by the length of the semitransverse axis a, that is, half the length of the transverse axis:

$$e = \frac{c}{a} \qquad (4)$$

Because the foci are located within the "wings" of the hyperbola, we see that $c > a$, so that $e > 1$. The eccentricity measures the "spread," or width, of the wings of a hyperbola. The larger the eccentricity, the wider the spread of its wings.

EXAMPLE 6 Determining the Eccentricity of a Hyperbola

Determine the eccentricity of the hyperbola that has the equation:

$$\frac{x^2}{4} - \frac{y^2}{9} = 1$$

Solution

In this example, we have $a = 2$ and $b = 3$. The focal length c is given by:

$$c^2 = a^2 + b^2$$
$$= 2^2 + 3^2$$
$$= 13$$
$$c = \sqrt{13}$$

Therefore, the eccentricity e is given by:

$$e = \frac{c}{a}$$
$$= \frac{\sqrt{13}}{2}$$
$$\approx 1.8028$$

We have defined the eccentricity of a hyperbola and an ellipse. Using analogous reasoning, we can define the eccentricity of a parabola. Namely, the eccentricity is the ratio of the focal length to the semiaxis. In the case of a parabola, the semiaxis is the line segment from the vertex to the focus. The eccentricity of a parabola is therefore 1.

We can classify the conics by their respective eccentricities as follows:

ECCENTRICITY OF A CONIC
Circle: $e = 0$
Ellipse: $0 < e < 1$
Parabola: $e = 1$
Hyperbola: $e > 1$

It can be proved, using calculus and elementary physics, that the orbit of a comet or an asteroid is always a conic section. The eccentricity of a comet can indicate the total energy of the comet. Roughly speaking, the faster the comet

Exercises 13.3

Match each equation in Exercises 1–6 with its graph in (a)–(f).

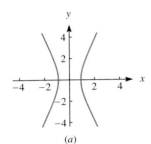
(a)

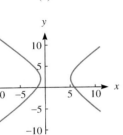
(b)

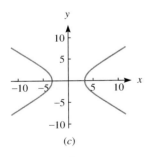
(c)

1. $x^2 - \dfrac{y^2}{4} = 1$

2. $\dfrac{x^2}{4} - \dfrac{y^2}{4} = 1$

3. $\dfrac{x^2}{12.25} - \dfrac{y^2}{6.25} = 1$

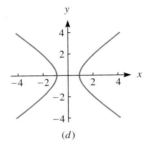
(d)

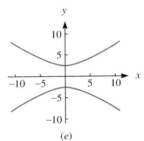
(e)

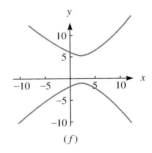
(f)

4. $\dfrac{y^2}{6.25} - \dfrac{x^2}{12.25} = 1$

5. $\dfrac{(x-2)^2}{9} - \dfrac{(y-2)^2}{16} = 1$

6. $\dfrac{(y-2)^2}{9} - \dfrac{(x-2)^2}{9} = 1$

Determine the center, vertices, foci, and asymptotes of the following hyperbolas. Sketch the graph with the aid of this information.

7. $x^2 - y^2 = 1$

8. $y^2 - x^2 = 1$

9. $\dfrac{x^2}{4} - y^2 = 1$

10. $\dfrac{y^2}{9} - \dfrac{x^2}{4} = 1$

11. $\dfrac{x^2}{100} - \dfrac{y^2}{36} = 1$

12. $y^2 - \dfrac{x^2}{3} = 1$

13. $(x-1)^2 - (y-2)^2 = 1$

14. $4(x+3)^2 - 9y^2 = 36$

15. $\dfrac{(x-4)^2}{9} - \dfrac{(y-1)^2}{25} = 1$

16. $\dfrac{x^2}{\frac{1}{3}} - \dfrac{y^2}{\frac{1}{2}} = 1$

17. $4x^2 - 9y^2 = 1$

18. $2x^2 - 3y^2 = 1$

19. $5(x+6)^2 - 3(y-4)^2 = 15$

20. $10y^2 - 3(x+2)^2 = 30$

21. $x^2 - 6x - 4y^2 + 12y = 30$

22. $2y^2 - 3x^2 + 12x - y = 0$

23. $(x+2)^2 - 3(y-4)^2 = 1$

24. $16x^2 - 48x - 2y^2 - 10 = 0$

25. $9y^2 + 18y - 4x^2 + 24x - 28 = 0$

26. $100x^2 + 200x - y^2 = 200$

Determine the equation of the hyperbola that has the following properties.

27. center: (0, 0)
    transverse axis: 4, horizontal
    conjugate axis: 2
28. center: (0, 0)
    transverse axis: 10, horizontal
    conjugate axis: 20
29. center: (1, 1)
    transverse axis: 2, vertical
    conjugate axis: 3
30. center: (−1, −2)
    transverse axis: 3, vertical
    conjugate axis: 1
31. vertices: (±3, 0)
    foci: (±5, 0)
32. vertices: (±8, 0)
    foci: (±12, 0)
33. vertices: (−5, 1), (3, 1)
    foci: (−7, 1), (5, 1)
34. vertices: (−2, −2), (−2, 4)
    foci: (−2, −5), (−2, 7)
35. vertices: (±2, 0)
    asymptotes: $y = \pm 2x$
36. vertices: (0, ±1)
    asymptotes: $y = \pm 2x$
37. vertices: (0, ±2)
    The hyperbola passes through the point (5, 3).
38. vertices: (1, −2), (1, 5)
    The hyperbola passes through the point (−1, 8).

Without completing the square, determine which conic is the graph of the equation.

39. $x^2 + 3y^2 + 12x - 14y - 30 = 0$
40. $x + 12y^2 - 13y + 128 = 0$
41. $y^2 = 3x^2 + 10$
42. $100x^2 - 300x + 100y^2 - 200x + 50 = 0$
43. $-30x = y^2 - 3y + 100$
44. $y^2 + 10y = 100 - 12x^2$

Applications

45. **Hyperbolic navigation.** Suppose that rescuers are searching a wooded area for a trapped camper. Two teams hear the camper's shouts. Suppose that the two rescue teams are 5 miles apart and one team hears the shouts of the camper 3 seconds after the other team hears them. Use a coordinate system that has its origin at the rescue team that hears the call first, and make the positive x-axis extend from the first rescue team to the second. Write an equation describing the possible location of the camper. Assume that sound travels at 1,100 feet per second.

46. **Hyperbolic navigation.** Refer to Exercise 45. Suppose that a third rescue team, located at (0, 3 miles), hears the camper 1 second after the first rescue team hears the camper. Locate the camper.

Technology

Use a graphing utility to graph the following hyperbolas. (You will need to graph each one as two separate equations obtained by solving for y in terms of x.)

47. $\dfrac{x^2}{2.25} - \dfrac{y^2}{6.25} = 1$

48. $3y^2 - 5x^2 = 1$

49. $\dfrac{(x-2)^2}{9} - \dfrac{(y+2)^2}{20.25} = 1$

50. $7(y-2)^2 - 6(x-2)^2 = 17$

51. Graph the hyperbolas that have a horizontal transverse axis and eccentricities 1, 2, 4, and 10. What happens as the eccentricities increase?

52. Graph the hyperbolas that have a vertical transverse axis and eccentricities 1, 2, 4, and 10. What happens as the eccentricities increase?

In your own words

53. Describe how the shape of a hyperbola changes as its eccentricity increases.

54. Describe how the shape of a hyperbola changes as the foci get closer to the vertices.

55. Suppose that the transverse axis and conjugate axis are both multiplied by the same factor. Explain why the eccentricity is unchanged.

56. Suppose that the transverse axis and the conjugate axis are both multiplied by the same factor. Explain why the asymptotes remain unchanged.

57. Explain why a hyperbola never crosses its asymptotes.

13.4 TRANSLATION AND ROTATION OF AXES

In Chapter 2, Section 2.1, we introduced the idea of a Cartesian coordinate system in a plane. The origin for the coordinate system was based on an arbitrary choice, as was the choice to have perpendicular coordinate axes meet at the origin. By either translating or rotating a given set of coordinate axes, however, we arrive at a new set of axes and a new origin. In this section, we study the relationship of coordinates with respect to new and original coordinate systems. The equation of a curve can often be written in a simpler form by choosing an appropriate new coordinate system.

Translation of Coordinate Axes

Suppose that we are given an x-y coordinate system. We can construct another coordinate system by translating the coordinate axes to another origin. Let the point $P = (h, k)$ denote the new origin, where the coordinates (h, k) are given with respect to the original coordinate system. (See Figure 1.) We designate coordinates in the new coordinate system by (X, Y) and coordinates in the original coordinate system by (x, y). Then the translation T of the coordinates with respect to the two systems is determined by the equations:

$$T: \begin{cases} x = X + h \\ y = Y + k \end{cases}$$

This translation moves the x-y axes h units in the horizontal direction and k units in the vertical direction to determine the X-Y axes.

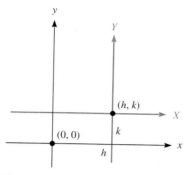

Figure 1

EXAMPLE 1
Calculating Translated Coordinates

Suppose that a new coordinate system is defined by the translation:

$$T: \begin{cases} x = X - 3 \\ y = Y + 4 \end{cases}$$

Calculate the coordinates of the point $P = (-1, 7)$ in the X-Y coordinate system.

Solution

To determine the new X-Y coordinates, we substitute $x = -1$ and $y = 7$ into the translation equations:

$$-1 = X - 3$$
$$7 = Y + 4$$

Solving for X and Y, we have:

$$X = 2$$
$$Y = 3$$

That is, with respect to the new coordinate system, the point P has coordinates $(X, Y) = (2, 3)$.

EXAMPLE 2
Equation of a Translated Curve

Suppose that a new coordinate system is defined by the translation:

$$T: \begin{cases} x = X + 1 \\ y = Y - 5 \end{cases}$$

A curve C is the graph of the equation $x^2 - 2xy + 3y = 1$. Determine the equation of C in the new coordinate system.

Solution

To obtain the equation in the new coordinate system, substitute in the equation of C the expressions $X + 1$ for x and $Y - 5$ for y to obtain the equation:

$$(X + 1)^2 - 2(X + 1)(Y - 5) + 3(Y - 5) = 1$$
$$(X^2 + 2X + 1) - (2XY - 10X + 2Y - 10) + (3Y - 15) = 1$$
$$X^2 + 12X - 2XY + Y = 5$$

Rotation of Coordinate Axes

Another method for obtaining a new coordinate system from a given one is by rotating the axes. Suppose we are given an x-y coordinate system. We can obtain another one by rotating the axes about the origin through an angle θ. (See Figure 2.)

The relationship between the coordinates in the original coordinate system and those in the rotated coordinate system is given by the following result:

ROTATION OF COORDINATE AXES

Suppose that an x-y coordinate system is rotated through an angle θ. The coordinates (X, Y) in the rotated coordinate system are related to the coordinates (x, y) in the original coordinate system by the equations:

$$T: \begin{cases} x = X \cos\theta - Y \sin\theta \\ y = X \sin\theta + Y \cos\theta \end{cases}$$

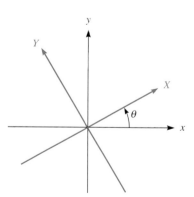

Figure 2

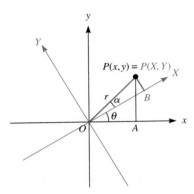

Figure 3

Proof Draw right triangles POA and POB, as shown in Figure 3, where $P(x, y) = P(X, Y)$. Denote by α the angle $\angle POB$. Then:

$$\angle POA = \angle POB + \angle BOA$$
$$= \alpha + \theta$$

Using the coordinates in the first coordinate system, we see that $\overline{OA} = x$ and $\overline{AP} = y$. Therefore, by the definition of the trigonometric functions, we have:

$$\cos(\theta + \alpha) = \frac{\overline{OA}}{\overline{OP}} = \frac{x}{r}$$

$$\sin(\theta + \alpha) = \frac{\overline{AP}}{\overline{OP}} = \frac{y}{r}$$

To these equations, apply the sum formulas for the sine and cosine functions to obtain:

$$x = r\cos(\theta + \alpha)$$
$$= r(\cos\theta \cos\alpha - \sin\theta \sin\alpha)$$
$$= r\cos\theta \cos\alpha - r\sin\theta \sin\alpha \qquad (1)$$
$$y = r\sin(\theta + \alpha)$$
$$= r(\sin\theta \cos\alpha + \cos\theta \sin\alpha)$$
$$= r\sin\theta \cos\alpha + r\cos\theta \sin\alpha \qquad (2)$$

Next, let's use the coordinates of P in the rotated coordinate system to obtain:

$$\cos\alpha = \frac{\overline{OB}}{\overline{OP}} = \frac{X}{r}$$

$$\sin\alpha = \frac{\overline{BP}}{\overline{OP}} = \frac{Y}{r}$$

Therefore, we have:

$$X = r\cos\alpha, \qquad Y = r\sin\alpha$$

Inserting these values for X and Y into equations (1) and (2), we have:

$$x = r\cos\alpha \cos\theta - r\sin\alpha \sin\theta$$
$$= X\cos\theta - Y\sin\theta$$
$$y = r\cos\alpha \sin\theta + r\sin\alpha \cos\theta$$
$$= X\sin\theta + Y\cos\theta$$

This completes the proof of the result.

EXAMPLE 3 Determining the Equation of a Rotated Ellipse

Suppose that the x-y coordinate system is rotated $45°$. Determine the equation of the ellipse

$$x^2 + 4xy + 4y^2 = 4$$

in the rotated coordinate system.

Solution

In this case $\theta = 45°$, so that the equations of the rotated axes read:

$$x = X\cos\theta - Y\sin\theta$$
$$= \frac{\sqrt{2}}{2}X - \frac{\sqrt{2}}{2}Y$$
$$y = X\sin\theta + Y\cos\theta$$
$$= \frac{\sqrt{2}}{2}X + \frac{\sqrt{2}}{2}Y$$

Substituting these expressions for x and y into the equation of the ellipse, we derive:

$$x^2 + 4xy + 4y^2 = 4$$

$$\left(\frac{\sqrt{2}}{2}X - \frac{\sqrt{2}}{2}Y\right)^2 + 4\left(\frac{\sqrt{2}}{2}X - \frac{\sqrt{2}}{2}Y\right)\left(\frac{\sqrt{2}}{2}X + \frac{\sqrt{2}}{2}Y\right)$$
$$+ 4\left(\frac{\sqrt{2}}{2}X + \frac{\sqrt{2}}{2}Y\right)^2 = 4$$

$$\left(\frac{1}{2}X^2 - XY + \frac{1}{2}Y^2\right) + 4\left(\frac{1}{2}X^2 - \frac{1}{2}Y^2\right) + 4\left(\frac{1}{2}X^2 + XY + \frac{1}{2}Y^2\right) = 4$$

$$\frac{9}{2}X^2 + 3XY + \frac{1}{2}Y^2 = 4$$

The last equation is the equation of the given ellipse in the rotated coordinate system.

EXAMPLE 4
Determining the Equation of a Rotated Hyperbola

Suppose that the hyperbola with equation

$$4x^2 - 9y^2 = 36$$

is rotated about the origin through an angle of $-30°$. Determine the equation of the rotated curve.

Solution

Let X-Y denote the rotated coordinate axes. Then the equations of rotation in this case read:

$$x = X\cos(-30°) - Y\sin(-30°)$$
$$= \frac{\sqrt{3}}{2}X + \frac{1}{2}Y$$
$$y = X\sin(-30°) + Y\cos(-30°)$$
$$= -\frac{1}{2}X + \frac{\sqrt{3}}{2}Y$$

Substituting these expressions for x and y into the equation of the hyperbola yields:
$$4x^2 - 9y^2 = 36$$
$$4\left(\frac{\sqrt{3}}{2}X + \frac{1}{2}Y\right)^2 - 9\left(-\frac{1}{2}X + \frac{\sqrt{3}}{2}Y\right)^2 = 36$$
$$4\left(\frac{3}{4}X^2 + \frac{\sqrt{3}}{2}XY + \frac{1}{4}Y^2\right) - 9\left(\frac{1}{4}X^2 - \frac{\sqrt{3}}{2}XY + \frac{3}{4}Y^2\right) = 36$$
$$\frac{3}{4}X^2 + \frac{13\sqrt{3}}{2}XY - \frac{23}{4}Y^2 = 36$$

Because it is actually the hyperbola that is to be rotated and not the axes, we replace X by x and Y by y in the last equation:
$$\frac{3}{4}x^2 + \frac{13\sqrt{3}}{2}xy - \frac{23}{4}y^2 = 36$$
This is the equation of the rotated hyperbola.

EXAMPLE 5
Eliminating the XY Term by Rotation

Rotate the x-y coordinate axes so that, in the new coordinate system, the equation
$$5x^2 + 3x + 4y^2 - xy + 10 = 0$$
has no XY term. Determine the angle θ of rotation.

Solution

Make the rotation specified by the equations:
$$x = X\cos\theta - Y\sin\theta$$
$$y = X\sin\theta + Y\cos\theta$$
In the new coordinate system, the equation becomes:
$$5x^2 + 3x + 4y^2 - xy + 10 = 0$$
$$5(X\cos\theta - Y\sin\theta)^2 + 3(X\cos\theta - Y\sin\theta) + 4(X\sin\theta + Y\cos\theta)^2$$
$$-(X\cos\theta - Y\sin\theta)(X\sin\theta + Y\cos\theta) + 10 = 0$$
$$-X^2\sin\theta\cos\theta + XY(2\sin\theta\cos\theta + \cos^2\theta - \sin^2\theta) + Y^2\sin\theta\cos\theta + 10 = 0$$
Expanding this equation, we see that the coefficient of the term XY equals:
$$2\sin\theta\cos\theta + \cos^2\theta - \sin^2\theta$$
We wish to choose θ so that this coefficient equals 0. That is, we wish to solve the equation:
$$2\sin\theta\cos\theta + \cos^2\theta - \sin^2\theta = 0$$
However, we can use the double-angle trigonometric identities we proved in Section 8.3.
$$\cos^2\theta - \sin^2\theta = \cos 2\theta$$
$$2\sin\theta\cos\theta = \sin 2\theta$$

SECTION 13.4 Translation and Rotation of Axes

Substituting, we have the equation:
$$\sin 2\theta + \cos 2\theta = 0$$
$$\frac{\sin 2\theta}{\cos 2\theta} = -1$$
$$\tan 2\theta = -1$$
$$2\theta = \tan^{-1}(-1)$$
$$2\theta = -45°$$
$$\theta = -22.5°$$

This choice of θ makes the coefficient of XY equal to 0.

Let's consider a generalization of the last example. Consider the curve with equation:
$$Ax^2 + Bxy + Cy^2 + Dx + Ey + F = 0$$

Let's determine an angle θ, where $0° < \theta < 90°$, such that a rotation through this angle eliminates the xy term in the above equation. If we make the substitution
$$T: \begin{cases} x = X\cos\theta - Y\sin\theta \\ y = X\sin\theta + Y\cos\theta \end{cases}$$
in the equation of the curve, and if we use trigonometric identities, then the coefficient of the XY term is:
$$-2A\sin\cos\theta + B(\cos^2\theta - \sin^2\theta) + 2C\sin\theta\cos\theta$$
$$= -A\sin 2\theta + B\cos 2\theta + C\sin 2\theta$$
$$= (-A + C)\sin 2\theta + B\cos 2\theta$$

Setting this coefficient equal to 0 gives us:
$$B\cos 2\theta = (A - C)\sin 2\theta$$
$$\frac{\cos 2\theta}{\sin 2\theta} = \frac{A - C}{B}$$
$$\cot 2\theta = \frac{A - C}{B}$$

That is, we have proved the following result:

ELIMINATING THE *xy* TERM USING ROTATION OF COORDINATE AXES

To eliminate the xy term in the equation
$$Ax^2 + Bxy + Cy^2 + Dx + Ey + F = 0$$
rotate the coordinate system through an angle θ such that:
$$\cot 2\theta = \frac{A - C}{B} \qquad (0° < \theta < 90°)$$

Exercises 13.4

A new, X-Y coordinate system is defined in each of the following exercises. Calculate the coordinates of the given point P in the new coordinate system.

1. $x = X - 2$, $y = Y + 2$,
   $P = (1, -3)$
2. $x = X + 3$, $y = Y - 2$,
   $P = (-1, 2)$
3. $x = X + 3$, $y = Y + 5$,
   $P = (-1, -2)$
4. $x = X - 2$, $y = Y - 5$,
   $P = (4, 1)$
5. $x = X + 3$, $y = Y - 1$,
   $P = (-3, +3)$
6. $x = X - 1$, $y = Y - 3$,
   $P = (-1, -4)$
7. $x = X - 5$, $y = Y - 3$,
   $P = (-3, 0)$
8. $x = X + 2$, $y = Y + 3$,
   $P = (0, 0)$
9. $x = 2X + 1$, $y = Y - 5$,
   $P = (-2, -3)$
10. $x = X + 5$, $y = 2Y - 1$,
    $P = (-3, 4)$

In the following exercises, a new, X-Y coordinate system is defined by the given translation. A curve C is the graph of the given equation. Determine the equation of C in the new coordinate system.

11. $x = X + 2$, $y = Y + 1$,
    $x^2 + 2xy - 3y = 1$
12. $x = X - 1$, $y = Y + 2$,
    $x^2 - 3x + 2y = 1$
13. $x = X - 2$, $y = Y - 3$,
    $x^2 + 4xy + 2y = 2$
14. $x = X - 5$, $y = Y - 1$,
    $2x^2 + 3xy - 2y = 4$
15. $x = X + 3$, $y = Y - 1$,
    $x^2 + 2xy + y^2 = 1$
16. $x = X + 2$, $y = Y - 3$,
    $x^2 - 3xy + y^2 = 0$
17. $x = X - 5$, $y = Y + 2$,
    $2x^2 - 3y^2 = 9$
18. $x = X + 3$, $y = Y + 2$,
    $x^2 + 3xy - y^2 = 4$
19. $x = 2X - 1$, $y = Y - 2$,
    $x + 2y = 6$
20. $x = X - 7$, $y = 3Y + 2$,
    $x - 3y = 6$

In the following exercises, a new, X-Y coordinate system is determined by rotating the x-y coordinates by the given angle θ. Determine the equation of the rotated curve in this new coordinate system.

21. $\theta = 45°$, $x^2 + y^2 = 9$
22. $\theta = 60°$, $x^2 + y^2 = 16$
23. $\theta = 30°$, $x^2 - y^2 = 1$
24. $\theta = 45°$, $4x^2 + 9y^2 = 36$
25. $\theta = 135°$, $x^2 + 4xy + 4y^2 = 4$
26. $\theta = -30°$, $y - 1 = \frac{1}{4}(x+2)^2$
27. $\theta = -60°$, $y = x^2 + 2x - 3$
28. $\theta = 210°$, $4x^2 + 4xy + y^2 = 4$
29. $\theta = -45°$, $16x^2 - 25y^2 = 400$
30. $\theta = -45°$, $y = 3x - 5$

Determine the angle of rotation θ such that the rotated equation has no XY term in the new coordinate system.

31. $xy = 1$
32. $2xy + 9 = 0$
33. $x^2 = xy - 3$
34. $2x^2 + 4xy - 3y^2 - 4y - 2 = 0$
35. $x^2 - 5xy + 3y^2 + 2y + 10 = 0$
36. $2x^2 + 5x + 4y^2 - 3xy + 12 = 0$
37. $x^2 + 2xy + y^2 = 10$
38. $4x^2 - 6xy - 9y^2 = 36$
39. $y = 4x^2 - 6xy + 2$
40. $(x - y)^2 + 4 = y$

In Exercises 41–44, rotate the axes so that the XY term in the following equation is eliminated. Describe and sketch the graph of the rotated curve.

41. $16x^2 - 24xy + 100 = 60x - 8y - 9y^2$
42. $4y^2 + 4xy = 18y\sqrt{5} - 6x\sqrt{5} - 45 - x^2$
43. $32x^2 + 53y^2 - 72xy - 80 = 0$
44. $5x^2 + 5y^2 - 8xy - 9 = 0$

In your own words

45. Describe how to obtain the equation of a curve if the coordinate system is translated. Use a concrete example for your explanation.

46. Describe how to obtain the equation of a curve if the coordinate system is rotated. Use a concrete example for your explanation.

47. Suppose you are given an equation of the second degree in two variables (in other words, the equation may have an xy term). Describe how to sketch its graph.

13.5 POLAR EQUATIONS OF CONICS

Definition of a Conic in Terms of Eccentricity

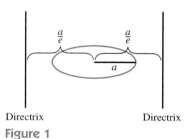

Directrix   Directrix

Figure 1

Recall that we defined the parabola in terms of a focus and a directrix. However, in defining an ellipse and a hyperbola, we did not make any mention of directrixes. Actually, all conic sections have directrixes. Let's now introduce them and use them to provide an alternate definition of the conics.

Consider first an ellipse with semimajor axis a and eccentricity e. The lines perpendicular to the major axis and at the distance a/e from the center are called the **directrixes of the ellipse**. (See Figure 1.) Because $e < 1$, we see that $a/e > a$, so that the directrices do not intersect the ellipse. Each focus has an associated directrix, namely the directrix that lies on the same side of the center as the focus.

It is possible to verify the following property of the ellipse (we omit the proof):

FOCUS-DIRECTRIX PROPERTY OF AN ELLIPSE

An ellipse is the set of all points P such that the distance from a point on the ellipse to the focus F is e times the distance from the same point to the associated directrix D.

Let us now consider a hyperbola with semitransverse axis a and eccentricity e. The lines perpendicular to the transverse axis and at the distance a/e from the center are called the **directrixes of the hyperbola**. (See Figure 2.) In the case of a hyperbola, $e > 1$, so that $a/e < a$. Therefore, the directrixes do not intersect the hyperbola. Each focus has an associated directrix, namely the directrix that lies on the same side of the center as the focus.

It is possible to verify the following property of the hyperbola (we omit the proof):

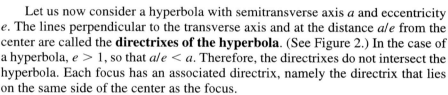

Directrix Directrix

Figure 2

FOCUS-DIRECTRIX PROPERTY OF A HYPERBOLA

The hyperbola is the set of all points P such that the distance from a point on the hyperbola to the focus F is e times the distance from that point to the associated directrix D.

Recall that the eccentricity of a parabola equals 1. From the definition of a parabola, we have:

FOCUS-DIRECTRIX PROPERTY OF A PARABOLA

A parabola is the set of all points P such that the distance from a point on the parabola to the focus F is e times the distance from that point to the directrix D.

CONICS IN TERMS OF FOCUS AND DIRECTRIX

Suppose that a conic has eccentricity e, focus F, and associated directrix D. Then the conic consists of all points P such that the distance from a point on the conic to F is e times the distance from that point to D.

Equation of a Conic in Polar Coordinates

It is possible to give an elegant form of the equation of a conic by using polar coordinates. This form of the equation makes use of the eccentricity and is independent of the type of conic involved. We have the following result:

POLAR EQUATIONS OF CONICS

Let e be the eccentricity of a conic, let r be the distance from a point on the conic to the focus, and let p be the directed distance from the focus F to the directrix D. Set up a polar coordinate system with the origin at the focus F.

1. If the directrix is vertical, then the equation of the conic is

$$r = \frac{ep}{1 \pm e \cos \theta}$$

where the $+$ sign holds if the directrix is to the right of the focus and the $-$ sign holds if the directrix is to the left of the focus.

2. If the directrix is horizontal, then the equation of the conic is

$$r = \frac{ep}{1 \pm e \sin \theta}$$

where the $+$ sign holds if the directrix is above the focus and the $-$ sign holds if the directrix is below the focus.

Conversely, any equation in form 1 or 2 has a conic as its graph.

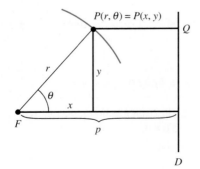

Figure 3

Proof Let's prove the first equation in the case $p > 0$. Draw the directrix and focus as in Figure 3. Let $P(r, \theta) = P(x, y)$ be a typical point on the conic, given in both polar and rectangular coordinates. Because p is positive, the directrix lies to the right of the focus as in the figure. According to the general definition of conics in terms of focus and directrix,

$$\text{distance to focus} = e \cdot \text{distance to directrix}$$

$$\overline{PF} = e\overline{PQ}$$

$$r = e(p - x)$$

$$= e(p - r \cos \theta) \quad \text{Definition of cosine}$$

SECTION 13.5 Polar Equations of Conics

Multiply out the parenthesis and solve for r:

$$r = ep - er\cos\theta$$
$$r + er\cos\theta = ep$$
$$r(1 + e\cos\theta) = ep$$
$$r = \frac{ep}{1 + e\cos\theta}$$

This gives us the equation of the conic in polar coordinates. The proof of the second equation is similar.

EXAMPLE 1
Sketching the Graph of a Polar Equation

Sketch the graph of the conic that has polar equation:

$$r = \frac{20}{5 + 4\cos\theta}$$

Solution

First put the equation in the form of one of the standard equations. To do so, divide numerator and denominator of the right side by 5 to obtain:

$$r = \frac{4}{1 + \frac{4}{5}\cos\theta}$$

We see that $e = \frac{4}{5}$. Therefore, the directrix is vertical and is located 5 units to the right of the focus. The conic is an ellipse with eccentricity $\frac{4}{5}$. Moreover,

$$ep = 4$$
$$p = \frac{4}{e} = \frac{4}{\frac{4}{5}} = 5$$

Because the major axis of the ellipse is horizontal, the vertices correspond to the points for which $\theta = 0, \pi$. That is, the vertices, in polar coordinates, are:

$$(r, \theta) = \left(\frac{20}{5 + 4\cos 0}, 0\right), \left(\frac{20}{5 + 4\cos\pi}, \pi\right)$$
$$= \left(\frac{20}{9}, 0\right), (20, \pi)$$

In particular, the major axis is $\frac{20}{9} + 20 = \frac{200}{9}$ and the semimajor axis is $a = \frac{100}{9}$. Because $c = ea$, we have:

$$b^2 = a^2 - c^2 = a^2 - (ea)^2 = a^2(1 - e^2)$$
$$= \frac{100}{9}\left(1 - \frac{4}{9}\right)$$
$$= \frac{500}{81}$$
$$b = \frac{10\sqrt{5}}{9}$$

The graph is shown in Figure 4.

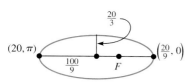

Figure 4

EXAMPLE 2
Sketching the Graph of a Polar Equation

Sketch the graph of the conic with polar equation:

$$r = \frac{10}{2 - 2\cos\theta}$$

Solution
Divide both numerator and denominator by 2 to obtain the equation:

$$r = \frac{\frac{5}{2}}{1 - \cos\theta}$$

According to our polar equation of conics, the directrix is vertical and lies to the left of the focus. Because $e = 1$, the graph is a parabola. And because the directrix is to the left of the focus, the parabola opens to the right. The point corresponding to $\theta = \pi$ is the vertex. This point is:

$$(r, \theta) = \left(\frac{10}{2 - 2\cos\pi}, \pi\right) = \left(\frac{5}{2}, \pi\right)$$

The graph is shown in Figure 5.

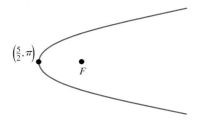

Figure 5

EXAMPLE 3
Determining the Polar Equation of a Conic

Give the polar form of the equation of an ellipse of eccentricity $\frac{1}{2}$ that has its focus at the origin and for which the directrix is the line $x = -3$.

Solution

The directrix is vertical and the distance from the focus to the directrix is 3. Therefore, the equation of the ellipse is:

$$r = \frac{ep}{1 - e\cos\theta} = \frac{(3)\frac{1}{2}}{1 - \frac{1}{2}\cos\theta} = \frac{3}{2 - \cos\theta}$$

We have used the minus sign in the denominator because the directrix lies to the left of the focus.

EXAMPLE 4
Determining the Polar Equation of a Conic

The orbit of the earth around the sun is an ellipse with the sun at the focus. The eccentricity of the orbit is 0.0167 and the major axis is 186 million miles. Find the equation of the orbit.

Solution

The semimajor axis of the ellipse is $\frac{186}{2} = 93$ million miles. Assume that the sun is located at the focus and that the directrixes are vertical. Then the equation is of the form:

$$r = \frac{ep}{1 + e\cos\theta}$$

We are given the information that $e = 0.0167$. Moreover, the vertices of the ellipse occur for $\theta = 0, \pi$. Therefore, if a denotes the semimajor axis, we have:

$$2a = \frac{0.0167p}{1 + 0.0167} + \frac{0.0167p}{1 - 0.0167}$$

$$\approx 0.033409p$$

$$p \approx \frac{2a}{0.033409} \approx 5567.4$$

Therefore, the desired equation is:

$$r = \frac{0.0167 \cdot 5567.4}{1 + 0.0167 \cos \theta} \approx \frac{92.976}{1 + 0.0167 \cos \theta}$$

Exercises 13.5

Determine an equation in polar form for the given conic that has a focus at the origin and has the following properties:

1. parabola; directrix: $x = 1$
2. parabola; directrix: $y = 2$
3. parabola; directrix: $x = -2$
4. parabola; directrix: $y = 1$
5. parabola; vertex: $(5, 0)$
6. parabola; vertex: $(3, \pi)$
7. parabola; vertex: $\left(4, \dfrac{\pi}{2}\right)$
8. parabola; vertex: $\left(3, \dfrac{3\pi}{2}\right)$
9. ellipse; $e = \tfrac{1}{3}$, directrix: $x = 5$
10. ellipse; $e = \tfrac{2}{3}$, directrix: $y = 3$
11. ellipse; vertices: $(4, 0), (2, \pi)$
12. ellipse; vertices: $\left(2, \dfrac{\pi}{2}\right), \left(3, \dfrac{3\pi}{2}\right)$
13. hyperbola; $e = 2$, directrix: $x = 3$
14. hyperbola; $e = 2$, directrix: $y = -1$
15. hyperbola; vertices: $(2, 0), (4, 0)$
16. hyperbola; vertices: $\left(4, \dfrac{3\pi}{2}\right), \left(1, \dfrac{3\pi}{2}\right)$

Sketch the graphs of the following polar equations.

17. $r = \dfrac{1}{1 + \cos \theta}$
18. $r = \dfrac{3}{1 - \cos \theta}$
19. $r = \dfrac{2}{1 + 2 \sin \theta}$
20. $r = \dfrac{5}{1 - 3 \cos \theta}$
21. $r = \dfrac{4}{1 + 10 \cos \theta}$
22. $r = \dfrac{3}{2 + \cos \theta}$
23. $r = \dfrac{4}{1 + \sin \theta}$
24. $r = \dfrac{6}{2 - 3 \sin \theta}$
25. $r = \dfrac{10}{25 + 6 \cos \theta}$
26. $r = \dfrac{5}{10 + 25 \sin \theta}$
27. $r = \dfrac{3}{1 - 3 \cos \theta}$
28. $r = \dfrac{1}{3 - 3 \sin \theta}$

Determine the polar equations of the conics that have the following equations in rectangular coordinates. (Set up the polar coordinate system so that one focus is at the origin.)

29. $x^2 - y^2 = 1$
30. $y^2 - 2x^2 = 1$
31. $\dfrac{x^2}{4} - \dfrac{y^2}{16} = 1$
32. $\dfrac{y^2}{16} - \dfrac{x^2}{25} = 1$

Applications

33. A satellite has an orbit with the earth as a focus and an eccentricity of 0.05. The closest distance of the orbit from the center of the earth is 12,000 miles. Determine the equation of the orbit.

34. Refer to Exercise 33. Determine the furthest distance of the orbit from the center of the earth.

In your own words

35. Give the definitions of focus, directrix, and eccentricity for each of the conic sections.

36. Describe a conic in terms of its focus, directrix, and eccentricity.

37. State the polar equations of conics. Illustrate each equation with several examples that show equations for each type of conic.

13.6 PARAMETRIC EQUATIONS

In Chapter 2, we introduced graphs of equations of the form $f(x, y) = 0$, where $f(x, y)$ is an expression in x and y. We have graphed many equations and arrived at an interesting collection of graphs, not the least of which are the conic sections considered in the preceding sections.

Rather than specifying a curve by an equation relating x and y, it is often expedient to relate both x and y to a third variable, which we will call t. In this case, both x and y are given as functions of t:

Definition
Parametric Equations of a Curve

A set of **parametric equations** in the **parameter** t is a pair of equations
$$x = f(t), \qquad y = g(t)$$
where t ranges over an interval (finite or infinite). Each value of t corresponds to a point $(x, y) = (f(t), g(t))$. As t ranges over the interval, the set of points (x, y) traces out a curve in the plane.

In many examples, the parameter has a physical interpretation. For instance, t can represent time, so that the point $(f(t), g(t))$ can be interpreted as the position of a point at time t.

A set of parametric equations for a curve gives much more information than just the set of points describing the curve. As t runs through a specified set of values, the parametric equations specify the speed with which the curve is traced out, the direction in which it is traced, and even if certain portions of the curve are traced over themselves more than once.

EXAMPLE 1
Motion of a Particle

The position of a particle at time t is given by the parametric equations:
$$x = 2t, \qquad y = 3t, \qquad 0 \le t \le 5$$
Describe the motion of the particle.

Solution

At time $t = 0$, the particle starts at the point $(x, y) = (2 \cdot 0, 3 \cdot 0) = (0, 0)$. The following table lists the location of the particle at various times t:

t	x	y
0	0	0
1	2	3
2	4	6
3	6	9
4	8	12
5	10	15

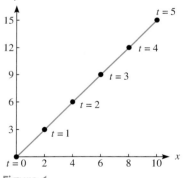
Figure 1

We plot the various points in Figure 1 and we see that the graph is a line. As time runs from 0 to 5, the point moves along the line from $(0, 0)$ to $(10, 15)$.

EXAMPLE 2
Graphing Parametric Equations

Graph the curve described by the parametric equations:
$$x = t, \qquad y = \frac{3}{2}t, \qquad 0 \le t \le 10$$

Solution

This curve is the same curve shown in Figure 1. However, the present curve is traced out at half the speed of the curve in Figure 1. That is, for any given time t,

SECTION 13.6 Parametric Equations

the location on the present curve is the same as the location on the previous curve at time $t/2$. This example shows that there is no single set of parametric equations to describe a particular curve.

EXAMPLE 3
Parametric Form of a Circle

Graph the curve described by the parametric equations:
$$x = 2\cos t, \qquad y = 2\sin t, \qquad 0 \le t \le 2\pi$$

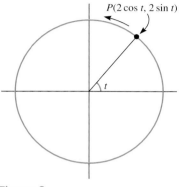

Figure 2

Solution

To solve this problem, let's pull a rabbit out of a hat. Draw a circle of radius 2 about the origin. From the definitions of the trigonometric functions sine and cosine, the coordinates of a point P on the circle determined by an angle t in standard position are $(\cos t, \sin t)$. So if we interpret the parameter t as the angle shown in Figure 2, we see that the parametric equations describe the point P. As t ranges over the interval $[0, 2\pi]$, the point described by the parametric equations traces out the circle shown, in the counterclockwise direction. This example illustrates how the parameter can often be interpreted in geometric terms, as an angle, length, or area.

In some instances, it is possible to solve the parametric equation for t in terms of either x or y and then eliminate the parameter t, thereby obtaining an equation for the curve in nonparametric form. The next two examples illustrate how this is done.

EXAMPLE 4
Translating Parametric Equations into Rectangular Coordinates

Write the following set of parametric equations in rectangular coordinates:
$$x = \sqrt{t}, \qquad y = 3t + 1, \qquad t \ge 0$$

Solution

We can solve the first equation for t in terms of x:
$$x = \sqrt{t}$$
$$t = x^2$$

Now substitute the expression for t into the equation for y:
$$y = 3t + 1 = 3x^2 + 1$$

From the first equation, the values of x are all nonnegative. In fact, each nonnegative real number is a value of x for some value of t. Therefore, when we graph the nonparametric equation $y = 3x^2 + 1$, we must subject x to the restriction $x \ge 0$, which is inherited from the parametric equations.

EXAMPLE 5
Translating Parametric Equations into Rectangular Coordinates

Write the following equations in rectangular coordinates:
$$x = e^t, \qquad y = t^2$$

Solution

Note that the value of x is positive. Solving the first equation for t in terms of x, we have:

$$t = \ln x$$

Substituting this value into the equation for y, we have:

$$y = t^2 = (\ln x)^2, \qquad x > 0$$

Using a Graphing Calculator to Graph Parametric Equations

Most graphing calculators allow us to graph parametric equations. For instance, the TI-82 can plot parametric equations when we enter the MODE menu and select **Par** instead of **Func**. We press ENTER and then Clear. Next, we press Y=. The screen now looks like the one shown in Figure 3. On this screen, we can enter up to three parametrically defined equations. For example, to graph

$$x = 3\cos t, \qquad y = 3\sin t, \qquad 0 \le t \le 2\pi$$

we enter $3\cos t$ for X_{1T} and $3\sin t$ for Y_{1T}. Then we graph the equation. We may need to adjust the viewing box. To do so, use the **Square** command in the ZOOM screen to get a graph that looks like Figure 4.

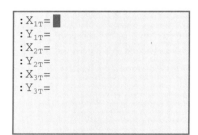
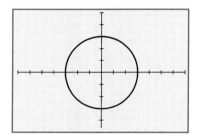

Figure 3 Figure 4

Exercises 13.6

Sketch the graphs of the following parametric equations.

1. $x = 3t, \quad y = 5t - 1, \quad 0 \le t \le 1$
2. $x = 2t - 1, \quad y = t + 1, \quad -1 \le t \le 1$
3. $x = 2t, \quad y = e^t, \quad t \ge 0$
4. $x = 2t, \quad y = \ln t, \quad 0 < t < 5$
5. $x = 2\cos\theta, \quad y = 2\sin\theta, \quad 0 \le \theta \le 2\pi$
6. $x = 3 + 2\sin\theta, \quad y = 1 + 2\cos\theta, \quad 0 \le \theta \le 2\pi$
7. $x = \sin\theta, \quad y = \cos\theta, \quad 0 \le \theta \le 2\pi$
8. $x = \sin 3t, \quad y = \cos 3t, \quad 0 \le t \le \pi$
9. $x = \cos\theta, \quad y = \sin\theta, \quad -\pi \le \theta \le \pi$
10. $x = 3\sin\theta, \quad y = -3\cos\theta, \quad 0 \le \theta \le 6\pi$
11. $x = 2t, \quad y = 3t, \quad t$ is in the interval $[0, 1]$
12. $x = 2 + \cos t, \quad y = -1 + 3\sin t, \quad 0 \le t \le 2\pi$
13. $x = \dfrac{1}{t}, \quad y = t, \quad t > 0$
14. $x = \dfrac{1}{t^2 + 1}, \quad y = \dfrac{t}{t^2 + 1}, \quad$ all real t
15. $x = \sqrt{t}, \quad y = \dfrac{1}{t}, \quad t > 0$
16. $x = \cos t, \quad y = \sin t, \quad 0 \le t \le \dfrac{\pi}{2}$

For each of the parametric equations in exercises 17–32, write the equation in nonparametric form. Be careful to state the correct domain in each case.

17. $x = 3t, \quad y = 5t - 1, \quad 0 \le t \le 1$
18. $x = 2t - 1, \quad y = t + 1, \quad -1 \le t \le 1$
19. $x = 2t, \quad y = e^t, \quad t \ge 0$
20. $x = 2t, \quad y = \ln t, \quad 0 < t < 5$

21. $x = 2\cos\theta$, $y = 2\sin\theta$, $0 \le \theta \le 2\pi$
22. $x = 3 + 2\sin\theta$, $y = 1 + 2\cos\theta$, $0 \le \theta \le 2\pi$
23. $x = \sin\theta$, $y = \cos\theta$, $0 \le \theta \le 2\pi$
24. $x = \sin 3t$, $y = \cos 3t$, $0 \le t \le \pi$
25. $x = \cos\theta$, $y = \sin\theta$, $-\pi \le \theta \le \pi$
26. $x = 3\sin\theta$, $y = -3\cos\theta$, $0 \le \theta \le 6\pi$
27. $x = 2t$, $y = 3t$, $t = 1$ to $t = 0$
28. $x = 2 + \cos t$, $y = -1 + 3\sin t$, $0 \le t \le 2\pi$
29. $x = \dfrac{1}{t}$, $y = t$, $t > 0$
30. $x = \dfrac{1}{t^2 + 1}$, $y = \dfrac{t}{t^2 + 1}$, all real t
31. $x = \sqrt{t}$, $y = \dfrac{1}{t}$, $t > 0$
32. $x = \cos t$, $y = \sin t$, $0 \le t \le \dfrac{\pi}{2}$

Show that the given conic section can be written in the following parametric form.

33. Circle: $(x - h)^2 + (y - k)^2 = r^2$
    $x = h + r\cos\theta$, $y = k + r\sin\theta$, $0 \le \theta \le 2\pi$
34. Ellipse: $\dfrac{(x - h)^2}{a^2} + \dfrac{(y - k)^2}{b^2} = 1$
    $x = h + a\cos\theta$, $y = k + b\sin\theta$, $0 \le \theta \le 2\pi$
35. Hyperbola: $\dfrac{(x - h)^2}{a^2} - \dfrac{(y - k)^2}{b^2} = 1$
    $x = h + a\sec\theta$, $y = k + b\tan\theta$,
    $\begin{cases} -\dfrac{\pi}{2} < \theta < \dfrac{\pi}{2} & \text{(left branch)} \\ \dfrac{\pi}{2} < \theta < \dfrac{3\pi}{2} & \text{(right branch)} \end{cases}$

36. Derive a set of parametric equations for a circle that has radius 2 and center $(3, -1)$.
37. Derive a set of parametric equations for the unit circle.
38. Derive a set of parametric equations for an ellipse that has equation $x^2/4 + y^2/25 = 1$, traced out twice in the clockwise direction starting from the point for which $x = 0$.
39. Derive a set of parametric equations for an ellipse that has its center at the origin, a horizontal semimajor axis of length 2, and a semiminor axis of length 1.
40. Derive a set of parametric equations for an ellipse that has its center at $(5, -2)$, a vertical semimajor axis of length 3, and a semiminor axis of length 2.
41. Derive a set of parametric equations for an ellipse that has foci at $(5, 1)$, $(-3, 1)$ and a semimajor axis of length 10.
42. Derive a set of parametric equations for the hyperbola $x^2 - y^2 = 1$.
43. Derive a set of parametric equations for the right branch of the hyperbola $(x - 1)^2/3 - (y - 1)^2/6 = 1$.
44. Prove that the line segment connecting (x_1, y_1) to (x_2, y_2) can be written parametrically in the form:
    $x = x_1 + t(x_2 - x_1)$, $y = y_1 + t(y_2 - y_1)$, $0 \le t \le 1$
45. Write parametric equations for a line segment that goes from $(2, 3)$ to $(-1, 4)$.
46. Describe the graph of the parametric equations:
    $x = 3 + 4t$, $y = -1 + 2t$, $0 \le t \le 1$
47. Describe the graph of the parametric equations obtained by replacing t by $1 - t$ in Exercise 46.

Suppose that $x = f(t)$ and $y = g(t)$, where $0 \le t \le 1$, is a set of parametric equations for a curve C. Describe the graph of the following equations.

48. $x = f(2t)$, $y = g(2t)$, $0 \le t \le \tfrac{1}{2}$
49. $x = f(1 - t)$, $y = g(1 - t)$, $0 \le t \le 1$
50. $x = g(t)$, $y = f(t)$, $0 \le t \le 1$
51. $x = f(t) + 1$, $y = g(t) + 2$, $0 \le t \le 1$

Technology

Use the parametric plotting features of your graphing utility and the formulas from Exercises 33–35 to graph the conic sections that have the following equations.

52. $\dfrac{x^2}{16} + \dfrac{y^2}{36} = 1$
53. $(x - 2)^2 + (y + 3)^2 = 11$
54. $\dfrac{(x - 2)^2}{49} + \dfrac{(y + 2)^2}{20} = 1$
55. $7(x - 2)^2 + 6(y - 2)^2 = 17$
56. $\dfrac{x^2}{2} - \dfrac{y^2}{9} = 1$
57. $7(y + 1)^2 - 5(x + 2)^2 = 13$
58. $\dfrac{(x - 2)^2}{9} - \dfrac{(y + 2)^2}{20} = 1$
59. $(y - 2)^2 + (x - 2)^2 = 17$

Mathematics and the World around Us—Fermat's Last Theorem

Mathematicians often come up with problems using experimentation or by means of analogy with other known results. Sometimes, even the simplest problem, however arrived at, can be incredibly difficult to solve. And this is certainly the case for the problem known as Fermat's Last Theorem.

Pierre de Fermat was a French lawyer of the seventeenth century who also had a deep interest in mathematics. He was especially fascinated by the branch of mathematics concerned with the properties of the integers. In his copy of the works of the ancient Greek mathematician Diophantus, he made a marginal note that said that if n is an integer at least 3, it is impossible to have nonzero integers x, y, and z that satisfy:

$$x^n + y^n = z^n$$

Furthermore, Fermat claimed to have a marvelous proof of this result, but that the margin was too small to contain it.

Fermat died without leaving any indication of his proof. However, the mathematical statement remains. Is it true or not?

For three and a half centuries, the finest mathematicians worked on this problem and failed to prove it one way or another. To be sure, many partial results were proved. For example, it was proven to be true for thousands of particular values of n. However, a general proof was elusive.

Because of the simplicity of its statement, as well as the glamor of a possible "lost" proof, Fermat's Last Theorem became a hotbed of activity for mathematical amateurs and cranks. Almost every professional mathematician who has been editor of a mathematics journal regularly receives alleged proofs of Fermat's Last Theorem. Almost without exception, these amateur attempts have logical flaws that can be found in a few minutes.

In 1989, Professor Kenneth Ribet of Berkeley showed that Fermat's Last Theorem would follow if you could prove a more general statement about mathematical objects called "elliptic curves." The connection between Fermat's Last Theorem and this other branch of mathematics was surprising. The statement about elliptic curves seemed just as intractable as Fermat's Last Theorem itself. However, in the summer of 1993, Professor Andrew Wiles of Princeton University announced a proof of the elliptic curve statement and, with it, a proof of Fermat's Last Theorem. Wiles' proof is over 200 pages long and is currently being subjected to intense scrutiny by experts. It looks hopeful that, after more than three centuries, mathematicians finally are able to answer the challenge created by Fermat.

 In your own words

60. Explain how parametric equations can be used to describe a curve.

61. Explain how to eliminate the parameter from a set of parametric equations.

62. Explain how a set of parametric equations contains more information than just the graph.

Chapter Review

Important Concepts—Chapter 13

- conic section; conic
- locus
- parabola
- directrix of a parabola
- focus of a parabola
- axis of a parabola
- vertex of a parabola
- focal length of a parabola
- standard forms of the equation of a parabola
- reflection property of the parabola
- ellipse
- foci of an ellipse
- vertices of an ellipse
- center of an ellipse
- major axis of an ellipse
- semimajor axis of an ellipse
- minor axis of an ellipse
- semiminor axis of an ellipse
- focal length of an ellipse
- standard forms of the equation of an ellipse
- eccentricity of an ellipse
- hyperbola
- foci of a hyperbola
- vertices of a hyperbola
- center of a hyperbola
- transverse axis of a hyperbola
- focal length of a hyperbola
- standard forms of the equation of a hyperbola
- asymptotes of a hyperbola
- conjugate axis of a hyperbola
- eccentricity of a hyperbola
- translation of coordinate axes
- rotation of coordinate axes
- eliminating the xy term using rotation of coordinate axes
- definition of conics in terms of focus and directrix
- polar equations of conics
- parametric equations of a curve
- translating parametric equations into rectangular coordinates

Important Results and Techniques—Chapter 13

Parabolas	A parabola with focus F and directrix D is the set of all points that are equidistant from F and D	p. 672, 673
	Standard Forms of the Equation of a Parabola: $(x-h)^2 = 4p(y-k), \quad p \neq 0, \quad$ axis is vertical $(y-k)^2 = 4p(x-h), \quad p \neq 0, \quad$ axis is horizontal where p is the focal length.	
Ellipses	An ellipse with foci F_1 and F_2 is the set of all points P such that the sum of the distances $d(P, F_1)$ and $d(P, F_2)$ is $2a$.	p. 680, 682
	Standard Forms of the Equation of an Ellipse: $$\frac{(x-h)^2}{a^2} + \frac{(y-k)^2}{b^2} = 1$$ If $a > b$, then the major axis is horizontal. If $a < b$, then the major axis is vertical.	
	If c is the focal length, then $$a^2 + b^2 = c^2$$ where a is the length of the semimajor axis, and b is the length of the semiminor axis.	

Hyperbolas	A hyperbola with foci F_1 and F_2 is the set of all points satisfying the equation $d(P, F_1) - d(P, F_2) = 2a$.	p. 689, 690, 692
	Standard Forms of the Equation of a Hyperbola: $\dfrac{(y-k)^2}{a^2} - \dfrac{(x-h)^2}{b^2} = 1,$ transverse axis is vertical $\dfrac{(x-h)^2}{a^2} - \dfrac{(y-k)^2}{b^2} = 1,$ transverse axis is horizontal	
	If c is the focal length, then $$c^2 = a^2 - b^2$$ where a is the semitransverse axis, and b is half the length of the conjugate axis.	
	Asymptotes are the lines: $y = k \pm \dfrac{b}{a}(x-h),$ horizontal transverse axis $y = k \pm \dfrac{a}{b}(x-h),$ vertical transverse axis	
Translation of coordinate axes	Suppose that the origin of the new X-Y coordinate system is the point $P = (h, k)$ of the original x-y coordinate system. Then the relationship between the coordinate systems is: $$T: \begin{cases} x = X + h \\ y = Y + k \end{cases}$$	p. 700
Rotation of coordinate axes	Suppose that an x-y coordinate system is rotated through an angle θ. The coordinates (X, Y) in the rotated coordinate system are related to the coordinates (x, y) in the original coordinate system by the equations: $$T: \begin{cases} x = X \cos \theta - Y \sin \theta \\ y = X \sin \theta + Y \cos \theta \end{cases}$$	p. 701
Polar form of equations for conics	Let e be the eccentricity of a conic, let r be the distance from a point on the conic to the focus, and let p be the distance from the focus F to the directrix D. Set up a polar coordinate system with the origin at the focus F. 1. If the directrix is vertical, then the equation of the conic is $$r = \dfrac{ep}{1 \pm e \cos \theta}$$ where the $+$ sign holds if the directrix is to the right of the focus and the $-$ sign holds if the directrix is to the left of the focus.	p. 708

	2. If the directrix is horizontal, then the equation of the conic is $$r = \frac{ep}{1 \pm e \sin \theta}$$ where the $+$ sign holds if the directrix is above the focus and the $-$ sign holds if the directrix is below the focus.	
Parametric equations of a curve	We have: $$x = f(t), \quad y = g(t)$$ where t ranges over a finite or an infinite interval.	p. 712

Review Exercises—Chapter 13

Sketch the graphs of the following conics. For each, determine the focus or foci, the directrix or directrixes, and the eccentricity. For ellipses, determine the semimajor and semiminor axes. For hyperbolas, determine the transverse and conjugate axes and the asymptotes.

1. $y^2 = 3x + 1$
2. $x^2 = 2y - 3$
3. $2x^2 - y^2 = 4$
4. $y^2 - 3x^2 = 10$
5. $x^2 + (y - 3)^2 = 10$
6. $2(x - 1)^2 + 5(y - 3)^2 = 25$
7. $\frac{x^2}{25} - y^2 = 6$
8. $\frac{(x - 1)^2}{3^3} + \frac{(y + 2)^2}{4^3} = 1$
9. $-x^2 + 3x + y = 1$
10. $x^2 = y^2 + 3$
11. $y^2 = -5x - 8$
12. $5x^2 - 3x + 2y^2 + 6y + 3 = 0$
13. $x^2 - 2y^2 + 3y = 0$
14. $(x + 1)^2 + 2(y - 3)^2 + 3x = 0$

A new, X-Y coordinate system is defined in each of the following exercises. Calculate the coordinates of the given point P in the new coordinate system.

15. $x = X + 2, \quad y = Y - 3, \quad P = (1, 4)$
16. $x = X - 3, \quad y = Y + 4, \quad P(-2, 3)$
17. $x = X - 4, \quad y = Y - 5, \quad P = (4, -2)$
18. $x = X + 5, \quad y = Y - 2, \quad P = (-1, -3)$
19. $x = 3X + 1, \quad y = Y + 2, \quad P = (-2, -4)$
20. $x = 4X - 2, \quad y = 2Y + 1, \quad P = (0, 3)$

A new, X-Y coordinate system is defined by the given translation. A curve C is the graph of the given equation. Determine the equation of C in the new coordinate system.

21. $x = X - 1, \quad y = Y + 2, \quad x^2 + 3xy - 2y = 2$
22. $x = X + 2, \quad y = Y - 3, \quad x^2 + 2xy - 2y = 3$
23. $x = X + 3, \quad y = Y - 1, \quad x^2 + 4xy + y^2 = 1$
24. $x = X - 3, \quad y = Y - 2, \quad x^2 - 3xy + y^2 = 4$
25. $x = X - 3, \quad y = Y - 2, \quad 2x + 3 = 6$
26. $x = X + 1, \quad y = Y - 3, \quad x^2 + 3xy = 6$

A new coordinate system is determined by rotating the x-y coordinates by the given angle θ. Determine an equation of the rotated curve in this new coordinate system.

27. $\theta = 60°; \quad 9x^2 - y^2 = 36$
28. $\theta = 45°; \quad x^2 + y^2 = 25$
29. $\theta = -45°; \quad x^2 + y^2 = 49$
30. $\theta = -30°; \quad 4x^2 - 9y^2 = 36$
31. $\theta = 210°; \quad y = x - 2$
32. $\theta = 135°; \quad x^2 - 4xy + 4y^2 = 4$

Determine the angle of rotation θ such that the given equation has no XY term in the new coordinate system.

33. $5x^2 - 8xy + 5y^2 = 9$
34. $xy = 4$
35. $5x^2 + 6xy\sqrt{3} - y^2 + 8x - 8y\sqrt{3} - 12 = 0$
36. $25x^2 4y^2 = 100$

Determine the polar equations of the following conics.

37. parabola; directrix: $x = -2$
38. ellipse; $e = \frac{1}{3}$, directrix: $z = 3$
39. hyperbola; vertices: $(1, 0), (6, 0)$
40. hyperbola; $e = 2$, directrix: $y = -3$

Graph the following conics.

41. $r = \dfrac{1}{1 + 2\cos\theta}$
42. $r = \dfrac{4}{2 - \cos\theta}$
43. $r = \dfrac{2}{1 - 2\sin\theta}$
44. $r = \dfrac{6}{1 + 3\cos\theta}$

Graph the following parametric equations.

45. $x = -t + 1, \quad y = 2t, \quad 0 \le t \le 4$
46. $x = t^2, \quad y = -t, \quad 0 \le t \le 1$
47. $x = -1 + 2\sin t, \quad y = 4 + 2\cos t, \quad 0 \le t \le \pi$
48. $x = \sin t, \quad y = t, \quad 0 \le t \le \pi$

Determine the rectangular-coordinate form of the following parametric equations.

49. $x = 3z - 1, \quad y = 2z + 3, \quad 0 \le z \le 5$
50. $x = t^2, \quad y = \dfrac{1}{t - 1}, \quad t > 1$
51. $x = t, \quad y = \sqrt{t - 1}, \quad t \ge 1$
52. $x = -1 + \cos t, \quad y = 5 + \cos t, \quad 0 \le t \le 2\pi$
53. $x = 4\sin t, \quad y = 3\cos t, \quad 0 \le t \le 2\pi$
54. $x = 5\ln(t - 1), \quad y = 3t + 1, \quad 1 < t \le 5$

Chapter Test

Sketch the graphs of the conics in problems 1–6. For parabolas, determine the focus, vertex, and directrix. For ellipses, determine the foci, center, and semimajor and semiminor axes. For hyperbolas, determine the foci, center, transverse and conjugate axes, and asymptotes.

1. $y^2 = 5x$
2. $x^2 + 6x = y - 5$
3. $25x^2 + 100y^2 = 5{,}000$
4. $2x^2 + 4x + 9y^2 + 54y = 300$
5. $x^2 - y^2 = 9$
6. $12x^2 - 48x - y^2 + 2y = 72$
7. By rotating the axes, graph the conic that has the following equation: $x^2 - xy + y^2 = 1$.

8. Here is a system of parametric equations:
$$x = 2t + 1$$
$$y = t^2 - 1$$
$(0 \le t \le 1)$

   a. Sketch the graph of the system.
   b. Determine the corresponding rectangular-coordinate equation.

9. The eccentricity of the orbit of a weather satellite is 0.025. The furthest distance of the satellite from the earth is 22,000 miles. What is the closest distance of the satellite from the earth?

10. The main span of a suspension bridge is 5,200 feet long. Suppose that the towers are 750 feet high and the roadway is 400 feet above the surface of the water. The cables are parabolic in shape. Using a coordinate system with the horizontal axis along the roadway and the vertical axis along the left tower, determine the equation of the parabola.

PHOTO CREDITS

CHAPTER 1
Page 7: Courtesy of Texas Instruments, Inc. All rights reserved.
Page 11: Courtesy of Idaho Forest Industries.
Page 12: Courtesy of Electrolux Corporation.
Page 13: Courtesy of The Document Company, Xerox.
Page 17: © Daemmrich/The Image Works.
Page 26: © Daemmrich/The Image Works.
Page 34: Courtesy of UPS.
Page 41: Courtesy of Cardinal Pool, Champaign, IL.
Page 42: © Daemmrich/The Image Works.

CHAPTER 2
Page 79: © M. Siluk/The Image Works.
Page 85: © Bruce Hands/The Image Works.
Page 95: Courtesy of International Business Machines Corporation.
Page 99: © Daemmrich/The Image Works.
Page 102: Courtesy of International Business Machines Corporation.
Page 113: © Daemmrich/The Image Works.
Page 120: © M. Douglas/The Image Works.
Page 121: Bryn Mawr College archives.
Page 122: © H. Dcatch/The Image Works.
Page 147: Courtesy of International Business Machines Corporation.
Page 148: © H. Gans/The Image Works.

CHAPTER 3
Page 163, drawing: From Lester M. Sdorow, *Psychology*, 3rd edition. Copyright ©1995 Wm. C. Brown Communications, Inc., Dubuque, Iowa. Reprinted by permission of Times Mirror Higher Education Group, Inc., Dubuque, Iowa. All rights reserved.
Page 163: © Granitsas/The Image Works.
Page 172: Courtesy of Intel Corporation.
Page 173, top: © Daemmrich/The Image Works.
Page 173, bottom: Courtesy of Pioneer Electronics.
Page 185: © Daemmrich/The Image Works.
Page 186: © Albert/The Image Works.
Page 193: © Bob Stern/The Hartford Courant/ The Image Works.
Page 195: Courtesy of Pat Martin.
Page 200, drawing: From William P. Crummett and Arthur B. Western, *University Physics* Copyright © 1994 Wm. C. Brown Communications, Inc., Dubuque, Iowa. Reprinted by permission of Times Mirror Higher Education Group, Inc., Dubuque, Iowa. All rights reserved.

CHAPTER 4
Page 211: © T. Arruza/The Image Works.
Page 222: © M. Siluk/The Image Works.
Page 230, #45: © David Wells/The Image Works.
Page 230, #49: © John Griffin/The Image Works.
Page 230, #50: © Suzanne Arms/The Image Works.
Page 243: The Granger Collection, New York.
Page 257, #57: Chairman photo, used by permission of Maxell Corporation of America.
Page 257, #59: © Crandall/The Image Works.
Page 258: The Granger Collection, New York.
Page 259: The Granger Collection, New York.
Page 264: © Daemmrich/The Image Works.
Page 210, drawing: From Lester M. Sdorow, *Psychology*, 3rd edition. Copyright ©1995 Wm. C. Brown Communications, Inc., Dubuque, Iowa. Reprinted by permission of Times Mirror Higher Education Group, Inc., Dubuque, Iowa. All rights reserved.

CHAPTER 5
Page 273: © Crandall/The Image Works.
Page 283: © M. Antman/The Image Works.
Page 289, #19: © Daemmrich/The Image Works.
Page 289, #20: © Daemmrich/The Image Works.
Page 289, #22: Smithsonian Institution photo #75–3965.
Page 290, #27: © John Eastcott/YVA Momatiuk/The Image Works.

Page 290, #30: © Granitsas/The Image Works.
Page 292: Animals/Animals/Earth Scenes/Stouffer Productions.

CHAPTER 6
Page 300: © Pat Watson/The Image Works.
Page 303: © Varian Associates, Inc.
Page 308: © Eastcott/The Image Works.
Page 311: © J. Pickerell/The Image Works.
Page 321: Courtesy of the Mexican Government Tourism Office.
Page 324: © Daemmrich/The Image Works.
Page 331, #53: Courtesy of the National Park Service.
Page 331, #54: © Daemmrich/The Image Works.
Page 331, #63: Courtesy of U.S. Geological Survey.
Page 342: © J. Greenburg/The Image Works.

CHAPTER 7
Page 352: Courtesy of John Deere.
Page 353, #74: © Lee Snider/The Image Works.
Page 353, #75: © L. Kolvoord/The Image Works.
Page 354: Courtesy of Schwinn Cycling & Fitness, Inc. © 1994.
Page 372: © Pedrick/The Image Works.
Page 373: The Granger Collection, New York.
Page 375, #23: © Daemmrich/The Image Works.
Page 375, drawing: From William P. Crummett and Arthur B. Western, *University Physics*. Copyright © 1994 Wm. C. Brown Communications, Inc., Dubuque, Iowa. Reprinted by permission of Times Mirror Higher Education Group, Inc., Dubuque, Iowa. All rights reserved.
Page 377: © T. K. Wanstall/The Image Works.
Page 391: © John Griffin/The Image Works.
Page 395, drawing: From William P. Crummett and Arthur B. Western, *University Physics* Copyright © 1994 Wm. C. Brown Communications, Inc., Dubuque, Iowa. Reprinted by permission of Times Mirror Higher Education Group, Inc., Dubuque, Iowa. All rights reserved.

CHAPTER 8
Page 461: Courtesy of National Oceanic and Atmospheric Administration.

CHAPTER 9
Page 476, #35: © R. Sidney/The Image Works.
Page 476, #36: USDA Forest Service.
Page 477: Courtesy AAR/Wayne R. Gaylord.
Page 483: Courtesy of U.S. Geological Survey.
Page 484: Courtesy of American Airlines.
Page 499: The Granger Collection, New York.
Page 510: Courtesy of James Young.
Page 513: © B. Bachman/The Image Works.
Page 518: © Topham/The Image Works.

CHAPTER 10
Page 528: © Lee Snider/The Image Works.
Page 533, #15: Courtesy of Cardinal Pool, Champaign, IL.
Page 533, #17: Courtesy of Greenview Companies.
Page 533, #18: © Alan Carey/The Image Works.
Page 533, #19: © D. Banks/The Image Works.
Page 555: © Wajnarowicz/The Image Works.
Page 556, #36: © M. Siluk/The Image Works.
Page 556, #39: © J. Griffin/The Image Works.
Page 556, #38: Courtesy of Otis Elevator.
Page 560: Courtesy of General Motors Corporation.

CHAPTER 11
Page 603: Courtesy of NASA.

CHAPTER 12
Page 617: The Granger Collection, New York.
Page 619, #62: Courtesy of International Business Machines Corporation.
Page 619, #61: © Mulvehill/The Image Works.
Page 636: © Daemmrich/The Image Works.
Page 645, #66: © M. Douglas/The Image Works.
Page 645, #68: © B. Bachman/The Image Works.
Page 645, #71: © Esbin-Anderson/The Image Works.
Page 649: Courtesy of Buick Motor Division.
Page 658: © Daemmrich/The Image Works.
Page 662: © Dion Ogust/The Image Works.
Page 665: From the Collections of The St. Louis Mercantile Library Association.
Page 669: © Lorraine Rorke/The Image Works.

CHAPTER 13
Page 677: © Daemmrich/The Image Works.
Page 679, #43: © Daemmrich/The Image Works.
Page 679, #41: © J. Sohm/The Image Works.
Page 680: © Daemmrich/The Image Works.
Page 687: Copyright Hansen Planetarium, Salt Lake City, Utah. Reproduced with permission.
Page 699: © John Topham Picture Library/The Image Works.

ANSWERS TO ODD-NUMBERED EXERCISES

Chapter 1

Section 1.1, page 11

1. $2x + y$
3. $0.3w$
5. $xz + 5$
7. $2 \cdot (-a)$
9. $x \leq y$
11. $x < 1.1y$
13. $|x| \leq 3$
15. $7x$
17. $3a + 2b + c$
19. $6x$
21. $24xy$
23. $5x$
25. $-5 + 3y$
27. $21 - 7y$
29. $7x - y$
31. $1 + x$
33. $11x + 5$
35. xy
37. $<$
39. $>$
41. $<$
43. $>$
45. $<$
47. $>$
49. $\frac{1}{4}$
51. 11
53. 10
55. 0
57. 3.162
59. 19
61. $>$
63. $<$
65. $\frac{x}{45}$
67. $60h$
69. $1.35W$
71. $0.8x$
73. $x - 4$
75. $|N - O| \leq 0.02O$, $N - O \leq 50{,}000$, where $N =$ new budget and $O =$ old budget.
77. $10{,}000 \leq O - N \leq 0.05O$, $N \geq 0.95O$ where $N =$ new budget and $O =$ old budget.
79. $|L - 120| > 0.15$, where $L =$ length of stud in inches.
81. $\$300$
83. 6.08
85. $380{,}353{,}132$
87. $2{,}870.8999$
89. 0.8272780833
91. 62.8075
93. -35.0464
95. True
97. True
99. $\$6{,}657.50$

Section 1.2, page 25

1. $\frac{1}{a^3}$
3. q^4
5. $-12x^{11}$
7. $\frac{36}{y^2}$
9. $9a^2b^8$
11. $6^4x^4y^4$
13. $\frac{1}{t^5}$
15. $\frac{a^5}{b^4}$
17. $\frac{b^8}{9x^4}$
19. $\frac{8b^3}{27a^3}$
21. $\frac{yz^8}{2x^3}$
23. $-\frac{25}{8}$
25. $\frac{1}{3}$
27. 81
29. $-\frac{1}{256}$
31. 5
33. 32
35. $y^{2/3}$
37. $\frac{1}{y^{17/6}}$
39. $\frac{1}{a^{16}}$
41. $x^{1/10}$
43. $\frac{1}{a^6}$
45. $7x$
47. $8(t + 1)$
49. $3\sqrt{2x}$
51. $2xy^2$
53. $\frac{7\sqrt{2}a}{b^3}$
55. $3xy\sqrt[3]{2y}$
57. 3
59. $2x^2\sqrt{x}$
61. $5ab\sqrt[3]{b^2}$
63. $2(x + 1)\sqrt{2(x + 1)}$

65. $\dfrac{3ac^2}{b}$
67. 2
69. a^{20}
71. 2×10^9
73. 6.672×10^{-11}
75. $12,000,000,000
77. 0.00000000000000000578
79. 1.29316×10^{19} miles
81. 2.387×10^5 miles
83. $8,358.84

85. $5,415.71
87. 8 years
89. $64,532.61
91. 5.76%
93. 86.16%
95. 0.015625 units
97. 66.779 ft
99. $\sqrt{12.506072 \times 10^7} \approx$ 11,183 m/sec
101. 11.355 sec

103. 346.34613 m/sec
105. 759,375
107. 802.64904
109. 6.4474196
111. 2.428373942
113. 1.710089937
115. 25.77424038
117. 43,928,742.73
119. 29.78148

Section 1.3, page 38

1. Yes, leading coefficient = 34, degree 5.
3. No, negative exponent.
5. No, negative exponent.
7. No, radicals not allowed.
9. $17x^3 - 7x^2 + 3x - 1$
11. $4x^2 + 2x^3y - 5x^2y^2 - 4xy^3 + y^4$
13. $4p^4 - 9p^3q^2 + 2p^2q^3 - 3q^4$
15. $-3x\sqrt{y} - 5y\sqrt{x} + 6\sqrt{xy}$
17. $8x^2 + 26x + 15$
19. $81a^2 - 4$
21. $a^3 - 8$
23. $6a^3b^4 + 4a^4b^3 - 15a^2b^3 - 10a^3b^2$
25. $25x^2 - 30xy + 9y^2$
27. $4a^2b^2 + 12ab^2c + 9b^2c^2$
29. $16x^4 - 56x^3y + 49x^2y^2$
31. $a^4 - 2a^2b^2 + b^4$
33. $a^2 + 2ab + b^2 - y^6$
35. $2x^2y^2 - 6\sqrt{2}x^2y + 9x^2$
37. $3a^2 - 2b^2$
39. $\dfrac{16}{25}x^2 - \dfrac{9}{49}y^2$
41. $\dfrac{4}{9}t^4 - \dfrac{4}{5}t^2y^3 + \dfrac{9}{25}y^6$
43. $\dfrac{1}{8}x^2 - x^3y - \dfrac{3}{8}x^2y^2 + 4xy^3 - \dfrac{1}{2}y^4$

45. $A^4 + 4A^3B + 6A^2B^2 + 4AB^3 + B^4$
47. $x^4 + 32x^3 + 384x^2 + 2,048x + 4,096$
49. $1,296x^8 + 864x^6 + 216x^4 + 24x^2 + 1$
51. $3x(-xy + 1)$
53. $(x - 7)(x + 3)$
55. $2(2x - 3)(5x - 2)$
57. $2ab(7a - 6b)$
59. $(y - w)(x - z)$
61. $(x - 2)(x + 2)(2x + 1)$
63. $(23 - 18x)(23 + 18x)$
65. $4x(2a - 3x + a^2 - 5)$
67. $(x - 3)(x^2 + 3x + 9)$
69. $(2x + 5)(3x + 7)$
71. $(0.5x - 0.7y)(0.5x + 0.7y)$
73. $(-2t + 3)(t - 4)$
75. $(3x^2 - 2y)(9x^4 + 6x^2y + 4y^2)$
77. $(a + 4)(a^2 + 2a + 4)$
79. $(x + y - 5)(x + y + 5)$
81. $3(x - \sqrt{2}y)(x + \sqrt{2}y)(x^4 + 2x^2y^2 + 4y^4)$
83. $\dfrac{1}{72}(4x - 3)(3x + 2)$

85. $9x^{2a} - 12x^a y^c + 4y^{2c}$
87. $20t^{2n} + 3t^n - 35$
89. $x^{2m^2 + 2mn}$
91. $(a^n - b^n)(a^n + b^n)$
93. $(6t^n - 5)(6t^n - 5)$
95. $\left(\dfrac{x^4}{10} + 1\right)\left(\dfrac{x^8}{100} - \dfrac{x^4}{10} + 1\right)$
97. $25r^2 + 100r + 100$
99. a. $3w + w^2$
    b. Yes.
    c. No.
101. a. $65p + 20c$
     b. $74p + 24c$
     c. $139p + 44c$
103. $A = x(x+3)$, $A = y(y-3)$
105. $V = x(x + 10)(x + 13)$, $V = y(y + 3)(y - 10)$
107. $P(x) = -\dfrac{1}{3}x^3 + 5x^2 + 11x - 100$
109. $78x - 1.5x^2$
111. 16.05404

Section 1.4, page 50

1. $\dfrac{(x-1)(x+1)}{(2x+1)(x+3)}$
3. a^2
5. $\dfrac{1}{r+s}$
7. $\dfrac{x+1}{x^2+x+1}$
9. $\dfrac{6x-10}{2x+1}$
11. $\dfrac{x-3y-7}{x-3}$
13. $\dfrac{3t^2-4t+7}{t^2+3t-5}$
15. $\dfrac{-2x^2+4x-6}{x^2-3x-8}$
17. $\dfrac{t-3}{(t+3)(t+1)}$
19. $\dfrac{8x-29}{(x+5)(x-3)}$
21. 1
23. $\dfrac{t-1}{t+1}$
25. $\dfrac{a^2+b^2}{a+b}$
27. $\dfrac{xy+yz+xz}{x+z}$
29. $-\dfrac{1}{x^2+xh}$
31. $-\dfrac{3x^2+3xh+h^2}{x^3(x+h)^3}$
33. 1
35. $2a+h$
37. $2x+3+h$
39. $\dfrac{\sqrt{a}+\sqrt{b}}{a-b}$
41. $\dfrac{(\sqrt{a}+\sqrt{2})(\sqrt{a}+\sqrt{b})}{a-b}$
43. $\dfrac{1}{\sqrt{a+h}+\sqrt{a}}$
45. $\dfrac{a-2}{a-2\sqrt{2a}+2}$
47. $\dfrac{1}{\sqrt{n}(\sqrt{n+1}+\sqrt{n})}$
49. $\dfrac{1}{1+\sqrt{x}}$
51. 0.5545
53. a. 2.2222×10^{-5}
    b. $L = 10.0533$ m
55. 0.0051231345
57. 4.406400318
59. 3.222222222

Section 1.5, page 60

1. $x = 13$
3. $t = 2$
5. $x = 35$
7. $y = 4.35$
9. $x = -\dfrac{83}{40}$
11. $x = -25$
13. $x = 5{,}000$
15. $y = 650$
17. $x = -8$
19. $x = 8$
21. $x = -7$
23. $x = 5$
25. $x = 1$
27. $x = -2$
29. $t = -\dfrac{600}{23}$
31. $x = -\dfrac{29}{3}$
33. $y = \dfrac{14}{17}$
35. Mid-1996.
37. Late 1989.
39. $x \approx -1.051413882$
41. $x \approx 0.70988$
43. $x \approx 2.0172711$

Chapter 1 Review, page 65

1. $\sqrt{x+y}$
3. $x - 0.37x$
5. $4^{5/2}, 17, 6^0, 9^{3/2}, 67.2\%, 0.892, 2.\overline{345}, \sqrt{3}$
7. $\sqrt[3]{-25}, \sqrt{3}$
9. Associative law of multiplication.
11. Additive inverse.
13. $<$
15. $<$
17. $6x^3 - x^2 - 10x - 3$
19. $\dfrac{6-t^2}{3t}$
21. $\dfrac{a-3}{2a-1}$
23. $2\sqrt[3]{x^2} + x\sqrt[3]{x^2} + 2x\sqrt{x}$
25. $\dfrac{x-9}{(x+9)(x+7)}$
27. $\dfrac{1}{9x^{74}}$
29. x^4
31. $\dfrac{b^{24}}{27a^{30}c^{18}}$
33. $|y|$
35. $t^{1/2}$
37. $\dfrac{1}{x^{25/6}}$
39. -4
41. $3{,}125$
43. $\dfrac{y^2+xy+x^2}{xy(x+y)}$

45. $\dfrac{x^4 - 4 + x^2}{x(x^4 - 4)}$

47. $2xz\sqrt{3yz}$

49. $5ab\sqrt[3]{b^2}$

51. $27a^3b^6$

53. $\dfrac{2x \cdot n^5}{(n+1)^5}$

55. $2(2x+1)(2x-3)$

57. $(5a+1)(25a^2 - 5a + 1)$

59. $(x^a + 3)(x^{2a} - 3x^a + 9)$

61. $2x(2x+1)(x+7)$

63. $-(x+11)(x-5)$

65. $(a+b)(x+y)^2$

67. $4(3t - 2m)(3t + 2m)$

69. $\left(\dfrac{a}{4} + \dfrac{b}{2} - \dfrac{ab}{3}\right)\left(\dfrac{a}{4} + \dfrac{b}{2} + \dfrac{ab}{3}\right)$

71. $\left(x - \sqrt[3]{7}\right)\left(x^2 + \sqrt[3]{7}x + \sqrt[3]{49}\right)$

73. $\dfrac{1 - 2\sqrt{a} + a}{1 - a}$

75. $\dfrac{\sqrt[3]{450}}{5}$

77. $\dfrac{1}{5\sqrt[4]{t^3}}$

79. $-\dfrac{n}{\sqrt{n^2 - n} + n}$

81. $\sqrt{1 - a^2}(1 + a^2)$

83. $\dfrac{\sqrt{1+x}}{\sqrt{1-x}}$

85. All real numbers.

87. $-\dfrac{16}{17}$

89. 2.467×10^3

91. 0.0648

93. 1.118 m

95. $\sqrt{3}x$

97. $a\sqrt{2}$

Chapter 1 Test, page 67

1. a. $2x - 11y$
   b. $-4x + 8$
   c. -3

2. a. x^{21}
   b. x^{16}
   c. $\dfrac{y^2}{x^4}$

3. a. x^6
   b. $\dfrac{1}{x^2 y}$
   c. x^2

4. 8

5. a. $>$
   b. $<$
   c. $=$

6. a. $4x^2 - 12xy + 9y^2$
   b. $3x^2 - 20x - 7$
   c. $x^3 - 1$

7. a. $(x+3)^2$
   b. $(x+1)(x-6)$
   c. $(xy - z)(x^2y^2 + xyz + z^2)$

8. a. $\dfrac{x+1}{x-1}$
   b. $\dfrac{3x}{x^2 + x + 1}$

9. a. $\dfrac{x^2 + 2x - 5}{x^2 - 1}$
   b. $\dfrac{2}{x-1}$

10. a. $\dfrac{x+2}{x^3}$
    b. $\dfrac{x+5}{x^4}$

11. $125r^2 + 500r + 500$

12. $x = -\dfrac{1}{5}$

Chapter 2

Section 2.1, page 74

1.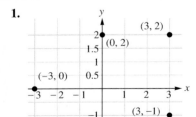

3. a. Quadrant I.
   b. Quadrant IV.
   c. y-axis.
   d. x-axis.

5. $\sqrt{61}$

7. $2\sqrt{2}$

9. 6

11. $\sqrt{9a^2 + 4}$

13. $\sqrt{a^2 + b^2}$

15. $\sqrt{a+b}$

17. 1

19. $\sqrt{a^2 + b^2}$

21. $\left(-1, -\frac{1}{2}\right)$ **29.** $\left(\frac{a}{2}, \frac{b}{2}\right)$

23. $(1, -4)$

25. $(1, 5)$ **31.** $\left(\frac{\sqrt{a}}{2}, \frac{\sqrt{b}}{2}\right)$

27. $\left(\frac{a}{2}, 4\right)$ **33.** 21 sq. units.

35. $\left(0, -\frac{61}{20}\right)$

37. No.

39. Yes, the Pythagorean theorem is satisfied.

41. $x - y = 2$

43. Assume that the triangle is positioned with one vertex at $(0, 0)$ and both legs along the axes. Let the two other vertices be $A(0, a)$ and $B(b, 0)$. The midpoint of the hypotenuse is $M(b/2, a/2)$ and

$$d(M, A) = \sqrt{(b/2 - 0)^2 + (a/2 - a)^2}$$
$$= \frac{1}{2}\sqrt{a^2 + b^2}$$

Similarly,

$$d(M, B) = \sqrt{(b/2 - b)^2 + (a/2 - 0)^2}$$
$$= \frac{1}{2}\sqrt{a^2 + b^2}$$

45. 3.3 mm

47.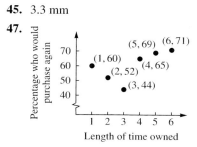

49. 7.898716415

51. (1,478, 1,892.5)

Section 2.2, page 84

1. Yes. **5.** Yes.

3. No. **7.** Yes.

9.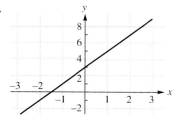

11.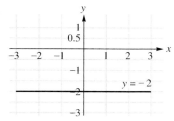

13.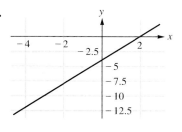

15.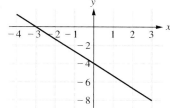

17.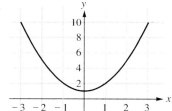

19.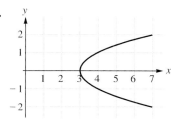

21.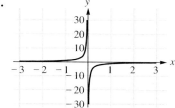

23.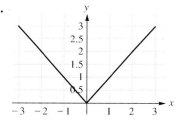

25.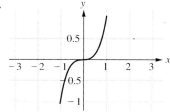

A-6 ANSWERS Section 2.2, page 84—Section 2.3, page 97

27.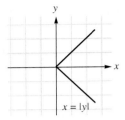

29.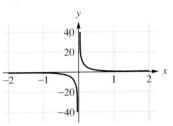

31.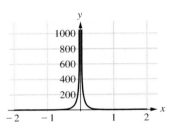

33.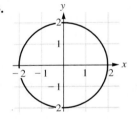

35.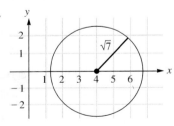

37.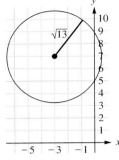

39.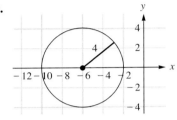

41.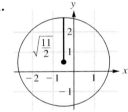

43. $(x + 6)^2 + (y - 1)^2 = 4$

45. $x^2 + y^2 = 39$

47. a.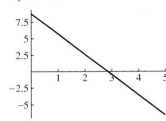

b. y-intercept: demand that occurs if the price is 0; x-intercept: price at which demand is 0.

49. a.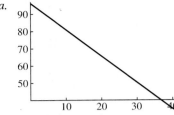

b. y-intercept: market share in 1980.

c. 50%

51. a.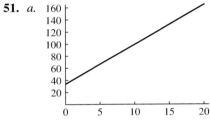

b. y-intercept: income of private legal offices in 1982.

53. a.

b. y-intercept: amount of itemized deductions if income is 0.

55. a.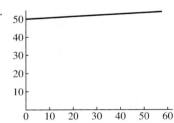

b. y-intercept: amount of itemized deductions for health when income is 0.

57. (5, 64.75)   **59.** (5, 0.1428571429)

Section 2.3, page 97

1. 0.3

3. −6

5. $\frac{1}{2}$

7. $\frac{2}{3}$

9. a. y decreases by 36.

b. y increases by 600.

11. 2   **13.** −2

15. 0

17. 2

19. $y = -3x - 1$

21. $x = 2$

23. $y = -\dfrac{1}{5}x - 1$

25. $y = -\dfrac{2}{3}x + \dfrac{13}{3}$

27. $x = 9$

29. $m = -\dfrac{2}{3};\ b = 1$

31. $m = \dfrac{5}{3};\ b = -3$

33. $m = 0;\ b = \dfrac{3}{2}$

35. Parallel.
37. Neither.
39. Neither.
41. Neither.

43. $y = -x - 5$

45. $y = x - 4$

47. $y = \dfrac{2}{3}x$

49. $y = -\dfrac{3}{2}x - 2$

51. $y = -x$

53. $x = -2$

55. $m_{AB} = \dfrac{1}{3}$; parallel line:
$y = \dfrac{1}{3}x - \dfrac{22}{3}$;
perpendicular line: $y = -3x + 6$.

57. $m_{AB} = -\dfrac{5}{2}$; midpoint $= (4, -2)$; equation of perpendicular bisector: $y = \dfrac{2}{5}x - \dfrac{18}{5}$.

59. $\dfrac{1}{2}$

61. To find the x-intercept, set $y = 0$ in the equation and solve for x:
$$\dfrac{x}{a} + \dfrac{0}{b} = 1$$
$$\dfrac{x}{a} = 1$$
$$x = a$$
So the x-intercept is a. A similar argument works for the y-intercept.

63. a. The hourly earnings increase by $0.27 per year.
    b. $1.35 per hour.

65. a. The slope is the change in the itemized deductions per dollar change in income.
    b. $1,543.48

67. a. Change in weekly orange juice sales per unit change in price.
    b. Sales increase by 3,000 cans.

69. a. $P = -\dfrac{12}{5}A + 82$
    b. 10%

71. The value the machine loses per year.

73. $S = 24{,}000 + 0.38x$; $S = $ salary, $x = $ sales

75. a. Rate of growth of cost per mile of driving per additional unit of time.
    b. $0.56 increase.

77. a. Rate of change of Fahrenheit temperature per unit change in Celsius temperature.
    b. $\dfrac{9}{5}$ degrees increase.

79. $m = -5.1,\ b = 2.3$

81. $m = 0.8,\ b = -2$

83. $y = -0.8539325843x + 6.252808989$

85. $y = 29{,}500x - 48{,}000$

Section 2.4, page 112

1. $c > 0$
3. $a \geq 0$
5. $m \geq -3$
7. $a \leq b$
9. $[-2, 3)$
11. $[-4, 5]$
13. $[-2, 3)$
15. $[-2, \infty)$
17. $\{x: -1 \leq x < 3\}$
19. $\{x: x > -2\}$
21. $\{x: x \leq 2\}$
23. $x \geq -1;\ [-1, \infty)$
25. $x < -\dfrac{3}{4};\ \left(-\infty, -\dfrac{3}{4}\right)$
27. $x \leq 4;\ (-\infty, 4]$
29. $x \geq -\dfrac{4}{5};\ \left[-\dfrac{4}{5}, \infty\right)$
31. $x \geq 4;\ [4, \infty)$
33. $x < -2.6;\ (-\infty, -2.6)$
35. $-4 \leq x \leq 5;\ [-4, 5]$
37. $1 \leq x \leq \dfrac{13}{3};\ \left[1, \dfrac{13}{3}\right]$
39. $-5 < x \leq -2;\ (-5, -2]$
41. $x \geq 25;\ [25, \infty)$
43. $-8 \leq x < \dfrac{14}{3};\ \left[-8, \dfrac{14}{3}\right)$
45. $x > \dfrac{14}{3};\ \left(\dfrac{14}{3}, \infty\right)$
47. $x < -\dfrac{3}{2};\ \left(-\infty, -\dfrac{3}{2}\right)$
49. $(-\infty, -5) \cup (3, \infty)$
51. $(-\infty, 3] \cup [4, \infty)$
53. $(-5, 5)$
55. $(-\infty, 0) \cup \left(\dfrac{2}{3}, \infty\right)$
57. $(-3, -1) \cup (2, \infty)$
59. $(-\infty, -5] \cup [0, 2]$
61. $(-\infty, -2) \cup (-1, 1)$
63. $(-\infty, -1) \cup (3, \infty)$
65. $\left[\dfrac{3}{2}, 4\right)$
67. 5,000 or more.
69. $I \geq 150$
71. $52 < S < 92$

A-8 ANSWERS Section 2.4, page 112—Section 2.6, page 136

73. $(-\infty, -4) \cup (-1, 2)$

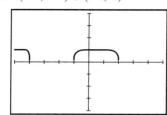

77. $(-5, 2]$

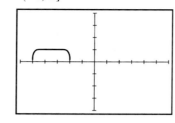

79. $[-8, 4.66)$

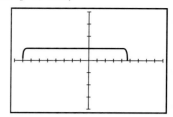

75. $(-3, -1) \cup (2, \infty)$

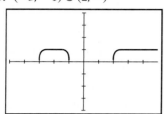

Section 2.5, page 121

1. -5
3. $-8a - 5$
5. 309
7. $3(a - h)^2 + (a - h) - 1$
9. 5
11. 3.56
13. 0 is not in the domain of g.
15. 11
17. -1
19. $-2h^3 + 9h^2 - 12h + 4$
21. 27
23. $(1 + a + h)^3$
25. 1
27. $4a^2 + 8ah + 4h^2 + 4a + 4h + 1$
29. 12
31. 12
33. domain = all real numbers; range = all real numbers
35. domain = all real numbers; range = $\{12\}$
37. domain = all real numbers except 3; range = all real numbers except 0
39. domain = $\left\{x: x \leq \dfrac{3}{2}\right\}$; range = set of all nonnegative real numbers
41. domain, range = all real numbers
43. domain = all real numbers; range = set of all nonnegative real numbers
45. domain = $\{x: -2 \leq x \leq 2\}$; range = $\{y: 0 \leq y \leq 2\}$
47. domain = $[1, \infty)$; range = $[0, \infty)$
49. 2
51. $2a - 5 + h$
53. $R(x) = -x^2 - 10x + 1{,}200$
55. $V(x) = x(20 - 2x)^2$
57. $C(r) = \dfrac{1}{5}\pi r^2 + 12\pi r$

Section 2.6, page 136

1. Yes.
3. No.
5. $0, 2, -2, 5$
7. $2, 4, 0, -2$
9. domain = $[-2, 5]$; range = $[-2, 5]$
11. domain = $[-2, \infty)$; range = $[-5, \infty)$
13. domain = $[-1, 1]$; range = $[-4, 12]$

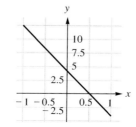

15. domain = $[-1, 1]$; range = $[0, 1]$

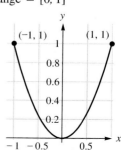

ANSWERS Section 2.6, page 136—Section 2.6, page 136 A-9

17. domain = $[-2, 2]$;
range = $[-3, 1]$

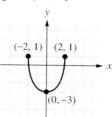

19. domain = all real numbers;
range = all real numbers

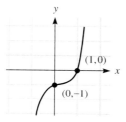

21. domain = $[-2, 2]$;
range = $[0, 2]$

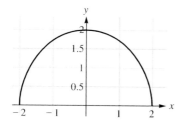

23. domain = all positive reals;
range = all positive reals

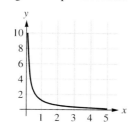

25.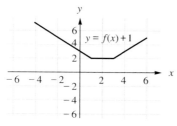

27.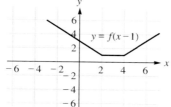

29.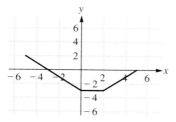

31.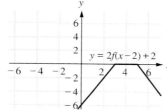

33.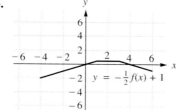

35. a.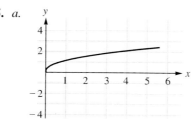

b.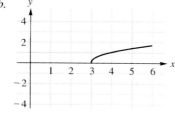

c.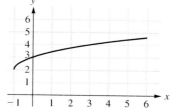

d.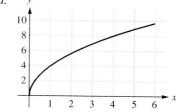

e.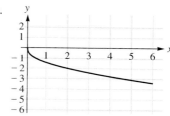

f.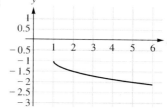

37. Neither.
39. Neither.
41. Even.
43. Neither.
45. Odd.
47. Even.
49.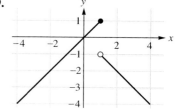

51.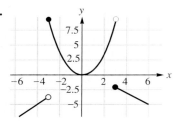

55.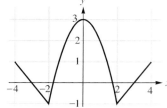

59.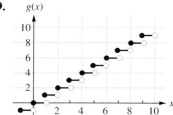

53.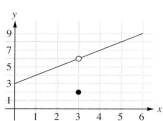

57.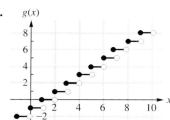

61. 0, 6, 2, 2

63.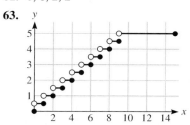

Section 2.7, page 143

1.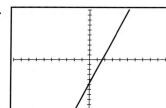

3.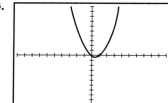

5.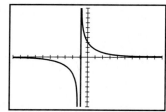

7.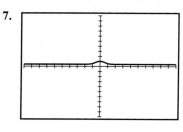

9. a. $(1.2, -2.160)$
b. $(2.1, -1.8935)$
c. $(-1, 4)$
d. $(0.001, -0.002999)$

11. $x = \pm 1.5275$

13. No x-intercepts

15.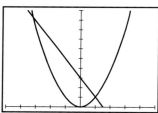

17.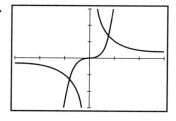

19. They all cross the horizontal axis at $x = 0$. If the vertex is on the horizontal axis, then a scaled graph will have its vertex at the same location.

21.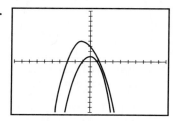

23. domain $= [0, 2]$; range $= [-2, 4]$

25. domain $= [1, 3]$; range $= [-1.25, 1]$

27. domain $=$ all reals except 1 and -1; range $= (-\infty, \infty)$

29. domain $= [-0.5, 0.5]$; range $= [0, 1]$

31. domain $= (-\infty, \infty)$; range $= [0, \infty)$

33. No solutions.

35. $t = 0, -0.5$

37. $x = -0.2, 0.5$

39. No solution.

41. $x = -4, 0.5$

43. $x = -2.1, 0.3$

45. $a = -13.9, 0.9$

47.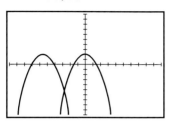

49.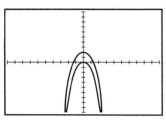

51.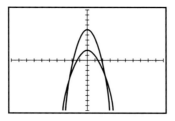

53.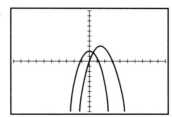

55. Even function by graph.
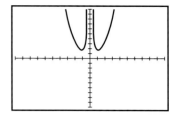

57. Not an even function.
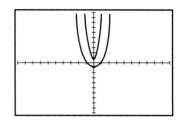

Section 2.8, page 147

1. $2x$

3. $-x + 3$

5. $4x - 1$

7. $-\dfrac{8}{3}x - 4$

9. $-\dfrac{5}{2}x + 10$

11. $-\dfrac{11}{4}x + 33$

13.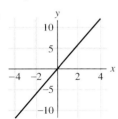

15.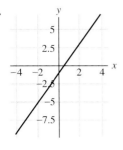

17.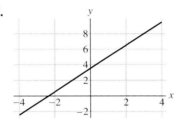

19.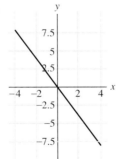

21. Translate graph of f down 1 unit.

23. Translate graph of f down 3 units.

25. Translate graph 1 unit to the left.

27. Translate graph right 1 unit.

29. Reflect graph of f in x-axis.

31. *a.*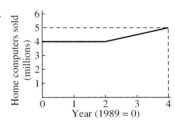

b. 4 million.

c. 4.5 million.

d. Demand was constant from 1989 through the end of 1990. Then demand began to increase.

33. *a.* $D(x) = 3.7 - \dfrac{7}{60}x$

b.

c. Equals the value of the building at year 0 ($3.7 million).
d. $1.37 million.
e. 23.143 years.

Chapter 2 Review, page 151

1. $\sqrt{34}$
3. $\frac{3}{5}$
5. slope = $\frac{1}{2}$, y-intercept = 4
7. Neither.
9. Neither.
11. $\sqrt{47}$
13.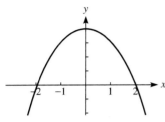
15.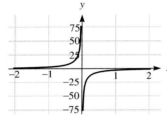
17.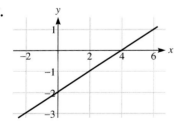

19. $(-5, \infty)$
21. $[-9, 9)$
23. $\left(-1, \frac{5}{4}\right)$
25. Circle, center (0, 3), radius 3.
27. Circle, center (2, −1), radius $2\sqrt{3}$.
29. $y = \frac{4}{3}x - \frac{11}{3}$
31. $y = -\frac{1}{3}x + \frac{7}{3}$
33. Yes.
35. No.
37. −6
39. −4
41. $(a - 1)^2 + 3(a - 1) - 4$
43. $2a + 3 + h$
45. 1
47. domain = all real numbers; range = all real numbers
49. domain = $\{x: x \le 6\}$; range = $[0, \infty)$
51. domain = $(-\infty, \infty)$; range = $\{56.7\}$
53. domain = $\{x: -8 \le x \le 8\}$; range = $[0, 8]$
55. domain = $(-\infty, 0) \cup (0, \infty)$
57. domain = $[-1, \infty)$
59. domain = $(-\infty, \infty)$

61.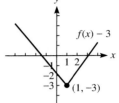
63.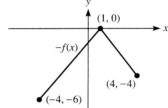

65. Even.
67. Even.
69. Even.
71. Neither.
73. Even.
75. Odd.
77. Even.
79. a. For each increase in Fahrenheit temperature, Celsius temperature increases $\frac{5}{9}$ of a degree.
b. Celsius temperature drops $\frac{5}{9}$ of a degree.

Chapter 2 Test, page 152

1. $5\sqrt{2}$
2. $\left(\frac{7}{2}, 4\right)$
3. a.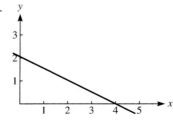
b.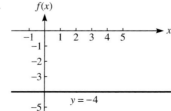
c.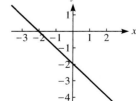

4. Slope = $-\frac{2}{3}$, y-intercept = $-\frac{19}{3}$
5. Decreases by 14.
6. a. $y = -\frac{1}{4}x - 1$
   b. $y + 2 = -3(x - 6)$
   c. $y = x - 5$
7. a. $x \geq \frac{1}{4}$
   b. $-\frac{3}{2} < x < \frac{3}{2}$

c. $-1 < x < 1$
8. 42.5%
9. a. 6
   b. 1
   c. $0, \frac{3}{2}$
10. a. All real numbers
    b. All nonzero real numbers
    c. $x \geq -1$

11.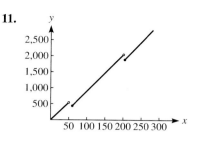

Chapter 3

Section 3.1, page 162

1. $V = \frac{F}{pq}$, $p = \frac{F}{qV}$
3. $y = -\frac{2x+5}{2}$, $x = -\frac{2y+5}{2}$
5. $x = \frac{144}{8 + 3\pi}$
7. $r = \sqrt{\frac{A}{\pi}}$
9. $b = \sqrt[3]{\frac{P}{a}}$
11. $d = \frac{k-p}{hq + \frac{v^2}{2}}$, $q = \frac{k - p - \frac{1}{2}dv^2}{hd}$
13. $a_1 = \frac{S(1-r)}{1 - r^n}$

15. $m_2 = \frac{m_1 v_1 - m_1 u_1}{u_2 - v_2}$,
    $v_1 = \frac{m_1 u_1 + m_2 u_2 - m_2 v_2}{m_1}$
17. $\lambda = \pm \frac{1}{2}$
19. $x = \pm \frac{2\sqrt{3}}{3} r$
21. $92
23. $1,500
25. Janitor: $11,500; president: $115,000; sales: $17,250.
27. 8%

29. 60,000/7 gallons of mixture A; 80,000/7 gallons of mixture B.
31. $6,500 at 6%, $5,500 at 9%.
33. 7,800
35. a. $V = 6,000 - \frac{2,350}{3}n$
    b. $6,000, $5,216.67, $4,433.33, $3,650, $2,083.33, $1,300
37. a. 68°F
    b. 100°C
    c. $-40°$

Section 3.2, page 174

1. $\pm\sqrt{3}$
3. $\pm 2\sqrt{2}$
5. No solutions.
7. 3 or 1
9. $\frac{2}{3}, -\frac{4}{3}$
11. $-\frac{5}{2} \pm \frac{\sqrt{65}}{2}$
13. No solutions.
15. $-\frac{2}{3} \pm \frac{\sqrt{7}}{3}$

17. $\frac{10}{3}, -1$
19. $0, -\frac{4}{3}$
21. No solutions.
23. $\frac{-1 \pm \sqrt{33}}{4}$
25. $0, \frac{5}{9}$
27. $\frac{3}{2}, -\frac{7}{3}$
29. No solutions.

31. $n = \frac{1 \pm \sqrt{1 + 8G}}{2}$
33. $t = -\frac{-v_0 \pm \sqrt{v_0^2 + 64(h-H)}}{32}$
35. $r = \frac{-h \pm \sqrt{h^2 + 2S/\pi}}{2}$
37. 5%
39. 8%
41. a. $h = \frac{\sqrt{3}a}{2}$
    b. $A = \frac{\sqrt{3}a^2}{4}$

   c. $P = 3a$
   d. 30 cm
43. $500 or $400
45. a. 1982
    b. $946,000
47. $x \approx 1.913136513, -1.269418785$
49. $x \approx 74.26128677, -75.26128677$

Section 3.3, page 183

1. (b)
3. (d)
5. (c)
7. $f(x) = \frac{5}{2}x^2$
9. $f(x) = -7x^2$
11. $y = -\frac{1}{3}(x-2)^2 + 4$
13. a. (0, 0)   b. Downward.
   c.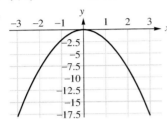
15. a. (2, 0)   b. Upward.
   c.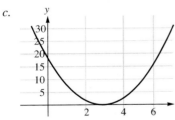
17. a. (3, 0)   b. Upward.
   c.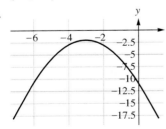
19. a. (−3, −2)   b. Downward
   c.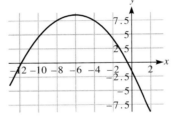
21. a. $f(x) = (x+3)^2 - 4$
    b. (−3, −4)
    c. Upward.
    d. min = −4
    e.
23. a. $f(x) = -\frac{1}{4}(x+6)^2 + 8$
    b. (−6, 8)
    c. Downward.
    d. max = 8
    e.
25. max = 0
27. min = 3
29. max = $\frac{121}{8}$
31. min = $-\frac{169}{12}$
33. a. $A(x) = -2x^2 + 150x$
    b. 37.5 yd
    c. 2,812.5 yd²
35. Dimensions are 18.5 ft × 18.5 ft; area = 342.25 ft².
37. 1,250
39. a. $R(x) = 1{,}200 - 10x - x^2$
    b. 0
    c. $1,200
41. a. $A(x) = \frac{4+\pi}{16\pi}x^2 - \frac{7}{2}x + 49$
    b. $\frac{28\pi}{4+\pi} \approx 12.32$ in.
    c. 27.44 in.²
43. a. 65   c. February
    b. 86.75   d. 64.82
45. Minimum value = 0.42
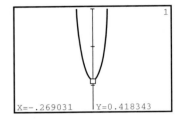
47. Minimum value = −160
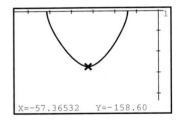
49. Minimum value = 1.8
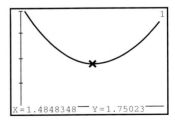

Section 3.4, page 194

1. No solution.
3. $\frac{43 \pm 5\sqrt{85}}{6}$
5. $\frac{17 \pm \sqrt{293}}{2}$
7. 100
9. No solution.
11. 9
13. $\frac{1 + \sqrt{13}}{2}$
15. −3, ±1
17. 0, 1
19. −1, 1
21. $\frac{35}{6}$
23. 0.7

ANSWERS Section 3.4, page 194—Chapter 3 Test, page 203   A-15

25. $\frac{16}{3}$
27. No solution.
29. No solution.
31. $\frac{1}{16}$
33. 19
35. 5
37. ± 2
39. 4, −1, 2, 1
41. 256, 225
43. $\frac{5}{6}, \frac{4}{3}$
45. $\frac{1}{4}$, 9
47. −4, 2, −1
49. 0, $\frac{6}{5}$
51. $\frac{15}{8}$ hr
53. $\frac{6}{5}$ hr
55. 36 hr
57. 7%
59. $x \approx 5.940339$
61. $x \approx -0.0622272$, 0.93199221

Section 3.5, page 199

1. ± 3
3. −5, −1
5. ± 4
7. 9, −6
9. $-23 \le x \le 23$; $[-23, 23]$
11. $x > 14$ or $x < -14$; $(-\infty, -14) \cup (14, \infty)$
13. $x < -2$ or $x > 6$; $(-\infty, -2) \cup (6, \infty)$
15. $-1 < x < 5$; $(-1, 5)$
17. $-6 \le x \le \frac{22}{3}$; $\left[-6, \frac{22}{3}\right]$
19. $-\frac{23}{12} < x < \frac{1}{4}$; $\left(-\frac{23}{12}, \frac{1}{4}\right)$
21. No solutions.
23. No solutions.
25. $|x - 5| \le r$
27. $|x - 2| \ge 6$
29. [$2,775,000, $3,225,000]
31. 6 in. $< d <$ 14 in.
33. (−1.193333333, 0.19333333333)
35. $(-\infty, -0.4) \cup (5.6, \infty)$

Chapter 3 Review, page 202

1. $\frac{-5 \pm \sqrt{41}}{8}$
3. $\frac{1 \pm \sqrt{4{,}001}}{20}$
5. $-\frac{1}{2}, -1$
7. 4, $\frac{2}{3}$
9. No solutions.
11. 16, 25, 0
13. −10
15. No solutions.
17. $\pm\sqrt{2}, \pm\sqrt{5}$
19. $\frac{4}{3}, -12$
21. $\pm 3, -4$
23. 6, −1
25. $(-\infty, -7] \cup [-1, \infty)$
27. $n = \pm\sqrt{\frac{Q}{m}}$
29. $L = \frac{T}{4v^2 - 1}$
31. $n = \frac{-1 \pm \sqrt{1 + 8S}}{2}$
33. a. $f(x) = \left(x - \frac{3}{2}\right)^2 - \frac{9}{4}$
    b. $\left(\frac{3}{2}, -\frac{9}{4}\right)$
    c. Upward.
    d. min $= -\frac{9}{4}$
    e.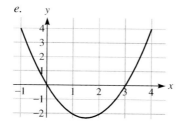
35. a. $f(x) = 2(x + 3)^2 - 16$
    b. $(-3, -16)$
    c. Upward.
    d. min $= -16$
    e.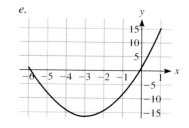
37. min $= -\frac{569}{8}$
39. $0.55x^2 + \frac{246.4}{x}$
41. 9%

Chapter 3 Test, page 203

1. $x = -3, \frac{1}{2}$
2. $x = 7 \pm 2\sqrt{14}$
3. a. $x = -\frac{2}{3}, 1$
   b. $x = -\frac{1}{3}$
4. $x = 2 \pm \sqrt{5}$
5. $x = 4$
6. $r = \sqrt{\frac{a - 3}{4\pi}}$
7. a. $-1 < x < 5$
   b. $x < -7$ or $x > -3$
8. −4
9. y-intercept −2; x-intercepts −2, $\frac{1}{3}$
10. a. $P(x) = -2x^2 + 80x - 200$
    b. 20
    c. $600

Chapter 4

Section 4.1, page 212

1. (d)

3. (a)

5. (c)

7.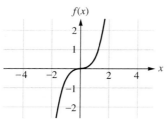

9.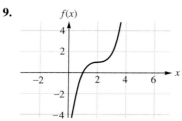

11.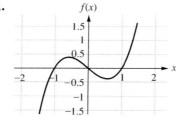

13.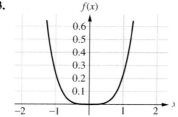

15.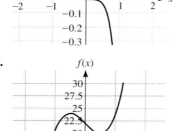

17.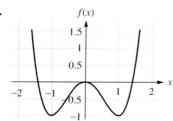

19.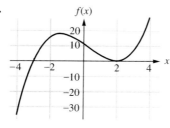

21.

23. a. $V(x) = 4x^3 - 32x^2 + 64x$

b.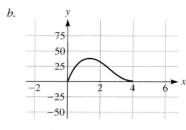

c. $x \approx \frac{4}{3}$ gives the approximate maximum; max ≈ 38

25. a. $P(x) = -\frac{1}{3}x^3 + 5x^2 + 11x - 100$

b.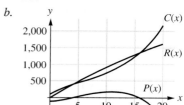

c. Max when $x = 11$; max $= 182.\overline{3}$.

27. 52

29. 2 and 24 weeks into the flu season.

31. $x = -10, -1, 14$

33. $x = -2, 0, 2$

35. a.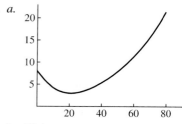

b. 19.4 years.

Section 4.2, page 221

1. $q(x) = x^4 + x^2 + 2x + 4$, $r(x) = 3$

3. $q(x) = x^3 + \frac{1}{3}x^2 + \frac{2}{9}x + \frac{22}{27}$, $r(x) = -\frac{118}{27}$

5. $q(x) = x^3 - 2x^2 + 3x - 4, r(x) = 6x - 13$
7. $f(x) = -\frac{5}{2}(x + 1)(x - 2)$
9. $f(x) = 4(x + 1)(x)(x - 1)$
11. $q(x) = 4x^4 - 8x^3 + 18x^2 - 36x + 71, r(x) = -137$
13. $q(x) = 2x^2 + 5x - 3, r(x) = 0$
15. $q(x) = x^2 - 5x + 25, r(x) = 0$
17. $q(x) = 3x^3 - 2x^2 - \frac{61}{3}x - \frac{50}{9}, r(x) = \frac{224}{27}$
19. $q(x) = x^5 + x^4y + x^3y^2 + x^2y^3 + xy^4 + y^5, r(x) = 0$
21. 0, 30, 402
23. 4, 0, 34
25. 41, -351, 2,353, $\frac{43}{27}$
27. -1, yes; 1, no.
29. Yes, yes.
31. $q(x) = x - 1, r(x) = 2$
33. $q(x) = 4(x + 1)^4(x - 5)^3 + 3(x + 1)(x - 5)^4, r(x) = 0$
35. $f(x) = -0.078125x^2 + 0.9375x + 5.5$

37. $2,492,500
39. a. $q(x) = x + 2; r(x) = -x + 1$
   b.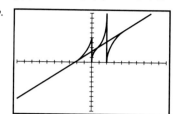
   c. After zooming out twice:
   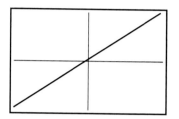
41. $q(x)$ is the oblique asymptote that the graph of $f(x)/g(x)$ approaches.

Section 4.3, page 229

1. $f(-1) = -1, f(0) = 1$
3. $f(2) = -1, f(3) = 7$
5. $f(0) = -2, f(1) = 9$
7. $f(-1) = -3, f(-2) = 9;$ $f(1) = -3, f(2) = 9$
9. -0.7
11. 2.2
13. 0.2
15. 1.4, -0.8
17. -0.7, 1.37, 1.8
19. 1, $\frac{2}{3}, \frac{1}{3}$
21. $\pm 1, \frac{2}{3}$
23. $\pm \frac{5}{2}$

25. Lower bound $= -1$; upper bound $= 2$.
27. Lower bound $= -4$; upper bound $= 1$.
29. Lower bound $= -2$; upper bound $= 1$.
31. Lower bound $= -3$; upper bound $= 3$.
33. $3, \frac{-3 \pm \sqrt{13}}{2}$
35. $-2, -2 \pm \sqrt{5}$
37. $\frac{-3 + \sqrt{5}}{4} \pm \sqrt{\frac{19 \pm 7\sqrt{5}}{2}},$

$\frac{-3 - \sqrt{5}}{4} \pm \sqrt{\frac{19 \pm 7\sqrt{5}}{2}}$
39. $\frac{1}{2}, -\frac{2}{3}$
41. $2, 4, -1, -\frac{1}{2}, \frac{3}{2}$
43. $\pm\sqrt{\frac{13 \pm \sqrt{145}}{2}}$
45. a. 25
    b. 13
47. 3.92, 15.89
49. $t = 0.64608, 3.9084$

Section 4.4, page 237

1. $\sqrt{3}i$
3. $9i$
5. $7i\sqrt{2}$
7. $-7i$
9. $4 - 2i\sqrt{15}$
11. $2i(1 + \sqrt{3})$
13. $-i$
15. 1
17. -1
19. i
21. $8 + 0i = 8$
23. $1 - 26i$
25. $0 + 0i = 0$
27. 0
29. 1
31. $5 - 8i$
33. $2 - \frac{\sqrt{6}}{2}i$
35. $\frac{4}{5}$
37. $5 + 6i$
39. 6

41. $2 + 4i$
43. $-7 + 10i$
45. $14 - 2i$
47. 29
49. $6 + 8i$
51. $21 - 20i$
53. $\frac{1}{10} + \frac{7}{10}i$
55. $\frac{1}{2} - \frac{\sqrt{3}}{2}i$
57. $-3 - 5i$
59. $-\frac{1}{2} + \frac{1}{2}i$
61. $\frac{2}{25} + \frac{11}{25}i$
63. $\frac{1}{5} + \frac{2}{5}i$
65. $\frac{1}{34} + \frac{2}{17}i$
67. $\frac{3}{5} + \frac{1}{5}i$
69. $-\frac{1}{2}i$
71. $\pm\sqrt{5}i$
73. $\pm 8i$
75. $(x - 2i)(x + 2i)$
77. $(x - \sqrt{3}i)(x + \sqrt{3}i)$
79. $(a - bi)(a + bi)$

Section 4.5, page 244

1. $6x^2\left[x - \dfrac{-5 + i\sqrt{95}}{12}\right]\left[x - \dfrac{-5 - i\sqrt{95}}{12}\right]$
3. $(x - i)(x + i)(x + 1 + \sqrt{2}i)(x + 1 - \sqrt{2}i)$
5. $(x - 1)\left(x - \dfrac{-1 + i\sqrt{3}}{2}\right)\left(x - \dfrac{-1 - i\sqrt{3}}{2}\right)$
7. $(x + 3)(x + \sqrt{5}i)(x - \sqrt{5}i)$
9. $\left(x - \dfrac{-1 + i\sqrt{11}}{2}\right)^2\left(x - \dfrac{-1 - i\sqrt{11}}{2}\right)^2$
11. 0, multiplicity $= 1$; 2, multiplicity $= 2$; $\frac{5}{2}$, multiplicity $= 3$
13. -2, multiplicity $= 5$; 2, multiplicity $= 2$
15. 2, multiplicity $= 3$; -1, multiplicity $= 1$
17. 0, multiplicity $= 1$; 3, multiplicity $= 2$; i, multiplicity $= 1$; $-i$, multiplicity $= 1$
19. 0, multiplicity $= 5$; $\dfrac{-1 + \sqrt{5}}{2}$, multiplicity $= 3$; $\dfrac{-1 - \sqrt{5}}{2}$, multiplicity $= 3$
21. $(x - 2)(x + 2)(x - 1 - \sqrt{3}i)(x - 1 + \sqrt{3}i) \times (x + 1 - \sqrt{3}i)(x + 1 + \sqrt{3}i)$
    Zeros are ± 2, $1 \pm \sqrt{3}i$, $-1 \pm \sqrt{3}i$.
23. $(x - 1)\left(x + \dfrac{1 + \sqrt{3}i}{2}\right)\left(x + \dfrac{1 - \sqrt{3}i}{2}\right)$
    Zeros are 1, $\dfrac{-1 \pm \sqrt{3}i}{2}$.
25. $(m + 2)(m - 1 + \sqrt{3}i)(m - 1 - \sqrt{3}i)$
    Zeros are -2, $1 \pm \sqrt{3}i$.
27. $(y - \sqrt{2})(y + \sqrt{2})(y + \sqrt{2}i)(y - \sqrt{2}i)$
    Zeros are $\pm\sqrt{2}$, $\pm\sqrt{2}i$.
29. $r^3(r + i)^2(r - i)^2$
    Zeros are 0, $\pm i$.
31. $(x + 2i)(x - 2i)\left(x - \dfrac{1 + \sqrt{5}}{2}\right)\left(x - \dfrac{1 - \sqrt{5}}{2}\right)$
    Zeros are $\pm 2i$, $\dfrac{1 \pm \sqrt{5}}{2}$.
33. $(x - \sqrt{3})(x + \sqrt{3})(x + i)(x - i)$
    Zeros are $\pm\sqrt{3}$, $\pm i$.
35. $(r - 2)^2(r + 3)$
    Zeros are 2, -3.
37. At most 2 positive; at most 2 negative.
39. At most 2 positive; at most 2 negative; at least 6 nonreal.
41. At most 2 positive; at most 2 negative.
43. At most 3 positive; at most 2 negative.
45. At most 2 positive; at most 2 negative; at least 6 nonreal.
47. At most 2 positive; at most 2 negative; 0 is a zero of multiplicity 2.
49. $-2 + i$, -1
51. 0, -1
53. 3
55. $1 - 2i$, 1, and 2

Section 4.6, page 255

1. (d)
3. (f)
5. (e)

7.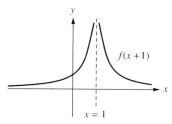

9.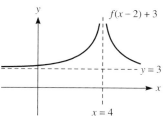

11.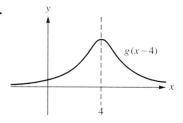

13. a. $x = 0, \quad y = 0$
    b.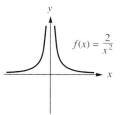

15. a. $x = 0, \quad y = 0$
    b.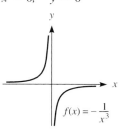

17. a. $x = -1, \quad y = 0$
    b.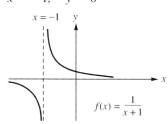

19. a. $x = -2, \quad y = 0$
    b.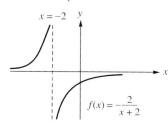

21. a. $x = -3, \quad y = 0$
    b.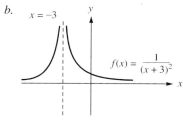

23. a. $x = -2, \quad y = -1$
    b.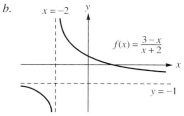

25. a. $x = -\dfrac{10}{3}, \quad y = -\dfrac{2}{3}$
    b.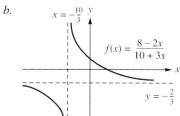

27. a. $y = 0$
    b.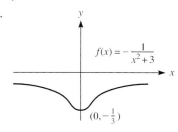

29. a. $x = \pm 3, \quad y = 0$
    b.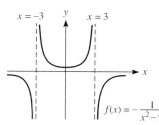

31. a. $x = \dfrac{1}{2}, \quad x = -3, \quad y = 0$
    b.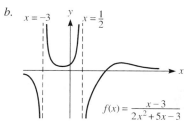

33. a. $x = 3, \quad y = x + 3$
    b.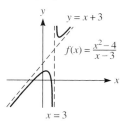

35. a. $x = 2, \quad x = -1, \quad y = 1$
    b.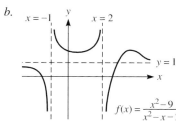

37. a. $x = -4, \quad y = x - 6$
    b.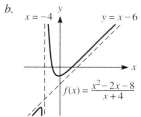

39. a. $x = 0$, $y = x$
b.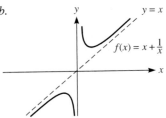

41. a. $x = \pm\dfrac{\sqrt{3}}{3}$, $y = \dfrac{1}{3}$
b.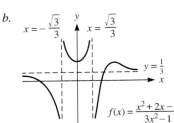

43. a. $x = 0$, $y = \dfrac{x^2}{4}$
b.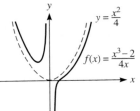

45. a. None.
b.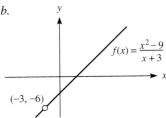

47. a. None.
b.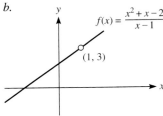

49. a. None.
b.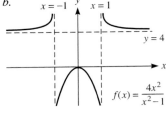

51. a. $x = \pm 1$, $y = 4$
b.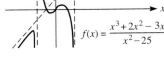

53. a. $x = \pm 5$, $y = x + 2$
b.

55. a.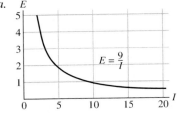
b.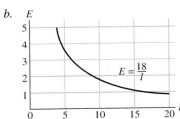
c.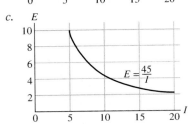

57.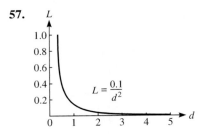

59. a. $W_h = 200\left(\dfrac{4{,}000}{4{,}000 + h}\right)^2$
b.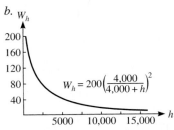

Chapter 4 Review, page 262

1. $q(x) = x^4 + 4x^3 + 4x^2 + 2x + 8$, $r(x) = 5$

3. $f(x) = -\dfrac{5}{2}(x + 2 - \sqrt{3}) \times (x + 2 + \sqrt{3})$

5. $q(x) = 4x^4 - 8x^3 + 16x^2 - 32x + 70$, $r(x) = -139$

7. $q(x) = \dfrac{1}{6}x - \dfrac{1}{3}$, $r(x) = 0$

9. $f(0) = -1$, $f(-1) = 4$, $f(i) = 0$

11. 2, no; 3, yes; -2, no.

13. $(x-5)(x^2+5x+25)$; zeros are $5, \dfrac{-5 \pm 5i\sqrt{3}}{2}$.

15. 0: multiplicity = 3; 4: multiplicity = 2; $-\dfrac{3}{2}$: multiplicity = 1.

17. 1 positive real zero, 1 negative real zero, 2 nonreal zeros.

19. $\pm\sqrt{\dfrac{3}{2}}, -1 + i\sqrt{2}$

21. $2 \pm 2i\sqrt{3}$

23. No rational zeros.

25. No rational zeros.

27. $1, \dfrac{-1 \pm i\sqrt{3}}{2}$

29. $1 \pm \sqrt{2}, 3$

31. $-2, 2 \pm \sqrt{3}$

33. $f(2) < 0, f(3) > 0$

35. 2.0946

37.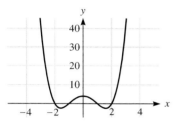

39.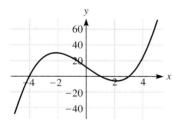

41.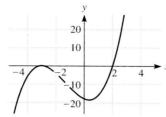

43.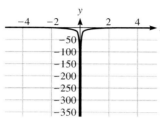

45.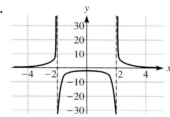

47.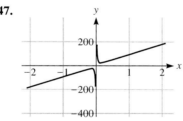

49.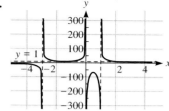

51.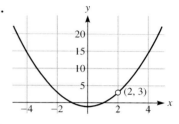

53.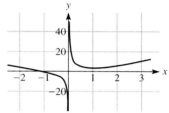

55. $2, \dfrac{21 - \sqrt{177}}{4}$

57.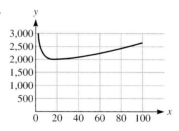

Chapter 4 Test, page 264

1. Quotient $= 5x^2 + \dfrac{5}{2}x + \dfrac{11}{4}$, remainder $= \dfrac{7}{4}$

2. Quotient $= x^4 - 4x^3 - x^2 - x - 2$, remainder $= 15$

3. -67

4. $10x^2 + \dfrac{5}{3}x - \dfrac{10}{3}$

5. 0 of multiplicity 4, 2 of multiplicity 3, i of multiplicity 2, $-i$ of multiplicity 2.

6. -1.167

7. 1 or 3 positive zeros, 2 or 0 negative zeros.

8. a. $8 + 2i$
   b. $-i$
   c. $26 - 95i$
   d. $1 - 2i$

9. $x = \pm i\sqrt{2}, \pm\sqrt{\dfrac{5}{2}}$

10. 1 of multiplicity 2, $-\dfrac{1 \pm \sqrt{3}i}{2}$ of multiplicity 2, $\pm i$ of multiplicity 1.

11. Vertical asymptote $x = 0$, oblique asymptote $y = x$

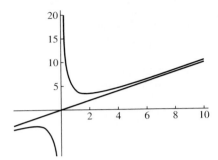

Chapter 5

Section 5.1, page 272

1. $f(x) = x - 4$ Domain (f) = all real numbers
   $g(x) = x + 5$ Domain (g) = all real numbers
   $(f + g)(x) = 2x + 1$ Domain $(f + g)$ = all real numbers
   $(f - g)(x) = -9$ Domain $(f - g)$ = all real numbers
   $(fg)(x) = x^2 + x - 20$ Domain (fg) = all real numbers
   $(ff)(x) = (x - 4)^2$ Domain (ff) = all real numbers
   $\left(\dfrac{f}{g}\right)(x) = \dfrac{x - 4}{x + 5}$ Domain (f/g) = all real numbers, $x \neq -5$
   $\left(\dfrac{g}{f}\right)(x) = \dfrac{x + 5}{x - 4}$ Domain (g/f) = all real numbers, $x \neq 4$
   $(f \circ g)(x) = x + 1$ Domain $(f \circ g)$ = all real numbers
   $(g \circ f)(x) = x + 1$ Domain $(g \circ f)$ = all real numbers

3. $f(x) = 2x^2 + x - 3$ Domain (f) = all real numbers
   $g(x) = x^3$ Domain (g) = all real numbers
   $(f + g)(x) = x^3 + 2x^2 + x - 3$ Domain $(f + g)$ = all real numbers
   $(f - g)(x) = -x^3 + 2x^2 + x - 3$ Domain $(f - g)$ = all real numbers
   $(fg)(x) = 2x^5 + x^4 - 3x^3$ Domain (fg) = all real numbers
   $(ff)(x) = (2x^2 + x - 3)^2$ Domain (ff) = all real numbers
   $\left(\dfrac{f}{g}\right)(x) = \dfrac{2}{x} + \dfrac{1}{x^2} - \dfrac{3}{x^3}$ Domain (f/g) = all real numbers, $x \neq 0$
   $\left(\dfrac{g}{f}\right)(x) = \dfrac{x^3}{2x^2 + x - 3}$ Domain (g/f) = all real numbers, $x \neq 1, -\dfrac{3}{2}$
   $(f \circ g)(x) = 2x^6 + x^3 - 3$ Domain $(f \circ g)$ = all real numbers
   $(g \circ f)(x) = (2x^2 + x - 3)^3$ Domain $(g \circ f)$ = all real numbers

5. -1

7. $x^2 + 2x + 2$

9. 72

11. Undefined.

13. Undefined.

15. $\dfrac{2x + 3}{x^2 - 1}$

17. $4x^2 + 12x + 8$

19. $2(2x + 3) + 3 = 4x + 9$

21. $f(x) = x^5, g(x) = 4x^3 - 1$

23. $f(x) = \dfrac{1}{x^4}, g(x) = x + 5$

25. $f(x) = \dfrac{x + 1}{x - 1}, g(x) = x^3$

27. $f(x) = \left(\dfrac{1+x}{1-x}\right)^4$, $g(x) = x^3$

29. $f(x) = \sqrt{x}$, $g(x) = \dfrac{x+2}{x-2}$

31. $f(x) = x^{63}$, $g(x) = x^5 + x^4 + x^3 - x^2 + x - 2$

33.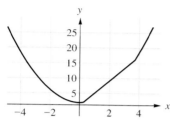

35.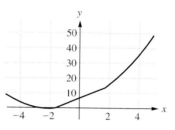

37.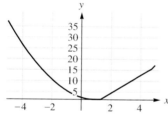

39.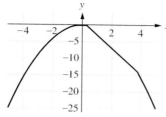

41.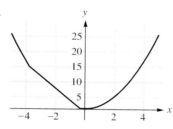

43.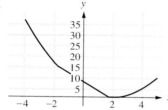

45. a. $r(t) = 10t$
b. $A(r) = \pi r^2$
c. $(A \circ r)(t) = 100\pi t^2$. This is the area of the spill at time t hours.

47.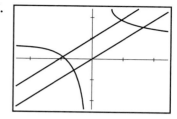

49.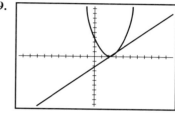

51.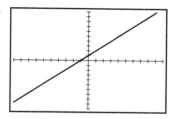

53.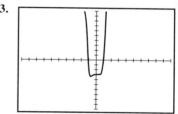

Section 5.2, page 282

1. No inverse. The function is not one-to-one.

3. No inverse. The function is not one-to-one.

5. No inverse. The function is not one-to-one.

7. $f^{-1}(x) = \dfrac{5}{8} - \dfrac{x}{8}$

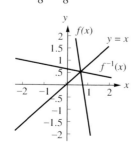

9. No inverse. The function is not one-to-one.

11. No inverse. The function is not one-to-one.

13. $f^{-1}(x) = \sqrt{4 - x^2}$; $0 \leq x \leq 2$

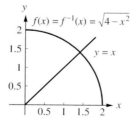

15. No inverse. The function is not one-to-one.

17. Yes.

19. Yes.

21. Yes.

23. No.

25. Yes.

27. No.

29. $f^{-1}(x) = \dfrac{1}{2}(x + 3)$

31. $f^{-1}(x) = \dfrac{3}{2}\left(x - \dfrac{3}{5}\right)$

33. $f^{-1}(x) = \sqrt{x}$, $x \geq 0$

35. $f^{-1}(x) = x^2 - 2$, $x \geq 0$

37. $f^{-1}(x) = \sqrt[5]{x}$

39. $f^{-1}(x) = \dfrac{2}{x}$

41. $f^{-1}(x) = \dfrac{x + 3}{2 - 3x}$

43. $f^{-1}(x) = -\sqrt{25 - x^2}$, $0 \leq x \leq 5$

45. $f^{-1}(x) = x^3 - 8$

47. $(f \circ f^{-1})(739) = 739$, $(f^{-1} \circ f)(5.00023) = 5.00023$

49. $f^{-1}(x) = \dfrac{x-b}{m}$

51. $F \circ C(x) = C \circ F(x) = x$

53. Since $f(x)$ is not one-to-one, it has no inverse function.

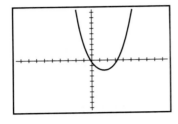

55. Since $f(x)$ is one-to-one, it has an inverse function.

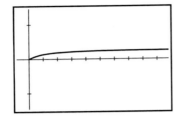

57. Since $f(x)$ is not one-to-one, it has no inverse function.

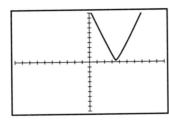

59. Function is one-to-one by horizontal line test.

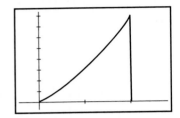

61. Function is one-to-one by horizontal line test.

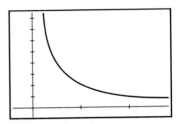

Section 5.3, page 288

1. a. $y = \dfrac{1}{70}x$
   b. $y = 5$

3. a. $y = 20x^2$
   b. $y = 273.8$

5. a. $y = \dfrac{0.98}{x^2}$
   b. $y = 98$

7. a. $y = \dfrac{xz}{w}(3.25)$
   b. $y = 62.4$

9. a. $y = \dfrac{xz}{w^2}(22.62857143)$
   b. $y = 15{,}713.28$

11. a. $y = \dfrac{z^3}{x^2}\left(\dfrac{2}{9}\right)$
   b. $y = \dfrac{1}{6}$

13. $A = \pi r^2$
Constant of variation is π.
A varies directly as r^2.

15. $d = 65t$
Constant of variation is 65.
d varies directly as t.

17. $22°C$

19. 28.16 g

21. $4\sqrt{2}$ ft

23. a. $s = \sqrt{30d}, d = 53.3333$ ft, $s = 65.03845$ mph
   b. $s = \sqrt{12d}, d = 252.08333$ ft, $s = 49.95998$ mph

25. $360°$

27. 66.3158 mph

29. $w = 1.14875 \times 10^{-38}$ m

31. -3.59375×10^{-13} dyn

Chapter 5 Review, page 293

1. All real numbers.

3. a. All real numbers, $x \neq -\tfrac{3}{2}$.
   b. $\left(\dfrac{f}{g}\right)(x) = \dfrac{x^2 - 1}{2x + 3}$

5. a. All real numbers.
   b. $(f + g)(x) = x^2 + 2x + 2$

7. a. All real numbers.
   b. $(g \circ f)(x) = 2x^2 + 1$

9. $(f \circ g)(x) = x; (g \circ f)(x) = x$

11. $(f \circ g)(x) = x; (g \circ f)(x) = x$

13. $f(x) = x^{11}$; $g(x) = 3x^2 - 5x + 1$

15. $f(x) = x^5 + x^3 - 2x^2 - 7$; $g(x) = x^3 + 2$

17. $D_{f \circ g}: \emptyset; D_{g \circ f}$: all real numbers, $x \neq 4$

19. Yes.

21. No.

ANSWERS Chapter 5 Review, page 293—Section 6.1, page 302   A-25

23. $f^{-1}(x) = \frac{1}{9}(x + 14)$

25. $f^{-1}(x) = \sqrt{16 - x^2}, 0 \leq x \leq 4$

27. $(f \circ f^{-1})(78.8999) = 78.8999$

29. $L = \frac{R}{d^2}$
The loudness decreases by a factor of $\frac{1}{9}$ if d is tripled.

31. a. $V = \frac{k}{P}$
  b. $k = 800; V = \frac{800}{15} = \frac{160}{3}$ ft^3

Chapter 5 Test, page 294

1. a. $(f + g)(x) = \sqrt{2x - 1} + x^2 - 3$, $x \geq \frac{1}{2}$
  b. $(fg)(x) = \sqrt{2x - 1}(x^2 - 3)$, $x \geq \frac{1}{2}$
  c. $\frac{f}{g}(x) = \frac{\sqrt{2x - 1}}{x^2 - 3}$, $x \geq \frac{1}{2}, x \neq \sqrt{3}$
  d. $(f \circ g)(x) = \sqrt{2(x^2 - 3) - 1}$, $|x| \geq \sqrt{\frac{7}{2}}$

2. $f^{-1}(x) = \sqrt{x + 1}, x \geq -1$

3. No. Since $f(x) = f(-x)$, the function is not one-to-one.

4. a. One-to-one;
  $f^{-1}(x) = \frac{1}{5}(x + 1)$
  b. Not one-to-one, since $f(x) = f(-x)$

  c. One-to-one;
  $f^{-1}(x) = \frac{x}{1 - x}, x \neq 1$
  d. Not one-to-one, since $f(x) = f(-x)$

5. 70 cm

6. $y = 52/x$

7. -12

Chapter 6

Section 6.1, page 302

1. a. $f(1) = 3; f(2) = 9; f(3) = 27;$
$f(0) = 1; f(-1) = \frac{1}{3};$
$f(-2) = \frac{1}{9}; f(-3) = \frac{1}{27};$
$f\left(\frac{1}{2}\right) = \sqrt{3}$

b.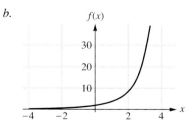

3.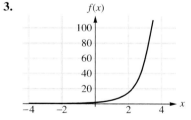

5.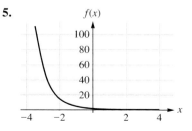

7.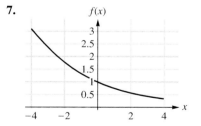

9.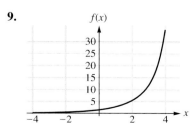

11.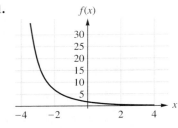

13.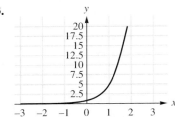

15.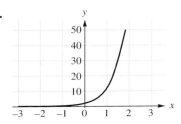

17.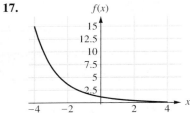

19.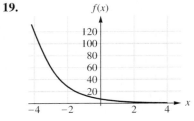

21.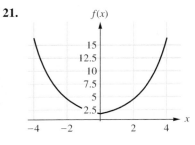

23.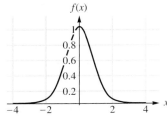

25.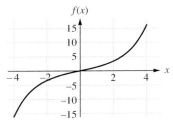

27.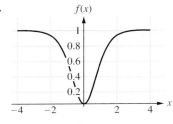

29. (d)

31. (h)

33. (g)

35. (j)

37. (b)

39. a. $A(t) = 10{,}000(1.08)^t$

b. 1 yr: $10,800
3 yr: $12,597.12
2 yr: $11,664
10 yr: $21,589.25

c.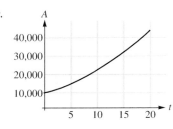

41. a. $Q(t) = A_0(0.99997)^t$

b. 1 yr: 399.988 kg
200 yr: 397.607 kg
2 yr: 399.976 kg

43. a. $N(t) = 16{,}370{,}000(1.005955)^t$

b. 10 yr: 22,127,445
15 yr: 25,726,019
23 yr: 32,740,000
46 yr: 65,480,000

45.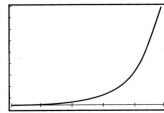

47.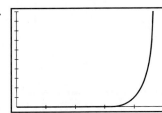

49.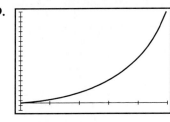

51. 1 point of intersection: (2.5, 6)

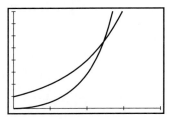

Section 6.2, page 310

1. $e^3 \approx 20.086$
3. $e^3 + 2e^7 \approx 2{,}213.4$
5. $\dfrac{e^{10} - 1}{e^{10} + 1} (= 0.9999092)$
7. $e^2 - 2e + 1 \approx 2.9525$
9. e^{4x}
11. $e^{2x} - 2 + e^{-2x}$
13. $e^{3x} - 3e^{-2x} + 3e^x - 1$
15. $4e^{2x-2} + 5e^{x-2}$
17. $e^{2x} - 9$
19. e^x
21. $\dfrac{2 + 2e^{-x}}{(1 + e^x)^2}$
23. $\dfrac{e^x(e^h - 1)}{h}$
25. $\dfrac{(e^x - e^{-x}e^{-h})(e^h - 1)}{2h}$
27. a.

t	$1 - e^{-0.18t}$
1	0.165
2	0.302
3	0.417
5	0.593
13	0.904

b. $P(t)$
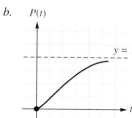

29. $P(t) = 0.5(1 - e^{-0.08t})$

a. 1: 0.0384
2: 0.0739
10: 0.2753
20: 0.399
50: 0.4908

b.
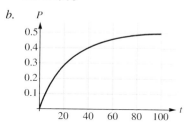

31. $P(t) = 100(1 - e^{-0.3t})$

a. 1: 25.918%
2: 45.119%
5: 77.687%
10: 95.021%
12: 97.268%

b.
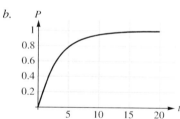

33. a. 135, 491, 499, 499

b.
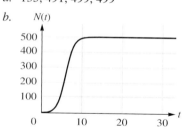

35. 1.276192609
37. $0 < y < 50$
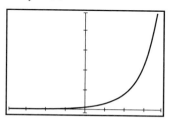

39. $.5 < y < 1.5$
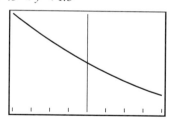

41. $0 < y < 1$
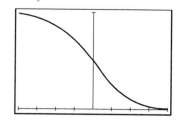

Section 6.3, page 320

1. $2^3 = 8$
3. $3^5 = 243$
5. $2^{-3} = \dfrac{1}{8}$
7. $x^{-12} = 3$
9. $\log_4 x = 2$
11. $\log_{10} 0.001 = -3$
13. $\log_4 2 = \dfrac{1}{2}$
15. $\log_x 0.01 = -3$
17. 1
19. 1
21. 0
23. 0
25. 4
27. -3
29. 0.8241
31. 3
33. $2x - 1$
35. x^4
37. $f(x)$
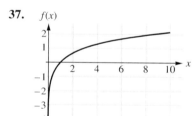

39.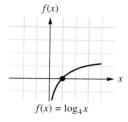

41.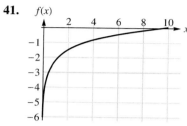

43.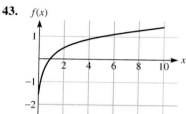

45.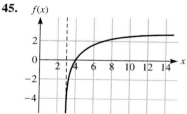

47.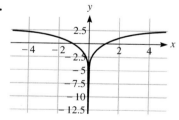

49.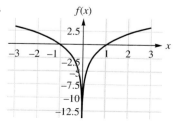

51. f(x) graph

53. $-\log_{10} 3$

55. $\log_a 3$

57. 1

59. $\log_a(x + 3)$

61. 1

63. $\log_a(x - y)$

65. $\log_b x + \frac{1}{2}\log_b y - 2\log_b z$

67. $2\log_2 x + 2\log_2 y + 4\log_2 z$

69. $4\log_3 x + \log_3 y - 2\log_3 z$

71. $2\log_{10} x + \frac{2}{3}\log_{10} y - 4\log_{10} z$

73. $\frac{3}{2}\log_{10} x + \frac{1}{4}\log_{10} y + \frac{5}{4}\log_{10} z$

75. $x = 5$

77. $x = 0$

79. $x = 4$

81. $x = 2{,}197$

83. $x = 6$

85. $x = \sqrt{3}$

87. $x = \pm\sqrt{5 + \log_5 14}$

89. $x = -\frac{1}{3}$

91. $x = \log_b 5$

93. $x = \log_5 7$

95. $x = -2.5$

97. $x = \frac{7}{4}$

99. $x = \frac{-1 \pm \sqrt{7}}{2}$

101. $x = \pm 3$

103. $f^{-1}(x) = \log_3 x$

105. $f^{-1}(x) = \log_2 \dfrac{\sqrt{x^2 - 4} + x}{2}$

107. $\dfrac{x + \sqrt{x^2 - 1}}{x - \sqrt{x^2 - 1}} \cdot \dfrac{x + \sqrt{x^2 - 1}}{x + \sqrt{x^2 - 1}} = \dfrac{(x + \sqrt{x^2 - 1})^2}{x^2 - (x^2 - 1)}$

$\qquad = (x + \sqrt{x^2 - 1})^2$

Thus:

$$\log_a\left(\frac{x + \sqrt{x^2-1}}{x - \sqrt{x^2-1}}\right) = \log_a(x + \sqrt{x^2-1})^2$$
$$= 2\log_a(x + \sqrt{x^2-1})$$

109. $H = \dfrac{-10}{\log_2(0.2466)}$ (≈ 4.95109 days)

111. $D = \dfrac{1}{\log_2\left(\dfrac{10{,}352{,}650}{10{,}000{,}000}\right)}$ (≈ 20 min)

113.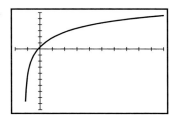

115.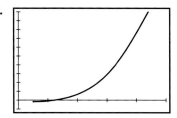

Section 6.4, page 329

1. 1
3. 3
5. -1
7. -3
9. 1
11. 2
13. 1
15. t
17. $64e^3$
19. 4
21. 66
23. $\dfrac{1}{\ln 100} \cdot (10)^{2x}$
25. $\dfrac{1}{2}$
27. x^2
29. $\ln 2$
31. $\dfrac{1}{2} \cdot 36 \cdot e - \dfrac{1}{2} \cdot 4 \cdot e = 16e$
33. $\ln 19$
35. (c)
37. (a)
39. ≈ 2.010382
41. ≈ -1.5058883
43. $\dfrac{\ln 4{,}578}{\ln 10}$
45. $2\dfrac{\ln x}{\ln 10}$
47. $\dfrac{\log 0.987}{\log e}$
49. $\dfrac{1}{2}\dfrac{\log y}{\log e}$
51. 130 decibels
53. 10^{-15} watts per cm^2
55. $D(I_1) = 10\log_{10}\left(\dfrac{I_1}{I_0}\right) = 10\log_{10} I_1 - 10\log_{10} I_0$

$D(I_2) = 10\log_{10}\left(\dfrac{I_2}{I_0}\right) = \log_{10} I_2 - 10\log_{10} I_0 10$

$D(I_2) - D(I_1) = 10\log_{10} I_2 - 10\log_{10} I_1 = 10\log_{10}\left(\dfrac{I_2}{I_1}\right)$

57. $I = I_0 \cdot 10^{D/10}$
59. a. 1   d. 4
    b. 2   e. 5
    c. 3   f. 8

61. The magnitude is 2 greater on the Richter scale.
63. $R(I_1) = \log_{10}\left(\dfrac{I_1}{I_0}\right) = \log_{10} I_1 - \log_{10} I_0$

$R(I_2) = \log_{10}\left(\dfrac{I_2}{I_0}\right) = \log_{10} I_2 - \log_{10} I_0$

$R(I_2) - R(I_1) = \log_{10} I_2 - \log_{10} I_1 = \log_{10}\dfrac{I_2}{I_1}$

65. $I = I_0 10^R$
67. a. 1.585×10^{-8} mol/l
    b. 6.3096×10^{-5} mol/l
69. $H^+ < 10^{-7}$
71. 2.523 days
73. a. 1995: 260,249,015; 2001: 274,688,828
    b. 77.0163534 yr
75. a. 0.0150964837
    b. $P(t) = 23{,}669{,}000 e^{0.015t}$
    c. 31,949,808
77. $1,582.02
79. 28.2%
81. a. 9.477121255
    b. 10.477121255
83. a. $80,000
    b. $1,465.25
85. 1.92153
87. 0.538573734
89. -0.33913
91. 6.828144
93. 49.9615
95. -3.0654
97. -22.9164
99. No solutions.
101. $x = -7.1, 3.3$

Section 6.5, page 336

1. $x = \pm\sqrt{\dfrac{\ln 4}{\ln 7}}$
3. $x = \dfrac{2}{5 - \ln 4}$
5. $x = \dfrac{1}{3e^6 - 1}$
7. $t = \ln 100$
9. $t = \ln 100$
11. $k = \dfrac{\ln 1.845}{10}$
13. $k = \dfrac{1}{10} \ln \dfrac{5}{3}$
15. $x = \dfrac{-3 \pm \sqrt{13}}{2}$
17. $x = \dfrac{2}{3}$
19. $x = 1$
21. $x = 2$
23. $x = \sqrt{41}$
25. $x = \dfrac{96}{31}$
27. $x = \dfrac{1}{8}, 4$
29. $x = \dfrac{1 + \sqrt{5}}{2}$
31. $x = 10^{1/4}, 10^{-2/3}$
33. $x = 5$
35. $x = 8$
37. $x = \ln 2$
39. No solutions.
41. $x = 9, \dfrac{1}{9}$
43. $x = 16$
45. $x = 1{,}000, \dfrac{1}{1{,}000}$
47. $x = 3^{1/(\log_2 3 - 1)}$
49. $t = \dfrac{1}{k} \ln \dfrac{P}{P_0}$
51. $t = -\dfrac{1}{k} \ln \dfrac{P - N}{AN}$
53. $t = \ln(x \pm \sqrt{x^2 - 1})$
55. $t = \ln \sqrt{\dfrac{1 + x}{1 - x}}$
57. $t = -\dfrac{L}{R} \ln\left(1 - \dfrac{RI}{E}\right)$
59. $f(x) = 5e^{x(3 - \ln 5)/2}$
61. $f(x) = \dfrac{1}{5} e^{x \ln 5}$
63. $f(x) = 10 e^{x \ln 100/10}$
65. 113.219669
67. No solution.
69. 0.0448656373

Chapter 6 Review, page 340

1. $e^{0.6931} = 2$
3. $(0.5)^2 = 0.25$
5. $\log_2 2 = 1$
7.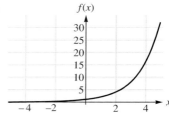
9.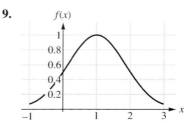
11.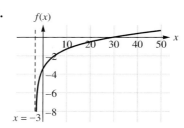
13.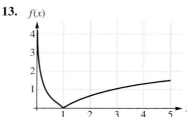
15. $x = 2$
17. $x = e^4$
19. $x = 2$
21. $x = 2$
23. $x = -\dfrac{5}{6}$
25. $x = -\dfrac{1}{2} \ln 0.01$
27. $x = 5$
29. $x = 4, \dfrac{1}{8}$
31. $x = 10^4, 10^{-4}$
33. $x = 16, \dfrac{1}{16}$
35. $5x - 2$
37. 1
39. 3
41. 4
43. $\log_a \dfrac{x^3 + x^2 y}{y^3}$
45. $\log_a \dfrac{\sqrt[3]{xv^3}}{\sqrt[6]{w}}$
47. $\log_3 x + \log_3(y + z)$

49. 2

51. b

53. $\frac{1}{3}$

55. $\frac{1}{2}$

57. 18

59. $t = \frac{1}{2} \ln\left(\frac{y+1}{y-1}\right)$

61. $t = \sqrt[n]{\frac{y}{a}}$

63. $t = \frac{-0.07}{\ln(0.02)}$ (≈ 0.0178936)

65. *a.* 0, 0; 1, 2.083317; 2, 3.9496; 5, 8.461004; 10, 13.343

b.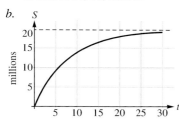

67. *a.* 50 watts

b. $W(30) = 44.346$ watts, $W(365) = 11.612$ watts

69. *a.* $N(0) = 711$

b.

t	$\dfrac{6{,}400}{1 + 8e^{-0.23t}}$
0	711
1	870
2	1,058
6	2,124
12	4,249
30	6,349
60	6,400

71. *a.* 1: 1.4816 mg 4: 0.60239 mg
  2: 1.0976 mg 6: 0.330598 mg

b. $t \approx 2.3105$ hr

73. $t \approx 40.3817$ days

75. 11.086186

77. *a.* 0.012

b. $P(t) = 2{,}286{,}000 e^{0.012t}$

c. 3,276,591

Chapter 6 Test, page 342

1. $3^{3/2} = \sqrt{27}$

2. $2 = \ln 7.3891$

3. *a.* 1

b. e^{6x-2}

c. e^{6x-5}

4. *a.*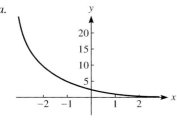

b.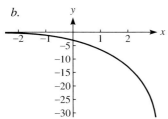

5. *a.* $-\dfrac{3}{2} \log x$

b. $\log \dfrac{xy^3}{z^4}$

c. $\ln(5\sqrt{x})$

6. *a.* 1

b. 2

c. 1

7. *a.*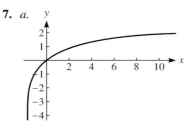

b.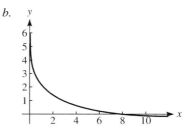

8. $x = (2 + \ln 3)/3 \approx 1.03287$

9. $x = \dfrac{1}{4}$

10. $x = \pm\sqrt{1+e}$

11. *a.* $N(t) = 20 \cdot 2^{t/23}$

b. 40 million, 80 million, 27.03414 million, 31.43069 million

c. 7.4 years

Chapter 7

Section 7.1, page 352

1.

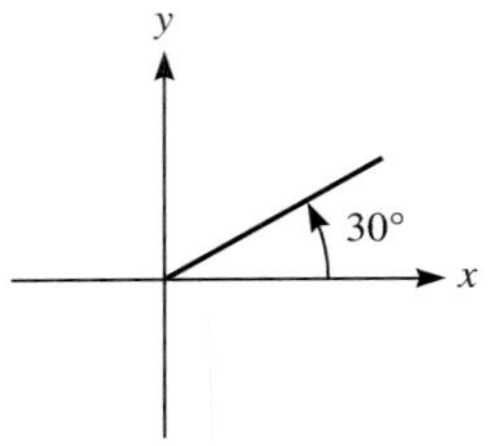

3.

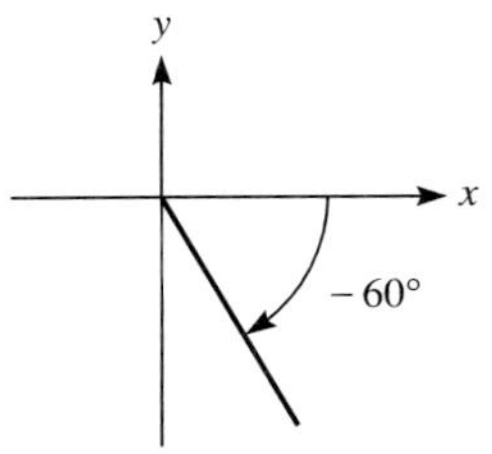

5.

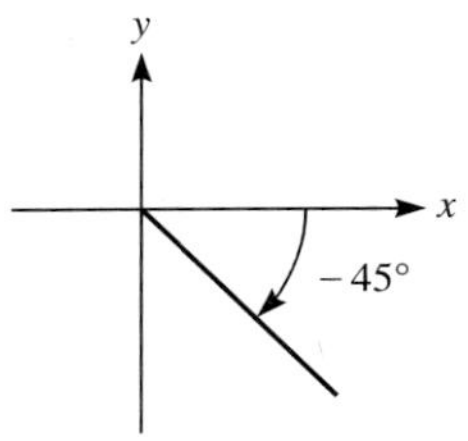

7.

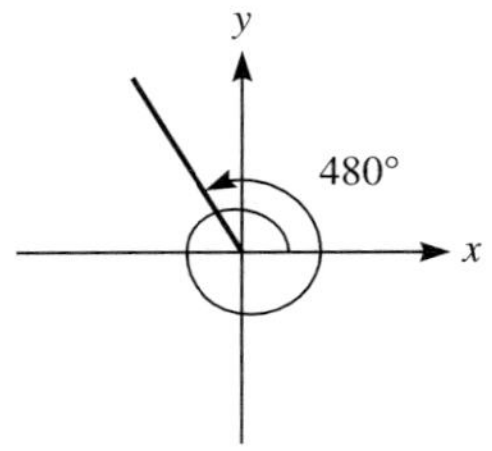

9. *a.* $53°36'$
b. $143°36'$

11. *a.* $33°25'7''$
b. $123°25'7''$

13. $36.4°$

15. $56.5814°$

17. $-35.8°$

19. $142.5667°$

21. $46°19'37''$

23. $-72°15'0''$

25. $364°2'42''$

27. $-1°39'20''$

29. $109°2'36.9912''$

31. $-114°35'29.61''$

33. $\frac{\pi}{5}$

35. $\frac{3\pi}{4}$

37. $-\frac{4\pi}{3}$

39. $\frac{1,528}{1,125}\pi$

41. 1.041855438 radians

43. −1.026544488 radians

45. $150°$

47. $-315°$

49. $201.73844°$

51. $-390°$

53. $135°, 45°$

55. $\frac{2\pi}{5}$ radians; complement measures $18° = \frac{\pi}{10}$ radians.

57. $\frac{5\pi}{9} \approx 1.745329252$

59. 7.5

61. $\frac{5\pi}{4}$

63. $\frac{5\pi}{2}$

65. $\frac{7\pi}{6}$

67. $45°, 45°, 90°; \frac{\pi}{4}, \frac{\pi}{4}, \frac{\pi}{2}$

69. 108.81 in.

71. 3.998 ft

73. 6.27 ft

75. 24.43 m

77. 73,107.692 radians/hr

79. $\frac{6,600}{\pi}$ rev/sec

Section 7.2, page 362

1.

$\sin\theta$	$\cos\theta$	$\tan\theta$	$\cot\theta$	$\sec\theta$	$\csc\theta$
$-\frac{\sqrt{2}}{2}$	$-\frac{\sqrt{2}}{2}$	1	1	$-\sqrt{2}$	$-\sqrt{2}$

3.

$\sin\theta$	$\cos\theta$	$\tan\theta$	$\cot\theta$	$\sec\theta$	$\csc\theta$
$-\frac{1}{2}$	$-\frac{\sqrt{3}}{2}$	$\frac{1}{\sqrt{3}}$	$\sqrt{3}$	$-\frac{2}{\sqrt{3}}$	-2

5.

$\sin\theta$	$\cos\theta$	$\tan\theta$	$\cot\theta$	$\sec\theta$	$\csc\theta$
$-\frac{\sqrt{3}}{2}$	$-\frac{1}{2}$	$\sqrt{3}$	$\frac{1}{\sqrt{3}}$	-2	$-\frac{2\sqrt{3}}{3}$

7.

$\sin\theta$	$\cos\theta$	$\tan\theta$	$\cot\theta$	$\sec\theta$	$\csc\theta$
$-\frac{\sqrt{2}}{2}$	$\frac{\sqrt{2}}{2}$	-1	-1	$\sqrt{2}$	$-\sqrt{2}$

ANSWERS Section 7.2, page 362—Section 7.3, page 368

9.

sin θ	cos θ	tan θ	cot θ	sec θ	csc θ
$-\frac{\sqrt{2}}{2}$	$-\frac{\sqrt{2}}{2}$	1	1	$-\sqrt{2}$	$-\sqrt{2}$

11.

sin θ	cos θ	tan θ	cot θ	sec θ	csc θ
$-\frac{\sqrt{3}}{2}$	$-\frac{1}{2}$	$\sqrt{3}$	$\frac{1}{\sqrt{3}}$	-2	$-\frac{2}{\sqrt{3}}$

13.

sin θ	cos θ	tan θ	cot θ	sec θ	csc θ
0	1	0	undef.	1	undef.

15.

sin θ	cos θ	tan θ	cot θ	sec θ	csc θ
$-\frac{\sqrt{3}}{2}$	$-\frac{1}{2}$	$\sqrt{3}$	$\frac{1}{\sqrt{3}}$	-2	$-\frac{2}{\sqrt{3}}$

17.

sin θ	cos θ	tan θ	cot θ	sec θ	csc θ
$-\frac{\sqrt{2}}{2}$	$\frac{\sqrt{2}}{2}$	-1	-1	$\sqrt{2}$	$-\sqrt{2}$

19.

sin θ	cos θ	tan θ	cot θ	sec θ	csc θ
1	0	undef.	0	undef.	1

21.

sin θ	cos θ	tan θ	cot θ	sec θ	csc θ
$-\frac{1}{2}$	$-\frac{\sqrt{3}}{2}$	$\frac{1}{\sqrt{3}}$	$\sqrt{3}$	$-\frac{2}{\sqrt{3}}$	-2

23.

sin θ	cos θ	tan θ	cot θ	sec θ	csc θ
$\frac{1}{2}$	$-\frac{\sqrt{3}}{2}$	$-\frac{1}{\sqrt{3}}$	$-\sqrt{3}$	$-\frac{2}{\sqrt{3}}$	2

25.

sin θ	cos θ	tan θ	cot θ	sec θ	csc θ
$-\frac{\sqrt{2}}{2}$	$\frac{\sqrt{2}}{2}$	-1	-1	$-\sqrt{2}$	$-\sqrt{2}$

27.

sin θ	cos θ	tan θ	cot θ	sec θ	csc θ
$-\frac{1}{2}$	$\frac{\sqrt{3}}{2}$	$-\frac{1}{\sqrt{3}}$	$-\sqrt{3}$	$\frac{2}{\sqrt{3}}$	-2

29.

sin θ	cos θ	tan θ	cot θ	sec θ	csc θ
0	-1	0	undef.	-1	undef.

31.

sin θ	cos θ	tan θ	cot θ	sec θ	csc θ
$\frac{4}{5}$	$\frac{3}{5}$	$\frac{4}{3}$	$\frac{3}{4}$	$\frac{5}{3}$	$\frac{5}{4}$

33.

sin θ	cos θ	tan θ	cot θ	sec θ	csc θ
$-\frac{12}{13}$	$\frac{5}{13}$	$-\frac{12}{5}$	$-\frac{5}{12}$	$\frac{13}{5}$	$-\frac{13}{12}$

35. 0.79864 **43.** 0.91656 **51.** 0.69109
37. 0.32557 **45.** 0.40015 **53.** 2.66079
39. 0.20466 **47.** 1.19674 **55.** undefined
41. 29.37111 **49.** 0.07567 **57.** 0.84147

Section 7.3, page 368

1.

sin θ	cos θ	tan θ	cot θ	sec θ	csc θ
$\frac{12}{13}$	$\frac{5}{13}$	$\frac{12}{5}$	$\frac{5}{12}$	$\frac{13}{5}$	$\frac{13}{12}$

3.

sin θ	cos θ	tan θ	cot θ	sec θ	csc θ
$\frac{7}{25}$	$\frac{24}{25}$	$\frac{7}{24}$	$\frac{24}{7}$	$\frac{25}{24}$	$\frac{25}{7}$

5.

sin θ	cos θ	tan θ	cot θ	sec θ	csc θ
$\frac{3}{\sqrt{34}}$	$\frac{5}{\sqrt{34}}$	$\frac{3}{5}$	$\frac{5}{3}$	$\frac{\sqrt{34}}{5}$	$\frac{\sqrt{34}}{3}$

7.

sin θ	cos θ	tan θ	cot θ	sec θ	csc θ
$\frac{12}{13}$	$\frac{5}{13}$	$\frac{12}{15}$	$\frac{5}{12}$	$\frac{13}{15}$	$\frac{13}{12}$

9.

sin θ	cos θ	tan θ	cot θ	sec θ	csc θ
$\frac{3}{\sqrt{13}}$	$\frac{2}{\sqrt{13}}$	$\frac{3}{2}$	$\frac{2}{3}$	$\frac{\sqrt{13}}{2}$	$\frac{\sqrt{13}}{3}$

11.

sin θ	cos θ	tan θ	cot θ	sec θ	csc θ
$\frac{a}{\sqrt{1+a^2}}$	$\frac{1}{\sqrt{1+a^2}}$	a	$\frac{1}{a}$	$\sqrt{1+a^2}$	$\frac{\sqrt{1+a^2}}{a}$

A-34 ANSWERS Section 7.3, page 368—Section 7.5, page 386

13.

$\sin\theta$	$\cos\theta$	$\tan\theta$	$\cot\theta$	$\sec\theta$	$\csc\theta$
$\frac{2}{3}$	$\frac{\sqrt{5}}{3}$	$\frac{2}{\sqrt{5}}$	$\frac{\sqrt{5}}{2}$	$\frac{3}{\sqrt{5}}$	$\frac{3}{2}$

15.

$\sin\theta$	$\cos\theta$	$\tan\theta$	$\cot\theta$	$\sec\theta$	$\csc\theta$
$\sqrt{\frac{2}{3}}$	$\frac{\sqrt{3}}{3}$	$\sqrt{2}$	$\frac{1}{\sqrt{2}}$	$\sqrt{3}$	$\sqrt{\frac{3}{2}}$

17.

$\sin\theta$	$\cos\theta$	$\tan\theta$	$\cot\theta$	$\sec\theta$	$\csc\theta$
$\frac{24}{25}$	$\frac{7}{25}$	$\frac{24}{7}$	$\frac{7}{24}$	$\frac{25}{7}$	$\frac{25}{24}$

19.

$\sin\theta$	$\cos\theta$	$\tan\theta$	$\cot\theta$	$\sec\theta$	$\csc\theta$
$\sqrt{1-a^2}$	a	$\frac{\sqrt{1-a^2}}{a}$	$\frac{a}{\sqrt{1-a^2}}$	$\frac{1}{a}$	$\frac{1}{\sqrt{1-a^2}}$

21. 2.34385688
23. 4.22618
25. 54.5698
27. 10.76506
29. 173.7196
31. 36.86990°
33. 39.86834°
35. 36.86990°
37. 17.10464°
39. 60°

41.

$\sin\theta$	$\cos\theta$	$\tan\theta$	$\cot\theta$	$\sec\theta$	$\csc\theta$
$\frac{12}{13}$	$\frac{5}{13}$	$\frac{12}{5}$	$\frac{5}{12}$	$\frac{13}{5}$	$\frac{13}{12}$

43. 79.8510°
45. 81.62185°
47. By approximation, 0.9803999; by calculation, 0.980066578.

Section 7.4, page 374

1. $\alpha = 37°, b = 19.9, c = 24.9$
3. $b = 7.5, \beta = 36.8699°, \alpha = 53.1301°$
5. $c = 7.647876, \beta = 62.765°, \gamma = 90°, \alpha = 27.235°$
7. $a = 30.25, \alpha = 52.75°, \beta = 37.25°, \gamma = 90°$
9. $\beta = 66°20', a = 0.6224, c = 1.5504, \gamma = 90°$
11. $\alpha = 33°52', b = 506.6098, c = 610.1258, \gamma = 90°$
13. $c = 51.522, \alpha = 38.54°, \beta = 51.46°, \gamma = 90°$
15. $\beta = 52°17', \gamma = 90°, b = 1616.34, c = 2,043.29$
17. $\alpha = 30.84°, \beta = 59.16°, \gamma = 90°, b = 54.44$
19. 95.395 ft
21. 272.56 ft
23. 1.11355 m
25. 26,469.4 ft
27. 41.44°
29. 485.57 ft
31. $c = 58.097, b = 43.061$
33. 7,498 m
35. 100.92 ft
37. $5.6 \times 6 = 33.6$ m
39. 10,404.7 ft
41. South: 100 miles, west: 173.2 miles.
43. 110°
45. 356.3 ft
47. 12 ft, 53.1301°

Section 7.5, page 386

1. (b)
3. (c)
5. (h)
7. (d)
9. 2π
11. π
13. 2π
15. a. 2π b. 1
17. a. 4π b. 1

c.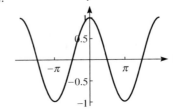

c.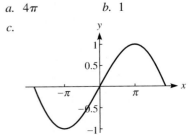

19. a. 1   b. 1
c.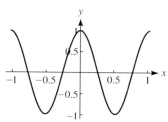

21. a. $\frac{2\pi}{3}$ b. 1
c.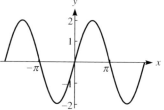

23. a. 2π b. 2
c.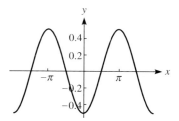

25. a. 2π b. $\frac{1}{2}$
c.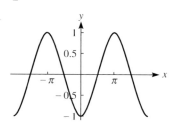

27. a. π b. 2
c.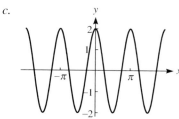

29. a. 1   b. 4
c.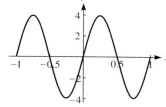

31. a. 2   b. 3
c.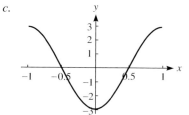

33. a. 3   b. 1
c.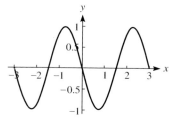

35. Periodic, $\frac{2}{3}$.

37. Periodic, 2π.   **45.** $2\sin 2\pi x$
39. Not periodic.   **47.** $\cos 2x$
41. $2\sin 2x$ **49.** $2\cos x$
43. $3\sin x$ **51.** $2\cos x$

53. a. 2
b. 10 inches
c. 5 and -5
d.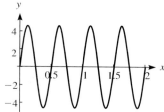

55. The first graph shows six cycles, the second three cycles, and the third two cycles.

57. The second has twice the amplitude of the first.

59. The second graph is the reflection of the first graph in the x-axis.

Section 7.6, page 393

1. a. 2π b. 1
c. $\frac{\pi}{2}$ right
d.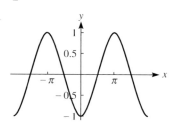

3. a. 2π b. 3
c. π left
d.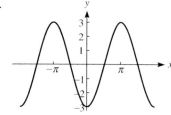

5. a. $\frac{\pi}{2}$ b. 2
c. $\frac{\pi}{4}$ left
d.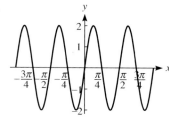

7. a. π
b. 3
c. $\frac{\pi}{6}$ right
d.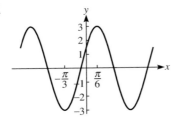

9. a. 1
b. $\frac{3}{2}$
c. $\frac{1}{2\pi}$ left
d.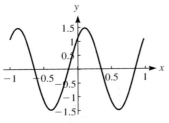

11. a. 2π
b. 1
c. None.
d.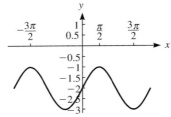

13. a. 2
b. 2
c. None.
d.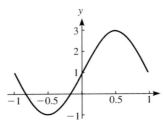

15. a. π
b. 6
c. $\frac{\pi}{2}$ right

17. a. $\frac{1}{2}$
b. 5
c. $\frac{1}{2\pi}$ left

19. a. $\frac{2\pi}{3}$
b. $\frac{2}{3}$
c. $\frac{\pi}{12}$ left

21. $y = 3\sin\left(\frac{\pi}{6}x\right)$

23. $y = 2\cos\left(2x + \frac{\pi}{2}\right)$

25. $y = 4.3\sin\left[\frac{2\pi}{7}x + \frac{20\pi}{7}\right]$

27. a.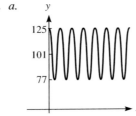
b. 80 beats/minute
c. 77 min, 125 max

29. a. $\frac{2\pi}{\pi/45} = 90$
b. 5,000 miles
c.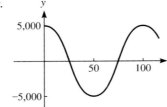

31. a.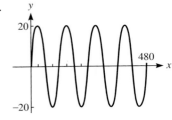
b. max = 20 m, min = −20 m
c. 120

33. a.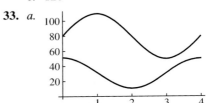
b. 110 hares/mile², 50 hares/mile²
c. 51 lynx/mile², 11 lynx/mile²

35. $p = \frac{2\pi}{\pi/4} = 8$, p.s.: 1 right

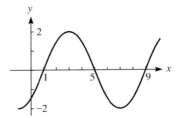

37. period = $\frac{8\pi}{3}$, amplitude = 4, phase shift = 0

39. period = 2π, amplitude = 1.6, phase shift = −1

41. The graph is the same as that of the line $y = 1$. It illustrates graphically the Pythagorean identity for sine and cosine.

43. −1.90, 0, 1.9

Section 7.7, page 402

1. (b)   3. (c)   5. (d)

7.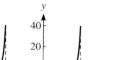

9.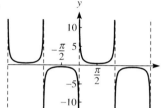

11.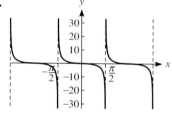

13.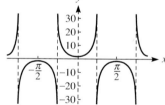

15.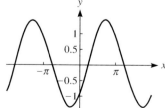

17.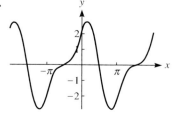

19.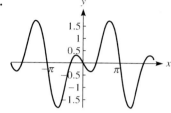

21.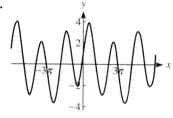

23.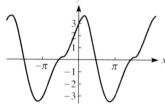

25.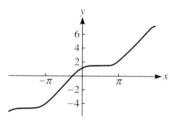

27.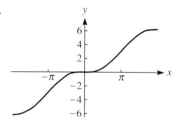

29.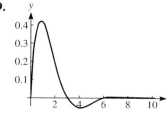

31.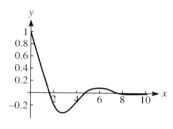

33.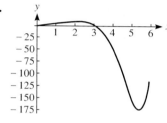

35.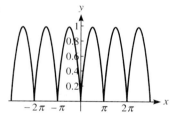

37.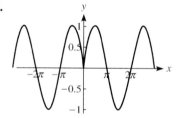

39.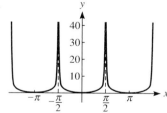

41.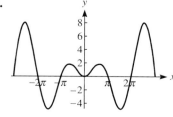

43.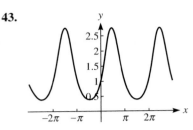

45. a.
$D(0) = 7$
$D(1) = 6.3339$
$D(2) = 5.7311$
$D(3) = 5.1857$
$D(4) = 4.6922$
$D(5) = 4.2457$
$D(6) = 3.8417$
$D(7) = 3.4761$
$D(8) = 3.1453$
$D(9) = 2.846$
$D(10) = 2.572$

b.

c. 7
d. -6.884397

47. a.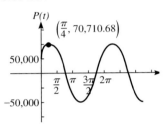

b. Revenue does not cover operating expenses.

49. $x = 3.14159$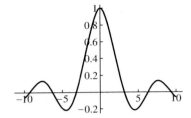

51. $x = 1.25331$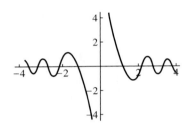

Section 7.8, page 409

1.

$\sin\theta$	$\cos\theta$	$\tan\theta$	$\cot\theta$	$\sec\theta$	$\csc\theta$
$-\dfrac{2}{5}$	$-\dfrac{\sqrt{21}}{5}$	$\dfrac{2}{\sqrt{21}}$	$\dfrac{\sqrt{21}}{2}$	$-\dfrac{5}{\sqrt{21}}$	$-\dfrac{5}{2}$

3.

$\sin\theta$	$\cos\theta$	$\tan\theta$	$\cot\theta$	$\sec\theta$	$\csc\theta$
$\dfrac{\sqrt{5}}{3}$	$-\dfrac{2}{3}$	$-\dfrac{\sqrt{5}}{2}$	$-\dfrac{2}{\sqrt{5}}$	$-\dfrac{3}{2}$	$\dfrac{3}{\sqrt{5}}$

5.

$\sin\theta$	$\cos\theta$	$\tan\theta$	$\cot\theta$	$\sec\theta$	$\csc\theta$
$-\dfrac{\sqrt{5}}{4}$	$\dfrac{\sqrt{11}}{4}$	$-\dfrac{\sqrt{5}}{\sqrt{11}}$	$-\dfrac{\sqrt{11}}{\sqrt{5}}$	$\dfrac{4}{\sqrt{11}}$	$-\dfrac{4}{\sqrt{5}}$

7.

$\sin\theta$	$\cos\theta$	$\tan\theta$	$\cot\theta$	$\sec\theta$	$\csc\theta$
$\dfrac{2}{\sqrt{13}}$	$-\dfrac{3}{\sqrt{13}}$	$-\dfrac{2}{3}$	$-\dfrac{3}{2}$	$-\dfrac{\sqrt{13}}{3}$	$\dfrac{\sqrt{13}}{2}$

9. $2\sin^2 t - 1$
11. $9 - 5\sin^2 t$
13. $2\cos^4 t - 2\cos^2 t + 1$
15. $1 - 2\cos^2 t$
17. $\sin^5 t - 2\sin^3 t + \sin t$
19. $\dfrac{2\cos^2 t - 1}{\cos^2 t - \cos^4 t}$
21. $\cot^4 t + \cot^2 t - 4\cot t - 3$
23. $\dfrac{1}{\tan^3 t} + \tan^3 t$
25. $\cos t + 1 - \cos^2 t$
27. $\left(\dfrac{\sin^2 t - 1}{\sin t}\right)^2$
29. $\tan t$
31. $\dfrac{7\sin^3\theta}{1 - \sin^2\theta}$
33. $\sin x - 1$
35. $\dfrac{1}{\cos\alpha + 5}$
37. 1
39. $\dfrac{\cos y}{5}$
41. 1
43. $\dfrac{1}{4}(7 - 3\cos^2 x)(\cos^2 x + 5)$
45. $\sec^8 x$
47. $1 - \sin^2 t + \sin^4 t$
49. Graph both sides of the identity as separate functions and observe that the two graphs are the same.
51. Graph both sides of the identity as separate functions and observe that the two graphs are the same.

Chapter 7 Review, page 412

1. a. $42°9'$
   b. $132°9'$
3. $47.85°$
5. $-124.5667°$
7. $56°47'20''$
9. $117°2'26''$

11. $\dfrac{5\pi}{12}$ radians
13. $-\dfrac{5}{6}\pi$ radians
15. $\left(\dfrac{180}{\pi}\right)° = 57.296°$
17. $-2{,}340°$

19.

sin θ	cos θ	tan θ	cot θ	sec θ	csc θ
$\dfrac{6}{2\sqrt{34}} = \dfrac{3}{\sqrt{34}}$	$-\dfrac{5}{\sqrt{34}}$	$-\dfrac{3}{5}$	$-\dfrac{5}{3}$	$-\dfrac{\sqrt{34}}{5}$	$\dfrac{\sqrt{34}}{3}$

21.

sin θ	cos θ	tan θ	cot θ	sec θ	csc θ
$\dfrac{2}{7}$	$\dfrac{3\sqrt{5}}{7}$	$\dfrac{2}{3\sqrt{5}}$	$\dfrac{3\sqrt{5}}{2}$	$\dfrac{7}{3\sqrt{5}}$	$\dfrac{7}{2}$

23.

sin θ	cos θ	tan θ	cot θ	sec θ	csc θ
$\dfrac{5}{\sqrt{26}}$	$\dfrac{1}{\sqrt{26}}$	5	$\dfrac{1}{5}$	$\sqrt{26}$	$\dfrac{\sqrt{26}}{5}$

25. 0.5973 radians, or $34.2219°$
27. 1.4706 radians, or $84.2608°$

29.

sin θ	cos θ	tan θ	cot θ	sec θ	csc θ
0	1	0	undef.	1	undef.

31.

sin θ	cos θ	tan θ	cot θ	sec θ	csc θ
$-\dfrac{1}{2}$	$-\dfrac{\sqrt{3}}{2}$	$\dfrac{1}{\sqrt{3}}$	$\sqrt{3}$	$-\dfrac{2}{\sqrt{3}}$	-2

33.

sin θ	cos θ	tan θ	cot θ	sec θ	csc θ
$\dfrac{\sqrt{3}}{2}$	$\dfrac{1}{2}$	$\sqrt{3}$	$\dfrac{1}{\sqrt{3}}$	2	$\dfrac{2}{\sqrt{3}}$

35.

sin θ	cos θ	tan θ	cot θ	sec θ	csc θ
$\dfrac{\sqrt{2}}{2}$	$-\dfrac{\sqrt{2}}{2}$	-1	-1	$-\dfrac{2}{\sqrt{2}}$	$\dfrac{2}{\sqrt{2}}$

37. $(2n-1)\pi$, n an integer.
39. $\dfrac{\pi}{6} + 2k\pi;\ \dfrac{5\pi}{6} + 2k\pi$, k an integer.
41. $20 + 9\cos t - 20\cos^2 t$
43. $\dfrac{\sin^4 t}{1 - \sin^2 t}$
45. $\dfrac{-1}{\sec t(\sec t + 1)}$
47. $\cos^2 x$
49. $\cos^2 \theta$
51. 0
53. $\cot^2 x$
55. $0,\ 1$
57. ± 1
59. 1
61. $\dfrac{\sqrt{25 - \cos^2 \theta}}{5 + \cos \theta}$
63. Domain of $\sin \theta$, $\cos \theta$: all real numbers.
    Domain of $\tan \theta$, $\sec \theta$: $x \neq \dfrac{2k+1}{2}\pi$, $k = 0, \pm 1, \pm 2, \ldots$
    Domain of $\csc \theta$, $\cot \theta$: $x \neq k\pi$, $k = 0, \pm 1, \pm 2, \ldots$
65. 1
67. a. 2π b. $\dfrac{\pi}{2}$ left   c. 3
    d.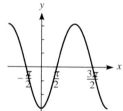

69. a. π b. $\dfrac{\pi}{4}$ right   c. 3
    d.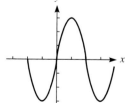

71.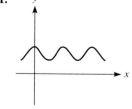

73.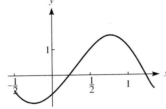

75.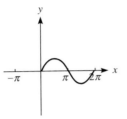

77. $y = -4\cos(\pi x)$

79. $y = 0.3\sin\left(\dfrac{\pi}{6}x + \dfrac{2\pi}{3}\right)$

81. $\dfrac{15}{\pi}$ cm

83. 0.785 radians

85. Angular speed is 248,914.29 radians/hr.

87. 28 ft

89. 70.5°

91. Distance west: 75 miles; distance north: 129.9 miles.

93. 1.11 miles

95. a.

t	$14 - 8\cos(39\pi t)$	t	$14 - 8\cos(39\pi t)$
0	6 in.	1	22 in.
0.1	6.392 in.	2	6 in.
0.2	7.528 in.	3	22 in.
0.4	11.528 in.	4	6 in.
0.5	14 in.	10	6 in.

b.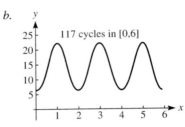

c. 14.25 in.

Chapter 7 Test, page 414

1. a. 41.58667°

b. $\dfrac{2\pi}{9}$

c. 220°

2. 32.46 radians

3. 7.85398 ft

4. $\sin t = \dfrac{1}{2},\ \cos t = -\dfrac{\sqrt{3}}{2},$
$\tan t = -\dfrac{\sqrt{3}}{3},\ \sec t = -\dfrac{2\sqrt{3}}{3},$
$\csc t = 2,\ \cot t = -\sqrt{3}$

5. $\cos t = \dfrac{2\sqrt{2}}{3},\ \tan t = \dfrac{\sqrt{2}}{4},\ \sec t = \dfrac{3\sqrt{2}}{4},\ \csc t = 3,\ \cot t = 2\sqrt{2}$

6. 132.27928 yd

7. Hypotenuse $= 13$, 22.61986°, 67.38014°

8. a.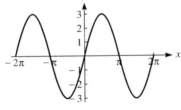

b.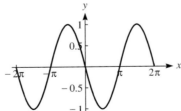

c.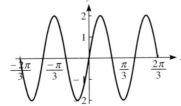

9. amplitude $= 5$, phase angle $= -\pi/3$

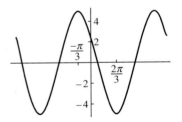

10. a. 2.5 bounces per second.
    b. 20 inches.
    c. 10 inches above and below the starting position.
    d.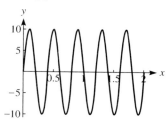

11. a.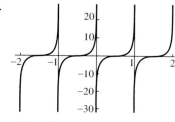
    b. All real numbers except for integers.
    c. All real numbers.

12. a.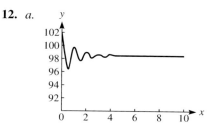
    b. 102.1°

Chapter 8

Section 8.1, page 423

1. 1
3. $\sin^2 x$
5. $\dfrac{1 - \sin^2 x}{\sin x}$
7. $\dfrac{1}{\sin^2 x}$
9. $\sin^2 x$
11. $\dfrac{1 - \cos^2 x}{\cos x}$
13. $\dfrac{1}{\cos^2 x}$
15. $\dfrac{1}{1 - \cos^2 x}$
17. $-\dfrac{1}{\cos^2 x}$
19. $\tan^2 x$
21. $\cot x$
23. $\cos x$
25. $|a| \sec x$
27. $\dfrac{5(1 - \cos^2 x)}{\cos x}$
29. $|a| \tan x$
31. $4 \sec x$

33. $\cos^2 a + \cos^2 a \tan^2 a = \cos^2 a + \cos^2 a \dfrac{\sin^2 a}{\cos^2 a}$
$= \cos^2 a + \sin^2 a$
$= 1$

35. $\cos^2 y - \sin^2 y = (1 - \sin^2 y) - \sin^2 y$
$= 1 - 2\sin^2 y$

37. $\dfrac{\tan a - 1}{\tan a + 1} = \dfrac{\dfrac{\sin a}{\cos a} - 1}{\dfrac{\sin a}{\cos a} + 1} = \dfrac{\sin a - \cos a}{\sin a + \cos a}$
$= \dfrac{\sin a - \cos a}{\sin a + \cos a} \cdot \dfrac{\dfrac{1}{\sin a}}{\dfrac{1}{\sin a}} = \dfrac{1 - \dfrac{\cos a}{\sin a}}{1 + \dfrac{\cos a}{\sin a}}$
$= \dfrac{1 - \cot a}{1 + \cot a}$

39. $\dfrac{\cos \alpha}{\cos \alpha - \sin \alpha} = \dfrac{\cos \alpha}{\cos \alpha - \sin \alpha} \cdot \dfrac{\dfrac{1}{\cos \alpha}}{\dfrac{1}{\cos \alpha}}$
$= \dfrac{1}{1 - \dfrac{\sin \alpha}{\cos \alpha}}$
$= \dfrac{1}{1 - \tan \alpha}$

41. $\dfrac{\cos \alpha + \sin \alpha}{\cos \alpha} = \dfrac{\cos \alpha}{\cos \alpha} + \dfrac{\sin \alpha}{\cos \alpha}$
$= 1 + \tan \alpha$

43. $\dfrac{1 + \csc x}{\cot x + \cos x} = \dfrac{1 + \dfrac{1}{\sin x}}{\dfrac{\cos x}{\sin x} + \cos x}$
$= \dfrac{\sin x + 1}{\cos x(1 + \sin x)}$
$= \dfrac{1}{\cos x}$
$= \sec x$

45. $\dfrac{\cos x}{\sec x - \tan x} = \dfrac{\cos x}{\dfrac{1}{\cos x} - \dfrac{\sin x}{\cos x}}$
$= \dfrac{\cos^2 x}{1 - \sin x}$
$= \dfrac{1 - \sin^2 x}{1 - \sin x}$
$= \dfrac{(1 + \sin x)(1 - \sin x)}{1 - \sin x}$
$= 1 + \sin x$

47. $\dfrac{1-\sin\alpha}{1+\sin\alpha} - \dfrac{1+\sin\alpha}{1-\sin\alpha} = \dfrac{(1-\sin\alpha)^2 - (1+\sin\alpha)^2}{1-\sin^2\alpha}$

$= \dfrac{1 - 2\sin\alpha + \sin^2\alpha - (1 + 2\sin\alpha + \sin^2\alpha)}{\cos^2\alpha}$

$= \dfrac{-4\sin\alpha}{\cos^2\alpha}$

$= -4 \dfrac{1}{\cos\alpha} \cdot \dfrac{\sin\alpha}{\cos\alpha}$

$= -4\sec\alpha\tan\alpha$

49. $\dfrac{\csc^2\theta - \cot^2\theta\,\csc^2\theta}{\cot^2\theta} = \dfrac{\dfrac{1}{\sin^2\theta} - \dfrac{\cos^2\theta}{\sin^2\theta} \cdot \dfrac{1}{\sin^2\theta}}{\dfrac{\cos^2\theta}{\sin^2\theta}}$

$= \dfrac{\sin^2\theta - \cos^2\theta}{\cos^2\theta \sin^2\theta}$

$= \dfrac{1}{\cos^2\theta} - \dfrac{1}{\sin^2\theta}$

$= \sec^2\theta - \csc^2\theta$

51. $\dfrac{1+\cos\theta}{\sin\theta} + \dfrac{\sin\theta}{1+\cos\theta} = \dfrac{(1+\cos\theta)^2 + \sin^2\theta}{\sin\theta(1+\cos\theta)}$

$= \dfrac{1 + 2\cos\theta + \cos^2\theta + \sin^2\theta}{\sin\theta(1+\cos\theta)}$

$= \dfrac{1 + 2\cos\theta + 1}{\sin\theta(1+\cos\theta)}$

$= \dfrac{2}{\sin\theta}$

$= 2\csc\theta$

53. $\dfrac{\csc(-x)}{\cot(-x)} = \dfrac{\dfrac{1}{\sin(-x)}}{\dfrac{\cos(-x)}{\sin(-x)}}$

$= \dfrac{-\dfrac{1}{\sin x}}{\dfrac{\cos x}{-\sin x}}$

$= \dfrac{1}{\cos x}$

55. $\tan^2\beta - \sin^2\beta = \dfrac{\sin^2\beta}{\cos^2\beta} - \sin^2\beta$

$= \sin^2\beta\left(\dfrac{1}{\cos^2\beta} - 1\right)$

$= \sin^2\beta\,\dfrac{1 - \cos^2\beta}{\cos^2\beta}$

$= \sin^2\beta\,\dfrac{\sin^2\beta}{\cos^2\beta}$

$= (\sin\beta\tan\beta)^2$

57. $\dfrac{\sin^2 x}{1-\cos x} = \dfrac{1-\cos^2 x}{1-\cos x} = 1 + \cos x$

59. $\sec^2 x - 2\sec x\cos x + \cos^2 x$

$= \dfrac{1}{\cos^2 x} - 2 \cdot \dfrac{1}{\cos x} \cdot \cos x + \cos^2 x$

$= \dfrac{1}{\cos^2 x} - 2 + 1 - \sin^2 x$

$= \dfrac{1 - \cos^2 x}{\cos^2 x} - \sin^2 x$

$= \dfrac{\sin^2 x}{\cos^2 x} - \sin^2 x$

$= \tan^2 x - \sin^2 x$

61. $\dfrac{\sin x}{1+\cos x} = \dfrac{\sin x(1-\cos x)}{1-\cos^2 x}$

$= \dfrac{\sin x(1-\cos x)}{\sin^2 x}$

$= \dfrac{1-\cos x}{\sin x}$

$= \dfrac{1}{\sin x} - \dfrac{\cos x}{\sin x}$

$= \csc x - \cot x$

63. No. $x = \dfrac{3\pi}{2}$

65. No. $x = \dfrac{\pi}{4}$

67. Yes.

69. Yes.

71. No. $x = \dfrac{\pi}{4}$

73. $2\sin x \cos x = \sin 2x$

75. $-4\sin^3 x + 3\sin x = \sin 3x$

Section 8.2, page 434

1. $\dfrac{\sqrt{2} - \sqrt{6}}{4}$

3. $-2 - \sqrt{3}$

5. $-\dfrac{\sqrt{2} + \sqrt{6}}{4}$

7. $\dfrac{\sqrt{2} + \sqrt{6}}{4}$

9. $2 - \sqrt{3}$

11. $\dfrac{\sqrt{2} - \sqrt{6}}{4}$

13. $-\sin\theta$

15. $\dfrac{\sqrt{3}}{2}\cos\theta + \dfrac{1}{2}\sin\theta$

17. $\dfrac{\tan\theta - 1}{\tan\theta + 1}$

19. $\dfrac{\sqrt{3}}{2}\cos\theta + \dfrac{1}{2}\sin\theta$

21. $\dfrac{\sqrt{2}}{2}\cos\theta - \dfrac{\sqrt{2}}{2}\sin\theta$

23. $\dfrac{\tan\theta + 1}{1 - \tan\theta}$

25. $-\dfrac{\sqrt{3}}{2}$

27. $\dfrac{\sqrt{2} + \sqrt{6}}{4}$ or $\dfrac{\sqrt{2} - \sqrt{6}}{4}$

29. 0

31. $-2 - \sqrt{3}$

33. $\dfrac{-\sqrt{6} - \sqrt{2}}{4}$

35. 0

37. $\dfrac{7}{25}$

39. $2 + \sqrt{3}$

41. 0

43. $-\dfrac{24}{25}$

45. $-\dfrac{120}{119}$

47. 7

49. Undefined.

51. $\dfrac{14}{3}$

53. 0.3080528 radians

55. $\tan\left(\dfrac{\pi}{2} - x\right) = \dfrac{\sin\left(\dfrac{\pi}{2} - x\right)}{\cos\left(\dfrac{\pi}{2} - x\right)}$
$= \dfrac{\cos x}{\sin x} = \cot x$

57. $\sec\left(\dfrac{\pi}{2} - x\right) = \dfrac{1}{\cos\left(\dfrac{\pi}{2} - x\right)}$
$= \dfrac{1}{\sin x} = \csc x$

59. $\sqrt{2}\sin\left(x + \dfrac{\pi}{4}\right)$

61. $2\sin\left(x + \dfrac{\pi}{3}\right)$

63. $\sqrt{194}\sin\left[x + \sin^{-1}\left(-\dfrac{13}{\sqrt{194}}\right)\right]$

65. $2\sin\left(\dfrac{\pi}{4}t + \dfrac{5\pi}{3}\right)$

67. $\dfrac{\sin(x + h) - \sin x}{h} = \dfrac{\sin x \cos h + \cos x \sin h - \sin x}{h}$
$= \dfrac{\sin x(\cos h - 1) + \cos x \sin h}{h}$

69. $\dfrac{\tan(x + h) - \tan x}{h} = \dfrac{\dfrac{\tan x + \tan h}{1 - \tan x \tan h} - \tan x}{h}$
$= \dfrac{\tan x + \tan h - \tan h - \tan^2 x \tan h}{h(1 - \tan x \tan h)}$
$= \dfrac{\tan h(1 - \tan^2 x)}{h(1 - \tan x \tan h)}$
$= \dfrac{\sin h}{h} \cdot \dfrac{\sec^2 h}{\cos h - \tan x \sin h}$

71. $2\cos\alpha\cos\theta - \cos(\alpha - \theta) = 2\cos\alpha\cos\theta -$
$(\cos\alpha\cos\theta + \sin\alpha\sin\theta)$
$= \cos\alpha\cos\theta - \sin\alpha\sin\theta$
$= \cos(\alpha + \theta)$

73. $\sin(x + y) - \sin(x - y) = \sin x \cos y + \sin y \cos x -$
$(\sin x \cos y - \sin y \cos x)$
$= 2\sin y \cos x$

75. $\cos(x + y) - \cos(x - y) = [\cos x \cos y - \sin x \sin y] -$
$[\cos x \cos y + \sin x \sin y]$
$= -2\sin x \sin y$

77. $\sin\left(\dfrac{\pi}{4} + x\right) - \sin\left(\dfrac{\pi}{4} - x\right) =$
$\sin\dfrac{\pi}{4}\cos x + \cos\dfrac{\pi}{4}\sin x - \left[\sin\dfrac{\pi}{4}\cos x - \cos\dfrac{\pi}{4}\sin x\right]$
$= 2\cos\dfrac{\pi}{4}\sin x$
$= \sqrt{2}\sin x$

Section 8.3, page 448

1. a. $\dfrac{\sqrt{3}}{2}$
   b. $\dfrac{1}{2}$
   c. $\sqrt{3}$

3. a. -0.96
   b. 0.28
   c. $-\dfrac{24}{7}$

5. a. $\dfrac{\sqrt{3}}{2}$
   b. $-\dfrac{1}{2}$
   c. $-\sqrt{3}$

7. a. -1
   b. 0
   c. Undefined.

9. $4\cos^3 x - 3\cos x$

11. $-4\sin x \cos x(-\cos^2 x + \sin^2 x)$

13. a. $\dfrac{1}{\sqrt{10}}$
    b. $\dfrac{3}{\sqrt{10}}$
    c. $\dfrac{1}{3}$

15. a. $\dfrac{\sqrt{2 + \sqrt{2}}}{2}$
    b. $-\dfrac{\sqrt{2 - \sqrt{2}}}{2}$
    c. $\dfrac{\sqrt{2}}{\sqrt{2} - 2}$

17. a. $\dfrac{\sqrt{10}}{4}$

b. $-\dfrac{\sqrt{6}}{4}$

c. $-\dfrac{\sqrt{15}}{3}$

19. a. $\sqrt{\dfrac{3-\sqrt{5}}{6}}$

b. $-\sqrt{\dfrac{3+\sqrt{5}}{6}}$

c. $-\dfrac{2}{3+\sqrt{5}}$

21. $-\dfrac{1}{2}$

23. $-\dfrac{\sqrt{7}}{4}$

25. $\cos 40°$

27. $\tan \dfrac{2\pi}{5}$

29. $\tan 70°$

31. $-\sin \dfrac{\pi}{8}$

33. $\cos 7x$

35. $\tan 135°$

37. 1

39. 1

41. $\sec x$

43. $\tan x \sin 2x = 1 - \cos 2x$
$\tan x \sin 2x = \dfrac{\sin x}{\cos x} \cdot 2\sin x \cos x = 2\sin^2 x + 1$
$= 1 - \cos 2x$

45. $\csc x - \tan \dfrac{x}{2} = \cot x$
$\csc x - \tan \dfrac{x}{2} = \dfrac{1}{\sin x} -$
$\dfrac{1-\cos x}{\sin x} = \dfrac{\cos x}{\sin x} = \cot x$

47. $\tan x + \dfrac{\sin 3x}{\cos x} = \dfrac{\sin x}{\cos x} + \dfrac{\sin(2x+x)}{\cos x}$
$= \dfrac{\sin x}{\cos x} + \dfrac{\sin 2x \cos x + \cos 2x \sin x}{\cos x}$
$= \sin 2x + \dfrac{\sin x(\cos 2x + 1)}{\cos x}$
$= \sin 2x + \dfrac{\sin x \cdot 2\cos^2 x}{\cos x}$
$= \sin 2x + \sin 2x$
$= 2\sin 2x$

49. $\dfrac{\sec^3 x}{2-\sec^2 x} = \sec 2x$
$\dfrac{\sec^2 x}{2-\sec^2 x} = \dfrac{\frac{1}{\cos^2 x}}{2-\frac{1}{\cos^2 x}} = \dfrac{1}{2\cos^2 x - 1} = \dfrac{1}{\cos(2x)}$
$= \sec(2x)$

51. $2\csc 2x = \tan x + \cot x \;\; \tan x + \cot x = \dfrac{\sin x}{\cos x} +$
$\dfrac{\cos x}{\sin x} = \dfrac{\sin^2 x + \cos^2 x}{\cos x \sin x} = \dfrac{1}{\cos x \sin x} = \dfrac{2}{2\cos x \sin x}$
$= 2\csc 2x$

53. $\cos^2\left(\dfrac{x}{2}\right) = \dfrac{\sin x + \tan x}{2\tan x}$
$\cos^2\left(\dfrac{x}{2}\right) = \dfrac{1+\cos x}{2} = \dfrac{\sin x + \tan x}{2\tan x}$

55. $\tan 2x = \dfrac{2}{\cot x \tan x}$
$\tan 2x = \dfrac{2\tan x}{1-\tan^2 x} = \dfrac{2}{\cot x - \tan x}$

57. $4\sin x \cos x \cos 2x = \sin 4x$
$\sin 4x = 2\sin 2x \cos 2x = 4\sin x \cos x \cos 2x$

59. $\tan \dfrac{x}{2} = \dfrac{\sin 2x - \sin x}{\cos 2x + \cos x}$
$\dfrac{\sin 2x - \sin x}{\cos 2x + \cos x} = \dfrac{2\sin x \cos x - \sin x}{2\cos^2 x - 1 + \cos x}$
$= \dfrac{\sin x(2\cos x - 1)}{(2\cos x - 1)(\cos x + 1)} = \tan \dfrac{x}{2}$

61. $\csc x = \cot x + \tan \dfrac{x}{2}$
$\cot x + \tan \dfrac{x}{2} = \dfrac{\cos x}{\sin x} + \dfrac{1-\cos x}{\sin x} = \dfrac{1}{\sin x} = \csc x$

63. $\dfrac{1-\sin^2 x}{1-\cos^2 x} = \cot^2 x$
$\dfrac{1-\sin^2 x}{1-\cos^2 x} = \dfrac{\cos^2 x}{\sin^2 x} = \cot^2 x$

65. $\csc x \sin 4x - \cos 3x = \sin 3x \cot x$
$\csc x \sin 4x - \cos 3x$
$= \csc x(\sin(3x+x)) - \cos 3x$
$= \csc x[\sin 3x \cos x + \cos 3x \sin x] - \cos 3x$
$= \sin 3x \cot x + \cos 3x - \cos 3x = \sin 3x \cot x$

67. $\dfrac{\sin 5x + \sin 3x}{\sin 5x - \sin 3x} = \dfrac{2\sin 4x \cos x}{2\sin x \cos 4x} = \dfrac{\tan 4x}{\tan x}$

69. $\dfrac{\sin 2x - \sin x}{\cos 2x \cos x} = \dfrac{2\sin\frac{x}{2}\cos\frac{3x}{2}}{2\cos\frac{3x}{2}\cos\frac{x}{2}} = \tan \dfrac{x}{2}$

71. $\dfrac{\sin 4x - \sin 2x}{\cos 4x + \cos 2x} = \dfrac{2\sin x \cos 3x}{2\cos 3x \cos x} = \tan x$

73. $\sin 2x \sin 6x = \dfrac{1}{2}[\cos(4x) - \cos(8x)] =$
$\dfrac{1}{2}[1 - 2\sin^2 2x - (1 - 2\sin^2 4x)] = \sin^2 4x - \sin^2 2x$

75. $\dfrac{\sin x}{\sec 2x} + \dfrac{\cos x}{\csc 2x} = \sin x \cos 2x + \cos x \sin 2x$
$= \sin(x+2x) = \sin 3x$

77. $-\dfrac{\cos x - \cos y}{\cos x + \cos y} = \dfrac{2\sin\left(\frac{x+y}{2}\right)\sin\left(\frac{x-y}{2}\right)}{2\cos\left(\frac{x+y}{2}\right)\cos\left(\frac{x-y}{2}\right)}$
$= \tan\left(\dfrac{x+y}{2}\right)\tan\left(\dfrac{x-y}{2}\right)$

79. $\dfrac{\cos 2x - \cos 6x}{\sin 6x - \sin 2x} = \dfrac{-2\sin 4x \sin(-2x)}{2\sin 2x \cos 4x}$
$= \dfrac{\sin 4x \sin 2x}{\cos 4x \sin 2x}$
$= \tan 4x$

81. $\dfrac{1}{2}[\sin 7x + \sin x]$

83. $\dfrac{1}{2}[\sin 2x - \sin 2y]$

85. $2\sin 8y \cos y$

87. $2\sin y \cos x$

89. $2\cos\dfrac{3x}{2}\cos\dfrac{x}{2}$

Section 8.4, page 454

1. $\dfrac{\pi}{6}$
3. $\dfrac{\pi}{4}$
5. $-\dfrac{\pi}{3}$
7. $\dfrac{\pi}{3}$
9. $\dfrac{\pi}{4}$
11. $\dfrac{5\pi}{6}$
13. $\dfrac{3\pi}{4}$
15. $\dfrac{\pi}{6}$
17. $\dfrac{2}{\sqrt{3}}$
19. $\sqrt{3}$
21. $\dfrac{\sqrt{3}}{2}$
23. $\dfrac{\sqrt{3}}{2}$
25. $\dfrac{\sqrt{34}}{3}$
27. $\dfrac{2\pi}{3}$
29. $\dfrac{1}{2}$

31. $\tan^{-1}(-x) = -\tan^{-1} x$
Let $\theta = \tan^{-1} x$. Then $x = \tan \theta$. Now $-x = \tan(-\theta) \Rightarrow \tan^{-1}(-x) = -\theta = -\tan^{-1} x$.

33. $\sin(\cos^{-1} x) = \sqrt{1 - x^2}$ for $|x| \le 1$.
Let $\theta = \cos^{-1} x$. Then $\cos\theta = x$, so $\sin\theta = \sin(\cos^{-1} x) = \sqrt{1 - x^2}$.

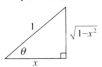

35. $\sin^{-1} x + \cos^{-1} x = \pi/2$ for $|x| \le 1$.
From Exercise 34, $\sin(\sin^{-1} x + \cos^{-1} x) = 1$, so $\sin^{-1}\sin(\sin^{-1} x + \cos^{-1} x) = \sin^{-1} 1 \Rightarrow \sin^{-1} x + \cos^{-1} x = \pi/2$.

37. $\cos[\sin^{-1}(-x)] = \cos(\sin^{-1} x)$ for $0 \le x \le 1$.
Let $\theta_1 = \sin^{-1} x$ and $\theta_2 = \sin^{-1}(-x) = -\theta_1$. Now $\cos(\theta_1) = \cos(-\theta_1) = \cos\theta_2 \Rightarrow \cos[\sin^{-1}(-x)] = \cos(\sin^{-1} x)$.

39. $\tan^{-1} x + \cot^{-1} x = \pi/2$ for $x \ge 0$.
Note $\sin(\tan^{-1} x + \cot^{-1} x)$
$= \sin(\theta_1 + \theta_2)$
$= \sin\theta_1 \cos\theta_2 + \sin\theta_2 \cos\theta_1$
$= \dfrac{x}{\sqrt{x^2+1}} \cdot \dfrac{x}{\sqrt{x^2+1}} + \dfrac{1}{\sqrt{x^2+1}} \cdot \dfrac{1}{\sqrt{x^2+1}}$
$= \dfrac{x^2+1}{x^2+1} = 1$
Now $\sin^{-1}\sin(\tan^{-1} x + \cot^{-1} x) = \tan^{-1} x + \cot^{-1} x = \sin^{-1} 1 = \pi/2$.

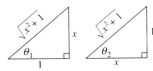

41. $\sin^{-1} x = \tan^{-1} \dfrac{x}{\sqrt{1-x^2}}$ for $|x| < 1$.
Let $\theta_1 = \sin^{-1} x$, $\theta_2 = \tan^{-1} \dfrac{x}{\sqrt{1-x^2}}$
Triangles are congruent; hence $\theta_1 = \theta_2$.

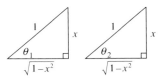

43. $\cot^{-1} x = \dfrac{\pi}{2} - \tan^{-1} x$. Clear from Exercise 39.

45. $\sin \dfrac{y}{2} - 1$
47. $1 - \tan 2y$
49. $\dfrac{\csc(-y/4) - 1}{2}$
51. $\cos\left(-\dfrac{3}{2} y\right) + 1$
53. $\dfrac{2 - \sec y}{3}$
55. $\dfrac{x}{\sqrt{4 + x^2}}$

57. $\dfrac{\sqrt{1 - x^2}}{x}$
59. $\dfrac{\sqrt{x^2 + 2}}{x}$
61. $\dfrac{3}{\sqrt{9 + x^2}}$
63. $\dfrac{\sqrt{x^2 - 1}}{x}$
65. $2x^2 - 1$
67. 0

69. $\sin y = -x \quad \cos y = \sqrt{1 - x^2} \quad \tan y = -\dfrac{x}{\sqrt{1 - x^2}}$
$\sec y = \dfrac{1}{\sqrt{1 - x^2}} \quad \csc y = -\dfrac{1}{x} \quad \cot y = -\dfrac{\sqrt{1 - x^2}}{x}$

71. $\sin y = \sqrt{1 - 4x^2} \quad \cos y = 2x \quad \tan y = \dfrac{\sqrt{1 - 4x^2}}{2x}$
$\sec y = \dfrac{1}{2x} \quad \csc y = \dfrac{1}{\sqrt{1 - 4x^2}} \quad \cot y = \dfrac{2x}{\sqrt{1 - 4x^2}}$

73. $\sin y = \dfrac{1}{x} \quad \cos y = \dfrac{\sqrt{x^2 - 1}}{x} \quad \tan y = \dfrac{1}{\sqrt{x^2 - 1}}$
$\sec y = \dfrac{x}{\sqrt{x^2 - 1}} \quad \csc y = x \quad \cot y = \sqrt{x^2 - 1}$

75. $\sin y = \dfrac{x}{\sqrt{4 + x^2}} \quad \cos y = \dfrac{2}{\sqrt{4 + x^2}} \quad \tan y = \dfrac{x}{2}$
$\sec y = \dfrac{\sqrt{4 + x^2}}{2} \quad \csc y = \dfrac{\sqrt{4 + x^2}}{x} \quad \cot y = \dfrac{2}{x}$

77.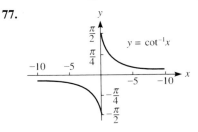

79.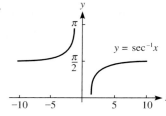

81. Set $x = 1$.

83. Set $x = 0$.

85. Set $x = \dfrac{1}{2}$.

87. Domain: $-1 \le x \le 1$; range: $-\dfrac{\pi}{2} \le y \le \dfrac{\pi}{2}$

89. Domain: all real numbers; range: $-\dfrac{\pi}{2} < y < \dfrac{\pi}{2}$

91. Domain: $-1 \le x \le 1$; range: $0 \le y \le \dfrac{\pi}{2}$

Section 8.5, page 459

1. $\{\pi k : k = 0, \pm 1, \pm 2, \pm 3, \ldots\}$

3. $\{\pi k : k = 0, \pm 1, \pm 2, \pm 3, \ldots\}$

5. No solution.

7. $\left\{\dfrac{3\pi}{2} + 2\pi k : k = 0, \pm 1, \pm 2, \pm 3, \ldots\right\}$

9. $\left\{\dfrac{\pi}{4} + \pi k : k = 0, \pm 1, \pm 2, \pm 3, \ldots\right\}$

11. $\left\{\dfrac{\pi}{3} + 2\pi k, \dfrac{5\pi}{3} + 2\pi k : k = 0, \pm 1, \pm 2, \ldots\right\}$

13. $\left\{\dfrac{5\pi}{6} + \pi k : k = 0, \pm 1, \pm 2, \ldots\right\}$

15. $\left\{\dfrac{2\pi}{3} + 2\pi k, \dfrac{4\pi}{3} + 2\pi k : k = 0, \pm 1, \pm 2, \ldots\right\}$

17. $\{1.01804 + \pi k : k = 0, \pm 1, \pm 2, \ldots\}$

19. $\{-0.1279965 + \pi k, 3.013596 + \pi k : k = 0, \pm 1, \pm 2, \ldots\}$

21. No solutions.

23. $\left\{-\dfrac{\pi}{6}, \dfrac{\pi}{6}\right\}$

25. $\left\{-\dfrac{\pi}{12}, \dfrac{\pi}{4}\right\}$

27. $\left\{-\dfrac{5\pi}{12}, -\dfrac{\pi}{12}\right\}$

29. No solutions.

31. $\left\{\dfrac{11\pi}{6}, \dfrac{\pi}{6}\right\}$

33. $\left\{0, \dfrac{\pi}{4}, \pi, \dfrac{5\pi}{4}\right\}$

35. $\left\{0, \dfrac{\pi}{2}, \dfrac{3\pi}{2}, \pi\right\}$

37. $\left\{\dfrac{2\pi}{3}, \dfrac{4\pi}{3}\right\}$

39. $\left\{\dfrac{\pi}{3}, \dfrac{5\pi}{3}, 0\right\}$

41. $\left\{\dfrac{\pi}{2}, \dfrac{7\pi}{6}, \dfrac{11\pi}{6}\right\}$

43. $\left\{\dfrac{3\pi}{2}\right\}$

45. $\left\{0, \dfrac{\pi}{3}, \dfrac{2\pi}{3}, \pi, \dfrac{4\pi}{3}, \dfrac{5\pi}{3}\right\}$

47. $\left\{\dfrac{\pi}{3}, \dfrac{5\pi}{3}\right\}$

49. $\left\{\dfrac{\pi}{12}, \dfrac{5\pi}{12}, \dfrac{13\pi}{12}, \dfrac{17\pi}{12}\right\}$

51. $\{0\}$

53. 0.98279, 4.12439, 1.76819, 4.90978

55. 1.41182, 4.87137

57. $0, \dfrac{\pi}{2}, \pi, \dfrac{3\pi}{2}$

59. $\dfrac{\pi}{6}, \dfrac{7\pi}{6}$

61. Equation is true for all x in $[0, 2\pi)$ except for $x = 0, \dfrac{\pi}{2}, \pi, \dfrac{3\pi}{2}$.

63. $\dfrac{\pi}{3}, \dfrac{\pi}{2}, \dfrac{3\pi}{2}, \dfrac{5\pi}{3}$

65. $0, \dfrac{\pi}{2}, \pi, \dfrac{3\pi}{2}$

67. $\dfrac{\pi}{2}, \dfrac{3\pi}{2}, \dfrac{\pi}{6}, \dfrac{5\pi}{6}$

69. $\left\{0, \dfrac{2\pi}{3}, \dfrac{4\pi}{3}, \dfrac{\pi}{6}, \dfrac{5\pi}{6}, \dfrac{3\pi}{2}\right\}$

71. $\dfrac{\pi}{6}, \dfrac{5\pi}{6}, \dfrac{7\pi}{6}, \dfrac{11\pi}{6}$

73. $\dfrac{7\pi}{6}, \dfrac{11\pi}{6}, \dfrac{\pi}{4}, \dfrac{5\pi}{4}$

75. $\dfrac{\pi}{4}, \dfrac{3\pi}{4}, \dfrac{5\pi}{4}, \dfrac{7\pi}{4}, \dfrac{\pi}{3}, \dfrac{2\pi}{3}, \dfrac{4\pi}{3}, \dfrac{5\pi}{3}$

Chapter 8 Review, page 464

1. $\cot x$

3. $\sec x$

5. $\sin x$

7.

sin x	cos x	tan x	cot x	sec x	csc x
$-\dfrac{1}{2}$	$\dfrac{\sqrt{3}}{2}$	$-\dfrac{1}{\sqrt{3}}$	$-\sqrt{3}$	$\dfrac{2}{\sqrt{3}}$	-2

9.

sin x	cos x	tan x	cot x	sec x	csc x
$\frac{4}{5}$	$\frac{3}{5}$	$\frac{4}{3}$	$\frac{3}{4}$	$\frac{5}{3}$	$\frac{5}{4}$

11. $3\cos x$

13. $\dfrac{\tan^2 x + 1}{\cot^2 x + 1} = \dfrac{\sec^2 x}{\csc^2 x} = \dfrac{\sin^2 x}{\cos^2 x} = \tan^2 x$

15. $\dfrac{1}{\sec x + \tan x} = \dfrac{1}{\dfrac{1}{\cos x} + \dfrac{\sin x}{\cos x}}$

$= \dfrac{\cos x}{1 + \sin x} \cdot \dfrac{1 - \sin x}{1 - \sin x}$

$= \dfrac{1 - \sin x}{\cos x}$

17. $\dfrac{\sqrt{3} + 1}{\sqrt{3} - 1}$

19. $\dfrac{1 + \sqrt{3}}{2^{3/2}}$

21. $\dfrac{\sqrt{3}}{2}\cos x + \dfrac{1}{2}\sin x$

23. $\dfrac{1 + \tan x}{\tan x - 1}$

25. 0

27. $-\dfrac{7}{17}$

29. 0

31. $\dfrac{1 - \sqrt{3}}{1 + \sqrt{3}}$

33. $\tan^{-1} \dfrac{7}{4} \approx 1.05165$

35. $\dfrac{\cos(x+y) - \cos x}{y} = \dfrac{\cos x \cos y - \sin x \sin y - \cos x}{y}$

$= \cos x \left(\dfrac{\cos y - 1}{y}\right) - \sin x \left(\dfrac{\sin y}{y}\right)$

37. $\cos\left(\dfrac{\pi}{4} + x\right) = \dfrac{1}{\tan\left(\dfrac{\pi}{4} + x\right)}$

$= \dfrac{1 - \tan\dfrac{\pi}{4}\tan x}{\tan\dfrac{\pi}{4} + \tan x}$

$= \dfrac{1 - \tan x}{1 + \tan x}$

39.

sin 2x	cos 2x	tan 2x	sin $\frac{x}{2}$	cos $\frac{x}{2}$	tan $\frac{x}{2}$
0.96	-0.28	$-\frac{24}{7}$	$\frac{1}{\sqrt{5}}$	$\frac{2}{\sqrt{5}}$	0.5

41.

sin 2x	cos 2x	tan 2x	sin $\frac{x}{2}$	cos $\frac{x}{2}$	tan $\frac{x}{2}$
$\frac{\sqrt{3}}{2}$	$-\frac{1}{2}$	$-\sqrt{3}$	$\frac{\sqrt{3}}{2}$	$-\frac{1}{2}$	$-\sqrt{3}$

43. $-4\sin^3 x + 3\sin x$ **45.** $1 - 2\sin^2 5x$

47. $\dfrac{1 - \cos 2x}{\sin 2x} = \dfrac{1 - (1 - 2\sin^2 x)}{2\sin x \cos x}$

$= \dfrac{\sin x}{\cos x} = \tan x$

49. $\tan 3x = \tan(2x + x)$

$= \dfrac{\tan 2x + \tan x}{1 - \tan 2x \tan x}$

$= \dfrac{\dfrac{2\tan x}{1 - \tan^2 x} + \tan x}{1 - \dfrac{2\tan x}{1 - \tan^2 x}\tan x}$

$= \dfrac{2\tan x + (\tan x - \tan^3 x)}{(1 - \tan^2 x) - 2\tan^2 x}$

$= \dfrac{3\tan x - \tan^3 x}{1 - 3\tan^2 x}$

51. $\tan\dfrac{x}{2} = \pm\sqrt{\dfrac{1 - \cos x}{1 + \cos x}}$

$= \pm\sqrt{\dfrac{(1 - \cos x)(1 + \cos x)}{(1 + \cos x)^2}}$

$= \pm\sqrt{\dfrac{1 - \cos^2 x}{(1 + \cos x)^2}}$

$= \pm\dfrac{\sqrt{\sin^2 x}}{1 + \cos x}$ since $1 + \cos x \geq 0$

$= \pm\dfrac{|\sin x|}{1 + \cos x}$

Since $\sin x$ and $\tan x/2$ always have the same sign and the denominator is positive, the $+$ sign always holds.

53. $-\dfrac{\sqrt{3}}{2}$

55. $-\dfrac{\sqrt{7}}{3}$

57. $\dfrac{\sqrt{2 + \sqrt{2}}}{2}$

59. $\dfrac{\sqrt{2 - \sqrt{2}}}{2}$

61. $\dfrac{1}{2}$

63. $\cos\dfrac{2\pi}{5}$

65. $\cos x$

67. $\sin 3x$

69. $\cos 8x - \cos 10x$

71. $\sin 8x - \sin 2x$

73. $2\sin 20° \cos 40°$

75. $-2\sin 2x \sin x$

77. $\dfrac{\cos 4x + \cos 2x}{\sin 4x - \sin 2x} = \dfrac{2\cos 3x \cos x}{2\sin x \cos 3x} = \cot x$

79. $\dfrac{\cos 2x - \cos 4x}{2\sin 3x} = \dfrac{-2\sin 3x \sin(-x)}{2\sin 3x} = \sin x$

81. $\dfrac{\pi}{6}$

83. $\dfrac{1}{2}$

85. $x = \sin\left(\dfrac{y}{3}\right) - 1$

87. $x = 1 - \cos\left(\dfrac{2y}{3}\right)$

89. $\dfrac{x}{\sqrt{x^2+9}}$

91. $\dfrac{\sqrt{25-x^2}}{5}$

93. $\sin y = x \quad \cos y = \sqrt{1-x^2} \quad \tan y = \dfrac{x}{\sqrt{1-x^2}}$
    $\sec y = \dfrac{1}{\sqrt{1-x^2}} \quad \csc y = \dfrac{1}{x} \quad \cot y = \dfrac{\sqrt{1-x^2}}{x}$

95. $\sin y = -\dfrac{1}{x} \quad \cos y = -\dfrac{\sqrt{x^2-1}}{x} \quad \tan y = \dfrac{1}{\sqrt{x^2-1}}$
    $\sec y = -\dfrac{x}{\sqrt{x^2-1}} \quad \csc y = -x$
    $\cot y = \sqrt{x^2-1}$

97. $\sin(x+y) = \dfrac{\sqrt{3}}{2} \quad \cos(x+y) = -\dfrac{1}{2} \quad \sin 2x = \dfrac{\sqrt{3}}{2}$
    $\cos 2x = -\dfrac{1}{2} \quad \tan 2x = -\sqrt{3}$

99. $x = \dfrac{\pi}{6} + 2\pi k$ or $x = 5\dfrac{\pi}{6} + 2\pi k : k = 0, \pm 1, \pm 2, \ldots$

101. $x = -1.0180407 + \pi k : k = 0, \pm 1, \pm 2, \ldots$

103. $-\dfrac{\pi}{6}, \dfrac{\pi}{6}$

105. $x = \dfrac{\pi}{3}, \dfrac{2\pi}{3}, \dfrac{4\pi}{3}, \dfrac{5\pi}{3}$

107. No solution.

109. No solution.

111. 1.77215, 4.51103, 1.318116, 4.96507

Chapter 8 Test, page 466

1. $4 - 4\cos^2 x$

2. $1 - \sec^2 x = \dfrac{1-\tan^2 x}{1-\cot^2 x}$
   $= \dfrac{1 - \dfrac{\sin^2 x}{\cos^2 x}}{1 - \dfrac{\cos^2 x}{\sin^2 x}}$
   $= -\dfrac{\dfrac{1}{\cos^2 x}}{\dfrac{1}{\sin^2 x}}$
   $= -\dfrac{\sin^2 x}{\cos^2 x}$
   $= -\tan^2 x$
   $= -(\sec^2 x - 1)$

3. $\dfrac{\sqrt{6}-\sqrt{2}}{4}$

4. $\dfrac{1}{2}\sin y + \dfrac{\sqrt{3}}{2}\cos y$

5. $8, -8$

6. $3\sin x - 4\sin^3 x$

7. $-\dfrac{3\sqrt{10}}{10}$

8. 1

9. $2\tan x = \dfrac{\sin x + \tan x}{\cos^2 \dfrac{x}{2}}$
   $= \dfrac{\sin x + \dfrac{\sin x}{\cos x}}{\dfrac{1+\cos x}{2}}$
   $= \dfrac{\dfrac{\sin x}{\cos x}[\cos x + 1]}{\dfrac{1+\cos x}{2}}$
   $= 2\dfrac{\sin x}{\cos x}$
   $= 2\tan x$

10. $\dfrac{1}{2}[\sin 8x - \sin 2x]$

11. $\dfrac{1}{\sqrt{x^2+1}}$

12. $\dfrac{1-x^2}{1+x^2}$

13. $x = \sin\dfrac{y}{5} - 1$

14. $x = \dfrac{1}{3}\left[1 + \dfrac{\pi}{6} + 2k\pi\right]$
    $= \dfrac{1}{3}\left[1 + \dfrac{5\pi}{6} + 2k\pi\right]$,
    $k = 0, \pm 1, \pm 2, \ldots$

Chapter 9

Section 9.1, page 475

1. 33.1°, 94.9°, 127.7

3. 48.6°, 101.4°, 156.84 or 131.4°, 18.6°, 51.03

5. 90°, 275.5, 165.8

7. $\angle B = 34.5°, c = 19.9, b = 22.6$

9. $\gamma = 75°, a = 9.66, c = 9.66$

11. $\beta = 126°, b = 302.39, c = 137.9$

13. $\angle B = 19.15°, \angle C = 60.6°, c = 42.5$

15. $\angle C = 154.31°, \angle A = 10.186°, c = 11.029$

17. $\gamma = 48.74°, \alpha = 21.26°, a = 48.23$

19. None.

21. None.

ANSWERS Section 9.1, page 475—Section 9.3, page 490

23. $\gamma = 90°, \beta = 60°, b = 34.64$
25. $\frac{1}{2}bc \sin \alpha$
27. 363.3 ft^2
29. 565.33
31. $\frac{a+b}{b}$
33. $\frac{a-b}{a+b}$
35. 20 km
37. 9.3 m
39. $0.33\overline{MN}$
41. 17.06 ft
43. 576.7 m
45. 32 ft
47. 9,609.4 m
49. 25.4 miles

Section 9.2, page 483

1. $c = 2.5, \angle B = 97.5°, \angle A = 52.5°$
3. $a = 2.6, \angle B = 40.9°, \angle C = 79.1°$
5. $a = 8.1, \angle B = 37.9°, \angle C = 7.1°$
7. $c = 8.9, \angle A = 39.1°, \angle B = 70.9°$
9. $a = 9.6, \angle B = 41.6°, \angle C = 85.4°$
11. $o = 44.1, \angle M = 12.5°, \angle O = 107.5°$
13. $c = 4.5, \angle A = 10.9°, \angle B = 153.7°$
15. $\angle A = 51°, \angle B = 87.3°, \angle C = 41.8°$
17. $\angle A = 29°, \angle B = 46.6°, \angle C = 104.5°$
19. $\angle A = 34°, \angle B = 44.4°, \angle C = 101.5°$
21. $\angle A = 26.4°, \angle B = 18.4°, \angle C = 135.2°$
23. $a = 22.1, \angle B = 33.4°, \angle C = 76.9°$
25. 30
27. 36.5
29. 168.4
31. Area $= \frac{1}{2}ab \sin C$
$\frac{\sin B}{b} = \frac{\sin A}{a}$ or $b = a\frac{\sin B}{\sin A}$
Therefore, Area $= \frac{1}{2}a^2 \frac{\sin B \sin C}{\sin A}$

33. From the law of cosines,
$$a^2 = b^2 + c^2 - 2bc \cos A$$
$$\frac{a^2 - b^2 - c^2}{-2bc} = \cos A$$
Thus,
$$1 + \cos A = \frac{a^2 - b^2 - c^2}{2bc} = \frac{2bc - a^2 + b^2 + c^2}{2bc}$$
$$= \frac{(b+c)^2 - a^2}{2bc}$$
$$= \frac{(b+c-a)(b+c+a)}{2bc}$$

35. $\overline{BC} = 45.9$ m
37. 85 m
39. 154 miles; course should change to 26.4°.
41. 514.4 miles
43. 257 m
45. 51.4 km
47. 4.9 miles, 11.3 miles

Section 9.3, page 490

1.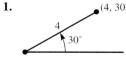

3.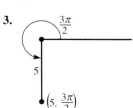

5.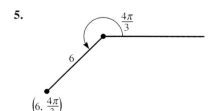

7.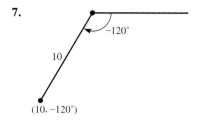

9.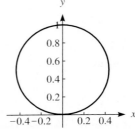

11.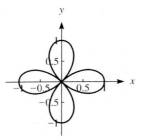

13.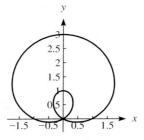

15.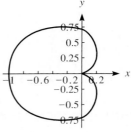

17.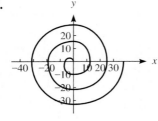

19.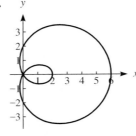

21.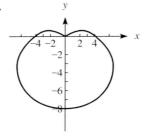

23.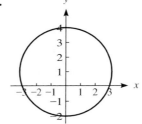

25.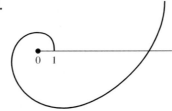

27.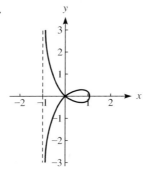

29.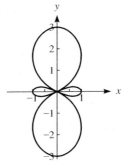

31.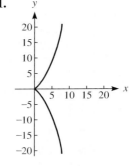

33. $(2^{5/2}, 45°)$

35. $(1, 120°)$

37. $(2, 30°)$

39. $(5, 53.1°)$

41. $(2^{3/2}, 225°)$

43. $(\sqrt{13}, 33.7°)$

45. $r = \dfrac{4}{3\cos\theta + 2\sin\theta}$

47. $r = 4$

49. $r = 6\csc\theta$

51. $r^2 = \dfrac{4}{\cos^2\theta - 4\sin^2\theta}$

53. $r^2 = \dfrac{36}{\cos^2\theta + 9\sin^2\theta}$

55. $x^2 + y^2 = 25$

57. $y = -2$

59. $x^2 + y^2 = 6y$

61. $y = x$

63. $x^2 + y^2 = 4x$

65. $(2, k\pi), k = 0, \pm 1, \pm 2, \ldots$

Section 9.4, page 496

1. $\sqrt{13}$
3. $\sqrt{3}$
5. 3
7. $2\sqrt{7}$
9. $2\sqrt{2}\left(\cos\frac{\pi}{4} + i\sin\frac{\pi}{4}\right)$
11. $4\left(\cos\frac{\pi}{2} + i\sin\frac{\pi}{2}\right)$
13. $3(\cos 0 + i\sin 0)$
15. $3\sqrt{2}\left(\cos\frac{7\pi}{4} + i\sin\frac{7\pi}{4}\right)$
17. $4\sqrt{2}\left(\cos\frac{\pi}{4} + i\sin\frac{\pi}{4}\right)$
19. $2\left(\cos\frac{11\pi}{6} + i\sin\frac{11\pi}{6}\right)$
21. $2\left(\cos\frac{\pi}{3} + i\sin\frac{\pi}{3}\right)$
23. $4\left(\cos\frac{7\pi}{6} + i\sin\frac{7\pi}{6}\right)$
25. $2(\cos\pi + i\sin\pi)$
27. $5\left(\cos\frac{\pi}{2} + i\sin\frac{\pi}{2}\right)$

29.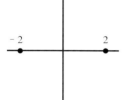

31.

33.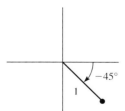

35.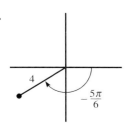

37.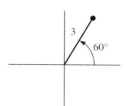

39.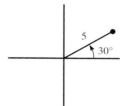

41.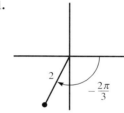

43.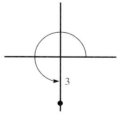

45. $10(\cos 225° + i\sin 225°)$, $\frac{5}{2}(\cos 135° + i\sin 125°)$
47. $4(\cos\pi + i\sin\pi)$, $\frac{1}{4}(\cos 2\pi + i\sin 2\pi)$
49. $6\left(\cos\frac{2\pi}{3} + i\sin\frac{2\pi}{3}\right)$, $\frac{2}{3}\left(\cos\frac{5\pi}{3} + i\sin\frac{5\pi}{3}\right)$
51. $10\left(\cos\frac{\pi}{12} + i\sin\frac{\pi}{12}\right)$, $\frac{5}{2}\left(\cos\frac{19\pi}{12} + i\sin\frac{19\pi}{12}\right)$
53. $12(\cos 150° + i\sin 150°)$, $3(\cos 330° + i\sin 330°)$
55. $27(\cos 75° + i\sin 75°)$, $3(\cos 195° + i\sin 195°)$
57. $32\left(\cos\frac{\pi}{3} + i\sin\frac{\pi}{3}\right)$, $\frac{1}{2}\left(\cos\frac{4\pi}{3} + i\sin\frac{4\pi}{3}\right)$
59. $-\sqrt{2} + \sqrt{2}i$
61. $-6i$

63. $-\dfrac{\sqrt{3}}{2} - \dfrac{1}{2}i$

65. $\dfrac{3}{2} - \dfrac{\sqrt{3}}{2}i$

67. $-5\sqrt{2} + 5\sqrt{2}i$

69. $15i$

71. $\left(2\sqrt{2} + \dfrac{3\sqrt{3}}{2}\right) + \left(-\dfrac{3}{2} + 2\sqrt{2}\right)i$

73. $4 + 2i$

75. Let $z = a + bi, w = c + di$. Then

$$\left|\dfrac{z}{w}\right| = \left|\dfrac{a + bi}{c + di}\right| = \left|\dfrac{(a + bi)(c - di)}{(c + di)(c - di)}\right|$$

$$= \left|\dfrac{ad - bdi^2 + i(bc - ad)}{c^2 + d^2}\right|$$

$$= \left|\dfrac{ad + bc}{c^2 + d^2} + i\dfrac{bc - ad}{c^2 + d^2}\right|$$

$$= \sqrt{\left(\dfrac{ac + bd}{c^2 + d^2}\right)^2 + \left(\dfrac{bc - ad}{c^2 + d^2}\right)^2}$$

$$= \dfrac{1}{c^2 + d^2}$$

$$\times \sqrt{a^2c^2 + 2acbd + b^2d^2 + b^2c^2 - 2acbd + a^2d^2}$$

$$= \dfrac{1}{c^2 + d^2} \sqrt{(a^2 + b^2)(c^2 + d^2)}$$

$$= \dfrac{\sqrt{a^2 + b^2}}{\sqrt{c^2 + d^2}}$$

$$= \dfrac{|z|}{|w|}$$

77. $zw = 6(\cos 170° + i\sin 170°)$
$\approx -5.909 + 1.042i$
$\dfrac{z}{w} = \dfrac{3}{2}[\cos(-90°) + i\sin(-90°)]$
$= -\dfrac{3}{2}i$

79. $zw = 10\left(\cos\dfrac{16\pi}{45} + i\sin\dfrac{16\pi}{45}\right) \approx 4.384 + 8.988i$
$\dfrac{z}{w} = \dfrac{5}{2}\left[\cos\left(-\dfrac{34\pi}{45}\right) + i\sin\left(-\dfrac{34\pi}{45}\right)\right]$
$\approx -1.798 - 1.737i$

Section 9.5, page 503

1. $-0.5 - \dfrac{\sqrt{3}}{2}i$

3. $1{,}024i$

5. $-125i$

7. -1

9. $-(5\sqrt{2})^{14}i$

11. $4^8\left(-\dfrac{1}{2} + \dfrac{\sqrt{3}}{2}i\right)$

13. $\dfrac{81}{2} + \dfrac{81}{2}\sqrt{3}i$

15. $-32 + 32\sqrt{3}i$

17. $4\sqrt{3} - 4i$

19. $-512\sqrt{2} - 512\sqrt{2}i$

21. $-2, 1 \pm \sqrt{3}i$

23. $t_j = 2^{1/10}\left[\cos\left(\dfrac{7\pi/6 + 2\pi j}{10}\right) + i\sin\left(\dfrac{7\pi/6 + 2\pi j}{10}\right)\right]$,
$(j = 0, 1, \ldots, 9)$

25. $\dfrac{\sqrt{3}}{2} + \dfrac{1}{2}i, -\dfrac{\sqrt{3}}{2} + \dfrac{1}{2}i, -i$

27. $\dfrac{3\sqrt{2}}{2} + \dfrac{3\sqrt{2}}{2}i, -\dfrac{3\sqrt{2}}{2} - \dfrac{3\sqrt{2}}{2}i$

29. $t_j = (\sqrt{2})^{1/5}\left[\cos\left(\dfrac{\pi/4 + 2\pi j}{5}\right) + i\sin\left(\dfrac{\pi/4 + 2\pi j}{5}\right)\right]$,
$(j = 0, 1, 2, 3, 4)$

31. $t_j = 2\left[\cos\left(\dfrac{5\pi/3 + 2\pi j}{4}\right) + i\sin\left(\dfrac{5\pi/3 + 2\pi j}{4}\right)\right]$,
$(j = 0, 1, 2, 3)$

33. $\dfrac{3}{2} + \dfrac{3\sqrt{3}}{2}i, \dfrac{3}{2} - \dfrac{3\sqrt{3}}{2}i, -3$

35. $\pm\left(\dfrac{3\sqrt{3}}{2} + \dfrac{3}{2}i\right), \pm\left(\dfrac{3}{2} - \dfrac{3\sqrt{3}}{2}i\right)$

37. $2\sqrt{2} + 2\sqrt{2}i, -2\sqrt{2} - 2\sqrt{2}i$

39. $3\left[\cos\dfrac{2\pi j}{5} + i\sin\dfrac{2\pi j}{5}\right]$, $(j = 0, 1, 2, 3, 4)$

41. $t_j = 3\left[\cos\left(\dfrac{3\pi/2 + 2\pi j}{4}\right) + i\sin\left(\dfrac{3\pi/2 + 2\pi j}{4}\right)\right]$,
$(j = 0, 1, 2, 3)$

43. $t_j = 2\left[\cos\left(\dfrac{5\pi/6 + 2\pi j}{4}\right) + i\sin\left(\dfrac{5\pi/6 + 2\pi j}{4}\right)\right]$,
$(j = 0, 1, 2, 3)$

45. $t_j = \cos\left(\dfrac{2\pi j}{5}\right) + i\sin\left(\dfrac{2\pi j}{5}\right)$, $(j = 0, 1, 2, 3, 4)$

47. $\dfrac{3}{2} + \dfrac{3\sqrt{3}}{2}i, \dfrac{3}{2} - \dfrac{3\sqrt{3}}{2}i, -3$

49. $\pm 5i$

51. $t_j = 2^{1/3}\left[\cos\left(\dfrac{5\pi/3 + 2\pi j}{3}\right) + i\sin\left(\dfrac{5\pi/3 + 2\pi j}{3}\right)\right]$,
$(j = 0, 1, 2)$

53. $t_j = 2^{1/5}\left[\cos\left(\dfrac{3\pi/4 + 2\pi j}{5}\right) + i\sin\left(\dfrac{3\pi/4 + 2\pi j}{5}\right)\right]$,
$(j = 0, 1, 2, 3, 4)$

55. $\dfrac{27}{4} + \dfrac{27}{4}i$

Section 9.6, page 512

57. $\dfrac{1+\sqrt{3}}{16} - \dfrac{1-\sqrt{3}}{16}i$

59. $16\sqrt{3} + 16 + (16 - 16\sqrt{3})i$

61. $\dfrac{9}{4}$

63. $-2\sqrt{2} - 2\sqrt{2}i$

1.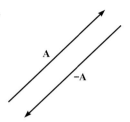

3.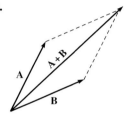

5.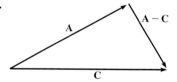

7.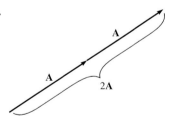

9.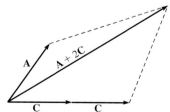

11. *a.* $\sqrt{2}$
  b. $\mathbf{i} - \mathbf{j}$

13. *a.* $3\sqrt{2}$
  b. $3\mathbf{i} + 3\mathbf{j}$

15. *a.* 2
  b. $-\mathbf{i} + \sqrt{3}\mathbf{j}$

17. $\langle 6, 4 \rangle$

19. $\langle 1, 6 \rangle$

21. $\langle 3, -2 \rangle$

23. $\langle -6, 13 \rangle$

25. $\langle 5, 6 \rangle$

27. $-3 - 5\sqrt{2}$

29. Let $\mathbf{v}_1 = \langle x_1, y_1 \rangle$, $\mathbf{v}_2 = \langle x_2, y_2 \rangle$. Then $c_1(\mathbf{v}_1 \cdot \mathbf{v}_2) = c_1(x_1 x_2 + y_1 y_2)$, whereas
$$(c_1 \mathbf{v}_1) \cdot \mathbf{v}_2 = \langle c_1 x_1, c_2 x_2 \rangle \cdot \langle x_2, y_2 \rangle$$
$$= c_1 x_1 x_2 + c_1 y_1 y_2$$
$$= c_1(\mathbf{v}_1 \cdot \mathbf{v}_2)$$

31. Let $\mathbf{v}_1 = \langle x_1, y_1 \rangle$, $\mathbf{v}_2 = \langle x_2, y_2 \rangle$, $\mathbf{v}_3 = \langle x_3, y_3 \rangle$.
$$\mathbf{v}_1 \cdot (\mathbf{v}_2 + \mathbf{v}_3) = \mathbf{v}_1 \cdot \langle x_2 + x_3, y_2 + y_3 \rangle$$
$$= x_1(x_2 + x_3) + y_1(y_2 + y_3)$$
$$= x_1 x_2 + x_1 x_3 + y_1 y_2 + y_1 y_3$$
$$= \langle x_1, y_1 \rangle \cdot \langle x_2, y_2 \rangle + \langle x_1, y_1 \rangle \cdot \langle x_3, y_3 \rangle$$

33. Yes.

35. No.

37. $\cos^{-1}\left(\dfrac{3}{\sqrt{10}}\right)$

39. $\cos^{-1}\left(\dfrac{\sqrt{5}}{5}\right)$

41. $10\sqrt{3} - \dfrac{15\sqrt{2}}{2} \approx 6.7$ knots

43. 136 miles, 172°

45. 466 mph, bearing 97.45°

47. 521.54 mph, bearing 175.6°

49. 3.92 mph

51. 44.54 miles, bearing 191.05°

53. 321.39 lb

55. $\theta \approx 18.2°$

57. $15 \sin 24° = 6.1$ lb

59. 6,815.66 lb

61. 13.1 lb

Chapter 9 Review, page 516

1. $c = 17.52$, $\angle B = 5.74°$, $\angle C = 144.26°$
3. $\angle B = 24.32006°$, $\angle C = 142.68°$, $c = 700.402$ or $\angle B = 155.7°$, $\angle C = 11.3°$, $c = 226.5$
5. $c = 8.62$, $\angle A = 55.1°$, $\angle B = 79.9°$
7. 21
9. No solution.
11. $\sqrt{5}$
13. 2
15. a. $2\left(\cos\frac{\pi}{2} + i\sin\frac{\pi}{2}\right)$
    b. 2
    c. $\frac{\pi}{2}$
17. a. $\sqrt{19}(\cos 203.4° + i\sin 203.4°)$
    b. $\sqrt{19}$
    c. $203.4°$
19.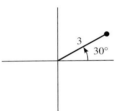
21.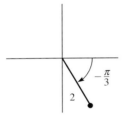
23. $zw = 10(\cos 135° + i\sin 135°)$, $\frac{z}{w} = \frac{5}{2}(\cos 45° + i\sin 45°)$
25. $zw = 6\left(\cos\frac{4\pi}{3} + i\sin\frac{4\pi}{3}\right)$, $\frac{z}{w} = \frac{3}{2}\left(\cos\frac{4\pi}{3} + i\sin\frac{4\pi}{3}\right)$
27. $1 + i$
29. 5
31. $\frac{1 - \sqrt{3}i}{2}$
33. $-2^{30}i$
35. $-2^{15} - \sqrt{3}\cdot 2^{15}i$
37. 3, $-\frac{3}{2} + \frac{3\sqrt{3}}{2}i$, $-\frac{3}{2} - \frac{3\sqrt{3}}{2}i$
39. $(2\sqrt{2})^{1/5}\left(\cos\frac{\pi}{20} + i\sin\frac{\pi}{20}\right)$, $(2\sqrt{2})^{1/5}\left(\cos\frac{9\pi}{20} + i\sin\frac{9\pi}{20}\right)$,
    $(2\sqrt{2})^{1/5}\left(\cos\frac{17\pi}{20} + i\sin\frac{17\pi}{20}\right)$,
    $(2\sqrt{2})^{1/5}\left(\cos\frac{5\pi}{4} + i\sin\frac{5\pi}{4}\right)$,
    $(2\sqrt{2})^{1/5}\left(\cos\frac{33\pi}{20} + i\sin\frac{33\pi}{20}\right)$
41. $5\sqrt{3} - 5i$, $-5\sqrt{3} + 5i$
43. $\cos\left(\frac{\pi + 2\pi j}{5}\right) + i\sin\left(\frac{\pi + 2\pi j}{5}\right)$, $(j = 0, 2, 3, 4)$
45. $2^{1/3}\left(\cos\frac{5\pi}{9} + i\sin\frac{5\pi}{9}\right)$, $2^{1/3}\left(\cos\frac{11\pi}{9} + i\sin\frac{11\pi}{9}\right)$, $2^{1/3}\left(\cos\frac{17\pi}{9} + i\sin\frac{17\pi}{9}\right)$
47.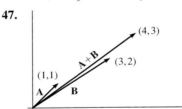
49.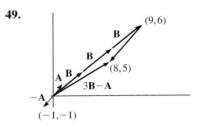
51. a. $2\sqrt{2}$
    b. $2\mathbf{i} + 2\mathbf{j}$
53. a. 2
    b. $-\sqrt{2}\mathbf{i} + \sqrt{2}\mathbf{j}$
55. $\langle 6, 15 \rangle$
57. $\langle -3, -4 \rangle$
59.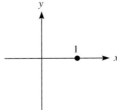

61.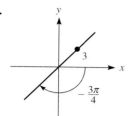

63.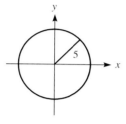

65.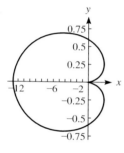

67. $\left(2\sqrt{2}, \dfrac{3\pi}{4}\right)$

69. $(6, -60°)$

71. $r\cos\theta = -3$

73. $r = 2$

75. $x^2 + y^2 = 4$

77. $x = 5$

79. $x^2 - y^2 = 1$

81. 50.8 ft

83. 13.26 ft

85. 256.87 ft

87. 203.96 mph; therefore, bearing is 94.94°.

89. 8.03848 mph

91. Centrifugal force = 57.74 lb
Force by cable = 115.48 lb

Chapter 9 Test, page 518

1. $a = 12.00958$, $c = 12.65595$, $\gamma = 82°$

2. a. 422.57 miles
b. 175.01882 miles

3. $\dfrac{a}{\sin\alpha} = \dfrac{b}{\sin\beta}$

$b = \dfrac{a\sin\beta}{\sin\alpha}$

$a + b = \dfrac{a\sin\beta}{\sin\alpha} + a$

$= a\left(\dfrac{\sin\beta + \sin\alpha}{\sin\alpha}\right)$

$\dfrac{a+b}{a} = \dfrac{\sin\beta + \sin\alpha}{\sin\alpha}$

$\dfrac{a}{a+b} = \dfrac{\sin\alpha}{\sin\beta + \sin\alpha}$

4. $a = 10.02441$,
$\beta = 49.43614°$,
$\gamma = 80.96386°$

5. 297.17218 m

6.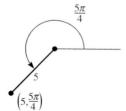

7. a. $x = -2.59808$, $y = 1.5$
b. $r = \sqrt{13}$, $\theta = 0.588$ radians

8.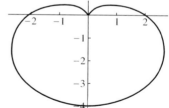

9. $5\sqrt{2}\left(\cos\dfrac{\pi}{4} + i\sin\dfrac{\pi}{4}\right)$

10. $-i$, $\dfrac{-\sqrt{3}+i}{2}$, $\dfrac{\sqrt{3}+i}{2}$

11. a. $5\mathbf{i}, 7\mathbf{j}$
b. $\langle -4, 13 \rangle$
c. $88.152°$

Chapter 10

Section 10.1, page 527

1. $x = -1, \quad y = \frac{1}{3}$
3. $x = 5, \quad y = -4$
5. $x = -8, \quad y = 14$
7. $(x, y) = \left(\dfrac{-6 + \sqrt{71}}{5}, \dfrac{3 + 2\sqrt{71}}{5}\right),$
   $\left(\dfrac{-6 - \sqrt{71}}{5}, \dfrac{3 - 2\sqrt{71}}{5}\right)$
9. $(3, 0), \left(-\dfrac{9}{5}, \dfrac{12}{5}\right)$
11. No real solutions.
13. $(x, y) = \left(\dfrac{-4 + \sqrt{26}}{5}, \dfrac{-2 + 3\sqrt{26}}{10}\right),$
    $\left(\dfrac{-4 - \sqrt{26}}{5}, \dfrac{-2 - 3\sqrt{26}}{10}\right)$
15. $(x, y) = (\pm \sqrt{11}, \pm 2\sqrt{3})$ (all four sign combinations)
17. $(x, y) = (-2, -4), (2, 4), (4, 2)$
19. Aluminum $= 13$, antimony $= 51$
21. 700 balcony tickets, 900 main floor tickets.
23. $4,000 at 12%, $1,000 at 6%
25. 20 cm $\times$ 34 cm
27. 0
29. a. 2.85 hours
    b. 26.9 mph = avg. speed of jet stream.
       511.2 mph = avg. speed of plane.
31. 614 or 423
33. No solutions.
35. $x = -10.3851648, \quad y = 1,068.1313$
    $x = 1.515512, \quad y = 3.2967765$
37. $x = 0.89432255, \quad y = 0.44742282$

Section 10.2, page 532

1. $x = 1, \quad y = 2$
3. $x = 7, \quad y = -6$
5. No solutions.
7. $x = -\dfrac{14}{3}, \quad y = \dfrac{9}{4}$
9. $x = \dfrac{26}{11}, \quad y = -\dfrac{28}{11}$
11. $x = \dfrac{5}{3}, \quad y = \dfrac{2}{5}$
13. $x = a + b, \quad y = a - b$
15. Length $= 24$ m, width $= 18$ m
17. Old mower $= 6$ hours, new mower $= 4$ hours
19. 2 mph
21. 100 @ $4.50, 350 @ $6
23. $x = 1.0049796,$
    $y = 1.9850611$
25. $x = 2.1706655,$
    $y = -2.377999$

Section 10.3, page 547

1. $x = 1, y = 2, \quad z = 1$
3. $x = 5, \quad y = -1, \quad z = -2$
5. $x = -4, \quad y = 2, \quad z = 3$
7. $\begin{bmatrix} -1 & 2 & 3 \\ 4 & 1 & 0 \\ 0 & -1 & 5 \end{bmatrix}$
9. $\begin{bmatrix} 4 & 1 & 0 \\ -\frac{1}{2} & 1 & \frac{3}{2} \\ 0 & -1 & 5 \end{bmatrix}$
11. $\begin{bmatrix} 4 & 1 & 0 \\ 9 & 0 & -3 \\ 0 & -1 & 5 \end{bmatrix}$
13. $\begin{bmatrix} 4 & 1 & 0 \\ -1 & 2 & 3 \\ -6 & 2 & 11 \end{bmatrix}$
15. $\begin{bmatrix} 1 & 2 & 0 & 3 \\ 0 & 1 & -\frac{1}{3} & 4 \\ 0 & 0 & 1 & -\frac{33}{7} \end{bmatrix}$
17. $\begin{bmatrix} 1 & 1 & 2 \\ 0 & 0 & 1 \end{bmatrix}$
19. $\begin{bmatrix} 1 & 3 & 2 & 1 \\ 0 & 1 & 1 & 0 \\ 0 & 0 & 1 & 1 \end{bmatrix}$
21. $\begin{bmatrix} 1 & 5 & 6 \\ 0 & 1 & 1 \end{bmatrix}$
23. $\begin{bmatrix} 1 & 2 & -3 & -15 \\ 0 & 1 & 2 & 11 \\ 0 & 0 & 1 & 5 \end{bmatrix}$
25. $\begin{bmatrix} 1 & 2 & 3 & 0 \\ 0 & 1 & 1 & 1 \\ 0 & 0 & 1 & -\frac{5}{3} \end{bmatrix}$
27. $x = -\dfrac{1}{3}, \quad y = \dfrac{8}{3}, \quad z = -\dfrac{5}{3}$
29. $x = 2, \quad y = -1, \quad z = 0$

ANSWERS Section 10.3, page 547—Section 10.4, page 555   A-57

31. $x = 2 - \dfrac{7z}{5}, \quad y = -1 + \dfrac{3z}{5},$
$z =$ any real number
33. $x = -1, \quad y = 2, \quad z = -2$
35. $x = 1, \quad y = -2, \quad z = 3,$
$w = -4$
37. $x = -5, y = -2, z = 3$
39. $x = 2y + 4, \quad y =$ any number
41. $x = 1, y = 2$

43. $x = 1, y = 0, z = -1, w = 2$
45. 150 mph, 30 mph
47. 30 sofas, 15 chairs
49.
```
rowSwap([A],1,3)
   [[-2  0  1 -1]
    [ 4  5  1  0]
    [ 1  2  0  3]]
```

51.
```
*row(-2,[A],2)
[[ 1   2   0  3]
 [-8 -10  -2  0]
 [-2   0   1 -1]]
[[ 1   2   0  3]
 [-8 -10  -2  0]
 [-2   0   1 -1]]
```

53.
```
[[ 1   2   0  3]
 [-4  -5  -1  0]
 [-2   0   1 -1]]
*row+(4,Ans,3,2)
[[ 1   2   0  3]
 [-12 -5   3 -4]
 [-2   0   1 -1]]
```

Section 10.4, page 555

1.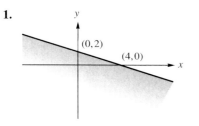

3.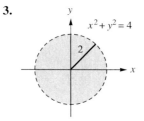

5.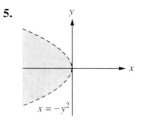

7.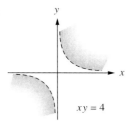

9.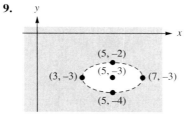

11.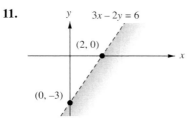

13.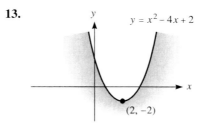

15.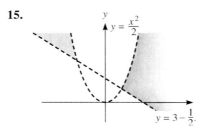

17.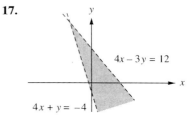

19.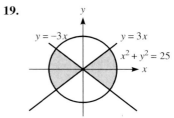

21.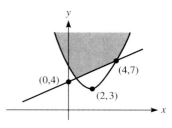

23.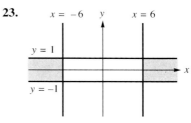

25.

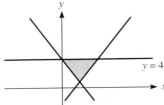

27.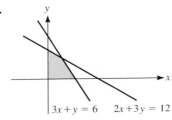

29.

31.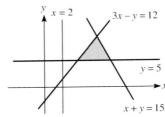

33. a. x = recreational skis, y = racing skis
$x \leq 60, y \leq 45, 3x + 4y \leq 240, x \geq 0, y \geq 0$

b.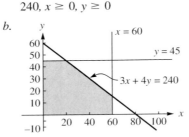

35. a. $x \geq 0, y \geq 0, x + y \leq 16, 3x + 6y \leq 60$

b.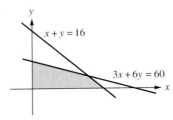

37. a. $0 \leq L \leq 74, 0 \leq w \leq 50$

b.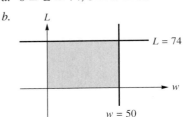

39. a. $2w + t \geq 60, t \geq 0, w \geq 0$

b.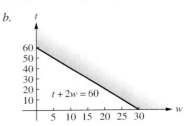

Section 10.5, page 559

1. $(x, y) = (4, 8)$ maximum value $= 52$
$(x, y) = (1, 1)$ minimum value $= 8$

3. $(x, y) = (0, 10)$ maximum value $= 30$
$(x, y) = (0, 0)$ minimum value $= 0$

5. $(x, y) = \left(\frac{5}{2}, 6\right)$ maximum value $= 97$
$(x, y) = \left(\frac{37}{7}, \frac{16}{7}\right)$ minimum value $= \frac{562}{7}$

7. $(x, y) = (7, 3)$ maximum value $= 5$

$(x, y) = (4, 6)$ minimum value $= -10$

9. $(x, y) = (5, 5)$ maximum value $= 30$
$(x, y) = (1, 2)$ minimum value $= 7$

11. $(x, y) = (8, 0)$ maximum value $= 64$
$(x, y) = (0, 0)$ minimum value $= 0$

13. Five 19-inch and four 13-inch televisions give a profit of $460.

15. 46¢ per bag with 0.6 kg bran and 0.4 kg rice.

17. Deluxe = 300, Top of the Line = 200; maximum profit = $1,300/week.

19. 134.5 acres in corn, 177.5 acres in beans, maximum profit = $50,248.50.

21. 818 ft²

23. 100 units of lumber, 300 units of plywood, maximum profit = $11,000.

Chapter 10 Review, page 564

1. $x = 4$, $y = -1$
3. $(x, y) = (4, -3), (-3, 4)$
5. $x = 7$, $y = 3$, $z = -1$
7. $(x, y) = (-1, 2)$
9.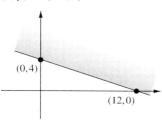

11.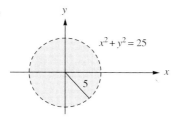

13.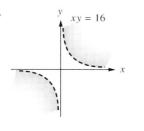

15. $y > x^2 + 16$

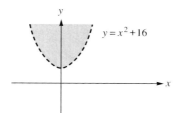

17.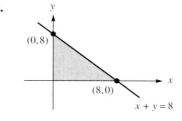

19.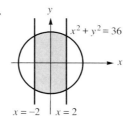

21.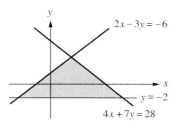

23.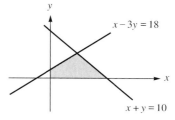

25. $(3, 1)$, minimum $= 9$; $(8, 1)$, maximum $= 19$.
27. Minimum value $\frac{9}{8}$ at $x = 0$, $y = \frac{3}{8}$; maximum value $\frac{107}{9}$ at $x = \frac{25}{9}$, $y = \frac{19}{9}$.
29. $(2, 6)$, maximum $= 26$; $(3, 1)$, minimum $= 7$.
31. $14,000 at 12.5% and $46,000 at 14%.
33. 835 at $20, $585 at $16.
35. 17 plain drawers, 13 fancy drawers.

Chapter 10 Test, page 566

1. $(x, y) = \left(-\frac{9}{5}, \frac{12}{5}\right), (3, 0)$
2. $(x, y, z) = (\sqrt{3}, 2\sqrt{3}, 3\sqrt{3})$, $(-\sqrt{3}, -2\sqrt{3}, 6 + 3\sqrt{3})$
3. $(x, y, z) = (0, 2, 3)$
4. $(x, y, z) = \left(-\frac{5}{3}, -\frac{65}{3}, -\frac{8}{3}\right)$
5. $y =$ any value, $x = -6 + \frac{3}{2}y$

6. a.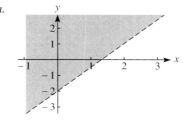

b.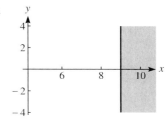

c.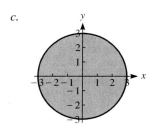

7.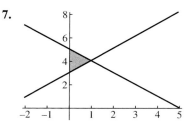

8. Minimum value 16 at (8, 0); maximum value 63 at (0, 9).
9. Width = 29 m, length = 41 m.
10. 24/5 plain desks and 84/5 fancy desks per day yield a maximum profit of $3,744 per day.

Chapter 11

Section 11.1, page 574

1. $\begin{bmatrix} 1 & 3 \\ 1 & 1 \end{bmatrix}$
3. $\begin{bmatrix} 2 & 6 \\ 4 & 12 \end{bmatrix}$
5. $\begin{bmatrix} -5 & 1 \\ 1 & -2 \end{bmatrix}$
7. $\begin{bmatrix} 7 & 0 \\ 2 & 9 \end{bmatrix}$
9. $\begin{bmatrix} 0 & 3 \\ 0 & -1 \end{bmatrix}$
11. $\begin{bmatrix} -3 & 3 \\ 0 & 0 \end{bmatrix}$
13. $\begin{bmatrix} 5 & -15 \\ 0 & 10 \end{bmatrix}$
15. $\begin{bmatrix} -2 & 6 \\ -1 & -3 \end{bmatrix}$
17. $\begin{bmatrix} 4 & 2 \\ -1 & 0 \end{bmatrix}$
19. $\begin{bmatrix} -14 & -4 \\ -1 & -2 \end{bmatrix}$
21. $\begin{bmatrix} 13 \\ 25 \end{bmatrix}$
23. $\begin{bmatrix} -10 & -7 & -9 \\ -2 & -19 & -14 \end{bmatrix}$
25. Undefined.
27. $\begin{bmatrix} -1 & -3 & 14 \\ 6 & 9 & 16 \end{bmatrix}$
29. Undefined.
31. $[\,-3 \quad 2 \quad 18\,]$
33. Undefined.
35. 3×5
37. Undefined.
39. 1×3
41. $x = 3, \quad y = 1, \quad z = 4$
43. $x = 18, \quad y = 3, \quad z = 3, \quad w = 3$
45. $\begin{bmatrix} 1 & 0 \\ 0 & 1 \end{bmatrix}\begin{bmatrix} a_{11} & a_{12} \\ a_{21} & a_{22} \end{bmatrix}$
$= \begin{bmatrix} a_{11} & a_{12} \\ a_{21} & a_{22} \end{bmatrix} = A$
$\begin{bmatrix} a_{11} & a_{12} \\ a_{21} & a_{22} \end{bmatrix}\begin{bmatrix} 1 & 0 \\ 0 & 1 \end{bmatrix}$
$= \begin{bmatrix} a_{11} & a_{12} \\ a_{21} & a_{22} \end{bmatrix} = A$
47. $k(A + B)$
$= k\begin{bmatrix} a_{11} + b_{11} & a_{12} + b_{12} \\ a_{21} + b_{21} & a_{22} + b_{22} \end{bmatrix}$
$= \begin{bmatrix} ka_{11} + kb_{11} & ka_{12} + kb_{12} \\ ka_{21} + kb_{21} & ka_{22} + kb_{22} \end{bmatrix}$
$= \begin{bmatrix} ka_{11} & ka_{12} \\ ka_{21} & ka_{22} \end{bmatrix} + \begin{bmatrix} kb_{11} & kb_{12} \\ kb_{21} & kb_{22} \end{bmatrix}$
$= kA + kB$

49. ```
[A]+[B]
[[5.87 4.31 5. ...
 [-.65 4.98 11 ...
 [-.95 -2.23 5. ...
```

51. ```
[A]*[B]
[[-1.735 1.508 ...
 [-.943  4.508 ...
 [15.297 14.754...
```

53. ```
[A]*[B]
[[2.0404 .9171 ...
 [5.8975 2.6083...
 [5.238 3.6003...
```

## Section 11.2, page 584

1. $-7$
3. $-5$
5. $-14$
7. $-13$
9. $-5$
11. $-14$
13. $8$
15. $-6$
17. $0$
19. $720$
21. $-410$
23. $9{,}072$
25. $-153$
27. $x = 3, \quad y = 2$
29. $x = -6, \quad y = 1$
31. $x = 1, \quad y = 2$
33. Zero determinant; Cramer's rule not applicable.
35. $x = -\frac{1}{3}, \quad y = \frac{8}{3}, \quad z = -\frac{5}{3}$
```

37. $x = 2$, $y = -1$, $z = 0$
39. 12
41. $a = \pm 4, -1$
43. $-10x - 2y - 4z$
45. $\begin{vmatrix} x_{11} & x_{12} & x_{13} & x_{14} \\ x_{21} & x_{22} & x_{23} & x_{24} \\ 0 & 0 & x_{33} & x_{34} \\ 0 & 0 & x_{43} & x_{44} \end{vmatrix}$

$= x_{11} \begin{vmatrix} x_{22} & x_{23} & x_{24} \\ 0 & x_{33} & x_{34} \\ 0 & x_{43} & x_{44} \end{vmatrix} - x_{21} \begin{vmatrix} x_{12} & x_{13} & x_{14} \\ 0 & x_{33} & x_{34} \\ 0 & x_{43} & x_{44} \end{vmatrix}$

$= x_{11} x_{22} \begin{vmatrix} x_{33} & x_{34} \\ x_{43} & x_{44} \end{vmatrix} - x_{21} x_{12} \begin{vmatrix} x_{33} & x_{34} \\ x_{43} & x_{44} \end{vmatrix}$

$= (x_{11}x_{22} - x_{21}x_{12}) \begin{vmatrix} x_{33} & x_{34} \\ x_{43} & x_{44} \end{vmatrix}$

$= \begin{vmatrix} x_{11} & x_{12} \\ x_{21} & x_{22} \end{vmatrix} \cdot \begin{vmatrix} x_{33} & x_{34} \\ x_{43} & x_{44} \end{vmatrix}$

47. det [A] 78

49. $x = 7.576419214$,
   $y = 0.8558951965$,
   $z = -1.152838428$,
   $w = -1.170305677$

51. $x = 0$, $y = 2$, $z = -2$, $w = 1$

Section 11.3, page 592

1. $\begin{bmatrix} \frac{7}{16} & \frac{5}{32} \\ -\frac{1}{16} & -\frac{3}{32} \end{bmatrix}$

3. $\begin{bmatrix} -1 & -2 \\ -\frac{1}{2} & -\frac{3}{2} \end{bmatrix}$

5. Inverse doesn't exist.

7. $\begin{bmatrix} \frac{7}{15} & -\frac{1}{5} & \frac{1}{15} \\ -\frac{2}{3} & 0 & \frac{1}{3} \\ -\frac{3}{5} & \frac{2}{5} & \frac{1}{5} \end{bmatrix}$

9. $\begin{bmatrix} 15 & 4 & -5 \\ -12 & -3 & 4 \\ -4 & -1 & 1 \end{bmatrix}$

11. Yes.
13. Yes.
15. Yes.

17. $\begin{bmatrix} x \\ y \end{bmatrix} = \begin{bmatrix} 4 \\ 1 \end{bmatrix}$

19. $\begin{bmatrix} x \\ y \end{bmatrix} = \begin{bmatrix} \frac{61}{19} \\ \frac{6}{19} \end{bmatrix}$

21. $\begin{bmatrix} x \\ y \end{bmatrix} = \begin{bmatrix} \frac{1}{2} \\ -1 \end{bmatrix}$

23. $\begin{bmatrix} x \\ y \end{bmatrix} = \begin{bmatrix} -2 \\ 5 \end{bmatrix}$

25. $A^{-1} = \begin{bmatrix} a^{-1} & 0 \\ 0 & b^{-1} \end{bmatrix}$ provided $ab \neq 0$.

27. Suppose that $A = \begin{bmatrix} a & b \\ c & d \end{bmatrix}$ and that $ad - bc = 0$. Suppose that $A^{-1} = \begin{bmatrix} x & y \\ z & w \end{bmatrix}$. Then

$$AA^{-1} = I$$

$\begin{bmatrix} ax + bz & ay + bw \\ cx + dz & cy + dw \end{bmatrix} = \begin{bmatrix} 1 & 0 \\ 0 & 1 \end{bmatrix}$

$\begin{cases} ax + bz = 1 \\ cx + dz = 0 \end{cases}$

Multiply the first equation by c and the second equation by a and subtract the first from the second to obtain:

$$(ad - bc)z = c$$
$$0 \cdot z = c$$
$$c = 0$$

Similarly, by multiplying the first equation by d and the second by b and then subtracting, we see that

$$d = 0$$

However, from the matrix equation, we have

$$cy + dw = 1$$

which is impossible. So A does not have an inverse.

29. $\begin{bmatrix} 0.375 & -0.25 & 0.125 \\ -0.125 & 0.75 & -0.375 \\ -0.25 & 0.5 & 0.25 \end{bmatrix}$

31. $\begin{bmatrix} 0.5 & 0.5 & -0.25 & 0.5 \\ -1 & 4 & -0.5 & -2 \\ -0.5 & 2.5 & -0.25 & -1.5 \\ 0.5 & -0.5 & 0.25 & 0.5 \end{bmatrix}$

33. $x = -2.25$, $y = 3.75$, $z = 10$

35. $x = 0$, $y = 1$, $z = -1$

Section 11.4, page 597

1. $\begin{bmatrix} 1.17 & 0.26 \\ 0.26 & 1.17 \end{bmatrix}$

3. $\begin{bmatrix} 1.14 & 0.11 & 0 \\ 0.23 & 1.02 & 0 \\ 0.03 & 0.11 & 1.11 \end{bmatrix}$

5. Semiconductor division $21.5 million, computer division $21 million, printer division $2.5 million.

7. $\begin{bmatrix} 0.03 & 0.15 & 0.20 \\ 0.15 & 0.10 & 0.09 \\ 0.08 & 0.18 & 0.02 \end{bmatrix}$

9. United States branch $62.1 million, the Canadian branch $29.4 million, and the Mexican branch $23.8 million.

11. $\begin{bmatrix} 0.0588312 & 0.0197403 \\ 0.0197403 & 0.0588312 \end{bmatrix}$

13. $\begin{bmatrix} 0.0263636 & 0.00363636 & 0 \\ -0.0127273 & 0.0227273 & 0 \\ 0.0252525 & 0.00363636 & 0.00111111 \end{bmatrix}$

Section 11.5, page 602

1. $\dfrac{2}{x-1} + \dfrac{4}{x+1}$

3. $\dfrac{1}{x} + \dfrac{2}{x+1} - \dfrac{3}{x+2}$

5. $\dfrac{\frac{4}{3}}{x-2} + \dfrac{\frac{13}{3}}{2x-1}$

7. $-\dfrac{1}{x-4} + \dfrac{2x-9}{x^2+1}$

9. $\dfrac{4}{x-3} + \dfrac{1}{3x+2}$

11. $1 + 3x + \dfrac{2}{x-1} + \dfrac{2}{2+x}$

13. $3x + 1 + \dfrac{4}{x-2} + \dfrac{2}{x-1}$

15. $\dfrac{4}{x-3} + \dfrac{1}{x+2}$

17. $\dfrac{2}{x} - \dfrac{4}{x-1} + \dfrac{7}{(x-1)^2}$

19. $\dfrac{3}{2x-1} - \dfrac{2}{x+2} + \dfrac{10}{(x+2)^2}$

21. $1 + 3x$

23. $\dfrac{x+1}{x^2+2} - \dfrac{2x}{(x^2+2)^2}$

25. $2x + 3 + \dfrac{5}{3x+1} + \dfrac{1}{x-1}$

27. $\dfrac{3}{x+2} - \dfrac{2}{(x+2)^2} + \dfrac{1}{(x+2)^3}$

29. $\dfrac{5}{x+1} - \dfrac{5}{x+2} - \dfrac{5}{(x+2)^2}$

31. $\dfrac{1}{4a^2(x+a)} + \dfrac{1}{4a^2(x-a)} - \dfrac{x}{2a^2(a^2+x^2)}$

33. $y = \dfrac{6}{x+1} - \dfrac{5}{x-1}$

35. $y = 2 - \dfrac{1}{x} + \dfrac{3}{x+1}$

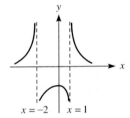

37. $x = -\dfrac{1}{9},\ y = \dfrac{1}{9},\ z = \dfrac{14}{9}$

Chapter 11 Review, page 606

1. $\begin{bmatrix} 23 & 13 & 26 \\ 4 & 15 & -1 \\ -5 & -5 & -6 \end{bmatrix}$

3. $\begin{bmatrix} 25 & 20 \\ 1 & 3 \end{bmatrix}$

5. $\begin{bmatrix} 18 & 13 & 19 \\ 1 & -7 & 4 \end{bmatrix}$

7. $\begin{bmatrix} 6 \\ -9 \end{bmatrix}$

9. $\begin{bmatrix} 0 & 0 & 0 \\ 12 & 3 & 0 \end{bmatrix}$

11. Not defined.

13. 1×1

15. $x = 3,\ y = 3,\ w = 7,\ z = 2$

17. 5

19. -13

21. -5

23. -85

25. -33

27. 157

29. 9

31. $\begin{bmatrix} 2 & -5 \\ -1 & 3 \end{bmatrix}$

33. No inverse exists.

35. Not inverses.
37. $x = 5, \quad y = -3$
39. $x = -1, \quad y = 5, \quad z = -3$
41. $-\dfrac{4}{a} + \dfrac{1}{a+1} - \dfrac{3}{(a+1)^2}$
43. $-\dfrac{4}{3x-1} + \dfrac{5}{2x-1}$

Chapter 11 Test, page 608

1. a. 3×7
   b. Undefined.
   c. 1×3
   d. 2×2
2. $\begin{bmatrix} 8 & -3 \\ 1 & 3 \end{bmatrix}$
3. $\begin{bmatrix} 15 & -15 \\ 1 & 5 \end{bmatrix}$
4. $\begin{bmatrix} 18 & -6 \\ 9 & 2 \end{bmatrix}$
5. 10
6. 7
7. -2
8. -7
9. $\begin{bmatrix} \frac{1}{3} & 0 & \frac{1}{3} \\ -\frac{2}{5} & \frac{2}{5} & \frac{1}{5} \\ \frac{2}{15} & \frac{1}{5} & -\frac{1}{15} \end{bmatrix}$
10. No inverse exists.
11. $x = 39, y = -94$
12. a. $\begin{bmatrix} 0.05 & 0.09 & 0.40 \\ 0.18 & 0.10 & 0.20 \\ 0.11 & 0.22 & 0.02 \end{bmatrix}$
    b. U.S. branch $818.12 million, Canadian branch $429.49 million, Mexican branch $596.41 million.

Chapter 12

Section 12.1, page 618

1. 1
3. 40,320
5. 15
7. 21
9. 6
11. 126
13. $35x^4y^3$
15. $-177,324,147t^5$
17. $\dfrac{42}{3,125} \cdot \dfrac{1}{a^2}$
19. $504t^{5/2}, -504\sqrt{2}t^2$
21. $a^4 + 4a^3b + 6a^2b^2 + 4ab^3 + b^4$
23. $4,096a^6 - 6,144a^5b + 3,840a^4b^2 - 1,280a^3b^3 + 240a^2b^4 - 24ab^5 + b^6$
25. $292 + 112\sqrt{6}$
27. $\dfrac{a^{12}}{b^6} - \dfrac{6a^{10}}{b^2} + 15a^8b^2 - 20a^6b^6 + 15a^4b^{10} - 6a^2b^{14} + b^{18}$
29. $243x^{10} + 810x^9y^3 + 1,080x^8y^6 + 720x^7y^9 + 240x^6y^{12} + 32x^5y^{15}$
31. $\dfrac{1}{x^3} + \dfrac{6}{x^2} + \dfrac{15}{x} + 20 + 15x + 6x^2 + x^3$
33. 0
35. 0
37. $a^6 + 6\sqrt{5}a^5 + 75a^4 + 100\sqrt{5}a^3 + 375a^2 + 150\sqrt{5}a + 125$
39. $1,024a^{10} - 5,120a^9b + 11,520a^8b^2 - 15,360a^7b^3$
41. $x^{-4} - 24x^{-3/2} + 252x$
43. $\dfrac{-5,940x^3}{y^9} + \dfrac{594x^2}{y^{10}} - \dfrac{36x}{y^{11}} + \dfrac{1}{y^{12}}$
45. $\binom{12}{8}\left(\dfrac{3a^{-2}}{2}\right)^4 \cdot \left(\dfrac{a}{3}\right)^8 = \dfrac{55}{144}$
47. $a^7 + 7a^6b + 21a^5b^2 + 35a^4b^3 + 35a^3b^4 + 21a^2b^5 + 7ab^6 + b^7$
49. $a^9 + 9a^8b + 36a^7b^2 + 84a^6b^3 + 126a^5b^4 + 126a^4b^5 + 84a^3b^6 + 36a^2b^7 + 9ab^8 + b^9$
51. $\dfrac{3y}{x^{5/3}}$
53. $nx^{n-1} + \binom{n}{2}x^{n-2}h + \cdots + nxh^{n-2} + h^{n-1}$
55. $1 - x + x^2 - \cdots - x^7$
57. $1 + \dfrac{1}{2}x - \dfrac{1}{8}x^2 + \dfrac{1}{16}x^3 - \dfrac{5}{128}x^4 + \dfrac{7}{256}x^5 - \dfrac{21}{1,024}x^6 + \dfrac{33}{2,048}x^7$
59. $\dfrac{2(n+2)^2}{n^2(n+1)}$

61. $\dfrac{100 \cdot 99 \cdot 98 \cdot 97}{4!}(0.91)^{96}(0.09)^4 \approx 0.03$

63. $x^{500}, 500x^{499}y, 124{,}750x^{498}y^2, 20{,}708{,}500x^{497}y^3,$
$2{,}573{,}031{,}125x^{496}y^4$

65. 3.548605186E11

67. Approximation, 1.18428; exact (10 places), 1.184304431.

Section 12.2, page 624

1. Prove: $1 + 5 + 9 + \cdots + (4n - 3) = 2n^2 - n$.
For $n = 1$: $4 - 3 = 1 = 2 - 1$. Assume true for n.
Consider: $1 + 5 + 9 + \cdots + (4n - 3) + 4(n + 1) - 3 = 2n^2 - n + 4(n + 1) - 3$
$$= 2n^2 + 3n + 1$$
$$= 2(n^2 + 2n + 1) - (n + 1)$$
$$= 2(n + 1)^2 - (n + 1)$$

3. Prove: $1 \cdot 2 + 3 \cdot 4 + 5 \cdot 6 + \cdots + (2n - 1)(2n) = \dfrac{n(n + 1)(4n - 1)}{3}$.
For $n = 1$: $(2 - 1)(2) = 2 = \dfrac{1(2)(3)}{3}$. Assume true for n.
Consider: $1 \cdot 2 + 3 \cdot 4 + 5 \cdot 6 + \cdots + (2n - 1)(2n) + [2(n + 1) - 1][2(n + 1)]$
$$= \dfrac{n(n + 1)(4n - 1)}{3} + \dfrac{6(2n + 1)(n + 1)}{3} = \dfrac{(n + 1)[4n^2 - n + 12n + 6]}{3}$$
$$= \dfrac{(n + 1)(4n + 3)(n + 2)}{3} = \dfrac{(n + 1)[(n + 1) + 1][4(n + 1) - 1]}{3}$$

5. Prove: $\dfrac{1}{2} + \dfrac{1}{4} + \dfrac{1}{8} + \cdots + \dfrac{1}{2^n} = \dfrac{2^n - 1}{2^n}$.
For $n = 1$: $\dfrac{1}{2} = \dfrac{2 - 1}{2}$. Assume true for n.
Consider: $\dfrac{1}{2} + \dfrac{1}{4} + \cdots + \dfrac{1}{2^n} + \dfrac{1}{2^{n+1}} = \dfrac{2^n - 1}{2^n} + \dfrac{1}{2^{n+1}} = \dfrac{2^{n+1} - 2 + 1}{2^{n+1}} = \dfrac{2^{n+1} - 1}{2^{n+1}}$.

7. Prove: $1^3 + 2^3 + 3^3 + \cdots + n^3 = \left[\dfrac{n(n + 1)}{2}\right]^2$.
For $n = 1$: $1 = \left(\dfrac{1(2)}{2}\right)^2$. Assume true for n.
Consider: $1^3 + 2^3 + \cdots + n^3 + (n + 1)^3 = \left[\dfrac{n(n + 1)}{2}\right]^2 + (n + 1)^3 = (n + 1)^2 \dfrac{[n^2 + 4(n + 1)]}{4} = \left[\dfrac{(n + 1)(n + 2)}{2}\right]^2$.

9. Prove: $1^2 + 2^2 + 3^2 + \cdots + n^2 = \dfrac{n(n + 1)(2n + 1)}{6}$.
For $n = 1$: $1 = \dfrac{1 \cdot 2 \cdot 3}{6}$. Assume true for n.
Consider: $1^2 + 2^2 + \cdots + n^2 + (n + 1)^2 = \dfrac{n(n + 1)(2n + 1)}{6} + (n + 1)^2$
$= \dfrac{(n + 1)[2n^2 + n + 6n + 6]}{6} = \dfrac{(n + 1)(2n + 3)(n + 2)}{6} = \dfrac{(n + 1)(n + 2)[2(n + 1) + 1]}{6}$.

11. Prove: $2 + 4 + 6 + \cdots + 2n = n^2 + n$.
For $n = 1$: $2 = 1 + 1$. Assume true for n.
Consider: $2 + 4 + 6 + \cdots + 2n + 2(n + 1) = n^2 + n + 2n + 2 = (n + 1)^2 + (n + 1)$.

13. Prove: $1 + 2 + 3 + \cdots + n = \dfrac{n(n + 1)}{2}$.
For $n = 1$: $1 = \dfrac{1 \cdot 2}{2}$. Assume true for n.
Consider: $1 + 2 + \cdots + n + n + 1 = \dfrac{n(n + 1)}{2} + (n + 1) = \dfrac{(n + 1)(n + 2)}{2}$.

15. Prove: $n(n^2 + 5)$ is a multiple of 6.
For $n = 1$: $1(1 + 5) = 6$. Assume true for n; that is, $n(n^2 + 5) = 6k$ where k is an integer. Is $(n + 1)[(n + 1)^2 + 5]$ a multiple of 6?
$(n + 1)[(n + 1)^2 + 5] = (n + 1)(n^2 + 2n + 6) = n^3 + 2n^2 + 6n + n^2 + 2n + 6 = (n^3 + 5n) + (3n^2 + 3n + 6)$
$= 6k + 3(n^2 + n + 2) = 6k + 3(2j) = 6(k + j)$
Therefore true for $n + 1$. Thus $n(n^2 + 5)$ is a multiple of 6.

17. Prove: 3 is a factor of $n^3 - n + 3$.
True for $n = 1$ since $1 - 1 + 3 = 3$. Assume true for n: that is, $n^3 - n + 3 = 3k$ where k is an integer.
Consider $(n + 1)^3 - (n + 1) + 3 = n^3 + 3n^2 + 3n + 1 - n - 1 + 3 = n^3 + 3n^2 + 2n + 3$
$= (n^3 - n + 3) + (3n^2 + 3n) = 3k + 3(n^2 + n) = 3(k + n^2 + n) \Leftarrow$ multiple of 3.

19. Prove $n^2 + n$ is divisible by 2.
Clear for n even. For n odd: n^2 is also odd, so $n^2 + n = $ odd + odd = even is divisible by 2.

21. Prove 7 is a factor of $11^n - 4^n$.
True for $n = 1$ since $11 - 4 = 7$. Assume true for n; i.e., $11^n - 4^n = 7k$ where k is an integer.
Consider: $11^{n+1} - 4^{n+1} = 11(11^n) - 4(4^n) = (7 + 4)11^n - 4(4^n)$
$= 7(11^n) + 4(11^n - 4^n) = 7(11^n) + 4 \cdot 7k = 7(11^n + 4k) \Leftarrow$ divisible by 7

23. Prove $4^n > n^4$ for $n \geq 5$.
True for $n = 5$ because $4^5 > 5^4$, that is, $1{,}024 > 625$. Assume true for n.
Show $4^{n+1} > (n + 1)^4 \Leftrightarrow 4 \cdot 4^n > n^4 + 4n^3 + 6n^2 + 4n + 1$
$\Leftrightarrow (3 + 1) \cdot 4^n > n^4 + 4n^3 + 6n^2 + 4n + 1$
$\Leftrightarrow 3 \cdot 4^n + 4^n > n^4 + 4n^3 + 6n^2 + 4n + 1$
We have assumed $4^n > n^4$ and it can be shown that $4^n \geq 4n^3, 4^n \geq 6n^2, 4^n \geq 4n + 1$.

25. Prove $n^2 < 2^n$ for $n > 4$.
True for $n = 5$ since $25 < 2^5 = 32$. Assume true for n.
Then $n^2 + 2n + 1 < 2^n + 2n + 1 \to (n + 1)^2 < 2^n + 2^n = 2^n \cdot 2 = 2^{n+1}$.
Justification of above: Show $2n + 1 < 2^n$ for $n > 4$. True for $n = 5$ since $11 < 2^5 = 32$. Assume true for n; i.e., $2n + 1 < 2^n$.
Now $2n + 3 < 2^n + 2 \to 2(n + 1) + 1 < 2^n + 2^n = 2 \cdot 2^n = 2^{n+1}$.

27. $f(n) = 1 + 2 + 4 + \cdots + 2^{n-1}$

a.
n	1	2	3	4	5	6
$f(n)$	1	3	7	15	31	63

b. $63 + 64 = 127 = f(7)$

c. $f(n) = 2^n - 1$

29. $f(n) = 2 + 9 + 16 + \cdots + (7n - 5)$

a.
n	1	2	3	4	5	6
$f(n)$	2	11	27	50	80	117

b. $f(7) = 161$

c. $f(n) = 2n + \dfrac{7}{2}n(n - 1)$

31. Prove that it is possible to pay any debt $4, $5, $6, \ldots, n by using only $2 and $5 bills. True for $4 = $2 + $2. Assume true for n; that is, $n = 2j + 5k$ where j and k are nonnegative integers. Consider $n + 1 = 2j + 5k + 1$. If $j \geq 2$, then:
$n + 1 = 2j + 5k + 5 - 4$
$= 2(j - 2) + 5(k + 1) \leftarrow$ true for $j \geq 2$

If $j = 0, 1$, then $k \geq 1$. So
$$n + 1 = 2j + 5k + 1 = 2j + 5(k - 1) + 6$$
$$= 2(j + 3) + 5(k - 1) \leftarrow \text{true here as well}$$

33. Prove that the number of diagonals of an n-sided convex polygon is $n(n - 3)/2$ for $n \geq 3$. n vertices can be connected in $_nC_2$ ways. Thus the number of diagonals is
$$_nC_2 - n = \frac{n!}{2(n-2)!} - n = \frac{n(n-1)}{2} - \frac{2n}{2} = \frac{(n-3)n}{2}$$

35. Prove $n + 5 < 2^n$ for all integers $n \geq 4$.
True for $n = 4$ since $9 < 2^4 = 16$. Assume true for n. Now,
$$n + 5 + 1 < 2^n + 1$$
$$(n + 1) + 5 < 2^n + 2^n \quad \text{(since } 2^n > 1 \text{ for } n \geq 4)$$
$$(n + 1) + 5 < 2 \cdot 2^n = 2^{n+1}, \quad \text{as was to be shown.}$$

37. Prove $3^n > 2^n + 20$ for $n \geq 4$.
True for $n = 4$ since $3^4 = 81 > 2^4 + 20 = 36$. Assume true for n. If $3^n > 2^n + 20$, then $3^n > 2^n + 10$. So,
$$2(2^n + 10) < 2 \cdot 3^n \rightarrow 2^{n+1} + 20 < 2 \cdot 3^n < 3 \cdot 3^n = 3^{n+1}.$$

39. P_n true for every integer $n \geq 18$.

41. Nothing.

43. P_n is true for $n \geq 1$.

Section 12.3, page 630

1. 3, 4, 5, 10
3. -2, undefined, 4, $\frac{3}{2}$
5. $\frac{1}{4}, \frac{1}{2}, 1, 32$
7. $a_1 = 1, \quad a_2 = \frac{1}{4}, \quad a_3 = \frac{1}{9}, \quad a_8 = \frac{1}{64}$
9. $-2, -2\sqrt{2}, -2\sqrt[3]{3}, -2\sqrt[8]{8}$
11. $-2, -6, -18, -4{,}374$
13. $0, \frac{3}{2}, \frac{8}{7}, \frac{63}{62}$
15. 2, 7, 12, 17, 22
17. $9, 3, 1, \frac{1}{3}, \frac{1}{9}$
19. $2, \frac{5}{2}, \frac{10}{3}, \frac{17}{4}, \frac{26}{5}$
21. $-\frac{1}{2}, \frac{1}{8}, -\frac{1}{24}, \frac{1}{64}, -\frac{1}{160}$
23. $\frac{3}{2}, \frac{3}{2}, \frac{3}{2}, \frac{3}{2}, \frac{3}{2}$
25. $3, \frac{1}{3}, 3, \frac{1}{3}, 3$
27. 2
29. 17
31. $\frac{13}{12}$
33. 56
35. 150
37. 10
39. $a_n = 2n - 1$
41. $a_n = (-1)^{1+n} 2^n$
43. $a_n = 2^{-(2n-1)}$
45. $\sum_{n=1}^{5} \frac{1}{n}$
47. $\sum_{n=1}^{10} n$
49. $\sum_{n=1}^{6} (-3)^n$
51. $x = 2$
53. $x = -46$
55. $x = 2$
57. $a_{10} = 48{,}828{,}125$
59. 5,120
61. 1,770
63. 2.928968254

Section 12.4, page 635

1. Difference is 2.
3. Difference is 3.
5. Difference is 1.
7. Difference is -2.
9. -25
11. 261
13. -19
15. -165

17. -16.5
19. $a_1 = -\frac{275}{4}, \quad d = \frac{19}{4}$
21. $a_1 = 114, \quad d = -7$
23. $a_1 = \frac{3}{5}, \quad d = \frac{3}{5}$
25. $a_1 = 8, \quad d = -3$
27. $a_1 = -10, \quad d = 4$
29. 5
31. -3
33. 21
35. -11
37. 30
39. 440
41. -532
43. 561
45. 35
47. $-1,005$
49. $4,185$
51. 136
53. $\$54.85$
55. $a_n = n^2, b_n = a_n - a_{n-1}$
$b_{n+1} = a_{n+1} - a_n = (n+1)^2 - n^2$
$b_{n+1} - b_n = (n+1)^2 - n^2 - [n^2 - (n-1)^2]$
$= n^2 + 2n + 1 - n^2 - [n^2 - n^2 + 2n - 1]$
$= 2$
Therefore, $b_1, b_2, b_3, \ldots$ is an arithmetic sequence.
57. 45.55
59. $59,900$

Section 12.5, page 644

1. $\frac{a_{n+1}}{a_n} = -\frac{1}{3}$
3. $\frac{a_{n+1}}{a_n} = 4$
5. $\frac{a_{n+1}}{a_n} = \frac{1}{3}$
7. $\frac{1}{1,024}$
9. 8
11. $\frac{256}{81}$
13. $6,144$
15. $6,144$
17. $-\frac{3}{10}$
19. $\frac{2}{81}$
21. 5
23. 4
25. $\frac{5}{3}, 15, -45, 135$; or $-\frac{5}{3}, -15, -45, -135$
27. $n = 4, \quad a_n = -40$
29. $n = 5, \quad S_n = 484$
31. $a_n = 324, \quad a_1 = 4$
33. $a_1 = 6, \quad S_n = 762$
35. $a_n = -\frac{2}{27}, \quad S_n = \frac{3,280}{27}$
37. $a_1 = 3, \quad n = 3$
39. $a_n = (12) \cdot \left(-\frac{1}{3}\right)^{n-1}$
41. $r = -\sqrt[3]{131}$,
$a_n = -\frac{50\left(-\sqrt[3]{131}\right)^n}{131}$
43. $\frac{2}{3}$
45. -5
47. 1
49. Sum does not exist.
51. $-\frac{1}{2}$
53. 30
55. 8
57. $\$28,809.07$
59. $\$11,865.23$
61. $1,024$
63. $322,961$
65. $\$101,578.77$
67. $\$1,489.35$
69. $93\frac{1}{3}$ ft
71. $\$543,179.08$
73. $\$28,146.97$
75. 1.300614744
77. 149.9548907

Section 12.6, page 650

1. $y_n = 1.1^n$
3. $y_n = 9 \cdot 2^n + 1$
5. $y_n = -490,000 \cdot 1.001^n + 500,000$
7. $y_n = 4,500 \cdot 1.01^n - 4,000$
9. 2
11. 2
13. $\$665.30$
15. $\$405.64$
17. $\$632.07$
19. $\$11,006.38$
21. $\$1,560.92$
23. $\$378,355.82$

25.
Month	Interest paid	Balance owed
1	$291.67	$50,291.67
2	293.37	50,585.04
3	295.08	50,880.12
4	296.80	51,176.92
5	298.53	51,475.45
6	300.27	51,775.72
7	302.03	52,077.75
8	303.79	52,381.54
9	305.56	52,687.10
10	307.34	52,994.44
11	309.13	53,303.57
12	310.94	53,614.51

Section 12.7, page 658

1. 24
3. 30
5. 3
7. $n(n-1)(n-2)(n-3)(n-4)$
9. 210
11. 56
13. 5
15. 45
17. n
19. 1
21. $n = 100$
23. $n = 7$
25. $_nP_4 - {_nP_3}$
    $= n(n-1)(n-2)(n-3) -$
    $\quad n(n-1)(n-2)$
    $= n(n-1)(n-2)(n-3-1)$
    $= (n-4)\,_nP_3$
27. 156
29. 20,000
31. 32
33. 720
35. 48
37. 3,024
39. 16
41. 311,875,200
43. 12
45. $\dfrac{12!}{5!3!2!2!} = 166{,}320$
47. 56
49. 24
51. 696,729,600
53. 16
55. a. 36
    b. 100
    c. 24
57. 4,800
59. 30,030
61. 2,162,160
63. 210

Section 12.8, page 663

1. {(H, 1), (H, 2), (H, 3), (H, 4), (H, 5), (H, 6), (T, 1), (T, 2), (T, 3), (T, 4), (T, 5), (T, 6)}
3. {MM, MF, FF, FM}
5. {2, 3, 4, 5, 6, 7, 8, 9, 10, 11, 12}
7. $_{52}C_5 = 2{,}598{,}960$ hands
9. 729
11. Ordered pairs of the numbers: {653, 654, 623, 624, 153, 154, 123, 124}; 64 elements in the sample space.
13. {WI, WL, WA, WM, IW, IL, II, IA, IM, LW, LI, LL, LA, LM, AW, AI, AL, AM, MW, MI, ML, MA}
15. {GGG, BGG, BBG, BBB}
17. {PN, PD, PQ, NP, ND, NQ, DN, DP, DQ, QP, QN, QD}
19. {3, 4, ..., 18}
21. $\dfrac{3}{5}$
23. $\dfrac{1}{8}$
25. $\dfrac{33}{16{,}660}$
27. $\dfrac{1}{9}$
29. $\dfrac{4}{465}$
31. $\dfrac{1}{15}$
33. $\dfrac{7}{30}$
35. $\dfrac{1}{200}$
37. $\dfrac{1}{312}$
39. $\dfrac{1}{5}$
41. $\dfrac{9}{10}$

Chapter 12 Review, page 667

1. $a^4 + 4a^3t + 6a^2t^2 + 4at^3 + t^4$
3. $2{,}187a^7b^{14} - 20{,}412a^8b^{13} + 81{,}648a^9b^{12} - 181{,}440a^{10}b^{11}$
   $+ 241{,}920a^{11}b^{10} - 193{,}536a^{12}b^9 + 86{,}016a^{13}b^8$
   $- 16{,}384a^{14}b^7$
5. -252
7. $2^3 = 8 = 2 \cdot 1^3 \cdot (1+1)^2 \rightarrow P_1$
   Assume $P_n: 2^3 + 4^3 + \cdots + (2n)^3 = 2n^2(n+1)^2$
   $2^3 + 4^3 + \cdots + (2n)^3 + (2n+2)^3$
   $\quad = 2n^2(n+1)^2 + (2n+2)^3$
   $\quad = 2n^2(n+1)^2 + 8(n+1)^3$
   $\quad = 2(n+1)^2(n^2 + 4(n+1))$
   $\quad = 2(n+1)^2(n+2)^2 \rightarrow P_{n+1}$

9. Prove $1 + 4 + 7 + \cdots + (3n - 2) = \frac{n}{2}(3n - 1)$

$$3 \cdot 1 - 2 = \frac{1}{2}(3 \cdot 1 - 1) \to P_1$$

Assume P_n:

$$1 + 4 + 7 + \cdots + (3n - 2) = \frac{n}{2}(3n - 1)$$

$$1 + 4 + 7 + \cdots + (3n - 2) + 3[(n + 1) - 2]$$

$$= \frac{n}{2}(3n - 1) + 3[(n + 1) - 2]$$

$$= \frac{n}{2}(3n - 1) + 3n + 1$$

$$= \frac{3n^2}{2} + \frac{5n}{2} + 1$$

$$= \frac{3n^2 + 5n + 2}{2}$$

$$= \frac{(n + 1)(3n + 2)}{2} \to P_{n+1}$$

11. a.

n	1	2	3	4	5	6	7	8
$f(n)$	2	7	15	26	40	57	77	100

   b. 126
   c. 126
   d. $f(n) = 2n + \frac{3}{2}n(n - 1)$

13. $-1, 0, 1, 8$

15. $2, \frac{4}{3}, 1, \frac{4}{11}$

17. $a_n = n(n + 1)$, $a_1 = 2$, $a_2 = 6$, $a_3 = 12$, $a_4 = 20$, $a_5 = 30$

19. $a_1 = 2$, $a_2 = \frac{3}{2}$, $a_3 = 2$, $a_4 = \frac{3}{2}$, $a_5 = 2$

21. 14

23. $\frac{5{,}269}{3{,}600}$

25. Common difference $= 1$.

27. Common difference $= 3$.

29. 17

31. -14

33. $d = 5$, $a_1 = -34$

35. $d = -3$, $a_1 = -5$

37. 16

39. 22

41. 150

43. $-1{,}120$

45. 1

47. 43,500

49. Common ratio $= \frac{1}{4}$.

51. Common ratio $= 2$.

53. 2^{19}

55. 243

57. 2^{n+1}

59. The fourth term.

61. $2,960.49

63. $a_1 = 4, S_5 = 484$

65. $a_n = (-2) \cdot (-5)^{n-1}$

67. $\frac{20}{3}$

69. 72

71. $46\frac{2}{3}$ ft

73. 24

75. 6,760,000

77. 6

79. $9! = 362{,}880$

81. 120

83. 900 (0 cannot be first digit)

85. 210

87. 84

89. 330

91. 27,405

93. $\{HHHHH, HHHHT, HHHTH, \ldots, TTTTT\}$, 32 sample points.

95. $\{BBB, BBR, BRB, RBB, BRR, RBR, RRB, RRR\}$

97. $\frac{4}{9}$

99. $\frac{3}{5}$

Chapter 12 Test, page 670

1. $a^4 - 12a^3b + 54a^2b^2 - 108ab^3 + 81b^4$

2. True for $n = 1$: $4 \cdot 1 - 3 = 2(1)^2 - 1$. Assume that $1 + 5 + 9 + 13 + \cdots + (4n - 3) = 2n^2 - n$. Add $4(n + 1) - 3 = 4n + 1$ to both sides:

   $1 + 5 + 9 + 13 + \cdots + (4n - 3) + 4((n + 1) - 4)$
   $= 2n^2 - n + (4n + 1)$
   $= 2n^2 + 3n + 1$
   $= 2(n^2 + 2n + 1) - n - 1$
   $= 2(n + 1)^2 - (n + 1)$

3. a. 7, 12, 17, 52
   b. $0.1, 0.01, 0.001, 0.0000000001$

4. 1,124

5. $a_1 = 16$, $d = -1$

6. 2,385

7. $a_1 = \frac{5}{8}$, $r = \frac{1}{2}$

8. $\frac{5{,}115}{4{,}096}$

9. 63 m

10. 24

11. 1,001

12. $374.13

Chapter 13

Section 13.1, page 678

1. (b)
3. (c)
5. (d)
7. Vertex: $(0, 0)$
   Focus: $\left(0, \frac{1}{36}\right)$
   Directrix: $y = -\frac{1}{36}$
9. Vertex: $(0, 0)$
   Focus: $\left(-\frac{1}{2}, 0\right)$
   Directrix: $x = \frac{1}{2}$
11. Vertex: $(1, -1)$
    Focus: $\left(1, -\frac{7}{8}\right)$
    Directrix: $y = -\frac{9}{8}$
13. Vertex: $\left(0, -\frac{1}{2}\right)$
    Focus: $(0, -1)$
    Directrix: $y = 0$
15. Vertex: $\left(-\frac{1}{3}, -3\right)$
    Focus: $\left(-\frac{1}{3}, -\frac{26}{9}\right)$
    Directrix: $y = -\frac{28}{9}$
17. Vertex: $\left(-\frac{41}{24}, -\frac{3}{2}\right)$
    Focus: $\left(-\frac{5}{24}, -\frac{3}{2}\right)$
    Directrix: $x = -\frac{77}{24}$
19. Vertex: $\left(-\frac{1}{5}, 4\right)$
    Focus: $\left(-\frac{1}{5}, \frac{401}{100}\right)$
    Directrix: $y = \frac{399}{100}$
21. $y^2 = -8x$
23. $(x - 2)^2 = 8(y - 1)$
25. $y^2 = 4x$
27. $(x - 3)^2 = 24(y + 2)$
29. $(x - 1)^2 = 8y$
31. $(y + 4)^2 = -8(x - 1)$
33. $\left(x - \frac{1}{4}\right)^2 = \frac{5}{4}\left(y + \frac{1}{20}\right)$
35. $\left(y - \frac{71}{6}\right)^2 = -\frac{140}{3}\left(x - \frac{5,041}{1,680}\right)$
37. $y - 4 = 4(x - 2) \Rightarrow y = 4x - 4$
39. $y = \frac{1}{8}x + \frac{45}{4}$
41. $y^2 = 8x$
43. 225 inches
45. $(x - 2,000)^2 = 8,000y$
47.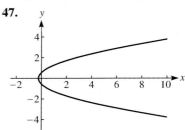
49.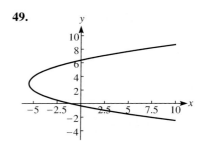

Section 13.2, page 688

1. (b)
3. (c)
5. (e)
7. $\frac{x^2}{5} + \frac{y^2}{9} = 1$
9. $\frac{(x + 1)^2}{5} + \frac{(y - 7)^2}{9} = 1$
11. $\frac{x^2}{9} + \frac{y^2}{49} = 1$
13. $\frac{(x - 1)^2}{4} + (y - 2)^2 = 1$
15. $\frac{(x - 5)^2}{21} + \frac{(y - 2)^2}{25} = 1$
17. $\frac{9x^2}{8} + y^2 = 1$
19. $\frac{(x - 10)^2}{25} + \frac{(y - 1)^2}{16} = 1$
21. $\frac{x^2}{4} + \frac{y^2}{16} = 1$
23. $\frac{x^2}{9} + \frac{5y^2}{9} = 1$

25. $a = 3$, $b = 2$, $c = \sqrt{5}$, $e = \dfrac{\sqrt{5}}{3}$, foci $(\pm\sqrt{5}, 0)$

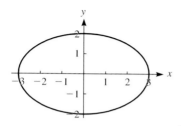

27. $a = 2\sqrt{2}$, $b = 2$, $c = 2$, $e = \dfrac{\sqrt{2}}{2}$, foci $(\pm 2, 0)$

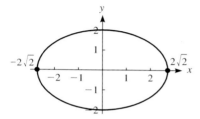

29. $a = 4$, $b = 2$, $c = 2\sqrt{3}$, $e = \dfrac{\sqrt{3}}{2}$, foci $(-2, 2 \pm 2\sqrt{3})$

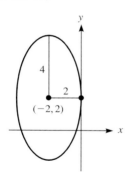

31. $a = 6$, $b = 3\sqrt{2}$, $c = 3\sqrt{2}$, $e = \dfrac{\sqrt{2}}{2}$, foci $(1 \pm 3\sqrt{2}, 2)$

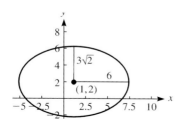

33. $a = 5$, $b = \sqrt{5}$, $c = 2\sqrt{5}$, $e = \dfrac{2\sqrt{5}}{5}$, foci $(2 \pm 2\sqrt{5}, -3)$

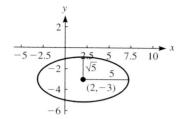

35. $a = 6$, $b = 2$, $c = 4\sqrt{2}$, $e = \dfrac{2\sqrt{2}}{3}$, foci $(0, \pm 4\sqrt{2})$

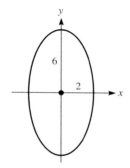

37. $a = 6$, $b = 2$, $c = 4\sqrt{2}$, $e = \dfrac{2\sqrt{2}}{3}$, foci $(3, 2 \pm 4\sqrt{2})$

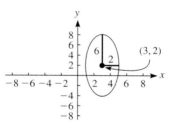

39. $a = 5$, $b = 4$, $c = 3$, $e = \dfrac{3}{5}$, foci $(-1, -1 \pm 3)$

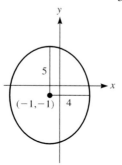

41. $a = b = 3$, $c = 0$, $e = 0$, center $(0, -3)$, focus $(0, -3)$

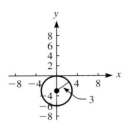

51.

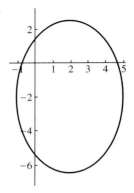

43. Endpoints of latera recta are $\left(\frac{4}{3}, \sqrt{5}\right), \left(\frac{4}{3}, -\sqrt{5}\right),$ $\left(-\frac{4}{3}, \sqrt{5}\right), \left(-\frac{4}{3}, -\sqrt{5}\right)$

45. Prove that $b^2 = a^2(1 - e^2)$.
$$a^2 = b^2 + c^2 \rightarrow b^2 = a^2 - c^2$$
$$\rightarrow b^2 = a^2\left(1 - \frac{c^2}{a^2}\right) = a^2(1 - e^2)$$

53. The ellipse becomes flatter in the vertical direction.

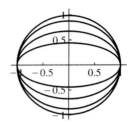

47. $\dfrac{x^2}{(24{,}000)^2} + \dfrac{y^2}{(22{,}000)^2} = 1$

49.

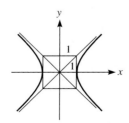

Section 13.3, page 698

1. (a)

3. (c)

5. (b)

7. center $(0, 0)$, $e = \sqrt{2}$, vertices $(\pm 1, 0)$, foci $(\pm \sqrt{2}, 0)$, asymptotes $y = \pm x$

9. center $(0, 0)$, $e = \sqrt{5}/2$, vertices $(\pm 2, 0)$, foci $(\pm \sqrt{5}, 0)$, asymptotes $y = \pm \dfrac{x}{2}$

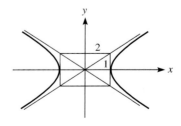

11. center $(0, 0)$, $e = \dfrac{\sqrt{34}}{5}$, vertices $(\pm 10, 0)$, foci $(\pm 2\sqrt{34}, 0)$, asymptotes $y = \pm\dfrac{3x}{5}$

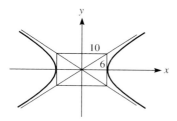

13. center $(1, 2)$, $c = \sqrt{2}$, $e = \sqrt{2}$, vertices $(1 \pm 1, 2)$, foci $(1 \pm \sqrt{2}, 2)$, asymptotes $y - 2 = \pm(x - 1)$

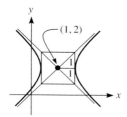

15. center $(4, 1)$, $e = \dfrac{\sqrt{34}}{3}$, vertices $(1, 1), (7, 1)$, foci $(4 \pm \sqrt{34}, 1)$, asymptotes $y - 1 = \pm\left[\dfrac{5}{3}(x - 4)\right]$

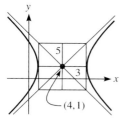

17. center $(0, 0)$, $e = \dfrac{\sqrt{13}}{3}$, vertices $\left(\dfrac{1}{2}, 0\right), \left(-\dfrac{1}{2}, 0\right)$, foci $\left(\pm\dfrac{\sqrt{13}}{6}, 0\right)$, asymptotes $y = \pm\dfrac{2x}{3}$

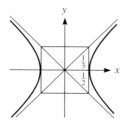

19. center $(-6, 4)$, $e = \dfrac{2\sqrt{6}}{3}$, vertices $(-6 \pm \sqrt{3}, 4)$, foci $(-6 \pm \sqrt{8}, 4)$, asymptotes $y - 4 = \pm\sqrt{\dfrac{5}{3}}(x + 6)$

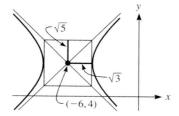

21. center $\left(3, \dfrac{3}{2}\right)$, $e = \dfrac{\sqrt{5}}{2}$, vertices $\left(3 \pm \sqrt{30}, \dfrac{3}{2}\right)$, foci $\left(3 \pm 5\dfrac{\sqrt{6}}{2}, \dfrac{3}{2}\right)$, asymptotes $y - \dfrac{3}{2} = \pm\dfrac{1}{2}(x - 3)$

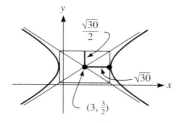

23. center $(-2, 4)$, $e = \dfrac{2\sqrt{3}}{3}$, vertices $(-2 \pm 1, 4)$, foci $\left(-2 \pm 2\dfrac{\sqrt{3}}{3}, 4\right)$, asymptotes $y - 4 = \pm\dfrac{1}{\sqrt{3}}(x + 2)$

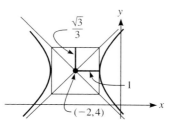

25. center $(3, -1)$, $e = \dfrac{\sqrt{13}}{2}$, vertices $\left(3, -1 \pm \dfrac{1}{3}\right)$, foci $\left(3, -1 \pm \dfrac{\sqrt{13}}{6}\right)$, asymptotes $y + 1 = \pm\dfrac{2}{3}(x - 3)$

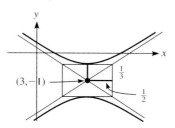

A-74 ANSWERS Section 13.3, page 698—Section 13.4, page 706

27. $\dfrac{x^2}{4} - \dfrac{y^2}{1} = 1$

29. $\dfrac{(y-1)^2}{1^2} - \dfrac{(x-1)^2}{(3/2)^2} = 1$

31. $\dfrac{x^2}{9} - \dfrac{y^2}{16} = 1$

33. $\dfrac{(x+1)^2}{16} - \dfrac{(y-1)^2}{20} = 1$

35. $\dfrac{x^2}{4} - \dfrac{y^2}{16} = 1$

37. $\dfrac{y^2}{4} - \dfrac{x^2}{20} = 1$

39. Ellipse

41. Hyperbola

43. Parabola

45. $\dfrac{256(x - \frac{5}{2})^2}{25} - \dfrac{256 y^2}{1{,}575} = 1$

47.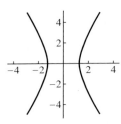

49.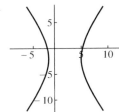

51. The graph of the hyperbola has a larger "opening."

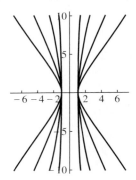

Section 13.4, page 706

1. $(3, -5)$
3. $(-4, -7)$
5. $(-6, 4)$
7. $(2, 3)$
9. $\left(-\dfrac{3}{2}, 2\right)$
11. $X^2 + 2XY + 6X + Y = -4$
13. $X^2 + 4XY - 16X - 6Y = -20$
15. $X^2 + 2XY + Y^2 + 4X + 4Y = -3$
17. $2X^2 - 3Y^2 - 20X - 12Y = -29$
19. $2X + 2Y = 11$
21. $X^2 + Y^2 = 9$
23. $X^2 - 2\sqrt{3}XY - Y^2 = 2$
25. $X^2 - 6XY + 9Y^2 = 8$
27. $X^2 + 2\sqrt{3}XY + 3Y^2 + \left(4 + 2\sqrt{3}\right)X + \left(4\sqrt{3} - 2\right)Y = 12$
29. $9X^2 - 82XY + 9Y^2 = -800$
31. $\theta = 45°$
33. $\theta = 22.5°$
35. $\theta \approx 34.1°$
37. $\theta = 45°$
39. $\theta \approx 61.845°$
41. $\theta \approx 53.13°$

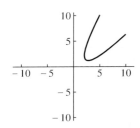

43. $\theta = 36.87°$

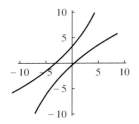

Section 13.5, page 711

1. $r = 1/(1 + \cos\theta)$
3. $r = 2/(1 - \cos\theta)$
5. $r = 10/(1 + \cos\theta)$
7. $r = 8/(1 + \sin\theta)$
9. $r = 5/(3 + \cos\theta)$
11. $r = 8/(3 - \cos\theta)$
13. $r = 6/(1 + 2\cos\theta)$
15. $r = 1/(1 + 3\cos\theta)$
17. Parabola, focus $(0,0)$, directrix $x = 1$
19. Hyperbola, focus $(0,0)$, directrix $y = 1$
21. Hyperbola, focus $(0,0)$, directrix $x = \frac{2}{5}$
23. Parabola, focus $(0,0)$, directrix $y = 4$
25. Ellipse, focus $(0,0)$, directrix $x = \frac{5}{3}$
27. Hyperbola, focus $(0,0)$, directrix $x = -1$
29. $r = 1 \big/ \left(1 + \sqrt{2}\cos\theta\right)$
31. $r = 8 \big/ \left(1 + \sqrt{5}\cos\theta\right)$
33. $r = 12{,}600/(1 + 0.05\cos\theta)$

Section 13.6, page 714

1.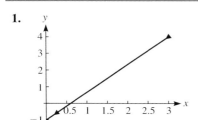

3.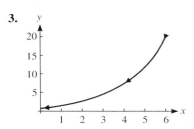

5.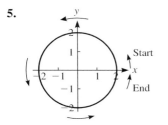

7.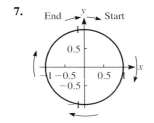

9.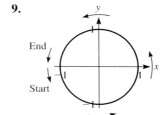

11.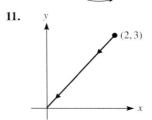

13.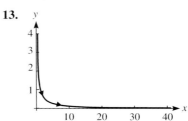

15.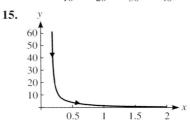

17. $y = \frac{5}{3}x - 1, 0 \le x \le 3$

19. $y = e^{x/2}, x \ge 0$

21. $x^2 + y^2 = 4, -2 \le x \le 2$

23. $x^2 + y^2 = 1, -1 \le x \le 1$

25. $x^2 + y^2 = 1, -1 \le x \le 1$

27. $y = \frac{3x}{2}, 0 \le x \le 2$

29. $y = \frac{1}{x}, x > 0$

31. $y = \frac{1}{x^2}, x > 0$

33. $x = h + r\cos\theta, y = k + r\sin\theta, 0 \le \theta \le 2\pi$
$$(x - h)^2 + (y - k)^2 = (r\cos\theta)^2 + (r\sin\theta)^2$$
$$= r^2(\sin^2\theta + \cos^2\theta)$$
$$= r^2$$

35. $x = h + a\sec\theta, y = k + b\tan\theta$
$$\begin{cases} -\frac{\pi}{2} < \theta < \frac{\pi}{2} & \text{left branch} \\ \frac{\pi}{2} < \theta < \frac{3\pi}{2} & \text{right branch} \end{cases}$$
$$\frac{(x - h)^2}{a^2} - \frac{(y - k)^2}{b^2} = \frac{a^2\sec^2\theta}{a^2} - \frac{b^2\tan^2\theta}{b^2}$$
$$= \sec^2\theta - \tan^2\theta = 1$$

37. $x = \cos t, y = \sin t, 0 \le t \le 2\pi$

39. $x = 2\cos t, y = \sin t, 0 \le t \le 2\pi$

41. $x = 1 + 10\cos t, y = 1 + \sqrt{84}\sin t, 0 \le t \le 2\pi$

43. $x = 1 + \sqrt{3}\sec t, y = 1 + \sqrt{6}\tan t, \frac{\pi}{2} \le t \le \frac{3\pi}{2}$

45. $x = 2 - 3t, y = 3 + t, 0 \le t \le 1$

47. Line segment from $(7, 1)$ to $(3, -1)$.

49. Same graph traced in the opposite direction.

51. Graph is translated horizontally 1 unit to the right and vertically 2 units upward.

Chapter 13 Review, page 719

1.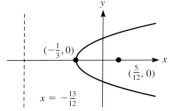

3.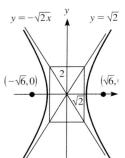

5.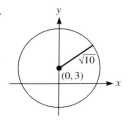

7.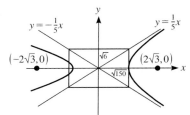

9.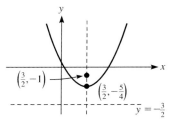

11.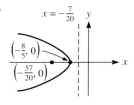

ANSWERS Chapter 13 Review, page 719—Chapter 13 Test, page 720 A-77

13.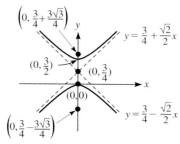

15. $(-1, 7)$
17. $(8, 3)$
19. $(-1, -6)$
21. $X^2 + 3XY + 4X - 5Y = 11$
23. $X^2 + 4XY + Y^2 + 2X + 10Y = 3$
25. $2X = 9$
27. $3X^2 - 10\sqrt{3}XY + 13Y^2 = 72$
29. $X^2 + Y^2 = 49$
31. $(\sqrt{3} - 1)X - (\sqrt{3} + 1)Y + 4 = 0$
33. $45°$
35. $30°$
37. $r = 2/(1 - \cos\theta)$
39. $r = 12/(5 + 7\cos\theta)$

41. Hyperbola, $e = 2$, focus at origin, directrix $x = \dfrac{1}{2}$
43. Hyperbola, $e = 2$, focus at origin, directrix $y = -1$
45.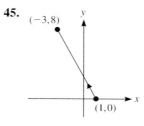

47.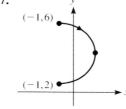

49. $y = \dfrac{2}{3}x + \dfrac{11}{3}, -1 \le x \le 14$
51. $y = \sqrt{x - 1}, x \ge 1$
53. $\dfrac{x^2}{16} + \dfrac{y^2}{9} = 1, -4 \le x \le 4$

Chapter 13 Test, page 720

1. Parabola
   Vertex $(0, 0)$
   Focus $\left(\dfrac{5}{4}, 0\right)$
   Directrix $x = -\dfrac{5}{4}$

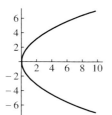

2. Parabola
   Vertex $(-3, -4)$
   Focus $\left(-3, -\dfrac{15}{4}\right)$
   Directrix $y = -\dfrac{17}{4}$

3. Ellipse
   Center $(0, 0)$
   Semimajor axis $10\sqrt{2}$
   Semiminor axis $5\sqrt{2}$
   Foci $(\pm 5\sqrt{6}, 0)$
   Eccentricity $2\sqrt{3}/3$

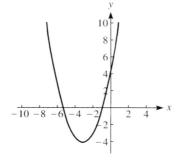

4. Ellipse
   Center $(-1, -3)$
   Semimajor axis $\sqrt{383/2}$
   Semiminor axis $\sqrt{383/9}$
   Foci $(-1 \pm \sqrt{5,362/6}, -3)$
   Eccentricity $\sqrt{7}/3$

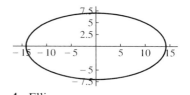

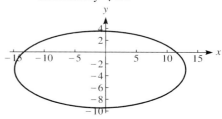

5. Hyperbola
Center (0, 0)
Transverse axis: 3, horizontal
Conjugate axis: 3, vertical
Foci $\left(\pm 3\sqrt{2}, 0\right)$
Eccentricity $\sqrt{2}$
Asymptotes $y = \pm x$

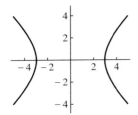

6. Hyperbola
Center (2, 1)
Transverse axis: $\sqrt{119/12}$, horizontal
Conjugate axis: $\sqrt{119}$, vertical
Foci $(2 \pm \sqrt{4{,}641}/6, 1)$
Eccentricity $\sqrt{13}$
Asymptotes $y - 1 = \pm 2\sqrt{3}(x - 2)$

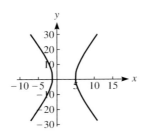

7. Rotate the axes $-45°$.

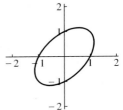

8. a.

b. $y = \left(\dfrac{x-1}{2}\right)^2 - 1$

9. 21,481 miles

10. $(x - 2{,}600)^2 = \dfrac{135{,}200}{7} y$

INDEX

A

Abscissa, 70
Absolute value, 9–11
  equations involving, 195–96
  inequalities involving, 197
Absolute Value Inequality, Fundamental, 197
Absolute value (modulus) of a complex number, 491–93
Abstract algebra, 121
Accrued interest, 119
Acute angle, 346
Addition
  of complex numbers, 232–33
  of rational expressions, 44–46
Addition and subtraction identities
  for the cosine function, 425–27
  for the sine function, 428–29
  for the tangent function, 431–33
Addition identities, 424–33
Additive identity, 3
  for complex numbers, 233
Additive inverse, 3
  of a complex number, 233
  of a matrix, 569
Adelman, L., 62
al-Khwarizmi, 1
Algebra, 1
  abstract, 121
  Fundamental Theorem of, 239
  history of, 1
Algebraic expression, 29
Algorithms, data compression, 562
Amplitude of sine and cosine graphs, 383
Analysis, break-even, 103–4, 225–26
Analysis, input–output, 593–97
Analytic geometry, 671–714
Analytic trigonometry, 416–64
Angle(s), 344–46
  reference, 357–58
  in standard position, 344
  of triangles, determining, 367–68
Angular speed, 351–52

Arc, length of an, 350–51
Arcsine. *See* Sine, function, inverse
Areas of triangles, 581–83
Arithmetic sequence, 631–35
Artin, Emil, 121
Associative law, 5
Associative law of addition for complex numbers, 233
Associative law of multiplication for complex numbers, 233
Asymptotes
  horizontal, 248–49
  of hyperbolas, 692
  oblique, 249–50
  of rational functions, 250–51
  vertical, 245–47
Attraction, electrical, 291
Augmented matrix of a system of equations, 539
Axes, 69
  real and imaginary, 491
  translation and rotation of, 700–705
Axis of a parabola, 177, 672

B

Back-substitution, 535
Barnsley, Michael, 201
Base, defined, 14
Binary number, 319
Binomial coefficient, 613
Binomial theorem, 610–617
  expansions using, 616–17
  first form, 611
  second form, 614
Biology, mathematical, 292
Bisection method for approximating real zeros of a polynomial, 224
Bounds for zeros of a polynomial, 228–29
Boyle's law, 294
Brahe, Tycho, 149
Break-even analysis, 103–4, 225–26
Briggs, Henry, 323
Byte, 562

C

Calculators
  angle calculations with, 349–50
  and a^x, 296
  and binomial coefficients, 615–16
  and common logarithms, 323
  [enter] and [=] keys, 7
  enter exponent ([EE]) key, 24
  entering negative numbers, 7
  and inequalities, 104–5
  matrix addition and multiplication with, 573
  and permutation and combination problems, 657
  and powers, 15
  quadratic formula implemented on, 170
  and row operations on matrices, 545–47
  and sine and cosine, 358–59
  translating a parabola with, 183
Calculators, graphing
  composition of functions on, 271
  and determinants, 583
  dots only mode, 695
  and ellipses, 686–87
  fundamentals of, 138–43
  and hyperbolas, 694–95
  and identities, 422
  and inequalities, 109–10
  and logarithmic equations, 336
  and matrix inverses, 591
  and the natural exponential function, 309
  operations on functions with, 268–69
  and optimization problems, 210
  Par, Func, and Square commands, 714
  and parabolas, 675–76
  and parametric equations, 714
  and polar coordinates, 490
  and rational functions, 254–55
  and sine and cosine functions, 380
  and zeros of a polynomial, 225–26
Canceling. *See* Reducing a rational expression to lowest terms

I-1

Celsius, 100
Center
  of a circle, 83
  of an ellipse, 681
  of a hyperbola, 690
Change of Base formula, 328
Chaos—Making a New Science, 201, 461
Chaos theory, 461
Charles's law, 99
Charles, Jacques, 99
Churchill, Winston, 258
Cipra, Barry, 410
Circle
  equation of, 83
  graphing, 83–84
  parametric form of, 713
  unit, 347
Clearing the denominators, 56
Closed intervals, 107
Closure, 2
Coefficient(s)
  binomial, 613
  of a monomial, 29
  of a polynomial, 29
Coefficient matrix of a system of equations, 539
Cofactor of a minor, 576
Cofactors, calculating a determinant by, 578
Cofunction identities, 424, 427–28
Coin flipping, 662–63
Collinearity, condition for, 582
Column matrix, 569
Combination(s), 654–57
  calculating number of, 655–57
  formula for, 655
  Fundamental Theorem of, 655
  of n objects taken r at a time, 654
Common (Briggsian) logarithms. *See* Logarithms, common
Commutative law of addition for complex numbers, 233
Commutative law of multiplication for complex numbers, 233
Commutative laws, 4
Complementary angle, 346
Completing the square, 166–67
Complex number(s), 231–37
  absolute value (modulus) of, 491–93
  additive identity and inverse for, 233
  algebraic properties of, 233–34
  arithmetic involving, 232–33
  conjugate of, 236
  division of, 237
  equality of, 231–32
  nth root of, 498–503
  powers of a, 499
  principal argument of, 495
  trigonometric (polar) form of, 491–96
Complex plane, 491

Complex roots (solutions) of quadratic equations, 235–36
Components of a vector, 508–9
Composition of functions, 269–71
Compound interest, 16–17, 161, 299
Conditional equation, 53
Conic section (Conic), 672–98
  definition of, in terms of focus and directrix, 708
  eccentricity of, 697
  polar equations of, 707–11
Conjugate axis of a hyperbola, 692
Conjugate of a complex number, 236
Conjugate Root theorem, 240
Constant polynomial, 31
Constant Sign theorem, 110
Constant term of a polynomial, 30
Constraints, 557
Coordinate representation of a vector, 505
Coordinate system, one-dimensional, 8
Coordinate system, two-dimensional, 69–74
Coordinates, converting between rectangular and polar, 486–88
Copernicus, Nicolaus, 149
Correspondence, 114–15
Cosecant (csc x), 399
  of an angle, 360
  as a ratio, 363
Cosine
  of an angle, 354–57
  function
    addition and subtraction identities for, 425–27
    alternate expressions for, 430–31
    period of, 381
    phase shift of, 389–90
    vertical translations of, 390–93
    zeros of, 388–89
  as a ratio, 363
Cosines, Law of, 478–82
  solving triangles using, 479–81
Cost of production, marginal, 92
Costs, fixed and variable, 92
Cotangent (cot x), 399
  of an angle, 359–61
  as a ratio, 363
Coterminal angles, 346
Counting numbers, defined, 2
Counting, Fundamental Principle of, 652
Cowling's rule, 264
Cramer's rule, 579–81
Cryptography, 258
Cubic function, 116
Cubic polynomial, 31
Curve fitting, 145
Curve, logistic, 308, 338
Cylinder, area and volume of, 120–21, 161

D

Damped sine wave, 401–2
Data compression, 562
Data, inferences from, 665
De Moivre's theorem, 498–500
Decay, exponential, 307
Decay, radioactive, 300
Decimal degrees, 347
Definition identities, 417
Degree of a monomial, 29
Degree of a polynomial, 29–30
Degrees, 344
  converting to radians, 348–49
Delta (Δ), 88
Demand matrices, 594
Denominator, 41
  rationalizing, 48–49
Dependent variable, 113
Depreciation
  defined, 13
  straight-line, 13, 158–59
Descartes's Rule of Signs, 242–44
Determinant, 575–83
  expanding about a given row or column, 577
  form of the equation of a line, 582
  formula for the area of a triangle, 582–83
  of an $n \times n$ matrix, 578
  of a square matrix, 575
  of a 3×3 matrix, 576–79
Dewey, John, 665
Dice, 662
Difference, 5
  of cubes, 36
  equation, 646–48
  of squares, 36
  of two functions, 266
Diophantus, 716
Direct variation, 284–87
Directed distance, 70
Directrix
  of an ellipse, 707
  of a hyperbola, 707
  of a parabola, 672
Discriminant of a quadratic equation, 168
Distance between points on the number line, 11
Distance formula, 71–72
Distributive law, 5
  for complex numbers, 233
Division
  algorithm for polynomials, 214
  of complex numbers, 237
  in polar form, 495–96
  of polynomials, 213–21
  of rational expressions, 46–47
  synthetic, 217–21
  by zero, 2, 42

Domain(s)
　of a function, 115, 116–17
　graphical interpretation of, 124–27
　of inverse trigonometric functions, 451, 464
　natural, 117
　of a quotient, 118
Dot product, 510–12
　criterion for perpendicularity, 511–12
Dots only mode, 695
Double-angle identities, 437–40
Drums, sounds made by, 410
Dummy variable, 629

E

E (exponent), 25
e (the natural number), 305–10
Eccentricity
　of a circle, 686–87
　of a conic, 697
　of an ellipse, 686–87
　of a hyperbola, 697–98
　of a parabola, 697
Echelon form, 534–36, 541
Economic output, 596–97
Electrical attraction or repulsion, 291
Element (of a set), 8
Elements (entries) of a matrix, 538
Ellipse(s), 680–87
　center of, 681
　directrix of, 707
　eccentricity of, 686–88
　focal length of, 681
　foci of, 680
　focus-directrix property of, 707
　major axis of, 681
　minor axis of, 681
　reducing the equation of to standard form, 684–85
　rotated, 702–3
　semimajor axis, 681
　semiminor axis, 681
　standard forms of the equation of, 682–85
　vertices of, 681
Encryption systems, 62
Enigma, 258
Equality of complex numbers, 231–32
Equality of polynomials, 31
Equation(s), 53–60
　in applications, 56–58
　of a circle, 83
　conditional, 53
　difference, 646–48
　equivalent, 53–54, 58–60
　exponential and logarithmic, 317–20, 334–36
　graphs of, 75–84
　inconsistent, 53
　linear. *See* Linear equations

of lines, 90, 94, 581–82
parametric, 712–14
quadratic. *See* Quadratic equations
radicals in, 192
raising to a power, 191–92
reducing to linear equations, 58
slope-intercept form of, 90
solving, 53, 76
substitution method for solving, 189–91
systems of. *See* Systems of equations
taking square roots of both sides of, 60
types of, 53
Error-correcting codes, 603
Escape velocity, 28
Euler's constant (e), 305–10
Even function, 133–34
Event, 660–61
Expansion, coefficient of linear, 52
Experiment, 660
Exponential equations, 317–20, 334–36
Exponential functions, 296–312
　applications of, 299–301
　with base a, 296–97
　graphs of, 297–98, 301
　natural, 305–10
Exponential growth and decay, 307
Exponents
　defined, 14–17
　eliminating negative and zero, 22
　laws of, 21–23
　rational, 19–20
　in scientific notation, 24
　zero, 20
Expressions, rational. *See* Rational expressions
Extraneous solutions, 59, 190–91
Extreme point, 207
Extreme point (value) of a quadratic function, 179–80

F

Factor theorem, 216–20
Factorial, 612–13
Factorial formula for binomial coefficients, 613
Factoring
　by grouping, 38
　identities (product to sum), for trigonometric functions, 446–48
　method for solving quadratic equations, 164–65
　polynomials, 35–38, 215–20
　a sum of squares, 233
Factorization identities, 36
Factorization of polynomials over the real numbers, 242
Fahrenheit, 100
Falling objects, 28

Feasible set (graph), 552–54
　determining, 553–54
　graphing, 553–54
　vertex of, 557
Fermat's Last Theorem, 716
Fermat, Pierre de, 716
Final demand matrix, 594
Finance, mathematics of, 646–50
Finite sequence, 626
Fitting a linear function to data, 145
Focal length
　of an ellipse, 681
　of a hyperbola, 690
　of a parabola, 673
Focus (foci)
　of an ellipse, 680
　of a hyperbola, 689
　of a parabola, 672
FOIL (First, Outside, Inside, Last), 32
Fractal, 201
Fractal image compression, 201
Fractals Everywhere, 201
Func command, 714
Function(s), 68–150, 265–93
　applications of, 118–21
　composition of, 269–71
　as correspondences, 115
　cubic, 115–16
　decreasing in an interval, 131–32
　defined, 113
　defined at x, 115
　domain and range of, 115, 116–17
　even and odd, 133–34
　exponential. *See* Exponential functions
　graphs of, 123–36, 205–11
　greatest integer, 134–35
　hyperbolic, 335
　increasing in an interval, 131–32
　inverse, 273–81, 292
　　ranges and domains of, 451
　inverse trigonometric, 450–54
　　ranges and domains of, 464
　linear, 144–47
　　graphing, 146
　logarithmic. *See* Logarithmic functions
　minimum and maximum values of, 179–80
　objective, 557
　one-to-one, 273–75
　operations on, 266–71, 291
　　using a graphing calculator, 268–69
　periodic, 379
　piecewise linear, 146–47
　piecewise-defined, 134
　polynomial, 116, 176, 204–44
　　of degree greater than two, 205–11
　quadratic. *See* Quadratic functions
　rational, 116, 245–55
　　graphing, 251–55
　trigonometric. *See* Trigonometric functions

Function(s)—*continued*
  undefined at x, 115
  values of, approaching infinity, 246–47
  values of, approaching zero, 246
Fund, college, 649–50
Fundamental Absolute Value Inequality, 197
Fundamental period, 381
Fundamental Theorem of Algebra, 239

G

Gas pressure, 289
Gauss elimination, 538–45
Gauss, Karl Friedrich, 235, 373, 542
Geometric sequence, 637–43
Geometry, analytic, 671–714
Gleick, James, 201, 461
Gordon, Carolyn, 410
Grade point average (GPA), 42–43
Graph(s)/graphing, 68–150
  absolute value equation, 80
  circle, 83–84
  cosecant (csc x), 399
  cosine (cos x), 378–82
  cotangent (cot x), 399
  damped sine waves, 401–2
  equation, 75–84
  equation by plotting points, 82
  equation that yields a pair of equations, 81
  equation with an undefined x-value, 81–82
  exponential functions, 297–98, 301
  feasible set, 552–54
  functions, 123–36
  the functions $y = ax^n$, 205–11
  hyperbolas, 691–95
  intersections and unions, 108–9
  inverse function, 278–81
  line using intercepts, 77
  linear function, 146
  logarithmic functions, 314–15
  $\log_a x$, 314
  parametric equations, 712–14
  by plotting points, 80–81
  polar equations, 488–90, 709–10
  quadratic equation, 80
  quadratic function, 176–79
  rational functions, 251–55
  secant (sec x), 398–99
  sine (sin x), 378–82
  solution set, 198
  solution sets of inequalities, 549–51
  sums that involve trigonometric functions, 399–401
  symmetry of, 133–34
  of tangent (tan x), 396–98
  of trigonometric functions, 378–85, 396–402
  vertical asymptotes, 247

Gravitation, Newton's law of universal, 24–25, 286–87
Gravity (falling objects), 28, 440
Greater than, 9
Greatest common factor, 35
Greatest integer function, 134–35
Grouping, factoring by, 38
Growth, exponential, 307, 338

H

Half-angle identities, 437, 441–46
Half-life of a radioactive element, 300
Half-open intervals, 107
Hamming, R. W., 603
Harmonic motion, 384–85
Hellman, Martin, 62
Heron's formula, 28, 481–82
History of algebra, 1
Hooke's law, 285
Horizontal line test for one-to-one functions, 274–75
Hyperbolas, 689–98
  applications of, 695–96
  asymptotes of, 692
  center of, 690
  conjugate axis of, 692
  directrix of, 707
  eccentricity of, 697–98
  eliminating the xy term using rotation of coordinate axes, 704–5
  focal length of, 690
  foci of, 689
  focus-directrix property of, 707
  graphing, 691–95
  rotated, 703–4
  standard forms of the equation of, 690
  transverse axis of, 690
  vertices of, 690
Hyperbolic function, 335

I

i, powers of, 234–35
Identities, 53
  addition and subtraction, 424–33
    for the cosine function, 425–27
    for the sine function, 428–29
    for the tangent function, 431–33
  additive and multiplicative, 3
  for complex numbers, 233–34
  cofunction, 424, 427–28
  definition, 417
  double-angle, 437–40
  factoring (product to sum), for trigonometric functions, 446–48
  factorization, 36
  fundamental trigonometric, 404–8
  half-angle, 437, 441–46
  multiple-angle, 437–40
  multiplicative, for complex numbers, 234

  parity, 408, 417
  product, 32
  product (sum to product), for trigonometric functions, 446–48
  Pythagorean, 417
  square, 437, 441
  symmetry, 424, 429–30
  for the tangent of the angle between two lines, 433–34
  trigonometric. *See* Trigonometric identities
  triple-angle, 438–39
Identity matrix, 587–88
  of size n, 573
Illumination, 289
Image compression, 201
Imaginary axis, 491
Imaginary numbers (imaginaries), 235
Imaginary part, 231
Imaginary unit, 231
Inconsistent equation, 53
Independent variable, 113
Induction, Principle of Mathematical, 620–24
Induction step, 621
Inequalities, 100–11
  absolute value, involving, 197–99
  cubic, 110–11
  linear, 101
  polynomial, 106, 110–11
  properties of, 101
  rational, 106–7
  satisfying, 549
  solution set of, 549
  solving, 101–7
    with a graphing calculator, 109–10
  system of. *See* System of inequalities
  three-termed, 105–6
Inequality symbols, 9
Inferences from data, 665
Infinite interval, 107–8
Infinite sequence, 626
Initial point of a vector, 505
Initial side of an angle, 344
Inner dimensions of a matrix, 571
Input-output analysis, 593–97
Input-output matrix, 593
Integers, defined, 2
Intercepts, determining, 77
Interest
  accrued, 119
  compound, 16–17, 161, 299
Intermediate Value theorem, 208
Internet, 62
Interpolating, 73–74
Intersection of sets, 105
  graphing, 108–9
Interval(s), 107–9
  infinite, 107–8
  open, closed, and half-open, 107
  test, 110

Inverse
  additive, of a complex number, 233
  functions, 273–81
  multiplicative, of a complex number, 236–37
  of a square matrix, 585–88
  sine function (arcsine), 450–51
  trigonometric functions, 450–54
    Fundamental Composition identities of, 452–54
  of a 2×2 matrix, 586–87
  variation. See Variation, inverse

J

Joint variation. See Variation, joint
Julia set, 609

K

Kac, Mark, 410
Kantorovic, V., 204
Kepler's law, 290
Kepler, Johannes, 149
Klein bottle, 671

L

Laws of exponents, 21–23
Laws of logarithms, 315–17
Laws of real exponents, 296
Leading coefficient of a polynomial, 30
Length of a circular arc, 350–51
Lenstra, A., 62
Leontieff, Vassily, 593
Less than, 9
Likelihood, 661
Line(s), 86–97
  equation of, 90, 92, 94
  determinant form of the equation of, 582
  parallel and perpendicular, 95–97
  vertical and horizontal, 78
Linear difference equation, 647–48
Linear equations, 54–56
  methods for solving, 54–56
  reducing equations to, 58
  in two variables, 77
Linear function, 115
Linear polynomial, 31
Linear programming, 557–59
Linear speed, 351–52
Linear system of equations, 522–24
  matrix solution of, 588–91
Loans, consumer, 648–49
Locus, 672
Logarithmic equations, 334–36
  solving, 317–20
Logarithmic function(s), 312–29
  with base a, 312–13
  common, 322–24
  natural, 324–28

Logarithms
  common (Briggsian), 322–24
  fundamental properties of, 313–14
  laws of, 315–17
  natural, 324–28
    function, 324–28
    laws of, 325
    properties of, 325
  to various bases, 328–29
Logistic curve, 308, 338
Long division of polynomials, 213–15
Lorenz, Edward, 461
Lotka, 292

M

Magnitude of a star, 333
Main diagonal of a matrix, 539
Major axis of an ellipse, 681
Mandelbrot, Benoit, 201
Mantissa (in scientific notation), 24
Mapping, 114–15
Marginal cost of production, 92
Matrix (Matrices), 538–45, 567–628
  additive inverse of, 569
  calculating a determinant by cofactors, 578
  column, 569
  determinant of a 2×2, 575
  determinant of a 3×3, 576–79
  determinant of an $n \times n$, 578
  in echelon form, 541
  elementary row operations on, 540
  elements (entries) of, 538
  final demand, 594
  identity, 587–88
    of size n, 573
  inner and outer dimensions of, 571
  input–output, 593
  inverse of a 2×2 matrix, 586–87
  main diagonal of, 539
  minor of a row or column in, 575
  minor of a 3×3, 575–76
  multiplication, 569–73
  multiplying by a real number, 568
  products of are defined, 571
  row, 569
  solution of a square linear system, 588–91
  square, 539
    determinant of, 575
    inverse of, 585–88
  subscript notation for elements of, 568
  sum of two, 568
  zero, 568
Maximum point of a graph, 207
Maximum point of a function, 180
Maximum value of a function, 180
Menachmus, 672
Midpoint formula, 72–73
Minimum point of a graph, 207
Minimum point of a function, 179

Minimum value of a function, 179
Minor
  cofactor of, 576
  of a row and a column in a matrix, 575
  of a 3×3 matrix, 575–76
Minor axis of an ellipse, 681
Minutes, 345
Mixture problem, 159–60
Modeling, simulation as a tool for, 514
Models, linear, 94
Models, population, 338
Modern Algebra, 121
Modulus of a complex number, 491–93
Monomial, 29
Monomial factors, 35
Mortgages, 648–49
Motion, harmonic, 384–85
Motion, particle, 712
Multiple-angle identities, trigonometric, 437–40
Multiplication
  of complex numbers, 232–33
    in polar form, 495–96
  of polynomials, 32–33
  of rational expressions, 46–47
  of a vector by a scalar, 506
Multiplicative identity, 3
  for complex numbers, 234
Multiplicative inverse, 3
  of a complex number, 236–37
Multiplicity of a zero, 239–42

N

Natural domain, 117
Natural exponential function, 305–10
Natural logarithms. See Logarithms, natural
Nautical mile, British, 439
Negative exponent, 20
Negative numbers, 8
Negatives, properties of, 5
Newton, Sir Isaac, 149
Newton's law of universal gravitation, 24–25, 286–87
Noether, Emmy, 121
Nonlinear system of equations, 523–24
Nonnegative number, 9
Normal probability curve, 310
Notation
  binomials, 613
  of logarithmic functions, 313
  matrix, 568
  scientific, 24
  set-builder, 8
  summation, 628
nth partial sum of a sequence, 627
nth root, 18–19
  of a complex number, 498–503
nth term of a geometric sequence, 638
nth term of an arithmetic sequence, 632

Number(s)
  binary, 319
  e, 305–10
  kinds of, 2–3
  line, 8
  negative, 8
  positive, 8
  theory, 62
  of Zeros theorem, 217
Numerator, 41
  rationalizing, 49–50

O

Objective function, 557
Obtuse angle, 346
Odd function, 133–34
One-dimensional coordinate system, 8
One-to-one correspondence, 8
One-to-one functions, 273–75
Open intervals, 107
Operations
  that generate equivalent equations, 54
  order of, 6, 14
  on real numbers, 3–4
Optimization, 176, 181–83
Orbits of planets, 149
Ordinate, 70
Origin, 8, 69
Oscillations, in a graph, 401–2
Outcome, 660
Outer dimensions of a matrix, 571
Output, economic, 596–97

P

Palladian windows, 172
Parabola(s), 177, 672–76
  axis of, 672
  directrix of, 672
  eccentricity, 697
  focal length of, 673
  focus of, 672
  focus-directrix property of, 707
  reducing the equation of to standard form, 675
  reflection property of, 676–77
  standard forms of the equation of, 673–75
  translating using a calculator, 183
  vertex of, 672
Parallel lines, 95
Parallelogram rule for adding vectors, 506–8
Parameter, 712
Parametric equations of a curve, 712–14
Parametric form of a circle, 713
Par command, 714
Parity bit, 603
Parity identities, 408, 417
Partial fraction, 598–602

Partial sum
  of an arithmetic sequence, 633–35
  of a geometric sequence, 641–43
  of a sequence, 627
Particle motion, 712
Pascal's triangle, 617
Pascal, Blaise, 617
Pendulum, period of, 289
Period, fundamental, 381
Periodic function, 379
Periods of general sine and cosine functions, 381
Permutation(s), 652–54
  Fundamental Theorem of, 653–54
  of n objects taken r at a time, 653
Perpendicular lines, 95
Perpendicularity, dot product criterion for, 511–12
Phase shift of a general sine or cosine function, 389–90
Piecewise linear function, 146–47
Piecewise-defined function, 134
Plato's Academy, 672
Point-slope formula, 93–94
Polar axis, 484
Polar coordinates, 484–90
  converting to rectangular, 486–88
Polar form of a complex number, 491–96
Pole, 484
Polynomial(s), 29–38
  addition and subtraction of, 31
  applications of, 33–35
  bisection method for approximating real zeros of, 224
  bounds for zeros of, 228–29
  calculating zeros of, 223–29
  coefficients of, 30
  combining like (similar) terms, 31
  constant, 31
  cubic, 31
  degree of, 29–30
  division algorithm for, 214
  division of, 213–21
  equality of two, 31
  factoring, 35–38, 215–20
  factorization over the real numbers, 242
  functions, 116, 176, 204–44
  inequalities, 106, 110–11
  linear, 31
  long division of, 213–15
  multiplication of, 32–33
  number of zeros of, 239–40
  quadratic, 31
  quartic, 31
  in several variables, 31
  terms of, 29–30
  zero of, 207
  zero, 31
Population models, 338
Positive numbers, 8

Power of a windmill, 290
Power(s)
  of a complex number, 499
  of i, 234–35
  raising numbers to. See Exponents
  of ten, 24
Pressure, gas, 289
Principal argument of a complex number, 495
Principal nth root, 18–19
Principal square root, 18
  of a negative real number, 235
Principle of horizontal translation, 129
Principle of reflection, 130
Principle of scaling, 130
Principle of vertical translation, 129
Probability, 660–63
  curve, normal, 310
Problems, applied, 154–199
  strategy for solving, 155–62
Product
  of a binomial sum and difference, 32
  of a matrix and a real number, 568
  of two binomials, 32
  of two functions, 267
Product identities, 32
Product identities (sum to product), for trigonometric functions, 446–48
Projecting, 94–95
Proof by mathematical induction, 620–24
Proportionality factor, 284
Ptolemy's theory of planetary motion, 149
Public key encryption systems, 62
Pythagorean equation for right triangles, 28
Pythagorean identities, 417
  for the cotangent and cosecant, 406–7
  for the sine and cosine, 404–6
  for the tangent and secant, 406

Q

Quadrants, 71
Quadratic equation(s), 164–74
  complex solutions of, 235–36
  discriminant of, 168
  equations that reduce to, 187–89
  graphing, 80
  solving, 168–69
    by completing the square, 166–67
    by factoring, 164–65
  using to solve other equations, 186–94
Quadratic formula, 167–74
  implemented on a calculator, 170
  summarized, 201
Quadratic function(s), 115, 176–83
  applications of, 179–81
  extreme point (value) of, 179–80
  graphing, 176–79

standard form of, 177–78
 summarized, 202
Quadratic polynomial, 31
Quartic polynomial, 31
Quotient
 domain of, 118
 of two functions, 267

R

R (the set of real numbers), 8
Radians, 347
 converting to degrees, 348–49
Radicals, 18–19
 properties of, 23
Radioactive decay, 300
Radius of a circle, 83
Range(s)
 of a function, 115, 116–17
 graphical interpretation of, 124–27
 of inverse trigonometric functions, 451, 464
Rate, 95
Ratio of a geometric sequence, 637
Rational exponents, 19–20
Rational expressions, 41–50
 addition and subtraction of, 44–46
 multiplication and division of, 46–47
 reducing to lowest terms, 43–44
 simplifying, 47–48
Rational functions, 116
 graphing, 251–55
Rational inequality, 106–7
Rationalizing the numerator and denominator, 48–50
Rational numbers, defined, 2
Rational Zero theorem, 226–27
Real axis, 491
Real exponents, 296
Real numbers, defined, 2–4
Real part, 231
Reciprocal, 3
Rectangular coordinates, converting to polar, 486–88
Recursive definition of the terms of a sequence, 627
Reducing a rational expression to lowest terms, 43–44
Reference angle, 357–58
Reflecting telescope, 677
Reflection of a function, 130
Reflection property of the parabola, 676–77
Remainder theorem, 215
Repulsion, electrical, 291
Resistance, electrical, 51, 286
Ribet, Kenneth, 716
Right angle, 346
Rise, 87
Rivest, R., 62
Root of an equation, 53
Roots, square and nth, 18–19

Rotation of coordinate axes, 700–705
Row matrix, 569
Rule of Signs, Descartes's, 242–44
Rules of arithmetic, 4–5
Run, 87

S

Sample space, 660–61
Satisfying an inequality, 549
Scalar product, 510–12
Scalar quantities, 505
Scaling a function, 128–31
Scientific American, 62
Scientific notation, 24
Secant (sec x), 398–99
 of an angle, 359–61
 as a ratio, 363
Seconds, 345
Semimajor axis of an ellipse, 681
Seminminor axis of an ellipse, 681
Sequence(s), 625–43
 arithmetic, 631–35
  nth term of, 632
  partial sum of, 633–35
 defining terms of recursively, 627
 finite, 626
 geometric, 637–43
  nth term of, 638
  partial sum of, 641–43
  ratio of, 637
 infinite, 626
 nth partial sum of, 627
 partial sum of, 627
 sum of an infinite geometric, 643
 term of, 626
Series, sum of an infinite geometric, 643
Set-builder notation, 8
Sets, 7
Shamir, A., 62
Shrinking a function, 130
Sides of triangles, determining, 366–67
Signs, Descartes's Rule of, 242–44
Similar polynomials, 31
Simulation as a modeling tool, 514
Sine(s)
 of an angle, 354–57
 and cosine functions, applications of, 384–85
 and cosine graphs, amplitude of, 383
 function
  addition and subtraction identities for, 428–29
  alternate expressions for, 430–31
  inverse (arcsine), 450–51
  period of, 381
  phase shift of, 389–90
  vertical translations of, 390–93
  zeros of, 388–89
 Law of, 669–87
  ambiguous case of, 472–74
  applications of, 474–75

 as a ratio, 363
 wave, damped, 401–2
Slope of a straight line, 87–90
 application of, 91–92
 geometric interpretation of, 91
 of parallel and perpendicular lines, 95–97
Slope-intercept form of the equation of a line, 90
Solution(s)
 of an equation, 53, 76
 of exponential and logarithmic equations, 317–20
 extraneous, 190–91
 of an inequality, 101–7
  using a calculator, 104–5
 of a system of equations, 521
 of a system of inequalities, 549, 552–54
Solution set, 53
 graphing, 198
 of an inequality, 105, 549
Sound, loudness of, 323–24
Sound, speed of, 20
Speed, linear versus angular, 351–52
Speed of boat, 119
Square
 of a binomial difference, 32
 of a binomial sum, 32
 command, 714
 identities, 437, 441
 matrix, 539
Square root, 18
 of a negative number, 235
Standard form(s)
 of the equation of a hyperbola, 690
 of the equation of a parabola, 673–75
 of the quadratic function, 177–78
Star, magnitude of, 333
Straight angle, 346
Straight-line depreciation, 158–59
Stretching a function, 130
Subscript notation for matrix elements, 568
Subset, 8
Substitution method for solving an equation, 189–91
Substitution, trigonometric, 422
Subtraction
 identities, 424–33
 of rational expressions, 44–46
 of real numbers, 5
Sum
 of cubes, 36
 of an infinite geometric sequence, 643
 of an infinite geometric series, 643
 of squares, factoring, 233
 of two functions, 266
 of two matrices, 568
Summation index, 628
Summation notation, 628
Supplementary angle, 346

Symbols
  absolute value, 9
  inequality, 9
Symmetry identities, 424, 429–30
Symmetry of graphs, 133–34
Synthetic division, 217–21
System of equations
  dependent, 530
  inconsistent, 530
  independent, 530
  linear, 522–24
    augmented matrix of, 539
    coefficient matrix of, 539
    elementary operations on, 536–37
    nonsquare, 543–45
    in any number of variables, 534–47
    reducing to triangular form, 536–38
    solving by substitution, 522–24
    square, 543
    in triangular form (echelon form), 534–36
  linear, in two variables, 521–32
    applications of, 526–27
    graphical analysis of, 524–25
    method of elimination for solving, 529–31
    solving with a graphing calculator, 525, 531–32
  nonlinear, 523–24
    solving by substitution, 523–24
  solution of, 521
System of inequalities, 552–54
  solution of, 549, 552–54
  in two variables, 549–54

T

Tangent (tan x), 396–98
  of an angle, 359–61
  function, addition and subtraction identities for, 431–34
  as a ratio, 363
Telescope, reflecting, 677
Temperature conversion, 100, 163
Term of a sequence, 626
Terminal point of a vector, 505
Terminal side of an angle, 344
Terminal velocity, 307
Terms of a polynomial, 29–31
Test interval, 110
Test point, 552
Test value, 208
Three-termed inequality, 105–6
Tracing, 139
Translating into rectangular coordinates, 713–14
Translation, horizontal and vertical, 128–31
Translation of coordinate axes, 700–705
Transverse axis of a hyperbola, 690

Triangle(s)
  area of, 480–81, 581–83
  solving, 371–72
  solving, using the Law of Cosines, 479–81
Triangular form, 534–36, 541
Trigonometric equations, solving, 419, 453–54, 456–59
Trigonometric (polar) form of a complex number, 491–96
Trigonometric functions, 343–411
  of acute angles, 363–68
    as ratios, 363–66
  graphs of, 378–85, 396–402
  signs of, 361
Trigonometric identities, 417–22
  fundamental, 404–8
  proving, 418
  table of, 462–63
  verifying, tips for, 420–22
Trigonometric substitution, 422
Trigonometry
  analytic, 416–64
  applications of, 371–73, 468–515
Triple-angle identity, 438–39
Truman, Harry, 665
Turing, Alan, 258
Turning point, 207

U

Union of sets, 106
Union, graphing, 108–9
Unit circle, 347

V

Value of a function, 113
van der Waerden, 121
Variables, dependent, 113
Variables, dummy, 629
Variables, independent, 113
Variation, 293
  direct, 284–87
  inverse, 285–86
  joint, 287
Vectors, 505–12
  adding, 506–8
  applications of, 509–10
  direction, 505
  magnitude, 505
  multiplication of by a scalar, 506
  resolution into components, 508–9
Velocity, 69, 95
  angular and linear, 352
  terminal, 307
Verhulst, P. F., 338
Vertex (vertices)
  of an angle, 344
  of an ellipse, 681

  of a feasible set, 557
  of a hyperbola, 690
  of a parabola, 177, 672
Vertical line test, 127
Vertical translations of sine and cosine functions, 390–93
Volterra, 292
Volume of a cylinder, 161

W

Wavelength, 290
Webb, David, 410
Wiles, Andrew, 716
Windmill, power of, 290
Wolpert, Scott, 410

X

x-axis, 69
x-coordinate, 70
x-intercept, 77

Y

y is inversely proportional to x^m, 285
y is proportional to x, 284
y is proportional to x and t, 287
y is proportional to x^m, 285
y varies directly as x, 284
y varies directly as x^m, 285
y varies inversely as x^m, 285
y varies jointly as x and t, 287
y-axis, 69
y-coordinate, 70
y-intercept, 77
Yeltsin, Boris, 258
Yield curve, 327–28
Young's rule, 264

Z

Zero matrix, 568
Zero polynomial, 31
Zero power (exponent), 20
Zero vector, 505
Zero, dividing by, 2, 42
Zeros
  of a polynomial, 207, 239
    approximating, using a graphing calculator, 225–26
    bounds for, 228–29
    calculating, 223–29
  of the general sine and cosine functions, 388–89
  number of, 217
  rational, 226–27
Zooming, 139

Graph of the Inverse of a Function

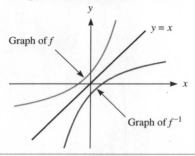

Graphs of Increasing Exponential Functions

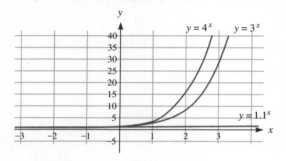

Graphs of Decreasing Exponential Functions

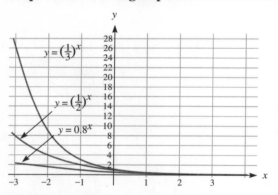

Graph of $\log_a x$

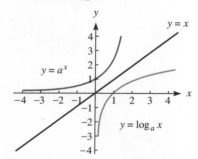

Method of Substitution for Solving Systems of Two Equations in Two Variables
1. Solve an equation for one of the variables in terms of the other.
2. Substitute the expression found in step 1 into the other equation.
3. Solve the resulting equation for the possible values of one variable.
4. For each value found in step 3, substitute into the expression of step 1 to determine the corresponding value of the other variable.

Elementary Row Operations on Matrices
1. E_{ij}. Interchange rows i and j.
2. kE_i. Multiply row i by a nonzero real number k.
3. $kE_i + E_j$. Add k times row i to row j.

Determinant of a Square Matrix
1. 2×2 matrix: If $A = \begin{bmatrix} a & b \\ c & d \end{bmatrix}$, then:
$$|A| = ad - bc$$
2. General $n \times n$ matrix:
$$|A| = a_{11}A_{11} + a_{12}A_{12} + \cdots + a_{1n}A_{1n}$$
where A_{ij} denotes the cofactor determined by eliminating the ith row and jth column.

Cramer's Rule
If a linear system of n equations in n variables $x, y, z, \ldots$ has a nonzero determinant $|A|$, then the system has a unique solution given by the formulas:
$$x = \frac{|A_1|}{|A|}, \quad y = \frac{|A_2|}{|A|}, \quad z = \frac{|A_3|}{|A|}, \quad \ldots$$

Computing the Inverse of a Square Matrix
1. Form the matrix:
$$\left[\begin{array}{cccc|cccc} a_{11} & a_{12} & \ldots & a_{1n} & 1 & 0 & \ldots & 0 \\ a_{21} & a_{22} & \ldots & a_{2n} & 0 & 1 & \ldots & 0 \\ \vdots & & & & \vdots & & & \\ a_{n1} & a_{n2} & \ldots & a_{nn} & 0 & 0 & \ldots & 1 \end{array}\right]$$
where the matrix on the right is the identity matrix I_n.
2. If possible, perform elementary row operations on this matrix to transform the left half into the identity matrix I_n.
3. After the transformation, the matrix in the right half is A^{-1}.